AF328002

Science of Sintering

New Directions for Materials Processing and Microstructural Control

Science of Sintering

New Directions for Materials Processing and Microstructural Control

Edited by

Dragan P. Uskoković
Serbian Academy of Sciences and Arts
Belgrade, Yugoslavia

Hayne Palmour III
North Carolina State University
Raleigh, North Carolina

and

Richard M. Spriggs
Alfred University
Alfred, New York

Plenum Press • New York and London

Library of Congress Cataloging-in-Publication Data

World Round Table Conference on Sintering (7th : 1989 : Herceg-Novi,
  Montenegro)
    Science of sintering : new directions for materials processing and
  microstructural control / edited by Dragan P. Uskoković, Hayne
  Palmour III, and Richard M. Spriggs.
        p.   cm.
    "Proceedings of the Seventh Round Table Conference on Sintering,
  held August 28-September 1, 1989, in Herceg-Novi, Yugoslavia"--Verso
  t.p.
    Includes bibliographical references and index.
    ISBN 0-306-43528-4
    1. Sintering--Congresses.   I. Uskoković, Dragan P., 1944-
  II. Palmour, Hayne.   III. Spriggs, Richard M.   IV. Title.
  TN695.W67   1989
  671.3'73--dc20   90-35888
                                                                  CIP

Proceedings of the Seventh Round Table Conference on Sintering,
held August 28–September 1, 1989, in Herceg-Novi, Yugoslavia

© 1989 Plenum Press, New York
A Division of Plenum Publishing Corporation
233 Spring Street, New York, N.Y. 10013

All rights reserved

No part of this book may be reproduced, stored in a retrieval system, or transmitted
in any form or by any means, electronic, mechanical, photocopying, microfilming,
recording, or otherwise, without written permission from the Publisher

Printed in the United States of America

TO GEORGE C. KUCZYNSKI

Sintering was a secret—not a science—before Dr. George C. Kuczynski developed the first laws and theory of sintering in 1949. We knew that sintering occurred, but we did not know how and why particles coalesced. Kuczynski's model provided the foundation for a scientific exploration of the sintering process.

While searching to process tungsten to build a better light bulb, Kuczynski's understanding of sintering shrinkage by volume diffusion and related grain growth processes occurred in the post-war laboratories of Sylvania Electric Co., Inc. As he was working with very pure copper crystals—a gift from his boss to encourage Kuczynski to pursue the secret of sintering—Kuczynski noticed something extraordinary in the commonplace occurrence when he heated the atomized copper in a copper crucible.

He examined the phenomenon more closely and heeded, he says, the advice of an oriental philosopher: "In science, you have to listen to the gentle murmurs of nature."

Kuczynski listened. He measured the neck that had formed with the particles as a function of time to give us the first laws of sintering on which so much subsequent work has depended. Kuczynski's broad classical education from his native Poland, his training in mathematics and physics, and his keen mind enabled him to cut through the complex to simple, clear, unambiguous truth. His publications, research and sintering conferences have encouraged new ideas and provided the scientific paradigm that has fostered pursuit of new theories and new understanding of sintering by other scientists and engineers. His work shed the first light on the path that has ultimately led us to grapple with the highest technology of sintered materials.

To Dr. George C. Kuczynski, we dedicate these published proceedings of the Seventh World Round Table Conference on Sintering with gratitude for his accomplishments that have made possible our own.

PREFACE

This volume, SCIENCE OF SINTERING: NEW DIRECTIONS FOR MATERIALS PROCESSING AND MICROSTRUCTURAL CONTROL, contains the edited Proceedings of the Seventh World Round Table Conference on Sintering, held in Herceg-Novi, Yugoslavia, Aug. 28 - Sept. 1, 1989. It was organized by the International Institute for the Science of Sintering (IISS), headquartered in Belgrade, Yugoslavia.

Every fourth year since 1969, the Institute has organized such a Round Table Conference on Sintering; each has taken place at some selected location within Yugoslavia. A separate series of IISS Topical Sintering Symposia (Summer Schools) have also been held at four year intervals, but they have been offset by about two years, so they occur between the main Conferences. As a rule, the Topical Sintering Symposia have been devoted to more specific topics and they also take place in different countries. The aim of these Conferences and their related "Summer Schools" has been to bring together scientists from all over the world who work in various fields of science and technology concerned with sintering and sintered materials. A total of seven IISS Conferences have been held over the period 1969-1989, and they have been supplemented by the four Topical Sintering Symposia held in Yugoslavia, Poland, India and Japan (in 1975, 1979, 1983 and 1987, respectively).

This most recent five day Conference addressed the fundamental scientific background as well as the technological state-of-the-art pertinent to science of sintering and high technology sintered materials. It encompassed many problems - and solutions - that are relevant to a wide variety of scientific concerns and industrial applications. In the editors' opinion, this Conference could best be characterized by (1) the high level and broad scope of the papers presented, and (2) the many nations (more than 20) that were represented by their respective authors.

It is also interesting to reflect upon some of the effects that are attributable to the cumulative influences of this whole series of Round Table Conferences on Sintering. Progression in the number counts alone have been impressive: this, for example, was by far the largest in total numbers, and in the numbers of nations from which participants have been drawn. The overall scientific quality and the real world relevance of the materials being presented and the issues being discussed have also shown steady improvement from meeting to meeting. In our opinion, in this most recent Conference, the interplay existing between quality, relevance and effective communication has been especially fruitful.

Nowhere has this kind of cumulative effect been more visible than
in the markedly increased participation evidenced by Yugoslavia's own
authors. In this most recent Conference, for example, there were more
significant papers by such authors, representing more well established
Yugoslav laboratories, than ever before. Nurtured in part through their
own hard work and in part through the insights, experiences and
exposures gained through their attendance at preceding Round Table
Conferences, it is evident that, collectively, this contingent of
Yugoslav sinterers was well prepared to present important scientific
results spanning a broad spectrum of relevant sintering issues.

Fifty-seven papers presented at the Conference were selected by
the editors for inclusion in this volume. Because of the significance
of their scientific contributions, two invited papers have also been
included even though, regrettably, it had not been possible to have
them presented at the Conference by their respective authors (G. S.
Upadhyaya and R. Pruemmer). Other papers presented at the Conference
will be published by the International Institute for the Science of
Sintering in its journal, "Science of Sintering".

Thematically, they are presented in the eleven sections which
comprise this book: (1) fundamentals of sintering, (2) controlled
preparation of ceramic powders, (3) preparation and sintering of
metallic powders, (4) sintering of multiphase systems, (5) pressure
sintering, (6) rate controlled sintering, (7) microstructure control,
(8) metals and composite processing, (9) sintering of oxide ceramics,
(10) high temperature superconductors, and (11) non-oxide ceramics.

The list of Contributors formally acknowledges the considerable
cooperation and assistance rendered to the General Secretary of the
International Institute for the Science of Sintering, Academician
Momčilo M. Ristić, and the President of the Organizing Committee,
Professor Dragan Uskoković, by (1) a distinguished International
Program Committee, (2) the several Session Chairmen, as well as (3) the
creative efforts of 129 distinguished contributing authors representing
many of the world's centers for sintering research. We extend our
personal thanks to all of them for their cooperative attitudes, timely
responses and many helpful suggestions which have characterized all our
relationship with them.

On behalf of the participants and the whole sintering community we
wish at this point to express our gratitude to the patron of the
International Institute for the Science of Sintering, the Serbian
Academy of Sciences and Arts, as well as to the Yugoslav Committee for
Electronics, Telecommunication, Automation and Nuclear Technology, the
Serbian Society for Powder Metallurgy, the Institute of Technical
Sciences of the Serbian Academy of Sciences and Arts and the Center for
Multidisciplinary Studies of Belgrade University for their support in
the organization of the Conference. Special thanks are due to Dr.
Branko Kostić, President of the Republic of Mentenegro and Academician
Dušan Kanazir, President of the Serbian Academy of Sciences and Arts,
who welcomed attendees to the Conference. We also note with gratitude
the considerable financial support for the Conference organization
provided by the Principal Sponsor "PRVI PARTIZAN" (Titovo Užice), the
Republic Council for Science, as well as other Yugoslav sponsors. We are
grateful to the U.S. sponsor, the American Ceramic Society and we would
like to emphasize the significance of the financial support for travel
provided by the AFOSR which enabled the participation of U.S.
scientists.

We wish to acknowledge our very special personal thanks to a modest group of persons who worked with dedication but largely behind the scenes: to Milivoj Jelačić for coordinating our Conference; to Mirjana Kosanović and the Department of International and Technical Relations for their secretarial assistance in preparing of materials before, during and after the Conference; to Aleksandar Kosanović and Dragan Tasić for serving as projectionists and providing other technical service; to Drs Stamenka M. Radić and Višnjica Vukčević for translation services; to Taisa Agaljcev, Miroslava Janković, Cveta Jakšić and Kiril Svinjarski for simultaneous translations during the Conference. We pay special tribute to Mirjana Stojanović for her special skills and experience in typing and/or revising the edited Proceedings.

It is also our pleasent duty to thank all our young collaborators: Rada Novaković, Olivera Milošević, Jelena Dumić, Petar Kostić, Čedomir Jovalekić and Ratomir Agatonović, who helped in performing various tasks during the Conference and in the preparation of these proceedings.

Last but not least, it is appropriate to acknowledge with real affection the patience, tolerance and moral support we have been accorded by our colleagues and our families through those extended periods of time we have had to pay attention to the organizing of this Conference and the editing of these Proceedings.

Belgrade   Dragan P. Uskoković
November 1989   Hayne Palmour III and
                                                  Richard M. Spriggs

CONTENTS

Part VI. RATE CONTROLLED SINTERING

Part VII. MICROSTRUCTURE CONTROL

Part VIII. METALS AND COMPOSITE PROCESSING

Part XI. NON-OXIDE CERAMICS

Part I. FUNDAMENTALS OF SINTERING

SINTERING OF CERAMICS

Carol A. Handwerker, John E. Blendell, and Robert L. Coble*

National Institute of Standards and Technology
Materials Science and Engineering Laboratory
Gaithersburg MD 20899 USA

*Massachusetts Institute of Technology
Cambridge MA 02139 USA

INTRODUCTION

The primary goal of sintering research is the controlled manipulation of microstructure. Out of the entire range of microstructures which are theoretically possible, each material system will be able to achieve only a subset of them, depending on the intrinsic material properties. Within these material constraints, the aim is to produce microstructures which enhance specific properties. Our understanding of the relationships among materials processing, microstructure, and properties is just beginning to emerge, and is producing unexpected results. For example, in a recent study of toughness in Al_2O_3 by Bennison and Lawn, microstructures with platy grains and a bimodal grain size distribution in undoped Al_2O_3 exhibited a greater resistance to crack propagation than did the more uniform microstructures in MgO-doped Al_2O_3 [1]. As a result of this emerging understanding, the focus of sintering science is changing from the modification of microstructures in incremental ways for correspondingly incremental improvement in properties to more effectual manipulation of microstructures to optimize properties. However, the production of the optimum microstructure will be dependent on both the material and the application and may require radically different processing routes for different materials. In this review paper, we have examined the research in sintering science over the past five years which has advanced the goal of microstructure manipulation.

THEORETICAL MODELLING AND MODEL MICROSTRUCTURES

Over the past five years, substantial progress has been made in developing more realistic sintering models and in designing materials to test these theories. The basis of the current theoretical advancement is that the specific set of coupled kinetic processes operating in certain systems can be separated and, thus, the enthusiasm for the modelling of competing mechanisms in sintering has been restored. When the sintering of a system cannot be categorized by a small subset of possible sintering processes, simplified geometries can be used to examine the roles of various processes in microstructure development. The recent advances in sintering experiments have been made possible by a revolution in the control of powder characteristics and processing. This revolution, fomented by H. K. Bowen, provides enormous flexibility in the design of specific microstructures [2]. In the following discussion, theories and their complementary model experiments are discussed together.

The simplest model whose use continues to provide insight is the two-sphere model. An example of recent results is the numerical simulation of neck growth by surface and grain boundary diffusion [3]. In agreement with previous studies [4], there exists a limiting ratio of grain boundary diffusivity to surface diffusivity above which the densification rate does not increase because surface diffusion is required to redistribute matter in the region of the neck.

The next step in geometrical complexity, is the sintering of a row of spheres. For this geometry, the number of concurrent processes which can be treated is large because each particle has only two neighbors and, if is assumed that there are no torques on the particles to rotate them out of the axially symmetrical position, the mathematical statement of the problem is simplified. An important new development is that differential densification has now been treated in a row of spheres by allowing a section of the chain to begin to densify before the rest of the chain [5,6]. It has been calculated that, when the other sections begin to densify the chain will either become uniform in its sintering or will break up catastrophically into individual segments depending on the relative rates of coarsening and densification and the dihedral angle.

This problem of differential densification and development of heterogeneous microstructures has recently been a topic of intense theoretical and experimental investigation [7-15]. The approach has been from two extremes in initial microstructures. In the first, mono-sized powders are packed and sintered, and the size of final defects are correlated with the initial amount of ordering [12,13]. In the second case, dense inclusions are placed into a powder compact and the effect of the non-densifying inclusions on the densification rate of the matrix is evaluated [14]. These two approaches reflect the extremes seen in the technology of powder processing with controlled composition and physical characteristics and of processing of ceramic-matrix composites. The types of heterogeneities examined over the last 5 years range from large cracks between regions of perfect packing to large spherical pores in a matrix containing smaller pores, to sintering of a porous matrix around a higher density spherical inclusion. The advances in sintering of composites are summarized in the next section.

A major result is that the maximum sintering stress is on the order of 0.1 - 2 MPa regardless of microstructure [7-11]. This result is important because the sintering stress determines the opposing tensile stress required to keep particles from densifying or, for higher tensile stresses, to pull particles apart. The size of the sintering stress is a major factor in specifying HIP'ing and hot pressing conditions to improve reliability of sintered composites.

Experiments on 2-D arrays of monosized or bimodally-sized particles have shown that crack-like defects develop in the imperfectly packed regions separating perfectly packed regions [12,13]. Liniger and Raj found that packing should be random to maximize final density and minimize final defect size. An additional important result is that differential densification between poorly packed and well-packed regions lead to the desintering of particle-particle necks, as predicted by the calculation for the row of sintering spheres. In addition, differential densification and desintering were found to be more important than asymmetrical neck growth in producing heterogeneous microstructure. The final result of this study is the suggestion that a *narrow but not mono-sized* particle size distribution provides the most homogeneous powder compact, the highest final density, and the lowest amount of sintering damage in the final sintered body.

As a result of the improvements in processing, primarily from the control of particle size and use of colloidal techniques, model uniform microstructures have now been fabricated in Al_2O_3 [16-19] ZrO_2 [20], Y_2O_3 [21,22], TiO_2[23,24], Si_3N_4 [25], and mullite [26,27]. In 1981, Rhodes' study of agglomerate effects on ZrO_2 sintering demonstrated that colloidal processing can be used to produce uniform, dense microstructures at low temperatures that show a resistance to discontinuous grain growth. A recent example is the use of colloidal processing of Al_2O_3 with a fine particle size (0.16 μm) and an extremely narrow particle size distribution to sinter Al_2O_3 to 99.5% of theoretical density at 1150°C [16,17].

The use of colloidal processing techniques for easily-sinterable materials, such as Al_2O_3 and TiO_2, creates microstructures which approach the simple geometrical models for sintering. With correct processing, microstructural parameters are representative of the microstructure at the scale of a few grains as well as the compact as a whole. These experimental developments permit a re-evaluation of 3-D sintering models based on uniform microstructures described by repeating unit cells [18,28-30]. The unit cell typically consists of a single grain of a space-filling geometry with pores along the boundary between grains. With these unit cell models, microstructure evolution during sintering is followed with the following features: competition between coarsening and densification and scaling laws for the sintering of a bimodal pore size distribution. The microstructural parameters needed experimentally to compare with the models are easily measurable parameters: density, average pore size, pore size distribution, grain size, grain size distribution, surface area, and so on.

There are many combinations of microstructural features that can be used to follow the competition between densification and coarsening, for example, pore size-grain size or surface area-boundary area [30,32-34]. The most reliably and easily measured are grain size and density. From plots of density versus grain size, changes in the slope of trajectories are due to changes in the ratio of coarsening rate to densification rate. These plots are of special utility because transitions between processes can be seen from changes in the slope for a given run and changes in microstructure with processing can be determined by comparing grain size at a different density [18,35]. For example, studies by Edelson and Glaeser and by Barringer showed that monosized TiO_2 follows the same density-grain size trajectory for the temperature range 1000° to 1160°C for a variety of processing conditions [23,24].

Systems which coarsen more than they densify have also been examined: TiO_2 in HCl [36], ZrO_2 in HCl [37], ZnO in H_2 [38], Fe_2O_3 [39]. Because each of these systems can be sintered close to theoretical density in air, the ratio of the coarsening rate to the densification rate can be varied by changes in atmosphere [40,41]. Studies of coarsening are important to our understanding of the generation of crack-like flaws by sintering. Specifically, coarsening processes maintain particle coordination without the generation of tensile stresses at particle contacts. When densification occurs, particles with lower coordination numbers will be under a tensile stress and these sintered necks will dedensify. If the necks sizes are increased by coarsening, dedensification and the generation of crack-like pores can be suppressed. These concepts have been demonstrated in studies of ZnO powders [42] and glass, alumina, and ZnO powders sintered to rigid substrates in which controlled amounts of coarsening at low temperature prior to high temperature densification led to denser final microstructures than with high temperature densification alone [43]. Grain growth during the intermediate stage of sintering has also been shown to increase homogeneity in pore size in a powder compact containing large pores by coalescence of smaller pores, while the larger pores grow less [44].

An additional refinement in sintering theory has been a consideration of how the free surface-grain boundary dihedral angle affects: the driving force for sintering [5,6,45], the breakup of cylindrical channels in the transition of intermediate to final stage sintering [46], the attachment of pores to grain boundaries during grain growth [47], and the transition to abnormal grain growth [48]. Using Kingery and Francois' model relating pore curvature, dihedral angle, and number of grains surrounding a pore [49], Lange suggested that grain growth may be necessary to lower the pore coordination for pore shrinkage to be possible thermodynamically. While that concept is valid, calculations by Zhao and Harmer [29] and by Evans and Hsueh [50] indicate that the sintering kinetics for large pores are so slow that even when a large pore can shrink thermodynamically, grain growth is ineffective in promoting densification. Other dihedral angle effects on sintering are discussed in more detail elsewhere [51]. These new models incorporating dihedral angle effects are especially timely since measurements of the free surface-grain boundary dihedral angle demonstrate that the average dihedral angle in undoped Al_2O_3, MgO-doped Al_2O_3, and MgO polycrystals is in the range 106° - 117°, not 150° as previously measured [52-53] Measurements of dihedral angle by Ikegami et al. show the same trends but the average values are slightly higher [54].

Our old workhorse, Al_2O_3, has been used to examine the effects of pore size, dopants, and liquid phases on pore-boundary separation and the orientation dependence of grain growth. Model experiments of a single crystal sapphire growing into a fine-grained matrix have been performed using sapphire crystals with c-axis, a-axis, and 60° off c-axis cuts in contact with dense polycrystalline samples [55-59]. The matrices examined are: with and without arrays of pores at the single crystal-matrix interface, with [55,59] and without [56-58] a silicate liquid, and with and without MgO. An important feature of some of these studies is the controlled etching of pore sizes and shapes in the sapphire surface using photolithographic techniques in order to measure accurately the effect of pore size and pore spacing on the conditions for pore-boundary separation [56-58]. The results include:

1. MgO doping to Al_2O_3 without a liquid appeared to increase D_s, thereby increasing pore mobility.

2. The pore spacing is a critical variable for breakaway with pores in undoped Al_2O_3 remaining attached to the growing sapphire (c-axis cut) for pore spacings less than 6 μm and for times up to 20 hrs.

3. In MgO-doped Al_2O_3, pore arrays moved with the growing interface at a faster velocity but pore-boundary separation occurred at an annealing time of 10 hrs, independent of pore spacing.

4. Without a liquid, c-axis sapphire grew into an MgO-matrix faster than into undoped Al_2O_3.

5. In the presence of a silicate liquid, c-axis sapphire grew more slowly than a-axis sapphire, with the difference in velocity between c-axis and a-axis increasing as the Ca to Si ratio increased. The authors suggested that this difference in velocity results from the process being interface controlled.

The latter notion of interface control conforms to Burke's suggestion that, since MgO is not found at grain boundaries in Al_2O_3 (but is so effective at suppressing breakaway grain growth) that the model for crystal growth on well-developed facets with steps and ledges and kinetics should be adopted for Al_2O_3 [60]. Thus, a very small amount of MgO at the steps/kinks could "poison" growth at special growth sites. However, the observation that MgO-additions increase the boundary mobility of c-axis sapphire without liquid or pores, summarized above, is in direct conflict with this notion. More work is required to address some of these discrepancies.

The effect of second phase particle pinning of grain boundaries during grain growth has been studied through experiments on Al_2O_3 with FeO particles [61], ZrO_2 with Al_2O_3 particles [62], and Y_2O_3 doped with La [63]. FeO and La-yttria phases appear to inhibit grain growth while Al_2O_3 particles in ZrO_2 have little effect on grain growth. A new evaluation of the Zener criteria for boundaries pinned by pores for various microstructures and theoretical formulations establishes the ranges over which different pinning equations are applicable [64].

It has become widely accepted that many ceramics earlier thought to be single phase bodies in fact contain liquids at the sintering temperature due to impurities in the powder or introduced during powder processing and sintering [65-71]. These liquids degrade the creep resistance at high temperatures and the uncontrolled amounts of liquid lead to wide variations in measured properties. For example, abnormal grain growth in undoped Al_2O_3[68] and Fe_2O_3 [65] was induced by unintentional silicate-based liquid phases. This has led to increased attention to the initial chemical powder composition and to control of contamination by clean-room processing [72-77]. In addition, organic inclusions introduced during processing may leave no chemical residue but may cause damage to the matrix during sintering or themselves comprise flaws after sintering that seriously degrade the mechanical properties at low temperatures.

Other recent theoretical studies have examined the kinetics and thermodynamics of particle rotation during sintering [78] and the contiguity of sintered structures [79]. Of particular interest from an engineering point of view are two studies on the non-destructive characterization of the progress of sintering using elastic property measurements [80] and SANS [81].

COMPOSITES

There is a natural link between sintering of single phase powders and composite sintering: any variability in the shape, particle size distribution, and degree of agglomeration in a powder may lead to inhomogeneous sintering resulting from heterogeneity stresses. Theoretical analyses have determined that inclusions with different densities than the matrix will retard the densification of the matrix by the creation of a hydrostatic tensile stress in the matrix of the order of 0.5 to 2 MPa [15,82-89]. For particulate composites, a rough estimate of the magnitude of the tensile stress is the volume fraction of the particles multiplied by the sintering stress, Σ. The magnitude of the heterogeneity stress was verified experimentally by Ostertag from the bending during sintering of Al_2O_3 compacts reinforced in an asymmetrical pattern with SiC fibers [90,91]. As determined both theoretically and experimentally, increases in the fraction of heterogeneities in the matrix or in the difference in initial density between heterogeneity and matrix lead to a increasing suppression of matrix sintering rate and lowering of the final density.

The number of different composite systems and experimental variables examined is impressive. Among the many studies of systems where the matrix and the dense inclusions are the same material are: dense MgO cylinders in MgO powder [92]; alumina agglomerates in alumina [93]; yttria agglomerates of various strengths in yttria powder [94]. Sintering of dissimilar materials has been examined in systems as diverse as SiC fibers, whiskers, or particulate in Al_2O_3 [95], Al_2O_3 agglomerates in TiO_2 [96], TiC-reinforced Al_2O_3 [97], BN-reinforced SiC [98], mullite produced by sol-gel processing reinforced with SiC whiskers [99-102], ZrO_2 particles in Al_2O_3 [103], B_4C-TiB_2 composites [104], and SiC-reinforced glass [105].

Most theoretical work has focused on identifying stress generation in the matrix resulting from the difference in densification between the matrix and the inclusion/agglomerate. The stress field generated by differential sintering around a spherical dense inclusion is composed of a tensile hoop stress and a radial compressive stress, with a tensile mean stress [82]. The magnitude of the stress depends on the relative rates of creep and matrix densification, and microstructural variables, such as the volume fraction and size of the inclusions. The creep rate is typically described in terms of a matrix viscosity which changes during densification [106- 113]. If the creep rate and densification rates are controlled by different mechanisms with different activation energies, the effect of dense inclusions may be changed by changing temperature. On the other hand, if same mechanism controls both creep and densification, then changes in temperature will produce no benefit.

In experiments on viscous matrices with dense inclusions, there appear to be three regimes of behavior as a function of inclusion volume fraction. At volume fractions ≤ 0.1, the densification rates of the composites can be described by a simple rule-of-mixtures. Only over a narrow range of inclusion volume fractions (between 0.1 and 0.12) are the results in agreement with the Scherer's theory for viscous sintering with rigid inclusions. However, for higher volume fractions, the densification rate deviates significantly from either of these models. In experiments on crystalline matrices with dense inclusions, the measured densification rates are significantly lower than predicted by theory over all ranges of inclusion volume fraction [86].

The explanation for this discrepancy between theory and experiment remains controversial. Scherer has argued that the stresses are of the same order as the sintering stress and cannot, therefore, explain the magnitude of the effect in polycrystals containing many heterogeneities/inclusions [111]. In addition, application of a hydrostatic pressure only slightly greater than the tensile hydrostatic stress is predicted to eliminate the effect of the inhomogeneities. However, retardation of matrix densification is observed for small volume fractions and for particles as small as the matrix particle size [86]. Bordia and Raj found that the application of a low quasi-hydrostatic pressure <10MPa would suppress damage formation in the matrix but does not restore the matrix densification rate to the rate without heterogeneities [114].

One plausible argument is that heterogeneities are imperfectly distributed so that heterogeneities form interconnected structures over short distances in the matrix. The percolation limit for a given particle shape and size distribution gives an estimate of the particle volume fraction when particles form a completely interconnected structure that cannot densify. For single-sized, spherical particles, the percolation limit is at a volume fraction of 0.16. At volume fractions smaller than the percolation limit, inhomogeneities will sinter together and, thereby, limit the density of matrix in the vicinity of the sintered heterogeneities. An additional explanation is that the powder packing density near a heterogeneity may be lower than in the matrix away from the inclusion and will, therefore, lead to a lower limiting density [45,84]. The magnitudes of these two effects must be calculated to assess their contributions to the observed suppression of matrix densification rate.

An alternate explanation proposed by Mataga is that the sintering in heterogeneous powder is not properly described by a linear, isotropic sintering model [115]. Mataga suggested that a non-linear material model may be required as a result of non-linear stress dependence on sintering rate, a creep threshold, or non-linear creep. He noted, however, that significant deviations from linear behavior are necessary to explain the experimental results. (Additional factors may be a change in the diffusion path due to the sintering stress [111] or a change in the coarsening/sintering trajectory in the presence of dense inclusions.)

In light of these theoretical and experimental results, Rahaman and Jeng have suggested practical guidelines for minimizing damage formation in composite systems [102]:

1. The packing in the matrix should be maximized.

2. The inclusions/agglomerates should be distributed *uniformly* not *randomly*, in the matrix to minimize inclusion/agglomerate interactions.

3. The optimum sinterability conditions for the unreinforced matrix should be used for sintering the reinforced composite.

4. If the powder is amorphous, the composite should be densified before crystallization occurs.

Of these four guidelines, (3) is the most provocative. What really is the proper sintering schedule for a composite as compared with the matrix without the reinforcing material? What is the effect of coarsening - to strengthen the necks before densification or to narrow the particle size distribution before densification and grain growth begin ? For materials that sinter easily under a variety of conditions, such as the classic model material Al_2O_3, the sintering heat treatment for composites may be very different from that which produces the highest final density of the matrix alone.

An alternative approach to preventing damage formation in composites is to allow the matrix to sinter to almost full density before it comes into contact with the fiber by creating a void space of controlled thickness around the inclusion. This void space can be produced by coating the dense inclusion with an organic which burns off before sintering. This approach also has the advantage that the bonding between the fiber and the matrix can be controlled by the initial gap thickness [106].

It should be noted that in the previous section, the systems chosen to examine stress generation in the matrix exhibit little or no intermediate phase formation between the matrix and the reinforcing materials. In many engineering situations the only possible materials for a specific application may react and it is the ceramic engineer's task to minimize the extent of reaction by careful, innovative processing.

An additional factor influencing the sinterability and final properties of composites is thermal expansion anisotropy when the same material is used for matrix and reinforcement and thermal expansion mismatch when the phases of the matrix and reinforcement are different. During heating of composites, neck formation begins at the same time that stresses develop. The stress generation during heating will then depend on the ratio of the creep rate required to relieve thermal expansion mismatch stresses and heterogeneity stresses to the densification rate.

LIQUID PHASE SINTERING

The progress in theory and experiment for solid state sintering summarized above has been matched by similar accomplishments in the field of liquid phase sintering. Before citing specific research topics, we would like to acknowledge the completion of a major work in liquid phase sintering: the book *Liquid Phase Sintering* by Randall M. German [116]. This textbook is an excellent general reference which is, as Prof. German states, truly materials-independent and appropriate for engineers and scientists with diverse technical backgrounds. We congratulate Prof. German on his accomplishment and recognize his significant contribution to the field.

The past five years have brought a better understanding of the processes occurring in the neck region where two solid particles are in contact with liquid, in assemblages of particles filled with liquid, and in the transient state when the liquid first becomes molten. The calculation of the equilibrium configuration of particles and liquid and the forces generated by a liquid meniscus was demonstrated by Heady and Cahn for liquid phase sintering of spherical and jagged particles [117,118]. The calculation is based on an energy minimization subject to certain conditions, such as volume conservation. Recent calculations by Park, Cho, and Yoon have used this approach to model the filling of isolated pores by liquid during liquid phase sintering [119]. Grains were assumed to maintain their equilibrium shape determined by the balance between the tendency of grains to become spherical ("sphering force") and the negative capillary pressure due to the liquid menisci. The filling of pores and contact flattening were found to depend on liquid volume fraction and grain size, thereby leading to greater pore filling as the grain size increases. This gradual pore filling as grain growth/coarsening proceeds explains the experimentally observed agglomerate formation during the early stages of liquid phase sintering. The thermodynamic criteria for sequential pore filling in different 2-D particle geometries were also examined by Shaw [120].

From Park, Cho, and Yoon, an additional result of special importance is that particles completely surrounded by liquid but having a vacuum-filled pore in the center of the liquid is an unstable equilibrium. An infinitesimal perturbation leads to filling of the spherical void. In contrast, if the pore is filled with an insoluble gas, the pore will reach a stable equilibrium size. This is contrary to the liquid phase sintering model previously derived by Kingery [121]. The observation of pores completely surrounded by liquid means that an entrapped gas is limiting the final density, and that removal of the gas by alternate processing techniques will allow the structure to achieve a higher density.

A major advancement in liquid phase sintering, as well as in high temperature materials properties, is the recognition that surface forces exist between two solid surfaces forming the neck region between two sintering particles. The surface forces have many sources, for example, electrostatic interactions, van der Waals attraction, or structural (steric) forces, and can lead to an equilibrium thickness of liquid separating the two solids. The measurement of surface forces between ceramic surfaces at room temperature is an active research area, made possible by the development of a technique for measurement of surface forces. For liquid phase sintering, the nature of these interactions will affect liquid redistribution as the volume fraction and/or grain size of the particle change and will, in many cases, determine the mechanical properties at high temperature.

Clarke has examined the equilibrium thickness of a thin intergranular where structure within the liquid determines the surface forces [122]. In a complementary experimental study by Greil and Weiss, the equilibrium thickness of thin liquid silicate layers in β-SiAlON was found to be constant for increases in the liquid volume fraction from 2% to 10% [123]. In addition, structure in a thin intergranular liquid phase was detected in an anorthite-based liquid in Al_2O_3 [124].

In most of the theoretical treatments described above, equilibrium configurations are the endpoints of the calculations. In liquid phase sintering experiments, the creation of equilibrium configurations is preceded by solid state sintering, disintegration of the sintered skeleton as melting occurs, mixed fluid and particle flow, particle dilatation, rearrangement, and finally to disintegration of the individual particles [125]. Grains will grow by grain growth/Ostwald ripening and coalescence processes. Grain shape may also change due to facetting transitions and/or growth rate anisotropy in the presence of a liquid phase. The identification of each stage is important in relating changes in processing variables to changes in microstructure. Selected examples of these processes are presented below.

As with solid state sintering, the magnitude of the sintering stress that provides the driving force for liquid phase sintering is an important parameter in determining sinterability. The magnitude of the sintering stress has been calculated by De Jonghe and coworkers for liquid phase systems, as well as for solid-state systems, from the ratio of the densification rate at no load to the creep rate with an applied uniaxial load. In the MgO-Bi_2O_3 system, De Jonghe and Srikanth found that the sintering stress was a constant throughout densification [126]. The calculated stress was low - 0.08 MPa - corresponding to a meniscus radius of curvature of about 6 μm, a value much larger than the MgO particle size. The low value of the sintering stress was attributed to formation of agglomerates by the sequential pore filling as suggested by Shaw and by Park, Cho, and Yoon.

In the final stages of sintering, grain growth/coarsening/Ostwald ripening, discontinuous grain growth, and densification occur simultaneously. Three recent studies illustrate the effect of the liquid phase on discontinuous grain growth and pore coalescence. In a study of grain growth in TiO_2-rich $BaTiO_3$, Hennings, Janssen, and Reynen found that the grain size of abnormally growing grains decreases with increasing additions of TiO_2-enriched seed grains[127]. Also, the grain size increases with increasing TiO_2 content to maximum at 2 mol% excess TiO_2. As noted above, studies of sintering and grain growth in undoped Al_2O_3 and Fe_2O_3 revealed that discontinuous grain growth can originate from regions containing a calcium-alumino-silicate second phase [65,68]. These results suggest that discontinuous grain growth is accelerated by a non-uniform distribution of liquid phase. A study of normal grain growth and pore coalescence in the MgO-$CaMgSiO_4$ system demonstrated that the coalescence of gas-filled pores is controlled by the rate of grain growth/Ostwald ripening of the solid grains [128].

An important feature of many liquid phase sintering systems is the facetting and shape change of the solid particles when they come into contact with the liquid. For example, pronounced facetting has been observed in anorthite- Al_2O_3 [59, 129, 130] and in β'-sialons [131]. The same growth phenomena identified for crystal growth in systems with facetted liquid-solid interfaces are expected to operate in these liquid phase sintering systems [132]. Among these phenomena are: (1) growth on a facet can be limited by the nucleation of a stable ledge, also known as "2-D nucleation limited growth"; (2) growth and dissolution shapes are not expected to be the same; (3) screw dislocations and special twin boundaries can serve as nucleation sites for growth in slow growing directions.

EFFECTS OF APPLIED PRESSURE: HOT PRESSING, HIP'ING, AND SINTER-FORGING

The industrial use of applied pressure to densify a wide variety of difficult-to-sinter materials has increased over the last several years. This has coincided with a increase in modelling of the effects of applied pressure on sintering [133-140]. These models fall into two different types. There are models based on the visco-elastic response of the porous compact to the applied stress. The constitutive equations for the entire body over the whole range of deformation are specified and then the response is numerically calculated [133-137]. An alternative approach is to divide the densification up into different regimes, as shown in hot pressing or hot isostatic pressing (HIP'ing) maps [138-140]. Within a regime a single mechanism will dominate, and the dividing lines between regimes are where the rates of densification are equal. The conditions required to obtain a desired microstructure can be estimated from trajectories on maps. The results of both techniques are consistent and provide insight into the densification process and the observed shape changes during hot isostatic pressing (HIP'ing). For example, a densification front moves inward during HIP'ing as densification proceeds. The main reason for the development of this densification front is a nonuniform temperature distribution during heating. The hotter material near the outer edge will densify faster. The heat conduction through the dense outer region can further change the temperature distribution. Also the dense outer region can support some of the load and thus reduce the pressure on the porous interior. These effects can lead to very anisotropic shape changes during HIP'ing. To study the densification in a systematic manner, model experiments have been preformed by Kaysser and co- workers [125]. The deformation of a single, polycrystalline particle under an applied uniaxial load was studied. In this technique they were able to measure changes in the geometry during densification.

The effect of nonuniform stresses during sintering is not clear. The addition of a shear stress, or uniaxial stress to a compact during sintering has yielded results which suggest that the creep rate may be control the rate of densification. The applied stress increases the creep rate significantly, so that it is no longer rate-controlling [141,142]. This effect is of importance only when creep is the rate limiting process which, for ceramics, occurs in a very limited number of systems. A more pronounced effect of shear stresses is an enhancement of the rearrangement during sintering. Large processing defects can be broken down during shear deformation, but would be stable during cold isostatic pressing [143]. Such large defects can also be removed during HIP'ing if the material is in a regime where extensive plastic deformation can occur, either by diffusion of dislocation motion [144-146].

A wide variety of materials which are difficult to densify at atmospheric pressure have been successfully sintered with the application of moderate to high pressures [147-150]. HIP'ing has been used to great advantage for removing defects in sintered metallic materials. It has been noted however, that near surface defects may increase in severity during post- sintering HIP'ing [151]. The development of HIP maps, and the underlying data base necessary for their use, has allowed the conditions to achieve high density to be estimated in advance of the experiment [152,153].

While experiments have shown the tremendous benefits of applied pressure on sintering, there are many limitations. The most severe is the bloating, or desintering problem associated with many HIP'ed or hot pressed materials. The bloating is due to two main effects. Oxidation of impurities which were incorporated during pressing produces a internal gas pressure which causes pore growth. This can be avoided by restricting use to inert environments, but this limits the usefulness of the materials. Another source of bloating is due to pores which remain in the material after pressing. These pores are due either to trapped residual gas or to thermodynamic stability of the pores as a result of the surface energy to boundary energy ratio [52,121]. While the applied pressure will reduce the size of such defects as compared to pressureless sintering, when the pressure is removed the defects are not in thermodynamic equilibrium and at elevated temperatures, they will grow until they are in equilibrium. This again limits the use temperature of a HIP'ed material. This effect implies that the minimum pressure possible should be used, and that the major benefit due to pressure will be an increase in the densification of large defects.

CASE STUDIES

Zirconia

Zirconia, one of the toughest ceramic materials, has mechanical properties which are dependent both on microstructure and chemistry. Additives are required to stabilize the tetragonal phase and give high toughness. Also, the grain size must be kept small to retain the metastable tetragonal phase at room temperature. For a detailed review of recent work in zirconia see Somiya et.al.[154]. Zirconia is a material which sinters well. Densities above 99% are achieved under a wide range of processing conditions and dopant concentrations. [1] Two important factors in determining the time and temperature required for densification are the starting powder particle size and the degree of powder agglomeration, as summarized below.

Ultrafine zirconia powders (10-20nm) forming 1-10 μm agglomerates can be produced by chemical precipitation [155,156] and these powders sinter to full density at 1400°C in 2 hrs. At the initial stage of sintering, density decreases with time [156] resulting from differential densification in the agglomerated powder. Density increases when grain growth begins. This coarsening at the initial stage of sintering is not disastrous in this system due to the extremely small grain size, even after grain growth. Rhodes [20] demonstrated that deagglomeration of ZrO_2 can reduce the time and temperature required to sinter a commercial power from 4 hours at 1500°C to 1 hour at 1100°C. An additional factor limiting final density is that gas evolution from residual organics used in washing the powders can cause bloating/desintering at higher

---

[1] One difficulty in interpreting the density data is the fact that the changes in the fraction of the different phases lead to changes in density without any change in porosity.

temperatures. Powders produced by electro-refining and grinding [157] have similar ultimate densities and properties as compared with chemically produced powders, but with somewhat reduced sintering rates. This allows a greater degree of control over the dimensional stability of the material.

Additions of Ce, Y, Ca, Mg, or Ti to stabilize the tetragonal phase do not reduce the sinterability of the powders [158-163] and, if Si is present as an impurity, may increase the densification by the formation of a silicate- based liquid phase. Also, the addition of 40% ZrO_2 to ZrC allowed the densification of the composite at 2000°C [164] if the green density was >55%.

Sinter forging of zirconia powders [165] determined that densification can occur by plastic flow at pressures and temperatures of 1400°C. Through hot forging of dense materials [146] it was found that fine- grained zirconia could be deformed but large grained zirconia could not. Diffusion control was thought to be limiting, probably due to the limited number of independent active slip systems. Thus for plasticity, diffusional accommodation is necessary and is only possible in fine-grained materials at the strain rates used. Sintering followed by HIP'ing has been used to remove large pores (30-40μm) and achieve full density at 1200°C [144], most likely due to plastic flow in addition to increased driving force for densification. It is necessary to achieve high density (>96%) before HIP'ing to prevent open pores from limiting the final density.

Dielectric ceramics

For production of dielectric ceramics by sintering, the critical issue is control of the microstructure. While many additives have been introduced to improve the electrical properties, recent work has focused on additives added to improve the sinterability and retard grain growth. Typically the additives promote the formation of a liquid phase and thus allow sintering to occur at low temperature [166-169], but may also work by reducing the vapor transport [170]. In the Pb-Mg-Nb-O system, avoidance of the pyrochlore phase is important, and the presence of liquid phases [171-173] controls not only the microstructure but also the phase content of the samples. With the formation of a liquid phase, the heating schedule becomes important: fast heating rates may lead to an inhomogeneous distribution of liquid phase [174].

Many dielectric materials have high vapor pressures at the sintering temperatures. This leads to weight loss and compositional changes during sintering. Some additives reduce the volatilization [175], but the problem is usually solved by adding an excess of the volatile components. Fast firing can reduce the problem as the material spends less time at temperature, and, therefore, has a finer grain size [176] than conventionally sintered PZT materials.

Different sintering atmospheres may also change the sintering behavior [177]. This can be for a variety of reasons. Often the coarsening rates can be reduced in specific atmospheres. In certain cases the phase relations may change as the atmosphere changes. Another possibility is that the defect concentrations change with atmosphere and thus the diffusion coefficients also change. All these effects are very system specific

[236]. Fe formed a liquid phase which aided densification and excess TiB_2 inhibited grain growth. In the Al_2O_3-TiC system, the reaction produces CO_2 which reduces densification [97]. However, rapid heating minimized this effect by reducing the time spent in a temperature range where the reaction occurred, and dense materials were obtained. In reactive sintering of ZrC-ZrO_2 composites, the amount of excess ZrO_2 was found to be important in determining the density of the sintered material [237]. In the case of Cr_2O_3 with TiO_2 additions, the densification depended on the atmosphere, with little densification in air and increasing as Po_2 decreased [238]. It was felt that this was due to a change in the defect chemistry rather then suppression of Cr volatilization. The effect of H_2O on the sintering of MgO is not well understood [239]. Closed pores were observed to shrink while the open pores coarsened, indicating that H_2O may activate surface or vapor transport. In the $MgAl_2O_4$-SiO_2 system, the sintering mechanism changed with temperature [240]. At low temperatures, viscous flow of SiO_2 dominated until the formation of crystobalite and then diffusion controlled the densification. At higher temperatures the formation of cordierite caused an expansion of the sample, and the grains were spinel cores surrounded by cordierite. Dense Al_2TiO_5 can be formed by reactive sintering of Al_2O_3 and TiO_2 [96,241], but careful control of the sintering temperature was required.

Very porous mullite gels (44% dense) can be sintered to 97% density but the resulting structure consists of elongated grains of $3Al_2O_3$- $2SiO2$ and fine grained Al_2O_3-$2SiO_2$ [101]. The structure of the gel may have an effect on the final microstructure produced [242]. Seeding of the gels to provide nucleation sites for the reaction proved effective in getting dense materials with a uniform grain size [243,244]. Seeding has also been effectively used for Al_2O_3 [244] and Al_2O_3-ZrO_2 composites [99].

SUMMARY

Significant advances in modelling of solid-state and liquid-phase sintering and grain growth have been made. The major driving force behind these advances has been the realization that fundamental thermodynamic concepts could be extended beyond what had been done for the two-sphere model. In particular, models of pore filling in multiparticle groups for liquid phase sintering and of the stability of a row of sintering spheres have provided new insight into the processes governing sintering. Improved powder processing has lead to the creation of closer-to-ideal microstructures with which to test theoretical predictions of models that assume uniform microstructures.

A major shift in research has been made to sintering in the presence of intentional and unintentional non-densifying inclusions. Modelling and experiment have examined the effects on the densification of a finer grained matrix by large particles (or agglomerates) of the matrix material and of dense particles, whiskers, or fibers of a different phase intended for reinforcement. Even for small volume fractions of inclusions, the matrix sinterability is seriously impeded. Strength-controlling defects can form in the matrix as a result of sintering in the presence of dense inclusions. Because of the severity of matrix damage during sintering, the analysis of damage formation during sintering and, in particular, the factors which inhibit crack formation are topics of wide interest. So far the theoretical estimates of the stresses produced by these heterogeneities do not seem to agree with experiment. However, there is a general recognition that application of constitutive laws is required for properly describing constrained sintering in single- and two-phase bodies.

For many composite systems, hot pressing may be required to produce high density materials. Unfortunately, applied pressure is not a panacea for composite systems: damage to high aspect ratio whiskers or fibers during pressure-sintering may degrade the mechanical properties of the composite and the applied pressure does not restore the sinterability to the level without inclusions. However, hot pressing and HIP'ing are viable techniques for producing ceramic parts. For example, a commercial cutting tool has been manufactured by hot pressing SiC whiskers in an Al_2O_3 matrix.

Of the recent research on hot pressing and HIP'ing, the generation of hot pressing and HIP'ing mechanism maps has the greatest impact on the use of pressure-sintering techniques in manufacturing. Although an extensive data base is required for the generating these maps, the data base is smaller than for the corresponding sintering maps/diagrams, since the number of mechanisms which dominate tends to be small when pressures are applied. These maps can also be used on-line in research and development to follow the densification process and to determine changes in processing conditions in real time.

When pressureless sintering, hot-pressing, and HIP'ing are not effective, or a special system presents unique opportunities, different techniques for processing should be used. The use of reactive sintering, shock activation, microwave sintering, self-combustion sintering and plasma sintering is of advantage only in the systems which allow the processing to exploit the special characteristics of these methods.

The case studies presented demonstrate the use of the fundamentals outlined in the beginning of the paper. Knowledge of the relationships between the observed microstructure and the processes which give rise to such microstructures has allowed the processing to be manipulated to achieve a specific microstructure. Without this understanding of the fundamentals of sintering, improvement of the properties of a specific material must occur by trial and error and in an incremental manner. Sintering has now progressed to the point where material development can be done more systematically to optimize properties.

ACKNOWLEDGEMENTS

The financial support of the U. S. Air Force Office of Scientific Research (Program Manager: Dr. Liselotte Schioler), the Max Planck Society, and the International Institute for the Science of Sintering are gratefully acknowledged.

REFERENCES

1. S. J. Bennison and B. R. Lawn, "Role of interfacial grain-bridging sliding friction in the crack-resistance and strength properties of non- transforming ceramics," submitted to Acta Metall.

2. H. K. Bowen, "Basic Research Needs on High-Temperature Ceramics for Energy Applications," Mat. Sci. Eng. 44 1-56 (1980).

3. H. E. Exner, "Neck Shape and Limiting GBD/SD Ratios in Solid State Sintering," Acta Metall. 35, 587-591 (1987).

4. W. S. Coblenz, J. M. Dynys, R. M. Cannon, and R. L. Coble, "Initial Stage Sintering Models: A Critical Assessment," in: *Sintering Processes*, ed. G. C. Kuczynski, Plenum Press, 1980.

5. W. C. Carter, R. M. Cannon ,"Sintering Microstructures: Instabilities and the Interdependence of Mass Transport Mechanisms," in: *Ceramic Transactions*, Vol. 7, Eds. C. A. Handwerker, J. E. Blendell, and W. A. Kaysser, American Ceramic Society, Westerville, Ohio, 1989, in press.

6. R. M. Cannon, W. C. Carter, "Interplay of Sintering Microstructures, Driving Forces, and Mass Transport Mechanisms," J. Amer. Ceram. Soc. 72 [8] 1550-1555 (1989).

7. K. R. Venkatachari, R. Raj, "Shear Deformation and Densification of Powder Compacts," 69 [6] 499-506 (1986).

8. M. N. Rahaman, L. C. De Jonghe, "Effect of Shear Stress on Sintering," J. Amer. Ceram. Soc. 69 [1] 53-58 (1986).

9. R. K. Bordia, *Sintering of Inhomogeneous or Constrained Powder Compacts: Modelling and Experiments*, Ph.D Thesis, Cornell University, Ithaca, NY, 1986.

10. T. Cheng, R. Raj, "Measurement of the Sintering Pressure in Ceramic Films," J. Amer. Ceram. Soc. 71, [4] 276-80 (1988).

11. L. C. DeJonghe, M. N. Rahaman, "Sintering stress of homogeneous and heterogeneous powder compacts," Acta Metall. 36 223-229 (1988).

12. E. G. Liniger, R. Raj, "Packing and Sintering of Two-Dimensional Structures Made from Bimodal Particle Size Distributions," J. Amer. Ceram. Soc. 70 [11] 843-849 (1987).

13. E. G. Liniger, "Spatial Variations in the Sintering Rate of Ordered and Disordered Particle Structures," J. Amer. Ceram. Soc. 71 [9] C-408- C-410 (1988).

14. M. W. Weiser, L. C. De Jonghe, " Rearrangement During Sintering in Two-Dimensional Arrays," J. Amer. Ceram. Soc. 69 [11] 822-26 (1986).

15. R. Raj, R. K. Bordia, "Sintering behavor of bi-modal powder compacts," Acta Metall. 32 1003-1020 (1984).

16. T.-S. Yeh, M. D. Sacks, "Effect of Particle Size Distribution on the Sintering of Alumina," J. Amer. Ceram. Soc. 71 [12] C-484-C-487 (1988).

17. T.-S. Yeh, M. D. Sacks, "Effect of Green Microstructure on Sintering of Alumina," in: *Ceramic Transactions*, Vol. 7, Eds. C. A. Handwerker, J. E. Blendell, and W. A. Kaysser, American Ceramic Society, Westerville, Ohio, 1989, in press.

18. M.A. Occhionero, J.W. Halloran, "The Influence of Green Density Upon Sintering," in: *Sintering and Heterogeneous Catalysis*, edited by G.C. Kuczynski, Albert E. Miller,, Gordon A. Sargent, Plenum Press (New York), Materials Science Research, Vol. 16.

19. C. P. Cameron, R. Raj, "Grain-Growth Transition During Sintering of Colloidally Prepared Alumina Powder Compacts," J. Amer. Ceram. Soc. 71 [12] 1031-35 (1988).

20. W.H. Rhodes, "Agglomerate and Particle Size Effects on Sintering Yttria-Stabilized Zirconia," J. Amer. Ceram. Soc. 64 [1] 19-22 (1981).

21. D. J. Sordelet, M. Akinc, "Sintering of Monosized, Spherical Yttria Powders," J. Amer. Ceram. Soc. 71 [12] 1148-53 (1988).

22. D. J. Sordelet, M. Akinc, "Sintering of Monosized, Spherical Y_2O_3 Powders," in: *Ceramic Transactions*, Vol. 7, Eds. C. A. Handwerker, J. E. Blendell, and W. A. Kaysser, American Ceramic Society, Westerville, Ohio, 1989, in press.

23. E. A. Barringer, H. K. Bowen, "Formation, Packing, and Sintering of Monodisperse TiO_2 Powders," J. Amer. Ceram. Soc. 65 [12] C-199-C-201 (1982).

24. L. H. Edelson, A. M. Glaeser, "Role of Particle Substructure in the Sintering of Monosized Titania," J. Amer. Ceram. Soc. 71 [4] 225-35 (1988).

25. T. M. Shaw, B. A. Pethica, "Preparation and Sintering of Homogeneous Silicon Nitride Green Compacts," J. Amer. Ceram. Soc. 69 [2] 88-93 (1986).

26. Y. Hirato, I. A. Aksay; "Colloidal Consolidation and Sintering Behavior of CVD-Processed Mullite Powders," Mat. Sci. Res. Vol. 21 *Ceramic Microstructures '86: Role of Interfaces* 6ll-621 (l988).

27. N. Otsuka, "Sintering of Monodisperse Particles, Seramikkusu, 22 473-8 (1987).

28. J. Zhao, M. P. Harmer, "Effect of Pore Distribution on Microstructure Development: II, Matrix Pores," J. Amer. Ceram. Soc. 71 [2] 113-20 (1988).

29. J. Zhao, M. P. Harmer, "Effect of Pore Distribution on Microstructure Development: II, First- and Second-Generation Pores," J. Amer. Ceram. Soc. 71 [7] 530-39 (1988).

30. T. Ikegami, "Microstructural Development during Intermediate- and Final-Stage Sintering," Acta Meta.. 35 667-675 (1987).

31. A. D. Rollett, D. J. Srolovitz, and M. P. Anderson,"Simulation and Theory of Abnormal Grain Growth-Anisotropic Grain Boundary Mobilities," in: *Ceramic Transactions*, Vol. 7, Eds. C. A. Handwerker, J. E. Blendell, and W. A. Kaysser, American Ceramic Society, Westerville, Ohio, 1989, in press.

32. N. J. Shaw, R. J. Brook, "Structure and Grain Coarseing During the Sintering of Alumina," J. Amer. Ceram. Soc. 69 107-110 (1986).

33. T. Yamaguchi, H. Kosha, "Sintering of Acicular Fe_2O_3 Powder," J. Amer. Ceram. Soc. 64 C-84 - C-85 (1981).

34. H. Kuno, M. Tsuchiya, Cumulative pore volume difference (CPVD) in the study of pore changes during compaction or sintering, Powder Technology, 52 187-92 (1987).

35. K. A. Berry, M. P. Harmer, "Effect of MgO Solute on Microstructure Development in Al_2O_3," J. Amer. Ceram. Soc. 69 [2] 143-49 (1986).

36. M. J. Readey, D. W. Readey, "Sintering TiO_2 in HCl Atmospheres," J. Amer. Ceram. Soc. 70, [12] C-358-C-361 (1987).

37. M. J. Readey, D. W. Readey, "Sintering of ZrO_2 in HCl Atmospheres," J. Amer. Ceram. Soc. 69 [7] 580-82 (1986).

38. T. Quadir, D. W. Readey, "Microstructure Development of Zinc Oxide in Hydrogen," J. Amer. Ceram. Soc. 72 [2] 297-302 (1989).

39. J. Lee, D.W. Readey, "Microstructure Development of Fe_2O_3 in HCl Vapor," in:*Sintering and Heterogeneous Catalysis*, edited by G.C. Kuczynski, Albert E. Miller, and Gordon A. Sargent, Plenum Press (New York), Materials Science Research, Vol. 16.

40. D.W. Readey, J. Lee and T. Quadir, "Vapor Transport and Sintering of Ceramics," in: *Sintering and Heterogeneous Catalysis*, edited by G.C. Kuczynski, Albert E. Miller, and Gordon A. Sargent, Plenum Press (New York), Materials Science Research, Vol. 16.

41. D. W. Readey ,"Vapor Transport and Sintering," in: *Ceramic Transactions.* Vol. 7, Eds. C. A. Handwerker, J. E. Blendell, and W. A. Kaysser, American Ceramic Society, Westerville, Ohio, 1989, in press.

42. L. C. DeJonghe, M. N. Rahaman, M.-Y. Chu, and R. J. Brook, "Effect of Heating Rate on Sintering and Coarsening," submitted to J. Amer. Ceram. Soc.

43. T. J. Garino, H. K. Bowen, "Deposition and Sintering of Particle Films on a Rigid Substrate," J. Amer. Ceram. Soc. 70 C315-C317 (1987).

44. L. C. De Jonghe, M. N. Rahaman, and M. Lin, "The Role of Powder Packing in Sintering," in : Ceramic Microstructures '86, Role of Interfaces, Ed. J. A. Pask and A. G. Evans, Plenum Press, New York, 1987, 447-454.

45. W. C. Carter, A. M. Glaeser, "The Morphological Stability of Continuous Intergranular Phases: Thermodynamic Considerations," Acta Metall. 35 237-45 (1987).

46. F.F. Lange, "Sinterability of Agglomerated Powders," J.Amer.Ceram.Soc. 67 83-89 (1984).

47. C. H. Hsueh, A. G. Evans, and R. L. Coble, "Microstructure development during final/intermediate stage sintering - I. Pore/Grain Boundary Separation," Acta Metall. 30 1269-1279 (1982).

48. A. D. Rollett, D. J. Srolovitz, and M. P. Anderson, "Simulation and theory of abnormal grain growth - aniostropic grain boundary energies and mobilities," Acta Metall. 37 [4] 1227-1240 (1989).

49. W. D. Kingery, B. Francois, "The sintering of crystalline oxides, I. Interactions between grain boundaries and pores," in: *Sintering and Related Phenomena*, Ed. G. C. Kuczynski, N. Hooten, and C. Gibbon, Gordon and Breach, NY (1967) 471-499.

50. A. G. Evans, C. H. Hsueh, "Behavior of Large Pores During Sintering and Hot Isostatic Pressing," J. Amer. Ceram. Soc. 69 [6] 444-48 (1986).

51. J. E. Blendell, C. A. Handwerker, "Effect of Chemical Composition on Sintering of Ceramics," J. Crystal Growth 75 138-160 (1986).

52. C. A. Handwerker, J. M. Dynys, R. M. Cannon, and R. L. Coble, "Dihedral Angles in MgO and Al_2O_3: Distributions from Surface Thermal Grooves," accepted by J. Amer. Ceram Soc.

53. C. A. Handwerker, J. M. Dynys, R. M. Cannon, and R. L. Coble, "Metal Reference Line Technique for Obtaining Dihedral Angles from Surface Thermal Grooves," accepted by J. Amer. Ceram. Soc.

54. T. Ikegami, K. Kotani, "Some Roles of MgO and TiO_2 in Densification of a Sinterable Alumina," J. Amer. Ceram. Soc. 70 [12] 885-90 (1987).

55. Y. Finkelstein, S. M. Wiederhorn, B. J. Hockey, C. A. Handwerker, and J. E. Blendell ,"Migration of Sapphire Interfaces into Vitreous Bonded Aluminum Oxide", in: *Ceramic Transactions*, Vol. 7, Eds. C. A. Handwerker, J. E. Blendell, and W. A. Kaysser, American Ceramic Society, Westerville, Ohio, 1989, in press.

56. J. Rödel, A. M. Glaeser,"Morphological Evolution of Pore Channels in Alumina", in: *Ceramic Transactions*, Vol. 7, Eds. C. A. Handwerker, J. E. Blendell, and W. A. Kaysser, American Ceramic Society, Westerville, Ohio, 1989, in press.

57. J. Rödel, A. M. Glaeser,"Pore Drag in Alumina," in: Ceramic Transactions, Vol. 7, Eds. C. A. Handwerker, J. E. Blendell, and W. A. Kaysser, American Ceramic Society, Westerville, Ohio, 1989, in press.

58. J. W. Rödel, *Application of Controlled Interfacial Pore Structures to Pore Perturbation and Pore Drag in Alumina*, Ph.D thesis, University of California, Berkeley, 1988 (LBL Publication No. LBL-26211).

59. W. A. Kaysser, M. Sprissler, C. A. Handwerker, and J. E. Blendell, "Effect of a Liquid Phase on the Morphology of Grain Growth in Alumina," J. Amer. Ceram. Soc. 70 [5] 339-343 (1987).

60. J. E. Burke,"Control of Grain Boundary Mobility," in: *Ceramic Transactions.* Vol. 7, Eds. C. A. Handwerker, J. E. Blendell, and W. A. Kaysser, American Ceramic Society, Westerville, Ohio, 1989, in press.

61. J. Zhao, M. P. Harmer, "Sintering of Ultra-High-Purity Alumina Doped Simultaneously with MgO and FeO," J. Amer. Ceram. Soc. 70 [12] 860-66 (1987).

62. F. F. Lange, T. Yamaguchi, B. I. Davis, P. E. D. Morgan, "Effect of ZrO_2 Inclusions on the Sinterability of Al_2O_3," J. Amer. Ceram. Soc. 71 [6] 446-48 (1988).

63. W.H. Rhodes, "Controlled Transient Solid Second-Phase Sintering of Yttria," J. Amer.Cer. Soc. 64 [1] 13-19 (1981).

64. D. L. Olgaard, B. Evans, "Effect of Second-Phase Particles on Grain Growth in Calcite, J. Amer. Ceram. Soc. 69 [11] C-272-C-277 (1986).

65. P. K. Gallagher, D. W. Johnson, Jr., and F. Schrey, "Some Effects of the Source and Calcination of Iron Oxide on Its Sintering Behavior," Bull. Amer. Ceram. Soc. 55 [6] 589-593 (1976).

66. S. J. Bennison, M. P. Harmer, "Effect of MgO Solute on the Kinetics of Grain Growth in Al_2O_3," J. Amer. Ceram. Soc. 66 [5] C90-C92 (1983).

67. S. J. Bennison, M. P. Harmer, "Grain-Growth Kinetics for Alumina in the Absence of a Liquid Phase," J. Amer. Ceram. Soc. 68 [1] C22-C24 (1985).

68. C. A. Handwerker, P. A. Morris, R. L. Coble, "Effects of Chemical Inhomogeneities on Grain Growth and Microstructure in Al_2O_3," J. Amer. Ceram. Soc. 72 [1] 130-36 (1989).

69. M. H. Drofenik, "Grain Growth During Sintering of Donor-Doped $BaTiO_3$," J. Amer. Ceram. Soc. Vol. 69, 1986, [1] C-8-C-9.

70. D. F. K. Hennings, R. Janssen, P. J. L. Reynen, "Control of Liquid- Phase-Enhanced Discontinuous Grain Growth in Barium Titanate," J. Amer. Ceram. Soc. Vol. 70, 1987, [1] 23-27.

71. D. Kolar ,"Discontinuous Grain Growth in Multiphase Ceramics," in: Ceramic Transactions, Vol. 7, Eds. C. A. Handwerker, J. E. Blendell, and W. A. Kaysser, The American Ceramic Society, Westerville, Ohio, 1989, in press.

72. C. A. Handwerker, *Sintering and Grain Growth of MgO*, Sc.D Thesis, M.I.T., Cambridge MA, 1983.

73. P. A. Morris ,"Impurities in Ceramics: Processing and Effects on Properties," in: *Ceramic Transactions*, Vol. 7, Eds. C. A. Handwerker, J. E. Blendell, and W. A. Kaysser, American Ceramic Society, Westerville, Ohio, 1989, in press.

74. P. A. Morris, *High-Purity Al_2O_3: Processing and Grain Boundary Structures*, Ph.D Thesis, M.I.T., Cambridge MA, June 1986.

75. P. A. Morris, R. H. French, R. L. Coble, F. N. Tebbe, U. Chowdhry, "Clean-Room and CO_2-Laser Processing of Ultra High-Purity Al_2O_3," in *Defect Properties and Processing of High- Technology Nonmetallic Materials*, Ed. Y. Chen, 60, Materials Research Society, Pittsburgh, PA, 1986, p.79.

76. S. J. Bennison, *The Effect of MgO on the Sintering of High Purity Alumina*, Ph.D Thesis, Lehigh University, 1987.

77. J. E. Blendell, H. K. Bowen, and R. L. Coble, "High Purity Alumina by Controlled Precipitation from Aluminum Sulfate Solutions," Bull. Amer. Ceram. Soc. 63 797-804 (1984).

78. J. W. Cahn ,"Grain Rotation in Sintering: An Examination of Driving Force Arguments," in: *Ceramic Transactions*, Vol. 7, Eds. C. A. Handwerker, J. E. Blendell, and W. A. Kaysser, American Ceramic Society, Westerville, Ohio, 1989, in press.

79. S. Prochazka ,"Surface Area, Average Mean Curvature and Chemical Potential in Porous Bodies," in: *Ceramic Transactions*, Vol. 7, Eds. C. A. Handwerker, J. E. Blendell, and W. A. Kaysser, American Ceramic Society, Westerville, Ohio, 1989, in press.

80. D. J. Green, C. Nader, and R. Brezny, "The Elastic Behavior of Partially-Sintered Alumina," in: *Ceramic Transactions*, Vol. 7, Eds. C. A. Handwerker, J. E. Blendell. and W. A. Kaysser, American Ceramic Society, Westerville, Ohio, 1989, in press.,

81. K. G. Frase, K. Hardman-Rhyne, "Porosity in Spinel Compacts Using Small-Angle Neutron Scattering," J. Amer. Ceram. Soc. 71 [1] 1-6 (1988).

82. R. K. Bordia, R. Raj, "Sintering of TiO_2-Al_2O_3 Composites: A Model Experimental Investigation," J. Amer. Ceram. Soc. 71 302-310 (1988).

83. J. P. Smith, G. L. Messing, "Sintering of Bi-modally Distributed Alumina Powders," J. Amer. Ceram. Soc. 67 238-42 (1984).

84. B. Kellett, F. F. Lange, "Stresses Induced by Differential Sintering in Powder Compacts," J. Amer. Ceram. Soc. 67 369-71 (1984).

85. L. C. DeJonghe, M. N. Rahaman, and C. H. Hsueh, "Transient Stresses in Bimodal Compacts During Sintering," Acta Metall. 34 1467-71 (1986).

86. M. W. Weiser, L. C. De Jonghe, "Inclusion Size and Sintering of Composite Powders," J. Amer. Ceram. Soc. 71 C125-127 (1988).

87. R. K. Bordia, G. W. Scherer,"On Constrained Sintering - Parts I, II, and III," Acta Metall. 36 2393-2416 (1988).

88. L. C. DeJonghe, M. N. Rahaman, "Sintering stress of homogeneous and heterogeneous powder compacts," Acta Metall. 36 223-229 (1988).

89. M. N. Rahaman, L. C. De Jonghe, "Effect of Rigid Inclusions on Sintering," in: *Ceramic Transactions*, Vol. 1, Ed. G. L. Messing, E. R. Fuller, and H. Hausner, American Ceramic Society, Westerville, Ohio, 1988, 887-896.

90. C. P. Ostertag ,"Reduction in Sintering Damage of Fiber-Reinforced Composites," in: *Ceramic Transactions*, Vol. 7, Eds. C. A. Handwerker, J. E. Blendell, and W. A. Kaysser, American Ceramic Society, Westerville, Ohio, 1989, in press.

91. C. P. Ostertag, "Technique for Measuring Stresses Which Occur During Sintering of a Fiber-Reinforced Ceramic Composite," J. Amer. Ceram. Soc. 70 C355-C357 (1987).

92. C. P. Ostertag, P. G. Charalambides and A. G. Evans ,"Observations and Analysis of Sintering Damage," in: *Ceramic Transactions*, Vol. 7, Eds. C. A. Handwerker, J. E. Blendell, and W. A. Kaysser, American Ceramic Society, Westerville, Ohio, 1989, in press.

93. W. H. Tuan, E. Gilbert, and R. J. Brook, "Sintering of heterogeneous ceramic compacts," J. Mat. Sci. 24 1062-68 (1980).

94. M. Ciftcioglu, M. Akinc, and L. Burkhart, " Effect of Agglomerate Strength on Sintered Density for Yttria Powders Containing Agglomerates of Monosize Spheres," J. Amer. Ceram. Soc. 70 C329-C334 (1987).

95. T. N. Tiegs, P. F. Becher, "Sintered Al_2O_3-SiC-Whisker Composites," Bull. Amer. Ceram. Soc. 66 339-42 (1987).

96. S. Kamiya, H. K. Bowen; "Microstructural Control of Al_2O_3-TiO_2 Composites by Cyclic Annealing," in: *Ceramic Transactions*, Vol. 1, Ed. G. L. Messing, E. R. Fuller, and H. Hausner, American Ceramic Society, Westerville, Ohio, 1988, 978-985.

97. M. P. Borom, M. Lee, "Effect of Heating Rate on Densification of Alumina-Titanium Carbide Composites," Adv. Ceram. Mat. 1 335-40 (1986).

98. G. Valentine, A. N. Palazotto, R. Ruh, and D. C. Larsen, "Thermal Shock Resistance of SiC-BN Composites," Adv. Ceram. Mat. 1 81-87 (1986).

99. G. L. Messing, M. Kumagai, "Low Temperature Sintering of Seeded Sol-Gel-Derived, ZrO_2-Toughened Al_2O_3 Composites," J. Amer. Ceram. Soc. 72 40-44 (1989).

100. M. Ishitsuka, T. Sato, T. Endo, M. Shimada, "Sintering and Mechanical Properties of Yttria-Doped Tetragonal ZrO_2 Polycrystal/Mullite Composites," J. Amer. Ceram. Soc. 70 [11] C342-C346 (1987).

101. M. N. Rahaman, L. C. DeJonghe, S. L. Shinde, P. H. Tewari; "Sintering and Microstructure of Mullite Aerogels," J. Amer. Ceram. Soc. 71 [7] C338-C341 (1988).

102. M. N. Rahaman, D.-Y. Jeng, "Sintering of Mullite and Mullite- Matrix Composites," in: *Ceramic Transactions*, Vol. 7, Eds. C. A. Handwerker. J. E. Blendell, and W. A. Kaysser, American Ceramic Society, Westerville, Ohio, 1989, in press.

103. Y.L. Tian, D. L. Johnson, and M. E. Brodwin, "Microwave Sintering of Al_2O_3-ZrO_2 Composites," in: *Ceramic Transactions*, Vol. 1, Ed. G. L. Messing, E. R. Fuller, and H. Hausner, American Ceramic Society, Westerville, Ohio, 1988, 933-938.

104. D. K. Kim, C. H. Kim, "Pressureless Sintering and Microstructural Development of B_4C-TiB_2 Composites," Adv. Ceram. Mat. 3 52-55 (1988).

105. M. N. Rahaman, L. C. De Jonghe, "Effect of Rigid Inclusions on Sintering of Glass Powder Compacts," J. Amer. Ceram. Soc. 70 C348-C351 (1987).

106. C.-H. Hsueh, A. G. Evans, R. M. Cannon, and R. J. Brook, "Viscoelastic Stresses and Sintering Damage in Heterogeneous Powder Compacts," Acta Metall. 34 927-36 (1986).

107. R. K. Bordia, G. W. Scherer, "Sintering of Composites: A Critique of Available Analyses," in: *Ceramic Transactions*, Vol. 1, Ed. G. L. Messing, E. R. Fuller, and H. Hausner, American Ceramic Society, Westerville, Ohio, 1988, 872-886.

108. G. W. Scherer, "Viscous Sintering with a Pore-Size Distribution and Rigid Inclusions," J. Amer. Ceram. Soc. 71 [10] C447-C448 (1988).

109. G. W. Scherer, "Sintering of Rigid Inclusions," J. Amer. Ceram. Soc. 70 719-725 (1987).

110. C.-H. Hsueh, "Comment on "Sintering with Rigid Inclusions"," J. Amer. Ceram. Soc. 71 C314-C315 (1988).

111. G. W. Scherer, "Reply," J. Amer. Ceram. Soc. 71 C315-316 (1988).

112. C.-H. Hsueh, "Sintering of Whisker-Reinforced Ceramics and Glasses," J. Amer. Ceram. Soc. 71 C441-C444 (1988).

113. Kurt R. Mikeska, George W. Scherer, and Rajendra K. Bordia, "Constitutive Behavior of Sintering Materials," in: *Ceramic Transactions*, Vol. 7, Eds. C. A. Handwerker, J. E. Blendell, and W. A. Kaysser, American Ceramic Society, Westerville, Ohio, 1989, in press.

114. R. K. Bordia, R. Raj, "Hot Isostatic Pressing of Ceramic/Ceramic Composites at Pressures < 10MPa," Adv. Ceram. Mat. 3 122-26 (1988).

115. P. A. Mataga ,"Retardation of Sintering in Heterogeneous Powder Compacts," in: *Ceramic Transactions*, Vol. 7, Eds. C. A. Handwerker, J. E. Blendell, and W. A. Kaysser, American Ceramic Society, Westerville, Ohio, 1989, in press.

116. R. M. German, *Liquid Phase Sintering*, Plenum Press, New York, 1985.

117. R. B. Heady and J. W. Cahn, "An Analysis of Capillary Force in Liquid Phase Sintering," Met. Trans. 1 185 (1970).

118. J. W. Cahn and R. B. Heady, "Analysis of Capillary Force in Liquid Phase Sintering of Jagged Particles," J. Amer. Ceram. Soc. 53 [7] 406 (1970).

119. H. Park, S. Cho, and D. N. Yoon, "Pore Filling During Liquid Phase Sintering," Met. Trans. A. 15A 1075-80 (1984).

120. T. M. Shaw, "Liquid Redistribution during Liquid Phase Sintering," J. Amer. Ceram. Soc. 69 [1] 27-34 (1986).

121. W. D. Kingery, "Densification during Sintering in the Presence of a Liquid Phase: I. Theory," J. Appl. Phys. 30 [3] 301-306 (1959).

122. D. R. Clarke, "On the Equilibruim Thickness of Intergranular Glass Phases," J. Amer. Ceram. Soc. 70 [1] 15-22 (1987).

123. P. Greil, J. Weiss, "Evaluation of Microstructure of β-SiAlON Solid solution Materials Containing Different Amounts of Amorphous Grain Boundary Phase," J. Mat. Sci. 17 1571 (1982).

124. J. E. Marion, C. H. Hsueh, and A. G. Evans, "Liquid Phase Sintering of Ceramics," J. Amer. Ceram. Soc. 70 [10] 708-13 (1987).

125. W. A. Kaysser, "Sintering and HIP with a Liquid Phase,"in: Ceramic Transactions, Vol. 1, Ed. G. L. Messing, E. R. Fuller, and H. Hausner, The American Ceramic Society, Westerville, Ohio, 1988, 955-968.

126. L. C. DeJonghe, V. Srikanth, "Liquid-Phase Sintering of MgO-Bi$_2$O$_3$," J. Amer. Ceram. Soc. 71 [7] C356-C358 (1988).

127. D. F. K. Hennings, R. Janssen, P. J. L. Reynen, "Control of Liquid-Phase-Enhanced Discontinuous Grain Growth in Barium Titanate," J. Amer. Ceram. Soc. 70 [1] 23-27 (1987).

128. U.-C. Oh, Y.-S. Chung, D.-Y. Kim, and D. N. Yoon, "Effect of Grain Growth on Pore Coalescence During the Liquid-Phase Sintering of MgO-CaMgSiO$_4$ Systems," J. Amer. Ceram. Soc. 71 [10] 854-57 (1988).

129. P.L. Flaitz, J.A. Pask, "Penetration of Polycrystalline Alumina by Glass at High Temperatures," J. Amer. Ceram. Soc. 70, 449-455 (1987).

130. H. Song, R. L. Coble, Liquid Phase Sintered Al$_2$O$_3$: I, Origin and growth kinetics of plate-like abnormal grains and II, Morphology of plate-like abnormal grains, submitted to J. Amer. Ceram. Soc.

131. D.-D. Lee, S.-J. L. Kang, and D. N. Yoon, "Mechanism of Grain Growth and $\alpha - \beta'$ Transformation During Liquid-Phase Sintering of β'-Sialon," J. Amer. Ceram. Soc. 71 [9] 803-808 (1988).

132. For example, B. Lewis, "Nucleation and crystal growth," in: *Crystal Growth*, Ed. B. Pamplin, Pergamon Press, New York, 1975, pp. 12-39.

133. A. Nohara, T. Nakagawa, T. Soh, T. Shinke, "Numerical Simulation of the Densification Behaviour of Metal Powder During Hot Isostatic Pressing," Inter. J. Numer. Methods in Engineering 25 213-225 (1988).

134. M. Oyane, S. Shima, Y. Kono, "Theory of Plasticity for Porous Metals," Bull. JSME 16 [99] 1254-1262 (1973).

135. B. Aren, E. Navara, "Modelling Shape Change of Parts Produced by Hot Isostatic Pressing of Powders," Powder Metallurgy 31 [2] 101-105 (1988).

136. M. Abouaf, J. L. Chenot, G. Raisson, P. Bauduin, "Finite Element Simulation of Hot Isostatic Pressing of Metal Powders," Inter. J. Numer. Methods in Engineering 25 191-212 (1988).

137. J. Besson, M. Abouaf, "Numerical Simulation of Hot Isostatic Pressing of Ceramic Powders, International Conference on Hot Isostatic Pressing of Materials, Antwerp 25-27 April, 1988.

138. W-B. Li, M. F. Ashby, K. E. Easterling, "On Densification and Shape Change During Hot Isostatic Pressing," Acta Metall. 35 [12] 2831-2842 (1987).

139. W. Li, K. E. Easterling, "Stresses Developed in the Hot Isostatic Pressing of Metals and Ceramics," Holanilulea Mechanical Report, 1287:48T, 1-21, 1987.

140. C. Brodhag, F. Thevenot, "Progress in Hot Pressing: Transitory Phenomena during Temperature Chenges," in: Ceramic Transactions, Vol. 1, Ed. G. L. Messing, E. R. Fuller, and H. Hausner, American Ceramic Society, Westerville, Ohio, 1988, 947-954.

141. M. M. Rahaman, L. C. De Jonghe, C. H. Hsueh, "Creep During Sintering of Porous Compacts," J. Amer. Ceram. Soc. 69 [1] 58-60 (1986).

142. M. Lin, M. N. Rahaman, L. C. DeJonghe, "Creep-Sintering and Microstructure Development of Heterogeneous MgO Compacts," J. Amer. Ceram. Soc. 70 [5] 360-366 (1987).

143. K. R. Venkatachari, R. Raj, "Enhancement of Strength through Sinter Forging" J. Amer. Ceram. Soc. 70 [7] 514-20 (1987).

144. A. P. Druschitz, "Processing Zirconia by Sintering/Hot Isostatic Pressing," Advanced Ceramic Materials, 3 [3] 254-56 (1988).

145. S. T. Lin, R. M. German, "Compressive Stress for Large-Pore Removal in Sintering," J. Amer. Ceram. Soc. 71 [10] C-432-C-433 (1988).

146. B. J. Kellett, F. F. Lange, "Hot Forging Characteristics of Fine Grained ZrO_2 and Al_2O_3/ZrO_2 Ceramics," J. Amer. Ceram. Soc. 69 [8] C-172-C-173 (1986).

147. T. Hattori, M. Yoshimura, S. Somiya, "High-Pressure Hot Isostatic Pressing of Synthetic Mica," J. Amer. Ceram. Soc. 69 [8] C-182-C-183 (1986).

148. B. K. Lograsso, D. A. Koss, "Densification of Titanium Powder During Hot Isostatic Pressing," Met. Trans. 19A 1767-1773 (1988).

149. James G. Schroth, Alan P. Druschitz,"Optimizing Unencapsulated Hot Isostatic Pressing of Al_2O_3," in: Ceramic Transactions, Vol. 7, Eds. C. A. Handwerker, J. E. Blendell, and W. A. Kaysser, American Ceramic Society, Westerville, Ohio, 1989, in press.

150. H. Schubert, W. A. Kaysser ,"HIPing of Alumina," in: Ceramic Transactions, Vol. 7, Eds. C. A. Handwerker, J. E. Blendell, and W. A. Kaysser, American Ceramic Society, Westerville, Ohio, 1989, in press.

151. B. J. Kellett, F. F. Lange, "Experiments on Pore Closure During Hot Isostatic Pressing and Forging," J. Amer. Ceram. Soc. 71 [1] 7-12 (1988).

152. A. S. Helle, K. E. Easterling, M. F. Ashby, "Hot-Isostatic Pressing Diagrams: New Developments," Acta Metall. 33 [12] 2163-2174 (1985).

153. S. Nair, J. K. Tien, "Densification Mechanism Maps for Hot Isostatic Pressing (HIP) of Unequal Sized Particles," Met. Trans. A, 18A (1987) 97-107.

154. *Science and Technology of Zirconia III*, Ed. S. Somiya, N. Yamamoto, H. Yanagida, (Advances in Ceramics, Vol 24), American Ceramic Society, Westerville, OH, 1988.

155. S. Somiya, M. Yoshimura, "Microstructure Development of Hydrothermal Powders and Ceramics," in: *Ceramic Microstructures '86 Role of Interfaces* (Materials Science Research), 21 464-475.

156. J.-M. Wu, C.-H. Wu, "Sintering Behavior of Highly Agglomerated Ultrafine Zirconia Powders," J. Mat. Sci. 23 3290-3299 (1988).

157. S. Blackburn, M. P. Hitchiner, C. R. Kerridge, "Green Density Characteristics and Densification Kinetics of PSZ Powders Produced by Electro-Refinning," in: Ceramic Transactions, Vol. 1, Ed. G. L. Messing, E. R. Fuller, and H. Hausner, American Ceramic Society, Westerville, Ohio, 1988, 864-865.

158. J.-G. Duh, H.-T. Dai, "Sintering, Microstructure, Hardness and Fracture Toughness Behavior of Y_2O_3-CeO_2-ZrO_2," J. Amer. Ceram. Soc. 71 [10] 813-19 (1988).

159. J.-G. Duh, H.-T. Dai, W.-Y. Hsu, "Synthesis and Sintering Behavior in CeO_2-ZrO_2 Ceramics," J. Mat. Sci. 23 2786-2791 (1988).

160. R. M. Dickerson, M. V. Swain, A. H. Heuer, "Microstructural Evolution in Ca-PSZ and the Room-Temperature Instability of Tetragonal ZrO_2," J. Amer. Ceram. Soc. 70 [4] 214-20 (1987).

161. C. L. Lin, D. Gan, P. Shen, "Stabilization of Zirconia Sintered with Titanium," J. Amer. Ceram. Soc. 71 [8] 624-29 (1988).

162. M. Ishitsuka, T. Sato, T. Endo, M. Shimada, "Sintering and Mechanical Properties of Yttria-Doped Tetragonal ZrO_2 Polycrystal/Mullite Composites," J. Amer. Ceram. Soc. 70 [11] C342-C346 (1987).

163. H.-Y. Lu, J.-S. Bow, "Effect of MgO Addition on the Microstructure Development of 3 mol% Y_2O_3-ZrO_2," J. Amer. Ceram. Soc. 72 [2] 228-31 (1989).

164. E. Min-Haga, W. D. Scott, "Sintering and Mechanical Properties of ZrC-ZrO_2 Composites," J. Mat. Sci. 23 2865-2870 (1988).

165. P. C. Panda, J. Wang, R. Raj, "Sinter-Forging Characteristics of Fine-Grained Zirconia," J. Amer. Ceram. Soc. 72 [12] C-507-C-509 (1988).

166. G. Zhilun, L. Longtu, G. Suhua, Z. Ziaowen, "Low Temperature Sintering of Lead-Based Piezoelectric Ceramics," J. Amer. Ceram. Soc. 72[3] 486-91 (1989).

167. J. P. Guha, H. U. Anderson, "Reaction During Sintering of Barium Titanate with Lithium Fluoride," J. Amer. Ceram. Soc. 69 [8] C193-C194 (1986).

168. G. M. Dynna, Y.-M. Chiang, "Mechanisms of Grain Growth Enhancement and Inhibition in Donor-Doped Barium Titanate," in: Ceramic Transactions, Vol. 7, Eds. C. A. Handwerker, J. E. Blendell, and W. A. Kaysser, American Ceramic Society, Westerville, Ohio, 1989, in press.

169. J.-S. Chen, R.-J. Young, T.-B. Wu, "Densification and Microstructural Development of $SrTiO_3$ sintered with V_2O_5," J. Amer. Ceram. Soc. 70 [10] C260-264 (1987).

170. V. L. Richards, "Agglomerate Size Effect on Sintering of Doped Lanthanum Chromite," in: Ceramic Transactions, Vol. 1, Ed. G. L. Messing, E. R. Fuller, and H. Hausner, American Ceramic Society, Westerville, Ohio, 1988, 897-898.

171. J. P. Guha, D. J. Hong, H. U. Anderson, "Effect of Excess PbO on the Sintering Characteristics and Dielectric Properties of $Pb(Mg_{1/3}Nb_{2/3})O_3$-Based Ceramics," J. Amer. Ceram. Soc. 71 [3] C152-C154 (1988).

172. J. P. Guha, H. U. Anderson, "Microstructural Inhomogeneity in Sintered $Pb(Mg_{1/3}Nb_{2/3})O_3$-$PbTiO_3$ Based Dielectrics," J. Amer. Ceram. Soc. 70 [3] C39-C40 (1987).

173. M. F. Yan, W. W. Rhodes, "Sintering, Microstructures and Dielectric Properties of $Pb(Mg_{1/3}Nb_{2/3})O_3$ Composition," in: Ceramic Transactions, Vol. 7, Eds. C. A. Handwerker, J. E. Blendell, and W. A. Kaysser, American Ceramic Society, Westerville, Ohio, 1989, in press.

174. Y.-S. Yoo, J.-J. Kim and D.-Yeon Kim, "Effect of Heating Rate on the Microstructural Evolution During Sintering of $BaTiO_3$ Ceramics," J. Amer. Ceram. Soc. 70 [11] C322-C324 (1987).

175. Z. S. Ahn, W. A. Schulze, "Conventionally Sintered $(Na_{0.5}, K_{0.5})NbO_3$ with Barium Additions," J. Amer. Ceram. Soc. 70 [1] C18-C21 (1987).

176. C. E. Baumgartner, "Fast Firing and Conventional Sintering of Lead Zirconate Titanate Ceramic," J. Amer. Ceram. Soc. 71 [7] C350-353 (1988).

177. B.-S. Chiou, S.-T. Lin, J.-G. Duh, "The Effect of Sintering Conditions on the Grain Growth of the $BaTiO_3$-Based GBBL Capacitors," J. Mat. Sci. 23 3889-3893 (1988).

178. P. Duran, J. F. F. Lozano, F. Capel, C. Moure, "Large Electromechanical Anisotropic Modified Lead Titanate Ceramics," Journal of Materials Science 23 4463-4469. (1988).

179. T. Kimura, T. Yoshimoto, FN. Lida, Y. Fujita, T. Yamaguchi, "Mechanism of Grain Orientation During Hot-Pressing of Bismuth Titanate," J. Amer. Ceram. Sco. 72[1] 85-89 (1989).

180. J. P. Gambino, W. D. Kingery, G. E. Pike, L. M. Levinson, H. R. Phillipp, "Effect of Heat Treatments on the Wetting Behavior of Bismuth- Rich Intergranular Phases in ZnO:Bi:Co Varistors," J. Amer. Ceram. Soc. 72 [4] 642-45 (1989).

181. J. Kim, T. Kimura, T. Yamaguchi, "Sintering of Sb_2O_3-doped ZnO," J. Mat. Sci. **24** (1989) 213-219 (1989).

182. J.E. Blendell, J.S. Wallace, M.J. Hill, "Effect of Po_2 on Microstructure of $Ba_2YCu_3O_{6+x}$," submitted to J. Am Ceram. Soc. 1989.

183. R.S. Roth, C.J. Rawn, F. Beech, J.D. Whitler, J.O. Anderson, "Phase Equilibria in the System Ba-Y-Cu-O-CO_2 in Air," Ceramics Superconductors II - Research Update, 243-251 (1988).

184. T. Aselage, K. Keefer, "Liquidus Relations in Y-Ba-Cu Oxides," J. Mat. Res. 3 [6] 1279-1291 (1988).

185. J.E. Ullman, R.W. McCallum, J.D. Verhoeven, "Effect of Atmosphere and Rare Earth on Liquidus Relations in RE-Ba-Cu Oxides," J. Mat. Res., 4 [4] 752-755 (1989).

186. K. Sadananda, A. K. Singh, M. A. Iman, M. Osofsky, V. Le Tourneau, L. E. Richards, "Effect of Hot Isostatic Pressing on $RBa_2Cu_3O_7$ Superconductors," Adv. Ceram. Mat. 3 [5] 524-26 (1988).

187. J. J. Rha, K. J. Yoon, S.-J. L. Kang, D. N. Yoon, "Rapid Calcination and Sintering of $YBa_2Cu_3O_x$ Superconductor Powder Mixture in Inert Atmosphere," J. Amer. Ceram. Soc. 71 [7] C328-C329 (1988).

188. P. Sarkar, T. B. Troczynski, K. J. Vaidya, P. S. Nicholson, "Reaction Sintering of $YBa_2Cu_3O_x$ in Different Oxygen partial Pressures," Ceramics Superconductors II - Research Update, 204-215 (1988).

189. J. S. Wallace, B. A. Bender, S. H. Lawrence, and D. J. Schrodt, "Reaction Sintering High-Density, Fine-Grained $Ba_2YCu_3O_{6.5+x}$ Superconductors Using $Ba(OH)_2H_2O$," Ceramics Superconductors II - Research Update, 243-251 (1988).

190. K. Sawano, A. Hayashi, T. Ando, T. Inuzuka, H. Kubo, "Processing of Superconducting Ceramics for High Critical Current Density," Ceramics Superconductors II - Research Update, 282-293 (1988).

191. P. Sainamthip, V. R. W. Amarakoon, "Role of Zinc Volatilization on the Microstructure Development of Manganese Zinc Ferrites," J. Amer. Ceram. Soc. 71 [8] 644-48 (1988).

192. J. T. Mullin, R. J. Willey, "Grain Growth of Ti-Substituted Mn-Zn Ferrites," Fourth International Conference on Ferrites Part 1 (Advances in Ceramics, Vol. 15), American Ceramic Society, Westerville, OH, 1985, 187- 191.

193. H. Rikukawa, I. Sasaki, "On the Sintering Atmosphere of Mn-Zn Ferrites," Fourth International Conference on Ferrites Part 1 (Advances in Ceramics, Vol. 15), American Ceramic Society, Westerville, OH, 1985, 215-219.

194. C. M. Srivastava, N. Venkataramani, R. Aiyar, "Studies on the Sintering Mechanism and on Texturization in the Hot-Pressed System (Mn, Zn, Fe)Fe_2O_4," Fourth International Conference on Ferrites Part 1 (Advances in Ceramics, Vol. 15), American Ceramic Society, Westerville, OH, 1985, 193-200.

195. *Fourth International Conference on Ferrites*, Ed. F.F.Y. Wang, Advances in Ceramics, Vol. 15, Amer. Ceramic Society, Westerville, OH, 1985.

196. K. Omatsu, T. Kumura, T. Yamaguchi, "Sintering of Acicular NiZn- Ferrite Powder," Ceramic Microstructures '86 - Role of Interfaces (Materials Science Research), 21 623-631 (1986).

197. S. Komarneni, E. Fregeau, E. Breval, R. Roy, "Hydrothermal Preparation of Ultrafine Ferrites and Their Sintering," J. Amer. Ceram. Soc. 71 [1] C26-C28 (1988).

198. F. J. C. M. Toolenaar, M. T. J. Verhees, "Reactive Sintering of Zinc Ferrite," J. Mat. Sci. **23** 856-861 (1988).

199. P. Kishan, D. R. Sagar, S. N. Chatterjee, J. K. Nagpaul, N. Kumar, K. K. Laroia, "Optimization of Bi_2O_3 Content and Its Role in Sintering of Lithium Ferrites," Fourth International Conference on Ferrites Part 1 (Advances in Ceramics Vol 15), Amer. Ceramic Society, Westerville, OH, 1985, 207-213.

200. C. Greskovich, S. Prochazka, "Selected Sintering Conditions for SiC and Si_3N_4 Ceramics," Ceramic Microstructures '86 - Role of Interfaces, Materials Science Research, 21 601-610 (1986).

201. T. M. Shaw, B. A. Pethica, "Preparation and Sintering of Homogeneous Silicon Nitride Green Compacts," J. Am Ceram. Soc. 69 [1] 88-93 (1986).

202. B. A. Bishop, M. S. Spotz, W. E. Rhine, H. K. Bowen, J. R. Fox, "Sintering of Silicon Carbide Prepared From a Polymeric Precursor," in: Ceramic Transactions, Vol. 1, Ed. G. L. Messing, E. R. Fuller, and H. Hausner, American Ceramic Society, Westerville, Ohio, 1988, 856-857.

203. William J. Hurley, Jr., Leonard V. Interrante, Roberto Garcia, and Robert H. Doremus ,"Sintering and Microstructural Studies of Nanosized Crystalline Si_3N_4 and Si_3N_4/AlN Powders Derived from Organometallic Precursors," in: Ceramic Transactions, Vol. 7, Eds. C. A. Handwerker, J. E. Blendell, and W. A. Kaysser, American Ceramic Society, Westerville, Ohio, 1989, in press.

204. O. J. Gregory, S.-B. Lee, R. C. Flagan, "Reaction Sintering of Submicrometer Silicon Powder," J. Amer. Ceram. Soc. 70 [3] C-52-C-55 (1987).

205. J. A. Palm, C. D. Greskovich, "Thermomechanical Properties of Hot- Pressed $Si_{2.9}Be_{0.1}N_{3.8}O_{0.2}$ Ceramic," Bull. Amer. Ceram. Soc. 59 [4] 447-452 (1980).

206. S. Bandyopadhyay, J. Mukerji, "Sintering and Properties of Sialons without Externally Added Liquid Phase," J. Amer. Ceram. Soc. 70 [10] C273- 277 (1987).

207. C. Greskovich, W. D. Pasco, G. D. Quinn, "Thermomechanical Properties of a New Composition of Sintered Si_3N_4," Bull. Amer. Ceram. Soc. 60 [9] 1165-1170 (1984).

208. W. A. Sanders, D. M. Mieskowski, "Strength and Microstructure of Si_3N_4 with ZrO_2 Additions," Adv. Ceram. Mat. 1 166-73 (1986).

209. M. H. Lewis, G. Leng-Ward, and C. Jasper, "Sintering Additive Chemistry in Controlling Microstructure and Properties of Nitride Ceramics,"in: Ceramic Transactions, Vol. 1, Ed. G. L. Messing, E. R. Fuller, and H. Hausner, The American Ceramic Society, Westerville, Ohio, 1988, 1019-1033.

210. P. K. Das, J. Mukerji, "Sintering Behavoir and Properties of Si_3N_4 Sintered with Nitrogen-Rich Liquid in the System Y_2O_3-AIN-SiO_2," Adv. Ceram. Mat. 3 [3] 238-43 (1988).

211. M. Omori, H. Takei, "Preparation of Pressureless-Sintered SiC- Y_2O_3-Al_2O_3," J. Mat. Sci. 23 3744-3749 (1988).

212. N. Hirosaki, A. Okada, K. Matoba, "Sintering of Si_3N_4 with the Addition of Rare-Earth Oxides," J. Amer. Ceram. Soc. 71 [3] C-144-C-147 (1988).

213. L. Cordrey, D. E. Niesz, and D. J. Shanefield ,"Sintering of Silicon Carbide with Rare-Earth Additions," in: Ceramic Transactions, Vol. 7, Eds. C. A. Handwerker, J. E. Blendell, and W. A. Kaysser, The American Ceramic Society, Westerville, Ohio, 1989, in press.

214. T. Takahashi ,"The Sintering Behavior of Y_2O_3-MgO-ZrO_2 Doped and Y_2O_3-Yb_2O_3 Doped Silicon Nitride," in: Ceramic Transactions, Vol. 7, Eds. C. A. Handwerker, J. E. Blendell, and W. A. Kaysser, American Ceramic Society, Westerville, Ohio, 1989, in press.

215. C. O'Meara, J. Sjoberg ,"The Development of Microstructure in Pressureless Sintered Si_2N_2O Bodies," in: Ceramic Transactions, Vol. 7, Eds. C. A. Handwerker, J. E. Blendell, and W. A. Kaysser, The American Ceramic Society, Westerville, Ohio, 1989, in press.

216. R. Jimbou, Y. Suzuki, R. Sugita, "Sintering of SiC-ZrB_2/AlN Heating Element by Hot-Pressing," in: Ceramic Transactions, Vol. 7, Eds. C. A. Handwerker, J. E. Blendell, and W. A. Kaysser, American Ceramic Society, Westerville, Ohio, 1989, in press.

217. W.A. Sanders, G. Baaklin, "Correlation of Processing and Sintering Variables with the Strength and Radiography of Silicon Nitride," Adv. Ceram. Mat. 3 [1] 88-94 (1988).

218. W.D. Carter, P.H. Holloway, C. White, R Clausing, "Boron Distribution in Sintered Silicon Carbide," Adv. Ceram. Mat. 3 [1] 62-65 (1988).

219. O. Yamada, Y. Miyamoto, M. Koizumi, "High-Pressure Self-Combustion Sintering of Titanium Carbide," J. Amer. Ceram. Soc. 70 [9] C-206-C-208 (1987).

220. M. Akaishi, S. Yamaoka, J. Tanaka, T. Ohsawa, O. Fukunaga, "Synthesis of Sintered Diamond with High Electrical Resistivity and Hardness," J. Amer. Ceram. Soc. 70 [10] C237-C239 (1987).

221. T. Akaishi, A. B. Sawaoka, "Dynamic Compaction of Cubic Boron Nitride Powders," J. Amer. Ceram. Soc. 69 [4] C-78-C-80 (1986).

222. M. A. Janney, H. D. Kimrey, "Microstructure Evolution in Microwave Sintered Alumina," in: Ceramic Transactions, Vol. 7, Eds. C. A. Handwerker. J. E. Blendell, and W. A. Kaysser, American Ceramic Society, Westerville, Ohio, 1989, in press.

223. M. A. Janney, H. D. Kimrey, "Microwave Sintering of Alumina at 28 GHz," in: Ceramic Transactions, Vol. 1, Ed. G. L. Messing, E. R. Fuller, and H. Hausner, American Ceramic Society, Westerville, Ohio, 1988, 919-924.

224. Y. L. Tian, H. S. Dewan, M. E. Brodwin, and D. Lynn Johnson, "Microwave Sintering Behavior of Alumina Ceramics," in: Ceramic Transactions, Vol. 7, Eds. C. A. Handwerker, J. E. Blendell, and W. A. Kaysser, American Ceramic Society, Westerville, Ohio, 1989, in press.

225. Y.L. Tian, D. L. Johnson, and M. E. Brodwin, "Microwave Sinteirng of Al_2O_3-ZrO_2 Composites," in: Ceramic Transactions, Vol. 1, Ed. G. L. Messing, E. R. Fuller, and H. Hausner, American Ceramic Society, Westerville, Ohio, 1988, 933-938.

226. J. Wilson, S. M. Kunz, "Microwave Sintering of Partially Stabilized Zirconia," J. Amer. Ceram. Soc. 71 [1] C40-C41 (1988).

227. F. Okada, S. Tashiro, M. Suzuki, "Microwave Sintering of Ferrites," Fourth International Conference on Ferrites Part 1, (Advances in Ceramics Vol 15), Amer. Ceramic Society, Westerville, OH, 1985, 201-205.

228. E. K. Beauchamp, M. J. Carr, R. A. Graham, "Hot-Pressing of Shock-Activated Aluminum Nitride," Advanced Ceramic Materials, 2 [1] 79-84 (1987).

229. R.F. Davis, Y. Horie, R.O. Scattergood, H. Palmore, III, "Defects Produced by Shock Conditioning: An Overview," Advances in Ceramics Vol 10, Structure and Properties of MgO and Al_2O_3 Ceramics, Ed. W.D. Kingery, American Ceramic Society, Westerville, OH, 1984.

230. D. Beruto, R. Botter, and A. W. Searcy, "The Influence of Thermal Cycling on Densification: Further Tests of a Theory," in: Ceramic Transactions, Vol. 1, Ed. G. L. Messing, E. R. Fuller, and H. Hausner, American Ceramic Society, Westerville, Ohio, 1988, 911-918.

231. D. Beruto, R. Botter, A. W. Searcy, "Influence of Temperature Gradients on Sintering: Experimental Tests of a Theory," J. Amer. Ceram. Soc. 72 [2] 232-35 (1989).

232. A. W. Searcy, "Theory for Sintering in Temperature Gradients: Role of Long-Range Mass Transport," J. Amer. Ceram. Soc. 70 [3] C61-C62 (1987).

233. D. Beruto, R. Botter, A. W. Searcy, "The Influence of Thermal Cycling on Densification: Further Tests of a Theory," in: Ceramic Transactions, Vol. 1, Ed. G. L. Messing, E. R. Fuller, and H. Hausner, American Ceramic Society, Westerville, Ohio, 1988, 911-912.

234. W. B. Johnson, T. D. Claar, G. H. Schiroky, "Preparation and Processing of Platelet Reinforced Ceramics by the Directed Reaction of Zirconium with Boron Carbide," to be published in in Ceramic Engineering and Science Proceedings; 9 [7-8] (1989).

235. T. D. Claar, W. B. Johnson, C. A. Andersson, G. H. Schiroky, "Microstructure and Properties of Platelet Reinforced Ceramics Formed by the Directed Reaction of Zirconium with Boron Carbide," to be published in: Ceramic Engineering and Science Proceedings, 9 [7-8] (1989).

236. D.K. Kim, C.H. Kim, "Pressureless Sintering and Microstructural Development of B_4C-TiB_2 Composites," Advanced Ceramic Materials, 3 [1] 52-55 (1988).

237. D. Barnier, F. Thevenot, "Fabrication of ZrC_xO_y and ZrC_xO_y-ZrO_2 Composite Ceramics," In: Ceramic Transactions: Ceramic Powder Science II, Vol. 1, Edited by Gary L. Messing and Edwin R. Fuller, American Ceramic Society, Westerville, OH, 1988.

238. H. Nagai, K. Ohbayashi, "Effect of TiO_2 on the Sintering and the Electrical Conductivity of Cr_2O_3," J. Amer. Soc. 72 [3] 400-03 (1989).

239. D. Beruto, R. Botter, "H_2O-Catalyzed Sintering of ≈ 2 nm-Cross-Section Particles of MgO," J. Amer. Ceram. Soc. 70 [3] 155-59 (1987).

240. R.J. Higgins, H.K. Bowen, "Preparation and Sintering Behavior of Fine-Grained $MgAl_2O_4$-SiO_2 Composites," Adv. in Ceram., Vol. 21, 691-98, 1987.

241. H. Okamura, E. A. Barringer, and H. K. Bowen, "Preparation and Sintering of Monosized Al_2O_3-TiO_2 Composite Powder," J. Amer. Ceram. Soc. 69 [2] (1986) C-22-C-24.

242. Y. Hirata, I.A. Aksay, "Colloidal Consolidation and Sintering Behavior of CVD-Processed Mullite Powders," In: Ceramic Microstructures '86: Role of Interfaces, Mat. Sci. Res. Vol. 21. Edited by J.A. Pask and A.G. Evans.

243. T.J. Mroz, Jr., J. W. Laughner, "Sintered Microstructures of Seeded Mullite Gels," in: Ceramic Transactions, Vol. 7, Eds. C. A. Handwerker, J. E. Blendell, and W. A. Kaysser, American Ceramic Society, Westerville, Ohio, 1989, in press.

244. T. J. Mroz, Jr., J. W. Laughner, "Microstructures of Mullite Sintered from Seeded Sol-Gels," J. Amer. Ceram. Soc. 72[3] 508-509 (1989).

245. J. L. McArdle, G. L. Messing, "Seeding with γ-Alumina for Transformation and Microstructure Control in Boehmite-Derived α-Alumina, " J. Amer. Ceram. Soc. 69 [5] C-98-C-101 (1986).

THE MAIN TRENDS IN STUDY AND QUANTITATIVE DESCRIPTION

OF THE SINTERING PROCESSES

V.V. Skorohod

Institute for Problems of Materials Science of the
Ukraynian Academy of Sciences, Kiev, USSR

ABSTRACT

A review of papers concerning the mechanism and kinetics of
sintering written in the last twenty years has shown that a significant
success was reached in the field of understanding, predicting and
quantitatively defining sintering. However, it has not been possible so
far to obtain a symbiosis of the whole sum of knowledge about sintering
and to establish a general physical and physico-chemical theory. An
attempt at such a synthesis is presented. Methods for the development
of the techniques of engineering calculation in the technology of
sintering are indicated.

INTRODUCTION

Sintered metals and ceramics play important roles in modern
technology. Many of the construction and engineering materials which
have facilitated technical progress in mechanical engineering,
energetics, metal-working industries, instrument making and
electronics, are produced by the sintering of different crystal
powders. Hence, the fundamental knowledge of the physical nature of
sintering process is connected to the prediction of the flow of these
processes and to the quantitative description of structure and property
changes necessary for the effective results in the powder metallurgy
and ceramics technology. So, the science of sintering has become one of
the most important fields in the modern material science.

The modern science of sintering is based on the classical works of
Frenkel,[1] Pines,[2] Kuczynski,[3] Mackenzie,[4] Herring,[5] Kingery,[6] Coble,[7]
and Geguzin.[8] At the end of the sixties and beginning of the seventies
the results in the field of science of sintering were reviewed in a few
monographs[8-12] and although they were written only by Soviet
scientists, they reflected the level of sintering science in the whole
world in that classical period of development (1945-1968). It was clear
at the end of this period that in spite of remarkable achievement in
the physics of sintering and in the phenomenology of macroscopic
kinetics of contact formation and densification, a certain gap between
scientists and engineers had developed. The divergence is not
surprising because based on the science of sintering it was possible to

explain much of the phenomena, but it was nevertheless difficult to predict or calculate final properties of the product. Such pragmatic needs have emphesised many questions of great practical interest including algorithms for the optimization of technological processes, the investigation of new more effective techniques of sintering of powders with different chemical bonds and the research in the field of engineering design of the linear and volume shrinkage during sintering of compacts with inhomogenious density and complex shape.

Owing to such requests the new era of development of the sintering science has begun. It may be symbolically identitified with the foundation of the International Institute for the Science of Sintering and with the first issue of the Journal Physics of Sintering (now Science of Sintering). The main topics in the Journal and in the Institute's areas of work were devoted to the scientific approach to the sintering of metal and ceramic powders and to the connection between theory and practice.[13]

THE MECHANISM OF FLOW AND MASS TRANSFER DURING SINTERING

Kinetic parameters of sintering are usually considered separately from the formation and growth of contacts between particles and the shrinkage of a porous disperse system influenced by capillary forces.[14,15]

From the physical point of view it was a new and important approach when the mechanism of dislocation formation in particle contacts was proposed. The most significant was the work of Schatt and his collaborators.[16-18] They have shown theoretically and experimentally that the dislocation density in interparticle contacts was increased during sintering of monocrystal particles on a monocrystal plate. The mechanism of the multiplication of dislocations may be based on sliding or creep influenced by the excess of non-equilibrium vacancies appearing as the result of the surface curvature. In the latter case the multiplication of dislocations has more mechanical nature than chemical, as it was first proposed by authors of dislocation-viscous flow during sintering of particles containing submicropores.[19]

From the mechanical point of view the formation of dislocations may be the result of fulfillment of the following condition

$$\tau_L \approx P_L = \gamma\left(\frac{1}{X} - \frac{1}{\rho}\right) \geqslant \tau_s \approx \frac{Gb}{X} \tag{1}$$

where τ_L is the maximal shear stress in the contact area appearing as the consequence of Laplace pressure gradient P_L, X is the radius of contact neck, ρ_2 is the minimal radius of curvature of the neck surface ($\rho = \frac{X^2}{4R}$, R is the particle radius), γ is the surface tension, τ_s is the critical shear stress at the beginning of dislocation source, G is the shear modulus, b is the Burgers vector and X is accepted as the mean length of dislocation sources. Hence, the multiplication of dislocations is possible if:

$$\frac{X}{R} \leqslant \frac{4\gamma}{Gb} \tag{2}$$

where if γ is 1.5×10^3 erg/cm^2, G is 10^{12} Dn/cm^2, b = 2.5×10^{-8} cm, X/R will be ≈ 0.2.[18] The absolute value of R has no influence on the generation of dislocations.

A similar result is obtained for the high-temperature process of helicoidal multiplication of dislocations caused by vacancy oversaturation. It means that the osmotic pressure acting on the

component of edge dislocation because of the presence of nonequillibrium
vacancy concentration C is[19]

$$P_{osm} = \frac{kT}{b^3} \ln \frac{C}{C_o} \qquad (3)$$

The concentration of excess vacancies is according to Thompson
relation:

$$C = C_o \exp \left[- \gamma \left(\frac{1}{X} - \frac{1}{\rho} \right) \frac{\Omega}{kT} \right],$$

where Ω is the atomic volume, γ is the surface tension. It means that P_{osm}
is identical to P_L given in equation (1).

The high-temperature mechanism of dislocation generation is
obviously prefereble since the diffusion fields of excess vacancies are
acting over longer distances than the fields of elastic stresses. It
must be also taken into account that dislocation movement can change
the value of the surface area because dislocation existing on the
surface may result in the formation of steps and in a increase of the
surface energy. The diminution of their surface energy is possible only
with a high surface mobility of atoms, enabling a rapid relaxation of
dislocation roughness and macroscopic changes of the system geometry.
The experiments which confirm the generation of dislocations during
sintering of a model system are authentic and are no longer subject to
doubt.[18]

However a fundamental question has appeared: do dislocations have
any significant influence on the kinetics of sintering? How can they
help to explain the increase in sintering rate in real systems by a few
levels of magnitude, which has been observed, but not explained by
means of diffusion theory?[1,8,11,12] None of the model experiments have
given results which could be used for the interpretation of mechanisms
of dislocation flow in the kinetics of contact formation during
sintering. It is an established fact that dislocations have a
quantitative influence on high-temperature sintering.[15,18] However, by
dislocation-diffusion theory is not possible to explain the observed
decrease of sintering temperature and the value of activation energy
which is much lower than in the case of volume diffusion.[20] Usually
experimental values of activation energy of sintering (contact
formation as well as volume shrinkage) are similar or less than the
activation energy of grain boundary and sometimes surface diffusion[11,20-22]
particularly in the case of sintering of fine powders of transition and
high-temperature metals (Fig. 1a,b).

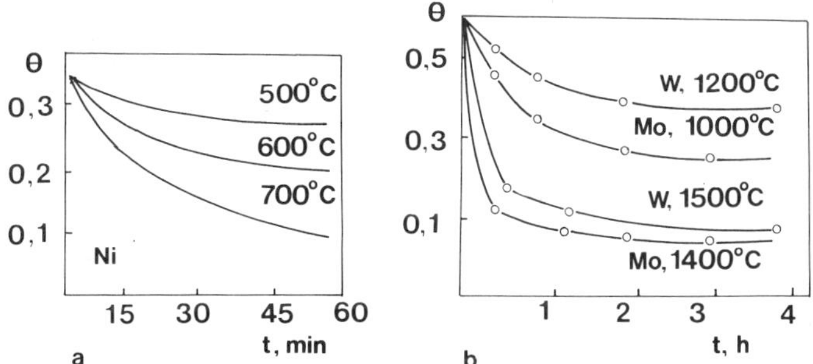

Fig. 1. Sintering kinetics of highly active powders
a) nickel; b) molybdenum tungsten.

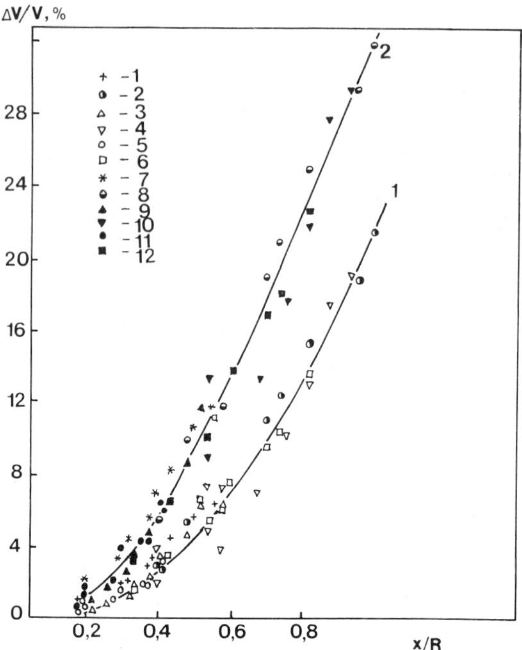

Fig. 2. Relationship between the volume shrinkage and interparticle
contacts of spherical powders of tungsten.

In the analyses of sintering as only a diffusion process, the
"activity" may be considered as the manifestation of grain boundary
diffusion. To Coble[7] belongs the credit of this idea and for many years
it has been the basis for quantitative estimation of sintering rate of
the disperse systems. However, the activation energy of sintering is
sometimes less than the activation energy of grain boundary diffusion,
approaching the activation energy of the surface diffusion.

It is well known from the theory and from model experiments that
surface diffusion alone cannot lead to the diminution of the distance
between the centers of particles and to the volume shrinkage during
sintering.[8] Recently, the kinetics of the growth of interparticle
contacts between compacted micron-size tungsten spheres in nearly real
conditions has been studied.[23,24] It was shown that there is a general
relationship between the mean size of interparticle contacts and the
linear shrinkage. It can be obtained from geometrical relation:

$$\frac{\Delta V}{V_o} = \frac{2}{3} \left[\left(\frac{X}{R}\right)^2 - \left(\frac{X_o}{R}\right)^2 \right]_{X>X_o} \tag{4}$$

if we accept that the contact growth is realized by the mass transport
from the center to the periphery of contact plane. The approach of
particle centers is defined by the volume of transfered material.[24] It
was experimentally verified (Fig. 2).

The kinetics of growth of interparticle contacts can be
qualitatively described by diffusion of vacancies along grain
boundaries:

$$\left(\frac{X}{R}\right)^6 - \left(\frac{X_o}{R}\right)^6 = \frac{96\gamma\Omega b D_{gb}}{R^4 kT} t, \tag{5}$$

where b_{gb} is the width of the boundary and D_{gb} is the coefficient of

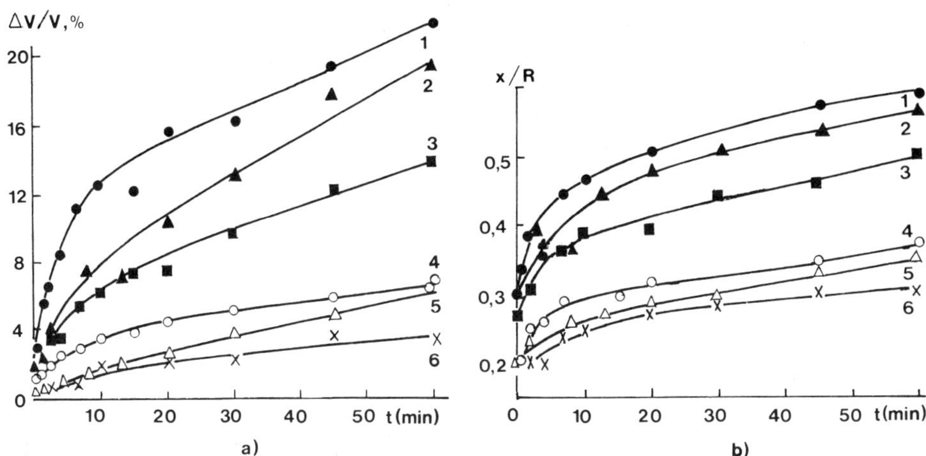

Fig. 3. Experimental and calculated curves of shrinkage kinetics
(a) and the growth of contacts; (b) of spherical tungsten
powders with particles size 4 and 10 μm.

grain boundary diffusion. Hence

$$\frac{\Delta V}{V_o} = \frac{2}{3} \left[\sqrt[3]{\left(\frac{X_o}{R}\right)^6 + \frac{96\gamma\Omega bD_{gb}\cdot t}{R^4 kT}} - \left(\frac{X_o}{R}\right)^2 \right] \tag{6}$$

Experimental data concerning the kinetics of contact growth and the
volume shrinkage are in accordance with the relations (5) and (6) if
the surface diffusion coefficients, taken from usual references,[26] are
put in these equations instead of the grain boundary diffusion
coefficient (Fig. 3a,b). If we take into account that, by simultaneous
action of two mechanisms (the surface diffusion and the grain boundary
diffusion), the contact growth is determined by more rapid mass
transport i.e. by the surface diffusion, then it is easy to understand
that in this case the shrinkage caused by the grain boundary diffusion
must be much less, than is given by equation (6). Decrease in the
driving force for the approach of particle centers is the result of the
diminution of neck curvature which is the consequence of the surface
diffusion. In this case the relation (4) is not valid. Both facts
confirm that the grain boundary diffusion is extremely intensified
compared to the diffusion along equilibrized grain boundaries in
polycrystals and it is similar to the surface diffusion.

In spite of the significance of diffusion theory in the
explanation of the "activity" during sintering, when one uses only this
theory, it is not possible to understand the influence of nature of
chemical bonds in crystal materials on their behaviour during
sintering. This question was studied most completely by Andrievski.[27]
The intensity of sintering as a function of relative sintering
temperature for a series of crystal materials is illustrated in Fig. 4.

We can explain this law if we accept that dislocations play a role
in the material flow caused by capillary processes. In ref. (27), a
comparison is given between the temperature ranges of the most
intensive increase in shrinkage and characteristic temperatures of
plastic deformation of crystals.[28] Such correlations undoubtly exist.

At the other side, during the last 15 years, ideas about the role
of diffusion along grain boundaries, i.e., the application of the
concept of structural superplasticity in sintering, has been

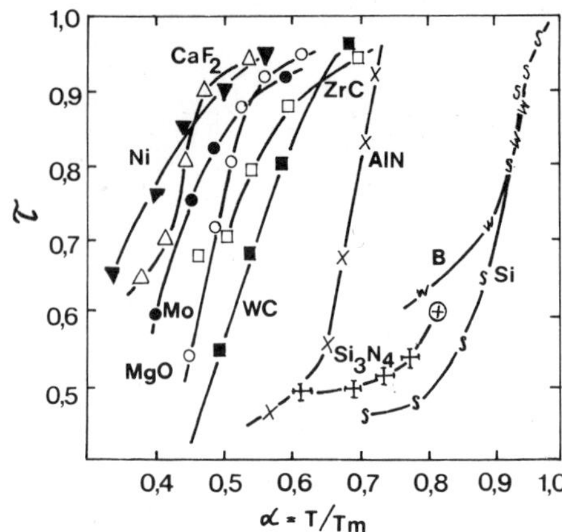

Fig. 4. Temperature relationship of the sinterability of crystal powders with different type of interatomic bonds.

extensively developed. Such ideas are based on the classical work of Ashby and Verral[29] and they have been completely developed in Geguzin's papers[30-32] and in ours.[33,34] From the quantitative point of view the concept of structural superplasticity i.e. quasi-viscous sliding along grain boundaries with the appropriate accomodation of grain shape[29] gives the results similar to Coble's theory,[7] although the numerical constant in the relation between the deformation rate and the stress is larger by one order of magnitude.

However, from a qualitative standpoint these ideas caused a significantly new approach in the theory of diffusion-viscous flow in polycrystals. First, if we take into account the effect of sliding, such a flow becomes stationary regarding the acting stress. This limit γ^* (with the accuracy on the level of the numerical constant) is:

$$\gamma^* \approx \frac{\gamma_{gb}}{L} \tag{7}$$

where γ_{gb} is the surface energy of the grain boundary and L is the mean grain size. Secondly, the flow becomes nonlinear if the accomodation of grain shape is realized by the dislocation mechanism. That possibility has been previously analysed.[33] The formation of dislocations and the appearance of plastic flow in the sintering process is conditioned by the localization of stress in the area of triple points between grains when the mean stress in the material is the linear superposition of viscous (grain boundary) and plastic flow:

$$\bar{\gamma} = \eta_{ef.}\,\dot{\varepsilon}\,f_v + \gamma_T f_p \tag{8}$$

where $\eta_{ef.}$ is the effective viscosity of polycrystal system without accomodation of grains, $\dot{\varepsilon}$ is the rate of flow resulted from sliding along grain boundaries, γ_T is the tension of plastic flow which slightly depends on the flow rate, f_v and f_p are statistically weighted surface cross-sections on which viscous and plastic flow is developed respectively ($f_v \gg f_p$).

An additional contribution to the formation of movable dislocations may appear in the form of mutual rotation of contacting crystal particles as a means of creating grain boundaries which possess

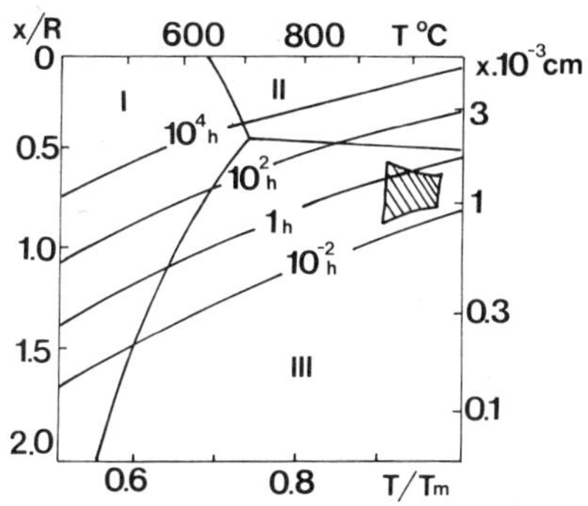

Fig. 5. Sintering map for copper powder.

a minimal energy.[34] The values of f_v and f_p may be changed during sintering process. Unfortunately these changes can not be quantitatively predicted. However, these analyses enable us to identify an important hypothesis - that "active" sintering is really followed by dislocation flow which may appear in latent form when the process rate is limited by the diffusion mechanism. Sometimes it may become a dominant mechanism of sintering, like in the case of covalent crystals. In such a way the dislocation flow of material during sintering is usually not the cause, but the consequence of high rates of sintering. It was often experimentally found[11] that this process acts as an inhibiting factor, slowing down the isothermal shrinkage during sintering. Hence, a significant influence of microstructural regularity on the sintering process must be expected.

Obviously, sintering of regularly packed spherical particles can be considered as a series of simultaneous processes of contact formation and approaching of particle centers. This process can be conditionally called coherent sintering. It is not followed by mutual motion of particles and the role of dislocation processes will be minimal. It is associated with the specific behaviour of monodisperse submicron oxide powders prepared by sol-gel method when they are sintered.[35] The study of sintering of regular monodisperse structures of micron and submicron size particles of covalent crystals is of special interest. Of course previous model experiments of the sintering of covalent crystals are needed. There are such data on ionic crystals.[36] Pure ionic crystals (LiF, NaF, CaF_2) and high-temperature oxides (MgO, TiO_2, ZrO_2) are analogous to metals[37] regarding their sintering behaviour, first of all regarding the role of dislocations, their formation in the contact zone, dislocation recovery and the possibility for very "active" sintering in a finely dispersed state.

SINTERING MAPS

It was Ashby's idea[38] to make sintering diagrams (maps). They are constructed on the basis of a chosen theoretical model of sintering. Usually it is the neck growth between spherical particles that is modeled and the proposed mechanism of mass transport is the diffusion. The goal of maps is to enable the estimation of area of temperature, time and sintering degree where some form of diffusion mechanism is dominant (e.g. volume diffusion, surface diffusion or grain boundary

diffusion). A definite value of the exponent in the kinetic
relationship between the relative neck growth and time of the
isothermal sintering[25] corresponds to each of the mentioned diffusion
mechanisms. A typical shape of such a sintering map can be seen in Fig.
5, illustrating sintering of copper particles. Lines of the identical
sintering time are shown on the map as well as the areas in which
experimental data for sintering of 100 μm particles can be found.

Wilkinson and Ashby have also proposed similar maps for the
estimation of the mechanism of pressure sintering.[41] The basic
parameter for their construction was the relative density of sintered
samples. In the case of these maps, besides the areas of dominant
volume diffusion and grain boundary diffusion, there are areas where
dislocation flow, realized in accordance to the exponential law, is the
basic mechanism of the mass transport. Further development and more
detailed study of sintering mechanisms by means of these maps was
contributed by Nikolić et al.[39,40] In their papers, algorithms for the
construction of maps are presented in detail, including the matrix
method (Fig. 6) and three-dimensional sintering maps as well (Fig. 7).
They have also applied sintering maps for the analyses of sintering
kinetics of real powders (for example carbonyl iron) using a
theoretical relation between the linear shrinkage and the neck size. In
the case of metals which have phase transformations (iron, titanium),
sintering maps are very sensitive regarding the structure of the
crystal lattice (influencing the value of diffusion coefficient). It is
of interest to emphasise significant contradictions existing between
conclusions about the basic sintering mechanism made by many authors of
previous experiments and by those who used these results for the
construction of sintering maps. It is specially characteristic for most
of the experimental data that they have been previously attributed to
volume diffusion, but now they are appearing in the area of surface

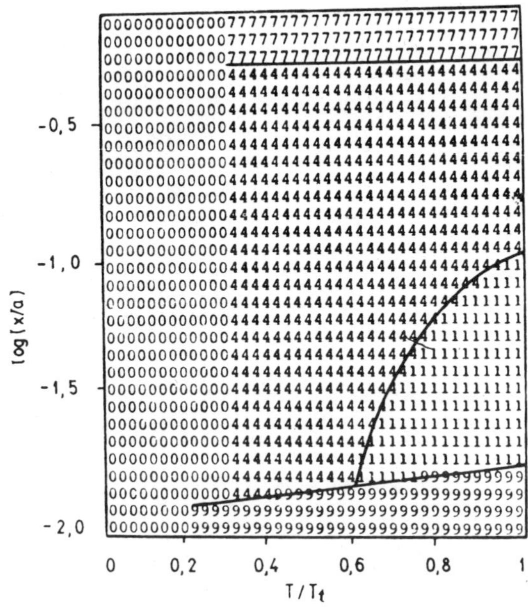

Fig. 6. Computed sintering maps for iron.

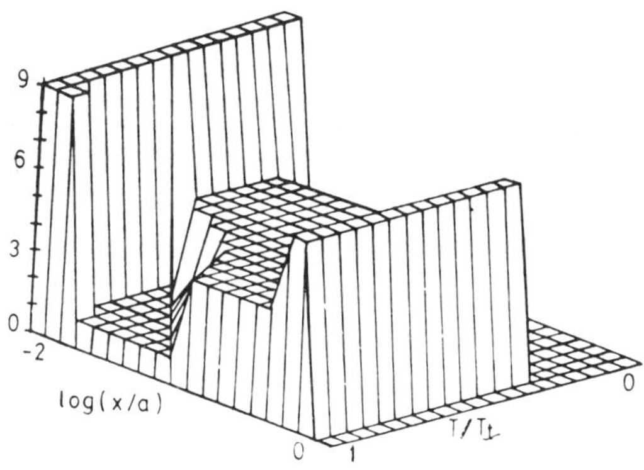

Fig. 7. Space (three dimensional) sintering map for iron.

diffusion on such sintering maps (including Kuczynski's classic work as well).[3]

We must point out that besides the logic of sintering maps, their application in the case of real systems require great caution and the basic method has to be developed more. First, it is necessary to take into account not only the areas where diffusion is the dominant mechanism of sintering, but to do something similar to the case of pressure sintering. The characteristics of the particulate assemblage like polydispersivity and the regularity of particle packing, should be included among already established parameters. As will be shown later, a simple relation between contact size and macroscopic shrinkage is possible only when there is not any local inhomogenious (differential) shrinkage, which is usully emphasised in the case of polydisperse powders

Generally, sintering maps should be connected to diagrams of the evolution and topological transformations of microstructures within a porous body during sintering, because they play a significant role in the formation of properties of sintered materials.

THE EVALUATION OF MICROSTRUCTURES DURING SINTERING

Interparticle contacts, grains (particles) and pores are the basic elements of the microstructure of sintered polycrystal materials. Interparticle contacts can be in one or more points. Pores are formed between particles (grains), "equiparticles", or may take the form of cracks. The classification of macrodefects in sintered bodies is given in Fig. 8.[42]

The evaluation of interparticle connections having various size and portion has been studied in detail in the case of sintered copper and iron obtained from atomized powders with large particles.[15,43] It has been shown that real contacts in compacts are not flattened (Fig. 9). Local areas of diffusion bonding in the particle contacts are developed during sintering resulting in a characteristic faceted structure. The authors have made a classification of facet types and they have shown that inside the contacts occur complex processes which result in the change of statistic distribution of structure elements. As the structure parameter which limits mechanical properties the authors have introduced the name "plain porosity" and have developed the method for its estimation by means of fractographic analyses. It

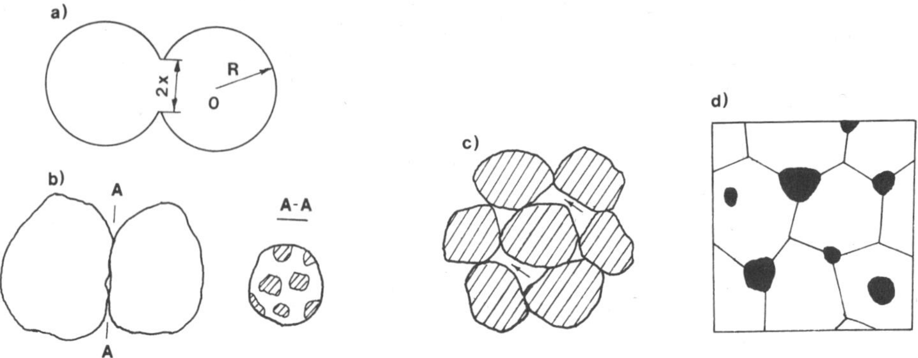

Fig. 8. Elements of the microstructure in dispersed body. a) with one
contact; b) with more contacts; c) interparticle pores; d) pores
between grains and inside grains.

has been shown that the fracture strength of analysed sintered
materials is practically linear (Fig. 10). The presence of quasisurface
macrodefects is of great influence on mechanical properties of porous
metals subjected to cold deformation processes. The anisotropy of
strengths was observed as well as some specific effects like the
diminution of the propagation rate of ultrasound waves, i.e., the
decrease of normal elastic moduli during extension of samples in the
direction of compaction, extending over a broad range, up to the point
of fracture (Fig. 11). [42] The temperature range of the plain porosity
decrease is far below the range of intensive shrinkage. [43] The
quantitative theory of this process has not been developed yet,
although it is probable that the plastic deformation has a very

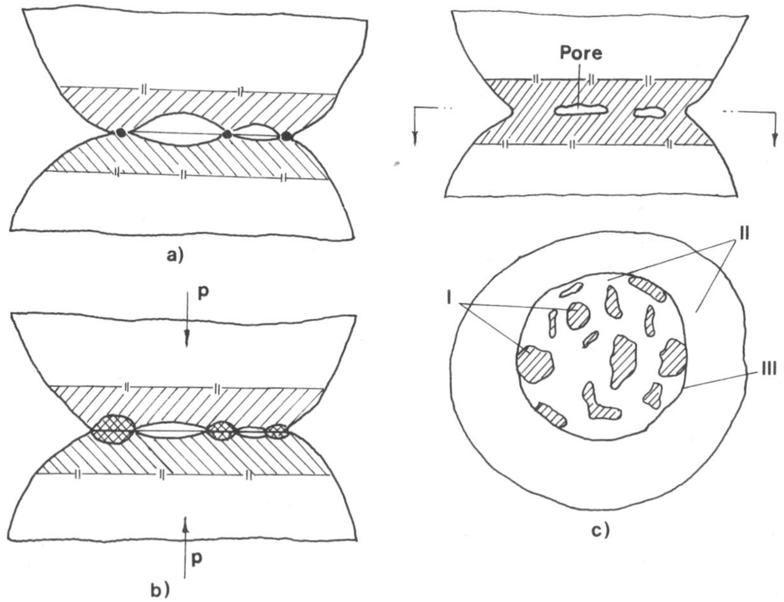

Fig. 9. The flowchart of the evolution of interparticle contacts
a) poured powder; b) compact; c) sintered sample.

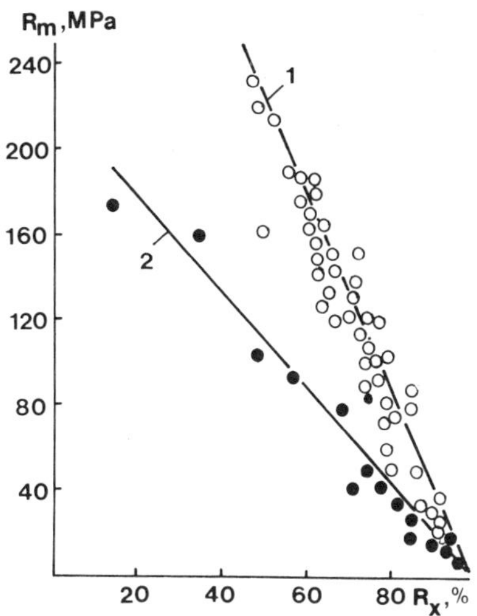

Fig. 10. Dependence of the fracture strength and tensile strength on
the plain porosity for iron (1) and copper (2).

important role, leading to dislocation clustering, within a contact
zone.

The second approach in the study of microstructural evolution in
sintering was to find the relationship of the mean grain size and pore
size to the total porosity. In this field, the most significant are
Kuczynski's papers[44] based on his statistical theory of sintering,
whose basic result was a simple relation between the mean particle
size, the pore size and the porosity:

$$\frac{4}{3} P \bar{a}^{-3} = K \, \overline{r^2} \bar{a}$$
(9)

where P is the porosity, $\bar{a}$ is the mean radius of grains, $\bar{r}$ is the mean
pore radius, K is a constant depending on the choice of polyhedral
shape of grains. It was assumed that the mechanism of recrystallization
and of the pore decrease is identical and that it is mainly attributed
to diffusion along grain boundaries. From the equation (9) follows a
qualitative conclusion that a decrease of porosity results in an
increase of mean grain size and of the dimensionless constant $\bar{a}/\bar{r}$
according to the hyperbolic law. Experimental data (Fig. 12)

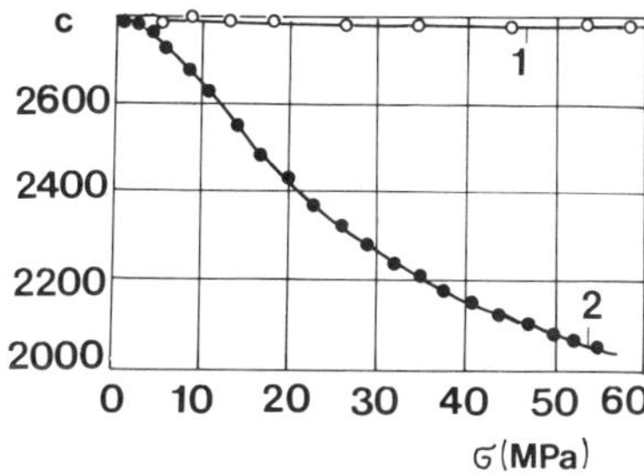

Fig. 11. Change of the propagation rate of ultrasound waves in iron
under tensile stress.

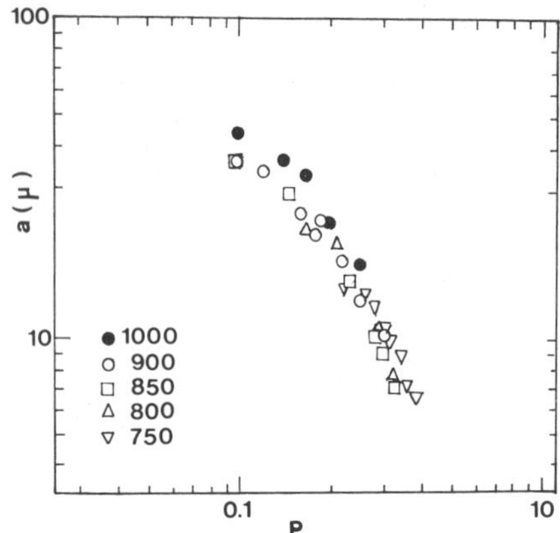

Fig. 12. Porosity vs. grain size of sintered copper.

qualitatively confirm the conclusions of the statistical theory.[44] The hyperbolic relationship plotted in log-log system gives a straight line. A few deviations in the range of low porosities indicate that there are some other inhibitors of grain boundary diffusion in the sintered body.

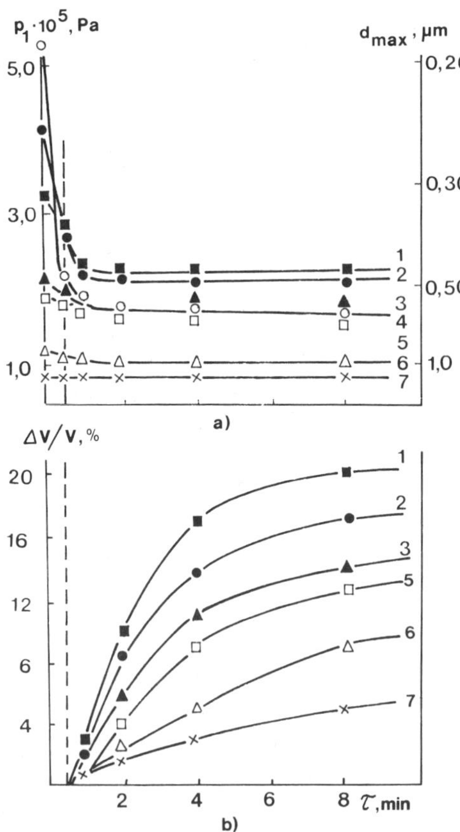

Fig. 13. Kinetics of the change of pore sizes (a), and volume shrinkage
(b) during sintering of tungsten powders of different
activity.

50

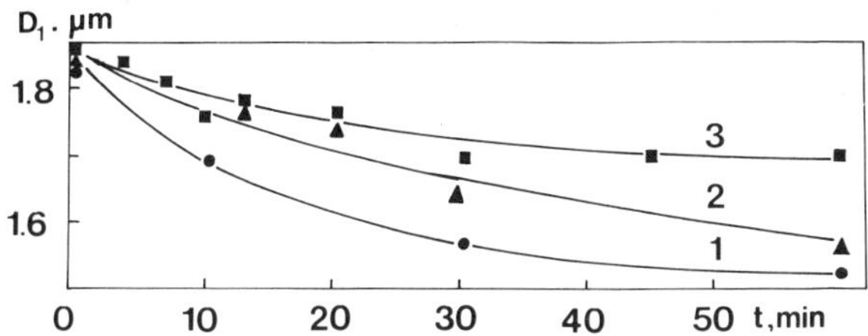

Fig. 14. Decrease in pore size during sintering of spherical tungsten
        powder.

The third direction of research is connected to the evolution of
porous structures. Kuczynski's theory postulated that the pore sizes
are decreased monotonously. In the early stages of sintering the mean
pore size is related to the mean particle size. When fine powders are
sintered these simple relations become complex. In such systems
differential sintering is characteristic. It results in a significant
increase of the mean size of the crossection of channel pores.[45,46]
Typical experimental data obtained by sintering of tungsten powders are
shown in Fig. 13. If tungsten powder is spherical, a more ordered
structure is formed during sintering, thereby accounting for a
monotonous decrease of the mean pore size (Fig. 14).[23]

Enlargement of pores in the early stages of sintering leads to the
diminution of driving force and to the formation of clusters of fine
particles, whose bonding results in large grains. Tendencies for the
onset of differential sintering become larger with the increase of the
degree of powder polydispersivity. In such a way, for powders with
wider particle size distribution,reduced shrinkage during sintering
under the same conditions can be explained.[47,48]

A simple geometrical consideration has resulted in the
quantitative relation between total V_t and local V_1 shrinkage for the
case of differential sintering:[46]

$$\frac{\Delta V}{V_1} = \theta_o \left[1 - \frac{R_o^3}{R_1^3} \left(1 - \frac{\Delta V}{V_t} \frac{1}{\theta_o} \right)^2 \right], \tag{10}$$

where θ_o is the initial porosity, R_o and R_1 are the initial and final
pore radius respectively. Quantitative estimation of the value of local
shrinkage is of great importance for both the theory (as previously
mentioned, the volume shrinkage is directly associated with the growth
of interparticle contacts) and for various practical applications,
because it gives potential opportunities for the optimization of
shrinkage during sintering.

CONCLUSION

Previously considered directions and perspectives of development
of the sintering process and analyses of its physical nature and its
quantitative definition have specific character and they have not yet
taken into account all problems of sintering. Neither have questions
concerning rate controlled sintering,[49] which is one of the highest
levels in the sintering technology, nor activated sintering, nor liquid

phase sintering, nor kinetics and mechanism of highly activated
sintering of ultrafine powders,nor pressure sintering been considered.
A separate analyses is needed for each of these fields.

The results obtained in the field of flow mechanisms and structure
formations in sintering, as well as a quantitative interpretation of
the sintering process, indicate that a general qualitative explanation
of sintering of monophase systems is now nearing an end. Of course,
quantitative definition of all stages of sintering, and of all
phenomena occuring in sintering, is still a challange for both
scientific and technological researches. It will require additional
experimental studies to find diffusion parameters and other kinetic
parameters in the case of real disperse systems. We must take care to
determine different influences of the microstructure and mass transport
processes and to develop calculation methods for the estimation and
prediction of physical and mechanical properties of sintered materials
using quantitative structure parameters.

All that will be necessary in order to reach the main goal of the
materials sciences – to synthesize materials with pre-defined structure
and properties. However, in solving these problems we will open a new
one – to develop an appropriate technology which will guarantee that
the requested structure is realized in the whole product which
sometimes has a complex shape and large size and owing to that, an
inhomogenious initial green density. Heating of such samples is
followed by higher temperature gradients in these parts thus causing
termal stresses. Inhomogenious density, and inhomogenious local
dispersivity in a compact results in inhomogenious volume shrinkage
during sintering. The resulting product can have an irregular shape,
the inhomogenious density and in the worst case, it may even be cracked
as the result of the stresses created. Besides, the shrinkage may be
influenced by gravitational forces, or by some other external
mechanical stresses. It is really a great challange for modern
engineers to calculate and design an overall sintering process, to
define the schedule of heating and cooling, the need for the
application of special equipment, etc. The unique theoretical base for
such engineering calculations in the sintering technology is the high-
temperature rheology of disperse systems. Its application in sintering
theory stems from Mackenzie's papers[4] and ours[12] although in the last
few years it has became possible to solve complex specific problems of
shrinkage of non-uniform areas using rheology and taking care about
boundary effects.[50-52] Further development in this field, and the
realization of the complete symbiosis of diffusional and rheological
approaches for the quantitative definition of sintering is an actual
problem in the application of sintering theory.

ACKNOWLEDGEMENT

The author is grateful to Prof. M.M.Ristić for the proposal to
prepare this review paper and for the opportunity to be informed about
results of the members of the International Institute for the Science
of Sintering.

REFERENCES

1. J. Frenkel, J. Physics USSR, 9(1945)385.
2. B. Ya. Pines, ZTF 16(1946)737; Usp.Fiz.Nauk, 52(1954)501.
3. G. Kuczynski, J.Metals 1(1949)169.
4. J. K. Mackenzie, Proc. Phys. Soc. (B) 63(1950)1.

5. C. Herring, J.Appl.Phys. 21(1950)347.

6. W. D. Kingery, J.Appl.Phys., 30(1959)300.

7. R. L. Coble, J.Appl.Phys., 32(1961)787; 34(1963)1679.

8. Ya. E. Geguzin, Physika Spekaniya, (Physics of Sintering, First Edition), M., Nauka, 1967,360.

9. V. N. Eremenko, Yu. V. Naydich, I. A. Lavrinenko, Spekanie v Prisustvii Zhidkoi Metallicheskoi Phasi, (Liquid Metal Phase Sintering) Kiev, Naukova Dumka, 1969,165.

10. A. I. Raichenko, "Diffuzioni Rascheti dlya Poroshkovih Smesei", Naukova dumka, (Diffusional Calculation in Particulate Systems) Kiev, 1969, 102.

11. V. A. Ivensen, Kinetika Uplotneniya Metallicheskih Poroshkov pri Spekanii, (Sintering Kinetics of Metal Powders Densification) Moscow, Metallurgiya, 1971, 265.

12. V. V. Skorohod, Reologicheskie Osnovi Teorii Spekaniya, (Rheological Basis of the Sintering Theory) Naukova dumka, Kiev, 1972, 149.

13. Sintering-Theory and Practice, Ed. M.M.Ristić, Phys. of Sintering 5(1973) 506.

14. Ya. E. Geguzin, Novie Issledovaniya v Oblasti Phisiki Spekaniya, (Recent Research in Physics of Sintering) in Poroshkovaya Metallurgiya-77, Kiev, Naukova dumka, 1977, 110.

15. Processi Masoperenosa pri Spekanii, Ed. V. V.Skorohod, (Mass Transfer Processes in Sintering) Naukova dumka, Kiev, 1987, 152.

16. W. Schatt, E. Friedrich, Planse. Pulver. 25(1977) 145.

17. E. Friedrich, W. Schatt, VII Intern.Powder Metal. Conf., Dresden, DDR, Sept. 1981, 121.

18. W. Schatt, E.Friedrich, K.P. Wieters, Rev. Powd. Met. and Phys. Ceram., 31(1986)1.

19. V. V. Skorohod, Phys. of Sintering, Spec. Issue, Sept. 1971, 53.

20. V. V. Skorohod, S. M Solonin, Physico-metallurgicheskie osnovi Spekaniya Poroshkov, (Fundamentals of Physics and Metallurgy of Sintering) Metallurgiya, Moscow, 1984, 159.

21. N. C. Kothari. Phys. of Sintering, Spec. Issue, Sept. 1971, 85

22. V. V. Skorohod, V. V. Panichkina, Yu. M. Solonin, I. V. Uvarova, Dispersnie Poroshki Tugoplavkih Metallov, (Fine Powders of High-Temperature Metals) Naukova dumka, Kiev, 1979, 171.

23. V. V. Skorohod, L. A. Vermenko, O. Getman et al., Porosh. Metal., No 5(1987)11 and No.6(1987)20.

24. V. V. Skorohod, VII Intern.Conf.Powd.Metal., Poland, Krakow, 1988.

25. G. C. Kuczynski, in Poroshkovaya Metallurgiya Specialnogo Naznacheniya, (Powder Metallurgy for High-Performance Applications) Eds. J. J. Burke, V. Weiss, Metallurgiya,Moscow, 1977, 88.

26. Ya. E. Geguzin, Yu. S. Kaganovski, Diffuzioni Processi na Poverhnosti Kristalov, (Diffusion Processes on Crystal Surfaces) Energoatomizdat, Moscow, 1984, 128.

27. R. A. Andrievski, J. Sci. Sintering 16(1984)3.

28. V. I. Trefilov, Yu. V. Milman, O. N. Grigoriev, Progres in Crystal Growth and Charact., 16(1988)225.

29. M. F. Ashby, R. A. Verral, Acta Met., 21(1973)53.

30. Ya. E. Geguzin, DAN SSSR, 229(1976)601.

31. Ya. E. Geguzin, Yu. I. Klinchuk, Porosh. Metal. No. 7(1976)17.

32. Ya. E. Geguzin, Physica Spekaniya, (Physics of Sintering, Second Edition) Nauka, Moscow, 1984, 312.

33. V. V. Skorohod, Porosh. Metal., No.5(1978) 34.

34. V. V. Skorohod, in Sintering-Theory and Practice, Eds. D. Kolar, S.Pejovnik, M.M. Ristić, Elsevier, 1982, 395.

35. E. A. Barringer, R. Brook, H. K. Bowen, Mat. Sci. Res. 16(1984)1.

36. Sintering of NaF, Eds. D.P. Uskoković, M.M.Ristić, Institute of
    Technical Sciences of SASA, Belgrade, Yugoslavia, 1976, 104.
37. V. V. Skorohod, V. A. Smirnov, A. F. Hrienko, D. P. Uskoković,
    Physics of Sintering 4(1972)47.
38. M. F. Ashby, Acta Metal., 22(1974)275.
39. D. Uskoković, Z. Nikolić, M. M. Ristić, J.Sci.Sintering,
    Spec.Issue, 11(1979)59.
40. M. M. Ristić, Z. S. Nikolić, The Theory of Sintering Diagrams with
    Elements of Sintering Physics, Serbian Academy of Sciences and
    Arts, Belgrade, Yugoslavia, 1987, 127.
41. D. S. Wilkinson, M. F. Ashby, J. Sci. Sintering 10(1978)67.
42. O. V. Roman, V. V. Skorohod, G. R. Fridman, Ultrazvukovoi i
    Rezistometricheskii Kontrol v Poroshkovoi Metallurgii,
    (Ultrasound and Resistometric Control in Powder Metallurgy)
    Vishaya shkola, Minsk, USSR, 1989, 182.
43. M. Slesar, M. Besterci, E. Dudrova, Pokroky Praskove Metal.
    No.2(1980)37.
44. G. C. Kuczynski, in Sintering-Theory and Practice, Eds. D. Kolar,
    S. Pejovnik, M.M.Ristić, Elsevier, 1982,37.
45. L. A. Vermenko, O. I. Getman, S. P. Rakitin, V. V. Skorohod,
    Porosh. Metal., No.11(1981)25 and No.9(1983)18.
46. V. V. Skorohod, Yu. M. Solonin, Porosh. Metal., No. 12(1983)25.
47. B. R. Patterson, V. D. Parkhe, J. A. Griffin, in Sintering'85,
    Eds. G.C.Kuczynski, D.P.Uskoković, H.Palmour III, M.M.Ristić,
    Plenum Press, New York, 1987, 43.
48. V. V. Skorohod, I. D. Fridberg, V. V. Panichkina, Porosh. Metal.,
    No. 9(1985)15.
49. H. Palmour III, T. M. Hare in Sintering'85, Eds. G.C.Kuczynski,
    D.P. Uskoković, H.Palmour III, M.M.Ristić, Plenum, New York,
    1987, 17.
50. A. G. Evans, J.Amer.Ceram.Soc., 65(1982)497.
51. B. A. Druyanov, K. B. Vartanov, Porosh. Metal., No.11(1984) 29.
52. K. B. Vartanov, Porosh. Metal., No. 2(1989)23.

STEREOLOGICAL THEORY OF SINTERING

R.T. DeHoff

Department of Materials Science and Engineering
University of Florida
Gainesville, Florida   U.S.A.

ABSTRACT

A geometrically general theory for microstructural evolution during solid state sintering is under development. The theory uses the global geometric properties of stereology to characterize the geometric state of the microstructure. The time evolution of these properties is related to the distribution of interface velocities on the pore-solid interface and the distribution of vacancy annihilation rates in grain boundaries through the geometrically general kinematic equations of stereology. The evaluation of local interface velocities via assumed mechanisms controlling mass transport completes the structure of the approach.

An example of a test of this theory is presented in an analysis of densification kinetics in the late stages of sintering of nickel powder.

INTRODUCTION

The evolution of understanding of the sintering process is characterized by the parallel development of two viewpoints. The path that focused upon the mechanisms that may operate to produce sintering [1 - 5] applied a sophisticated understanding of atomic processes to rudimentary models for the geometry through which these processes operate. The path that focused upon the rigorous description of the evolution of the geometry of the microstructure during sintering [6 - 9] combined a sophisticated understanding of the microstructure with a rudimentary view of the mechanisms that operate. During four decades of development of the theory of sintering the geometry of the mechanism-centered models became increasingly complex [10 -12], while the adaptation of the understanding of mechanistic operations to the geometry-centered approach proceeded slowly [13].

The time is ripe to lay out a description of sintering that combines a rigorous treatment of both mechanism and

Science of Sintering
Edited by D. P. Uskoković *et al.*
Plenum Press, New York

geometry. In order to be general this theory must be expressed
in terms of those geometric properties that have unambiguous
meaning for real, three dimensional microstructures. The basic
descriptors that satisfy this condition are precisely those
geometric properties that are experimentally accessible through
the methods of stereology [14,15]: volume, V, surface area,
S, total curvature, M, line length, L, connectivity, C and
number of separate parts, N. This stereological theory of
sintering is the subject of this paper.

COMPONENTS OF THE THEORY

The theory attempts to describe the rate of change of
Three of the stereological properties: volume, V, surface area,
S and total curvature, M*. Changes in these properties derive
from two contributions:

1. Local displacements of the pore-solid interface
   described by a local interface velocity, v; and
2. Annihilation of vacancies at grain boundaries in the
   system.

Formulation of a kinetic expression for the local velocity of
the pore-solid interface is complicated by the operation in
parallel of a number of mechanisms, each of which brings into
play unique aspects of the geometry of the structure.
Nonetheless, kinetic equations can be formulated for each
operating mechanism, and combined into a single general
interface velocity equation. Changes in the geometric
properties associated with these interface displacements are
then derived by applying the kinematic equations of stereology
[16,17]. The application is not straightforward; it is
complicated by the vacancy annihilation processes that are
responsible for densification. Incorporation of the vacancy
annihilation contribution requires development of a cell model
for the sintering microstructure.

The resulting equations relate rates of change of volume,
surface area and total curvature to the governing material and
microstructural parameters. Many of the required geometric
properties are not yet accessible to stereological measurement.
The value of these results is the rigorous identification of
the geometric properties that are operative in describing these
microstructural changes. Thus, approximate estimates may be
devised with some understanding of the magnitude and direction
of their expected deviation from the quantities rigorously
required by the theory.

LOCAL INTERFACE VELOCITIES

Complexities in the analysis of the sintering process
arise because a variety of mechanisms of mass transport may

----

*If κ_1 and κ_2 are the local principal normal curvatures at an
element of surface, then H = $(\kappa_1+\kappa_2)/2$ is the local mean
curvature. The total curvature is the integral of the mean
curvature over the surface of the structure (the pore surface
in this case): M = $\iint$ HdS = $\iint$ dM. This quantity is
accessible through the feature count or the tangent count in
stereology [14,15].

operate simultaneously to connect a variety of sources and sinks; in the process, the driving forces for these processes are altered as the geometry changes. The sources and sinks lie on the pore solid interface and on the grain boundary network. For each mechanism a specifically definable subset of these surfaces operates as source and sink; these regions on the surfaces are separated by a boundary that has definable geometric properties.

Consider an element of surface on the pore-solid interface. The geometry of this element is described by its local principal normal curvatures, κ_1 and κ_2, its mean curvature, $H = (\kappa_1+\kappa_2)/2$, and its spherical image, $d\Omega$**. Adopt the following conventions: curvatures are positive if directed into the solid phase; velocities and fluxes are positive if directed toward the pore phase; a concentration gradient at the surface is positive if concentration increases in passing toward the vapor phase. Interface accommodation equations (local velocity equations) for each of the assumed operating mechanisms are reviewed in this section.

<u>Vapor Transport</u>. The governing mechanism is assumed to be the local rate of evaporation at an element of surface. The local velocity of an element of surface is given by

$$v_{v.t.} = - V^O J_G = - V^O k [P - P*] \tag{1}$$

where V^O is the atomic volume, JG is the flux in the vapor phase at the surface, k is an evaporation rate constant, P* is the average vapor pressure in the pore volume, determined by the average mean surface curvature on the pore surface, and P is the local vapor pressure at the surface element, determined by the local mean curvature. Assuming local equilibrium at each element of surface, so that $P = P_O + 2\gamma H$, equation (1) becomes:

$$v_{v.t.} = 2V^O \cdot k \cdot \gamma \cdot [<H>_S - H] \tag{2}$$

where $<H>_S$ is the (surface area-weighted) average mean surface curvature for the interface and γ is the specific interfacial free energy. The pore surface is thus subdivided into domains by the set of lines defined by $H = <H>_S$, at which the local interface velocity is zero. Surface elements that lie on the side of this boundary for which $H > <H>_S$ are sources for vapor transport; on the other side, where $H < <H>_S$, each element is a sink.

<u>Surface Diffusion</u>. A thin layer of material at the surface has an enhanced rate of diffusion described by the surface diffusion coefficient, D_s. Material in this region obeys Fick's laws. The displacement of the surface normal to itself results from the accumulation of atoms locally because this surface flux diverges. The local interface velocity is given by

$$v_{s.d.} = - V^O \cdot \delta \cdot \text{div } J_s \tag{3}$$

**The spherical image of an element of surface is the area of a map of its normals on the unit sphere; it is thus the solid angle subtended by these normals [15].

57

where δ is the thickness of the zone of enhanced diffusivity and J_s is the surface diffusion flux. The flux is related to the local concentration gradient, which is in turn related to the gradient in mean surface curvature, so that:

$$v_{s.d.} = \delta \cdot D_s \cdot \Gamma \cdot \nabla^2 H \qquad (4)$$

where $\Gamma = 2\gamma V^{\circ}/kT$ is the capillarity shift coefficient, and $\nabla^2 H$ is the Laplacian of the mean surface curvature. The boundary that separates positive accumulation of surface layer atoms from negative is the condition $\nabla^2 H = 0$. On the side of this boundary for which $\nabla^2 H < 0$, the rate of accumulation is negative; elements in this region may be thought of as being sources for material flow. Regions for which $\nabla^2 H > 0$ are thus material sinks.

Volume <u>Diffusion</u>. The contribution to the displacement of a local element of interface arising from volume diffusion may be given by:

$$v_{v.d.} = V^{\circ} \cdot J_L = - V^{\circ} \cdot J_v \qquad (5)$$

where J_L is the atom flux in the lattice, and J_v the corresponding vacancy flux. Invoke Fick's first law:

$$v_{v.d.} = V^{\circ} \cdot D_v \nabla C \qquad (6)$$

D_v is the vacancy diffusion coefficient and ∇C is the concentration gradient of vacancies in the volume element adjacent to the moving surface element. On the basis of this formulation it is again possible to visualize a subdivision of the pore surface into two classes of regions bounded by curves on the surface defined by the condition $\nabla C = 0$. Where the vacancy concentration gradient normal to the surface is positive, by the chosen convention, vacancies flow into the solid (matter flows out), the velocity is positive, and the surface element is a source of vacancies or a sink for atoms. Where $\nabla C < 0$, the opposite condition holds.

It is useful to devise a second important subdivision of the pore-solid surface in this context. Focus on the region for which $\nabla C > 0$, i.e., the parts of the surface that act as a source for vacancies. Assuming that the vacancy concentration field is continuous in the solid, isoconcentration contours define the vector field of concentration gradients which in turn identify flux lines that fill the solid, connecting sources and sinks. Since no sinks exist within the solid, at any instant in time each element of pore surface within the vacancy source domains has its associated flux line which connects ultimately to its unique element of vacancy sink, called its communicating neighbor. If the communicating neighbor sink also lies on the pore surface, then transport between the two elements contributes to a change in the geometry of that boundary, but does not contribute to densification, i.e., a decrease in the total volume of the system. If and only if the communicating neighbor sink lies on a grain boundary will the vacancy source on the pore surface contribute to densification. Thus, there exists a boundary within the domain for which $\nabla C > 0$, defined, not by the local geometry, but by a change in the nature of the remote vacancy sink, that divides that domain into regions that contribute to densification, and those that do not.

An expression for the concentration gradient at a surface element may be formulated:

$$\nabla C = [C_n - C(H)]/\lambda \tag{7}$$

where $C(H) = C_o(1 + \Gamma H)$ is the vacancy concentration at the reference element, C_n is the concentration at its communicating neighbor, and λ is a diffusion length scale, which to a first approximation is the length of the flux line connecting the communicating neighbors. Thus, the lattice diffusion contribution to the local interface velocity depends, not only upon the local geometry (H), but upon the distance to and properties of its communicating neighbor.

Grain Boundary Diffusion. This mechanism can contribute to the displacement of only those elements on the pore surface that are adjacent to the triple line formed by the intersection of internal grain boundaries with pore surface. The local velocity of such a surface element that is due to flow down the grain boundary is

$$v_{g.b.} = - V^o J_b = V^o \cdot D_b \nabla C_b \tag{8}$$

where J_b is the vacancy flux into the solid at the pore surface adjacent to the grain boundary, D_b is the diffusion coefficient of vacancies in the grain boundary and ∇C_b is the concentration gradient of vacancies in the element of volume adjacent to the pore surface at the grain boundary. An equation analogous to equation (7) may be used to evaluate the concentration gradient.

The total velocity of any element of pore surface not adjacent to the grain boundary is given by the sum:

$$v = v_{v.t.} + v_{s.d.} + v_{v.d.} \tag{9}$$

For elements adjacent to the grain boundary,

$$v = v_{v.t.} + v_{s.d.} + v_{g.b.} \tag{10}$$

Elements for which $v > 0$ are net sinks for matter transport, while those for which $v < 0$ are net sources.

THE BIPYRAMID CELL MODEL FOR POROUS MICROSTRUCTURES

Consider a partially sintered polycrystal. The microstructure consists of a collection of crystals that are in contact at grain boundaries and a pore structure which may vary from a single, multiply connected network to a collection of simple isolated pores as sintering proceeds. Visualize the subdivision of this three dimensional microstructure into a collection of space filling cells, Figure 1, such that:

a. Each cell contains one grain and its "associated porosity"; and
b. Each cell face contains the grain boundary between a pair of contacting particles.

In the early stages of the process there may be some ambiguity in the construction of this cell structure; some faces in the structure may not contain any grain boundary area. This ambiguity will disappear after sufficient densification has occurred to bring together all potentially contacting grains.

THE VOLUME CHANGE

Focus upon a cell face with its associated grain boundary, Figure 2. Identify the centroids of the contacting particles and construct lines from these points to the vertices of the cell face. The resulting bipyramid has a face area A_j and a pair of altitudes, p_{1j} and p_{2j}, as shown in Figure 2. Let A_j^b be the area of the grain boundary on this face. In time dt the altitudes change their lengths by amounts dp_{1j} and dp_{2j}; let $dp_j = dp_{1j} + dp_{2j}$. The change in volume associated with this bipyramid is

$$dV_j = A_j \, dp_j \tag{11}$$

This volume change is the result of the removal of a layer of solid of thickness dp_j and area A_j^b. Define n_{Aj} to be the rate of annihilation of vacancies (number per square centimeter per second) on the jth boundary. (Forces operate to insure that this rate will be essentially uniform over the grain boundary face at any instant.) Then the volume removed may be written:

$$A_j^b \, dp_j = - v^o \, A_j^b \, n_{Aj} \cdot dt \tag{12}$$

Note that

$$\frac{dp_j}{dt} = - v^o \, \dot{n}_{Aj} \tag{13}$$

Equation (11) may now be expressed in terms of the vacancy annihilation rate:

$$dV_j = f_j \cdot A_j^b \cdot dp_j = - f_j \cdot v^o \cdot A_j^b \cdot \dot{n}_{Aj} \cdot dt \tag{14}$$

which introduces $f_j = A_j/A_j^b$, the efficiency factor for vacancy annihilation on the boundary. This quantity reports the ratio of the volume shrinkage of the bipyramid to the volume of vacancies annihilated. It is a very large number early in sintering when the grain boundary occupies a small fraction of the cell face, and decreases steadily toward 1 as the grain face progressively fills up the cell face during densification.

Since the collection of bipyramids is space filling, the change in volume for the system in time dt is the sum of the contributions from each of the cells:

$$dV = \sum_j dV_j = - v^o \sum_j f_j \cdot A_j b \cdot \dot{n}_{Aj} \cdot dt$$

This result may be written:

$$\frac{dV}{dt} = - v^o \cdot \langle f \rangle \cdot \langle \dot{n}_A \rangle \cdot A^b \tag{15}$$

where $\langle f \rangle$ is a peculiarly weighted average efficiency factor:

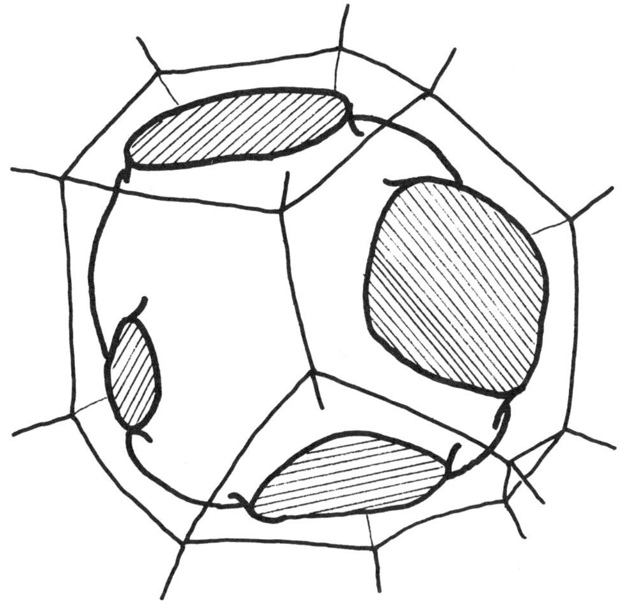

Figure 1. Subdivision of a partially sintered structure into space filling cells, each containing one grain and its associated porosity.

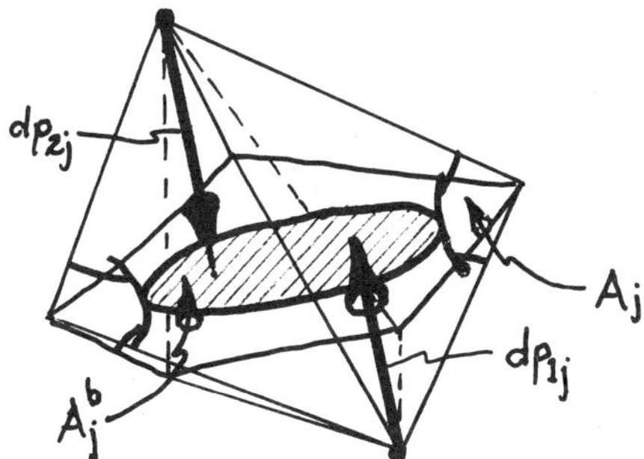

Figure 2. Bipyramid structure associated with each particle pair and their contacting grain boundary facet.

$$\langle f \rangle = [\ \sum_j f_j \cdot A_j^{\ b} \cdot \dot{n}_{Aj}\ ]/[\ \sum_j A_j^{\ b} \cdot \dot{n}_{Aj}\ ] \tag{16}$$

$\langle \dot{n}_A \rangle$ is the average rate of annihilation of vacancies in the system:

$$\langle \dot{n}_A \rangle = [\ \sum_j A_j^{\ b} \cdot \dot{n}_{Aj}\ ]/[\ \sum_j A_j^{\ b}\ ] \tag{17}$$

and A^b is the total grain boundary area in the structure:

$$A^b = \sum_j A_j^{\ b} \tag{18}$$

The number of vacancies annihilated in time dt in the system is given by

$$dN\ =\ \sum_j A_j^{\ b} \cdot \dot{n}_{Aj} \cdot dt\ =\ \langle \dot{n}_A \rangle \cdot A^b \cdot dt \tag{19}$$

These vacancies are supplied from those elements of the pore surface that have communicating neighbors on the grain boundaries; those adjacent to the triple line contribute by grain boundary diffusion, and those on the remainder of the subregion connected to the grain boundary by flux lines contribute by vacancy diffusion through the lattice. However, the domain of integration may be extended to the whole pore surface since the integrated flux over the remaining elements is zero. The domain of integration for the grain boundary source is the length of the triple line.

$$dN\ =\ \iint_S J_v \cdot dS \cdot dt\ +\ \int_L J_b \cdot \delta \cdot dL \cdot dt \tag{20}$$

Apply Fick's Law:

$$dN\ =\ \iint_S (-D_v \nabla C) \cdot dS \cdot dt\ +\ \int_L (-D_b \nabla C_b) \cdot \delta \cdot dL \cdot dt$$

Since no vacancies are accumulated in the solid, the expressions for dN in equations (19) and (21) are equal, and

$$\langle \dot{n}_A \rangle \cdot A^b\ =\ \iint_S (-D_v \nabla C) \cdot dS\ +\ \int_L (-D_b \nabla C_b) \cdot \delta \cdot dL \tag{21}$$

Equation (15) relates the left hand side of equation (21) to the rate of densification:

$$\frac{dV}{dt}\ =\ -\ V^o \cdot \langle f \rangle \cdot \{ \iint_S (-D_v \nabla C) \cdot dS\ +\int_L (-D_b \nabla C_b) \cdot \delta \cdot dL \} \tag{22}$$

Introduce the "surface area-weighted average vacancy concentration gradient over the pore surface":

$$\langle \nabla C \rangle_s\ =\ \iint_S \nabla C \cdot dS / \iint_S dS \tag{23}$$

and the "line length-weighted average vacancy concentration gradient on elements at the triple line":

$$<\nabla c>_b = \int_L \nabla c_b \cdot dL / \int_L dL \tag{24}$$

With these definitions, equation (22) may be written:

$$\frac{dV}{dt} = V^o \cdot <f> \cdot \{D_v <\nabla c>_s \cdot S + D_b <\nabla c>_b \cdot \delta \cdot L\} \tag{25}$$

This result may also be written:

$$\frac{dV}{dt} = V^o \cdot <f> \cdot D_v <\nabla c>_s \cdot S\{1 + \frac{D_b \cdot \delta L <\nabla c>_b}{D_v \cdot S <\nabla c>_s}\} \tag{26}$$

This result is similar to that obtained by Berrin and Johnson (5). The second factor in brackets is a quantitative measure of the competition between volume diffusion and grain boundary diffusion to control the process. The key stereological parameters are the surface area of the pore-solid interface and the length of the triple line at which grain boundaries emerge on the pore surface.

THE CHANGE IN SURFACE AREA

The kinematic equations of stereology [17] provide an estimate of the change in surface area due to the displacement of the pore surface described in equation (9):

$$[\frac{dS}{dt}]_S = 2 \iint_S v \cdot H dS = 2 \iint_M v \cdot dM \tag{27}$$

However, this is not the only contribution to the changing surface area in the system. As material is removed by vacancy annihilation at grain boundaries and the body densifies, surface area is lost where this removed material is incident on the pore surface, i.e., along the triple line.

Surface Area Loss Due to Shrinkage

Focus again upon the jth bipyramid cell. The perimeter of grain boundary region in the cell face is a segment of the triple line; let its length be L_j. In time dt a layer of thickness dp_j is removed by vacancy annihilation. The area of the segment of pore-solid interface bounding this layer is $L_j \cdot dp_j$. Multiply and divide this expression by the grain boundary area on the face:

$$dS_j = (L_j / A_j^{\;b}) \cdot A_j^{\;b} \cdot dp_j \tag{28}$$

Substitute from equation (12)

$$dS_j = (L_j / A_j^{\;b}) \cdot (-V^o \cdot A_j^{\;b} \cdot \dot{n}_{Aj} \cdot dt) \tag{29}$$

The total change in surface area associated with this contribution in time dt is

$$(dS)_{gb} = \sum_j dS_j = \sum_j (L_j / A_j^{\;b}) \cdot (-V^o \cdot A_j^{\;b} \cdot n_{Aj} \cdot dt) \tag{30}$$

The rate of change of surface area may be written:

$$\left[\frac{dS}{dt}\right]_{gb} = - V^o \ \langle \tfrac{L}{A}b \rangle \cdot \langle \dot{n}_A \rangle \cdot A^b \tag{31}$$

where

$$\langle \tfrac{L}{A}b \rangle = \sum_j (L_j/A_j^{\ b}) \cdot A_j^{\ b} \cdot n_{Aj} / \sum_j A_j^{\ b} \cdot n_{Aj} \tag{32}$$

is a peculiarly weighted average of the perimeter to area ratio for the grain boundary faces. The other factors on the right side of equation (31) are identical with those that appear in equation (15). Applying equation (15) to evaluate these properties yields

$$\left(\frac{dS}{dt}\right)_{gb} = \langle \tfrac{L}{A}b \rangle \cdot \frac{1}{\langle f \rangle} \cdot \frac{dV}{dt} \tag{33}$$

The total change in surface area is the sum of the contributions due to moving pore surface and shrinkage associated with vacancy annihilation, equations (27) and (33):

$$\frac{dS}{dt} = \left(\frac{dS}{dt}\right)_S + \left(\frac{dS}{dt}\right)_{gb}$$

$$= 2\iint_M v \cdot dM + \langle \tfrac{L}{A}b \rangle \cdot \frac{1}{\langle f \rangle} \cdot \frac{dV}{dt} \tag{34}$$

Since the velocity in the first term has three terms, see equation (9), the expression for the rate of change of surface area has four contributions.

<u>Vapor Transport</u>

$$\left(\frac{dS}{dt}\right)_{v.t.} = 2\iint_S 2V^o k \cdot \gamma \cdot [\langle H \rangle_S - H] \cdot H dS$$

$$= 4V^o k \cdot \gamma \cdot \{ \langle H \rangle_S \iint_S H dS - \iint_S H^2 \ dS \}$$

$$\left(\frac{dS}{dt}\right)_{v.t.} = 4V^o k \cdot \gamma \cdot [\langle H \rangle_S^2 - \langle H^2 \rangle_S] \cdot S \tag{35}$$

The quantity in brackets is the difference between the square of the mean and the mean square of the mean curvature distribution over the area of the pore surface.

<u>Surface Diffusion</u>

$$\left(\frac{dS}{dt}\right)_{s.d.} = 2\iint_M \delta \cdot \Gamma \cdot D_s \nabla^2 H \cdot dM$$

$$\left(\frac{dS}{dt}\right)_{s.d.} = 2\delta \cdot \Gamma \cdot D_s \cdot \langle \nabla^2 H \rangle_M \cdot M \tag{36}$$

The geometric property defined in this equation is

$$\langle \nabla^2 H \rangle_M = \iint_M \nabla^2 H \cdot dM / \iint_M dM \tag{37}$$

which is the M-weighted average of the Laplacian of the mean surface curvature on the pore surface.

<u>Volume Diffusion</u>

$$\left(\frac{dS}{dt}\right)_{v.d} = \iint_M v^o \cdot D_v \nabla c \cdot dM$$

$$\left(\frac{dS}{dt}\right)_{v.d} = v^o \cdot D_v \cdot \langle \nabla c \rangle_M \cdot M \tag{38}$$

The term in brackets is an M-weighted average of the vacancy concentration gradient over the pore surface. The components of this quantity may be further clarified with aid of equation (7):

$$\iint_S \frac{1}{\lambda} [C_n - C(H)] \cdot H dS = \langle \frac{1}{\lambda} \rangle_M \cdot [\langle C_n \rangle_M - C_o \Gamma \frac{\langle H^2 \rangle_S}{\langle H \rangle_S}] \cdot M \tag{39}$$

This factor involves appropriately weighted averages of the harmonic mean of the diffusion length scale, the communicating neighbor concentration, and the mean square of the surface curvature distribution.

The rate of change of the surface area of a sintering structure is the sum of the four expressions on the right hand sides of equations (33), (35), (36) and (38). This intractable result is made more palatable with the observation that in a given situation only one (or at most, two) of these mechanisms make an overwhelming contribution to the process, so that the rest may be neglected.

It is important to note that in a given circumstance the rate of change of surface area may be dominated by a different mechanism from that which controls densification. Thus, for example, surface diffusion may dominate in the determination of the rate of change of surface area, but cannot contribute to changing the volume of the system; volume diffusion or grain boundary diffusion may operate to produce densification.

THE CHANGE IN TOTAL CURVATURE

The development of equations that describe the rate of change of the integral mean curvature, M, parallels that in the last section for the surface area. The fundamental stereological equation for the rate of change of total curvature is [17]

$$\frac{dM}{dt} = \iint_S v \cdot K dS = \iint_\Omega v \cdot d\Omega \tag{40}$$

where the domain of integration is over the spherical image of the pore surface. As in the case for surface area, v is the local velocity of a surface element, given by equation (9).

Also, there is an additional contribution the change in M associated with the annihilation of vacancies and the associated disappearance of elements of surface elements adjacent to the triple line in the structure.

<u>Total Curvature Loss Due to Shrinkage</u>

Focus again upon the jth bipyramid cell. The perimeter of grain boundary region in the cell face is a segment of the triple line. Focus further upon an element of length of this triple line, dL. The surface adjacent to this line element has mean curvature, H. In time dt a layer of thickness dp_j is removed by vacancy annihilation. In the process an area $dLdp_j$ with its mean curvature H is removed from the system. The change in integral mean curvature associated with this loss of surface area is $HdLdp_j$. Over the perimeter of the grain boundary on the jth face this loss totals

$$dM_j = -\int_{L_j} HdL \cdot dp_j = - <H>_j \cdot L_j \cdot dp_j \tag{41}$$

where

$$<H>_j = \int_{L_j} HdL / \int_{L_j} dL \tag{42}$$

is the average mean curvature of the surface adjacent to the triple line on the grain boundary in the jth bipyramid. For the system,

$$dM = -\sum_j <H>_j \cdot L_j \cdot dp_j = - \sum_j <H>_j \cdot L_j \cdot (-v^o \cdot n_{Aj} \cdot dt)$$

so that the rate of change,

$$\left(\frac{dM}{dt}\right)_{g.b.} = - v^o \sum_j <H>_j \cdot (L_j / A_j^{\ b}) \cdot A_j^{\ b} \cdot n_{Aj}$$

$$\left(\frac{dM}{dt}\right)_{g.b.} = - v^o \cdot <H>_b \cdot <\frac{L}{A}b> \cdot <\dot{n}_A> \cdot A^b \tag{43}$$

where $<H>_b$ is an appropriately weighted average of the mean curvature of surface elements that are adjacent to the triple line on the pore surface. The other averages in this equation are identical with those defined previously in equations (32) and (17). Invoke equation (15):

$$\left(\frac{dM}{dt}\right)_{g.b.} = <H>_b \cdot <\frac{L}{A}b> \cdot \frac{1}{<f>} \cdot \frac{dV}{dt} \tag{44}$$

Comparison with equation (33) yields

$$\left(\frac{dM}{dt}\right)_{g.b.} = <H>_b \cdot \left(\frac{dS}{dt}\right)_{g.b.} \tag{45}$$

The derivations of the contributions to the rate of change of M due to displacements of the pore-solid interface parallel those presented for the rates of change of surface area, equations (35), (36) and (38). These results are obtained by inserting the velocity equations for each contribution, equations (2), (4) and (6), into the kinematic equation for M, equation (40). Differences arise because these velocities are weighted according to the local spherical image of the moving surface element, $d\Omega$. The results are:

<u>Vapor Transport</u>

$$(\frac{dM}{dt})_{v.t.} = 2V^{o}k\cdot\gamma\cdot\ [<H>_{s} - <H>_{\Omega}]\cdot\Omega \qquad (46)$$

where

$$<H>_{\Omega} = \iint_{\Omega} H\cdot d\Omega / \iint_{\Omega} d\Omega \qquad (47)$$

is a spherical image-weighted average of the mean curvature on the pore surface.

<u>Surface Diffusion</u>

$$(\frac{dM}{dt})_{s.d.} = \delta\cdot D_{s}\cdot\Gamma\cdot<\nabla^{2}H>_{\Omega}\cdot\Omega \qquad (48)$$

where

$$<\nabla^{2}H>_{\Omega} = \iint_{\Omega}\nabla^{2}H\cdot d\Omega / \iint_{\Omega} d\Omega \qquad (49)$$

is a spherical image-weighted average of the Laplacian of the mean surface curvature distribution.

<u>Volume Diffusion</u>

$$(\frac{dM}{dt})_{v.d.} = V^{o}\cdot D_{v}\cdot<\nabla C>_{\Omega}\cdot\Omega \qquad (50)$$

in which

$$<\nabla C>_{\Omega} = \iint_{\Omega}\nabla C\cdot d\Omega / \iint_{\Omega} d\Omega \qquad (51)$$

is a spherical image-weighted average of the vacancy concentration gradient in the solid adjacent to the pore surface. Application of the communicating neighbor description of the concentration gradient, equation (7), clarifies the components involved in this quantity.

$$\iint_{\Omega}\nabla C\cdot d\Omega = \iint_{\Omega}\frac{1}{\lambda}[C_{n} - C(H)]\cdot d\Omega = \iint_{\Omega}\frac{1}{\lambda}[C_{n} - C_{o}\Gamma H]\cdot d\Omega$$

$$= <\frac{1}{\lambda}>_{\Omega}\cdot[<C_{n}>_{\Omega} - C_{o}^{\Gamma}<H>_{\Omega}]\cdot\Omega \qquad (52)$$

Thus, this quantity contains spherical image-weighted averages of the reciprocal of the diffusion length scale distribution, the communicating neighbor concentration distribution and the distribution of mean curvature values on the pore surface.

The rate of change of total curvature is the sum of the four contributions presented in equations (45), (46), (48) and

(50). Fortunately, in any given sintering situation it is
expected that one, or at most two, of these contributions will
be important, and the others may be neglected.

DISCUSSION

It is evident that a rigorous formulation of the
description of the evolution of microstructure during sintering
evokes a bewildering array of geometric properties that require
evaluation. In addition to the global geometric properties
that appear in these equations (volume V, surface area S, total
curvature M and spherical image Ω) that are stereologically
accessible, a large number of geometric properties that are
(frequently peculiarly weighted) averages emerge. Properties
that appear include: the average efficiency factor, $<f>$;
certain averages of the mean curvature distribution, $<H>_S$,
$<H^2>_S$, $<H>_\Omega$; the distribution of values of the Laplacian of the
mean curvature, $<\nabla^2 H>_M$, $<\nabla^2 H>_\Omega$; the distribution of
communicating neighbor concentrations, $<C_n>_M$, $<C_n>_\Omega$; and the
reciprocal of the diffusion length scale distribution
associated with volume diffusion, $<1/\lambda>_M$, $<1/\lambda>_\Omega$. Most of
these geometric factors can not be rigorously estimated at the
current state of the stereological apparatus. The theory
identifies the properties that are required; methods for
estimating or approximating those properties, and assessing
those approximations, remain to be developed.

It has been pointed out that the relative importance of
the operating mechanisms in a given situation may be different
for the different properties, V, S and M. Ratios that combine
physical and geometric properties that determine these relative
importances may be obtained by applying the strategy that
yielded equation (26). In many cases this approach may yield
combinations of physical parameters that differ from 1 by
orders of magnitude. In these cases it may only be necessary
to obtain order of magnitude estimates of the ratios of the
unmeasurable geometric factors in the relationship.

Relationship of this Analysis to Other Applications

Many of the equations presented in this comprehensive
development have appeared in earlier analyses of sintering and
related processes. For example, those equations predicting
behavior contributed by vapor transport, equations (35) and
(46), are identical to those that apply to coarsening processes
controlled by an interface reaction rate [18]. Equations
describing the contribution from surface diffusion, equations
(36) and (48), were presented in an earlier paper that focused
only on this mechanism [19]. Equations (38) and (50)
describing geometric changes associated with volume diffusion
are similar to, but not identical with, equations describing
diffusion controlled coarsening [20]. Thus, the peculiar
geometric parameters needed to analyze geometric evolution
during sintering have other applications in materials science.
This observation provides further justification for the
theoretical study of this class of properties.

Path of Surface Area Change

One of the more puzzling experimental observations in the
literature is the linear relationship that appears to exist

between the pore surface area and pore volume fraction during loose stack sintering [7,21-23]. This result appears to be pervasive, having been observed in the sintering of metal powders [7,21,22], ceramics [23] and even sol gels [24]. It is seen in structures that have characteristic length scales in the range from 50 microns to a few nanometers. The analysis of the rate of change of surface area presented in this development provides a possible explanation for this observation. If most of the surface smoothing in the system is completed by the end of the first (neck growth) stage of sintering, then it may be possible that contributions to the change in surface area due to the local displacement of surface elements may be negligible during subsequent evolution. Thus, the rate of change of surface area arises from the annihilation of vacancies at grain boundaries, equation (33):

$$(\frac{dS}{dt})_{gb} = <\frac{L}{A}b> \cdot \frac{1}{<f>} \cdot \frac{dV}{dt} \tag{33}$$

The path of surface area change (variation of S with V) is then described by

$$(\frac{dS}{dV}) = <\frac{L}{A}b> \cdot \frac{1}{<f>} \tag{53}$$

The major change in the efficiency factor occurs in the first stage of sintering during which the ratio $f_j = A_j/A_j^b$ decreases rapidly from its initial very large value. During subsequent sintering this quantity slowly decreases toward 1. Its reciprocal thus increases slowly toward 1. The quantity $<L/A_j^b>$ is a peculiarly weighted average of the ratio of L_j to A_j within each cell. As a first approximation, it may be estimated as the ratio of the total triple line length to the grain boundary area in the system. The triple line length increases rapidly in first stage sintering, then peaks and gradually decreases as the grain boundary network forms, ultimately reaching zero when the pore phase disappears. The grain boundary area increases slowly as necks grow, then decreases when a sufficient fraction of the grain boundary network has formed to permit grain growth to begin.

Thus, L, A^b and $<f>$ all decrease slowly during the second and third stages of sintering. It is thus at least plausible that the right hand side of equation (53) does not change much during this part of the process. If there were no change, dS/dV would be constant, and a plot ofS versus V would be linear, with a slope equal to the right hand side of equation (53). Rough estimates show that the magnitude of this slope agrees with values that are determined experimentally.

<u>Experimental Test of the Densification Equation</u>

Her [25] has provided a definitive test of the densification equation (15). Stereological measurements were performed on a sequence of samples prepared by loose stack sintering nickel powder. Best estimates of the vacancy diffusion coefficient, surface energy and atomic volume were combined with measurements of the grain boundary area, triple line length, fraction of pores on the grain boundary network and rates of densification. By taking proper account of the fraction of the pore surface area communicating with grain boundary sinks, it was possible to obtain agreement for the whole series of samples to within ten percent.

Attempts to apply equation (15) to earlier stages of sintering, where the geometry of the structure is topologically more complicated, met with less success. Evidently more rigorous estimates of the geometric properties involved are required. These must await additional theoretical studies of the nature of the geometric parameters involved, and their possible relation to stereological properties that can be experimentally determined.

SUMMARY

A geometrically general stereological theory of sintering has been devised, based upon the kinematic equations of stereology and kinetic formulations of the interface velocities involved. The contribution to geometric change derived from vacancy annihilation at grain boundaries must be included in this description. The results appear largely intractable at the present state of development of the tools for microstructural characterization. Implementation of this formalism evidently will require significant advances in our understanding of the geometry of surfaces.

ACKNOWLEDGEMENT

The Defense Advanced Projects Research Agency (DARPA) provided funding for this theoretical research.

REFERENCES

1. G.C. Kuczynski. <u>Trans AIME</u>, Vol. 185, 169-178 (1949).
2. G. C. Kuczynski. <u>Journal Appl. Phys.</u>, Vol. 20, 1160 (1949).
3. W. D. Kingery and M. Berg. <u>J. Appl. Phys.</u> Vol.26, 1205 (1955).
4. R. L. Coble. <u>J. Appl. Phys.</u> Vol. 32, 787 (1961).
5. L. Berrin and D. L. Johnson. In Sintering and Related Phenomena, G.C.Kuczynski, N.A. Hooton and C. F. Gibbon, Eds, Gordon and Breach, New York (1967) 369-392.
6. F. N. Rhines. Plansee Proceedings, 3rd Seminar Ruette, Tyrol,(1958) 38.
7. E. H. Aigeltinger and R. T. DeHoff. <u>Met. Trans.</u> Vol. 6A 1853 (1975).
8. F. N. Rhines and R. T. DeHoff. In Modern Developments in Powder Metallurgy, Plenum Press, New York, (1971) 193.
9. F. N. Rhines and R. T. DeHoff, "Channel network decay in sintering", in Proc. Sixth Intl Conf. on Sintering and Related Phenomena, Including Catalysis, G.C. Kuczynski, Ed., Plenum Press (1984) 43.
10.W. Beere'. <u>Acta Met.</u> Vol. 22, 131 (1975).
11.R. L. Eadie, G. C. Weatherly and K.T. Aust. <u>Acta Met.</u> Vol. 26, 759 (1978).
12.G. C. Kuczynski. In Ceramic Microstructures - '76, R. M. Fulrath and J. A. Pask, Eds, Westview Press, Boulder CO, (1977) 233-245.

13. R. T. DeHoff, " A cell model for microstructural evolution in sintering", in Proc. Sixth Intl Conf. on Sintering and Related Phenomena, Including Catalysis, G.C. Kuczynski, Ed., Plenum Press (1984) 97.
14. E. E. Underwood. Quantitative Stereology, Addison Wesley, Reading, MS (1970).
15. R. T. DeHoff and F. N. Rhines. Quantitative Microscopy, McGraw-Hill, New York, (1968).
16. R. T. DeHoff. <u>Metals Forum</u>. Vol. 5, 4 (1982).
17. R. T. DeHoff. In: Encyclopedia of Materials Science. Edited by M. Bever. Pergamon Press, Oxford; 1982b; 1127-1129
18. R. T. DeHoff. <u>Acta Met</u>. Vol. 32, 43 (1984).
19. R. T. DeHoff. <u>Sci. of sintering</u>, Vol. 16, 97 (1984).
20. R. T. DeHoff. <u>Acta Met.</u> (1989) In press.
21. F. N. Rhines, R. A. Rummel, H. P. LaBuff and R. T. DeHoff. In Modern Developments in Powder Metallurgy, Volume 1. Plenum Press (1966) 53.
22. N. J. Shaw and R. J. Brook. <u>J. Am. Ceram. Soc.</u>, Vol. 69, 107 (1986).
23. W. Vasconcelos, Doctoral Dissertation, University of Florida (1989).
24. S. M. Her, Doctoral Dissertation, University of Florida (1987).

A STEREOLOGY - BASED EQUATION FOR ISOTROPIC SHRINKAGE DURING SINTERING BY VISCOUS FLOW

Hans Eckart Exner* and Edward A. Giess

IBM Thomas J. Watson Research Center
Yorktown Heights, NY 10598, USA

*On leave from Max-Planck Institut für Metallforschung
Institut für Werkstoffwissenschaften, D-7000 Stuttgart, FRG

A phenomenological shrinkage equation for compacts of a viscous powder is derived using stereological equations to describe the geometry of the pore/solid interface. The geometric assumptions conform to the pore structure observed experimentally over an extended period of shrinkage. The shrinkage equation does not contain adjustable fitting parameters and has the form

$$\ln\left(1 - \frac{\Delta L}{\Delta L_f}\right) = -\frac{2\pi}{3}\ \frac{\gamma}{\eta}\ \frac{1}{\lambda_0}\ \frac{1}{\Delta L_f/L_0}\ (t - t_0)$$

where ΔL and ΔL_f are the linear shrinkage values after isothermal sintering for a period $t - t_0$ and to full density, respectively, of a compact of length L_0. γ and η are the surface energy and the viscosity, respectively, and λ_0 is the mean linear intercept of the solid in the compact. This equation describes results for the shrinkage of cordierite-type glass powder compacts reported earlier.

INTRODUCTION

Equations to describe shrinkage (linear shrinkage, volume shrinkage, or decrease of porosity) of viscous materials have been derived by numerous authors[1-8]. In most of the early derivations[1-4], simplified geometries are assumed, e.g. spherical particles or spherical and cylindrical pores. These simplifications allow analytical solutions for the linear shrinkage, the closure of a pore and the coalescence of two particles. Obviously, only the very early and the very late stages of sintering can be described in this way. The effects of particle shape and size distribution, particle packing and compaction are usually neglected.

A very advanced treatment is due to Scherer[5-8], who developed a microstructural model consisting of a regular arrangement of the solid and the pore space (cylinder model). This model describes the effects of the more complex parameters of powder stacking like bimodal pore size distributions, varying green density or anisotropy. Here we use an alternative approach based upon a global sterological description of the pore/solid interface. An equation is derived which predicts the linear shrinkage of an isotropic compact of viscous powder particles over a wide range of porosities and sintering temperatures from parameters that can rigorously be measured experimentally.

THE MODEL

As Frenkel[1] proposed in his pioneering work of 1945, we relate the frictional work dissipated during viscous flow to the work done by surface forces. In order to find quantitative relationships between these types of work and the geometry of the pore space in a sintering compact, we consider a thin slice cut from the porous compact by two parallel planes as shown in Fig. 1.

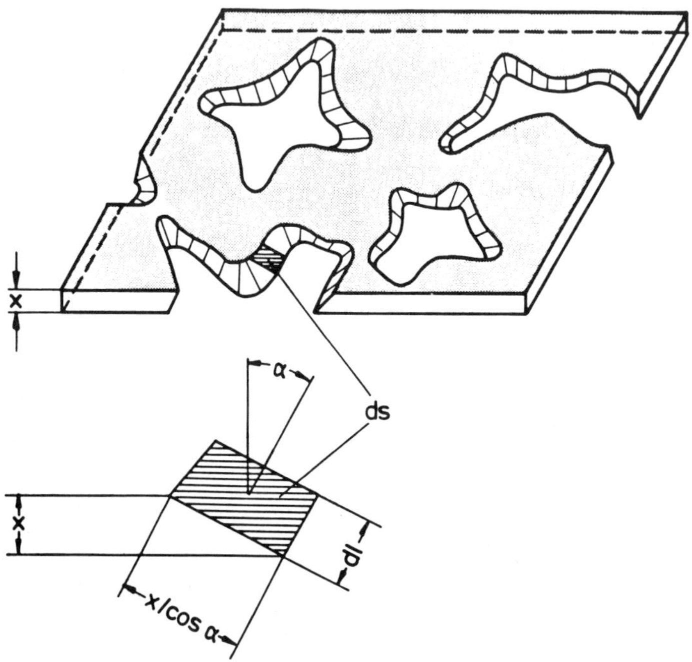

Fig. 1. Slice of a porous compact of height x and cross sectional area A.
A small element of the pore/solid interface inclined to the height direction by an angle α is indicated. In section, the pore/solid interface appears as a solid line of total length l.

We now make the following assumptions: (i) The pore/solid interface is isotropic and, within statistical fluctuations, homogeneously distributed throughout the porous compact. This assumption can be satisfied by careful powder processing and controlled compaction. (ii) The strain rate during an incremental time step, dt, can be treated independently in each of the three orthogonal directions x, y and z. The influence of shrinkage in the two other directions is considered with respect to the change of microstructural geometry. This assumption is based on the idea that linear shrinkage in each direction can be described by the instantaneous geometry of the pore space. (iii) Deformation in each direction can be modelled by a plain strain situation where the rate of relative height change of any slice is identical throughout the compact. (iv) Finally, we assume that the slice can be divided into prismatic segments, each having a hole (a pore section). For a compact made of spherical glass powder, this resembles the geometry in the early and intermediate stages of a sintering very well, as shown in Fig. 2 .

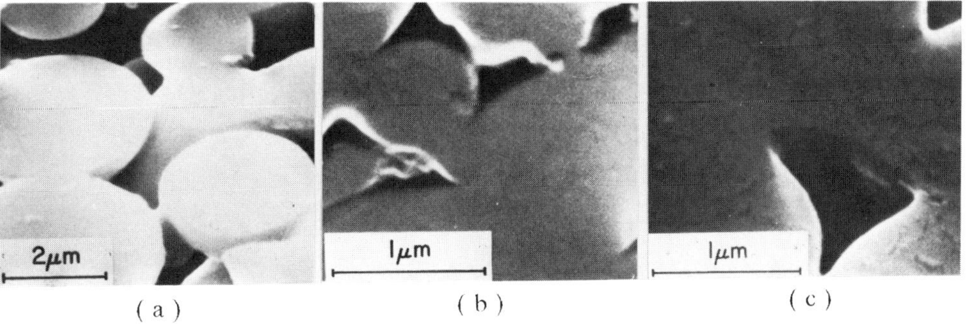

Fig. 2. Pore shapes in sintered glass powders.
(Note that little pore rounding takes place and pore shape is similar in compacts produced from different powders sintered for different times.)
(a) spheroidized glass powder, mean particle size about 6 μm, sintered 6 h at 840° C (relative density 0.89)
(b) milled glass powder, mean size about 3.5 μm, sintered 45 min at 840°C (relative density 0.83)
(c) as in (b) but sintered 150 min (relative density 0.98)

Even in the late stages, some parts of the pore/solid interface are convex with respect to the solid (Fig. 2 c). This situation may be better represented by flat prisms with irregular vertical holes. We did not succeed, however, to find a more general solution to this problem (see also the section on Discussion below).

Under these conditions, the linear shrinkage of the compact can be described by the relative height shrinkage of the slice which is derived as follows: The rate of energy dissipated during viscous flow of the slice, $\dot{e}_s$ (see, for example, ref (8)) is

$$\dot{e}_s = 3\eta v_s \dot{\varepsilon}^2 = 3\eta A_s x \dot{\varepsilon}^2 \tag{1}$$

(where v_s is the solid volume in the slice, ε is the rate of deformation in the direction vertical to the plane of the slice with a basal area occupied by the solid, A_s, and a height x, and η is the viscosity).

For an isotropic sample of height $L_o = nx$ made out of n slices of height x, the total solid volume in the sample, V_s (which does not change during sintering) is

$$V_s = nv_s = nA_s x = A_s L_o \tag{2}$$

and

$$\dot{E}_f = n\dot{e}_f = 3\eta V_s \dot{\varepsilon}^2 = 3\eta V_s \left(\frac{dL}{L_o dt}\right)^2 \tag{3}$$

where dL/L_o is the relative length change during a time step dt.

For calculating the rate of surface energy change during sintering, we again start with the slice shown in Fig. 1. In the sectioning plane the pore/solid interface appears as a line of total length l. The area of a surface element, Δs, inclined by an angle α with respect to the x-axis (see Fig. 1) is

$$\Delta s = \Delta l x / \cos \alpha \tag{4}$$

Here we are interested in the projection of the surface element in the direction of shrinkage $\Delta s \cos \alpha$. The related surface energy is

$$\Delta e_s = \gamma \Delta l x \tag{5}$$

where γ is the specific surface energy of the viscous solid. The rate of surface energy change due to linear shrinkage of the slice in the height direction is

$$\dot{e}_s = \gamma \frac{ds}{dt} = \gamma (l \frac{dx}{dt} + x \frac{dl}{dt}) \tag{6}$$

where $l = \Sigma \Delta l$ is the length of the intersection line of the pore/solid interface in the cross section of the slice. In the limits of statistical fluctuations, l takes the same value in each cross section of the sample. Thus, again for a sample of height $L = nx$, the total rate of surface energy produced by the shrinkage in the height direction is

$$\dot{E}_s = n\dot{e}_s = \gamma (\bar{l} \frac{dL}{dt} + L \frac{d\bar{l}}{dt}) \tag{7}$$

For evaluating $\frac{d\bar{l}}{dt}$, we have to consider only the contribution of shrinkage normal to the direction of the slice plane while for $\bar{l}$ the shrinkage in the plane of the slice must be considered as well. The quantitative assessment of $\frac{d\bar{l}}{dt}$ as a function of $\frac{dL}{dt}$ poses a problem.

For a solid cylinder and for any other convex body, the terms in the bracket of eq. (2) are of opposite sign while for a circular hole or any other concave part the two terms will add up. For the complicated shape of the pore/solid interface including positively and negatively curved parts, we have not succeeded in solving the problem rigorously. Since the reduction of $\bar{l}$ proceeds by the closing of pore channels (and, in the final stage, by the closing of individual pores), the relative reduction of the intersection length $\bar{l}$ is directly related to the relative shrinkage, i.e., $\frac{d\bar{l}}{\bar{l}dt} = \frac{dL}{Ldt}$, and we get

$$\dot{E}_s = 2\gamma \bar{l} \frac{dL}{dt} \tag{8}$$

The average length of the pore/solid intersection line, $\bar{l}$, for any arbitrary sintering time can be related to the average intersection line in the unsintered sample, $\bar{l}_0$, and the linear shrinkage (which, for the isotropic case considered here, is identical normal to and in the slice plane) in the following way. When the centers of the particles in contact approach each other during sintering, all surface elements are displaced in direction of the normal to the surface. Thus, the cross sections of the pores change retaining shape as shown in Fig. 3.

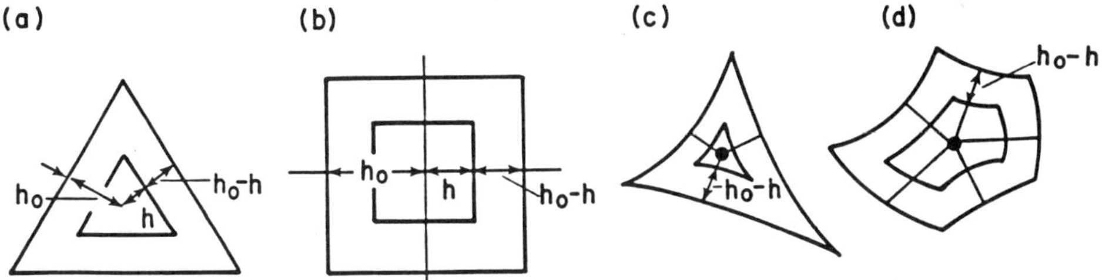

Fig. 3. Model for change of pore geometry during sintering.
(a) and (b) are idealized pore shapes showing the relationship between shrinkage
of the pore and its perimeter (eq. 10). (c) and (d) show that the same relationship
applies for more realistic pore shapes.

(The shape of the cross sections will change only if one pore breaks up into two or more
sections or if local material flow from mildly to sharply curved portions of the pore/solid
interface takes place. In the following we neglect shape changes, since, as we deal with
ratios and averages, most of these statistical variations do not effect the results markedly).
For any planar figure not changing shape, there is a simple relationship between the distance of parallel shift of its sides toward the center and the perimeter length following from
self-similarity rules (see Fig. 3)

$$\frac{l}{l_0} = \frac{h}{h_0} \tag{9}$$

where l_0 and h_0 are the perimeter length and the radius of the inscribed circle of an individual pore section in the unsintered sample and where l and h are these parameters after
sintering time t when a shift of the edges by the distance $h_0 - h$ has taken place. We now
average h_0 and h for all pores and add up the perimeter of all pore sections in the cross
section of the sample and get

$$\bar{l} = \bar{l}_0 \, \frac{\bar{h}}{\bar{h}_0} \tag{10}$$

With the average distance from the center of each particle to the point of contact with
its neighbors $\bar{d}_0$, the relative shrinkage to full density corresponds to $\bar{h}_0 / \bar{d}_0$

$$\frac{\bar{h}_0}{\bar{d}_0} = \frac{\Delta L_f}{L_0} = \frac{L_0 - L_f}{L_0} \tag{11}$$

where ΔL_f and L_f are the final length change and the final length of the (fully dense) sample
and L_0 is the initial (unsintered) sample length. Similarly, the relative shrinkage at any time
is related to the interface shift by

$$\frac{\bar{h}_0 - \bar{h}}{\bar{d}_0} = \frac{\Delta L}{L_0} = \frac{L_0 - L}{L_0} \tag{12}$$

where ΔL and L are the length change and the sample length at time t, and we get

$$\bar{l} = \bar{l}_0 \frac{L - L_f}{L_0 - L_f} \tag{13}$$

A stereological relationship relates the line density in a cross section to the surface density of a sample. This equation, which is generally valid for isotropic interfaces[9,10], yields the value of L_0 from other geometric parameters of the unsintered sample, i.e. the cross section area A_0, the pore/solid interface S_0, and the total volume, V_0 (including the pore volume)

$$\frac{\bar{l}_0}{A_0} = \frac{\pi}{4} \frac{S_0}{V_0} \tag{14}$$

Since $V_0 = A_0 L_0$, we get by combining equations (13) and (14)

$$\bar{l} = \frac{\pi}{4} \frac{S_0}{V_s} \frac{L - L_f}{L_0 - L_f} \tag{15}$$

and substituting $\bar{l}$ in eq.(8)

$$\dot{E}_s = \frac{\pi}{2} \gamma \frac{S_0}{V_s} \frac{L - L_f}{L_0 - L_f} \frac{dL}{dt} \tag{16}$$

We now relate the rate of work produced by the surface energy change $\dot{E}_s$ (eq.16) to the rate of energy dissipation by viscous friction $\dot{E}_F$ (eq.3) and, after rearranging, get

$$\frac{dL}{L_0 dt} = - \frac{\pi}{6} \frac{\gamma}{\eta} \frac{S_0}{V_s} \frac{L - L_f}{L_0 - L_f} \tag{17}$$

Separating the variables yields

$$\int_{l_0}^{L} \frac{\delta L}{L - L_f} = - K \int_{t_0}^{t} \delta t \tag{18}$$

where

$$K = \frac{\pi}{6} \frac{\gamma}{\eta} \frac{S_0}{V_s} \frac{L_0}{L_0 - L_f}$$

Integration of the left hand side is carried out by observing that, if $z = \dfrac{L}{L_f}$

$$\int \frac{\delta z}{z - 1} = \ln (z - 1) + C \tag{19}$$

and noting that $L = L_0$ at $t = t_0$:

$$\ln \frac{L - L_f}{L_0 - L_f} = - K(t - t_0) \tag{20}$$

Rearranging yields

$$\ln\left(1 - \frac{\Delta L/L_0}{\Delta L_f/L_0}\right) = -\frac{\pi}{6}\,\frac{\gamma}{\eta}\,\frac{S_0}{V_s}\,\frac{1}{\Delta L_f/L_0}\,(t - t_0) \tag{21}$$

t_0 is some initial time for $L = L_0$ or $\Delta L = 0$ correcting for the heating up period and other uncertainties of time measurement and initial adjustment of particle arrangement. The final relative shrinkage, $\Delta L_f/L_0$, can easily be estimated from initial porosity, P_0, or from the relative green density, ρ_0/ρ_{th}, according to

$$\Delta L_f/L_0 = 1 - (1 - P_0)^{1/3} = 1 - (\rho_0/\rho_{th})^{1/3} \tag{22}$$

The initial surface area of the sample per unit volume of the solid, S_0/V_s, can be determined by one of the methods for surface area measurement of powders (e.g. , by gas adsorption or permeability). The appropriate method, however, is quantitative microscopy. In a cross section of the compacted sample, the mean linear intercept λ_0 is determined by linear analysis[10]. Here λ_0 is related to the surface/solid volume ratio by the equation

$$\lambda_0 = 4\,\frac{V_s}{S_0} \tag{23}$$

which holds for any shape and size distribution of the powder particles. With eq. (23), we get from eq.(21)

$$\ln\left(1 - \Delta L/L_f\right) = -\frac{2\pi}{3}\,\frac{\gamma}{\eta}\,\frac{1}{\lambda_0}\,\frac{1}{\Delta L_f/L_0}\,(t - t_0) \tag{24}$$

Thus, a shrinkage equation is derived which contains only the physical constants γ and η and parameters, which can readily be obtained by simple measurements.

COMPARISON WITH EXPERIMENTS

In order to compare eq. (24) with experimental results, we use data for crushed cordierite-type glass powders presented in an earlier paper[11]. A pronounced anisotropy of shrinkage rates was found in these experiments with the radial shrinkage being higher than that in the axial direction of the cylindrical sample. A suitable average of the length shrinkage is obtained from the relative volume shrinkage, $\Delta V/V_0$, for each sample according to

$$\Delta L/L_0 = 1 - \left(1 - \frac{\Delta V}{V_0}\right)^{1/3} = 1 - \left[\left(1 - \frac{\Delta D}{D_0}\right)^2\left(1 - \frac{\Delta H}{H_0}\right)\right]^{1/3} \tag{25}$$

where $\Delta D/D_0$ and $\Delta H/H_0$ are the relative diameter and height shrinkages. The final length shrinkage, $\Delta L_f/L_0 = 0.155$, is obtained from the dimensions of a sample sintered to full density according to eq. (22). Taking the viscosity values from a study of similar glasses[12] given by

$$\eta = 1.15 \times 10^{-24}(\exp 85830/T) \tag{26}$$

(T is in K) and estimating the time correction t_0 for the heating period to be about 2.5 min, $\ln(1 - \Delta L/\Delta L_f)$ is plotted vs. $(t - t_0)/\eta$ in Fig. 4. As demonstrated earlier[11] using a semi-empirical Avrami type equation, which is exactly the same form as eq. (24), an excellent fit is obtained. The data points very closely follow a single straight line as expected from eq. (21) or (24) since the temperature dependence of the surface energy γ is negligible compared to that of η (one percent vs. two orders of magnitude over the temperature range considered here).

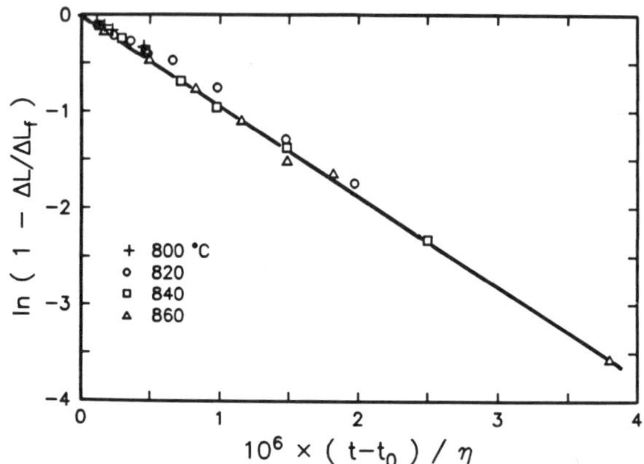

Fig. 4. Plot of shrinkage data calculated from results for a crushed glass powder for four temperatures. t is in s and η is in Pa.s.

The slope of the line in Fig. 4 is

$$\beta_{meas} = -0.94 \times 10^6 \ Pa^{-1} \tag{27}$$

In order to compare this value with that predicted by eq. (24), we calculate a slope β_{calc} from data used in reference (11), i.e. the surface energy of the glass, $\gamma = 0.36 N/m$, and the average particle radius of the powder $r_0 = 3.5\mu m$. r_0 can be related to the surface/solid volume ratio or to the mean linear intercept of the solid in the compact for particles of uniform simple shape and uniform size. For N spherical particles we get $S_0 = N \cdot 4\pi r_0^2$ and $V_s = N \cdot 4\pi r_0^3 /3$, and with eq. (23)

$$r_0 \, (\, spheres \,) \; = \; \frac{3}{4} \, \lambda_0 \tag{28}$$

Substituting λ_0 in eq.(24), we can calculate the slope

$$\beta_{calc} \; = \; - \; \frac{\pi}{2} \; \frac{\gamma}{r_0} \; \frac{1}{\Delta L_f/L_0} \; = \; - \; 1.04 \times 10^6 \ Pa^{-1} \tag{29}$$

The experimental and the calculated slope differ by only 10 percent. Considering the non-spherical shape and the averaging procedure for the mean particle radius using the distribution measured[11,13,14], this agreement seems more than satisfactory. Only a small correction of the value estimated for r_0 (3.9 μm instead of 3.5 μm) or using a slightly higher numerical factor in eq.(28) (5/6 = 0.83 instead of 3/4 = 0.75) would make the experimental and the calculated average rate constant coincide. A slight systematic decrease of the slope is noticeable from Fig. 4 for data taken at different temperatures, indicating that the viscosity values may not be adequately represented by eq. (26). Using a slightly lower exponential constant (80500 instead of 85830) and adjusting the preexponential constant to 1.73×10^{-22}, this inconsistency is eliminated and the fit of the experimental rate constants with the prediction of eq. (24) becomes better than 5% using the original values of the particle radius (3.5μm) and of the numerical constant.

DISCUSSION

It is not only the good agreement of the physical contants and of the experimental shrinkage curves that makes the shrinkage equation derived in this paper appropriate but also the fact that the derivation does not require the assumption of a regular geometry of the pore space.

Obviously, the most restrictive assumption in this derivation is that leading to eq. (8). With idealized geometric assumptions, a rigorous derivation based on volume constancy as a boundary condition would be possible. Of course, the same assumption on the geometry of the pore space would need to be made in the derivation of eq. (1). Our attempts to generalize the derivation using only microstructural dynamics based entirely on stereological concepts, as proposed by De Hoff[5], were not successful. During these attempts, we convinced ourselves that both the energy dissipated during viscous flow and that produced by the change of surface are related to the ratio of solid volume fraction / pore volume fraction (which increases rapidly at the later stages of sintering). Lacking a simple reasoning or a quantative derivation of these factors, we would like to leave this problem to future work.

The only other assumption is that of constant pore shape which we took from experimental observations (see Fig. 2, compare also ref. (16)). Compared to earlier treatments[2,3,5], this assumption seems less severe with respect to the microstructure of powder compacts during sintering. The particle size distribution and details of particle shape do not enter either the derivation of the shrinkage equation or the comparison with the experimental results, provided that the mean linear intercept of the pore space or the density of the pore/solid interface in the unsintered compact are assessed by quantitative microscopy. Uniform spherical pore shape, as assumed by Mackenzie and Shuttleworth[2], is hardly ever observed except in the very late stages (> 0.95 relative density) and extrapolation of the Frenkel approach[3] is restricted to monosized spherical particles and to small neck sizes (and therefore to very small amounts of linear shrinkage ($< 1\%$)). Since dramatic changes of pore shape only take place in the very late sintering stage, as shown in Fig. 3, eq.(21) describes shrinkage during a much larger portion of the period of sintering without any specific assumption about the pore space geometry.

Previous authors[13] and earlier experimental results[11,16] have shown that irregular particle shape leads to a significant acceleration of shrinkage rate compared to the rate for a spherical shape. This observation can probably be quantitatively understood if the larger surface to volume ratio of irregular particles is taken into account. In the very detailed treatment of this problem, Scherer[6,7] showed that the effect of pore size distribution is not pronounced unless the distribution is very broad or bimodal. We would expect that the effects of the particle size distribution and irregular particle shape are accounted for by taking the mean linear intercept λ_0 as the suitable average for the real powder compact.

ACKNOWLEDGEMENTS

The authors thank G. W. Scherer and M. W. Shafer for helpful discussions. C. Aliotta ably did the SEM work. Sponsorship of a visiting scientist position for one of the authors (H.E.E.) by IBM Germany and IBM EMEA is gratefully acknowledged.

REFERENCES

1. J. Frenkel, "Viscous Flow of Crystalline Bodies Under the Action of Surface Tension," J. Physics (Moscow), 9: {5} 385-91 (1945).

2. J.K. Mackenzie and R. Shuttleworth, "Phenomenological Theory of Sintering," Proc. Phys. Soc. London, 62: {12B} 838-52 (1949).

3. W.D. Kingery and M. Berg, "Study of the Initial Sintering Stages of Sintering Solids by Viscous Flow, Evaporation - Condensation, and Self-Diffusion," J. Appl. Phys., 26: {10} 1205-12 (1955).

4. N.V. Solomin and G.M. Tomilov, "Sintering Kinetics of Vitreous Silica," Inorg. Mater. (Engl. Transl. of Neorg. Mater., Moscow), 6: {10} 1631-34 (1970).

5. G.W. Scherer, "Sintering of Low Density Glasses: I, Theory," J. Am. Ceram. Soc., 60 {5-6} 236-39 (1977).

6. G.W. Scherer, "Sintering of Low-Density Glasses: III, Effect of a Distribution of Pore Sizes," ibid. pp 243-46.

7. G.W. Scherer, "Viscous Sintering of a Bimodal Pore-Size Distribution," J.Am. Ceram. Soc., 67: {11} 709-15 (1984).

8. G.W. Scherer, "Viscous Sintering of Inorganic Gels", in: Surface and Colloid Science, Vol. 14, Plenum Publ. Corp., New York, London (1987) pp. 265-300.

9. E.H. Aigeltinger and H.E. Exner, "Stereological Characterization of the Interaction between Interfaces and its Application to the Sintering Process," Metall. Trans. A {3} 421-24 (1977).

10. E. E. Underwood, Quantitative Stereology, Addison Wesley, Reading, 1970.

11. E.A. Giess, J.P. Fletcher and L.W. Herron, "Isothermal Sintering of Cordierite-Type Glass Powders," J. Am. Ceram. Soc., 77: {8} 549-52 (1984).

12. E.A. Giess and S.H. Knickerbocker, "Viscosity of $MgO - Al_2O_3 - SiO_2 - B_2O_3 - P_2O_5$ Cordierite-Type Glasses", J. Mat. Sci. Lett., 4: {7} 835-37 (1985).

13. I.B. Cutler and R.E. Henrichsen, "Effect of Particle Shape on the Kinetics of Sintering of Glass," J. Am. Ceram. Soc., 51: {10} 604-5 (1968).

14. E.A. Giess, C.F. Guerci, G.F. Walker and S.H. Wen, "Isothermal Sintering of Spheroidized Cordierite-Type Glass Powder," Comm. Am. Ceram. Soc., 68: {12} C-328-29 (1985).

15. R.T. De Hoff, " The Dynamics of Microstructural Changes", in: Treatise on Materials Science and Technology, Vol. 1 (Ed. H. Herman). Academic Press, New York, London (1972) pp. 247 - 292.

16. H.E. Exner, "Principles of Single Phase Sintering", Rev. Powder Metall. Phys. Ceram., 1: {1-4} 7-251 (1979).

A NEW MODEL FOR INITIAL SINTERING OF OXIDE POWDERS UNDER CONTROLLED GASEOUS ATMOSPHERE

Michèle Pijolat and Michel Soustelle

Ecole Nationale Supérieure des Mines, 158 cours
Fauriel - 42023 Saint-Etienne Cedex, France

INTRODUCTION

Initial sintering of oxide powders annealed at moderate temperatures corresponds to a surface area reduction or particle growth without increase in density and mechanical toughness of the material. Applications of ultrafine oxide powders in the field of catalysis requires a good thermal stability. More often such materials do not exhibit the appropriate high surface area during a long period at temperatures in the range 800-1200 K. The decrease in surface area as a function of time has already been studied by several authors, as for example, in the case of MgO[1,2], ZnO[3] and transition alumina[4].

It is well known that the rate of initial sintering of oxide powders may be influenced by two factors: impurities (native defects or incorporated as additives) and the nature of the gaseous atmosphere. In particular, water vapour has an enhancing effect on magnesia initial sintering[2,4,5].

A general property of the surface of oxide powders is to interact with the molecules of water contained in the ambient air. As a consequence, superficial hydroxyl ions can still be observed after treatments at temperatures as high as about 1000 K. The accelerating effect of water vapour may thus be understood from the basis of an interaction between H_2O molecules and the surface giving hydroxyl ions.

In the present work we shall propose a mechanistic model for initial sintering of oxide powders in the presence of water vapour. Then results of kinetics of surface area decrease during isothermal annealing under controlled partial pressures of water vapour and oxygen will be reported. Finally, the comparison between theoretical and experimental kinetic laws will be finally discussed.

```
O   M
M < >                                    (H₂O)g
O   M < >                                    :
M   O   M < >                                :
O   M   O   M   O   M   O   M < > M   O   M
M   O   M   O   M   O   M   O   M   O   M   O


O   M                              step i
M < >
O   M < >
M   O   M < >                            H       H
O   M   O   M   O   M   O   M   O   M   O   M
M   O   M   O   M   O   M   O   M   O   M   O


O   M                    step ii
M  OH
O   M < >
M   O   M  OH
O   M   O   M   O   M   O   M < > M < > M
M   O   M   O   M   O   M   O   M   O   M   O


O   M                    step iii
M < >
O   M  OH  □ < >
M   O   M  OH
O   M   O   M   O   M   O   M < > M < > M
M   O   M   O   M   O   M   O   M   O   M   O


O   M              (H₂O)g   step iv
M < >          ↑
O   M < > □ < >
M   O   M   O
O   M   O   M   O   M   O   M < > M < > M
M   O   M   O   M   O   M   O   M   O   M   O


O   M                    step v
M < >
O   M < > M < >
M   O   M   O
O   M   O   M   O   M   O   M < > M < > □
M   O   M   O   M   O   M   O   M   O   M   O


O   M                    step vi
M < >
O   M < > M < >
M   O   M   O
O   M   O   M   O   M   O   M < > M
M   O   M   O   M   O   M   O   M   O   M   O
```

FIG. 1. Model for initial sintering of an oxide MO.

A GENERAL MODEL FOR INITIAL SINTERING

1.Description of the model

It must be pointed out that the model described in this section, what we have called "initial sintering", concerns only the process responsible for surface area reduction, which may result from particle growth or elimination of external porosity.

Previous models for surface area reduction are based on the conventional approach which relies on geometrical assumptions[6]. Surface diffusion is generally considered to be the dominant mechanism for material transport. However these models do not account for the influence of water vapour or impurities on the sintering rate. Water vapour is usually considered to enhance surface diffusion but, to our knowledge, there does not exist a mechanistic model describing the intrinsic course of the process at an atomic level.

The model is detailed for an oxide having the general molecular formula MO. Then it will be easy to write similar equations in the case of any particular oxide.

As usually depicted, the driving force for initial sintering induces neck growth between particles. The neck region can be illustrated at a microscopic level, as in Figure 1 using a two-dimension illustration. We assume that the major point defects of the crystal MO are anionic vacancies. The symbol < > represents an oxygen vacancy and the symbol □ a cation vacancy. More oxygen vacancies are represented in the neck region, notated "R<0" because of the sign of its apparent curvature radius, than in the "R>0" region. This comes from the variation of the equilibrium constant with the curvature radius. Initial sintering may be described like the repeated transport of a building unit "MO" from the R>0 region towards the R<0 region.

Let us suppose first that a molecule of water dissociates at the R>0 surface creating two hydroxyl ions (step i). Then hydroxyl ions are assumed to diffuse in the direction of the R<0 surface (step ii). The movement of two hydroxyl ions to neighbouring sites may be considered like a step of vacancy creation (step iii). Then water desorption occurs at the R<0 surface due to the two interacting hydroxyl ions (step iv), followed by cation diffusion (step v). The last step will be vacancy annihilation at the R>0 surface (step vi).

These six elementary steps can be written as follows, using Kröger's notation:

$$\text{(i)} \quad H_2O_g + (O_O^x)_{R>0} + (V_O^{\cdot\cdot})_{R>0} = 2\,(OH_O^{\cdot})_{R>0}$$

$$\text{(ii)} \quad (OH_O^{\cdot})_{R>0} + (V_O^{\cdot\cdot})_{R<0} \rightarrow (OH_O^{\cdot})_{R<0} + (V_O^{\cdot\cdot})_{R>0}$$

$$\text{(iii)} \quad (OH_O^{\cdot})_{R<0} + (OH_O^{\cdot})_{R<0} = 2\,(OH_O^{\cdot})_{R<0} + (V_M^{\prime\prime})_{R<0} + (V_O^{\cdot\cdot})_{R<0}$$

$$\text{(iv)} \quad 2\,(OH_O^{\cdot})_{R<0} = H_2O_g + (O_O^x)_{R<0} + (V_O^{\cdot\cdot})_{R<0}$$

(v) $\qquad (M_M^x)_{R>0} + (V_M'')_{R<0} \rightarrow (M_M^x)_{R<0} + (V_M'')_{R>0}$

(vi) $\qquad (V_M'')_{R>0} + (V_O^{\cdot\cdot})_{R>0} = 0$

This mechanism points out a catalytic effect of water vapour. Effectively, the overall process coresponds to the diffusion of a cation and an anion without creation or consumption of any other species.

2. Rate equations

The rate can be calculated using the approximation of the rate-limiting step. By this way all the reactions but one, the rate-limiting step, are considered to be at equilibrium. Diffusion steps are evaluated according to Fick's first law within the Wagner approximation (constant gradient). The calculations are greatly simplified by using the equation of electroneutrality for the crystal. As it can be shown in the various steps of the model, hydroxyl ions must be considered like point defects. Then the various terms in the electroneutrality equation can be alternatively neglected and, according to Brouwer[7], several limiting conditions may be obtained.

Finally, the rate equation can be calculated for a given surface area, that is to say, for a given geometry of the system, and it takes the expression of the product of a constant and the partial pressure of water vapour raised to some power.

The calculations have been done in the case of two oxide powders: TiO_2 anatase[8] and γ-Al_2O_3 transition alumina[9].

The results are reported in Tables I and II respectively.

TABLE I. Predicted rate equation of surface area decrease for TiO_2 anatase.

Rate-lim. step	$[(OH_O^-)^{\cdot}] \gg [(V_O)^{\cdot\cdot}]$	$[(V_O)^{\cdot\cdot}] \gg [(OH_O^-)^{\cdot}]$
1	$P(H_2O)^{1/2}\ P(O_2)^{-1/4}$	$P(H_2O)\ P(O_2)^{-1/6}$
2	$P(H_2O)^{1/4}\ P(O_2)^{-1/8}$	$P(H_2O)^{1/2}\ P(O_2)^{-1/12}$
3	$P(H_2O)\ P(O_2)^{-1/2}$	$P(H_2O)^{2}\ P(O_2)^{-1/6}$
4	$P(H_2O)^{1/2}\ P(O_2)^{-1/4}$	$P(H_2O)\ P(O_2)^{-1/6}$
5	$P(H_2O)\ P(O_2)^{1/2}$	$P(O_2)^{1/3}$
6	1	1

TABLE II. Predicted rate equation of surface area
decrease for γ-Al$_2$O$_3$.

Rate-lim. step	$[(OH_O^-)^\cdot] \gg [(V_O)^{\cdot\cdot}]$	$[(V_O)^{\cdot\cdot}] \gg [(OH_O^-)^\cdot]$
1	1	$P(H_2O)$
2	1	$P(H_2O)^{1/2}$
3	1	$P(H_2O)^3$
4	1	$P(H_2O)$
5	$P(H_2O)^2$ or $P(H_2O)^4$	1
6	1	1

APPLICATIONS TO ANATASE AND TRANSITION ALUMINA

1. Experimental procedure

The preparation and characterization of TiO$_2$ (anatase)
and of transition alumina powders have been described
previously in previous papers[8,10]. Their initial surface
areas are equal to 100 m^2.g^{-1}.

The samples of anatase are placed into a preheated
furnace and annealed at 823 K. Once the ambient air has
been readily evacuated, oxygen and water vapour are
introduced at appropriate partial pressures. The overall
pressure is lower than the atmospheric pressure. The
temperature is chosen in order to avoid the phase
transformation of anatase into rutile.

The samples of transition alumina are placed into a
preheated furnace and annealed at 1378 K. After evacuation
an appropriate mixture of oxygen, water vapour and helium
is flowed over the sample. The overall pressure is equal
to the atmospheric pressure. The transformation of
transition alumina into α-Al$_2$O$_3$ induces a loss in surface
area but this contribution will be omitted in the
following analysis.

Samples are briefly cooled after the desired annealing
period.

Surface areas are measured by nitrogen adsorption at
77 K using the BET equation.

2. Rate of surface area decrease

The surface area variation with time of annealing can
be obtained for various partial pressures of water vapour
and oxygen. As previously described, it is possible to fit
this variation with a mathematical function of the
following form[8]:
$$S = S_0 (1+at)^n$$
where S_0 is the surface area at time t, S_0 is the initial

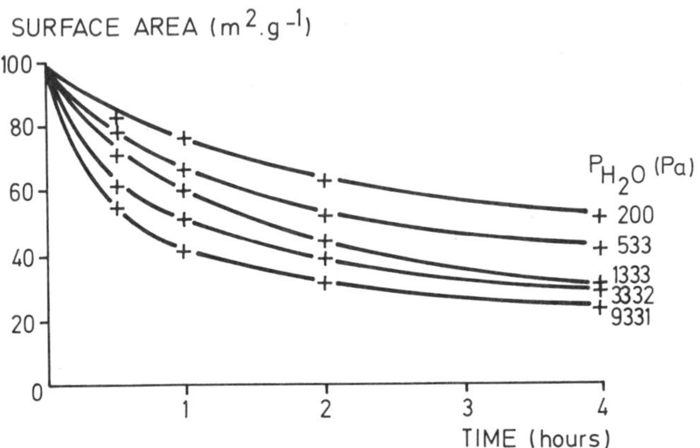

FIG. 2. Variation of surface area of TiO_2 anatase with time of annealing for various partial pressures of water vapour and 133 Pa of oxygen.

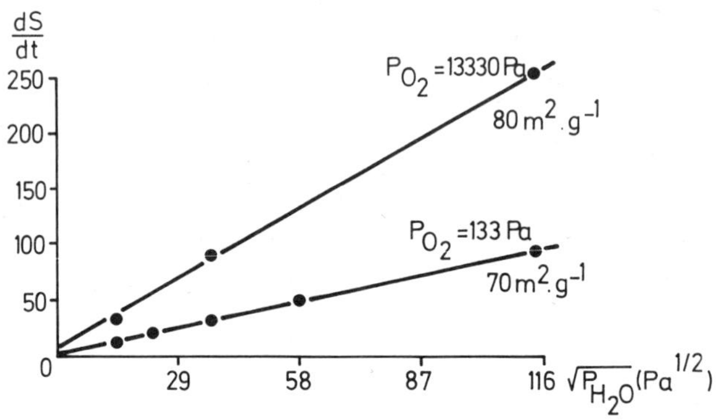

FIG. 3. Relation between the rate of surface area decrease and the partial pressure of water vapour raised at the power 1/2 for TiO_2 anatase powder.

surface area, a and n are adjusting parameters. As an example, Figure 2 illustrates the curves for surface area variation of anatase with time for various partial pressures of water vapour and fixed oxygen partial pressure.

By this way the experimental rate of surface area reduction, dS/dt, can be calculated. Finally, we obtain the law of variation of the rate with the partial pressure of water vapour for any constant value of the surface area S. The power exponent of water vapour is calculated by linear regression. The results are reported in Figures 3 and 4 for the case of anatase and transition alumina, respectively, as a function of $P(H_2O)^{1/2}$.

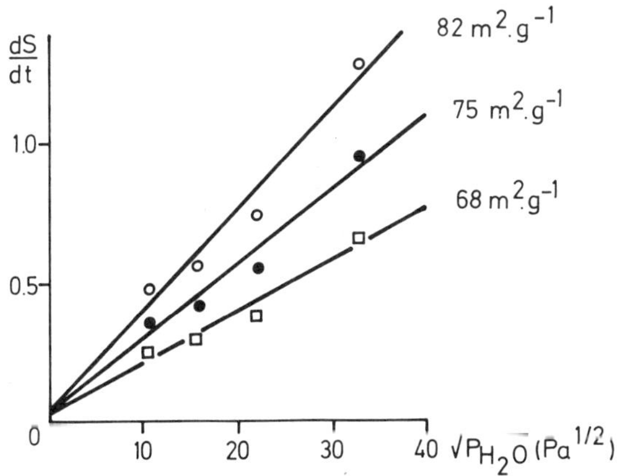

FIG. 4. Variation of the rate of surface area decrease with the partial pressure of water vapour raised to the power 1/2 for transition alumina powder.

3. Comparison with the model

At first it is interesting to compare the experimental and predicted rates for a fixed pressure of oxygen. The experiments give a rate proportionnal to $P(H_2O)^{1/2}$ for both anatase and transition alumina. By refering to Table II, one can see that there is only one rate-limiting step in agreement with experiment in the case of transition alumina: hydroxyl ions diffusion. The related Brouwer's case indicates that their concentration is negligible in comparison with the oxygen vacancies. This agrees with the formulation for transition alumina given in a previous paper[10]. In the case of anatase, several rate-limiting steps are found to be in agreement with $P(H_2O)^{1/2}$, as it can be seen in Table I. Among these, only one can be accepted when the influence of oxygen is taken under consideration[8]. The rate-limiting step is again the diffusion of hydroxyl ions, the same as Brouwer's case.

CONCLUSION

The model proposed to explain the influence of gaseous atmosphere on the surface area decrease of ultrafine oxide powders has been successfully applied to anatase and transition alumina. The major interests of this model are to not depend of the geometrical shape of the particles, but to determine the rate-limiting step of the process. These investigations have shown that hydroxyl ions diffusion is the rate-limiting step of the initial sintering of both anatase and transition alumina.

REFERENCES

1. P.F. Eastman and I.B. Cutler, J. Am. Ceram. Soc, 49 (1966) 526.
2. D. Beruto, R. Botter and A.W. Searcy, J. Am. Ceram. Soc., 70(1987)155.
3. D. Louër, R. Vargas and J.P. Auffredic, J. Am. Ceram. Soc, 67(1984)136.
4. H. Schaper, E.B.M. Doesburg, P.H.M. De Korte and L. Van Reijen, Solid State Ionics, 16(1985)261.
5. O.J. Whittemore and J. A. Varela, Advances in Ceramics, 10(1985)583.
6. R.M. German and Z.A. Munir, J. Am. Ceram. Soc, 59(1976) 379.
7. G. Brouwer, Philips Res. Rep., 9(1954)366.
8. J.L. Hébrard, P. Nortier, M. Pijolat and M. Soustelle, accepted for publication in J. Am. Ceram. Soc.
9. M. Dauzat, M. Pijolat and M. Soustelle, to be published.
10. P. Burtin, J. P. Brunelle, M. Pijolat and M. Soustelle, Applied Catal., 34(1987)239.

MECHANISM OF ANISOTROPIC DIMENSIONAL CHANGES

DURING SINTERING OF METAL POWDER COMPACT

Hidenori Kuroki and Masahiro Hiraishi

Department of Mechanical Engineering
Hiroshima University, Higashi-hiroshima City
724 Japan

INTRODUCTION

The dimensional change of a metal powder compact during sintering has
been considered as one of the major parameters with which the mechanism of
sintering can be described quantitatively. The changes, however, are anisotropic in a cold pressed compact. The value depends on the direction in
which it is measured, i.e., the compacting one or the lateral one.

E. H. Exner[1] has counted anisotropy as one of the fundamental problems which has not yet been solved. Various types of mechanisms have been
proposed and discussed for the occurrence of anisotropy since the days of
H. H. Hausner and F. V. Lenel[2]. One of the present authors has proposed
two mechanisms. One mechanism is the expansion of flat pores in the direction of thickness or the pressing direction of the compact by the internal
gas pressure. This has been well known for the hydrogen disease of copper
compacts. The other mechanism is the larger recovery in the pressing direction of the smallest size which has been attained during pressing and
ejecting from larger sizes before sintering of the ejected or dewaxed
compact. This mechanism has been proposed for iron powder compacts[3].

The dimensional changes of a compact in the latter mechanism can be
attributed to the changes of distances between particles. Thus, a quantitative explanation of the larger dimensional changes in the pressing direction would be possible by counting the number of the particle boundaries in that direction, and then multiplying the number by the mean change
of the distances between particles. In the present work, some experiments
are made on wire-wound model compacts and on powder compacts to evaluate
the changes of distances between particles during sintering.

EXPERIMENTAL PROCEDURE

Preparation of Wire-wound Model Compacts

Structure: A wire, 0.16mm in diameter, of austenitic steel 304(18Cr-
8Ni), is wound in 5 layers and 17 or 18 turns in a layer on a shaft, and
pressed in the axial direction of the shaft (Fig. 1). A movable flange is
fixed with a small bolt on the shaft. Fastening the bolt maintains the
tight contact condition attained by pressing.

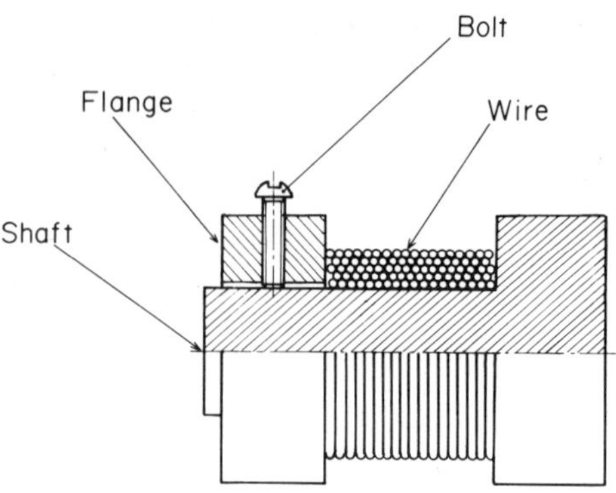

Fig. 1. Structure of wire-wound model compact.

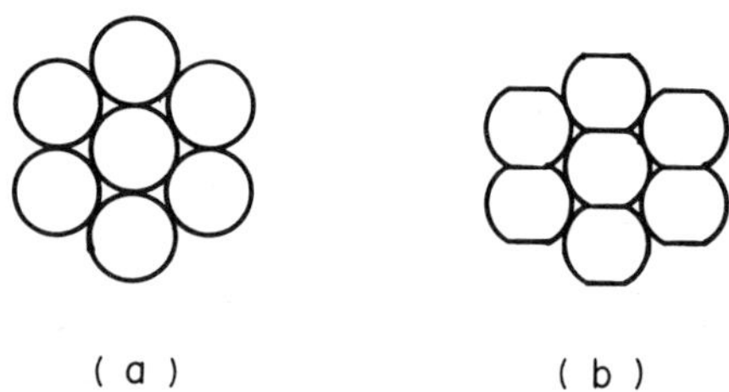

(a)   (b)

Fig. 2. Contacts between wires. (a) before
pressing; (b) after pressing.

Pressing: Flat contacts having a width of about one third of the
diameter are developed between the neighboring wires by the pressing operation under 3330N (Fig. 2).

Sintering: The model compacts are sintered in a furnace held at 1263-
1463K for 5 minutes–17 hours in a hydrogen atmosphere.

Evaluation of Sintered Model Compacts

A section of a pressed model compact shows narrow gaps of about 0.5

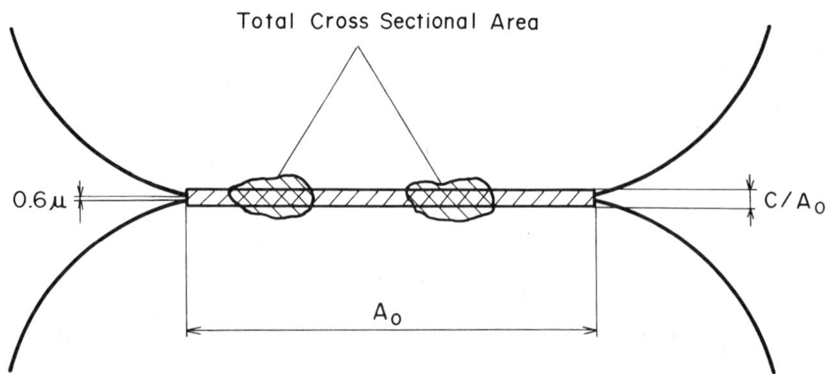

Fig. 3. Determination of mean gap between wires.

microns at flat contact areas between neighboring wires. The gaps change
into rows of micropores during sintering as many microwelds develop bridging the flat surfaces of the neighboring wires[4]. The size of the gaps is
given by the following equation, as shown in Fig. 3.

$$\text{mean gap} = C/A_o \tag{1}$$

where,

A_o: The width of a macrocontact with the gap not larger than 0.6
microns (the maximum value measured in a non-sintered compact) between a
pair of flat surfaces. Since the values of A_o are difficult to measure
before sintering, they are measured after sintering on polished sections
and confirmed to be unchanged from the respective values of the same contacts before sintering[4].

C: The total area of the gaps and pores of a sintered macrocontact on the polished section.
The measurement is carried out with the third layer.

Preparation of Powder Compacts

Compacts for counting particle boundaries: Cubic compacts, 10mm in
size, are compacted under 600MPa of electrolytic iron powder (-100mesh) by
admixing 0.7wt% of zinc stearate. One of the side faces of each compact is
impregnated with a cyano-acrylate adhesive and polished in order to count
the number of the particle boundaries in the compacting direction along
the total length of the compact (10.1mm).

Compacts for measuring particle movement: Each of a pair of compacts
prepared as above is ground on a side face with 1500grit abrasive paper.
Subsequently, the finely-ground faces are ground with a much coarser 400
grit paper so as to make marks for measuring; on one compact in the compacting direction (Fig. 4(a)), and on the other in the lateral direction
(Fig. 4(b)).

After the measurement of distances as described below between several
pairs of marks on each compact before sintering, the compact is heated at
the rate of 10K/min to sinter up to 1223K beyond the Ac_3 point of iron,
and then moved to the cooling zone of the furnace. The treatment corresponds to finishing the initial stage of sintering[3]. On sintered compacts,
the distances between pairs of marks measured before are measured again so
as to define the movement of particles from the changes of those distances.

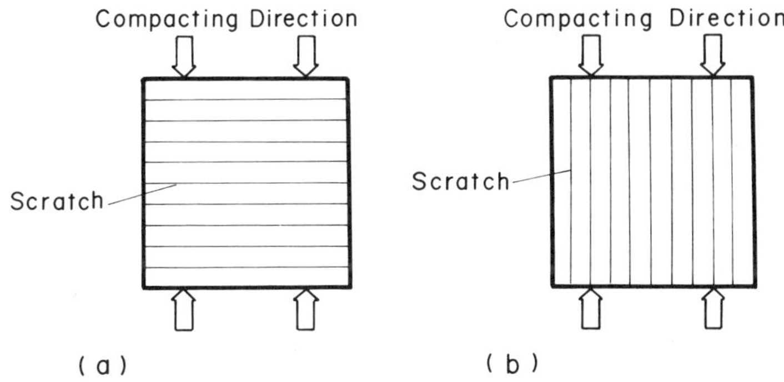

Fig. 4. Scratches as marks for measuring the movement
in different directions of powder compacts.
(a) for the compacting direction; (b) for the
lateral direction.

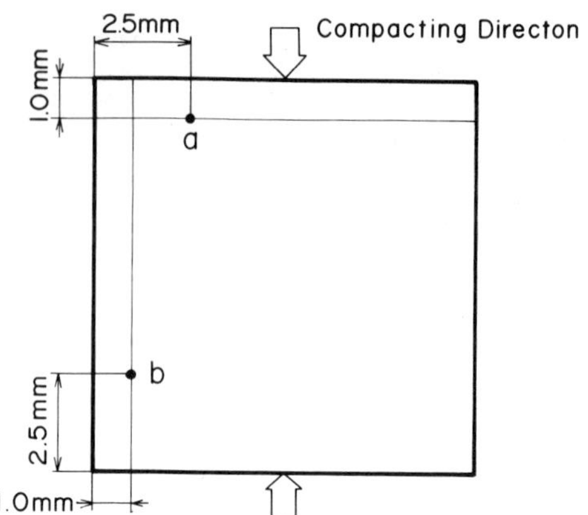

Fig. 5. Points on powder compacts photographed
by SEM before and after sintering.
(a) near the pressed face;
(b) near the side face.

Measurement of the Particle Movements in Powder Compacts

On each compact, several SEM photomicrographs are taken at points
indicated in Fig. 5 before and after sintering to show areas including the
particle boundary between the pair of marks. Thus, the particle movement
is defined as follows;

$$\text{particle movement} = X_1 - X_2 \tag{2}$$

where X_1 and X_2 are the distances between the pair of marks before and
after sintering.

RESULTS

Sintering Phenomena in Wire-wound Model Compacts

The pressed wire forms many microwelds between the pair of flat faces
of the contact, subsequently leading to the growth and the coalescence of
those welds and to the disappearance of micropores between the welds, and
finally resulting in a few welds occupying most of the contact area, as
reported elsewhere[4] and illustrated in Fig. 6.

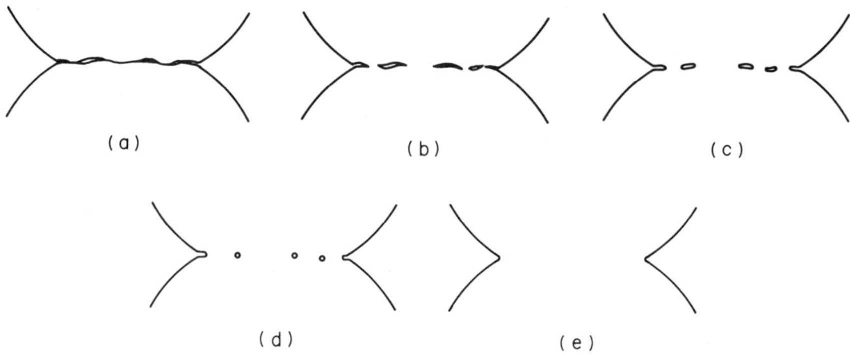

Fig. 6. A typical sintering sequence; (a) through (e).

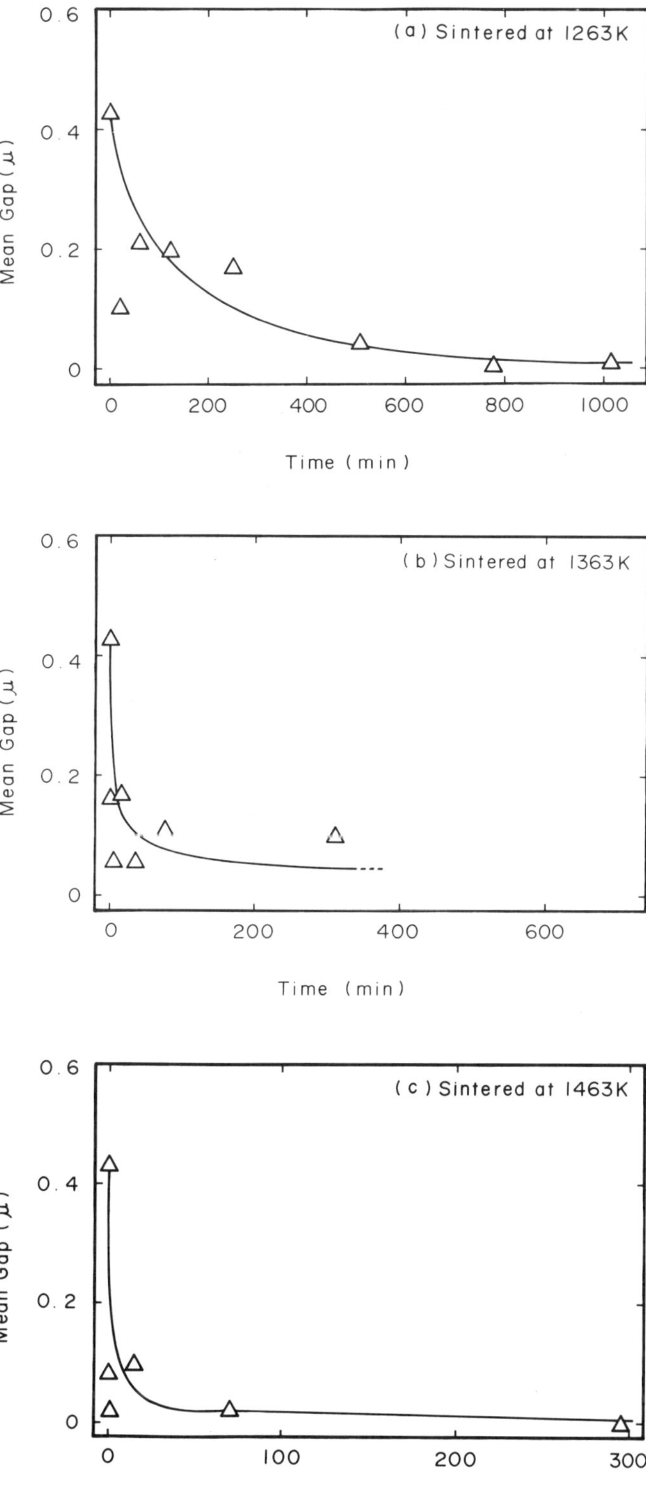

Fig. 7. Change of the mean gap between flat surfaces with time.

<u>Gaps between Contact Surfaces of the Wire</u>

Atoms forming the microwelds are supposed to have moved from unstable positions near the welds on the flat contact surfaces in the initial stage of sintering. When the sintering proceeds and the gaps and pores disappear, wires may move closer, resulting in a decrease of the center-to-center distance between the wires. Fig. 7 shows the changes of the mean gaps between the flat surfaces with time. It also shows that the distance decreases during the initial sintering at each temperature, and that the decrement is about 0.4 microns per contact interface or particle boundary.

<u>Particle Movements in Powder Compacts during Sintering</u>

Powder compacts often show anisotropic dimensional change during sintering, especially in the initial stage of sintering, resulting in larger shrinkage in the compacting direction than in the lateral direction. This shrinkage is supposed to be a kind of phenomena of the particle movement. The particles may contact each other at the deformed surfaces by pressing, separate by spring-back during unloading and ejecting, and move closer together again leading to the disappearance of the gap during sintering[3]. To confirm the supposition, the compacts are marked with a grinding paper to measure the changes of distances between the marks or to detect the approach of particles during the initial stage of sintering.

The movements of particles perpendicular to the contact surfaces of the particles themselves at areas near the pressed and side faces are shown in Table 1. Although the divergences of the data are not small, the mean values show the larger aproach of particles in the compacting direction rather than in the lateral direction, especially in the side faces.

DISCUSSION

<u>Progress of Sintering</u>

Appearance and disappearance of micropores and the changes in the values of the mean gap during sintering indicate that the sintering between the pressed and contacted faces proceeds as follows.

At first sight, contact surfaces deformed by pressing appear flat to form a perfect contact with no micropores. However, the contact is not flat microscopically. Rather it may have microwelds and micropores developed during the initial stage of sintering as has already been reported for powder compacts sintered for a short time[5]. Atom movement may start to reduce the real surface area near the microcontact when the compact is

Table 1. Movements of Particles Perpendicular to
Their Contact Surfaces.

point	direction	measured value	average
near pressed face	compacting	0.53, 0.25	0.39
	lateral	0.22, −0.04, −0.44	−0.09
near side face	compacting	1.02, 0.51, 0.09	0.54
	lateral	0.55, 0.26, 0.02	0.28

heated to the sintering temperature, forming microwelds in a short time.
From the viewpoint of pores, the phenomenon may be the spherodization and
the disappearance of pores between microwelds due to the surface tension.

<u>Comparison of Shrinkage between the Model and the Powder Compacts</u>

If the initial shrinkage during sintering in the compacting direction
is caused by the approach of the particles, the shrinkage per particle
boundary may be deduced to be equal to the value of the mean gap. Thus,
the value of the mean gap multiplied by the number of particle boundaries
may be equal to the initial shrinkage of the compact. From this viewpoint,
the measured and theoretical or calculated values of the shrinkage are
compared for an electrolytic iron powder compact. The measured values of
the shrinkage, mostly accomplished during the initial sintering stage,
have been obtained with compacts pressed at 4, 6 and 8t/cm^2 of the iron
powder mixed with 0.7wt% of zinc stearate and sintered at 1373K for 60
minutes in dissociated ammonia[3]. By interpolation with those values, approximate shrinkage of 0.4% is derived for the compacting pressure of
600MPa (6.12t/cm^2).

Remembering the above-described boundary number of 208/10.1mm in the
compacting direction, the theoretical shrinkage may be equal to the value
of the gap, 0.4 microns, multiplied by 208.

$$\frac{0.4 \times 208}{10.1 \times 1000} \times 100 = 0.82\% \tag{3}$$

This theoretical value is twice as large as the measured value of 0.4%.
The difference will be caused by the smaller value of the real mean gap
between the particles in a powder compact rather than between the wires in
a model compact. Thus, the disappearance of micropores is reason enough to
possibly cause the shrinkage of an iron powder compact in the initial
stage of sintering.

<u>Particle Movements in Powder Compacts during Sintering</u>

If the particle movement during the initial stage of sintering causes
the major part of the shrinkage of a whole compact in the compacting direction, the shrinkage for the whole length can be derived from the particle movement, 0.39 microns near the pressed surface, multiplied by the
number of the boundaries, 208.

$$0.39 \text{ microns} \times 208 = 81 \text{ microns}$$

Thus, the shrinkage in percentage is,

$$\frac{81}{10.1 \times 1000} \times 100 = 0.80\% \tag{4}$$

This is almost the same result as that in Equation (3) and, again, twice
as large as the measured value of about 0.4% shown above. In this calculation, the main reason for the differences in the results from the measured
values will be the fact that the measurement of the particle movement is
possible only with particle boundaries which clearly show the existence of
the gaps before sintering. Unclear boundaries may be narrow gaps, which
may cause minor shrinkage during sintering. If all boundaries are measured
for the particle movement, the mean values of the movement or the movement
per boundary will decrease. Thus, the theoretical values derived as above
will approach the measured values.

The anisotropy of the particle movement, the larger movement in the compacting direction rather than in the lateral direction, is observed at both the areas near the pressed surface and the side face during the initial sintering stage. The observation coincides with the larger shrinkage in the compacting direction than in the lateral direction of the compacts made of the same powder[3]. This also indicates the above-mentioned relation between the initial shrinkage of the iron powder compacts and the disappearance of the particle boundary gaps or the approach of the particles again.

CONCLUSION

Wire-wound model compacts of stainless steel are pressed along the core axis of the compacts to form flat contacts between the neighboring wires. The development of metallurgically welded necks during sintering is observed at the contacts in polished and etched sections of samples.

In most cases, the process of sintering may start with the rapid formation of welds at the microscopic real contacts. This is followed by the spherodization and the disappearance of small pores one micron or less in size originated from the microscopic unevenness on the wire surfaces.

The total area of the sections of the micropores which have disappeared during the initial stage of sintering is divided by the width of the mechanical contact to calculate an average change of distances between wires, which can be the shrinkage at that stage. The values have the same order as the shrinkage per particle boundary of an iron powder compact has at similar stages.

The anisotropy in the shrinkage during sintering of pressed powder compacts may be attributed partly to the difference between the movement of particles in the compacting and the lateral directions.

REFERENCES

1. H. E. Exner, Solid-state sintering: critical assessment of theoretical concepts and experimental methods, _Powder Metallurgy_ 23:203 (1980).
2. F. V. Lenel, H. H. Hausner, E. Hayashi and G. S. Ansell, Some Observations on the Shrinkage Behaviour of Copper Compacts and of Loose Powder Aggregates, _Powder Metallurgy_ No.8: p.25 (1961).
3. H. Kuroki, Anisotropy in the Dimensional Change of Metal Powder Compacts during Sintering, _in_: "Sintering '87", S. Somiya, M. Shimada, M. Yoshimura and R. Watanabe, ed., Elsevier Applied Science, London, p.387 (1988).
4. H. Kuroki, M. Hiraishi and T. Kimura, to be published.
5. H. Kuroki and T. Honda, Powder Particle Boundary in Sintered Iron, _The International Journal of Powder Metallurgy and Powder Technology_ 19 :269 (1983).

Part II. CONTROLLED PREPARATION OF CERAMIC POWDERS

PREPARATION OF WELL DEFINED CERAMIC POWDERS

Egon Matijević

Department of Chemistry
Clarkson University
Potsdam, NY 13676, USA

INTRODUCTION

Sufficient evidence has become available to demonstrate that the sintering processes can be greatly affected by the physical and chemical properties of the starting powders. Specifically, the temperature of sintering and the quality of the final products will depend on the size, shape, and uniformity of the original particles. As far as the chemical aspects are concerned, in addition to the obvious expectation that the bulk composition should be essential, the surface characteristics in terms of active sites and charge can play an important role in the powder processing.

For these reasons it is desirable, and in some cases essential, that "monodispersed" particles, varying in modal size from several nanometers to several micrometers, be available. It is also necessary to produce such dispersed matter of different shapes, especially of uniform spheres, rods, or platelets. Finally, depending on application, we need chemically simple, composite, or coated particles, either amorphous or crystalline.

In view of the importance of finely dispersed powders in general, and in ceramics in particular, it is of no surprise that much effort is presently devoted to the development of techniques for the preparation and characterization of such materials. Indeed, many uniform systems have been described in the literature, including elements, inorganic or organic compounds of varying degrees of complexity, and cores covered with chemically different coatings.

This article describes a number of well defined dispersions obtained by precipitation and aerosol techniques. The availability of these materials has been instrumental in the development of a better understanding of the mechanisms of the formation of solids from homogeneous solutions. Monodispersed colloids make it possible to relate various properties (optical, magnetic, adsorptive, adhesive, etc.) to particle shape and size.

Science of Sintering
Edited by D. P. Uskoković *et al.*
Plenum Press, New York

While many problems in fine particle science have been addressed
and solved with the aid of systems of controlled properties, new more
sophisticated problems wait to be resolved. Some of these questions
are raised in this paper.

PRECIPITATION

In principle, it is very easy to obtain solids from homogeneous
solutions. The only condition is to exceed the solubility product of
the reactants and a precipitate will form (occasionally after a given
induction period). However, to produce well defined dispersed matter,
it is necessary to conduct the process in a specific manner, which may
be based on relatively simple principles. First, it should be recognized
that not all constituents of a solution need to be directly involved
in the precipitation process. The chemical natures of the actual
reactants depend on the system: they may be merely hydrated ions, or
species of varying degrees of complexity. To generate uniform
dispersions, the *particle forming solutes* must be generated in a
kinetically controlled manner to produce a short burst of nuclei, which
should then grow uniformly (1-3). For example, in the case of metal
(hydrous) oxides this requirement can be met by deprotonation of hydrated
metal ions at an appropriate solution pH at elevated temperatures, the
process described as "forced hydrolysis" (3-6). Alternately, one can
introduce hydroxide ions within the solution in a steady manner by
decomposition of certain compounds, such as formamide or urea (3,7,8).
Finally, metal ions can be released into a basic solution, if complexes
(e.g., metal chelates) are broken up *in situ* to liberate the cations
(9,10).

Based on the same principles, uniform particles of metal sulfides
may be produced by continuous release of sulfide ions from thioaceta-
mide in metal salt solutions (11-13), while selenides are obtained by
decomposition of selenourea (14).

1. Spherical particles

Dispersions of spherical particles are of particular interest for
several reasons. Because of simple symmetry considerations, most
theoretical analyses are based on the sphere as the model.

Systems of spherical particles of narrow size distribution offer
significant advantages in quantitative evaluation of their properties.
For example, light scattering techniques can be used to determine the
size distribution and optical characteristics *in situ*, which makes it
possible to follow particle growth and/or chemical changes in the course
of the precipitation process. The stability of dispersions, consisting
of either identical or unlike particles, can also be established and
interpreted as a function of time in the presence of different additives.
Spherical particles are also essential in the quantitative evaluation
of adhesion phenomena. Furthermore, one can compare geometric specific
surface areas of solids to those obtained by gas adsorption. Finally,
in ceramics, powders of uniform spheres should help resolve many questions
related to powder compaction and sintering processes.

It is now possible to prepare finely dispersed spherical particles
of a large number of inorganic compounds of simple or composite nature,
either amorphous or displaying crystalline properties of known minerals.
These findings have raised some fundamental questions with respect to
chemical and physical mechanisms of their formation, as illustrated
in the following examples.

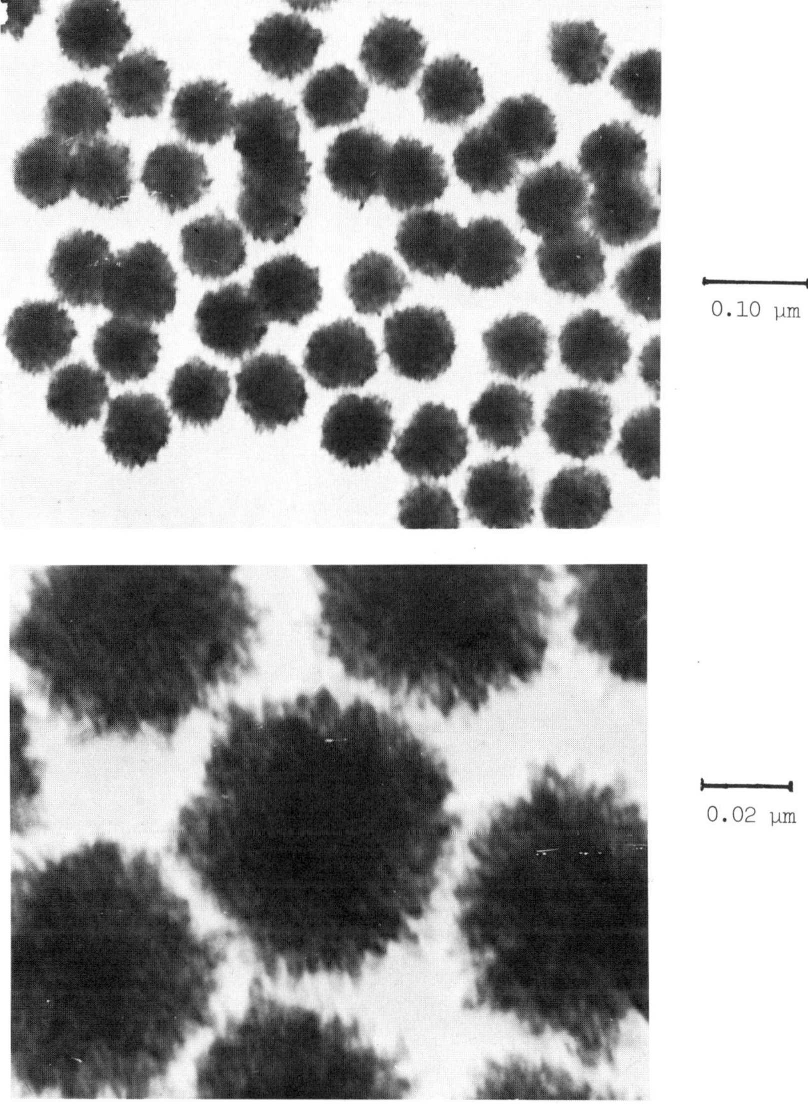

Fig. 1. Transmission electron micrographs (TEM) at two
different magnifications of SnO_2 particles obtained
by homogeneous precipitation by aging at 100°C
for 2 hr a solution 0.0030 mol dm^{-3} in $SnCl_4$ and
0.30 mol dm^{-3} in HCl (15).

The first case deals with the precipitation of SnO_2. The electron
micrograph in Fig. 1a shows that rather uniform, nearly spherical
particles are obtained by aging at 100°C for 2 hr a solution 0.0030
mol dm^{-3} in $SnCl_4$ and 0.30 mol dm^{-3} in HCl (15). The high acidity
is needed to control the hydrolysis of the tin(IV) ion. The X-ray
diffraction pattern of the resulting powder is characteristic of
cassiterite, SnO_2. It is intriguing that spherical, yet crystalline
particles can form by precipitation from homogeneous solutions. The
larger magnification of the same powder (Fig. 1b) clearly indicates
these solids to be internally composite; i.e., they consist of a large
number of much smaller subunits. Indeed, it has been documented in

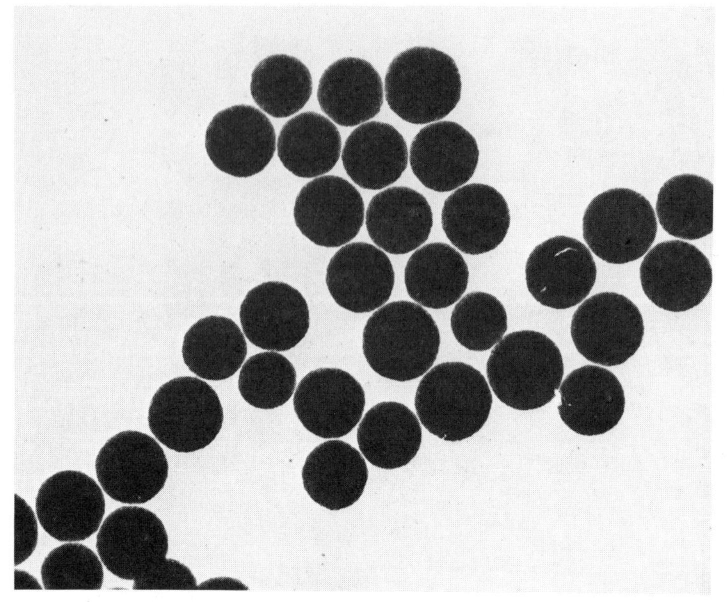

Fig. 2. TEM of manganese phosphate particles obtained by
aging at 80°C for 1 hr a solution 0.0050 mol dm^{-3}
in MnSO$_4$, 0.0050 mol dm^{-3} in H$_3$PO$_4$, 1.0 mol dm^{-3}
in urea, and 0.010 mol dm^{-3} in sodium dodecylsulfate (SDS) (17).

a number of cases that finely dispersed spheres, obtained
by precipitation, with distinct X-ray spectra of known crystalline matter,
are internally inhomogeneous (4,16).

It should be noted that many metal oxides and other inorganic
compounds, prepared by analogous processes, appear as perfectly smooth
uniform amorphous spheres. Figure 2 illustrates manganese phosphate
particles, generated by aging at 80°C for 1 hr a solution of 0.0050
mol dm^{-3} MnSO$_4$, 0.0050 mol dm^{-3} H$_3$PO$_4$, 1.0 mol dm^{-3} urea, and 0.010
mol dm^{-3} scdium dodecylsulfate (SDS) (17). The chemical analysis of
this powder showed it to be hydrated manganese phosphate, Mn$_3$PO$_4 \cdot$H$_2$O.

Amorphous spherical particles of metal (hydrous) oxides were also
obtained by forced hydrolysis; e.g., in solutions of aluminum (6),
chromium (5), and titanium (18) salts. Similarly, spheres of metal
basic carbonates were generated by aging aqueous lanthanide salts at
elevated temperatures in the presence of urea (4,8) (See Fig. 8, upper).

Whether spherical particles, obtained by precipitation from
homogeneous solutions, will appear amorphous or exhibit crystalline
characteristics, depends on the nature of the solute precursors. Complex
species that tend to polymerize in solution before and during the
precipitation process will largely yield amorphous solids. In contrast,
well defined monomeric or polynuclear complex ions will separate out
as tiny crystalline particles, which subsequently agglomerate into larger
spheres.

By coprecipitation from solutions of mixed metal salts, it is
possible to produce internally composite spherical particles. As the
example, Fig. 3 shows an Y(III)/Ce(III) basic carbonate powder, prepared

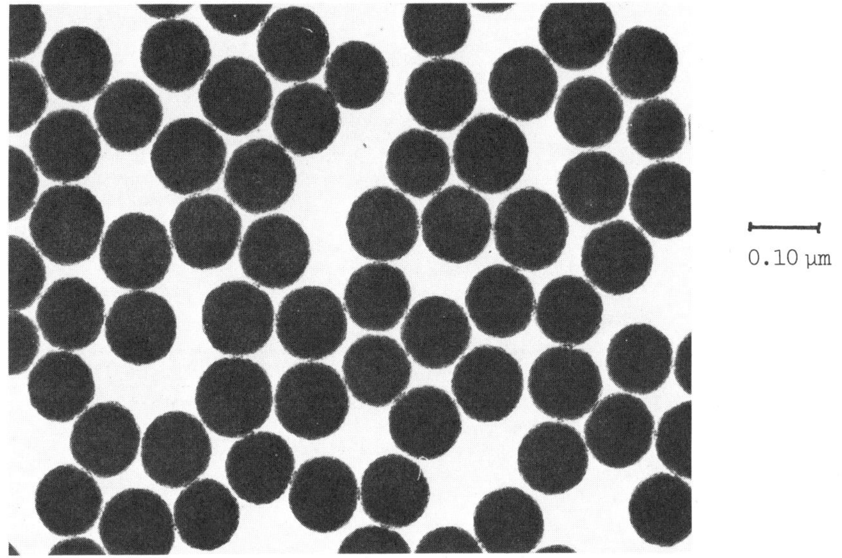

Fig. 3. TEM of internally mixed Y(III)/Ce(III) basic carbonate
particles obtained by aging at 90°C for 2 hr a
solution 0.0050 mol dm^{-3} in Y(NO$_3$)$_3$, 0.015 mol
dm^{-3} in Ce(NO$_3$)$_3$, 0.5 mol dm^{-3} in urea (8).

as described in the legend (8). One of the questions that need to be
resolved is the relationship of the composition of the solid phase
relative to that of the solution in which the particles are formed.
In this particular example, the analysis showed that the ratio
[Y(III)]/[Ce(III)] was rather close in both phases (8). However, this
finding does not imply that the particles are internally homogeneous,
i.e., that the two metal ions are uniformly distributed throughout the
solids. Indeed, it was established with mixed systems of Cu(II)/Y(III)-
and Cu(II)/La(III)-basic carbonates, that the composition of the particles
in terms of the constituent metal ions changes from the center to the
periphery (19). Once the solids reach the final size, the molar ratios
of cations in the solution and in the powder are approximately the same.
On calcination, stoichiometrically defined binary oxides have been
identified (19).

Finally, uniform spherical particles of complex composition and
exact stoichiometry, such as barium- or lead titanates, have now been
prepared (20,21).

2. Particles of Other Morphologies

The control of the morphology of precipitated solids remains a
vexing problem of fine particle science. With the availability of
"monodispersed" colloids, it has been possible to demonstrate that the
particle shape can be varied in many different ways, which are seldom
predictable. However, once the conditions that yield powders of a given
morphology are established, the materials are readily reproduced. The
factors that affect particle geometry include concentrations of the
reactants, the nature of anions with salts of the same cation, temperature
and duration of aging, pH, and the method of mixing. In some instances,
the variations in morphology are associated with the change in the

chemical composition of the precipitate, while in other cases, the same compound appears as particles of different shapes.

Figure 4 shows a rod-like crystalline SnO_2 powder obtained by using the same reactants as before (Fig. 1), except that a more highly acidic solution was aged for a longer period of time in the presence of formamide. These particles have the same main X-ray characteristics as the spheres displayed in Fig. 1.

Another example of diverse morphology of powders of the same chemical composition is offered in the scanning electron micrographs of Fig. 5. The three samples consist of ZnO obtained by aging solutions of $Zn(NO_3)_2$ in the presence of different bases (22). The X-ray diffraction pattern (Fig. 6) corresponds to zincite (ZnO) and is characteristic of all of these powders.

Anions may have a significant effect on the particle composition and shape in solutions of the same metal ion. They may be incorporated in a stoichiometric ratio of known crystalline compounds such as is the case of the ferric basic sulfates (alunites), $Fe_3(SO_4)_2(OH)_5 \cdot 2H_2O$ and $Fe_4(SO_4)(OH)_{10}$ (23,24), cerium(IV) basic sulfate, $Ce(OH)_2SC_4$ (4), or yttrium amine carbonate, $Y_2(CO_3)_3 \cdot NH_3 \cdot 3H_2O$ (8). In some other instances, anions affect the particle morphology, and are present in the precipitated solids as contaminants in varying amounts. Such occluded species can usually be readily removed by dialysis or other procedures, without changing the particle shape (5,6).

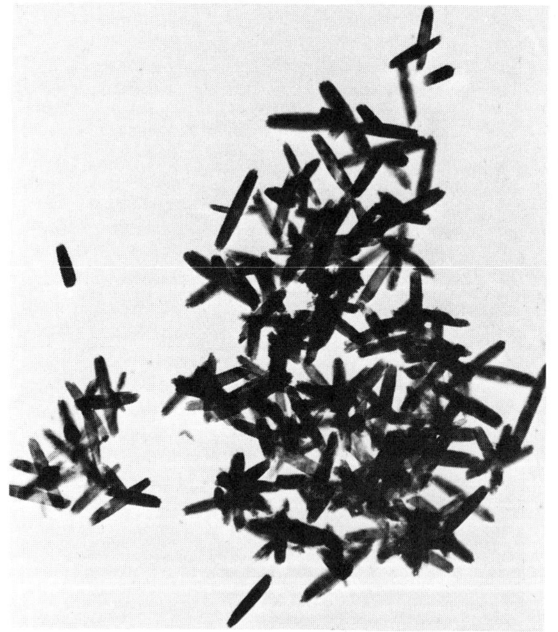

Fig. 4. TEM of SnO_2 particles obtained by aging at 100°C for one week a solution 0.0010 mol dm^{-3} in $SnCl_4$, 0.60 mol dm^{-3} in HCl and 0.15 mol dm^{-3} in formamide.

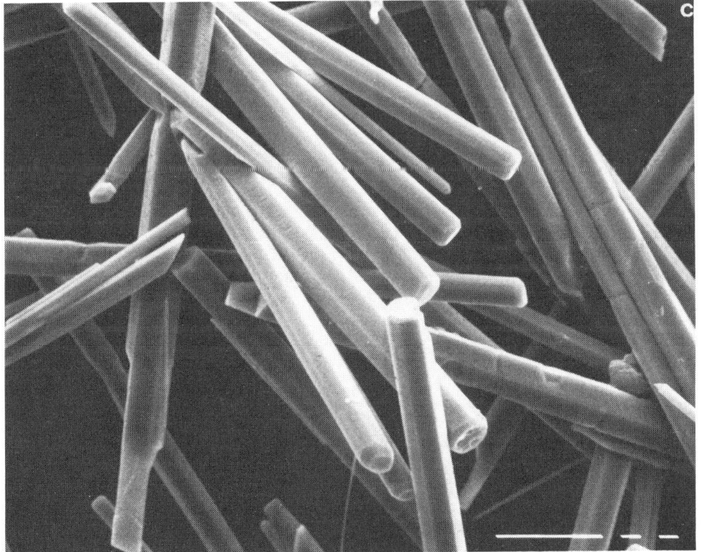

Fig. 5. Scanning electron micrographs (SEM) of ZnO particles
         obtained by aging solutions
         (a) 0.010 mol dm^{-3} in Zn(NO$_3$)$_2$ and 0.10 mol dm^{-3}
             in triethanolamine (TEA) at 90°C for 1 hr,
         (b) 0.0010 mol dm^{-3} in Zn(NO$_3$)$_2$, 0.0032 mol dm^{-3}
             in ethylenediamine at 90°C for 9 hr,
         (c) 0.040 mol dm^{-3} in Zn(NO$_3$)$_2$, 0.20 mol dm^{-3} in
             TEA and 1.2 mol dm^{-3} in NaOH at 150°C for 2 hr (20).
         The longest bar in each case corresponds to 10 μm.

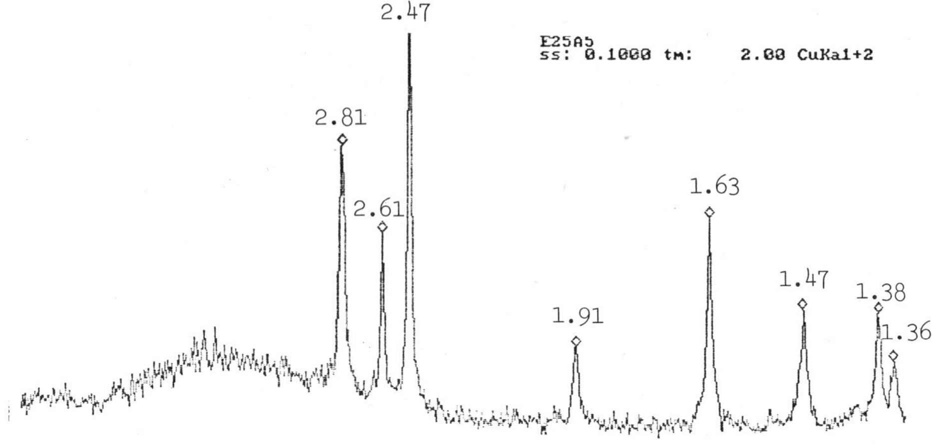

Fig. 6. X-ray diffraction pattern of particles illustrated
in Fig. 5.

Finally, small amounts of added anions can produce particles of a
given geometry without being detected in the final product. Spindle-type
hematite (α-Fe$_2$O$_3$) particles, as obtained when solutions of ferric
chloride were aged in the presence of a small amount of Na$_2$HPO$_4$, (25) are
displayed in Fig. 7.

The above examples are offered to demonstrate that particles of a
variety of shapes can now be produced; they are selected from a much
larger collection of materials presently available. It is also obvious
that much more work needs to be done in order to develop an understanding
of parameters that affect the particle morphology.

Fig. 7. TEM of hematite (α-Fe$_2$O$_3$) particles obtained by
aging at 100°C for 2 days a solution 0.020 mol
dm^{-3} in FeCl$_3$ and 0.0045 mol dm^{-3} in NaH$_2$PO$_4$ (23).

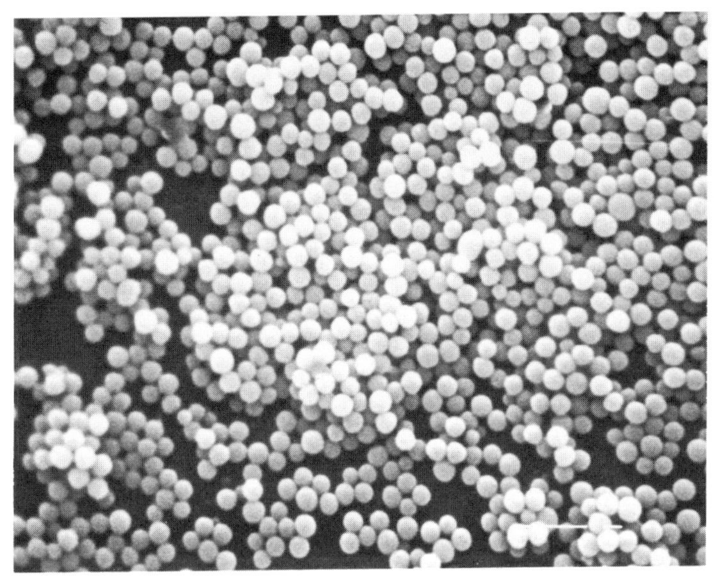

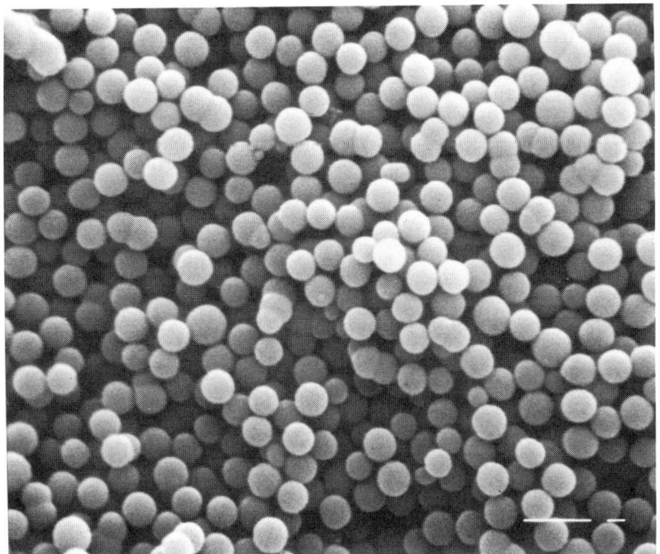

Fig. 8. Upper: SEM of lanthanum basic carbonate, $La(OH)CO_3$ particles obtained by aging at 90°C for 4 hr a solution 0.0080 mol dm^{-3} in $La(NO_3)_3$ and 0.20 mol dm^{-3} in urea.
Lower: The same particles coated with $Cu(OH)CO_3$, prepared by dispersing 22 mg of the above powder in 50 cm^3 of a solution 0.0080 mol dm^{-3} in $Cu(NO_3)_2$ and 0.20 mol dm^{-3} in urea and then aged the entire system at 90°C for 2 hr (25).
The longer bar corresponds to 1 μm.

3. Coated Particles

The properties of a powder can be altered by deposition of a given material of another chemical composition on the cores. This procedure may be used to change surface reactivity or to coat a precious solid onto a less valuable or inert material. Finally, one can achieve a desired particle morphology, which may not be possible with the coating material alone.

The precipitation process lends itself to this purpose, if the generation of the second phase is carried out in the presence of a dispersed core material (26). Again, forced hydrolysis or other techniques discussed above can be used for this purpose, as long as there is a proper ratio between the concentration of particles and the solutes that will precipitate onto them. Unless sufficient surface area is available, the coating material may separate as a fine dispersion, in addition to coating the cores, thus yielding an externally mixed system.

In Fig. 8 (upper) is shown the scanning electron micrograph of a lanthanum basic carbonate powder, whereas the lower SEM illustrates the same particles coated with copper(II) basic carbonate (27). It is quite obvious that the particles have become larger in the course of this process. Also, the color of the powder has changed from white (the pure core material) to green. One may still question whether in the latter case all original lanthanum basic carbonate particles are covered, since copper basic carbonate alone precipitates as spheres by the same procedure.

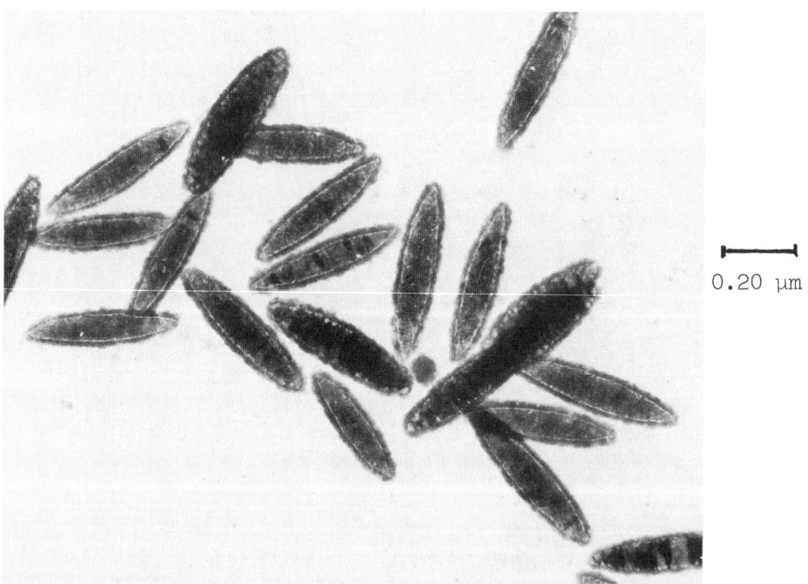

Fig. 9 . TEM of hematite particles coated with zirconium
hydrous oxide obtained by aging at 70°C for 2 hr
an aqueous dispersion containing 600 mg dm^{-3} spindle-
type α-Fe$_2$O$_3$ particles in the presence of 0.0050
mol dm^{-3} zirconium sulfate, 5 vol% formamide, and
0.5 wt% polyvinylpyrrolidone (26).

For this reason, studies have been carried out with cores of different shapes. As an example, Fig. 9 shows ellipsoidal hematite particles coated with zirconium hydrous oxide (28). Microcalorimetry curves (Fig. 10) of pure zirconium hydroxide and hematite particles covered with the latter are quite similar, indicating that the behavior of coated particles resembles that of the coating material. Using analogous procedures, it has been possible to precipitate various solids onto particles of different chemical compositions (26,29,30).

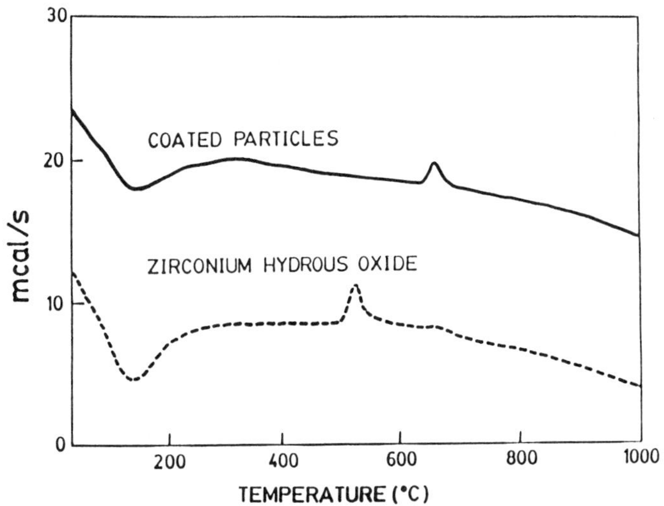

Fig. 10. Microcalorimetry curves of the zirconium (hydrous) oxide powder (-----), and of hematite coated with zirconium hydrous oxide (———) (16).

AEROSOLS

The use of aerosols in the preparation of fine particles is conceptually rather simple. Droplets of one reactant are contacted with a gas of a coreactant, to yield a desired product. The resulting particles are, as a rule, spherical and their size may be adjusted by controlling the droplet diameter. Since the liquid aerosols can be generated by condensation of vapors, exceedingly pure products are obtained. Finally, either by co-condensation of different vapors or by nebulizing mixtures of liquids, it is possible to produce internally composite solids of known ratios of constituents.

The technique described has been employed to generate a number of powders of interest in ceramics, such as titanium oxide (31), aluminum oxide (32), titanium silicate (33), or titania coated with silica.

Figure 11 illustrates a powder of amorphous titania particles prepared by reacting droplets of titanium(IV) ethoxide with water vapor (31). The entire process is completed in less than one second (34). On calcination, these solids crystallize to anatase without changing their morphology.

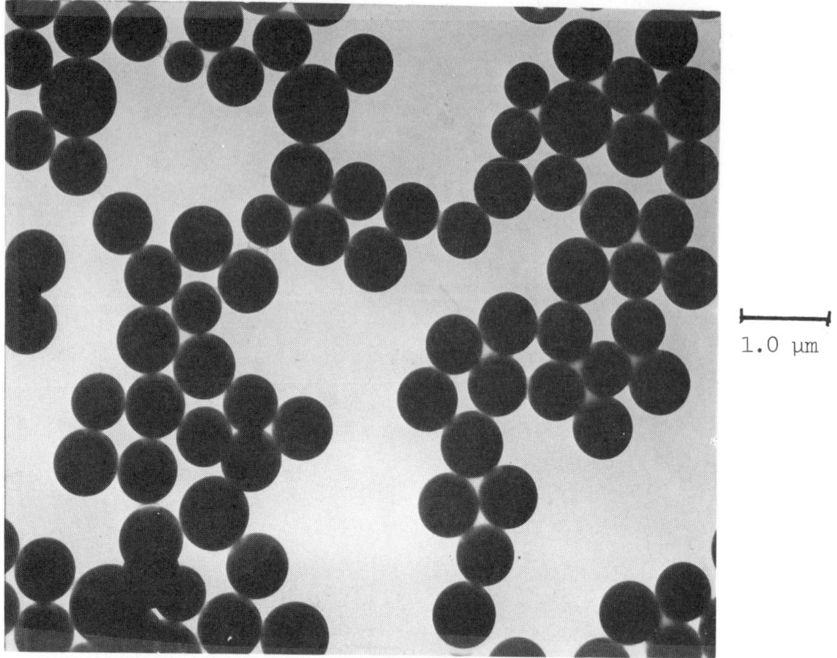

Fig. 11. TEM of titania particles obtained by interaction
of titanium(IV) ethoxide droplets with water vapor
(29).

More recently, particles of much more complex chemical composition
have been obtained by the aerosol technique. A mixed aluminum-magne-
sium-silicon ethoxide was nebulized in the generator schematically shown
in Fig. 12. The chemical stoichiometry of this alkoxide, in terms of
Al, Mg, and Si, compared well with that of cordierite. The droplets
were then brought into contact with water vapor (as a humid gas) at
temperatures varying between 95 and 240°C (35). The resulting powders
were calcined and their composition depended on the conditions in the
aerosol reactor and during calcination. Figure 13 gives X-ray diffraction patterns of spherical particles obtained at two different temperatures, then calcined at 900°C, which clearly show the crystalline character of the powder.

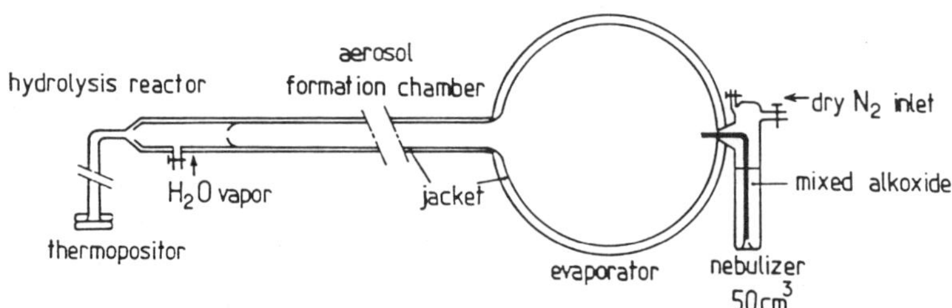

Fig. 12. Schematic diagram of the apparatus for the generation
of the alkoxide aerosols and subsequent hydrolysis.

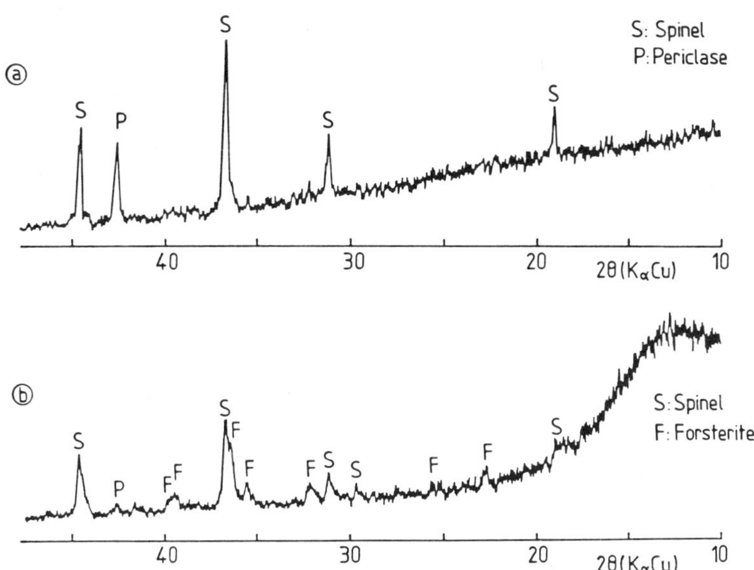

Fig. 13. X-ray diffraction patterns of two powders consisting
of spherical particles prepared by reacting Al-Mg-Si-ethoxide
droplets with water vapor at 130 and 240°C, respectively,
and subsequent calcination at 900°C.

While this technique is still principally in the exploratory state,
it offers definite possibilities for the preparation of ceramic powders
needed in special applications.

CONCLUDING REMARKS

This short review demonstrates that a large number of inorganic
powders of simple and mixed composition can now be prepared, consisting
of uniform particles of different shapes and of narrow size distributions. While the general principles of the precipitation and aerosol
techniques are reasonably well understood, there are still a number
of unresolved questions which need to be addressed.

From the applications point of view, it is also important that
engineering problems in scaling up and continuous processing be solved,
so that such powders can be produced economically and in quantity.

ACKNOWLEDGMENTS

This work has been supported by U.S. Air Force Contract
49620-85-C-0142 and by NSF Grants CHE-8619509 and CBT-8420786. The
author is greatly indebted to many of his collaborators and students,
who have made major contributions to this program. Their names appear
in the listed references.

REFERENCES

1. Matijević, E., 1988, Colloid Science of Ceramic Powders,
    Pure Appl. Chem., 60:1479.
2. Matijević, E., 1986, Monodispersed Colloids: Art and Science, Langmuir,
    2:12.
3. Matijević, E., 1985, Production of Monodispersed Colloidal Particles,
    Annu. Rev. Materials Sci., 15:483.
4. Hsu, W.P., Rönnquist, L., and Matijević, E., 1988, Preparation
    and Properties of Monodispersed Colloidal Particles of Lanthanide
    Compounds. II. Cerium(IV), Langmuir, 4:31.
5. Demchak, R. and Matijević, E., 1969, Preparation and Particle Size
    Analysis of Chromium Hydroxide Sols of Narrow Size Distributions,
    J. Colloid Interface Sci., 31:257.
6. Brace, R. and Matijević, E., 1973, Aluminum Hydrous Oxide Sols.
    I. Spherical Particles of Narrow Size Distribution,
    J. Inorg. Nucl. Chem., 35:3691.
7. Matijević, E. and Hsu, W.P., 1987, Preparation and Properties of
    Monodispersed Colloidal Particles of Lanthanide Compounds. I.
    Gadolinium, Europium, Terbium, Samarium, and Cerium(III), J. Colloid
    Interface Sci., 118:506.
8. Aiken, B., Hsu, W.P. and Matijević, E., 1988, Preparation and Properties
    of Monodispersed Colloidal Particles of Lanthanide Compounds.
    III. Y(III) and Mixed Y(III)/Ce(III) Systems,
    J. Amer. Ceramic Soc., 71:845.
9. Sapieszko, R.S. and Matijević, E., 1980, Preparation of Well Defined
    Colloidal Particles by Thermal Decomposition of Metal Chelates.
    II. Cobalt and Nickel, Corrosion, 36:522.
10. Sapieszko, R.S. and Matijević, E., 1980, Preparation of Well Defined
    Colloidal Particles by Thermal Decomposition of Metal Chelates.
    I. Iron Oxides, J. Colloid Interface Sci., 74:405.
11. Matijević, E. and Murphy-Wilhelmy, D., 1982, Preparation and Properties
    of Monodispersed Spherical Colloidal Particles of Cadmium Sulfide,
    J. Colloid Interface Sci., 86:476.
12. Murphy-Wilhelmy, D. and Matijević, E., 1984, Preparation and Properties
    of Monodispersed Spherical Colloidal Particles of Zinc Sulphide,
    J. Chem. Soc., Faraday Trans. I, 80:563.
13. Murphy-Wilhelmy, D. and Matijević, E., 1985, Preparation of Uniform
    Colloidal Particles of lead Sulfide and of Mixed Sulfides of Cadmium
    and Zinc and Cadmium and Lead, Colloids Surf. 16:1.
14. Gobet, J. and Matijević, E., 1984, Preparation of Monodispersed
    Colloidal Cadmium and Lead Selenides, J. Colloid Interface Sci.,
    100:555.
15. Ocana, M. and Matijević, E., Well Defined Colloidal Tin(IV) Oxide
    Particles, J. Materials Res. (submitted).
16. Matijević, E., The Science of Ultrafine Powders: Well Defined Composite
    and Coated Particles, J. Non-Crystalline Solids, in press.
17. Ishikawa, T. and Matijević, E., 1988, Preparation and Properties
    of Uniform Colloidal Metal Phosphates. III. Cobalt(II) Phosphate,
    J. Colloid Interface Sci., 123:122.
18. Matijević, E., Budnik, M., and Meites, L., 1977, Preparation and
    Mechanism of Formation of Titanium Dioxide Hydrosols of Narrow
    Size Distribution, J. Colloid Interface Sci., 61:302.
19. Ribot, F., Kratohvil, S., and Matijević, E., Preparation and Properties
    of Uniform Mixed Colloidal Particles: VI, Copper(II)-
    Yttrium(III) and Copper(II)-Lanthanum(III) Compounds, J. Mater.
    Res., in press.
20. Gherardi, P. and Matijević, E., 1988, Homogeneous Precipitation
    of Spherical Colloidal Barium Titanate Particles, Colloids Surf.,
    32:257.

21. Kim, M.J. and Matijević, E., Preparation and Characterization of Uniform Submicronic Lead Titanate Particles. <u>Chem. Materials,</u> in press.

22. Chittofrati, A. and Matijević, E., Uniform Particles of Zinc Oxide of Different Morphologies, <u>Colloids Surf.</u>, submitted.

23. Matijević, E., Sapieszko, R.S. and Melville, J.B., 1975, Ferric Hydrous Oxide Sols. I. Monodispersed Basic Iron(III) Sulfate Particles, <u>J. Colloid Interface Sci.</u>, 50:567.

24. Sapieszko, R.S., Patel, R.C. and Matijević, E., 1977, Ferric Hydrous oxide Sols. II. Thermodynamics of Aqueous Hyroxo and Sulfato Ferric Complexes, <u>J. Phys. Chem.</u> 81:1061.

25. Ozaki, M., Kratohvil, S. and Matijević, E., 1984, Formation of Monodispersed Spindle-type Hematite Particles, <u>J. Colloid Interface Sci.</u>, 102:146.

26. Kratohvil, S. and Matijević, E., 1987, Preparation and Properties of Coated Uniform Colloidal Particles. I. Aluminum (Hydrous) Oxide on Hematite, Chromia, and Titania, <u>Adv. Ceramic Mater.</u>, 2:798.

27. Kratohvil, S. and Matijević, E., to be published.

28. Garg, A. and Matijević, E., 1988, Preparation and Properties of Uniform Coated Colloidal Particles. III. Zirconium Hydrous Oxide on Hematite, <u>J. Colloid Interface Sci.</u>, 126:243.

29. Aiken, B. and Matijević, E., 1988, Preparation and Properties of Uniform Coated Colloidal Particles. IV. Yttrium Basic Carbonate and Oxide on Hematite, <u>J. Colloid Interface Sci.</u>, 126:645.

30. Garg, A. and Matijević, E., 1988, Preparation and Properties of Uniform Coated Colloidal Particles. II. Chromium Hydrous Oxide on Hematite, <u>Langmuir,</u> 4:38.

31. Visca, M. and Matijević, E., 1979, Preparation of Uniform Colloidal Dispersions by Chemical Reaction in Aerosols. I. Spherical Particles of Titanium Dioxide, <u>J. Colloid Interface Sci.</u>, 68:308.

32. Ingebrethsen, B.J. and Matijević, E., 1980, Preparation of Uniform Colloidal Dispersions by Chemical Reactions in Aerosols. 2. Spherical Particles of Aluminum Hydrous Oxide, <u>J. Aerosol Sci.</u>, 11:271.

33. Balboa, A., Partch, R.E. and Matijević, E., 1987, Preparation of Uniform Colloidal Dispersions by Chemical Reactions in Aerosols. IV. Mixed Silica/Titania Particles, <u>Colloids Surf.</u>, 17:123.

34. Ingebrethsen, B.J. and Matijević, E., 1984, Kinetics of Hydrolysis of Metal Alkoxide Aerosol Droplets in the Presence of Water Vapor, <u>J. Colloid Interface Sci.</u>, 100:1.

35. Salmon, R. and Matijević, E., Preparation of Colloidal Magnesium-Aluminum-Silicates by Hydrolysis of a Mixed Alkoxide, <u>Ceramics Intern.</u>, in press.

PROGRESS IN PREPARATION OF ZnO BASED VARISTOR CERAMICS

O. Milošević, D. Vasović, D. Poleti, Lj. Karanović,
V. Petrović and D. Uskoković

Institute of Technical Sciences of the Serbian Academy of
Sciences and Arts and Belgrade University, Yugoslavia

ABSTRACT

Chemical preparation methods were used to fabricate ZnO-based
varistor ceramics having nonlinearity coefficients between 35-45,
breakdown fields ranging from 400-1000 kV/m and leakage currents far
less than 10^{-2} A/m². The processing steps employed include (1) chemical
synthesis of powders having a complex initial composition of the form
(100-y) ZnO+y (additives) with y selected from the group of soluble
salts of Bi, Sb, Co, Mn, Ni and Cr, then (2) calcination of as-prepared
powders within the temperature interval from 373 to 1473K and (3)
sintering at temperatures from 1373 to 1573K. Two different methods for
powder preparation were used: 1) evaporation of ZnO suspensions with
additive solutions and 2) coprecipitation of additive solution and
afterwards by adding ZnO.

Various analytical methods such as optical and scanning electron
microscopy, DTA, TGA, X-ray and EDAX were used in characterizing the
powders and the resulting ceramics. Processes of formation of the main
phases as well as their influence on resulting non-Ohmic behaviour, was
investigated.

INTRODUCTION

ZnO-based varistors represent a class of multiphase ceramic
semiconductive devices which exhibit high nonlinear current-voltage
characterics.[1] These nonlinear properties are generally described by
the relation $J = C\,K^{\alpha}$, where J and K are the current density and the
electric field, respectively, α is the nonlinearity coefficient defined
as d(log J)/d(log K), and C is a proportionality constant. To meet
these special needs, the processing of ZnO varistors typically involves
searching for optimum phase content by means of high temperature solid
and liquid phase reactions: resultant microstructure is comprised of
semiconducting n-type ZnO grains surrounded by insulating barriers at
the ZnO grain boundaries.[2]

The as-formed gross ceramic microstructure, containing ZnO grains,
a spinel phase and an intergranular layer, does not itself play a major
role in the highly non-linear conduction process. Rather the electrical

properties are determined solely by the behaviour of the individual
grain-grain junctions.[2] However, when subjected to alternating or
pulsed currents, localized small differences in the number of grain
boundaries, due to inhomogeneous distribution of the main phases, may
result in such erratically distributed current densities as to cause
different local degradation and thermal breakdown processes. For these
reasons, the microstructure of varistors need to be closely controlled.

The well known fact is that these conditions can be hardly
fulfilled when using conventioanl processing of ZnO varistors, e.g., by
relaying mechanical blending of oxides as precursors. Fine reactive
powders of high purity and controlled composition and particle shape
are needed. Preparation of such powders by means of the homogeneous
precipitation method,[3-5] sol-gel process[6] and evaporation of solutions
and suspensions,[7-9] have prove to be the most important step in the
production of varistors with highly reliable ceramic and electric
properties.

The purpose of this paper is to describe an improved varistor
production procedure that includes chemical preparation of powders
having a complex initial composition by means of 1) evaporation of ZnO
with additive solution and 2) coprecipitation. The conditions for
decomposition and consolidation of these powders were thoroughly
investigated in order to obtain nonlinear ceramics with a high
coefficient of nonlinearity and low leakage currents.

EXPERIMENTAL

In order to obtain powder with complex initial composition (100-y)
ZnO + y (additives), where y=1.8-7.2 mol%, the following methods were
used: (1) evaporation of ZnO suspension with additives being introduced
from solutions (Powder I); (2) evaporation of ZnO suspension with
additives prepared through coprecipitation (Powder II).

The following additive salts were used: $C_O(NO_3)_2 \cdot 6H_2O$; $Ni(NO_3)_2$
$6H_2O$; $Cr(NO_3)_2 \cdot 9H_2O$; $Bi(NO_3)_3 \cdot 5H_2O$; $Mn(CH_3COO)_2 \cdot 4H_2O$ and $SbCl_3$. Details
concerning powder preparation were presented in a previous paper,[10] the
various steps involved being shown in Fig. 1.

Powders were dried at a temperature of 378K, then were heat
treated isothermally within a temperature interval from 423-1473K, for
2h. Both grinding and homogenization of calcined powders were done in a
high speed, agate planetary ball mill with an acceleration of 12 g
("Retsch") for periods of 4-16h.

Homogenized powders I and II were compacted into samples having a
diameter of $8 \cdot 10^{-3}$m and a heigh of $1-5 \cdot 10^{-3}$m at an uniaxial pressing
pressure of 80 MPa. Some samples of powder I having a diameter of $4 \cdot 10^{-2}$
m and a height of $1.5 \cdot 10^{-2}$m were formed by cold isostatic pressing at
140 MPa in an Autoclave engineering apparatus.

The $8 \cdot 10^{-3}$m dia samples were isothermally sintered at temperatures
from 1073-1573K, held for 0-240 min, and air quenched. The $4 \cdot 10^{-2}$m dia
samples were sintered at a temperature of 1473K held for 180 min;
heating and cooling rates were controlled at 5°/min. A sintering
processes were performed in air, in a furnace for which the controls
were precise within ±2 C.

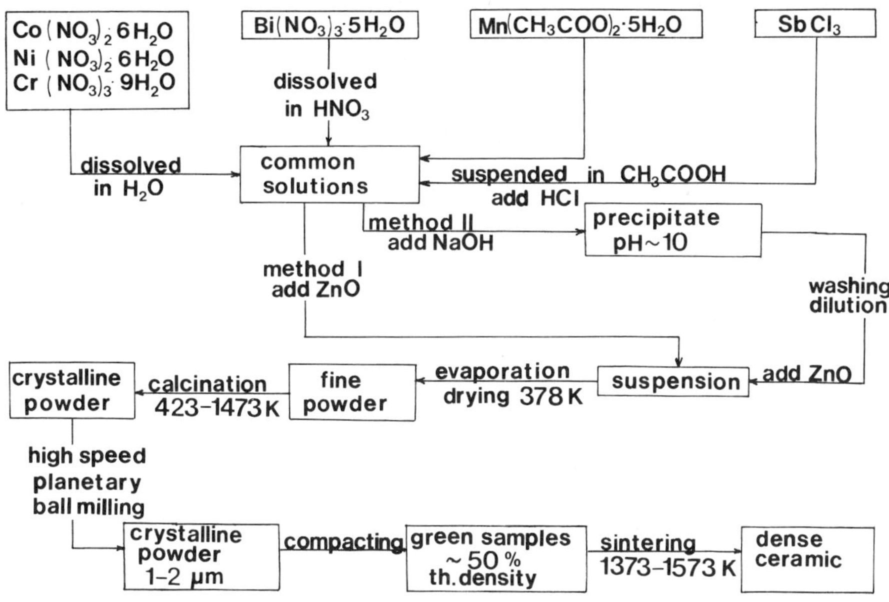

Fig. 1. Schematic of the chemical powder routes for processing of varistor ceramics.

DTA and TGA were performed in air at heating rates of 10°C/min ("Linseis", Pt-Pt+Rd thermocouple, sensitivity 0.1 mV). Development of crystal phases in powders was studied with "Phillips 1051" diffractometer using CuK_α radiation with graphite monohromater. X-ray diffraction was also utilized for crystal phase detection in the resulting ceramics. For such analyses, samples ~ $6\cdot10^{-3}$m dia. and $1.5\cdot10^{-3}$m in height were used.

Microstructural investigations were carried out by optical microscopy (apparatus "Reighert") on free, polished and chemically etched fracture sample surfaces. Chemical etching was done with HNO_3 solution for 7-10 s. Free fracture surfaces were also examined by scanning electron microscopy.

Energy dispersive X-ray analysis (EDAX) was used for qualitative analysis of constitutive phases, using polished and chemically etched fracture surfaces of $\sim 6\cdot10^{-3}$m dia samples.

Electric measurements were registered within interval from 1-100 A/m^2 by a dc power supply $\sim 6\cdot10^{-3}$m dia samples and by ac power suplly (50 Hz) within the low current region up to $10_3 A/m^2$ for $\sim 3.5\cdot10^{-2}$ m dia samples. Within the high current region ($<4\ 10^3$ KA/m^2) the K-J measurements were done by a pulse technique (8x20 μs current pulses) for $\sim 3.5\cdot10^{-2}$m dia samples. With this purpose, a generator with impact currents of 66 KV, 24 KJ and a high voltage apparatus of 6 KV, 150 mA were used. The $3.5\cdot10^{-2}$m dia samples were measured by electrodes and insulation. The nonlinearity coefficients were determined within the range from 1-10 A/m^2 (α_1) and 10-100 A/m^2 (α_2), the breakdown field, Kc, was determined at $10\ A/m^2$ and the leakage current was measured at a voltage of 0.8 Kc.

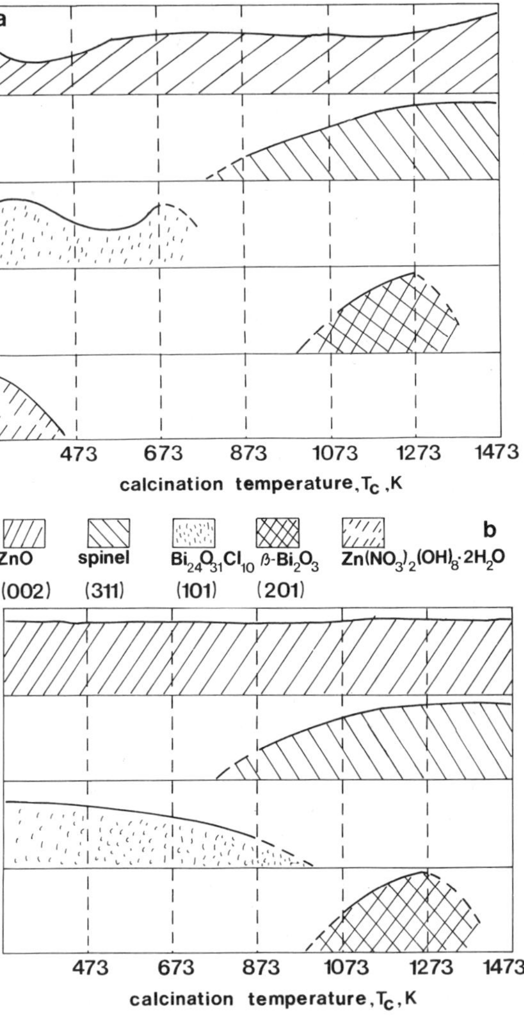

Fig. 2. Effect of calcination temperature on the crystal phase formation in (a) type I and (b) type II powders.

RESULTS AND DISCUSSION

Phase formation

According to TGA results, powder I decomposed in the temperature interval from 323–1093K, with a maximum weight loss of 39%. It was shown by DTA that decomposition proved to be endothermic up to 473K, corresponding to dehydratation reactions and the decomposition of nitrates.[10] Decomposition above this temperature is exothermic, indicative of formation of crystal phases; the phases thus formed proved to be typical for varistor ceramics.

In powder I, dried for 20 h at 378K, the $Bi_{24}O_{31}Cl_{10}$ phase, having a relatively low degree of crystalinity, was the most representative one. Together with this phase, also appeared ZnO and a phase of $Zn_5(NO_3)_2(OH)_8 \cdot 2H_2O$ type, formed during reaction of initial ZnO with anions from common acidic solutions.[10] In the course of calcination of powder I at 426K (Fig. 2a), the phase of $Bi_{24}O_{31}Cl_{10}$ type diminished somewhat, while the ZnO content increased. At 473K no evidence of the

zinc-hydroxyl-nitrate phase was detected; the system proved to be a
two-phase system. In it, the presence of ZnO phase was confirmed, the
crystallinity of which gradually increased with temperature. At 573K, a
loss of the $Bi_{24}O_{31}Cl_{10}$ phase was observed, as well as a further
increament in the ZnO content. At 673K both phases were well
crystallized. At 873K, ZnO prevailed and the formation of a spinel
phase began. At 1073K, the ZnO phase predominates; the content and the
degree, of crystallinity of a spinel phase increased: diffraction
maxima appeared at 3.26, 3.22, 2.88 and 2.78 Å indicates the presence
of tetragonal β-Bi_2O_3 and/or of phase $Bi_{20}M_0O_3$ type. During calcination
at 1273K the content of spinel increased, the content of ZnO phase
remained unchanged and diffraction peaks at 3.2 and 2.74 Å pointed to
gradual re-arrangement of β-Bi_2O_3 lattice. The phases of ZnO and spinel
were only identified in the sample of powder I calcined at 1473K.

Powder II attained its maximum weight loss (5%) at a temperature
of 1173K.[10] This decomposition proved to be endothermic up to 773K,
corresponding to dehydratation reactions. Above this temperature,
decomposition showed exothermic character. In powder II, dried for 17 h
at 378K, a well-crystallized ZnO phase was observed, as well as a small
amount of a less well-crystallized $Bi_{24}O_{31}Cl_{10}$ phase. No change of
phase composition was detected at temperature up to 870K, when poorly
crystallized spinel phase appeared (Fig. 2b). At 1073K, the phase
composition was similar to that of powder I at the same temperature. At
1273K the contents of both spinel and β-Bi_2O_3 increased. At 1473K, the
β-Bi_2O_3 phase was not identified any more.

For bulk ceramics obtained by either of the methods and sintered
at 1573K, the XRD patterns yielded an intense pattern identifiable as
hexagonal ZnO, with lattice parameters similar to those observed in
conventionally prepared samples (a=3.246 Å, c=5.199 Å).[11] ZnO grains
displeyed polygonal cross sections, and often showed twin boundaries.
When samples obtained from powder I are investigated, EDAX confirmed
that the ZnO phase dissolves a small quantity of Co, Ni and Mn ions.[10]
In the case of Powder II, only cobalt ions were present in the ZnO
grains. The spinel phase, with a main structure $Zn_7Sb_2O_{12}$, typically
contained Cr, Mn, Co and Ni in substitutional solid solution. It
appeared in the form of grain-like uniaxial structured along the ZnO
grain boundaries. When examined by SEM, fractured surfaces showed a
much higher content of uniformly dispersed spinel phases along ZnO
grain boundaries than did those obtained by conventional processing.[11]
It was also noticed that the distribution of the ZnO grains, and that
of the spinel phases, was much more uniform in the chemically prepared
ceramics.

Initial powders obtained by either of the methods yield as highly
dispersed homogeneous microparticles due to a dehydratation process as
well as decomposition of nitrates during calcination. This cause the
solid and liquid phase reactions took place at far lower temperatures
when compared with temperatures of phase formation in varistor systems
that were obtained by conventional synthesis.[12,13] This certainly
accounts for the observed differences in phase composition and the
extremely uniform microstructures with higher contents of a spinel
phase obtained in those samples produced by chemical synthesis (I and
II).

When sintered at 1573K, the intergranular layer was found to
consist mainly of β-Bi_2O_3, with lattice parameters of a=7.7425 Å,
c=5.6313 Å (JCPDS 22-515). The degree of crystallization proved to be
higher in mixture II. Besides the principal Bi-component, intergranular
layer also contained Zn, Sb and Cr.

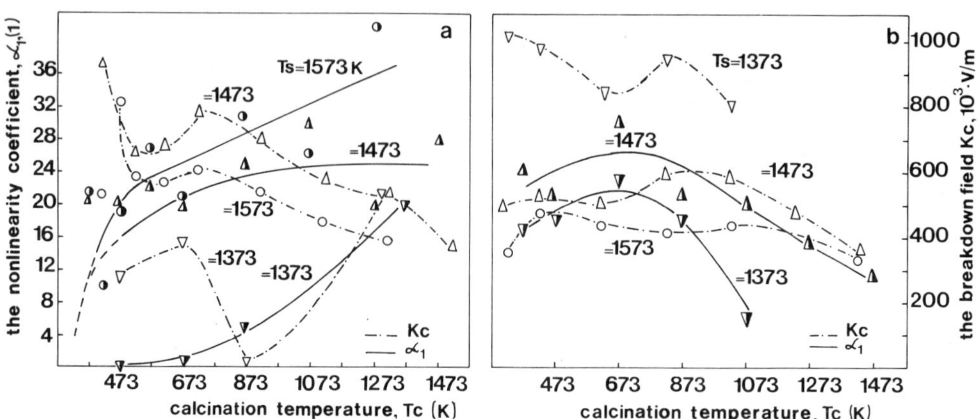

Fig. 3. The nonlinearity coefficients, α_1, and the breakdown fields, K_c, vs calcination temperature for powders I (a) and II (b), sintered at 1373K, 1473K and 1573K, respectively.

Nonlinear properties

The nonlinearity coefficients and the breakdown fields observed for samples sintered at 1373, 1473 and 1573K, respectively, are given in Fig. 3, plotted as functions of calcination temperature for powders I and II. In case of powder I densified by sintering at 1373K (Fig. 3a), modest nonlinear properties are formed, but only when calcined at higher temperatures (<1273 K). When sintered at 1473K, the nonlinearity coefficients >12 are evident already for powder calcined as low as 423 K, but will exceed values of 20 when the temperature of calcination reaches 873K or higher. The highest values for the nonlinearity coefficients ($\alpha_1 > 20$ with maximum of 36) are realized by sintering of powders at 1573K, when the initial calcination temperature had been 473K. For both powders I and II, the values of the breakdown fields slightly decreased with the rise of calcination temperature, but the sintering temperature proved to be the most influental parameter.

Because of the higher sinterability of powder II[10], its nonlinear properties are noticed already after sintering at 1373K. The nonlinearity coefficients reached their maximum values at calcination temperature of 673K (22, when sintered at 1373K, and 26, when sintered at 1473K). Above this calcination temperature, the values for α_1 decreased.

The nonlinearity coefficients, α_1 and α_2, are given as functions of isothermal sintering time at 1573K in Fig. 4. The corresponding values approximately range from 37–42 (α_1) and from 32–35 (α_2) when samples of powder I were isothermally sintered at 1573K for 5–120 min. For samples of powder II, the nonlinearity coefficients range from 30–35 (α_1) and to 30 maximum (α_2). The abrupt fall of nonlinearity, noticed after 120 min at temperature of 1573K, resulted either from the evaporative loss of, or from transformation of, the β-Bi_2O_3 phase.[15] Note that the average breakdown voltages of individual grain boundaries, given by the slope of the curve $K = f(1/D)$,[10] as plotted in Fig. 4, are 5–6 V, for both powders.

For powder II, the lowest values of the nonlinearity coefficients were obtained over a wide range of breakdown fields. However, the

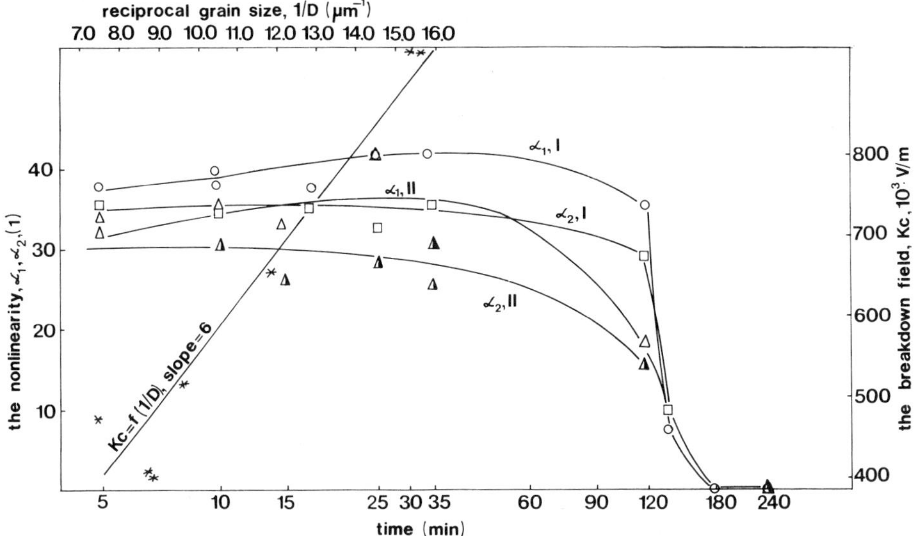

Fig. 4. The nonlinearity coefficients α_1 and α_2 vs sintering time for samples of powder I and II, isothermally sintered at 1573K.

observed leakage currents were also higher (10^3 –10^4 μA/m^2). This certainly resulted from insufficient doping of ZnO grains when method II is employed. The Mn^{2+} ion, which is the performance substituent of ZnO grain influencing the height and formation of a potential barrier in the vicinity of the ZnO grain boundaries,[16] was not identified in ZnO grains by means of EDAX in case of method II. A higher doping of ZnO grains in case of method I resulted from the fact that during the calcination step a transformation of the ZnO component occured, thus enabling it to be more reactive. Apart from this, the presence of NaOH, an agent of precipitation, can also inhibit substitution in the ZnO lattice in case of method II.[17]

The current-voltage characteristics within the region from 10^{-2} – 10^7 A/m^2 for samples obtained by conventional synthesis[11] and those by chemical process of method I are given in Fig. 5. The values of nonlinearity coefficients for samples obtained by chemical synthesis (powder I) with factor 2, are higher in the whole breakdown region for the current densities within 10^{-1}– 10^4 A/m^2. For the low current region, the corresponding values are 30.3 (Powder I) vs. 5.84 (powder obtained by conventional method) within the currents densities from $2.86 \cdot 10^{-3}$– 10 A/m^2, and 38.9 (powder I) vs 24.6 (powder obtained by conventional method) within the current densities from 1–10 A/m^2. As the shape of the current-voltage characteristics is a function of the grain boundary uniformity within the breakdown region, far higher values of the nonlinearity coefficients were obtained in the system obtained by chemical synthesis (method I); these improvements resulted from a more uniform distribution of the main microstructural phases.

In the upturn region, a higher slope of the K–J curve is evident for samples obtained by chemical synthesis (method I). This certainly limits the preformance of ZnO varistors in surge suppression applications. Within this region, at frequencies of >10^8 Hz, the equivalent impendance of a varistor is reduced to the resistance of ZnO grains and therefore the slope of the K–J curve in this region is controlled by the concentration of donor atoms in ZnO grains. This

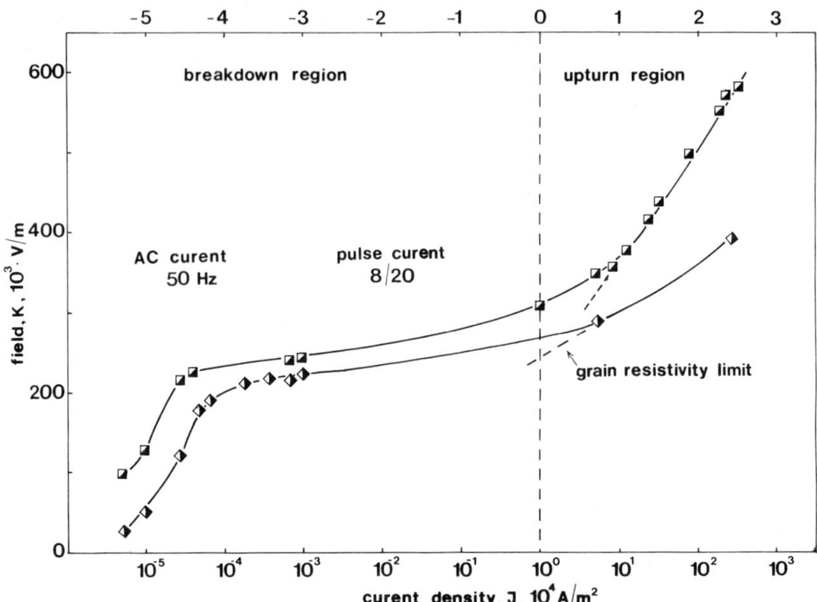

Fig. 5. K-J characteristics of the samples obtained by means of chemical synthesis (method I, ▨) and by conventional processing (◈).

observation certainly reinforces the need for controlling the doping level of ZnO grains when chemically prepared samples are employed.

CONCLUSION

Two chemical preparation methods: (1) evaporation of ZnO suspensions with additive common solutions, and (2) coprecipitation of additive solutions by NaOH, and afterwards adding of ZnO component, were used in order to obtain ZnO-based varistor ceramics. A summary of phase development in powders during calcination has been presented. It was established that during calcination of as-prepared powders within temperature interval from 373-1473K the following phases are formed: ZnO phase, existing throughout the whole temperature interval; $Bi_{24}O_{31}Cl_{10}$, appearing up to 673K in case of the first method and up to 873K in the second one; spinel phase, with the main composition $Zn_7Sb_2O_{12}$; and $\beta-Bi_2O_3$, when temperatures of 873K and 1073K respectively, are exceeded. The $Zn_5(NO_3)_2(OH)_8 \cdot 2H_2O$ phase exists up to temperature of 473K when the first method is employed; it results from ZnO transformation during the preparation procedure. This facilitates higher activity of Zn-component, causing a higher doping level of ZnO grains in the sintered ceramics, and corresponding higher values of the nonlinearity coefficients.

The current-voltage characteristics for the whole current region in a sample obtained by the first method is presented, together with those obtained by conventional processing. Having compared these two characteristics, the following can be noticed:

(i)   the liquid and solid phase reactions and their development began at lower temperatures due to the higher activity of chemically prepared powders. This enables one to obtain fine grained ceramics with the corresponding higher values of the breakdown fields;

(ii) the nonlinearity coefficients are higher in the breakdown region
     for chemically prepared samples (method I) due to a much more
     uniform distribution of the main microstructural constituents. In
     the upturn region where the slope of K-J curve is limited by the
     resistivity of ZnO grains, the surge protection capability is
     reduced in the case of chemically prepared samples (method I) as a
     result of insufficient doping of ZnO grains.

ACKNOWLEDGEMENT

This paper was financially supported by the Republic Committee for
Sciences of Serbia through the Project "Physical Chemistry of Condensed
Systems".

REFERENCES

1. M. Matsuoka, "Nonohmic properties of zinc oxide ceramics", J.J.
    Appl. Phys., 10(6),(1971)736.
2. H. R. Philipp, "Grain resistivity and conduction in metal oxide
    varistors", in "Tailoring multiphase and composite ceramics",
    ed. R. E. Tressler, G. L. Messing, C. G. Pantano and R. E.
    Newnham, Mat. Sci. Res., 20, Plenum Press, New York and
    London, 1987, 481.
3. E. Sonder, T. C. Quinby and D. L. Kinser, "ZnO varistors made from
    powders produced using a urea process", Am. Ceram. Soc. Bull.
    65(4), (1986), 665.
4. R. G. Dosch, "The effects of processing chemistry on electrical
    properties of high-field ZnO varistors", in "Science of
    ceramics chemical processing", ed., L. L. Hench and D. R.
    Ulrich, Wiley & Sons, (1986),311.
5. S. Hishita, Y. Yao and S. Shirasaki, "Zinc oxide varistors made
    from powders prepared by amine processing", J. Am Ceram. Soc.
    72(2), (1989), 338.
6. R. J. Lauf and W. D. Bond, "Fabrication of high field zinc oxide
    varistors by sol-gel processing", Am. Ceram. Soc. Bull., 63(2),
    (1984), 278.
7. K. Seitz, E. Ivers - Tiffee, H. Thomann and A. Weiss, "Influence
    of zinc acetate and nitrate salts on the characteristics of
    undoped ZnO powders", Proceeding of the VI World Conf. High.
    Techn. Ceram., Milano, Italy, June 1986, Elsevier, ed. P.
    Vincenzini, 1987.
8. E. Ivers-Tiffee and K. Seitz, "Characterisation of varistor type
    raw materials prepared by the evaporative decomposition of
    solutions technique", Am. Ceram. Soc. Bull., 66(9), (1987),
    1384.
9. F. C. Palilla, "Process for the preparation of homogeneous metal
    oxide varistors", U.S. Patent 4,575,440, Mar. 11, 1986.
10. O. Milošević, D. Vasovic, D. Poleti, Lj. Karanović, V. Petrović
    and D. Uskoković, "Microstructural and electrical properties of
    ZnO varistors prepared by coprecipitation and evaporation of
    suspensions and solutions", Proceedings of Second Varistor
    Conference, Schnenectady, USA, Decembar 1988 (in Press).
11. O. Milošević, P. Kostić, V. Petrović, Lj. Trontelj and D.
    Uskoković, "Crystal phases and electrical properties in
    nonohmic ZnO ceramics", Proceeding of the 14th Conference on
    Silicate Science, Budapest, 6-10 May, 1985.
12. M. Inada, "Crystal phases of nonohmic zinc oxide ceramics", J. J.
    Appl. Phys. 17(1) (1978) 1.
13. V. F. Katkov, A. I. Ivon, V. O. Makarov and I. M. Cernenko,

"Formirovaniye struktury oksidno-cinkovoi keramiki" Neorganicheskiye materialy, <u>24</u>, 8(1988) 1358.
14. J. Wong, "Barrier voltage measurement in metal oxide varistors", J. Appl. Phys., 47 (11)(1976) 4971.
15. O. Milošević, D. Vasović, D. Poleti, Lj. Karanović, V. Petrović and D. Uskoković, "Development of crystal phases and nonlinear properties in chemically prepared varistor ceramics", Paper presented at ECerS '89, Maastricht, 18-23 Jun 1989, Holland (to be published in Conference Proceeding).
16. R. Eizinger, "Grain boundary phenomena in ZnO varistors", in "Grain boundaries in semiconductors", North-Holland, Amsterdam, 1982, 343.
17. M. V. Vlasova, N. G. Kakazey, O. Milošević, D. Poleti, D. Vasović and D. Uskoković, "Electronic paramagnetic resonance study of the structure of ZnO varistors prepared by various chemical methods", J. Mater. Sci., 1989 (Accepted for publication).

CHEMICAL PREPARATION OF ALUMINA-ZIRCONIA POWDERS FOR

LOW TEMPERATURE SINTERING AND PARTICULATE COMPOSITES

Jean-Luc Rehspringer, Sami Dick and Marc Daire

Ecole Européenne des Hautes Etudes des Industries Chimiques
de Strasbourg - Institut de Physique et Chimie des
Matériaux - BP 296 - F-67008 Strasbourg-cedex

INTRODUCTION

As a solution to difficulties encountered in the development of-fiber
reinforced materials, the use of particulates as toughening constituents
appears more near-term, and to a certain extent is economically rather
appealing, because of simplicity of processes. When synthesizing with ZrO_2
these tougher materials, it has been found that retention of the tetragonal
structure at room temperature is critically dependent upon the dimensions
of the microstructure. Peculiarly, a critical grain or inclusion size
exists, below which the high temperature tetragonal phase can persist, and
above which retention is not observed (1).

For example, for a composite with 2 mol% Y_2O_3, the retention of the
tetragonal Zirconia phase was obtained with an overage size of the ZrO_2
powder grain lower than 0,2 um (2). But the sintering process needs a high
temperature (1500°C -1600°C), where grain growth occurs and induces a
decrease of the amount of the ZrO_2 tetragonal phase. Thus hot pressing
processes are achieved to lower the sintering temperature and to limit the
grain growth.

Another way to obtain submicron tetragonal ZrO_2 particles in the
alumina matrix is to experiment submicronic powders of high sinterability.
Several methods of elaboration for such Al_2O_3-ZrO_2 powders have been
reported in the literature. Several investigators used aqueous solutions
(or sols) of aluminum or zirconium salts to produce powders by
coprecipitation (3), spray drying (4) or the so-called spray-ICP technique
— which involves spraying the solution into an inductively coupled plasma
at 5000K (5). At the same time,a new preparation technique of zirconia-
alumina powder was proposed (6), involving hydrolysis of a zirconium
alkoxide in an   -alumina dispersion. Compositionally homogeneous powders
can be obtained by these preparation methods, nevertheless they do not
display the needed excellent capabilities for sintering.

The main objective of the present work was to study the sinterability
of two powders obtained from very different processes, theoretically
guaranteeing a good homogeneity in composition and a narrow distribution of
particle sizes. The first passes through the coprecipitation in an aqueous
solution at constant pH, and the other is a sol-gel method improved by a
modification of alkoxides.

Science of Sintering
Edited by D. P. Uskoković *et al.*
Plenum Press, New York

PREPARATION and EXPERIMENTAL PROCEDURES

Aqueous Precipitation

The mineral salts ($Al(NO_3)_3$, $9H_2O$ and $ZrOCl_2$, $8H_2O$) are dissolved in water, with a concentration of 0.5 mol/liter. In another vessel, an aqueous ammonia solution at 33 mol % is maintained under vigourous stirring on a magnetic stirrer. The aluminum-zirconium salt solution is pourred rapidly into the ammonia solution, and a white gel appears. Classically, we have chosen to introduce the salt into the basic solution in order to preserve a quite constant pH from the beginning of the coprecipitation. PH measurements before and after indicate a small decrease during precipitation, from 10 to 9 pH units. It is well known that in this range of pH, aluminum and zirconium cations precipitate into hydroxyde types, namely $Al(OH)_3$ and $Zr(OH)_4$ (metastable). If the reverse proceeding is used (introduction of ammonia into Al-Zr solution), the increase of pH, from acid to neutral to basic, leads to the intermediate formation of AlO^+ cations. Thus it can cause orientation of grain crystallization during coprecipitation, and threafter promote a bimodal distribution of the grains, contrary to our requirements of homogeneous composition and grain size.

The hydrous gel is then thoroughly washed with distilled water, in order to rid it of all chloride anions, which can hinder the sintering process. A final wash is carried out with ethanol or other organic solvent; with the aim to decrease the inter-particle water of hydration, which participate, during the drying step, in the floculation of powders and the cementation of hydroxyde particles into hard agglomerates.

The gel is filtered, dried at 110 °C for 12 hours, then raised to 600 °C in one hour. After an annealing at this temperature for an additional hour, the material is ready for sintering.

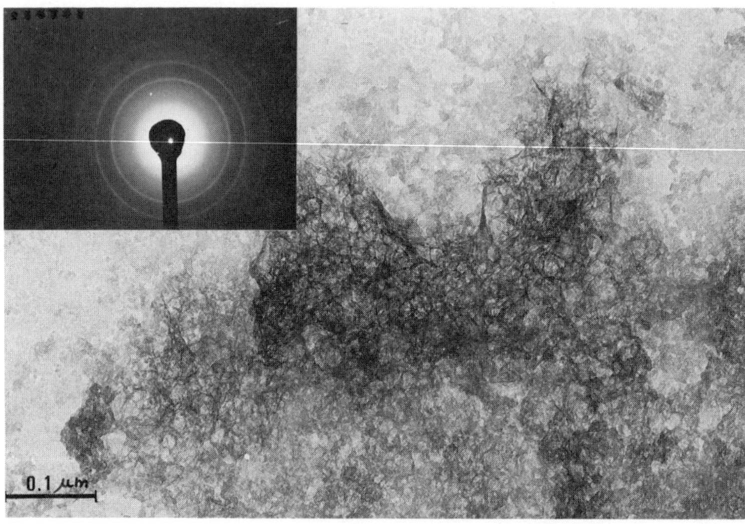

Fig. 1. TEM observation of an alcohol washed hydrous gel dried at 110°C.

Sol-gel Precipitation

Several sol-gel processes have been developed that depend upon the resources of chemistry (6-7). Eventhough theses experimenters obtained alumina-zirconia powders, they could not adequately control the hydrolysis process and thereby regulate crystallization and growth. In sintering science it has been shown (8) that a regular morphology often induces a high sinterability of powders for oxide systems. The sol-gel preparation route also allows one to use highly reactive alkoxides in alcoholic solution. Hydrolysis by water, in alcoholic medium, occurs within a very short time and for this reasonthe crystal growth cannot be controlled. An alternative procedure involves the modification *in situ* of the alkoxides in order to obtain compounds that are less reactive with water.

Previous results (9) have pointed out that spherical particles can be obtained according to the following method. Aluminum butoxide and Zirconium i-propoxide are dissolved in a suitable solvent (about 1 mol/l). An organic acid (propionic acid) is rapidly pourred into the vessel containing the alkoxides, while the medium is stirred vigorously. An alcoholic solution of water is then added, causing the precipitates to develope slowly. After filtration, the gel is heattreated in the same manner as before.

Physical tests

The grain size distribution is obtained using a Malvern Autosizer, with samples ultrasonically treated for 30 minutes in alcohol. The porosity is measured with a Carlo Erba microporosimeter.

TGA is performed with a Setaram thermogravimeter. The expansion is studied at the rate of 300 °C per hour, on an Adamel apparatus with pellets (diameter and length about 7 mm) prepared from annealed powders by uniaxial pressing at 5 t/cm^2.

MORPHOLOGY OF POWDERS

Hydrous Coprecipitation

Transmission electron microscopy (Fig. 1) reveals a bimodal distribution of the powder particles, made up of small needles and very small particles.. These views were taken after the drying step. Electron diffraction patterns show a set of crystallization rays which cannot be indexed; but it can be reasonably proposed that the observed needles are composed of aluminum hydroxide, and the very small particles of zirconium hydrate or hydroxide. Furthermore no morphological difference has been demonstrated for samples washed respectively with alcohol, acetone or pure water.

The distribution of the grain size is obtained after annealling. The representative example given in Fig. 2 (54 mol % Al_2O_3 - 46 mol % ZrO_2) shows a very narrow distribution of particle sizes in the case of washing with alcohol; we note also that the powder de-agglomerates very easily. In contrast, washing with water induces a vary broad distribution that cannot be dispersed by an ultrasonic treatment and is reliably causes cementation of particles into hard agglomerates. Moreover porosity measurements verify that in the range 0.5 - 10 um, the amount of pores in the water-washed powder is ten times the quantity in the alcohol-washed precipitate.

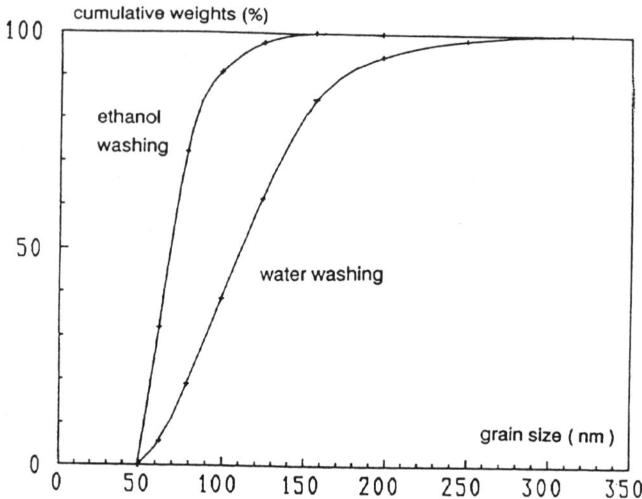

Fig. 2. Particle size distribution of hydrous gel washed with ethanol or water and heated at 600°C.

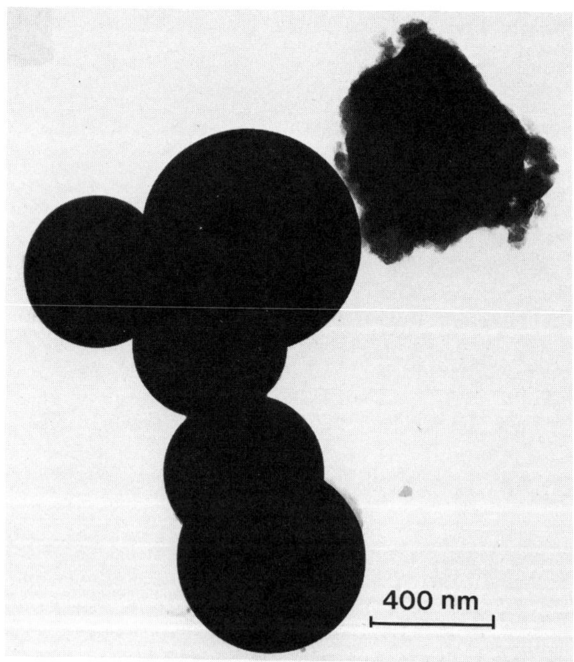

Fig. 3. TEM observation of sol-gel powders heated at 600°C.

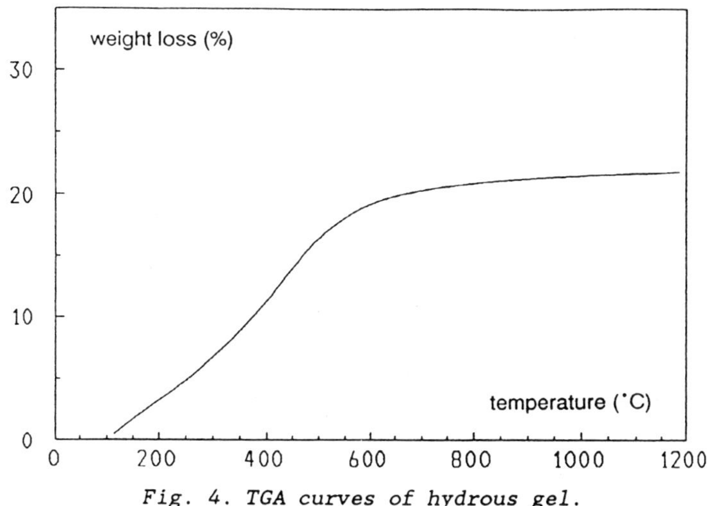

Fig. 4. TGA curves of hydrous gel.

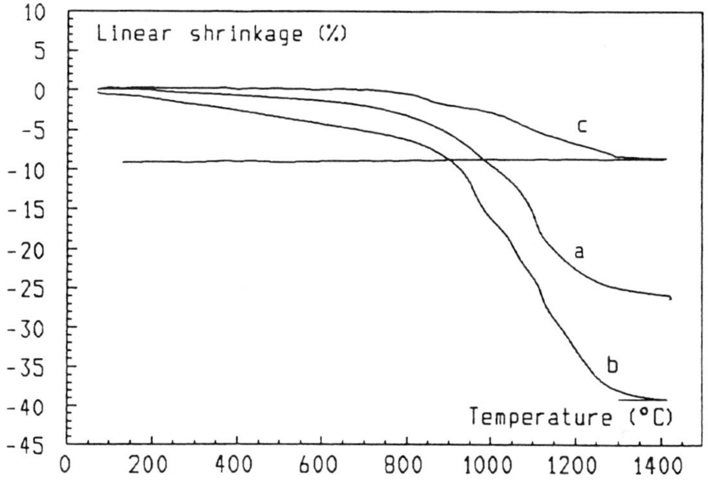

Fig. 5. Dilatometric measurements on hydrous gel washed with:
a. ethanol; b. acetone-toluene; c. water.

<u>Sol-gel Powders</u>

TEM examinations show a very singular crystallization (Fig. 3), with polyhedral and typical well formed spherical particles, leading systematically to a bimodal distributed population. The dried powders vary in dimension from 30 to 80 nm, regardless of the form they may have. Their regular shapes denote some degree of organization, and is quite different from the various facets built by the coprecipitation process.

SINTERING CAPABILILITIES

Before conducting dilatometric experiments, it is necessary to analyse the weight behaviour of dried gels as a function of temperature. TGA curves (Fig. 4) give evidence of an important loss of weight due to dehydration. At 600°C it has almost ended, but a weak rate phenomenon persists, as some water losses are still observed until 1200°C. At 600°C ZrO_2 has crystallized, and calculations from the TGA diagrams demonstrate that AlO(OH) appears at 400°C and leads to a hydrated alumina.

<u>Powders from Hydrous Coprecipitation</u>

For the composition 54 % Al_2O_3 - 46 % ZrO_2, the effect of the final washing step for gels is compared. For washing with an organic solvent (Fig.5), we notice a high rate sintering process between 700°C and 1200°C, i.e in a low temperature regime. When the wash-medium is pure water, the linear shrinkage is much reduced, and two steps of sintering can be clearly distinguished, leading one are to look for two successive mechanisms of dehydration and/or sintering. For the other compositions we chose to wash with alcohol, which is easier to use than acetone-toluene mixtures and especially so since that latter medium results in low green density.

<u>Powders from Sol-Gel Modified Process</u>

As shown in Fig. 3 these powders lead one to expect promising results during densification. However, sintering studies (Fig. 6) demonstrate a very low shrinkage. The green density of these samples was also very poor (1.97 g/cm^3); it attained only a value of 2.56 g/cm^3 (i.e., 64 % of the theoretical density) as a final density after heat treatment at 1400°C.

Tables I recapitulates the sintering capabilities of various powders observed after a heat treatment at 1270°C in the case of alcohol-washed coprecipitates and 1400°C for the tested sol-gel powder.

CONCLUSIONS

For the first process investigated in this work we saw the basic influence of the steps of washing. The organic final washing has a strong influence on the desagglomeration *in situ* of the small particles. It seems to be clearly established that the critical path for low temperature powder preparation is the drying process. During this process, we have very small particles,highly surface charged, and thus ready to respond to any and all; attractive or repulsive actions.

In water especially, particles group themselves in so-called "flocs" which constitute large networks of small particles or colloïdal particles, bonded by coulombian and capillary forces. After a part of

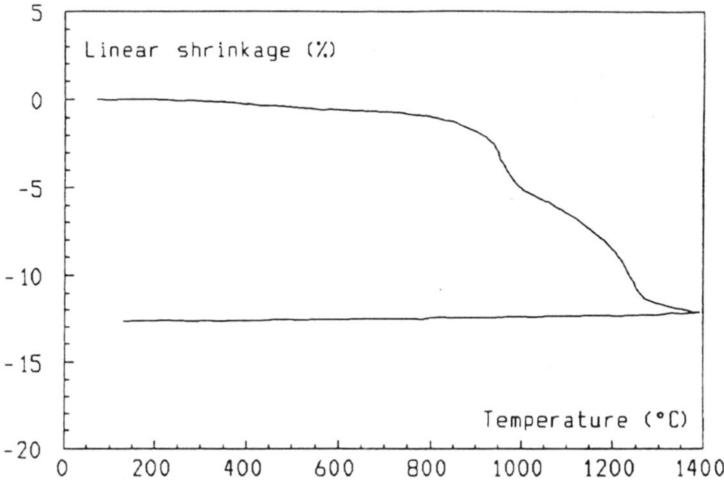

Fig. 6. Dilatometric measurments on sol-gel powder.

this water has been taken off by vaporisation, particles can be joined
together with already present soluble salts - among them hydroxides;
a partial dissolution of the particles can occur during the drying step,
supplying the inter-particle solution in sufficient raw material to
produce a gathering together of the disorganized matter into a hard
cement during vaporisation stage. The resultant grains made up during
this first step (including simultaneous vaporisation, presintering
and beginning of shrinkage), are then less reactive and thus do not
sinter very well.

A partial removal or substitution of this unfavorable liquid medium,
by a solvent in which hydroxides are less soluble, will therefore cause the
cementation effect to vanish. Consequently the de-agglomeration will be
easier to achieve with narrow size distribution, and hence the porosity
will be better suited for an efficient sintering.

In the case of acetone-toluene washing, the powder has been found in
an intermediary stage of de-agglomeration between water and alcohol. This
can be attributed to acetone , which is a solvant having a dielectric
constant higher than alcohol; it is able to dissove partially, but in les
quantity, the indesirable hydroxides.

The results about the densification of the powders obtained from the
sol-gel modified process seem disappointing, since at 1400°C the density
was only 64% of the theoritical value. In fact we have two phenomena:
joining of hard agglomerates and sintering of resulting grains. As seen in
the dilatometric curves, the former corresponds to the first shrinkage due
to agglomeration; the latter to the shrinkage due to sintering *per se*.
At this stage, large grains have appeared, and the coordination number
of pores has decreased. According to the energetical concepts of F.F.
Lange (9), this change in pore coordination leads to a decrease of the
ratio of grain boundary energy to surface energy. Mass transport thus
decreases, and coarsening kinetics of grains can dominate the further
sintering process.

Table I. Experimental density of several compositions after sintering process at 1270°C (300°C/h)

Preparation method	Al_2O_3-ZrO_2 composition	Theoretical density	Experiment dens at 1270°C		green density
Coprecip.	46/54	4.7	4.34 -	(91.0 % dth)	1.73
"	54/46	4.66	4.40 -	(94.4 % dth)	1.72
"	60/40	4.56	4.23 -	(92.8 % dth)	1.71
"	70/30	4.40	3.98 -	(90.5 % dth)	1.71
"	80/20	4.25	3.78 -	(88.9 % dth)	1.70
Sol-gel	93/17	4.20	2.56 -	(61.0 % dth)	1.97

REFERENCES

1. F. F. Lange, Transformation toughening part 1, J. Mat. Science, 17:225 (1982).
2. F. F. Lange, Transformation toughening part 2, J. Mat. Science, 17:240 (1982).
3. Y. Murase, E. Kato and K. Diamon, Stability of tetragonal ZrO_2-Al_2O_3 mixtures, J. Amer. Ceram. Soc., 69:83 (1986).
4. A. H. Heuer, N. Claussen, W. M. Kriven and M. Ruhle, Stability of ultra fine tetragonal ZrO_2 particles in ceramic matrices, J. Amer. Ceram. Soc., 65:642 (1982).
5. M. Kagawa, M. Kikuchi, Y. Syono and T. Nagae, Stability of ultrafine tetragonal ZrO_2 coprecipitated with Al_2O_3 by the spray-ICP technique, J. Amer. Ceram. Soc., 66:751 (1983).
6. B. Fegley, P. White and H. K. Bowen ,Preparation of Zirconia-Alumina Powders by Zirconium Alkoxide Hydrolysis, J. Amer. Ceram. Soc., 68,.[2],.C60 (1985).
7. J. C. Debsikdar, Influence of synthesis chemistry on alumina-zirconia powder characteristics, J. Mat. Science, 22:2237 (1987).
8. E. A. Barringer and H. K. Bowen, Formation, packing and sintering of monodisperse TiO_2 powders, J. Amer. Ceram. Soc., 65:C199 (1989).
9. S. Dick, C. Suhr, J.L. Rehspringer, and M. Daire, Mat. Sc. Eng., A 109: 227 (1989).
10. F. F. Lange, Powder processing science and technology for increased releability, J. Amer. Ceram. Soc., 72 [1]:3 (1989).

EVOLUTION OF MULLITE FROM A SOLGEL PRECURSOR

G. Klassen, J. Laughner, and G. Fischman

NYS College of Ceramics
Alfred University
Alfred, NY 14802

Abstract

The usage of a solgel precursor to make dense
mullite has been shown to be effective. Work on the
characterization of the process energetically and
microstructurally has started. This paper presents data
on the synthesis of mullite from the precursor, some
possible mechanisms for that formation will be
discussed.

Mullite, an oxide with strongly bonded oxygens, is an
important material in systems which call for moderately high
temperatures. It is stable in many gaseous environments and has
good engineering properties to about $1600^{\circ}C$.

Stability is a double edged sword; on one hand, it is
necessary for applications under extreme environments but to
retain temperature stability, the material must have a strong
bonding character. The result is generally a material swith a
high surface energy which tends to diffuse poorly at "reasonable"
temperatures. Mullite has been difficult to form from the raw
powders because of its slow diffusion properties. Incomplete
reaction of the raw materials to a true mullite often occurs.

To make a better mullite, smaller silica and alumina
particles «that are» well -mixed is a very promising possibility.
This system has been used previously and has been shown to be a
very good system for microstructural control[1]. Simpler forming
methods yielding a high density mullite is a positive effect of
this method, but improved mullite material is the driving force.

Sample Preparation

Samples were prepared from raw materials mixed on a colloidal
scale with the view that this size realm would be sufficient for
diffusion and reaction to form mullite in a fast firing process.
The raw materials were Ludox - an amorphous silica suspension
and Disperal, a colloidal pseudoboehmite.

The samples were batched to form the stoichiometric 3 to 2 mixture. the mixture was suspended in water and then peptized at a pH of 1.0 - 1.5. After the gel had formed, the pH was increased to about 3.0 so that the sample could be spray dried without causing damage to the spray dryer.

After Spray drying in a Bowen* spray dryer, some sample was used for Simultaneous Thermal Analysis and the rest was reserved for sintering studies.

Simultaneous Thermal Analysis was done using a Setaram TAG 24**. Differential Thermal Analysis and Thermogravimetric Analysis was done on samples simultaneously to 1650C at two ramp rates: 5°C/min and 15°C/min.

The rest of the material was pressed using a Carver Press and a Cold Isostatic Press to a final pressure of 140 MPa. These samples were then soaked at temperatures and times determined by the results of the thermal analysis. X-Ray diffraction was used to determine the phases present.

Results

Figures 1 and 2 are the results of the thermal analysis for the samples. Both figures represent the active run with a blank run subtracted for determination of baseline. Figure 1 was done with a sample of mass 24.7 mg and a ramp rate of 5°C/min, Figure 2 was done with 21.9 mg of sample and a ramp rate of 15°C/min.

An analysis of Fig. 1 shows two major regions. At low temperature there is a region where gas evolves. In the differential thermal gravimetric curve, this evolution occurs in three distinct steps. Furthermore, a glance at the DTA curve shows that the evolutions yield endotherms sitting atop a large, diffuse exotherm. The exotherm and the endotherms are all reproducible in this system. These are dehydration and decompositional reactions which will be discussed in further detail in a later paper. The goal of this paper is to concentrate on the high temperature regime. In this figure there is a large amount of thermal activity between 1000°C and 1500°C that occurs without a further change in mass; this suggests reconstructive and solid state reactive processes. Three peaks are identifiable: the lowest temperature peak has a shoulder on its low temperature side suggesting another possible peak; the second peak is very small; and the third peak is the sharp peak clearly indicative of the transition to mullite. The first two peaks sit atop an exotherm and their is an exotherm following the mullite peak which is questionable due to its closeness to the range at which the Argon gas thermal conductivity gives rise to noise in the system. Given these peaks, it was decided that microstructure studies would be done at 1100, 1200, 1300 and 1400°C with a ramp rate of 5°C/min and no hold to bracket these peaks.

Moving on to Fig. 2, the low temperature curves are broadened but still existent. More important, all the sample peaks save the

---

* Bowen Engineering, North Branch, NJ
** Setaram, Caluire, Cedex, France

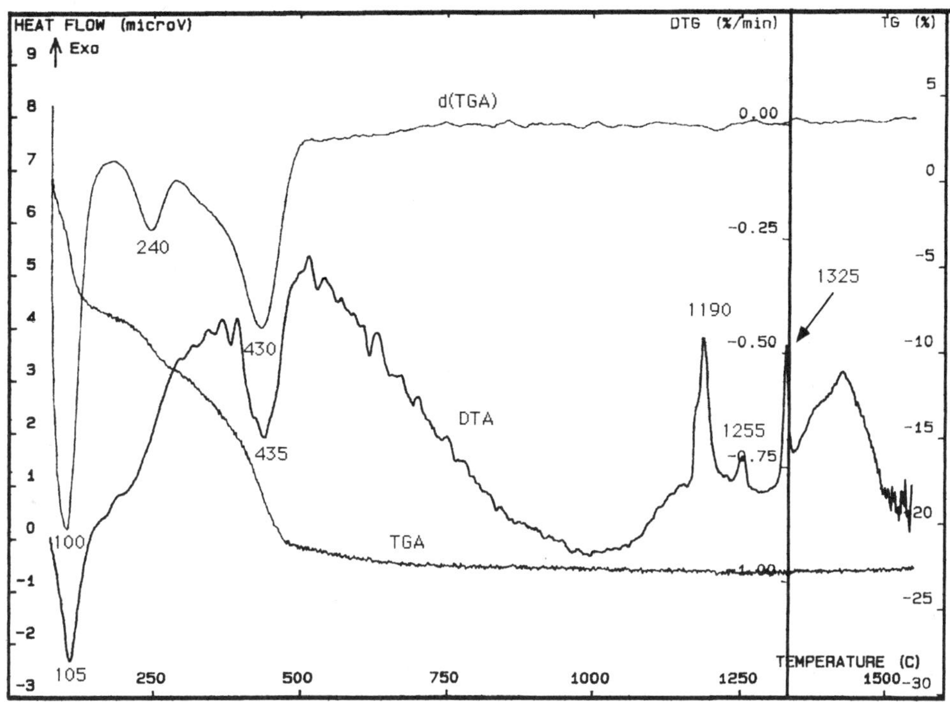

Figure 1. Simultaneous thermal analysis results for a heating rate of 5° C/min.

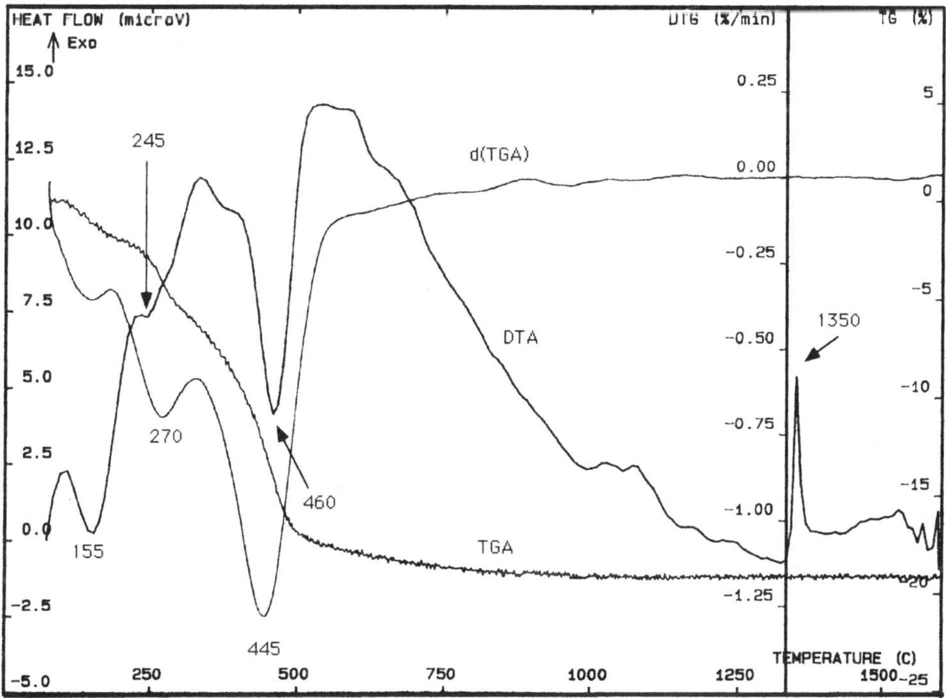

Figure 2. Simultaneous thermal analysis results for a heating rate of 15°C/min.

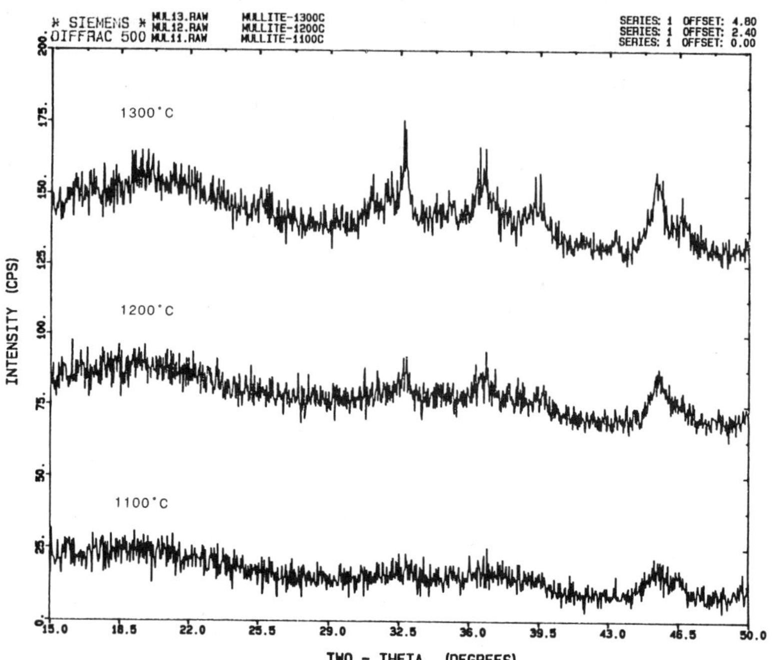

Figure 3. X-ray Diffraction patterns for samples heated to 1100, 1200 and 1300°C at 5°C/min with no high temperature hold. The formation of δ-alumina is noted.

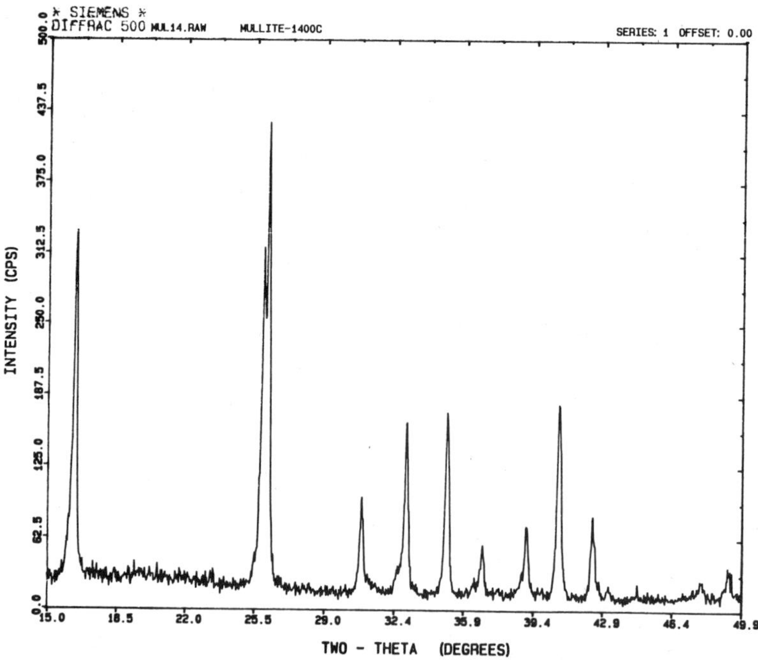

Figure 4. X-ray Diffraction pattern for a sample heated to 1400°C at 5° C/min with no high temperature hold. The pattern is characteristic of mullite.

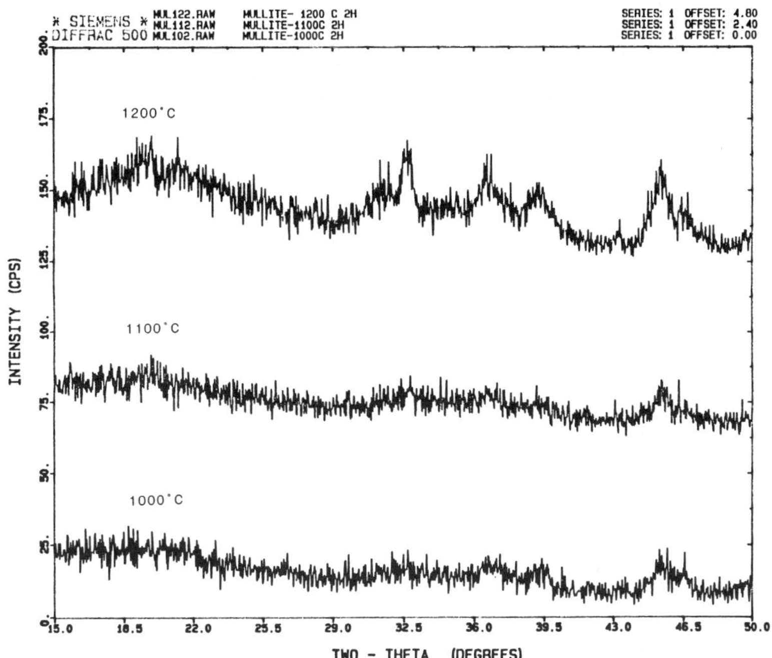

Figure 5. X-ray Diffraction pattern for samples heated to 1000, 1100 and 1200°C at 5°C/min with a 2 hour high temperature hold.

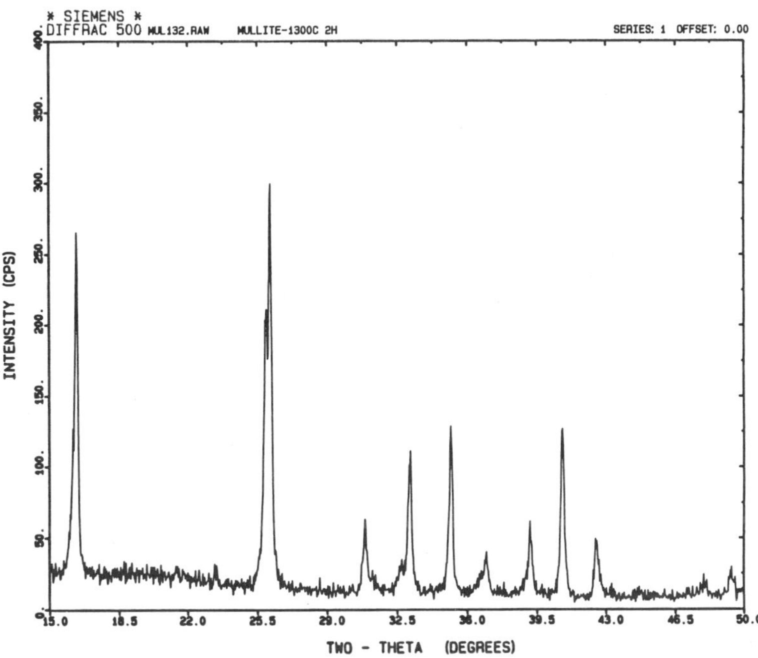

Figure 6. X-ray Diffraction pattern for a sample heated to 1300°C at 5 C/min with a 2 hour high temperature hold.

sharp mullite peak are replaced by a small diffuse exotherm and
the mullite peak itself has shifted to a higher temperature. The
dampening of the preliminary peaks strongly suggests a diffusion
process and because of this, another set of microstructural
studies were started at 1000, 1100, 1200 and 1300°C at a ramp rate
of 5°C/min but with a two hour hold.

X-ray diffraction from the runs described in the two
preceding paragraphs are shown in Figs. 3-6. Figures 3 and 4 are
for the samples that were treated without a hold at the final
temperature. Figures 5 and 6 were treated with the two hour
temperature hold at the temperature described.

Looking at Figs.3 and 4 together, the evolution of δ-Al_2O_3
from the solgel is apparent. At 1100^{δ}, the material is, for the
most part, amorphous with but a hint of structure showing through
in curve undulations and a very diffuse peak in the range of about
45° for 2Θ. As the final temperature is increased to 1300°C and
1400°C, we see the formation of a δ phase. Upon an increase past
the mullite transformation peak found in thermal analysis we find
the transition of this material to mullite (Fig. 4).

Comparing Figs. 5 and 6 to Figs. 3 and 4 we see that these
figures are close to identical. The story given for Figs. 3 and 4
holds for the samples used to determine Figs. 5 and 6. The
difference is that each of corresponding curves in Figs. 5 and 6
were done at a temperature of 100°C lower than for the counterpart
in Figs. 3 and 4 with a 2 hour hold for each of these in the
former case. This strongly suggests that even at the size of
these starting materials, diffusion still plays an important role
in the formation of the mullite.

<u>Conclusion</u>

In this study we have done a preliminary evaluation of the
high temperature regime for mullite formation from a colloidal sol-
gel. To date, this study has allowed us to form the following
conclusions.

1.   Mullite crystallized between 1300° and 1400°C depending on
     firing conditions. Since this peak shifts to higher
     temperatures at higher ramp rates, it is probable that
     this is still a diffusion controlled situation. It is
     quite probable that the peak temperature can be reduced
     further by lowering the temperature ramp rate.

2.   Thermal activity between 1100° and 1200°C involves the
     formation of a δ-alumina. The dampening of the peaks in
     the 15°C/min thermal analysis run suggests that this
     transformation is diffusion controlled.

Reference

1. T.J. Mroz and J.W. Laughner, "Microstructures of Mullite
Sintered from Seeded Sol-Gels," <u>J Amer Ceram Soc 72</u>, [3] 508-509
(1989).

THE ROLE OF POWDER CALCINATION CONDITIONS IN THE SINTERING BEHAVIOUR OF
CALCINED ZrO_2 POWDERS

T. Kosmač, D. Kolar, V. Kraševec, and R. Gopalakrishnan

"J. Stefan" Institute
University of Ljubljana
Ljubljana, Yugoslavia

1. INTRODUCTION

Precipitation of hydrous zirconia and its thermal decomposition
represent the simplest method for the preparation of ultrafine zirconia
powders. These powders are now widely used either in unstabilized form
for toughening various ceramic matrices (so-called zirconia toughened
ceramics) (1), or they are used for the production of tetragonal zirconia
ceramics (2), containing a few % of Y_2O_3 or CeO_2 as stabilizing agent,
which also reduces ZrO_2 grain growth during sintering. Therefore the
crystallization of hydrous zirconia as well as the occurence and stability
of the transitional metastable tetragonal (mt) phase, have been rather
extensively investigated, but they are still not completely understood
(3-10). Besides, the relationship between powder processing and powder
characteristics, as well as their influence on the densification behaviour
and microstructure of sintered material seem to be of great importance.

In the first part of the present paper a short summary of our
crystallization studies of undoped hydrous zirconia precipitates during
calcination in air and in vacuum is given. In the second part of the
paper, we discuss the sintering behaviour of differently calcined
ultrafine zirconia powders.

2. EXPERIMENTAL PROCEDURE

Undoped amorphous hydrous zirconia, used for the present
investigation, was supplied by Hüls Troisdorf GmbH (FRG). It represents an
intermediate product in the production line of commercial Dynazirkon F
zirconia powder.

The precipitate was calcined under various atmospheres such as
vacuum, air and flowing (dry or wet) H_2 and O_2. Dehydration on heating was
studied using a standard DTA-TGA apparatus (Netzsch), and the crystalline
products were analysed by XRD, TEM and BET.

For sintering experiments the calcined powders were cold
isostatically compacted into pellets (ϕ 6 mm, $\ell = \phi$ 7 mm). A standard
dilatometer (Bähr Gerätebau, FRG)) was used to follow the densification
behaviour (i.e. relative shrinkage and shrinkage rate vs. temperature)

during heating at various heating rates. Microstructural development and grain size were estimated from SEM micrographs of fracture surfaces. The ratio of tetragonal (t) to monoclinic (m) ZrO_2 was determined by XRD using the integrated intensities of the tetragonal (111) and the monoclinic (111) peaks and (111) peaks (11).

3. RESULTS AND DISCUSSION

3.1. Crystallization of hydrous zirconia and characteristics of the calcined powders

The results of our crystallization study of hydrous zirconia precipitates (12,13) indicated that too rapid removal of water at lower temperatures during calcination in a vacuum prevents the required ordering of zirconium and oxygen atoms, forming Zr-O-Zr bonds, up to the temperature of explosive crystallization. The latter is reflected in a very sharp exothermic effect - the so-called glow effect. If, however, the dehydration rate is moderate (air calcination) or slowed down (calcination under a humidified atmosphere or under increased water vapour pressure during autoclaving), the retained water promotes the ordering of Zr and O atoms and thus the nucleation of crystalline ZrO_2 at temperatures below the glow effect. It was further supposed that the number of active nucleation sites for crystallization strongly depends on the presence of water during calcination. This assumption was based on *in situ* observation of the crystallization of an amorphous agglomerate in a high vacuum TEM chamber (14), which revealed that the crystallization started from one single nucleation site and spread almost abruptly over the whole aggregate, forming one nearly single crystalline region - the so-called domain. A more detailed description of the evolution and defect structure of mt ZrO_2 domains can be found elsewhere (14,15).

As seen from TEM images of air and vacuum calcined powders (Fig.1), aggregates of the former are composed of randomly oriented crystallites 50 nm in size, whereas agglomerates of the latter consists of one or a few 200 - 1000 nm sized domains, which are composed of individual crystallites 50 nm in size with clearly visible low angle boundaries in between.

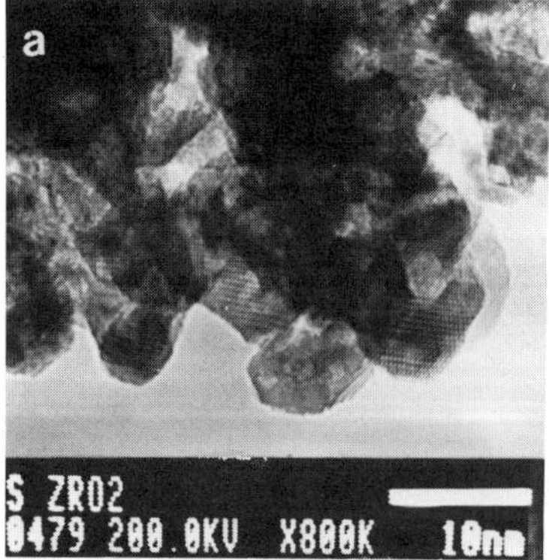

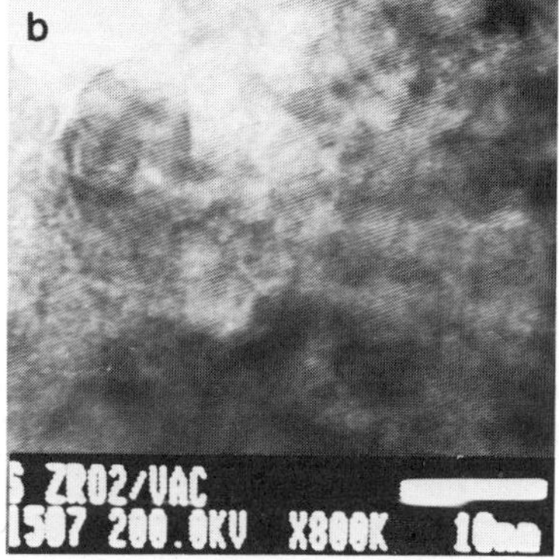

Fig. 1. TEM images of a) air calcined and b) vacuum calcined ZrO_2 powders

TABLE I. Characteristics of ZrO2 powders after calcination at 460°C for
3 hours in air or in vacuum

	Air	Vacuum
Crystallinity	100%	90%
Crystal structure	50% m, 50% mt	100 mt
Crystallite size	10-30 nm	100 nm
Spec. surface area (BET)	110 m 2 /g	19 m 2 /g
Green density	46%	51%

It is supposed that the highly distorted domain structure of the vacuum
calcined powder increases the high stability of the metastable tetragonal
ZrO2 phase towards mt → m transformation.

The formation of large nearly single crystalline domains during
vacuum crystallization also resulted in a drastic reduction of the
specific surface area of the powder as evident from Table I, where the
characteristics of air and vacuum calcined powders are summarized.

3.2. <u>Sintering behaviour</u>

Dilatometric curves of air and vacuum calcined powders are given in
Fig.2. The considerable retardation in the onset of shrinkage, as well as
the somewhat lower final amount of shrinkage of the vacuum calcined powder
are assumed to result from the essentially lower specific surface area of
this powder, and the consequently slightly higher green density of the
compacted sample. It is, however, interesting to note that, in spite of a
considerable shift in the onset of densification, the temperature of the
t → m transformation, reflected in a sharp expansion of the sample on
cooling, occurs at nearly the same temperature for both specimens. Since
the t → m transformation temperature of unstabilized ZrO2 mainly depends
on the grain size, it appears that the average final grain size of the two
samples does not differ significantly.

Further information on the sintering behaviour of the two powders was
obtained by analysing the corresponding densification rate curves (Fig.3).
It is worth noting that both peak densification rates are almost identical
although they occur at different temperatures. Both densification curves
also show two additional more or less pronounced maxima: in the case

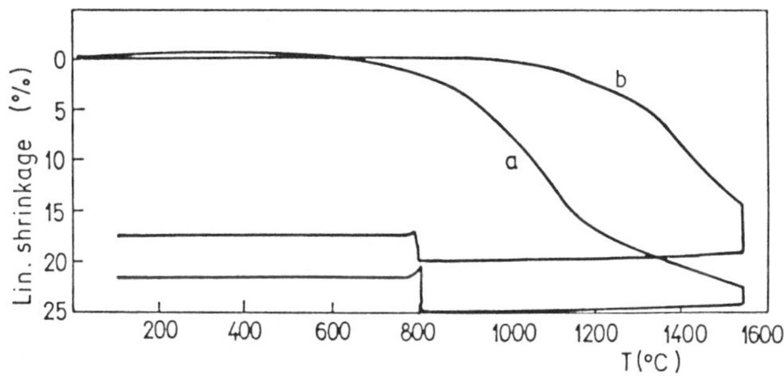

Fig. 2. Dilatometric curves of a) air calcined and b) vacuum calcined
sample, heating rate 5°K/min.

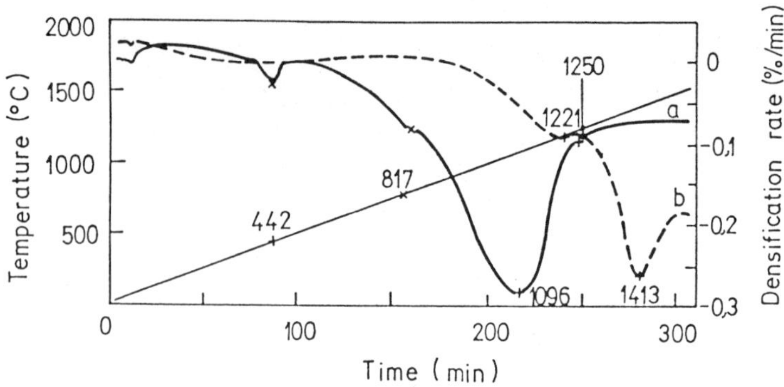

Fig. 3. Densification rate curves of a) air calcined and b) vacuum calcined sample during heating at constant rate (5°K/min).

of the air calcined powder they occur at   820°C and   1250oC, whereas in the case of the vacuum calcined powder the corresponding temperatures are 440°C and 1220°C, respectively. The low temperature maximum in the shrinkage rate of the vacuum calcined sample at 440°C results from the completion of crystallization, whereas the origin of the other discontinuities in the densification rate in both samples is less unambiguous: they may result from differential sintering of agglomerated powders, but they may also arise from the phase transformations of ZrO_2. Namely, both powders undergo two phase transformations on heating: the irreversible diffusional mt → m transformation is followed by the reversible martensitic m → t one at higher temperature. The latter

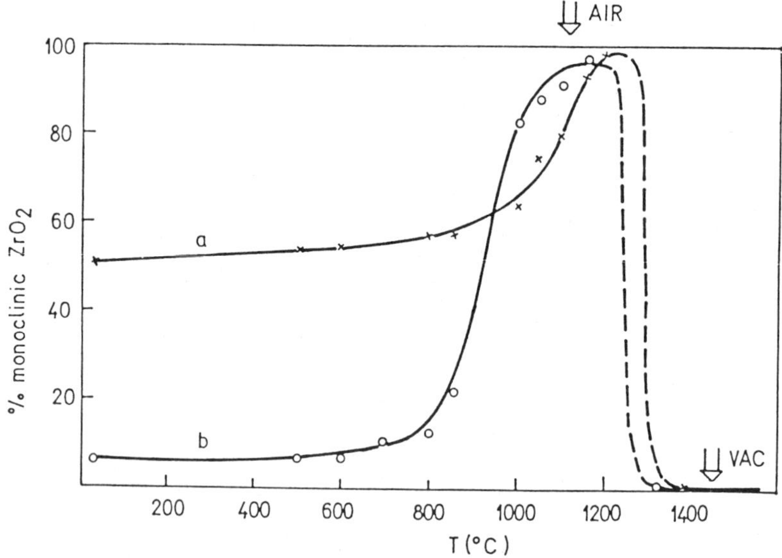

Fig. 4. Relative monoclinic ZrO_2 content vs. temperature as determined by XRD analysis after rapid cooling to room temperature. (m = monoclinic ZrO_2, mt = metastable tetragonal ZrO_2, t = tetragonal ZrO_2). The arrows denote maximum densification rate of air and vacuum calcined samples.

is well defined and occurs above 1150°C, whereas the temperature of the former transformation varies over a broad temperature range, mainly depending on the crystallization conditions, impurity content, defect structure, etc. (3-10).

In order to establish the possible relationship between ZrO2, phase transformations and sintering behaviour, the variation in the relative amount of monoclinic ZrO2 phase with temperature was estimated by XRD analysis of the samples on rapid cooling from selected temperatures (Fig.4). The comparison between XRD measurements and densification rate curves revealed that the maximum densification rate in the air calcined sample coincides with the temperature of intense mt → m transformation (i.e. in the predominantly monoclinic ZrO2 region), whereas in the case of the vacuum calcined sample the peak densification rate is reached at 1450°C, i.e. in the tetragonal ZrO2 region. Further, the discontinuity in the shrinkage rate at 1220°C in the vacuum calcined sample as well as that at 1250°C in the air calcined sample coincide with the m → t transformation of ZrO2. Since the discontinuity in the densification rate at 1220°C of the vacuum calcined sample remains clearly visible irrespective of the preheating treatment of the powder at 1300oC (i.e. above the temperature of its occurence), it seems that the change in the densification rate at 1220°C arises due to the m → t transformation of ZrO2, rather than from differential sintering.

In contrast, preheating of the air calcined powder at 850°C prior to sintering resulted in complete disappearance of the discontinuity at 820°C. The preheating caused a considerable reduction of the specific surface area of the powder (from 110 to 45 m^2/g) but did not noticeably alter the content of the mt ZrO2 phase, thus indicating that the mt → m transformation of agglomerated ultrafine ZrO2 powder is less important for the densification behaviour than agglomeration itself. Also, the mt → m transformation does not result in any measurable change in the dilatometric curve nor in the densification rate curve of the vacuum calcined powder, thus giving support to the above assumption.

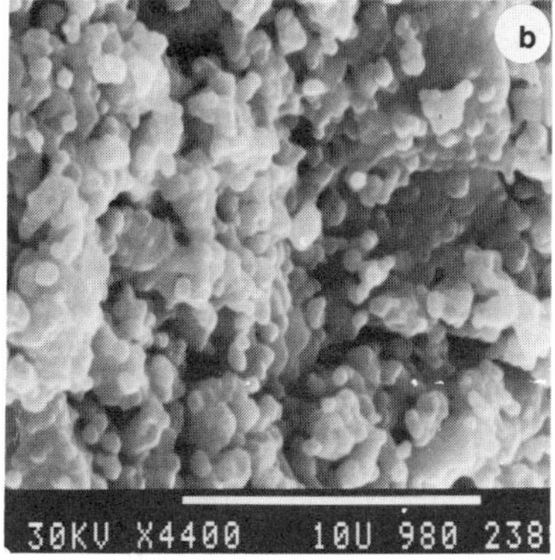

Fig. 5. SEM micrographs of fracture surfaces of a) air calcined and b) vacuum calcined sample after sintering up to 1400°C.

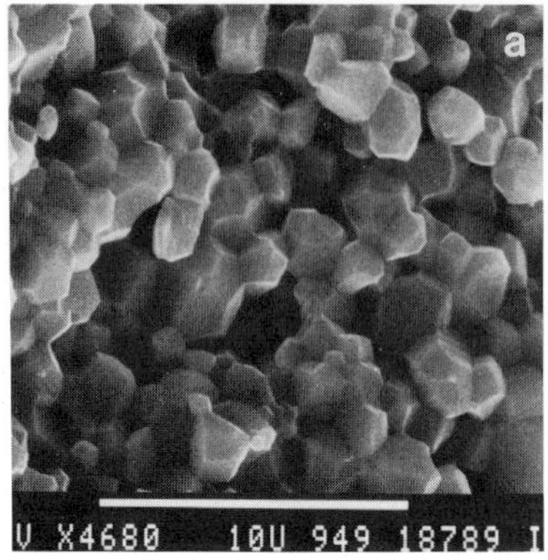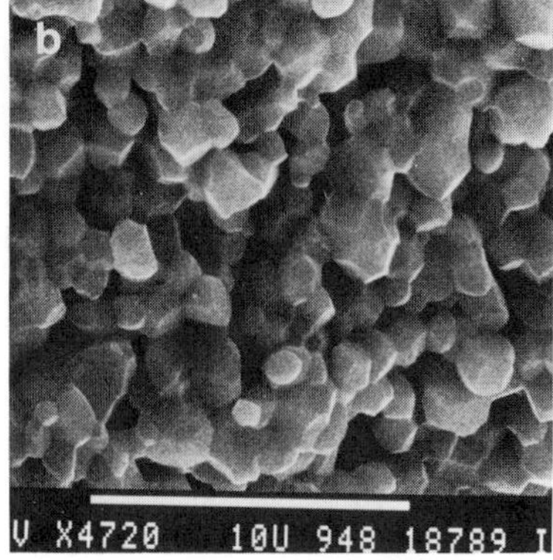

Fig. 6. SEM micrographs of fracture surfaces of a) air calcined and b) vacuum calcined sample after sintering at 1500°C for 1 hour.

An attempt was also made to follow the grain growth during densification of both differently calcined powders. Two sintering temperatures were chosen for these purposes: 1400°C, at which the peak densification rate in the air calcined sample is exceeded whereas in the vacuum calcined sample it is not yet reached, and 1500°C, which lies above the temperature of the maximum densification rate of the latter sample as well. SEM micrographs of fracture surfaces of samples sintered at these two temperatures are represented in Figs. 5 and 6. After sintering at 1400°C the microstructure of the air calcined sample is highly heterogeneous, showing regions of 2 μm large grains with well developed morphology within fine grained (ca 0.1 μm) matrix, whereas almost no grain growth has taken place in the vacuum calcined sample at this sintering temperature. After sintering at 1500°C, however, both fracture surfaces are very homogeneous and nearly identical, at least as far as grain size and grain size distribution are concerned.

This observation is in good agreement with the results of dilatometric measurements showing approximately the same temperature of the t $\rightarrow$ m transformation on cooling from the sintering temperature for both powders.

4. CONCLUSIONS

Calcination conditions during crystallization of amorphous hydrous zirconia – especially the rate of water removal – result in essential differences in powder characteristics such as crystallinity, crystal structure, crystallite size, morphology and specific surface area. These differences are mainly reflected in the temperatures at which densification starts and reaches its maximum shrinkage rate. Thus the air calcined powder starts sintering below the temperature of the diffusional mt $\rightarrow$ m transformation of ZrO_2, and reaches its maximum densification rate in the predominantly monoclinic ZrO_2 region. The influence of the mt $\rightarrow$ m transformation on the densification rate seems to be overshadowed by the differential sintering of agglomerated powder. The densification of the vacuum calcined powder starts at a higher temperature in the mt + m region and reaches its maximum rate in the tetragonal ZrO_2 region. The discontinuity in the densification rate before its maximum value is

reached is supposed to result from the m → t phase transformation of
ZrO2. Despite all the above listed differences between air and vacuum
calcined zirconia powders, the maximum shrinkage rate is almost identical
for both powders.

The grains in the air calcined sample start to grow at lower
temperatures but once the maximum densification rate in the vacuum
calcined sample is reached, the difference in the grain size after
sintering becomes less pronounced until it almost vanishes.

5. REFERENCES

1. N.Claussen, Microstructural Design of Zirconia Toughened Ceramics
   (ZTC), Advances in Ceramics, Vol 12, 325-351 (1984)
2. K.Tsukuma, Y. Kubota, T. Tsukidate, Thermal and Mechanical Properties
   of Y2O3-Stabilized Tetragonal Zirconia   Polycrystals, :ibid. Vol 12,
   382-390 (1984)
3. R.Cypres, R. Wollast and J. Raucq, "Polymorphic Conversion of Pure
   Zirconia", Ber. DKG, 40, 527-532 (1963)
4. J.Livage, K. Doi and C. Mazieres, "Nature and Thermal Evolution of
   Amorphous Hydrated Zirconium Oxide", J. Amer. Ceram. Soc.,51, 349-353
   (1968)
5. V.G.Keramidas and W. B. White, "Raman Scattering Study of the
   Crystallization and Phase Transformations of ZrO2, J. Amer. Ceram.
   Soc. 57, 22-24 (1974)
6. R.C.Garvie, "The Occurance of Metastable Tetragonal Zirconia as a
   Crystallite Size Effect", J. Phys. Chem., 69, 1238-1243 (1965)
7. R.C.Garvie, "Stabilization of the Tetragonal Structure in Zirconia
   Microcrystals", J. Phys. Chem. 82, 218-224 (1978)
8. R.C.Garvie and M. F. Goss, "Intrinsic Size Dependence of the Phase
   Transformation Temperature in Zirconia Microcrystals", J. Mater. Sci.,
   21, 1253-1257 (1986)
9. T.Mitsuhashi, M. Tenihara and U. Tatsuke, "Characterization and
   Stabilization of Metastable Tetragonal ZrO2", J. Amer. Ceram. Soc. 57,
   97-101 (1974)
10. M.J.Torralvo, M. A. Alario, "Crystallization Behaviour of Zirconium
    Oxide Gels", J. Catal., 86, 473-476 (1984)
11. R.C.Garvie, P.S. Nicholson, Phase Analysis in Zirconia Systems, J. Am.
    Ceram. Soc. 55, 303-305 (1972)
12. T.Kosmač, V. Kraševec, R. Gopalakrishnan, Influence of Water Removal
    Rate During Calcination on the Crystallization of ZrO2 from Amorphous
    Hydrous Precipitates, Porc. IX German-Yugoslav Meeting on Materials
    Sciences and Development, Hirsau 1989, Eds. W.A. Kaysser, G. Petzow
13. R.Gopalakrishnan. M. Komac, V. Kraševec, T. Kosmač, Preparation of
    Zirconia Fine Powders by Critical Solvent Evaporation, Brit. Ceram.
    Soc. Proc., No 38, 37-48 (1986)
14. T.Kosmač, V. Kraševec, R. Gopalakrishnan, M. Komac, Crystallization of
    ZrO2 from Amorphous Precipitates and its Influence on Microstructure of
    Sintered Ceramics, Advances in Ceramics, Vol. 24 , 167-172 (1989)
15. V.Kraševec, T. Kosmač, R. Gopalakrishnan, An HREM and ED Study of
    Calcined Precipitated Zirconia Powder, Inst. Phys. Conf. Ser. No. 93,
    Vol 2, Chapter 8, 329-330 (1988)

SURFACE EFFECT OF MgO ADDITION ON GAMMA Al_2O_3 SINTERING

J.Katanić-Popović and Lj.Kostić-Gvozdenović[*]

Boris Kidrič Institute P.O.B.522, 11001 Belgrade
[*]Faculty of Technology and Metallurgy, Karnegijeva 4
11000 Belgrade, Yugoslavia

INTRODUCTION

In order to inhibit the sintering process of gamma alumina, magnesium oxide is mostly used[1-4] as additive for catalyst support preparation.

This paper deals with the effect of magnesia distribution and its quantity in the range of 0.02 to 0.06 of Mg/Al atom ratio on specific surface area and porosity changes in the temperature interval 600-800°C.

EXPERIMENTAL

Pseudoboehmite (I), gamma alumina (II) as well as coprecipitated pseudoboehmite and magnesium hydroxide (III) powders were used for sample preparations. These starting powders were obtained by precipitation of aluminum and magnesium chlorides with ammonium hydroxide.[3] Reagent grade salts: $MgCl_2.6H_2O$, $AlCl_3.6H_2O$ and $MgNO_3.6H_2O$ were used for this purpose. Gamma alumina was prepared through pseudoboehmite calcination at 500°C for 2 h. Specific surface area of this powder (210 m^2g^{-1}) was taken as the initial value (S_o) for the interpretation of results.

Magnesium nitrate and magnesium hydroxide were used as precursors of the doping agent. These magnesium compounds were chosen because their melting points and decomposition temperatures were low.

Magnesia and gamma alumina derived from these starting materials during calcination form spinel at temperatures above 500°C. The reaction of spinel formation is more pronounced in the coprecipitated powder (III), than in the mixture of pseudoboehmite and magnesium nitrate (I) with the Mg/Al atom ratio being 0.5 in both powders (Fig.1).

The additive concentration was always maintained in the range from 0.02 to 0.06 of Mg/Al atom ratio. The surface distribution of additive was realized by wetting of pseudoboehmite or gamma alumina with corresponding quantities of magnesium nitrate solutions. The volume distribution of magnesia was achieved by coprecipitation of magnesium hydroxide and pseudoboehmite (III). To deaglomerate the powder particles, they were wet ground in an alumina ball mill. Cylindrical samples were pressed at 50 MPa. The compacts were sintered from 600 to 800°C for 2 and 24h in air.

Science of Sintering
Edited by D. P. Uskoković *et al.*
Plenum Press, New York

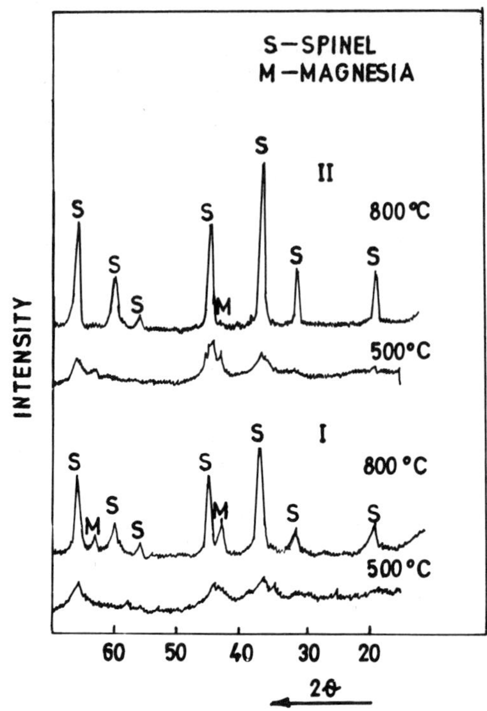

Fig.1. X-Ray diffractogrames of calcined powders with Mg/Al atom ratio
0.5, obtained from: the mixture of pseudoboehmite and magnesium
nitrate (I) and coprecipitated pseudoboehmite and magnesium
hydroxide (III).

The samples were characterized by several well known methods. Included
were: density measurement (immersion in xylene); determination of pore size
distribution (mercury porosimeter); specific surface area (areameter with
nitrogen adsorption) and structure determination (X-ray diffractometer).

RESULTS AND DISCUSSION

Changes in sintering temperature over the range 600 to 800°C lead to
small porosity increases and to larger specific surface area decreases
(Table I). The observed reduction of specific surface area ispresumably
due to intensive surface diffusion occurring during sintering.

In the composition range studied the addition of surface-distributed
magnesia increases the specific surface area of sintered samples when
compared to pure gamma alumina (Table I and Figs. 2 and 3). These results
confirm the hypothesis[1,2] that surface diffusion in gamma alumina is
suppressed in some degree by the presence of magnesia, or the reaction
product, spinel, on the particle surfaces. The most evident effect of
magnesia can be best evaluated at 800°C, where the reaction of spinel formation is more pronounced (Fig. 1), resulting in a less mobile surface layer.

Table I. Porosity (P) and specific surface area (S) of samples sintered
at 800°C (24h) and their relative changes between 600° and 800°C

No.	Starting Materials	Additive	Mg/Al ratio	P (%)	S (m^2g^{-1})	Relative changes	
						$\Delta P(\%)$	$-\Delta S(\%)$
1.	I	–	–	78.4	118	1.7	22.9
2.	I	$Mg(NO_3)_2$	0.02	75.9	130	1.8	21.0
3.	III	$Mg(OH)_2$	0.02	74.2	100	2.1	24.2
4.	I	$Mg(NO_3)_2$	0.06	73.6	129	1.9	19.9
5.	III	$Mg(OH)_2$	0.06	76.4	125	2.8	26.0
6.	II	–	–	82.7	116	1.8	23.0
7.	II	$Mg(NO_3)_2$	0.02	81.3	128	1.2	19.0
8.	II	$Mg(NO_3)_2$	0.06	80.5	135	1.2	18.7

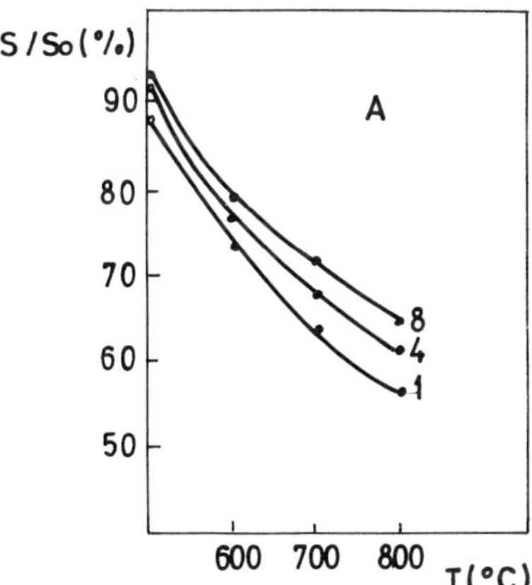
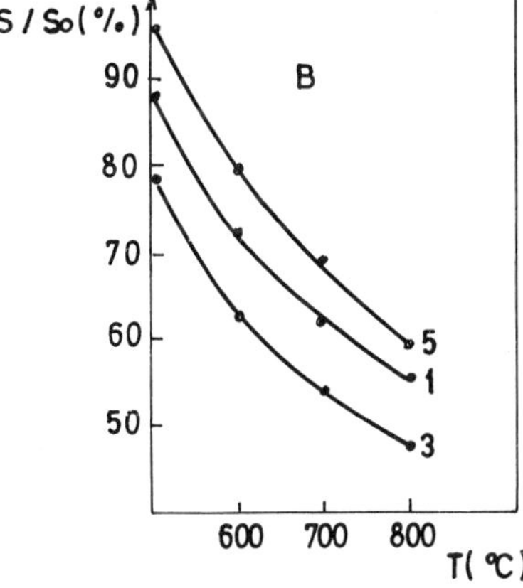

Fig.2. Temperature effect on relative changes of specific surface area
for sintered samples (24h) derived from: undoped pseudoboehmite
(1); doped pseudoboehmite (4) and gamma alumina (8) for 0.06 Mg/Al
ratio; coprecipitated pseudoboehmite and magnesium hydroxide with
ratios 0.02 (3) and 0.06 (5).

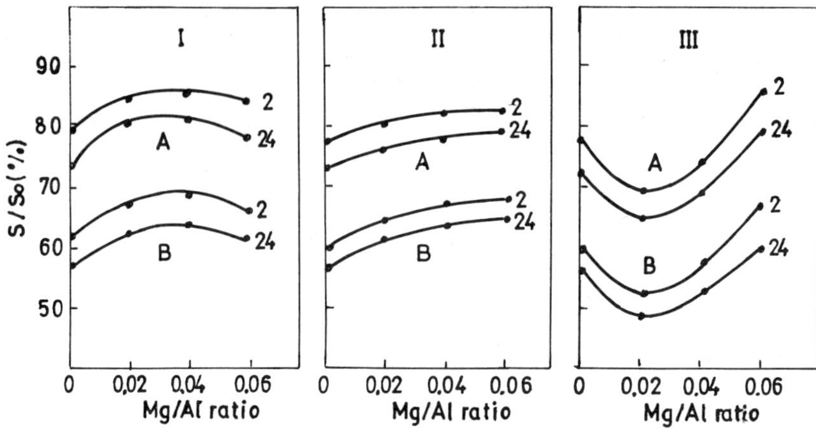

Fig.3. Additive effect on relative specific surface area changes of sin
tered samples obtained at 600°C (A) and 800°C (B) for times of 2h
and 24h:materials derived from pseudoboehmite (I) and gamma Al_2O_3
(II) doped with $Mg(NO_3)_3$ as well as coprecipitated pseudoboehmite
with $Mg(OH)_2$ (III).

A different behaviour was observed only in the case of a doped sample
having a _volume_ distribution of magnesia (Figs. 2 (B) and 3 (III)). The
obvious decrease of specific surface area for the composition up to 0.02
Mg/Al atom ratio is probably caused by some solubility of magnesia. Gamma
alumina has the spinel structure wherein the oxygen sublattice is fairly
well ordered, whereas the aluminum lattice is strongly disordered.[5] There
are numerous cation vacancies in the tetrahedral position, which are
therefore available for magnesium ions. The increase of specific surface
area in these samples for compositions above 0.02 Mg/Al atom ratio (Fig.
3 III) is ascribed to the dominant effect of the insoluble fraction of
magnesia which is deposited on the surface.

Specific surface area is more effected by sintering temperature than by
isothermal heating time. At the same time,the additive concentration in
crease has a less pronounced effect on the specific surface area (Table
I and Fig.3).This is in agreement with observed better crystallinity of the
gamma phase at higher temperature (800°C) and an unavoidable reducion of
the large internal surface area as a consequence of structure ordering and
crystal growth.

However the expected proportionality between porosity and specific
surface area was not observed. The surface distribution of the additive
can be accounted for indirectly through comparisons of the change of pore
size distribution of the doped and undoped samples (Fig.4). Magnesium
nitrate addition results in improving both the green and sintered sample
densities. Besides, the open porosity of doped samples is decreased because
the doping agent is deposited on the particle surface. The pore size
distribution in sintered samples derived from the doped pseudoboehmite
(Curve 4D) has the same form as the one for the undoped samples (Curve 1D).
Some differences in the pore size distribution curves of doped and undoped
sintered samples obtained from gamma alumina were observed (Curves 6D and
8D) mainly in the region of pore size from 10nm to 100 nm. A slight

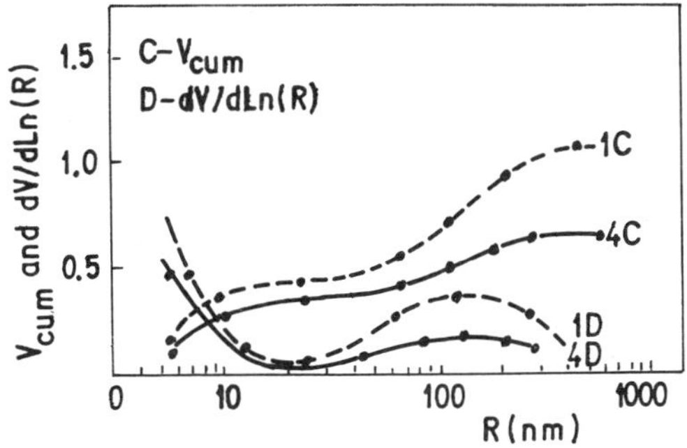

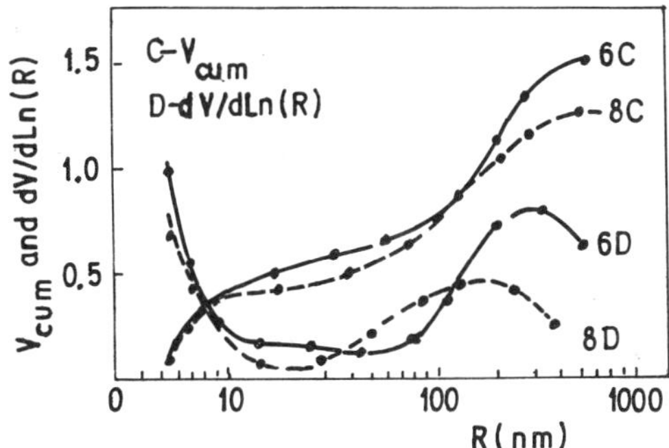

Fig.4. Pore size distribution for samples sintered at 800°C, 24h
obtained from: undoped pseudoboehmite (1) and gamma Al_2O_3
(6) as well as doped with magnesium nitrate, Mg/Al ratio
0.06 (4) and (6) respectively.

decrease of specific surface area, noticed for sintered samples derived
from doped pseudoboehmite with 0.06 Mg/Al atom ratio (Fig.3 I), is probably
caused by a filling up of pores with excess magnesia.

CONCLUSIONS

An exponential relation between sintering temperature and surface
area change was observed. Sintering temperature has a relatively small
influence on the open porosity. An expected proportionality of porosity
and specific surface area was not observed.

With additive concentrations up to 0.06 Mg/Al ratio, an increase of
specific surface area was found for sintered samples containing surface
distributions of additive. This positive change of specific surface area
with increased additive quantity is more pronounced at higher temperature
(800°C), where magnesium aluminate formation is more intense.

An obvious decrease of specific surface area of sintered samples
having volume distributions of additive in the composition range up to
0.02 Mg/Al ratio is the consequence of solid solubility of magnesia within
the gamma alumina structure. At higher concentrations, further increases

of specific surface area are due to the dominant effect of the insoluble fraction of magnesia, which deposits on the surface.

REFERENCES

1. J.Katanić-Popović and S.Zec, "Influence of additive on surface and structure stabilization of gamma alumina", In Proc. of Ist. European Conference of Ceramics, held in Holland, June, 1989 (to be published).
2. H.Shaper and L.L. Van Reijen, "Influence of dopes on the stability of gamma alumina catalyst supports", In Sintering Theory and Practice, D.Kolar and S.Pejovnik, eds., Elsevier Sci. Publ. 1982, p.173.
3. J.Katanić-Popović, M.Gašić and Lj.Kostić-Gvozdenović, "Formation and sintering of gamma alumina with magnesia addition from coprecipitated hydrates, J.Physique Coll C1, 47:131, 1986.
4. P.Reynen, "The impact of sintering theory on practical powder metallurgy" in Sintering Processes, G.C.Kuczynski ed., Material Science Research Vol.13, Plenum Press, New York and London, (1979) p.355.
5. K.Wefers and G.M.Bell, "Oxides and hydroxides of alumina", Tech. Paper No.19, Alcoa, 1972.

Part III. PREPARATION AND SINTERING OF METALLIC POWDERS

PRINCIPLES OF ATOMIZATION

W.A.Kaysser and K.Rzesnitzek

Max–Planck–Institut für Metallforschung
Pulvermetallurgisches Laboratorium, PML
Heisenbergstraße 5, 7000 Stuttgart 80, FRG

ABSTRACT

Basic principles of materials science and physics which apply to the formation and solidification of small droplets during gas and water atomization of metals and alloys are described. Several mechanisms of melt stream disintegration are presented, including melt stripping and subsequent disintegration due to Taylor instabilities. The cooling rates of the droplets and the extent of undercooling are estimated. Some aspects of homogeneous and heterogeneous nucleation as well as the microstructural variations caused by different solidification front velocities are outlined and compared with experiments with Cu–Sn alloys.

INTRODUCTION

The etymology of atomization leads back to the Greek word atomos, which means uncut or undivisible. In the modern English language, the meaning of atomization is less strict and includes the reduction of a bulk melt to fine particles as a spray. The reasons to prefer atomization to a large number of other physical, chemical and physico–chemical powder production methods are different for individual powders and processing routes. High compressibility and compactability at low cost are essential requirements for iron and steel powders used for PM structural parts.[1] Those powders are obtained by water atomization if the oxides forming during atomization can be tolerated or removed. In the case of complex multiphase alloys, atomization minimizes the segregation of the alloying elements and refines the microstructure compared to ingot castings, e.g. of high speed steel[2]. Inert gas atomization is chosen when the oxygen content of a powder must be low. These powders are usually hot consolidated by HIP or hot extrusion to semi–finished materials or parts[3]. Centrifugal atomization is used for high–melting alloys or when the reactivity of the melt requires crucible–free melting during atomization.

The impulse which atomizes (disintegrates) the melt stream or pool can be transmitted by a solid substrate (e.g. in the rotating electrode process) or more frequently by a fluid medium (e.g. gas or water atomization). The fragmentation determines the economics of the process and the quality of the atomized powders in regard to size, shape and purity. The microstructural development during solidification of the small droplets is partly based on stochastic events and, therefore, is still inadequately understood. Major problems in interpretation and prediction of rapidly solidified microstructures arise from the combined effects of cooling rate and undercooling.[4] In this paper, some recent

Science of Sintering
Edited by D. P. Uskoković *et al.*
Plenum Press, New York

investigations with gas atomized, water atomized and splat cooled Cu–Sn alloys are described. The experiments gave some insight of the microstructural response to the relative gas velocity, the water vapour shell around hot liquid droplets and the interaction of droplets with substrates. Undercooling of the droplets, which can change the heat flow conditition during solidification considerably, is discussed in detail. It is shown that undercooling can be widely varied by adding or eliminating heterogeneous nuclei, resulting in microstructures with featureless, fine dendritic and cellular areas.

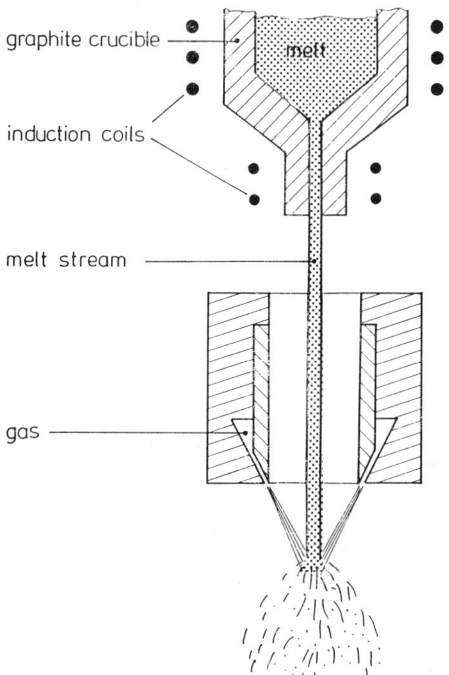

Fig. 1.   Principal arrangement of an atomizer. Water can be used instead of gas as atomizing medium.

PRINCIPLES OF FRAGMENTATION

Fragmentation during Water Atomization

Fig. 1 shows the general layout of a water or gas atomization facility. The molten metal is poured either directly or by means of a runner into a tundish which supplies a constant flow of melt through an orifice. The atomizing nozzle system provides high pressure water jets which impinge on the melt stream. The melt stream is disintegrated into many small droplets which solidify rapidly. Various fragmentation mechanisms can be thought of. As extremes, purely thermally induced fragmentation and purely hydrodynamic fragmentation are distinguished. In the first case, temperature differences yield atomization, while in the latter case, relative velocities between the metal melt and the water and, hence, accelerations are effective.

Thermal fragmentation can be caused by violent boiling processes leading to local pressure buildup during vapour bubble growth and to impinging water jets due to vapour bubble collapse.[5] In the latter case, the jets may produce fragments simply due to the impact causing splashing events, or the jets may penetrate into the molten material and evaporate there in an explosive manner. Non–uniform heat transfer through the vapour

158

shell around a molten droplet together with the inertia of the water suffices to explain pressure fluctuation along the melt surface, which may cause fragmentation.[6]

Hydrodynamic fragmentation mechanisms are based on the relative velocities of the adjacent fluid media. Firstly, the component of the water jet parallel to the melt stream either gives rise to parallel flow instabilities with subsequent stripping of the crests of the growing waves or provides a direct shearing off of a boundary layer of the melt. Fragmentation in relatively rough fragments also results from the jet component transverse to the melt stream. Fine fragments may be obtained due to additional Taylor instabilities (Figs. 2 and 5).[7,8] The impinging water jets are often assumed to be disintegrated into a dispersion of water droplets before hitting the metal stream, allowing cratering and splashing effects due to the impact of this spray on the melt stream as fragmentation modes (Fig. 5).

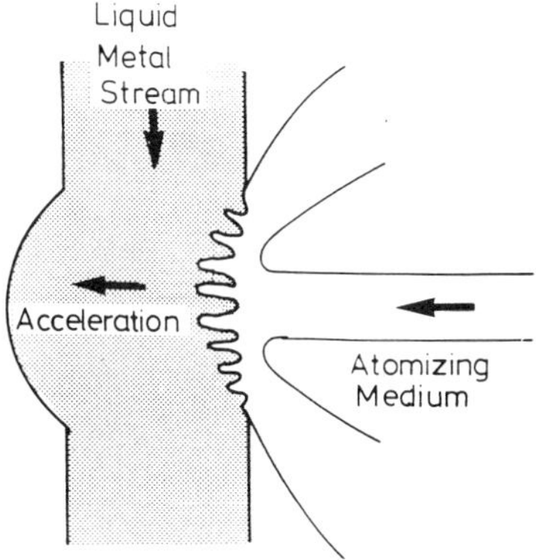

Fig. 2.   Hydrodynamic fragmentation mechanism caused by perpendicular impact of gas or water jets on a liquid metal surface. The Taylor instabilities are induced by sideways acceleration of the liquid metal stream by the less dense gas or water jet (after[6]).

In principle, two types of processes may be distinguished: 1) processes with a primary breakup of the metal stream into larger melt fragments which subsequently undergo a secondary breakup into fine fragments in the relative flow of the atomizing medium and 2) processes with a continuous fine fragmentation of the metal stream without an intermediate rough primary breakup (Fig. 5). Detailed analysis of the size distributions of water atomized Sn showed a frequency peak at particle sizes, d, between 25 and 30 μm and a shoulder between 80 and 100 μm. When superheating was low, the shoulder developed into a second peak (Fig. 3).[9] A similar size distribution was reported by[10]. The magnitudes of the peaks changed with different atomization parameters (superheating of the melt, metal flow rate, water jet pressure) but the peak position remained unchanged. The separation of the size distribution into two composing log–normal distributions gave a satisfying fit with a lower and higher size peak at d_l and d_h, the deviation of the peaks, σ_l and σ_h, and the proportion of the lower to the higher peak, p_l. Fig. 3 shows two examples. After separation, the total particle size distribution is represented in the form of

$$f = (2\pi p_l/\sigma_l) \bullet \exp \left(\frac{\ln d - \ln d_l}{\sigma_l}\right) + \frac{2\pi(1 - p_l)}{\sigma_h} \bullet \exp \left(\frac{\ln d - \ln d_h}{\sigma_h}\right) \qquad /1/$$

This suggests that the water atomized powder consists of two types of particles with different size distributions which reflect different physical disintegration mechanisms. Assuming thermal fragmentation causing the secondary fragmentation of initial coarse fragments into fine droplets, a pronounced increase in the fraction of fine powders is expected with increasing superheating of the Sn melt. In contrast, only a slight increase of the proportion of the finer peak was measured, indicating thermal fragmentation was not the determining disintegration mechanism. Following the interpretation of Halada et al., the liquid metal flow is destabilized by the ambient gas flow induced by the water jet and broken up into coarser droplets with a diameter of approx. 100 μm. In a second stage, disintegration into finer fragments occurs from the hydrodynamic interaction of the coarse droplets with the water jet.[9]

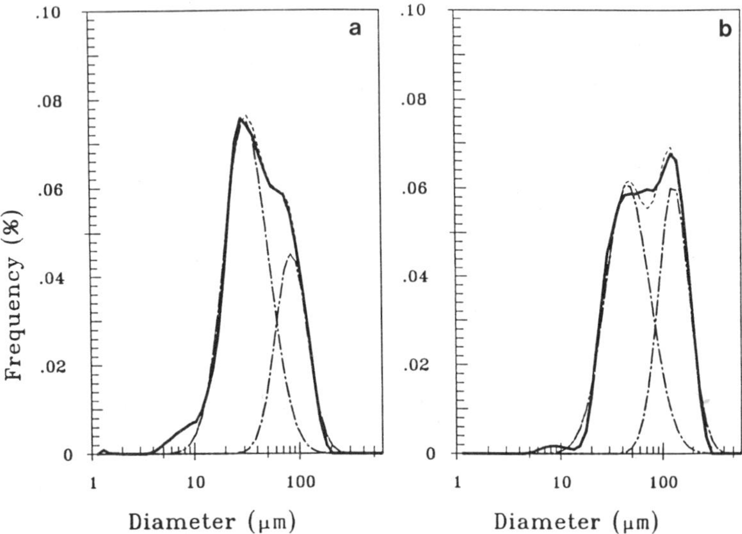

Fig. 3. Separation of particle size distributions into two log–normal distributions. Sn powder, water atomized a) at 12 MPa, atomization temperature 1100°C and b) at 8.5 MPa, atomization temperature 700°C.

Secondary atomization leads to spherical droplets whenever time is sufficient to form a droplet out of irregular melt fragments by surface tension before solidification. The time interval required for spheroidization is given by[11]

$$t_{sph} = \frac{3\pi^2 \cdot \eta_m}{4V \cdot \sigma_m} \cdot (R^4 - r^4), \qquad /2/$$

where η_m is the viscosity of the metal, V is the volume of the droplet, σ_m is the surface tension of the metal, R is the fragment radius before transformation to spherical shape, and r is the droplet radius after transformation to spherical shape. The time interval between disintegration and the start of solidification depends on superheating, undercooling and cooling rate.

If solidification occurs immediately after a disintegration event, hence in the more turbulent zones, then irregular droplets solidify into irregular shaped particles. Higher cooling rates during water atomization, sometimes in combination with the shape stabilizing event of oxide or hydroxide reaction layers forming at the surfaces, result in irregular shaped particles compared with the more spherical particles provided by gas atomization.

Principally in Japan, attempts have been made to obtain particles with smaller sizes by high–pressure water atomization. When the water pressure exceeds 20 MPa, the particle size decreases rapidly with increasing pressure (Fig.4). It is thought that high pressure water jets provide suffcent velocity differences, v_{rel}, to primarily disintegrate droplets and to produce much finer droplets during the secondary fragmentation. A half–empirical estimate is possible by the Weber criterion[12]

$$We_d = \rho \bullet v_{rel}^2 \bullet d / \gamma_{lv}$$ /3/

where ρ is the density of the flow field (water–gas mixture), γ_{lv} is the surface tension, v_{rel} is the relative velocity between the atomizing flow field and the droplet when the drop is first exposed to the flow field. As a rule of thumb, a metal melt droplet of initial size, d, may disintegrate if the Weber number, We_d, is above 10. Higher Weber numbers can lead to various other fragmentation mechanisms which are described later.

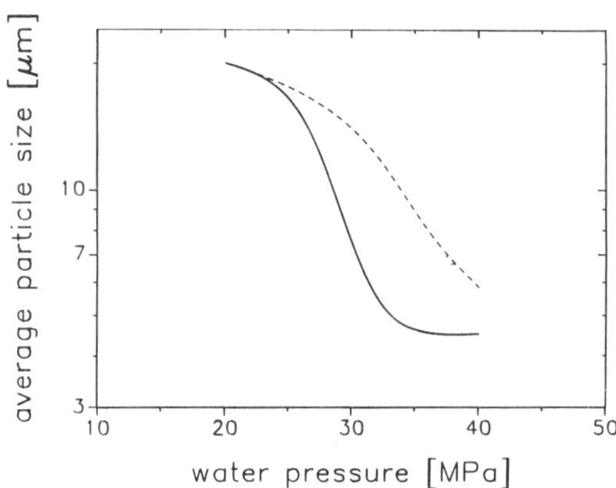

Fig. 4.   Influence of pressure during water atomization on average particle size.[13] Dotted line: undoped water; full line: polyethylene oxide–doped water.

Fragmentation during Gas Atomization

Nozzles for gas atomization usually take the form of either an annulus, which is concentric around the metal stream, or of several discrete jets. Nozzles are available in 'free–fall' and 'confined' designs. In the 'free fall' design, the melt falls by gravity a distance of up to 0.3 m and then interacts with the high pressure gas jets (Fig.1). The impinging gas leaves the nozzle at high speed but slows down during its travel to the metal impingement point. The amount of this attenuation increases with increasing jet distance, yielding coarser powders. With 'confined' designs, the atomization occurs at the orifice of the tundish nozzle. The energy transfer from gas to metal is then very efficient and uniform due to prefilming of the molten metal over the end of the tundish nozzle and due to the short distance between gas and metal stream[16,17,18]. As in water atomization, primary and secondary fragmentation mechanisms also have to be distingiushed in gas atomization.

A low pressure area is created above the impingement point of the atomizing gas, causing the liquid metal to form a hollow cone or umbrella shaped pattern (Fig. 5). This prefilm or sheet is then further primarily disintegrated into fragments, ligaments and drops. Primary breakup may also occur above the impingement point of the gas jets, due to inherent melt stream instabilities (i.e. of varicose or sinuous type[14]). Another fragmentation mechanism which may act as a primary (or secondary mechanism) is wave stripping due to parallel shear flow instabilities (Fig. 5). In the extensive work by Bürger[19] et al., the formulation by Bradley[20] was used with a logarithmic velocity profile (in recognition of the turbulent flow field) of the gas flow above the wavy liquid metal surface according to an approach by Miles.[49] After sufficient wave growth, the detachment of wave crests caused by the pressure forces exerted from the gas flow was calculated by an energy balance at the wave crest. The fragments and ligaments formed by stripping are assumed to undergo further secondary break up by surface tension.[20] Fig. 6 shows part of the sequence of fragmentation proposed by Bürger et al.[19] for gas atomization, which combines several breakup mechanisms. Gas atomization experiments with a parallel flow nozzle showed satisfying agreement between theoretically predicted and experimentally measured particle size distributions.[21,48]

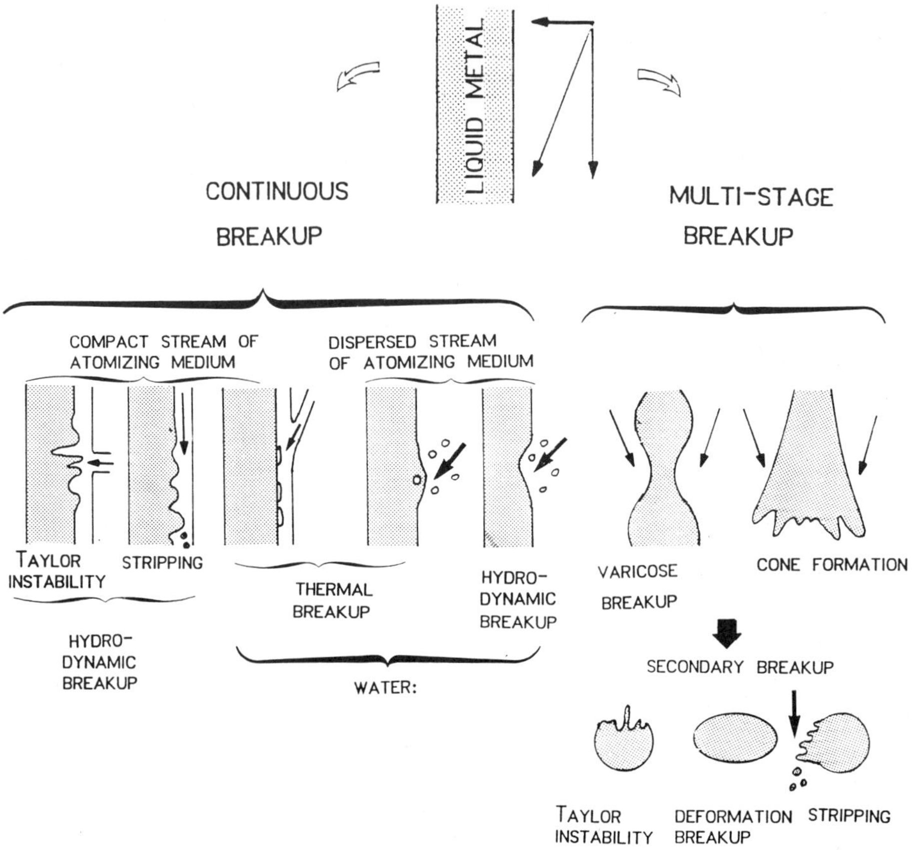

Fig. 5.   Stages of liquid metal stream fragmentation during gas and water atomization.

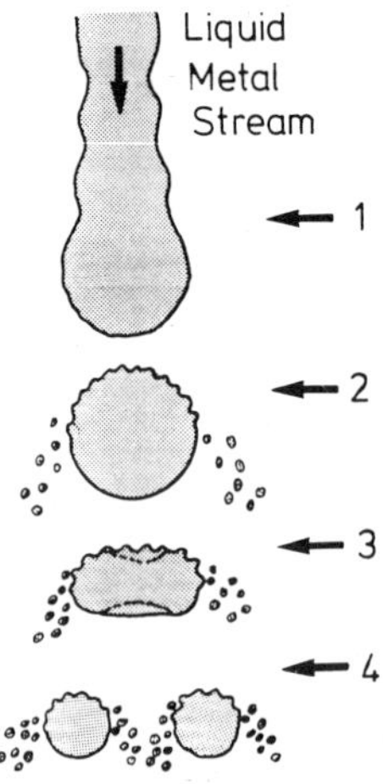

Fig. 6.   Combination of coarse breakup mechanisms (jet breakup due to varicose mechanism), drop breakup due to deformation and fine fragmentation due to parallel flow instability. 1 varicose breakup, 2 onset of high velocity relative flow, 3 wave crest stripping and deformation, 4 deformation breakup (after[21]).

A more general view of the breakup of a drop due to acceleration by a gas fluid flow field gives the following mechanisms, listed in the order of increasing Weber number[22]: Vibrational breakup: The drop develops into a flat disk by acceleration forces and starts to oscillate. When the surface tension force is exceeded by the dynamic pressure, two or more smaller fragments are formed. Bag breakup: A hollow bag is formed spanning a ring of fluid. The bag bursts and the ring breaks up into many droplets. Bag and stamen breakup: Similar to bag breakup but with the formation of a central club prior to disintegration. Stripping breakup: The drop flattens and a mist of droplets is stripped away from the circumference. Catastrophic breakup: The drop is penetrated by large amplitude surface waves. Smaller particles are formed and undergo the stripping mechanism.

Similar to solidification during water atomization, spheroidization time of the droplets has to be short compared to the solidification time to obtain essentially spherical particles. Due to the lower cooling rates provided by gas atomization compared to water atomization, solidification usually starts when the droplets are far from the turbulent impingement zone and when the relative velocities between droplet and gas are small (see chapter on cooling rate). This means that the Nusselt number is close to 2 and calculations can be made with the approximation of a zero relative velocity.

Fragmentation during Centrifugal Processes

In the centrifugal processes, a thin liquid film on the rotating electrode or disc is fragmented due to the instabilities caused by centrifugal forces. The surface tension due to the specific surface energy and the curvature of the liquid surface are restoring forces. Higher centrifugal forces and a slower melt supply lead to finer powders. The average particle size is much coarser than is commonly achieved by gas atomization but the size distribution can be held within narrow ranges.[23]

MICROSTRUCTURAL DEVELOPMENT DURING RAPID SOLIDIFICATION

Microstructural refinement, formation of metastable phases and extended solid solutions have all been often demonstrated by techniques which are designed primarily to maximize the heat extraction rate (e.g. splat cooling, melt spinning).[24] Rapid solidification processing techniques which are more suited to commerial production (e.g. high pressure gas atomization) often yield disappointing results with respect to the obtainable microstructures due to the lower heat transfer coefficents associated with the latter techniques. It will be shown that inferior heat extraction rates associated with atomization can be compensated for by an optimized use of the low nucleation potential and the high undercooling capbility of the technique. Methods of efficient heat extraction can be exploited to produce microstructures which display extended solid solubilities.

Cooling Rate during Gas Atomization

Calculations of the cooling rate on the basis of thermophysical data require some assumptions. We consider a spherical droplet in a surrounding gas field. There are two limiting cases for the cooling of this droplet. The Newtonian case, where no temperature gradients within the droplet are present and the ideal cooling case, where steep thermal gradients are present. The decision whether Newtonian conditions can be assumed is made with the help of the dimensionless Biot number. The Biot number indicates whether the internal heat transport is fast enough to keep the thermal gradients extremely small and is given by[25,26]

$$Bi = \frac{h \cdot L}{K} \qquad\qquad /4/$$

where h is the heat transfer coefficient, L is a typical length (e.g. the droplet radius) and K is the thermal conductivity. If the Biot number is smaller than 0.1, the Newtonian conditions are valid. It is well accepted now that during cooling of atomized droplets, the Biot number is much smaller than 0.1 and the cooling is mainly controlled by the convective and radiative heat transfer between the droplet and the atomizing gas.[25] The Newtonian model is, therefore, a useful approximation which depends on those factors affecting the cooling rate. The decrease of the droplet temperature in a gas stream due to heat transfer by forced convection and by radiation is described by[27]

$$\frac{\Delta T_p}{\Delta t} = \frac{6K \bullet Nu \bullet (T_p - T_g)}{d^2 \bullet \rho \bullet C_p} + \frac{6\epsilon \bullet \sigma \bullet (T_p{}^4 - T_w{}^4)}{d \bullet \rho \bullet C_p} \, , \qquad /5/$$

where Nu is the Nusselt number, T_p is the temperature of the droplet, T_g is the temperature of the gas, d is the droplet diameter, ρ is the density of the droplet, C_p is the specific heat of the metal, ϵ is the emissivity factor and σ is the Stefan—Boltzmann constant. The important factors affecting the quench rate in the first term (covering the convective heat transfer) of eq./5/ are the following, in order of decreasing influence

— droplet diameter
— thermal conductivity of the quenching medium
— temperature difference between droplet and medium,
    especially for lower melting alloys
— relative velocity between droplet and medium

In the temperature range up to 1400°C, the radiative heat transfer is only of minor importance and can be neglected in the calculation. On the other hand, the velocity difference between gas and particles is an important factor that can determine the cooling rate considerably. Often, the velocity difference at the beginning of the atomization, when the gas stream impacts the melt, is incorrectly used to calculate the Nusselt number, which determines the heat transfer coefficient. In reality, the droplets are accelerated by the gas stream, which decreases the velocity difference and the heat transfer coefficient. The acceleration of the droplets depends on their diameter. The smaller the droplet, the faster it will approach the velocity of the surrounding gas flow field. Solidification in a state of small relative velocity difference between the droplet and the gas is then most likely. Larger droplets are more slowly accelerated and leave the zone of rapid gas flow before the onset of solidification. Again, solidification without the influence of the gas flow is likely to occur. In both cases, the starting gas velocity plays no dominant role for the heat transfer during the solidification. Therefore, the cooling rate is mainly determined by the thermal properties of the gas and by the droplet diameter.[4] The velocity difference between gas and droplet during the droplet's flight and at the beginning of the solidification is shown in Fig. 7. The heat transfer coefficients until the end of solidification for different particle diameters are shown in Fig. 8. Fig. 9 shows the thermal history of a Cu—7wt.%Sn droplet calculated with the assumptions and heat transfer data shown in Fig. 8.

If the effect of the velocity difference of the gas jet and the droplet at the critical instant of crystallization is assumed to be negligible, the heat transfer coefficient h becomes proportional to 1/d. The cooling rate is then[25,26,29]

$$dT/dt \propto h/d \propto d^{-2} \qquad /6/$$

The velocity of the solidification front during cooling without initial undercooling can then easily be calculated. If we assume that the droplet is isothermal at any instant, the solidification front velocity, V_E, is given by a simple heat balance, controlled by the external heat flow[40]

$$V_E = \frac{6h_i \bullet \Delta T}{L} \qquad /7/$$

where h_i is the heat transfer coefficient, L the latent heat of fusion and ΔT the temperature difference between the cooling gas and the alloy droplet.

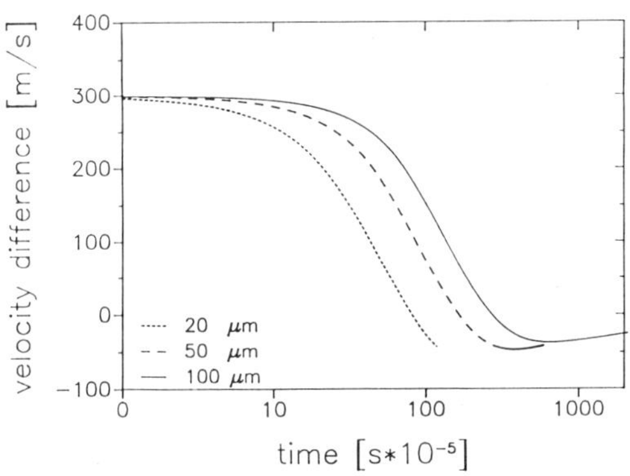

Fig. 7.   Velocity difference between gas and droplet during atomization (the parameters are the droplet diameters; initial velocity difference is 300 m/s).

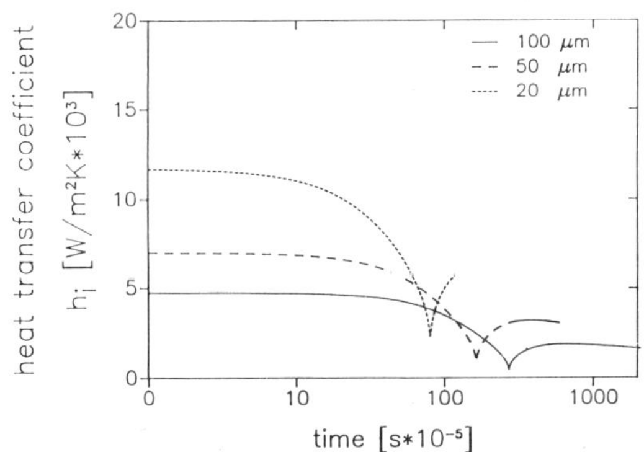

Fig. 8.   Heat transfer coefficient during atomization of Cu–7wt.%Sn droplets in Ar (velocity conditions as shown in Fig.7). The droplet was initially superheated by 150 K and subject to crystallization by a single nucleation event at an undercooling of 100 K (after[28]).

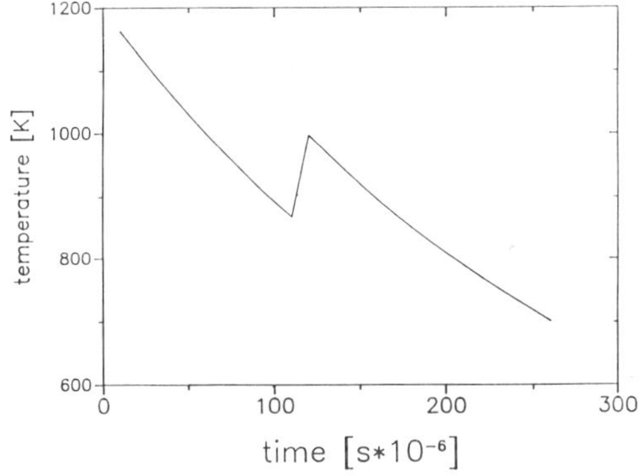

Fig. 9.   Predicted thermal history of a 10 μm diameter Cu–7wt.%Sn droplet during Ar–atomization.

Cooling Rate and Microstructural Development

An empirical relation, known for many years, relates the secondary dendritic arm spacing, SDAS, with the cooling rate, dT/dt. From this relation

$$SDAS = K_c \bullet (dT/dt)^{-n} \qquad\qquad /8/$$

we obtain

$$\log SDAS = \log K_c - n\log(dT/dt) \qquad\qquad /9/$$

where K_c and n are empirical constants which depend on the system and its particular composition. Fig. 10 shows the validity of these relations for a gas atomized NiCrBSi alloy. The reduced cooling rate was calculated after eq./6/ and normalized to the cooling rate of a droplet 315 μm in diameter. The straght line indicates that the cooling rate has controlled the solidification rate.[30] The effect of undercooling was clearly negligible.[31,32]

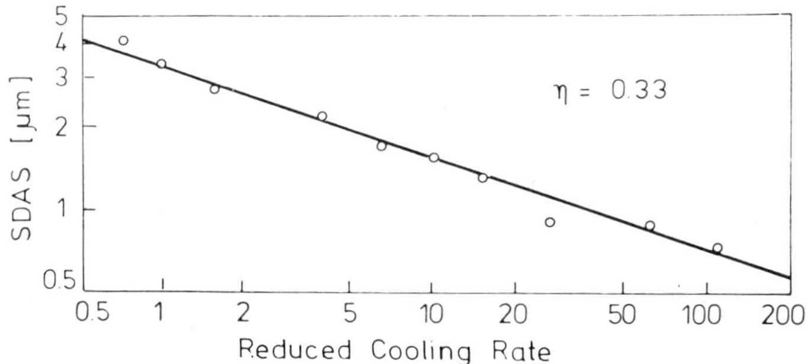

Fig. 10   SDAS and reduced cooling rate of a Ni hardfacing alloy.

The relationship of interdendritic spacing and the cooling rate or more precisely the local solidification rate can be calculated by an approximation of Feurer and Wunderlin which is based on the Ostwald ripening model of Kattamis[33], with

$$dT/dt = (5.5/\lambda_2)^3 \bullet M \bullet \Delta T \qquad\qquad /10a/$$

and

$$M = \frac{\gamma_{sl} \bullet D \bullet \ln(c_1/c_0)}{\Delta S^m \bullet m \bullet (1-k_0) \bullet (c_0 - c_1)} \qquad\qquad /10b/$$

Here, λ_2 is the secondary dendrite arm spacing, γ_{sl} is the solid/liquid interface energy, c_1 is the composition of the last liquid, M the coarsening factor, k_0 the equilibrium partition coefficient, D the diffusion coefficient in the melt, m the slope of the liquidus line, ΔS^m the entropy of fusion and c_0 the initial melt composition. The approximation is based only on thermodynamic data.

Fig. 11 shows the microstructure of Cu–7wt.%Sn powders of different size. The SDAS of these powders and of the furnace cooled and splat cooled samples of the same alloy are plotted in Fig. 12 vs measured and calculated local solidification rates. The line represents the calculation of the local solidification rate after eq./10a/. The agreement is good.

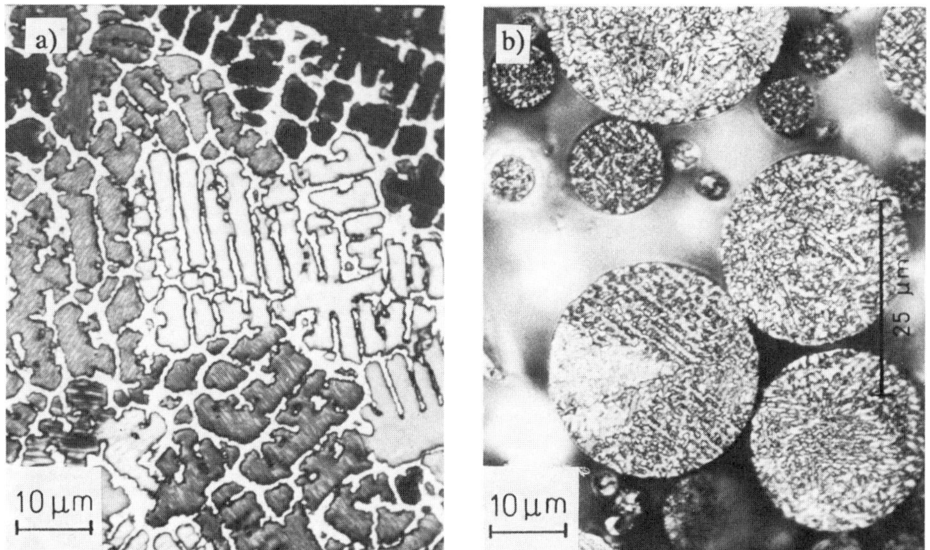

Fig. 11. Microstructure of Cu–7wt.%Sn. a) Water quenched droplet (diameter 4 mm). b) Gas atomized droplets.

The extrapolation of a plot as shown in Fig. 12, over some orders of magnitude requires a careful estimate of a variety of errors which are inherent to the measurements. After all, those plots are valid only for one particular alloy. This is partly due to the dendrite coarsening effect during the particular time intervals between the beginning and the end of solidification, which are often refered to as 'local solidification time'. Different local solidification times clearly lead to different time intervals available for coarsening of the secondary dendrite arms.

It is also clear that the local solidification times usually will depend on both the cooling rate and the prior undercooling. A case of suppressed undercooling was shown in Fig. 10. Cases of considerable undercooling are discussed in the following chapter.

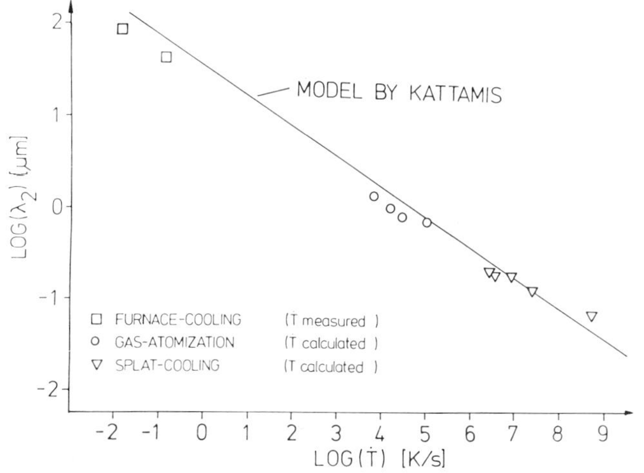

Fig. 12. Secondary dendrite arm spacing vs cooling rate. ——— Calculated after.[33]

Solidification of a bulk liquid metal usually requires only small undercooling[34] due to heterogeneous nucleation. Many experiments were accomplished to minimize the influence of heterogeneous nucleation and to achieve a high undercooling. Perepezko[35], for example, subdivided a liquid metal sample into a stabilized dispersion of fine droplets in an inert liquid. The droplet size was between 5 and 20 μm. Undercooling between 0.3 and 0.4 T_m (T_m is the melting temperature) was measured. A similar disintegration of a larger liquid metal pool into fine droplets occurs during atomization. If the number of particles produced from the melt pool is higher than the number of potent heterogeneous nuclei which had been present in the pool, the smallest droplets have a high probability of being free of potent nuclei. These particles will undercool by several hundred K before solidification initiates.

Significant undercooling of metal systems prior to crystalline solidification can affect both the dendritic structure and the solute distribution. The influence of undercooling on the dendritic structure was demonstrated on bulk specimens of Ni and Fe base alloys[36]. At small undercooling, the usual dendritic structure formed. The secondary dendritic arm spacing decreased with increasing undercooling. At moderate undercooling of appr. 60 to 170 K, the dendritic structure consisted of cylindrical arms elongated along three perpendicular <100> directions. Beyond a critical degree of undercooling of more than 170 K, the dendritic microstructure changed into a microstructure with equiaxed grains. Fig. 13 shows the microstructure of Cu–7wt.%Sn alloys after different undercooling. Similar microstructures were found by Caesar[37] after solidifying levitationally melted droplets of Cu–Ni alloys.

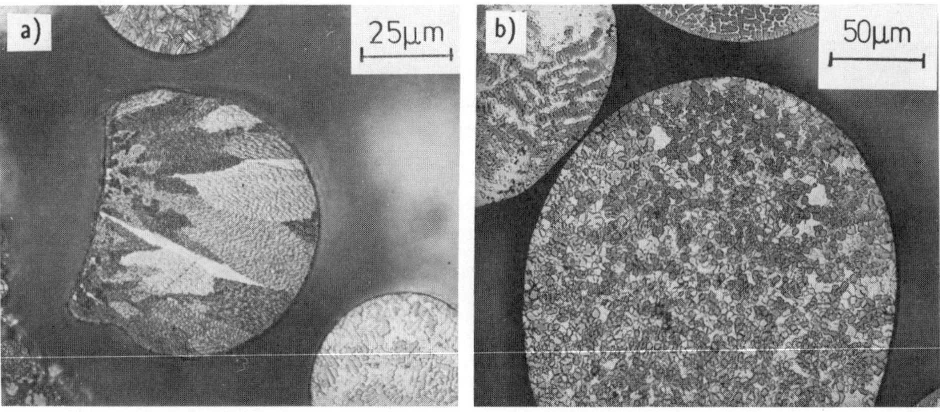

Fig. 13.   Microstructure of Cu–7wt.%Sn alloys after different undercooling. a) Undercooling = 180 K. b) Undercooling = 260 K

In hypercooled melts, the increase of the temperature due to recalescence is insufficient to heat up the material to the solidus temperature again. For a segregation–free solid solution of the melt composition this condition is identical with undercooling below the T_0 temperature, where the free energy of the melt and the solid are equal. On the other hand the hypercooling condition is a necessary thermodynamic requirement for the formation of a segregation free solid solution (homogeneous solid). It is stated once more that due to high undercooling, rapidly solidified structures are obtained during atomization without extremely high cooling rates.[38]

The interpretation and modeling of microstructures resulting from undercooled melts is still in its infancy and needs considerably further experimental and computer simulation efforts. In the case of large undercooling, the external heat extraction becomes less important. In the extreme case, solidification can be finished adiabatically. Here maximum solidification front velocities are obtained. The amount of undercooling is controlled by the onset of nucleation and, as a second effect, by the competition between the release of the latent heat of fusion and the external heat flow.

The nucleation temperature describes the instant when the first stable nucleus is formed. Nuclei can form by two different processes. The theoretically cited homogeneous nucleation temperature yields the maximum undercooling prior to crystalline solidification. This type of nucleation is unlikely to occur in the practice of an atomization process, but the models of homogeneous nucleation are easy to calculate and provide the lower bound of possible undercooling.[44] Heterogeneous nucleation is of more practical significance. In this case, the nucleation temperature is mainly controlled by the wetting angle between the heterogeneous nuclei and the melt. The first stable nucleation event starts the release of latent heat of fusion. For a short time interval, the droplet cools further as the heat loss due to external heat flow exceeds the latent heat released by the growth of the nuclei. The maximum undercooling is reached when the rate of latent heat release balances the external heat flow. Additional nuclei are formed as long as the droplet is sufficiently undercooled. When, during subsequent reheating by the latent heat release, a critical temperature is exeeded, the generation of new nuclei stops. The temperature of the droplet increases further and the solidification continues with the growth of those active nuclei which had been present at the critical temperature. Theoretically, the number of grains in the solidified droplet equals the number of active nuclei present at the critical temperature, if coarsening during further cooling is suppressed.

The release rate of the latent heat is determined by the number and the growth velocity of the nuclei. Two different assumptions on the growth controlling mechanism exist[45]: diffusion controlled and interface controlled growth. The diffusion controlled growth mechanism requires sufficient time for the atoms in the melt to diffuse to appropriate sites on the crystal lattice in contact with the melt. The maximum velocity is limited by the diffusion rate. In this case, maximum velocities of some m/s can be obtained. The second mechanism is controlled by collisions between atoms in the melt and the growing crystal. Maximum growth velocities are predicted to be in the range of the speed of sound of the melt[46], which is several thousand m/s. Since not every hitting atom fits to the solid structure, an efficiency factor is necessary, which reduces the predicted growth velocities. Unfortunately, values for the efficiency factor are largely unknown.

Few models exist in the literature which describe the nucleation behaviour during rapid solidification, taking account of the nucleation and growth problems mentioned afore[47]. For convenience, pure metallic systems and homogeneous nucleation are generally considered. Application of the models to the atomization of pure copper yields predicted grain sizes according to the assumption of homogeneous nucleation which are much larger than the experimentally determined grain sizes (Fig. 14). The assumption of homogeneous nucleation leads to high undercooling and the prediction of very few nucleation sites. When the first stable nucleus is predicted to form, the large undercooling results in extremely high growth velocities. The corresponding fast release rate of latent heat would warm the droplet too fast to allow for the fomation of additional critical nuclei. Therefore, the number of predicted critical nuclei is extremely small.

More realistically, heterogeneous nucleation conditions must be applied. It is then necessary to know the wetting angle between nuclei and the melt, a value which can only be deduced from theoretical considerations or approximated from a limited number of experiments. In experiments with atomized copper alloys, droplets with a diameter of 50 μm showed an undercooling of more than 80 K prior to solidification. Using this value and the number of grains measured from the microstructures allows the calculation of the wetting angle. Using this wetting angle for the calculation of grain sizes in atomized Cu gives much better results than the homogenous nucleation apprach. Nevertheless, the agreement is not completely satisfying. The calculated number of generated nuclei and, consequently, the number of grains in the final particle is still smaller than the real value (Fig. 14). This indicates that the temperature gradients within the solidifying droplet must be introduced into the theoretical model to allow a more exact description of the physical situation during the time interval of nucleation in atomized droplets.

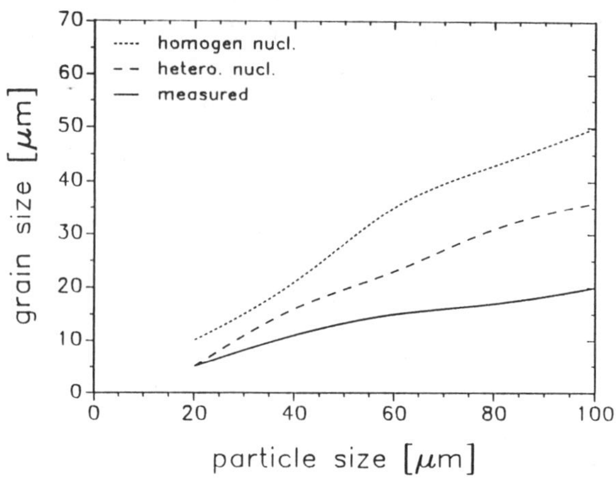

Fig. 14.   Grain size in atomized Cu powder. Comparision between calculations based on heterogeneus and homogeneous nucleation and experimental data from Rzesnitzek.[48]

SOLIDIFICATION OF Cu–Sn DURING ATOMIZATION

In the subsequent chapters experiments with Cu–Sn alloys are presented which show some applications of the principles outlined up to this point. Major emphasis is put on the problems of undercooling and the manipulation of the density of heterogeneous nuclei.

Seeded Melts

Heterogeneous nuclei are always present in the melt. Different heterogeneous nuclei are, however, of different nucleation potential. Especially, the metal oxides have a low nucleation potential compared to some other nuclei that are used in casting as seeding materials. Cu–7wt.%Sn ingots show, for example, a significant grain refinement and a smaller degree of undercooling when seeded with Fe and Zr. The addition of 0.1wt.% Fe and 0.1wt.% Zr results in ingots solidified after nearly zero undercooling with grain sizes about 30 times smaller than those of the untreated material.[39] In untreated melts, disintegration during atomization into small droplets makes the fine droplets particularly prone to the statistical absence of potent nuclei, leading to non– deterministic undercooling.[40] In contrast, in our experiments with seeded Cu–7wt.%Sn, undercooling was kept small in all droplets due to the presence of a sufficient number of potent heterogeneous nuclei. The solidification rate was only controlled by the external heat flow. Powder from seeded melts showed similar microstructures in particles of different sizes even for the smallest ones which were below 5 μm.

Untreated melts

Atomization was done with unpurified materials. The atmosphere in the atom–ization chamber contained sufficient oxygen to form oxides in the melt that were able to act as nuclei. Under usual conditions, the larger particles always showed multiple nucleation. This is indicated by the different growth directions of the dendrites in Fig. 15a. After a certain undercooling, multiple nuclei provide the formation of several crystals in one particle. If nucleation occurs closer to the solidus temperature by a single more potent nucleus, the whole particle forms a single crystal. This case is shown in Fig. 15b.

Here, the liquid droplet was hit during flight by an already solidified small particle. This particle, a so called "satellite", caused nucleation at once and subsequent solidification of the bigger particle. The undercooling of the big particle was smaller than in the case of multiple nucleation.

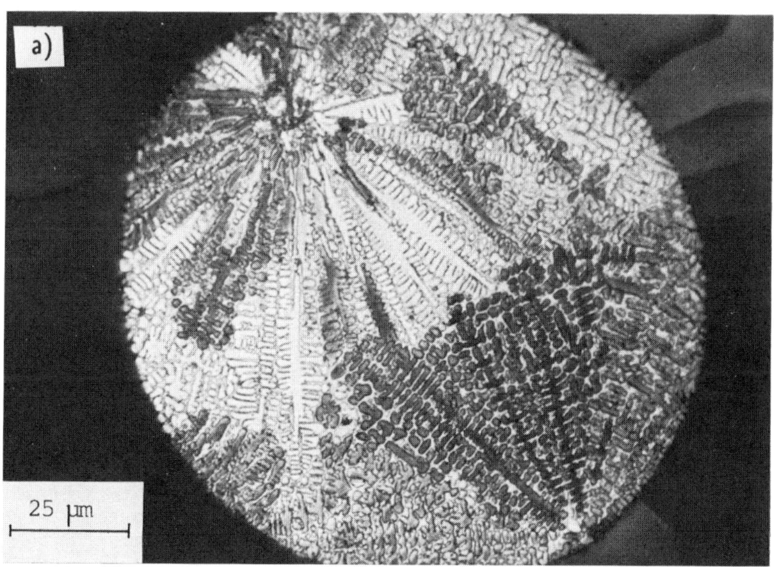

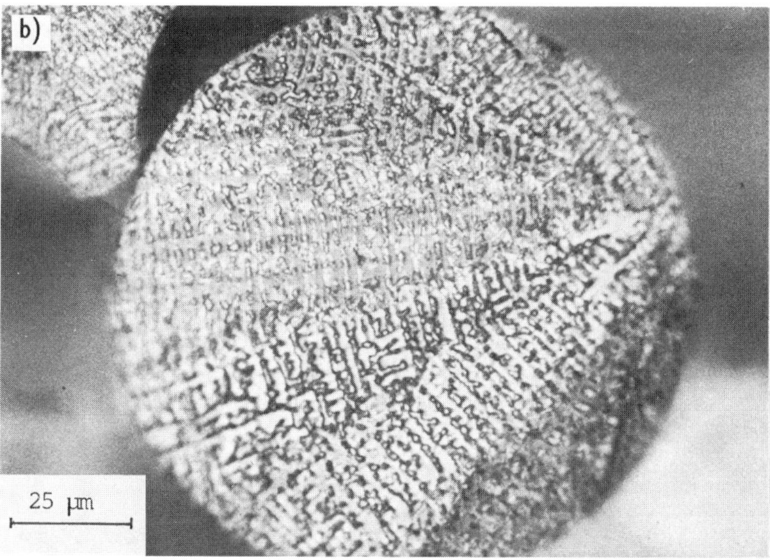

Fig. 15.   Microstructure of Cu–7wt.%Sn particles atomized from untreated melts. a) Multiple heterogeneous nucleation. b) Single nucleation event due to a satellite.

Using particular alloy compositions, the nuclei densities can be estimated easily. For example, the Cu—20wt.%Sn alloy (Fig. 16) shows a typical difference in the phases present in the large and the small particle size fractions. The nuclei density follows the Poisson distribution

$$X = \exp(-M \bullet v) \qquad\qquad /11/$$

where X is the nuclei free droplet fraction, M is average number of nuclei per volume element and v is the droplet volume. By quantifying the existent phases in the size fractions, the density of the nuclei can be calculated.[41] The most potent nucleus leads to the formation of the α—phase. Particles containing the ß—phase exist only in the size fraction smaller 50 μm. The nuclei density for the formation of the α—phase is $5 \cdot 10^{15}$ m^{-3}. The development of the ß—phase requires an undercooling of at least 80 K.

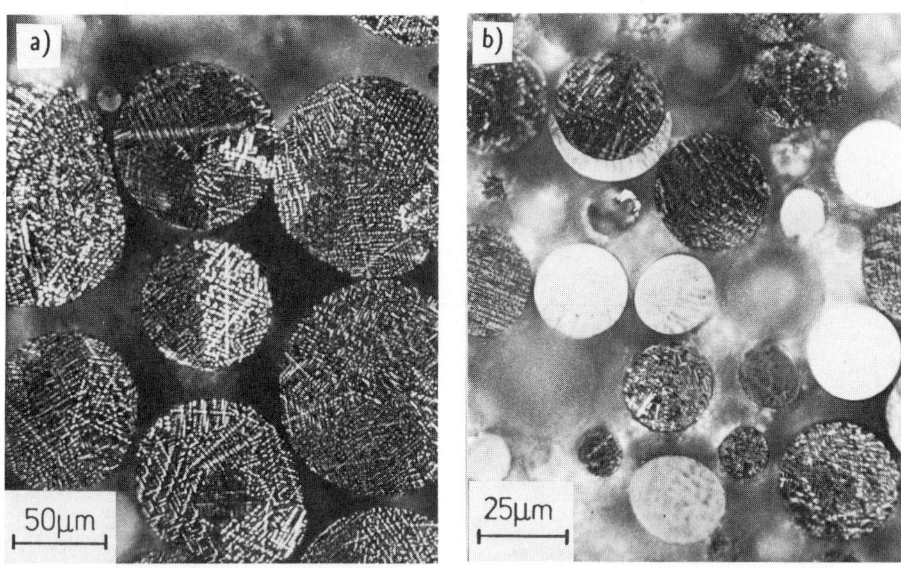

Fig. 16.   Microstructure of atomized Cu—20wt.%Sn. α = dendritic, ß = light phase
a) Particle size = 100 μm. b) Particle size $\leq$ 30 μm.

Refined Melts

Oxidic nuclei can be eliminated. Pronounced undercooling and high solidification front velocities result. Oxidic nuclei in a Cu—7wt.%Sn alloy were eliminated by the addition of metallic Al to the bulk melt. The high oxygen affinity of Al led to the reduction of most other metal oxides. Fine Al_2O_3 particles formed, which are known not to act as heterogeneous nuclei in this alloy. Therefore, the number of nuclei decreased and the possibility and magnitude of large undercooling increased. With an initial undercooling large enough to allow the particles to be their own heat sink, very rapid solidification velocities are realized. If the particle reaches the solidus temperature again before solidification is completed, due to the release of the latent heat of fusion, the remaining liquid solidifies by the normal external heat flow. The particle shown in Fig. 17 was strongly undercooled before the onset of nucleation. Before the solidus temperature was reached, part of the melt solidified in a segregation—free manner (triangles in Fig. 16b). The rest of the melt solidified at the normal external heat flow, as indicated by the usual dendritic microstructure. The change in the morphology of the solidifying microstructure was caused by a decrease of the solidification front velocity. The growth velocities are described as a function of recalescence by a mathematical model.[42] It was shown theoretically that the growth velocity may differ by a factor of 25 between the start and the end of solidification of a strongly undercooled droplet. The usual solidification front velocity in our particles cooling by convective heat transfer without initial undercooling is estimated after equation /7/ to be $V_E = 2.4 \cdot 10^{-2}$ m/s. The growth velocity during the segregation–free solidification must have exceeded 4 m/s.

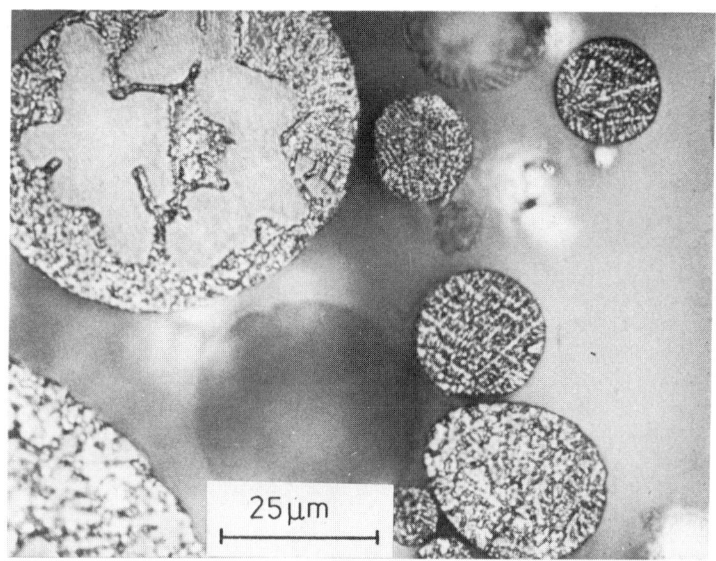

Fig. 17.   Refined Cu–7wt.%Sn with segregation–free solidified areas.

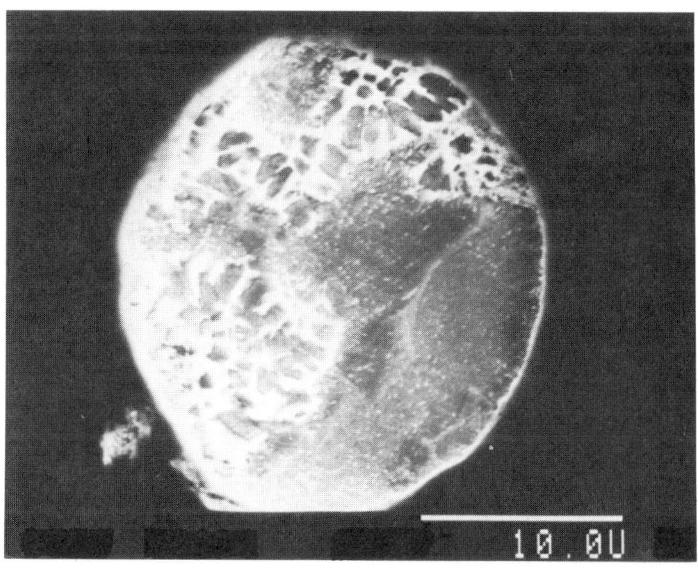

Fig. 18.   SEM micrograph of Cu–7wt.%Sn with segregation–free areas. The white dots are entrapped Al_2O_3 particles.

Another effect of the high growth velocities is the entrapment of the Al_2O_3 particles in segregation—free solidified alloy regions. Fig. 18 shows an example. The white dots in the matrix are the Al_2O_3 particles. The velocity of the growth front determines whether these particles are entrapped or not. In the powders without segregation—free solidification, the Al_2O_3 particles are always swept into the interdendritic space. The velocity necessary to entrap the particles in the solidification front can be calculated[43] and thus the entrapment of those particles in the solidification front would provide some information on the minimum growth velocity. Unfortunately, a lack of exact interface energy data for $Cu-Sn-Al_2O_3$ makes the calculation impossible at present.

CONCLUSIONS

Several mechanisms of melt stream disintegration are possible. One of the most important mechanisms is melt stripping and subsequent disintegration due to Taylor instabilities. During water atomization, the liquid metal flow is proposed to be initially destabilized by the ambient gas flow induced by the water jet. The metal flow then breaks up into coarser droplets with a diameter of approx. 100 μm. In the second stage, secondary disintegration into finer particles occurs by the hydrodynamic interaction of the coarse droplets with the water jet.

The cooling rate during gas atomization depends on the reciprocal square of the droplet diameter. During solidification of the particles, the velocity differences between the gas and the droplets were found to be small and independent from the initial gas velocity. Hence, the cooling rate during solidification can not be increased by an increased gas velocity at the nozzle outlet.

The coarsening of the secondary dendrite arms occurs during the time interval when the initially formed dendrites and the melt are in contact. A model of Kattamis describes the dendrite ripening based on thermodynamic data. Our present calculations of the cooling rate from thermal data and the measured secondary dendrite arm spacing are in good agreement with the model.

The solidification front velocities of particles with small initial undercooling are in the range of some 10^{-2} m/s. They result in a refinement of the microstructure compared to cast materials. The production of segregation—free solidified powders, however, requires growth velocities in the order of some m/s. In gas atomization, these high velocities can only be attained by undercooling. Undercooling is mainly controlled by the nucleation conditions within the melt droplet. Usually a large number of heterogeneous nuclei exist in a particle and multiple nucleation already occurs at medium undercooling.

The density of nuclei can be decreased or increased by a treatment of the melt. Metal oxides in the melt that act as heterogeneous nuclei can be transformed into inert oxide particles. In the case of Cu—7wt.%Sn, aluminum was used to reduce the potent metal oxides. The generated Al_2O_3 particles did not act as nuclei and larger undercooling was observed.

Existing models which describe nucleation during rapid solidification neglect temperature gradients within the solidifying droplet. Therefore, the calculated recalescence is too fast to allow further nucleation events. The number of finally predicted nuclei is always smaller than the experimentally observed. These models should be improved.

ACKNOWLEDGEMENT

This work was supported by Deutsche Forschungsgemeinschaft, Internationales Büro der KFA Jülich und Bundesministerium für Forschung und Technologie. We thank Profs. G.Petzow, R.M.German, Gh.Mattei, Drs. K.Halada and J.Wachter, Dipl.Ings. R.Laag and J.Murray for fruitful discussions and support.

REFERENCES

/1/   S. Takayo, Innovations in Ferrous Powders and their Production, in "Powder Metallurgy — State of the Art", (eds.) W.J.Huppmann, W.A.Kaysser and G.Petzow, Schmid—Verlag, Freiburg ,7—28(1986).

/2/   P.Hellmann and P.Billgren, Commercial Gas Atomization, in "Atomization and Processes: Current and Future", (eds.) A.Lawley and E.Klar, MPIF, Princeton, 33—50(1984).

/3/   E.Klar, Gas and Water Atomization, in "Metals Handbook, Ninth Edition, Volume 7, Powder Metallurgy", (coordinator) E.Klar, American Society for Metals, Metals Park, Ohio 44073, 25—39(1984).

/4/   W.A.Kaysser, K.Rzesnitzek and G.Petzow, Atomized Powders with Reduced Microstructure Variations, in "Modern Developments in Powder Metallurgy", (eds.) P.U.Gummeson and D.A.Gustavson, MPIF, Princeton, NJ, 18—21,(1988).

/5/   M.D.Oh and M.L.Corradini, A Propagation/Expansion Model for Large Scale Vapor Explosions, Nuclear Science and Engineering, 95,225—240(1987).

/6/   E.Berg und G.Fröhlich, Modell zur Beschreibung von Entrapment—Interaktionen, IKE 2—80, Universität Stuttgart, April (1987).

/7/   G.I.Taylor, The Instability of Liquid Surface when Accelerated in a Direction Perpendicular to their Planes I., Proc.Roy.Soc., 202A,192—196(1950).

/8/   H.W.Emmons, C.T.Chang, and B.C.Watson, Taylor Instabilities of Finite Surface Waves, J.Fluid Mech., 7,177—193(1960).

/9/   K.Halada, K.D.Rzesnitzek, W.A.Kaysser, and G.Petzow, Effect of Superheating on Water Atomization, Powder Metallurgy International, 21,17—21(1989).

/10/   G.Citran, Powder Metallurgy, 29,277—280(1986).

/11/   B.See and G.Johnston, Interactions between Nitrogen Jets and Liquid Lead and Tin Streams, Powder Technology, 21,119—133(1978).

/12/   W.Reinecke and G.Waldman, Shock Layer Shattering of Cloud Drops in Reentry Flight, AIAA 13th Aerospace Sciences Meeting, AIAA Paper, 75—152(1973).

/13/   T.Takeda, in "Proceedings of Lecture Meeting at National Research Institute for Metals", Japan, 13th Nov. 1985, cited in /1/.

/14/   G. de Jarlais, An Experimental Study of Inverted Annular Flow Hydrodynamics Utilizing an Adiabatic Simulation, NUREG/CR3339, ANL—83—44, Argonne NL, Illinois, March 1983.

/15/   A.Lawley, Atomization of Speciality Alloy Powders, J.Metalls, 33,13—17(1981).

/16/   S.A. Miller, Close Coupled Gas Atomization of Metal Alloys, in "Horizons of Powder Metallurgy", (eds.) W.A.Kaysser and W.J.Huppmann, Verlag Schmid, Freiburg, 29—32(1986).

/17/   I.E. Anderson, J.D. Ayers, R.G. Hughes and W.P. Robey, Gas Flow Effects on Atomization Performance, presented at 1984 AIME meeting, Los Angeles (1984).

/18/   G.Rai, E.Laverina and N.J.Grant, Powder Size and Distribution in Ultrasonic Gas Atomization, Journal of Metals, 37,22—26(1985).

/19/   M.Bürger, E.von Berg, S.H.Cho and A.Schatz, Analysis of Fragmentation Processes in Gas and Water Atomization Plants for Process Optimization Purposes; Part I: Discussion of the Main Fragmentation Processes in Gas and Water Atomization Procedures, Powder Metallurgy International, 21(1989), in print.

/20/   D.Bradley, On the Atomization of Liquids by High—Velocity Gases, J.Appl.Phys, D6,1724—1736(1973); D6,2267—2272(1973).

/21/   M.Bürger, E.von Berg, S.H.Cho and A.Schatz, Development of Optimized Atomization Procedures for Metal Powders by Coupling of Numerical Modelling and Experiments, presented at ILASS Europe '89, 5th Annual Conference, July 3—4, Bremen.

/22/   M.Pilch, C.Erdman and A.Reynolds, Acceleration Induced Fragmentation of Liquid Drops, U.S. Nuclear Regulatory Commission,1981, Report number: NUREG/CR—2247.

/23/   P.R.Roberts, Rotating Electrode Process, in "Metals Handbook, Ninth Edition, Volume 7, Powder Metallurgy", (coordinator) E.Klar, American Society for Metals, Metals Park, Ohio 44073, 39—41(1984).

/24/ P.Duwez, R.H.Willens and W.Klement, <u>J.Appl.Phys.</u>, 31,1136–1137(1960).

/25/ T.W.Clyne, R.A.Ricks and P.J.Goodhew, The Production of Rapidly–Solidified Aluminum Powder by Ultrsonic Gas Atomization. Part I: Heat and Fluid Flow, <u>Int.Jour.Rapid Solidification</u>, 1, 59–80(1984).

/26/ H.S.Carlslaw and J.C.Jaeger, Heat of Conduction in Solids, Oxford Univ. Press, (1959).

/27/ R.Patterson II, The Modelling of Rapidly Solidified Powders.

/28/ W.A. Kaysser, K.Rzesnitzek, R.Laag, J.Wachter, and G. Petzow, Rapid Solidification by Optimized Gas Atomization, <u>in</u> "Horizons of Powder Metallurgy", (eds.) W.A.Kaysser and W.J.Huppmann, Verlag Schmid, Freiburg, 84–88(1986).

/29/ W.H.McAdams, "Heat Transmission", McGraw–Hill, Series in Chemical Engineering, New York (1954).

/30/ Gh.Matei, E.Bicsak, Z.Sparchez, and W.A. Kaysser, Melt Disintegration During Ar–Atomization of Ni–Cr–B–Si Alloys, <u>in</u> "Horizons of Powder Metallurgy", (eds.) W.A.Kaysser and W.J.Huppmann, Verlag Schmid, Freiburg, 33–36(1986).

/31/ I.E. Anderson, R.A.Masumura, B.B.Rath, and C.L.Vold, <u>in</u> "Rapid Solidification Processing, Principles and Technologies,III", (ed.) R.Mehrabian, NBS, Washington DC, 178(1983).

/32/ S.K.Bhattacharyya, J.H.Perepezko, T.B.Massalski, Nucleation during Continuos Cooling — Application to Massive Transformations, <u>Acta Metall.</u>, 22,879(1974).

/33/ I.Dustin and W.Kurz, Modeling of Cooling Curves and Microstructures During Equiaxed Dendritic Solidification, <u>Z.Metallkde.</u>, 5,265–273(1986).

/34/ M.Flemings, "Solidification Processing", McGraw–Hill Series in Materials Science and Engineering, New York, 1974.

/35/ J.H. Perepezko, Nucleation in Metallic Melts, <u>in</u> "Keimbildung, Schnelle Erstarrung, Nucleation, Rapid Solidification", (ed.) P.R. Sahm, Gießerei Institut RWTH–Aachen, 14/15 März 1983, 9–33(1983).

/36/ R.Mehrabian, Relationship of Heat Flow to Structure in Rapid Solidification Processing, <u>in</u> "Rapid Solidification Processing", Int. Conference on Rapid Solidification Processing, November 13–16, Reston, USA (1987).

/37/ C.Caesar, U.Köster, R.Willnecker and D.M.Herlach, Comparison of Microstructures and Solidification Behaviour of Melt–Spun Cu–Ni Ribbons and Bulk Undercooled Cu–Ni Alloys, <u>Mat.Sci.Eng.</u> 98,339–342(1988).

/38/ J.C.Baker and J.W.Cahn, Thermodynamics of Solidification, <u>in</u> "Solidification", ASM publication (1971).

/39/ Y.Aoumer, Einfluß der Kornfeinung von G–CuSn10 auf den Erstarrungsablauf und die mechanischen Eigenschaften", Dissertation, TU Berlin (1984).

/40/ W.J. Boettinger, J.H.Perepezko, Fundamentals of Rapid Solidification, <u>in</u> "Rapidly Solidified Crystalline Alloys", (eds.) S.K.Das, B.H.Kear and C.M.Adam, Metallurgical Society, 21–58(1985).

/41/ I.E.Anderson and M.P.Kemppainen, Undercooling Effects in Gas Atomized Powders, <u>in</u> "Undercooled Alloy Phases", (eds.) E.W.Collins and C.C.Koch, Metallurgical Society, 269–285(1986).

/42/ R. Mehrabian, Rapid Solidification, <u>Int.Met.Rev.</u>, 27,185–208(1982).

/43/ J.Poetschke, Spacelab–Nutzung, Status–Seminar 1980 des BMFT, <u>DFVLR–Bericht 80–02</u>, 141–168(KFI–UB 025/80).

/44/ G.M.Pound, Perspectives on Nucleation, <u>Met.Trans.</u>, 16A,487–502(1985).

/45/ J.S.Huang and E.N.Kaufmann, Nucleation, Growth, and Glass Formation in an Electron–Beam Surface–Processed $Cu_{47}Zr_{53}$ Alloy, <u>J.Mater.Res.</u> 3(2),238–247(1988).

/46/ S.R.Coriell and D.Turnbull, Relative Roles of Heat Transport and Interface Rearrangement Rates in the Rapid Growth of Crystals in Undercooled Melts, <u>Acta. Metall.</u> 30,2135–2139(1982).

/47/ C.G.Levi, The Evolution of Microcrystalline Structures in Supercooled Metal Powders, <u>Met.Trans.</u>, 19A,699–708(1988).

/48/ K.Rzesnitzek, Einfluß von Abkühlgeschwindigkeit und Unterkühlung auf das Gefüge von verdüstem Cu– und Cu–Sn–Legierungen, <u>Ph.D.Thesis,</u>TU Berlin (1989).

/49/ J.W.Miles, On the Generation of Surface Waves by Shear Flow, <u>J.Fluid Mech.</u>, 3,185–204(1957); 6,568–582(1958); 7,422–448(1960); 13,433–448(1961).

DISPERSION HARDENING OF A RAPIDLY

SOLIDIFIED COPPER BASED MATRIX

D.Božić and M.Mitkov

Boris Kidrič Institute
Materials Department
P.O. Box 511
11001 Beograd, Yugoslavia

INTRODUCTION

Copper is widely used in industry because of its high electrical and
thermal conductivites, outstanding corrosion resistance, and ease of fabrication. In the pure form, however, it shows a relatively low yield strength and fatique resistance.

Precipitation-hardened alloys offer higher strength, while retaining
good electrical and thermal characteristics at room temperature. The fine
particles of a hard phase are dispersed uniformly in a matrix of copper
with a desirably low content in solution of atoms. These dispersoids act
as barriers to the motion of dislocations, thereby increasing the hardness
of the material. Cu-Be, Cu-Cr, Cu-Co-Be, Cu-Ni-Sn dispersion hardened alloys
are of commercial importance, whereas Cu-Ti, Cu-Ti-B alloys are prospective
ones due mainly to new RSP processes.

Besides there being a variety of RSP techniques, they are all based
on the same general fundamental principles. Basically, rapid solidification
processes include the fast overheating of a melt produced from solid mixtures of the components (the solubility of the alloying elements in the melt
increases significantly with temperature) and subsequent quenching. The
solidification process is characterised by undercooling of the molten phase which results in some desirable microstructural effects such as: grain
size refinement, extremly narrow microsegregation ranges, if any, and the
formation of metastable crystalline and amorphous phases as well.

The very fast oxidation of titanium which has been the main problem
in the Cu-Ti alloy production, has been satisfactorily solved lately. The
annealing of quenched alloys in the two phase regions leads to the formation
of a modular structure and metastable β'-phase precipitation. The mechanical
and electrical properties of some commercial (or promising) Cu-based alloys
are given in Fig.1.[1] The potential characteristics of Cu-Ti alloys are close
to those of Cu-Be alloys, which are mostly used. The addition of Ti leads
to a significant hardening of Cu-based alloys.[2-5] It was the purpose of the
present paper to study the possibility of gas-atomization to produce these
alloys as well as the hardening effects.

Science of Sintering
Edited by D. P. Uskoković *et al.*
Plenum Press, New York

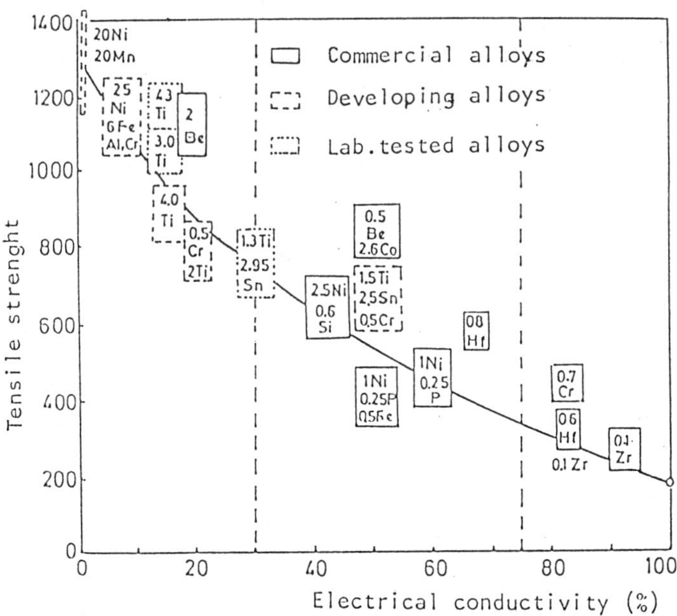

Fig.1. Mechanical and electrical properties of precipitation hardened copper alloys[1].

EXPERIMENTAL

Prealloys, as a starting materials for the atomization process, were vacuum induction melted, with a heating and cooling rates of about 100 K/min. and 200 K/min. respectively. The prealloys were cast in an Ar atmosphere into a copper mold, in order to assure good homogenization of the alloying elements in the master matrix. Figs.2 and 3 show the micro-structures of a homogenized Cu-Ti and Cu-B prealloys.

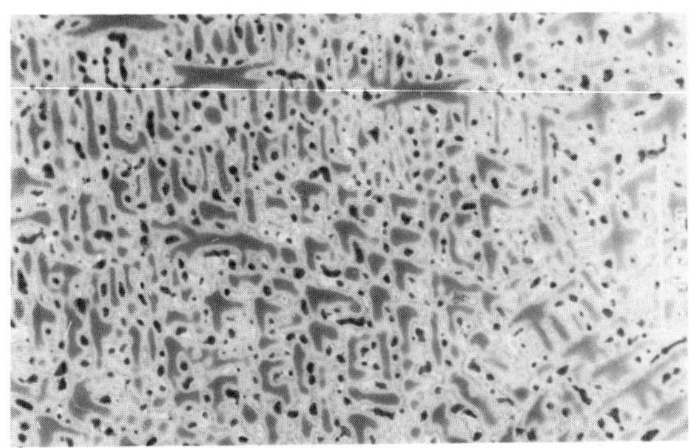

Fig.2. Microstructure of Cu-4%wt. Ti, as cast.

The prealloys were heated at 200 K/min. to 1400°C, homogenized for 5 min, and then Ar atomized 30 bar with a cooling rate between 10^3 K/s and 10^5 K/s.

The atomized powders were thermally treated in the temperature inter-

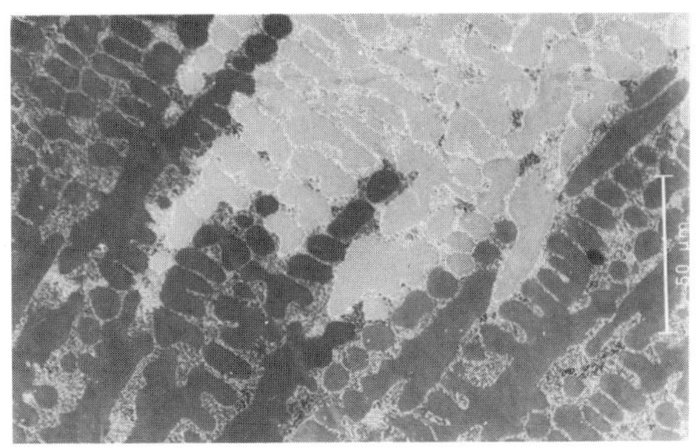

Fig.3. Microstructure of Cu-1,4%wt. B, as cast.

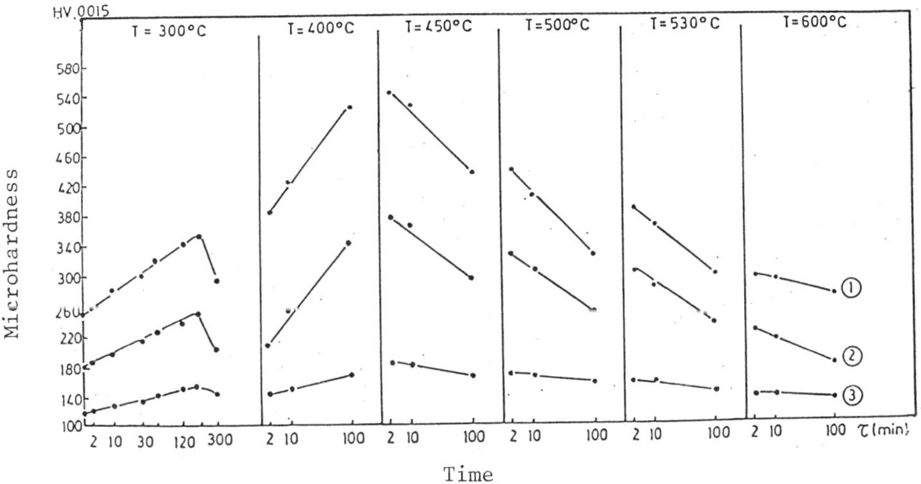

Fig.4. Microhardness of Cu-Ti and Cu-Ti-B alloy powders after thermal
treatment. 1) Cu-1,2Ti-0,01B ($2,9TiB_2$); 2) Cu-3,2Ti; 3) Cu-1,2Ti.

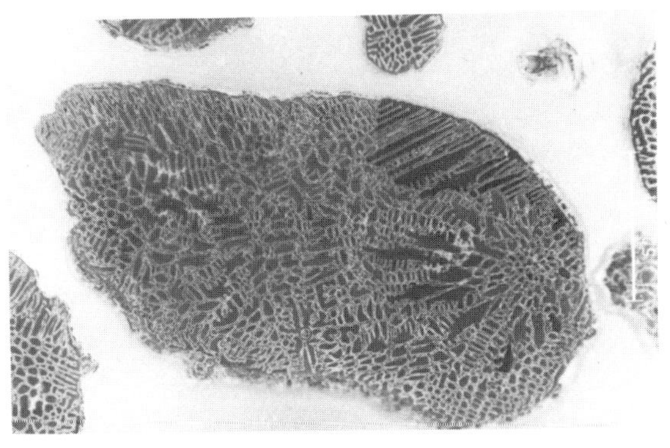

Fig.5. Cu-3,2%wt. Ti powder, as atomized.

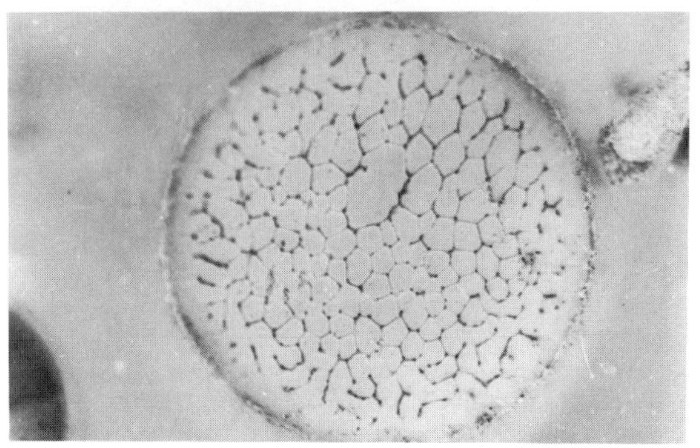

Fig.6. Cu-1,2%wt. Ti powder, as atomized.

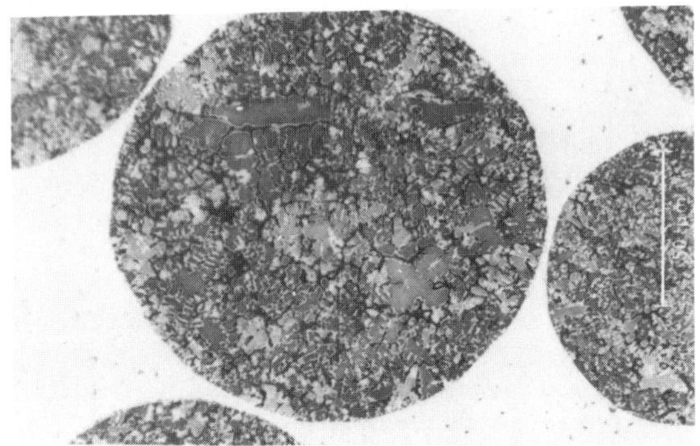

Fig.7. Cu-1,2%wt Ti-0,01%wt B 2,9%wt. TiB$_2$ (Thermal treatment 300°C, 300 min.)

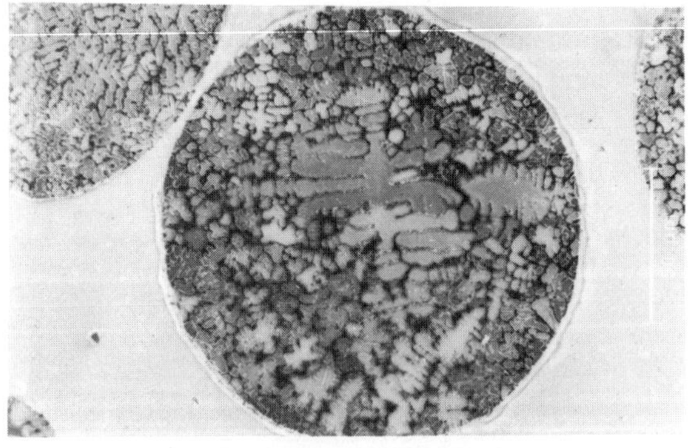

Fig.8. Cu-1,2%wt Ti-0,01%wt B powder 2,9%wt. TiB$_2$ (Thermal treatment 450°C, 100 min.)

val from $300^{\circ}C$ to $600^{\circ}C$ for different times in hydrogen. The microhardness of the thermally treated alloys is shown on Fig.4.

Figs.5-8. Microstructures of Cu-Ti and Cu-Ti-B powders.

DISCUSSION

The atomization of Cu-Ti-B alloys results in powders with a dendritic structure (Figs.2 and 3) with enriched solid solutions of Cu-Ti and Cu-B between the dendritic arms. The microstructure is characterised by cluster free homogeneous distribution of the alloying elements. The Cu-Ti prealloy shows the presence of numerous second phase particles, due to the small cooling rate with presumably Cu_4Ti. The dendritic morphology dominates the structure which is combined in some cases with cellular type areas (Fig. 6). The primary and secondary formed dendritic arms coarsen during ageing (Figs. 7-8) due to diffusion processes. The enlargement of the areas consisting of Cu rich solid solution occurs simultaneously with the formation of a second phase at about $420^{\circ}C$ which is presumably Cu_4Ti.

The given hardening curves, showing most likely the effect of spinodal decomposition, are typical for Cu-Ti containing up to 5% of Ti. The general trend of the curves is maintained if B is additionaly present (Fig.4). The lower the titanium concentration in the alloy, the smaller the hardening effect due to the lower quantity of precipitated phase in the interdendritic and intercellular regions.

The first maximum appearing at $300^{\circ}C$ is most likely due to the preasence of a modulated structure. As shown by,[5] spinodal decomposition involves the continous transformation of a disordered to an ordered Cu_4Ti_m phase, which is in equilibrium whith the fcc Cu solid solution. According to Cahn[4] the hardening effect in the early stage is due to the effect of the internal stresses, the self energy gradient, and the chemical hardening as well. According to Cahn, Kato and Ardell[4], the motion of combined dislocations in the periodic deformations field is strongly dependent on the statistical configuration of obstacles present in the system.

The structure ordering in the Ti-rich regions causes the disapperence of antiphase boundaries resulting from the lattice misfit. The easier dislocation motion through such ordered structure leads to a hardness decrease. This stage is rather short in higher alloyed alloys, which show, again, at $400^{\circ}C$ the hardening effect due to intensified coherent precipitation of Cu_4Ti particles, extending the antiphase boundary formation.[5]. The decrease of microhardness in the investigated alloys, after annealing at $450^{\circ}C$ appears as a consequence of precipitate coarsening and eventualy the discontinuous precipitation of Cu_4Ti.[2,3]

Our experimental results (Fig.9) show higher microhardness values than that reported in the literature[2,3], even for materials subjected to cold plastic deformation.

In the Cu-Ti-B system the modular structure (I maximum), precipitation of Cu_4Tim particles (II maximum) and primary TiB_2 dispersed precipitates contribute to powder particle hardening. The higher microhardness remains in the system, even after a subsequent thermal treatment (Fig.4), independent of the Ti-content.

The similarity of the age hardening curves of both systems (binary and ternary) indicates the absence of secondary formation of fine TiB_2 dispersoids in the Cu-matrix.

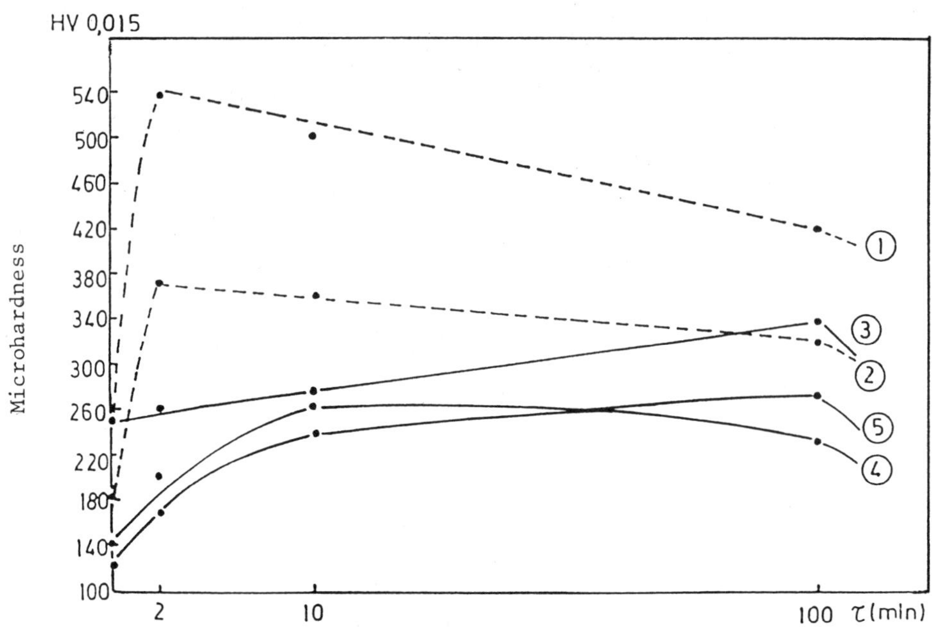

Time

Fig.9. The age hardening effects in Cu-Ti system
1). Cu-1,2Ti-0,01B (2,9TiB$_2$)   2). Cu-3,2Ti
    T= 450°C   T= 450°C

3). Cu-4,3Ti (Zwicker)
    T= 450°C, 70% plastic deformation

4). Cu-4,3Ti (Zwicker)   5). Cu-5 Ti (Heubner)
    T= 550°C   T= 500°C

CONCLUSIONS

1. Supersaturated solid solutions as well as primary TiB$_2$ dispersoids
   in the powder particles are obtained by rapid solidification.

2. The presence of both B and Ti in the solid solution results in
   higher hardening than an equivalent quantity of Ti only.

3. The hardening effects are due to the presence of a modular
   structure of Cu$_4$Ti and TiB$_2$.

4. The hardness properties of atomized Cu-Ti and Cu-Ti-B powders are
   superior compared to the other Cu-Ti systems investigated.

REFERENCES

1. W.Schatt: "Neure Aushärtbare Cu-Ti Legierungen", Planseeberchte
    fur Pulvermetallurgie, Bd. 23, 186-194 p. (1975).

2. U.Zwicker: "Aushärtung und mechanische Eigenschaften von Kupfer-
    Titan-Legierungen", Z.Metallkde, 53, 709-714 p. (1962).

3. U.Heubner, G.Wasserman: "Untersuchungen über das Ausscheidungs-
    und Aushärtungsverhalten übersättigter Kupfer-Titan-Mischkristal-
    le", Z.Metallkde. 50, 152-161 p. (1962).

4. A.J.Ardell: "Precipitation Hardening", Met.Trans. 16A, 2131-2165
    p. (1985).

5. P.Kratochvil, M.Saxlova, J.Pesicka: "The Age Hardening in Copper
    with Low Concentration of Titanium", in "Strength of Metals and
    Alloys", eds. P.Haasen, V.Gerold, G.Kostorz, Proceedings of the
    5th International Conference, Aachen, 687-692 p. (1979).

PROBLEMS OF SINTERING METALLIC ULTRAFINE POWDERS

L.I. Trusov, V.N. Lapovok, and V.I. Novikov

NPO "Krasnaya Zvezda"
Moscow, USSR

It seems that ultrafine media (UFM), i.e., the macroscopic
ensembles of small particles (SP) with size of about 10 nm, have been
systematically studied only for the last twenty years. The results
obtained for these objects testify to the fact that their physical
properties differ essentially, sometimes qualitatively, from the
properties of the same materials in the bulk state I. This distinction
is caused by the dominant role of the surface which noticeably
contributes to the net balance of the system free energy at these
sizes. This contribution becomes comparable with that of the volume,
the reconstruction of the surface atomic structure which initiates the
drastic change in the atomic structure type of a whole SP, the re-
building of elementary excitation spectra in SPs. In addition, the
typical lengths of most of the transfer processes (electroconductivity,
thermoconductivity, galvanomagnetic effects, etc.) become comparable
with the particle sizes in this particular range of dimensions, which
brings to life interesting "size" anomalies in transport processes.
First of all, these processes are structure sensitive, while the defect
types and their densities are specific for UFM. Dislocations were
observed in UFM neither by the electron microscopy, nor by the positron
annihilation method. It is of no surprise, since the sizes of the
dislocation loops for typical metals are one order-of-magnitude larger
than the particle size. In contrast, SPs are liable to twinning and
reveal different stacking faults. This fact is important, since most of
the processes analogous to recrystallization and sintering develop in
the foreground of the initial defect structure present in the system.
In particular, this implies that the primary recrystallization is not
typical of UFMs.

The distinction of UFMs from the usual pressed pellets and
polycrystals is characterized by the fact that in UFMs the ratio of the
total area of the boundaries to the UFM volume is 4-5 orders of
magnitude larger than, say, in a polycrystal with the grain size of $\sim 10^2 \mu m$.
Thus, the boundaries represent the main defect type in UFMs.

It is interesting, that in UFMs the boundaries between SPs
analogous to the grain boundaries in polycrystals are rather irregular,[1]
so that both the volume fractions of the free surface and the integrain
boundaries are comparable and the difference between the boundary and
surface energies is reduced to zero. The effects of zone isolation[2]

that influence the ratio between the densities of the boundaries of different types are aggravated by the fact that besides the statistical causes in UFM there are dynamic causes ensuring direct attraction of SPs[3] thus bringing about aggregate formations. There is experimental evidence[4] that in UFMs of some metals with SP sizes $1 \sim 10$ nm there are aggregates with $L \sim 1$ μm. The pore size in these powder media is of the same order. Since all the fundamental relations of the physics of sintering contain the initial particle size, one may expect the decrease of this dimension to cause the intensification or, on the contrary, the suppression of the corresponding mass flows in UFMs compared to the coarse-grain systems by many orders of magnitude. Besides, the type of the dependence on the particle size is different for different mass-transport mechanisms. Therefore, in principle, for one and the same material and identical external conditions the change of the mass-transfer mechanism is possible only with the change of dispersion. We shall give some experimental examples in confirmation of this assumption. Nonetheless, note that of greatest interest are the size effects which manifest themselves specifically only in the size range typical of UFMs.

One of the most important qualitative distinctions of UFMs from the coarse-grain systems lies in the fact that in UFMs the recrystallization and sintering processes are mutually stipulated and develop simultaneously at the temperature of $0.1 - 0.3$ T_{melt}, whereas in the coarse-grain systems collective recrystallization with a notable rate occurs at the temperature higher than 0.7 T_{melt}, and the most unstable range of sintering temperatures belongs to the interval $0.5-0.7$ T_{melt}.[2] Such a regularity has lately been experimentally proved by the authors for a number of different UFMs of metals.[5-8] When investigating these processes it was experimentally shown that the recrystallization migration of particles in UFMs is accompanied by the formation of essentially non-equilibrium vacancies, whose concentration in UFMs with the typical grain size of ~ 10 nm in the short time period ($\sim 10^2$ s) comprises more than 10^{-4} relative units.[5]

The effect of the formation of vacancies by migrating boundaries[11] in the case of UFMs having grain sizes less than the average diffusion free path of vacancy may reveal specific features of the collective size effect, which may probably relate to the development of the relaxation processes accompanying the recrystallization. In UFMs there are two types of boundaries: the boundaries between SPs in clusters (i) which are analogous to the intergrain boundaries and free boundaries (ii), i.e., the pore boundaries. Two simultaneous processes are connected with the relaxation of these boundaries in UFMs which are determined by the boundary and surface energy-recrystallization and sintering, respectively. Therefore, each SP in an aggregate represents the "recrystallization seed", while the pores may serve as efficient outlets for recrystallization vacancies.

The concentration of non-equilibrium vacancies is determined by the conditions of the self-consistency of the formation of vacancies by the migrating boundaries of type (i) in the course of the recrystallization process, their drainage to the boundaries of type (ii) and the activation of the process of the type (i) boundary migration by the recrystallization vacancies produced by different cluster boundaries. Evidently, such a self-activation of migrating boundaries is possible if the rate of the vacancy drainage is not too large compared to the intensity of their formation, so that a rather high level of the non-equilibrium vacancy concentration is ensured under steady state conditions.

In the conditions of the experiments being discussed the time non-equilibrium vacancies spend in the SP volume $\tau_v \simeq 1^2/D_v$ (where D_v is the coefficient of the vacancy diffusion) comprises $\sim 10^{-3}-10^{-1}$ s at $D_v \sim 10^{-12}-10^{-15} m^2/s$. Insert parameter S describing the length of the type (i) boundaries $S = a/1$, where a is the SP lattice constant. The rate of the boundary length decrease ($dS/dt \equiv \dot{S}$) in the recrystallization process is $\dot{S} = -\alpha V S^2$, where $\alpha \simeq 1/a$. If γ_1 is the probability of the vacancy generation by the migrating boundary normalized per a unit atomic cell, then the rate of generation is $\gamma_1 \dot{S}$. The vacancies with the momentary volume concentration c will drain to the type (ii) boundaries at the rate of $c/\tau_v = \gamma_2 c S^2$, where $\gamma_2 = \frac{D_v}{a^2} K^2$, $K = \frac{1}{L}$.

The density of the recrystallization vacancies in the self-activation mode reaches value c_m

$$c_m = c_0 + \frac{\gamma^2}{\alpha\beta} \left(\ln \frac{\gamma_2}{\alpha\beta} - 1\right) \tag{1}$$

at $S_m = \frac{\gamma_2}{\alpha\beta\gamma_1}$, i.e., at the certain grain size value $1_m = \frac{\alpha}{S_m}$.

Eqns (2) and (6) allow an easy estimation of the time required to get the maximum vacancy density

$$t_m \simeq \frac{1}{\alpha\beta C_m S_m^2} \tag{2}$$

If we make use of the typical parameters values given above for numerical estimations we find $C_m \approx 10^{-4}$, $S_m \approx 10^{-3}$, $t_m \approx 10^2$ s. At $1 > 1_m$ the vacancies are generated beginning with temperatures $\sim 0.3 T_{melt}$ as the result of a mass-scale intensive recrystallization boundary migration, the concentration of these vacancies reaching for a short time interval Δt, which is of the order of several minutes, the values corresponding to the premelting level $c_m \sim 10^{-3} - 10^{-4}$. This state is achieved due to a condition of self-activation which has the effect of increasing the boundary migration rate under the conditions of its interaction with the vacancy flows.

At $1 > 1_m$ the main role is played by the known inclination for the excessive point defect-vacancies toward rapid relaxation. A rather low (compared to the range $1 \sim 1_m$) density of boundaries along with the lower energetical stimulus for the recrystallization do not provide a sufficient rate of the vacancy formation and the self-activation regime is not achieved. This corresponds to the usual situation where the recrystallization develops as the relaxation process in which the average length of the boundaries and the concentration of non-equilibrium vacancies decrease monotonously.

The qualitative change in the kinetics of non-equilibrium vacancies in the recrystallization in UFMs (from monotonous kinetics to an extreme one) occurs at some typical grain size 1_m (for instance, in copper this size is $1_m \simeq 7 \cdot 10^{-8}$ m). This change in the kinetics is the size effect. It is natural that in these non-equilibrium conditions the kinetics of various activation processes may change considerably, in particular for those of mass-transfer processes. So, the non-equilibrium coefficient of self-diffusion may be evaluated as $D_n = cD_v$ which, say, for copper at T = 300 °C yields $D_n = \sim 10^{-4} \cdot 10^{-12} m^2/s \simeq 10^{-16}$ m^2/s. In this case the typical time of merging of the contacting particle is

$$t_n \simeq \frac{kT1^3}{D_n \sigma\omega} \sim 1 \div 10^2 s \tag{3}$$

where ω is the atomic volume, i.e., in the porous structures this may bring about the fast processes of shrinkage or zone isolation. The typical viscosity scale of such a medium impregnated with non-equilibrium vacancies may be estimated using the ideas of the diffusive-viscous flow theory of the polycrystalline materials[14] by means of the self-consistence change of the contacting particles shape

$$\eta = \frac{1^2 kT}{A \omega D_n} \simeq 10^7 \div 10^9 \text{ poise}, \quad A = 10^2 \tag{4}$$

Let's note that the migrating boundaries essentially facilitate the mutual rearrangement of the particles in UFMs, in particular in the process of sintering of porous materials. This is important, since in the under-threshold range where vacancy concentration $c < c_m$ the volume diffusion may not ensure the shrinkage at a sufficiently high rate $(\frac{d}{dt} (\frac{\Delta V}{V}) \sim 10^{-3} \text{ s}^{-1})$. In this case the mutual slippage of powder grains under the action of the surface tension forces will be the main mechanism of active sintering.[15] An important peculiarity of this mechanism's implementation in UFMs is that this process, possessing the features of a collective phenomenon,[16] occurs under the conditions when the migrating boundaries become rather slippery. Taking into account the premelting level of the vacancy concentration in the vicinity of the boundaries we may note the presence of a liquids phase.[1] When the phases are formed in the process of the mutual diffusion of components the time typical of UFMs comprises

$$t_m \simeq \frac{1^2}{D_n} \simeq 1 \div 10s \tag{5}$$

Other mass-transfer processes, e.g., the formation of solid solutions, chemical compounds and pore formation, should also develop in accordance with the essentially non-equilibrium diffusion coefficients values in the time scale of $\Delta t_m \sim 10^2$ s. It was such a kinetics that was typical for the recrystallization sintering of UFMs which was observed in[4-8] (Fig. 1).

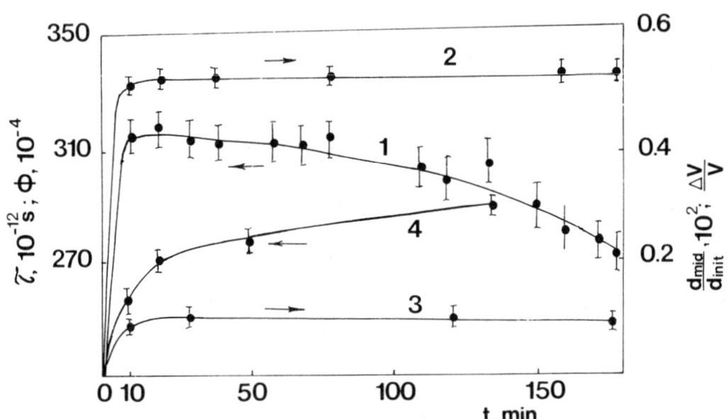

Fig. 1. Dependences of average positron lifetime $\bar{\tau}$ (1), shrinkage $\Delta V/V$ (2), relative particle enlargement d_{aver}/d_{init} (3) upon the duration of isothermal (T= 300°C) annealing of nickel (15 nm) UFM. The dependence of parameter $\emptyset$ ($\frac{2\theta t - 2\theta_{init}}{2\theta_{init}}$) of the peak profile (311) of nickel UFM (4) on the duration of isothermal (T = 300°C) annealing of the nickel (15 nm)-copper (50 nm) mixture UFM.

From the viewpoints of the ideas developed here all the entirety
of experimental data on solid phase transformations in UFMs limited by
the mass-transfer processes becomes physically clear. The process of
the UFM sintering in the given situation develops according to two
possible mechanisms of mass-transfer leading to shrinkage-mutual
slippage of particles (coagulation and liquid-like coalescence).[9] Since
according to the experimental data (Fig.1) the active stage of
recrystallization takes about several minutes, one can hardly expect
any noticeable contribution of the volume self-diffusion processes to
the shrinkage of UFMs in comparison with the values D traditional for
the coarse-grain systems. However, when the concentration of excessive
vacancies reaches values $c \geqslant 10^{-4}$, there appears the possibility of
liquid like merging (coalescence) of particles into larger ones.
Liquid-like coalescence as well as slippage causes the UFM shrinkage.

Experimental results on sintering copper and nickel UFMs testify
to the presence of two stages of the shrinkage process (Fig. 2). The
initial stage of sintering (stage 1) corresponds to the most active
recrystallization (recrystallization self-activation) at which the
stationary concentration of excessive vacancies reaches its peak value
$c \geqslant 10^{-4}$ (see Fig. 1). In the given period the shrinkage develops
simultaneously along the lines of two previously discussed mechanisms-
grain-boundary slippage and liquid-like coalescence. This circumstance
brings about the fact that the sintering activation energy during the
first stage is some effective value determined by the simultaneous flow
of coagulation and powder particle coalescence processes. The energies
of the sintering activation for copper and nickel UFMs during the first
stage were found to be 14·4.18 kJ/gram-atom-for Ni (70 nm), 40·4.18
kJ/gram-atom-for Ni (15 nm), 25·4.18 kJ/gram-atom-for Cu (70 nm) and 48·
4.18 kJ/gram-atom-for Cu (20 nm). The more dispersive the powder, the
more significant the coalescence contribution to the shinkage and,

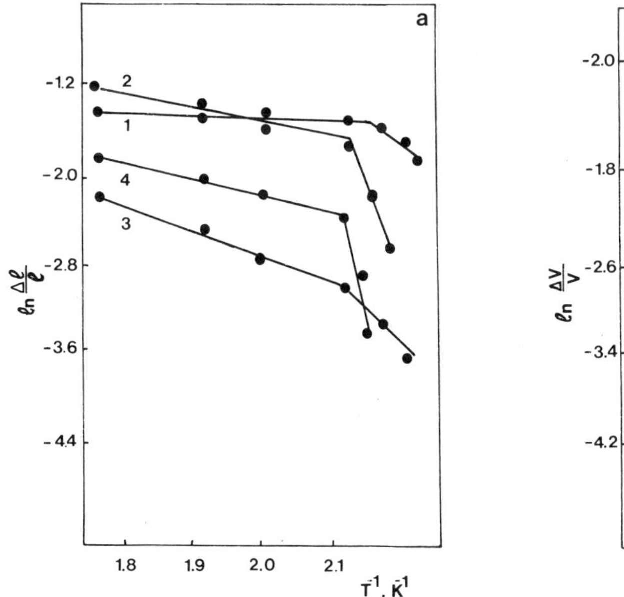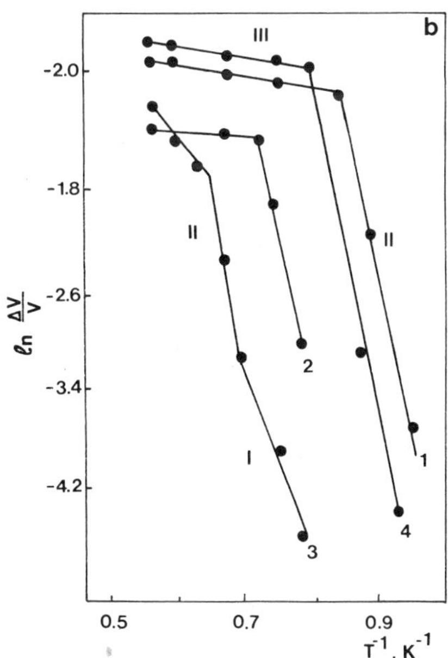

Fig. 2. Temperature dependence of the relative shrinkage of UFM pressed
pellets at sintering: a) (1) –Cu (70 nm); (2) –Cu (20 nm); (3)
–Ni (70 nm), b) (1) –W (40 nm) clad Ni; (2) –W (2 μm) clad Ni;
(3) –W (40 nm); (4) –W (40 nm) with the 0.5 weight %Ni (15 nm)
addition.

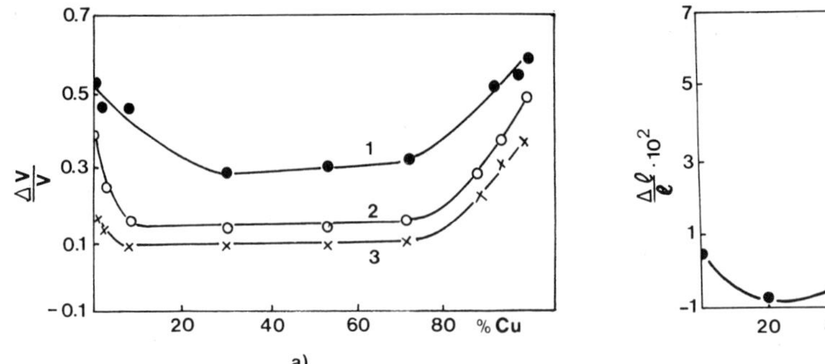
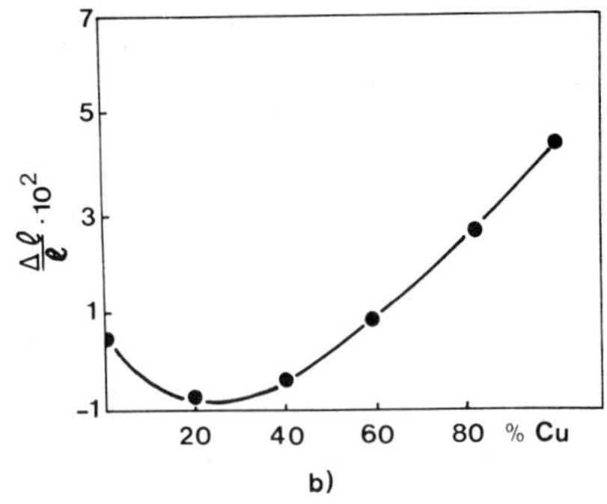

Fig. 3. The concentrational dependence of the shrinkage of Cu and Ni powder mixture: a) UFM (70 nm), (1) $-T = 500°C$; (2) $-T = 300°C$; (3) $-T = 200°C$; t = 20 min., b) coarse-grain powders (50 μm, T = 1000°C, t = 15 min.

hence, the higher the effective values of the sintering activation energy.

Under certain sintering conditions the coalescence may turn out to be the only mechanism that determines the shrinkage (the contribution of the grain-boundary slippage being negligible). In this case the activation energy of the sintering process corresponds to the process of volume self-diffusion, as is the case for the sintering of copper (20 nm) UFMs. Another extreme is posible when the level of the excessive vacancy concentration does not provide the conditions for liquid-like coalescence and, hence, only the coagulation shrinkage process takes place. The analysis of the experimental data on sintering refractory tungsten (40 nm) UFM (Fig. 2b) has shown that a similar situation occurs in the temperature interval 900–1200°C (stage I-32 4.18 kJ/gram-atom). Within the given temperature interval the dominating mechanism of the tungsten UFM shrinkage is the coagulation densification, which is definitely shown by the microstructural research on pressed pellets. And only the transition to a higher sintering temperature T ~ 1200°C brings about a considerable particle growth (~ 5 μm associated with the contribution of the volume diffusion processes (coalescence) (stage II-76·4.18 kJ/gram-atom). A purely coagulation shrinkage was also observed in nickel UFM and in copper-nickel mixture UFM at temperature ~200°C (Fig. 3,4). The study of the kinetics of the UFM sintering process in the case of copper and nickel has shown that after 2-3 minutes of isothermal exposure the shitch-over in the mechanism of densification occurs.[7] In the non-isothermal conditions of sintering the switch-over of the densification mechanism in the copper and nickel UFMs corresponds to the temperature level 220°C while for the tungsten UFM-above 1400°C (stage III). An extremely low energy of the process activation is characteristic of the final stage of sintering. Thus, for the copper and nickel UFMs it comprises 3-5·4.18 kJ/gram-atom (stage II), while for the tungsten UFM – 11·4.18 kJ/gram-atom (stage III).

The sintering mechanism switch-over in the process of the UFM annealing is caused by the change in the regimes of boundary migration. During the first stage period the motion of the boundaries occurs in the mode of secondary recrystallization when the imbalance of the grain-boundary tension of contacting particles serves as the motive force. As mentioned secondary recrystallization is of self-

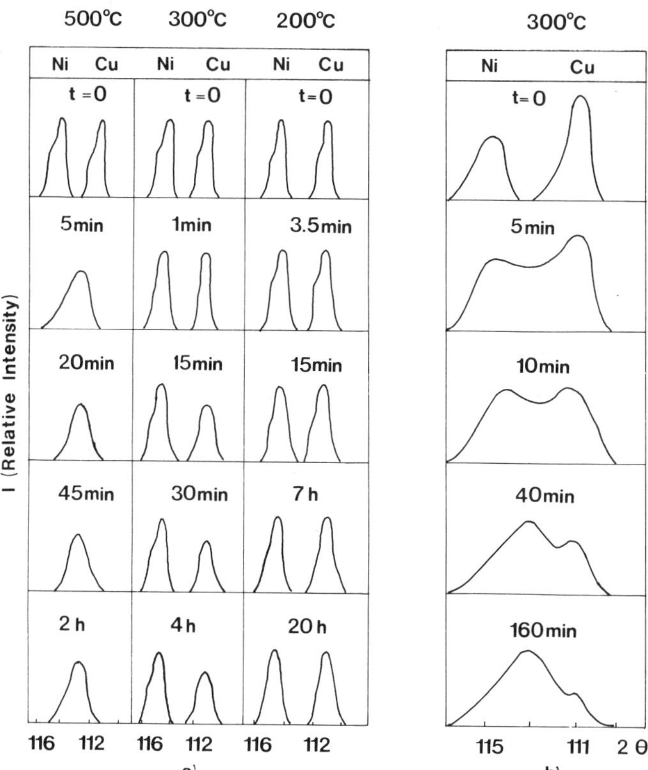

Fig. 4. Diffraction peaks (311) C_oK_α of copper, nickel and their alloys at various times of hydrogen annealing of UFM mixture: a) Cu (70 nm)-Ni (70 nm); b) Cu (50 nm)-Ni(15 nm).

activation type that leads to the possibility of momentary essential oversaturation of UFM with excessive vacancies ($c \geqslant 10^{-4}$). The decrease in the activation energy of the sintering process during the final stage is associated with the annealing of vacancy clusters formed during the first stage which lead to the considerable over-saturation of boundaries with excessive vacancies and to the decrease of the activation energy of their migration. At longer annealing times ($t \geqslant 20$-30 min) the UFM shrinkage is absent under the isothermal sintering conditions, which corresponds to the complete annealing of clusters prior to the given time.[8]

The above results show[17-20] that in UFMs a certain "rehabilitation" of the diffusive mass-transfer is possible, since a powerful source of dynamic vacancies, that acts during a period of time sufficient for the transformations limited by the mass-transfer rate, appears in these objects.

REFERENCES

1. I. D. Morokhov, L. I. Trusov, V. N. Lapovok, Physical phenomena in ultrafine media, Moscow, "Energoatomizdat", 1984, 224 p.
2. V. V. Skorohod, S.M. Solonin, Physical-metallurgical fundamentals of powder sintering, Moscow, "Metallurgia", 1984, 159 p.

3. I. D. Morokhov, L. I. Trusov, S. P. Tchizhik, Ultrafine metallic
    media, Moscow, "Atomizdat", 1977, 214 p.
4. L. I. Trusov, V. I. Novikov, Yu. A. Lopukhov, V. N. Lapovok,
    Recrystallization in ultrafine systems, in book: Physics and
    chemistry in ultrafine systems, Moscow, "Nauka", 1987, pp. 67
    -74.
5. V. N. Lapovok, V. I. Novikov, S. V. Svirida, A. N. Semenikhin, L.
    I. Trusov, Formation of non-equilibrium vacancies in the
    recrystallization process of ultrafine nickel powder, Solid
    state physics, 1983, $\underline{25}$,v.6,pp.1846-1848.
6. V. I. Novikov, L. I. Trusov, V. N. Lapovok, T. N. Heleischvili, On
    the mechanism of the low-temperature diffusion activated by a
    migrating boundary, Solid State Physics, 1983, $\underline{25}$,v.12,pp.3696
    -3698.
7. V. I. Novikov, L. I. Trusov, V. N. Lapovok, T. P. Heleischvili,
    Recrystallization mechanism of ultrafine powder sintering,
    "Poroshkovaya metallurgia",1984, no.5, pp.28-34.
8. V. I. Novikov, L. I. Trusov, V. N. Lapovok, T. P. Heleischvili,
    Peculiarities of mass-transfer processes in ultrafine media,
    "Poroshkovaya metallurgia", 1983, no.7, pp.39-46.
9. V. I. Novikov, L. I. Trusov, V. N. Lapovok, T. P. Heleischvili,
    Peculiarities of growth of ultrafine powder particles at
    sintering, "Poroshkovaya metallurgia", 1984, no.3, pp.30-35.
10. J. Estrin, K. Lucke, Grain boundary motion-II. The effect of
    vacancy production on steady state grain motion, Acta Met.,
    1981, $\underline{29}$, pp.781-789.
11. V. G. Gryaznov, V. I. Novikov, L. I. Trusov, Recrystallization
    size effect, "Poverkhnost.", 1986, no.1,pp.133-139.
12. M. A. Schtremel, Alloy strength, part 1, Lattice defects, Moscow,
    "Metallurgia", 1982, 227 p.
13. Ya. E. Geguzin, Yu. S. Kaganovski, Diffusion processes on crystal
    surface, Moscow, "Energoatomizdat", 1984, 128 p.
14. F. R. N. Nabarro, Report on a Conf. Strength of Solids Phys.,
    London, 1948, p. 75.
15. Ya. E. Geguzin, Yu. I. Klintchuk, Mechanism and kinetics of
    initial stage of solid-phase sintering of pressed pellets from
    the powders of crystalline materials ("activity" at sintering),
    "Poroshkovaya metallurgia", 1976, no.7, pp.17-25.
16. V. G. Gryaznov, L. I. Trusov, V. N. Lapovok, V. I. Novikov, E. V.
    Knyazev, T. P. Heleischvili, M. V. Kvernadze, Collective
    effects at diffusive interaction in ensembles of small metallic
    particles, Solid State Physics, 1983, $\underline{25}$, v. 8, pp. 2290-2298.
17. V. N. Lapovok, V. I. Novikov, S. V. Svirida, A. N. Semenikhin, L.
    I. Trusov, Formation of non-equilibrium vacancies in nickel
    ultrafine powder at plastic flow under pressure, "Fizika
    metallov i metallovedenie", 1984, $\underline{57}$, v.4, p.718-721.
18. V. I. Novikov, S. V. Svirida, L. I. Trusov, V. N. Lapovok, V. G.
    Gryaznov, T. P. Heleischvili, Activation of diffusion processes
    and phase transformations in ultrafine media at plastic
    deformation, "Metallofizika", 1984, 6, no.3, 114-115.
19. V. I. Novikov, P. F. Relushko, L. I. Trusov, Yu. A. Lopukhov,
    Mössbauer study of polymorphous transformations in ultrafine
    ferrous powder at plastic deformation, Solid State Physics,
    1984, $\underline{26}$, v. 11, pp.3447-3449.
20. V. I. Novikov, S. V. Svirida, L. I. Trusov, V. N. Lapovok, V. P.
    Fedotov, Yu. A. Voskresenski, T. P. Heleischvili, Initiating
    diffusive mass-transfer and phase transformations in ultrafine
    media at recrystallization, "Metallofizika" 1984, 6, no.4,
    pp.97-99.

SINTERING OF COPPER ULTRAFINE POWDERS

Yoshio Sakka, Tetsuo Uchikoshi and Eiichi Ozawa*

National Research Institute for Metals
3-12, Nakameguro-2, Meguro-ku, Tokyo 153, Japan

* Present Address: K. K. L'Air Liquide Laboratories
5-9-9 Tokodai, Tsukuba-shi, Ibaraki, 300-26, Japan

ABSTRACT

Sintering characteristics were examined for the Oxidized (subjected to slow oxidation treatment) Cu and the Fresh (not subjected to slow oxidation treatment) Cu ultrafine powders(UFPs). Gas desorption, reduction and densification of two kinds of Cu UFPs during sintering in a vacuum or in a hydrogen atmosphere were discussed.

INTRODUCTION

Metal ultrafine powders (UFPs) have been expected as one of the advanced materials because of their extremely large specific surface area and high reactivity. Although a great deal of efforts have been given to prepare ultrafine particles with uniform particle size,[1,2,3] the studies of the utilization of metal UFPs are limited. The fabrication of porous materials by consolidating UFPs seems to be one of the best utilizations of their large specific surface area. In this case, the initial sintering characteristics of UFP are important factors which control the structure of the porous materials.

Metal UFPs are so reactive that slow oxidation treatment is generally adopted to avoid the rapid oxidation reaction and to stabilize the surface of them.[4,5] Exposure of metal UFP to open air is accompanied by appreciable oxidation and by adsorption of various gases.[6] Sintering characteristics are considered to be greatly affected by the adsorbed gases and oxide phases.

In the present study, to clarify the effect of slow oxidation treatment on the metal UFP characteristics, sintering process, gas desorption behavior and structure changes were examined by various techniques.

Science of Sintering
Edited by D. P. Uskoković *et al.*
Plenum Press, New York

EXPERIMENTAL

Two kinds of Cu UFPs were prepared by Vacuum Metallurgical Co. Ltd. using a gas evaporation method. One is subjected to slow oxidation treatment (hereafter referred to as Oxidized Cu UFP) and the other is not subjected to such treatment (hereafter referred to as Fresh Cu UFP). The impurity contents of oxygen and carbon are summarized in Table 1 together with the specific surface area. Photo 1 shows the SEM photographs of the as-received Cu UFPs.

Table 1. Oxygen and Carbon impurity contents and specific surface areas of the Oxidized and Fresh Cu UFPs.

Sample	Oxygen (mass%)	Carbon (mass%)	Specific surface areas (m^2/g)
Oxidized Cu	9.9	0.10	6.7
Fresh Cu	1.7	0.19-0.25	5.1

Photo 1. SEM photographs as received (a)Oxidized Cu UFPs and (b) Fresh Cu UFPs.

Specific surface area measurement, continuous dimensional change measurement and scanning electron microscopy (SEM) observation were carried out to examine sintering characteristics.[7,11] The both powders were pressed (300 kg/cm^2) into a disc of approximately 6 mm dia. and 4 mm thick. The green density was approximately 30 % of the theoretical density. The continuous dimensional change of the pellet was measured by using a dilatometer in a vacuum (ca. 10^{-2}Pa) and in a stream of hydrogen (flow rate: 800 ml/min) at a constant heating rate of 5 K/min. The Cu UFPs were treated in a stream of nitrogen or argon gas to avoid contamination by contact with air.

The desorption of the adsorbed gases was monitored with a quadrupole mass spectrometer.[12,13,14] Temperature programmed desorption measurements were carried out after degassing the powder in a high vacuum ($<10^{-6}$ Pa). The thermogravimetry (TG), differential thermal analysis (DTA) and X-ray diffraction measurements were also performed to examine the structure changes of the Oxidized and the Fresh Cu UFP.

RESULTS

Fig. 1(a) shows the linear shrinkage of the Cu UFPs in a vacuum and a hydrogen atmosphere at heating rate of 5 K/min. Great difference in shrinkage curves is observed depending on atmosphere and on surface conditions of Cu UFPs. In a vacuum, the Oxidized Cu UFPs slightly shrank at approximately 450 K and expanded at temperatures between 500 and 700 K , and began to densify at temperature above 700 K. However, the Fresh Cu UFPs began to densify at a temperature above 540 K after expandingslightly at temperatures between 500 and 540 K. On the other hand, in a hydrogen atmosphere, the Oxidized Cu UFPs densify at a temperature above 440 K, and the Fresh Cu UFPs at a temperature above 410 K. These differences come from the oxidation levels of Cu UFPs as discussed later. Fig. 1(b) shows the dimensional changes after isothermal holding at fixed temperatures for 7.2 ks. And Photo 2 shows the typical SEM photographs of the fractured surface of the sintered Cu UFPs.

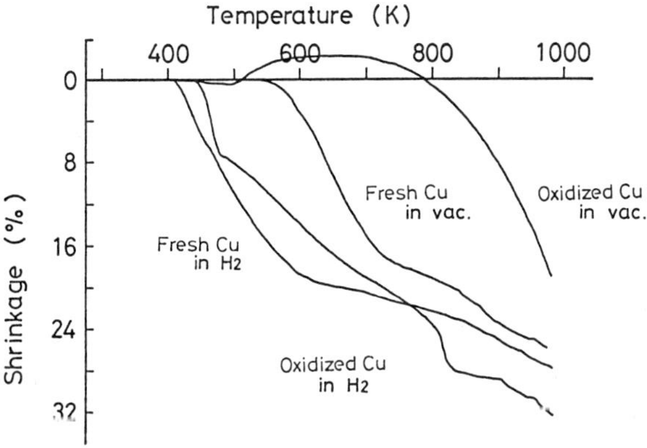

Fig. 1(a) Linear shrinkage of the Oxidized and the Fresh Cu UFPs at heating rate of 5 K/min in a vacuum and in an H_2 atmosphere.

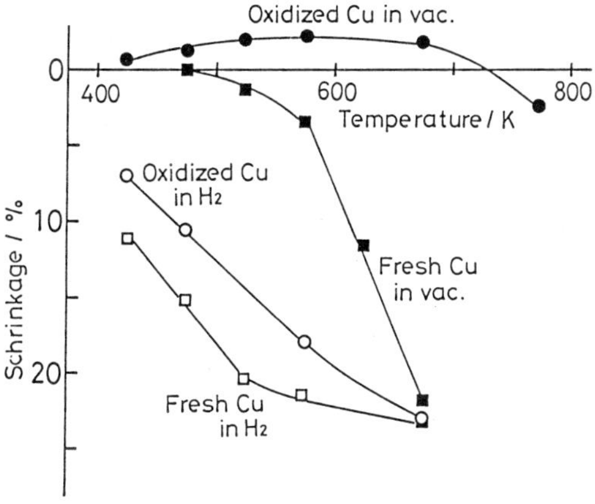

Fig. 1(b) Linear shrinkage of the Oxidized and the Fresh Cu UFPs after isothermal holding at fixed temperatures for 7.2 ks.

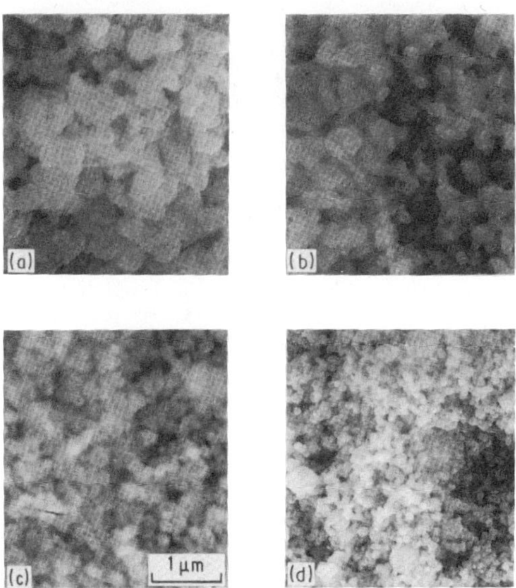

Photo 2. SEM photographs after heating for 7.2 ks at 473 K in a hydrogen atmosphere for (a) Oxidized Cu UFPs and (b) Fresh Cu UFPs; (c)at 673 K in a
vacuum for Oxidized Cu UFPs; and (d) at 573 K in a vacuum for Fresh Cu UFPs.

    Fig. 2 shows the specific surface areas and weight losses after out-
gassing at a fixed temperature for 7.2 ks. Here, the decrease in weight
after outgassing at room temperature is considered to result from the
desorption of physisorbed gases. The specific surface area of the Oxidized
Cu UFPs had a maximum value at approximately 425 K and decreased at temperatures above 425 K. The specific surface area of the Fresh Cu UFPs decreased
significantly at temperatures above 530 K.

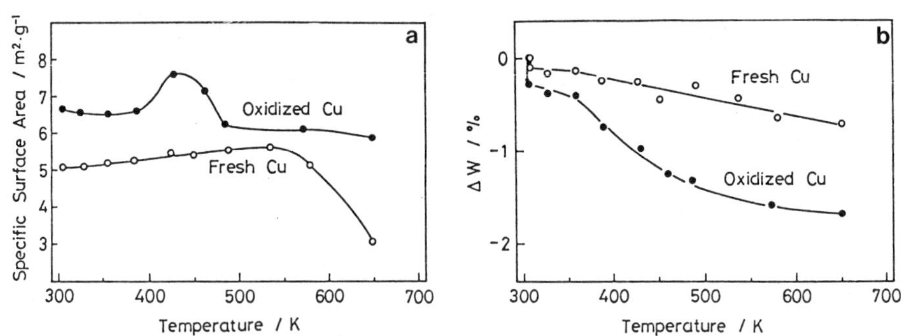

Fig. 2 Temperature dependences of the specific surface areas (a), and
  the weight losses (b) after outgassing at the fixed temperatures
  for 7.2 ks. The closed circles denote the data of the Oxidized Cu UFPs
  and the open circles denote the data of the Fresh Cu UFPs.

The desorbed gases detected by quadrupole mass spectrometer were mainly H_2O and CO_2 gases. Fig. 3 shows the thermal desorption spectra of H_2O and CO_2 gases at a heating rate of 5 K/min in a high vacuum. The amount of gas desorption of the Oxidized Cu UFPs is extremely larger than that of the Fresh Cu UFPs. This result corresponds to the weight losses shown in Fig. 2(b). One peak is observed of H_2O at approximately 400 K for the Fresh Cu UFPs, while another one peak is observed at approximately 440 K for the Oxidized Cu UFPs. The desorption spectra of CO_2 have several peaks, which are considered to be due to surface reactions. The interpretation of these results has proved difficult, largely because of the complexity of the oxidation process[8,9,10] as described later.

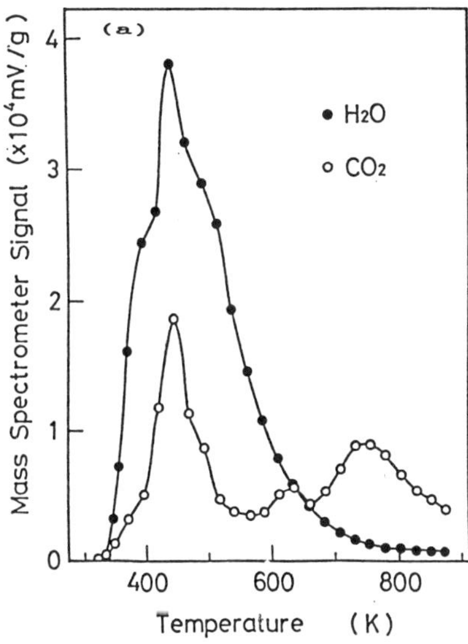

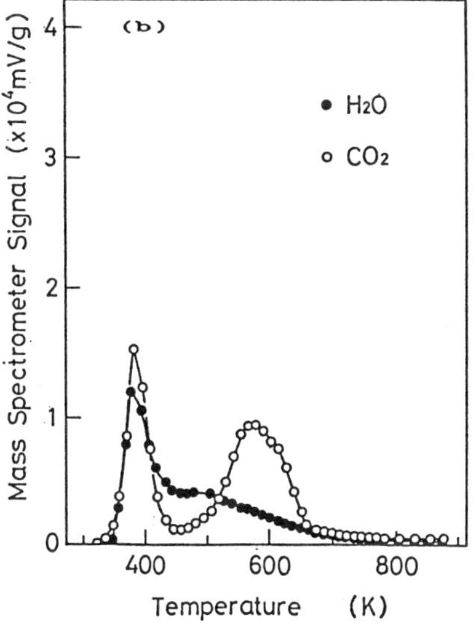

Fig. 3 Thermal desorption spectra of H_2O and CO_2 gases at a constant heating rate of 5 K/min.
(a) Oxidized Cu UFPs.
(b) Fresh CU UFPs.

Fig. 4 shows the TG and DTA curves of the Oxidized Cu UFPs in a He and an H_2 atmosphere. Here, the sold lines show the data obtained when the samples were heated at 5 K/min in an H_2 atmosphere and the dotted lines show the data when the samples were heated at 10 K/min in a He atmosphere. DTA curves show the broad exothermic peak at approximately 530 K in a He atmosphere, while large exothermic peak appears at approximately at 440 K in an H_2 atmosphere. The former exothermic reaction corresponds to the oxidation reaction, while the latter exothermic reaction corresponds to the reduction process as seen from the following X-ray diffraction patterns.

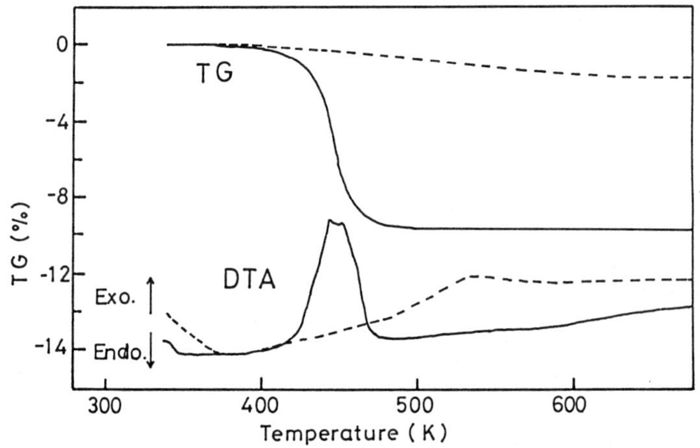

Fig. 4 TG and DTA curves of the Oxidized Cu UFPs. The solid lines denote the data heating at 5 K/min in an H_2 atmosphere and the dotted lines denote the data heating at 10 K/min in a He atmosphere.

As-received Cu UFPs contain small amounts of Cu_2O and CuO phases, as shown in Fig. 5. The Cu_2O phase increases as heating at temperatures above 450 K in a vacuum, especially for the Oxidized Cu UFPs. Since the Cu UFPs heated at temperature below 673 K in an H_2 atmosphere were very reactive, the slow oxidation treatment was carried out before exposed in air. Fig. 6 shows the X-ray intensity ratio of $Cu_2O(111)$ to Cu(100) as a function of heating temperature in a vacuum, where the open circles denote the data for the Oxidized Cu UFPs and the closed circles denote those for the Fresh Cu UFPs. A large amount of Cu_2O is produced when Cu UFPs are heated at temperatures above 423 K, especially for the Oxidized Cu UFPs.

DISCUSSION AND CONCLUSIONS

Sintering in a vacuum

The sintering in a vacuum can be classified into three regions. In Fig. 1(a), we define a region where the slight shrinkage is observed as the first region, a region where the expansion observed as the second region,

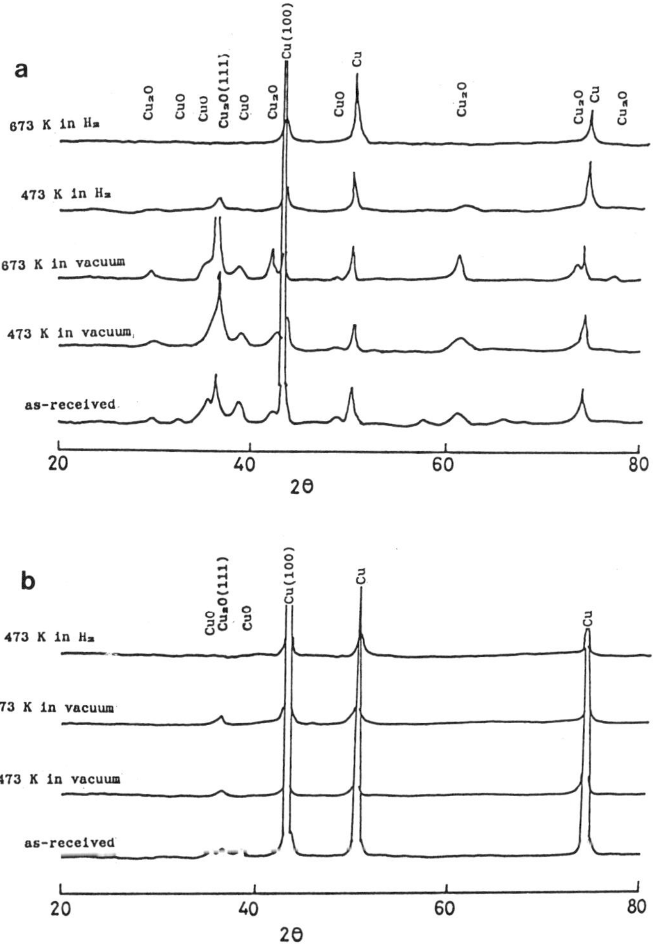

Fig. 5 X-ray powder diffraction patterns of (a) the Oxidized Cu UFPs and (b) the Fresh Cu UFPs.

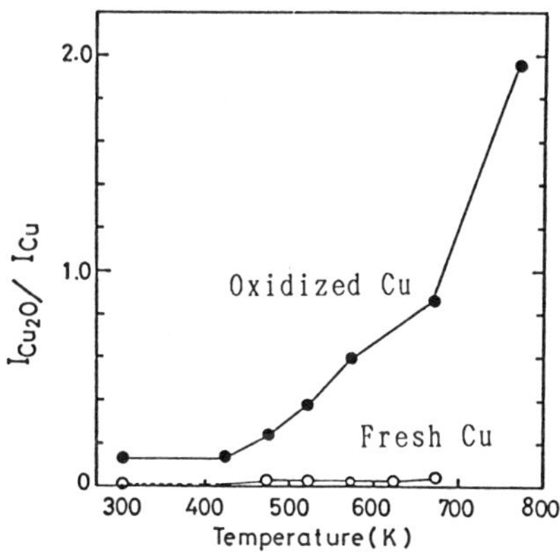

Fig. 6 X-ray intensity ratio of $Cu_2O(111)$ peak to $Cu(100)$ peak after heating the fixed temperature for 7.2 ks in a vacuum. The closed marks denote the data of the Oxidized Cu UFPs and the open marks denote the data of the Fresh Cu UFPs.

and a region where densification observed as the third region. The temperature ranges of these regions are different between the Fresh and the Oxidized Cu UFPs, as shown in Table 2, where the temperatures were the values obtained at heating rate of 5 K/min.

Table 2.   Temperature range of three regions.

Cu UFP	First Region (K)	Second Region (K)	Third Region (K)
Oxidized	500	500-700	700
Fresh	500	500-540	540

In the first region, the specific surface area increases for the Oxidized Cu UFPs, as seen in Fig. 2(a). The cause of the increase in specific surface area of metal powders is generally considered to be due to gas desorption or oxidation. Since the amount of Cu_2O is constant at the temperature range from 300 K to 423 K as seen in Fig. 6, the cause of increase in surface area can not be explained by the oxidation. As seen in Fig. 3, the amount of desorption of H_2O and CO_2 gases is observed to be large especially for the Oxidized Cu UFPs. Therefore, the increase in the surface area is considered to result from the gas desorption. This consideration may be supported by the increase in number of micropores.[7] It may be considered that the shrinkage in Fig. 1 is not due to sintering but to the rearrangement of the UFPs accompanied with the gas desorption. This is because the specific surface area did not decrease in the temperature range where this shrinkage occurred.

In the second region, remarkable expansion is observed for the Oxidized Cu UFPs, whereas slight expansion for the Fresh Cu UFPs as seen in Fig. 1. This region corresponds to a broad exothermic peak observed in Fig. 4. As seen in Fig. 6, the Cu_2O phase increases with increasing temperature. This leads us to consider that the expansion is due to the oxidation reaction of Cu UFPs. The Oxidized Cu UFPs expand remarkably owing to a great amount of Cu_2O produced. On the other hand, since the Fresh Cu UFPs have a small amount of Cu_2O, the expansion is slight. The origin of oxygen to cause the oxidation reaction is considered to be supplied by the adsorbed gases (OH(ad), O(ad), etc.) on the surface of Cu UFP. The water is adsorbed as OH on the surface of Cu UFP. The desorption of water is expressed as the following reaction,

$$2(OH)(ad) \rightarrow H_2O(gas) + O(ad).$$

The produced O(ad) accompanied with H_2O desorption is considered to further contribute to the oxidation reaction on the surface of Cu UFP.

In the third region, decrease in the specific surface area and shrinkage simultaneously occurs. Therefore, sintering by lattice diffusion or grain-boundary diffusion occurs in this region. The temperature at which densification occurs for the Fresh Cu UFPs is lower than that for the Oxidized Cu UFPs. This difference arises from the difference of oxidation level. The Oxidized Cu UFPs have a large amount of oxide phase (Cu_2O) in comparison with the Fresh Cu UFPs, as seen in Fig. 6. In Cu UFPs, the densification rate is retarded by the oxide phase. The grain growth is also hindered by the oxide phases. The apparent grain growth is observed for the Fresh Cu UFPs heated at 623 K for 7.2 ks, while this is not observed significantly for the Oxidized Cu UFPs heated at 673 K for 7.2 ks in a vacuum as shown in Photo 2.

Sintering in an H_2 atmosphere

The densification in an H_2 atmosphere occurs at much lower temperature than that in a vacuum, as seen in Fig. 1. It is also seen that the densification of the Fresh Cu UFPs occurs at lower temperature than that of the Oxidized Cu UFPs. These results show that the oxide phases depress the sintering of metal UFPs significantly. Almost all oxide phases are reduced into metal on heating at temperatures above 473 K in an H_2 atmosphere, as seen in comparison of TG data in Fig. 4 with the oxygen content shown in Table 1. The large exothermic reaction is considered to occur according to the following reactions,

$$Cu_2O \rightarrow 2Cu + 1/2O_2,$$

$$1/2O_2 + H_2 \rightarrow H_2O.$$

Apparent grain growth was observed for both Cu UFPs when they are heated at 473 K for 7.2 ks in H_2 atmosphere as shown in Photo 2. It is reported that the densification and grain growth of Au UFPs occurs at 383 K.[12] In this way, densification and grain growth in the clean metal UFPs is considered to occur at lower temperatures.

ACKNOWLEDGMENTS

The authors wish to thank S. Ohno, K. Halada, Y. Muramatsu and A. Oguchi for their helpful suggestions and discussion.

REFERENCES

1. S. Yatsuya, K. Mihama and R. Uyeda, A New Technique for Preparation of Extremely Fine Metal Particles, Japan J. Appl. Phys., **13**: 749 (1974).

2. E. Fuchita, M. Oda and S. Kashu, Properties and Applications of Ultra Fine Powders, Proc. 7th Int. Conf. Vacuum Metal. The Iron and Steel Institute of Japan, Tokyo, p973 (1982).

3. S. Ohno and M. Uda, Generation Rate of Ultrafine Metal Particles in "Hydrogen Plasma-Metal" Reaction, Nippon Kinzoku Gakkaishi, **48**:640 (1984).

4. S. Ohno and M. Uda, Room Temperature Oxidation of Ultrafine Iron Particles at Low Oxygen Partial Pressures, Nippon Kagaku Kaishi, 924 (1984).

5. S. Ohno, Y. Sakka, H. Okuyama, K. Honma and M. Ozawa, Synthesis and Characterization of the Mixed Ultrafine Particles of Ni and TiN, Proc. MRS International Meeting on Advanced Materials, Vol. 3: (1989)(in press).

6. Y. Sakka, T. Uchikoshi and E. Ozawa, Effect of the Slow Oxidation Treatment on the Characteristics of the Ultrafine Metal Powders, ibid., (in press).

7. Y. Sakka, T. Uchikoshi and E. Ozawa, Gas Evolution and Densification during Sintering of Ultrafine Copper Powders, <u>Nippon Kinzoku Gakkaishi</u>. **53**:422 (1989).

8. P. Hollins and J. Pritchard, Interaction of CO Molecules Adsorbed on Oxidized Cu(111) and Cu(110), <u>Surface Sci.</u>, **134**: 91 (1983).

9. A. Spilzer and H. Lueth, The Adsorption of Water on Clean and Oxygen Covered Cu(110), <u>ibid.</u>, **120**: 376 (1982).

10. K. Bange, D. E. Grider, T. E. Madey and J. K. Sass, The Surface Chemistry Of H_2O on Clean and Oxygen-Covered Cu(110), <u>ibid</u>, **136**: 38 (1984).

11. Y. Sakka, Reduction and Sintering of Ultrafine Copper Powders, <u>J. Mater. Sci. Letters</u>, **8**: 273 (1989).

12. Y. Sakka, T. Uchikoshi and E. Ozawa, Low-Temperature Sintering and Gas Desorption of Gold Ultrafine Powders, <u>J. Less-Common Metals</u>, **147**: 89 (1989).

13. T. Uchikoshi, Y. Sakka and E. Ozawa, Sintering and Gas Release of Ag Ultrafine Powders, <u>Nippon Kinzoku Gakkaishi</u>, **53**(6):(1989)(in press).

14. Y. Sakka, T. Uchikoshi and S. Ohno, Gas Desorption Characteristics of Ag Ultrafine Powders Synthesized by Different Methods, <u>Proc. Japan Congress on Materials Research</u>, **32**: (1989)(in press).

SYNTHESIS AND CHARACTERIZATION OF THE MIXED AND THE COMPOSITE
Ni-TiN UTRAFINE PARTICLES

Yoshio Sakka, Satoru Ohno, Hideo Okuyama, and Masaya Ozawa

National Research Institute for Metals

Advanced Powder Processing Group

3-12, Nakameguro-2, Meguro-ku

Tokyo 153, Japan

ABSTRACT

Mixed Ni-TiN ultrafine particles (UFPs) were synthesized by an arc melting of Ni and Ti, separately, in an N_2-H_2-Ar atmosphere, and mixing the produced Ni and TiN UFPs in the gas phase. Homogeneous distribution of spherical Ni UFPs and cubical TiN UFPs were confirmed by TEM photographs. Composite Ni-TiN UFPs were synthesized by an arc melting Ni-Ti alloy in an N_2-H_2-Ar atmosphere. The composite Ni TiN UFPs were in the shape of the dumbbell which consisted of Ni UFPs on both sides. Sintering and gas desorption characteristics of the mixed and the composite Ni-TiN UFPs were examined by means of BET specific surface area measurements, dilatometric measurements, and temperature programmed desorption measurements. The results were discussed by comparison of those of the Ni or the TiN UFPs.

INTRODUCTION

There has been increasing interest in the possibilities of ultrafine particles (UFPs) for many fields of application because of their unique chemical and physical properties. A great number of efforts have been made the production of metal or ceramic UFPs. Uda and the authors have been demonstrating an active plasma-metal reaction method for preparing metal or ceramic UFPs.[1,2,3]

Since Ni fine particles have been used as catalysts, Ni UFPs are expected to be an excellent calalyst. Ni UFPs are very reactive and sintering of Ni UFPs occurs at lower temperatures.[3,4] Therefore, Ni UFPs should be immobilized with stable materials. Supported metal catalysts have been used by wet processes, where another post-treatment is needed.[5] Although a lot of metal UFPs has been produced by dry processes, the

synthesis   of   the metal UFPs   immobilized   with   stable   materials   is
limited.[3]

In   the   present   study,   the   authors   demonstrate   two   types   of   the
catalysts synthesized   by   an   active   plasma-metal   reaction   method. One is
the mixed Ni-TiN UFPs, where   Ni UFPs and stable TiN UFPs are distributed
homogeneouly.   The   other   is   the   composite   Ni-TiN   UFPs,   where   Ni   UFPs
combine   with   TiN   UFPs   in   the   shape   of   dumbbells.The   synthetic   conditions,
sintering   and   gas   desorption   characteristcs   of   the   mixed   and   the
composite   Ni-TiN   UFPs   were   investigated   in   comparison   with   Ni   or   TiN
UFPs.

SAMPLE PREPARATION

Apparatus

Fig.   1   shows   the   apparatus   for   preparing   UFP   by   an   active   plasma
method.   It   consists   of   an   arc   melting   part,   a   collecting   room   and   a
treatment   room.   The   melting   part   consists   of   a   non-consumable   tungsten
electrode   fixed   in   a   water   cooled   copper   cathode,   and   a   shallow   plate-
like   water   cooled   copper   anode   where   the   metal   (Ni,   Ti   or   Ni-Ti   alloy)
block   is   placed.   When   the   arc-flame   of   nitrogen   or   hydrogen   strikes   the
metal,   UFPs   are   subsequently   generated   around   the   arc   spot   area   of   the
molten   metal   droplet.   The   prepared   UFPs   were   treated   in   a   glove   box   or
under   flowing N_2 atmosphere   to   avoid   oxidation   reaction.   A   slow   oxidation
was   applied   to   some   UFPs   before   being   exposed   to   air.[3,6]

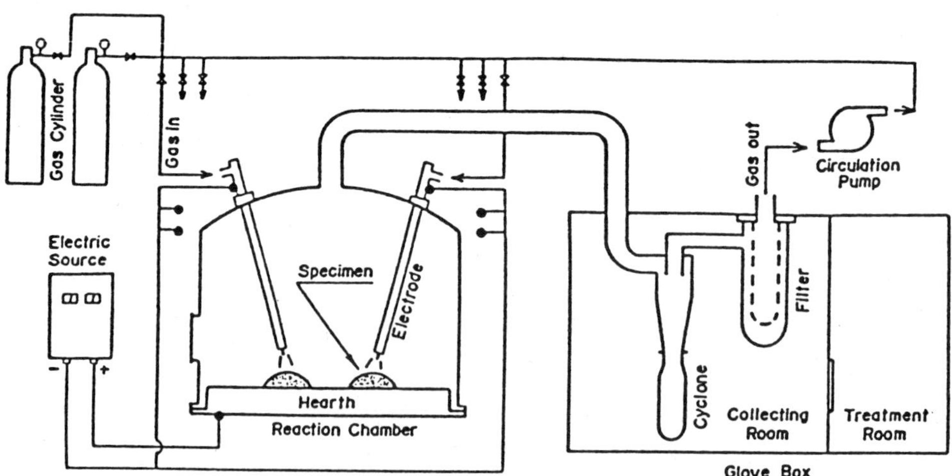

Fig. 1 Schematic diagram of apparatus for the synthesis of UFPs
by means of an active plasma-metal reaction.

<u>Ni UFP and TiN UFP</u>

Ni or TiN UFPs were synthesized by arc melting Ni or Ti metal, respectively, in an N_2-H_2-Ar atmosphere. The variations of concentration ratios of N_2 and H_2 had no effect on the nitrogen and titanium contents of the produced TiN UFPs.[7] X-ray diffraction measurements showed that the crystal structure of the TiN UFPs was a single phase of NaCl structure and that lattice constant was 0.4238+0.0002nm which was close to that of JCPDS's data (6-0642: 0.4240nm). Figs. 2 and 3 show the typical TEM photographs of spherical Ni UFPs and cubical TiN UFPs, respectively.

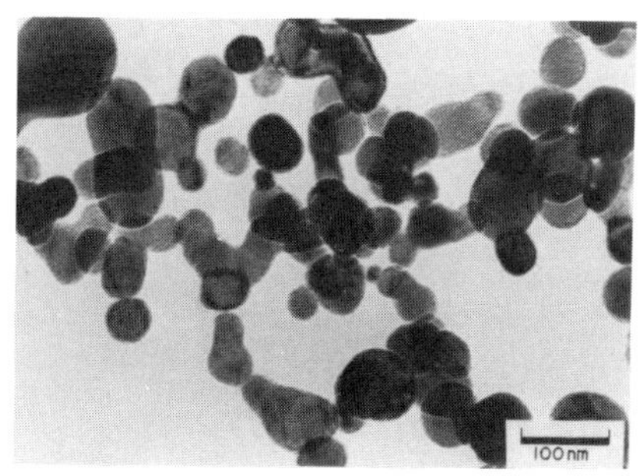

Fig. 2 TEM photograph of
the Ni UFPs.

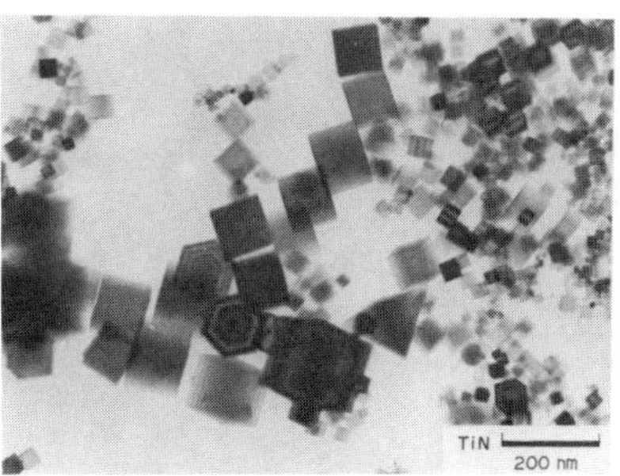

Fig. 3 TEM photograph of
the TiN UFPs.

<u>Mixed Ni-TiN UFP</u>

The mixed Ni-TiN UFPs were synthesized by arc melting of Ni and Ti, separately, at the same time in the same chamber shown in Fig. 1, and mixing the produced Ni UFPs and TiN UFPs in the circulating gas phase. The preparation conditions of the mixed Ni-TiN UFPs were as follows.

Table 1. Synthetic conditions of mixed Ni-TiN UFPs

Sample: Ni(99.95% purity), Ti(99.9% purity)
Atmosphere: N_2-H_2-Ar (N_2:5-10%, H_2:40-48%)
Total Pressure: 0.1MPa
Arc Current: 100-170A
Arc Voltage: 20-30V

Fig. 4 shows the TEM photograph of the mixed Ni-TiN UFPs. The spherical Ni UFPs and the cubical TiN UFPs were distributed homogeneously. X-ray diffraction measurements showed that only Ni and TiN phases were present and that the lattice parameters were close to the undoped ones.

Since TiN begins to dissolve into Ni phase at temperatures above 1173K,[8] it is considered that the mixing in the gas phase occurred at temperatures below 1173K.

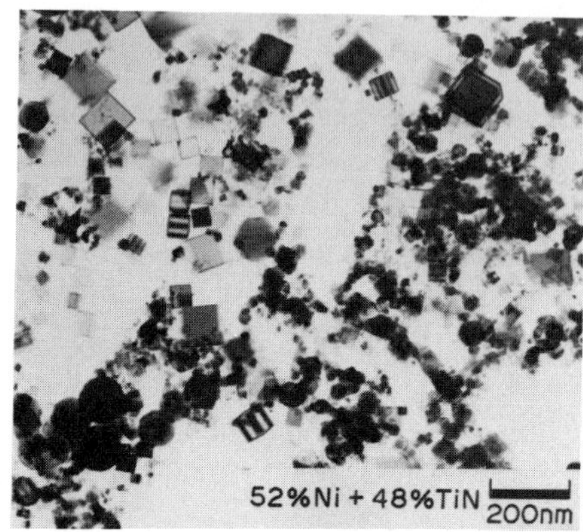

Fig. 4 TEM photograph of
the mixed UFPs.

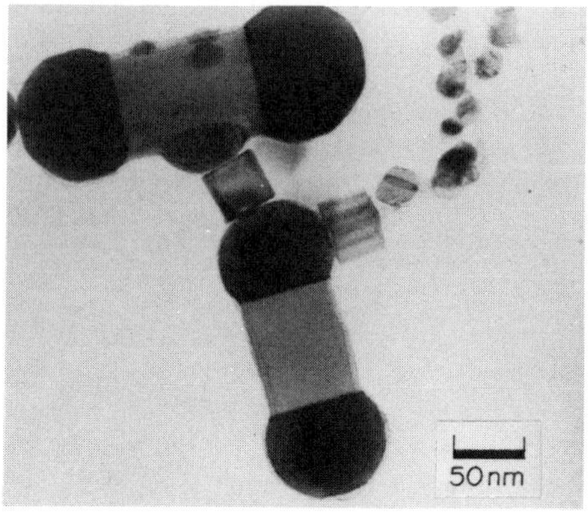

Fig. 5 TEM photograph of
the composite UFPs.

Composite Ni-TiN UFP

Fig. 5 shows the typical TEM photograph of the composite Ni-TiN UFPs. The composite Ni-TiN UFPs were synthesized obviously by arc melting Ni-Ti alloy of which composition exceeded 80 at% Ti under the following condition.[9]

Table 2. Synthetic condition of composite Ni-TiN UFPs

Atmosphere: $46\%H_2-7\%N_2-Ar$
Total Pressure: 0.1MPa
Arc Current: 260A
Arc Voltage: 25-30V

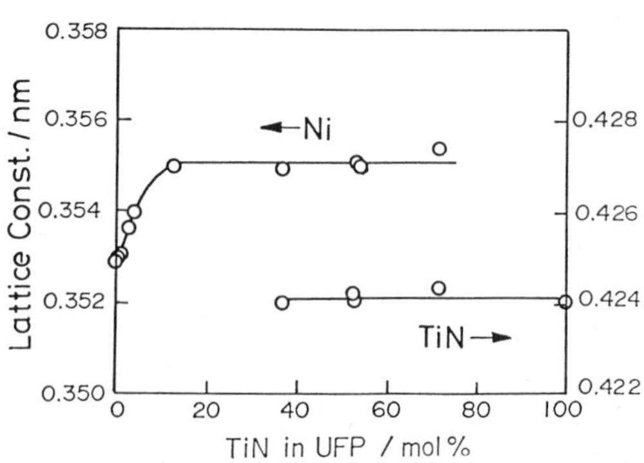

Fig. 6 Lattice constants of Ni and TiN as a function of
TiN mol% in UFPs.

X-ray diffraction measurements showed Ni, TiN and small amounts of unknown (probably Ni_3Ti) phases. Fig. 6 shows the lattice constants of Ni and TiN phases as a function of TiN mol% in UFPs. The increase of the lattice constant of Ni means that Ti or TiN dissolved into Ni. Kato et al. reported that the growth of TiN whiskers on Ni started by the VLS (vapor-liquid-solid) mechanism via a melt containing Ni at temperature of 1373 to 1523K.[10,11] Similar mechanism is considered to occur in the growth of the composite Ni-TiN UFPs as the following steps. The nucleation and growth of TiN particles occurred at first. The vaporized Ni collided with TiN and deposited on the surface of TiN at lower temperatures. Here, it is noted that Ni film was not formed because TiN is not wetted by the molten Ni(the angle of contact between TiN and molten Ni is 113° [12]). The growth of TiN particles occurred via molten Ni at temperatures of 1373 to 1523K because at temperatures above 1523K evaporation of Ni was predominated.[10,11]

SINTERING CHARACTERISTICS

To clarify the sintering characteristics of Ni UFPs, three kinds of Ni UFPs were used. Ni-1 was prepared by an active plasma-metal method, and Ni-2 and Ni-3 were prepared by a gas evaporation method. Ni-2 and Ni-3 were subjected to slow oxidation treatment and Ni-1 was not subjected to such treatment. The oxygen impurity contents and specific surface areas are summarized in Table 3.

Table 3. Oxygen impurity content and specific surface area of Ni UFPs

Ni UFP	Oxygen(mass%)	Specific surface area(m^2/g)	Green Density (g/cm^3)
Ni-1	0.69-0.80	13	2.6
Ni-2	9.0	40	1.6
Ni-3	1.3	5.0	3.7

These powders were pressed ($300Kg/cm^2$) into a disc. Here, the green densities shown in Table 3 varied with the specific surface areas of Ni UFPs. The continuous dimensional changes of the pellets were measured by using a dilatometer. Fig. 7 shows the linear shrinkage of Ni UFPs in a vacuum and in a stream of H_2 at a constant heating rate of 5K/min. Great

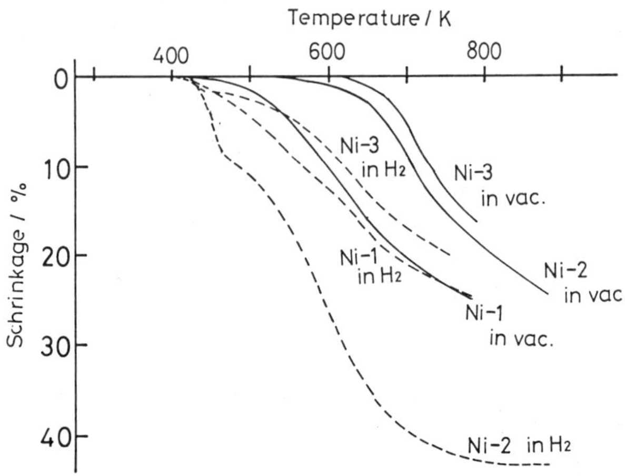

Fig. 7 Linear shrinkages of Ni UFPs in a vaccum and in an H₂ atmosphere at a constant heating rate of 5 K/min.

Table 4. Composition, specific surface areas and green densities of the UFPs.

Sample Name	Composition (mass%)	Specific surface area (m²/g)	Green Density (g/cm³)
Ni-1	Ni	13	2.6
TiN	TiN	106	1.1
Mixed	52%Ni-TiN	23	1.9
Composite	28%Ni-TiN	21	2.1

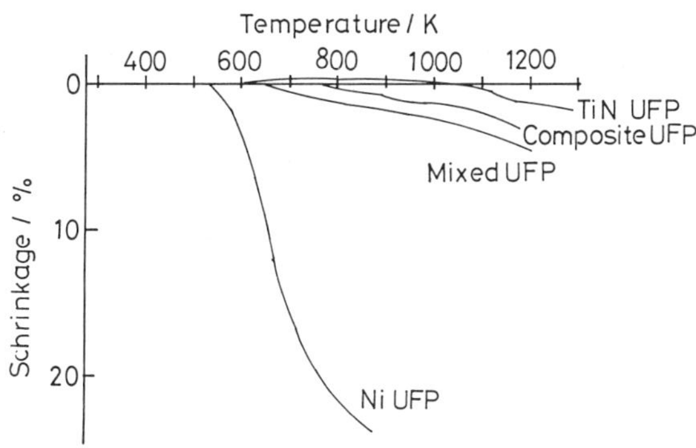

Fig. 8 Linear shrinkages of UFPs in an N_2 atmosphere
at a constant heating rate of 5 K/min.

difference in shrinkage curves is observed depending on atmosphere and oxidation levels. The shrinkage of Ni-2 and Ni-3 in an H_2 at temperatures around 423 K is considered to result from the reduction of the Ni UFPs. Here, it is noted that the clean Ni UFP shrinked at lower temperatures. Densification rate was retarded significantly by the adsorbed gases and oxide phase. Therefore, sintering characteristics were examined for the mixed and composite UFPs which were relatively clean.

Fig. 8 shows the effect of TiN on the linear shrinkage of Ni UFP in a stream of N_2 at heating rate of 5K/min. The composition and specific surface areas of the UFPs are summarized in Table 4 together with green densities of the pellets. The densification of Fresh Ni UFPs (Ni-1) started at approximately 500 K, while that of the mixed and the composite UFPs started at temperatures above 600 K. Similar behavior was observed by measurements of specific surface areas of the UFPs.[3,9] These phenomena imply that Ni UFPs are surrounded by TiN UFPs homogeneously. Therefore, Ni UFPs were immobilized by mixing or combination with TiN UFPs.

GAS DESORPTION CHARACTERISTICS

The desorption of the adsorbed gases was monitored with a quadrupole mass spectrometer, where temperature-programmed desorption (TPD) measurements were carried out after degassing the UFPs in a high vacuum(less than 10^{-6} Pa) at room temperature.[13] Fig. 9 shows the TPD spectra of four kinds of UFPs which had been exposed to air and adsorbed much gases. H_2, NH_3, H_2O, CO, N_2 and CO_2 gases were mainly desorbed depending on the heating temperature. The origin of these gases except N_2 is considered to be from adsorbed gases onto the surface of the UFPs or produced by surface reaction. The desorption of NH_3 at around 500 K for the TiN, the

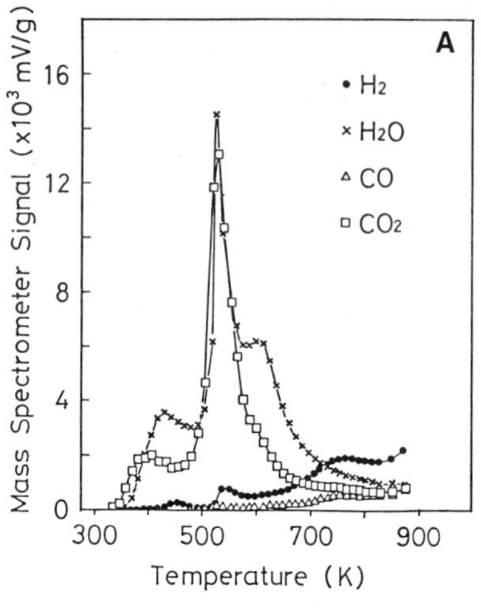
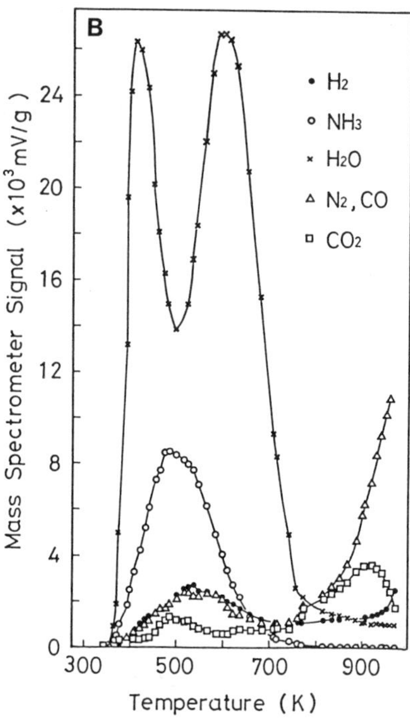
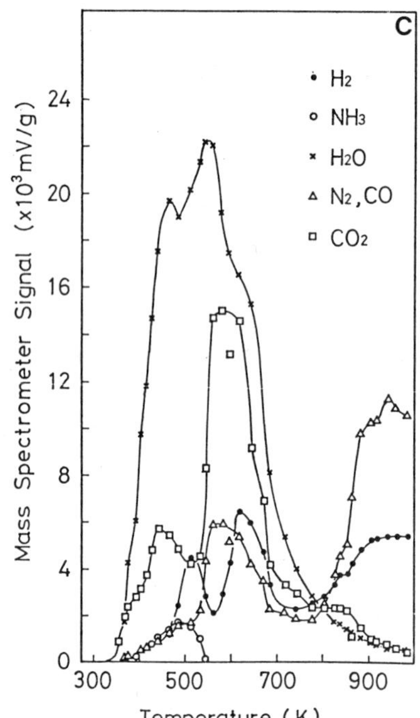
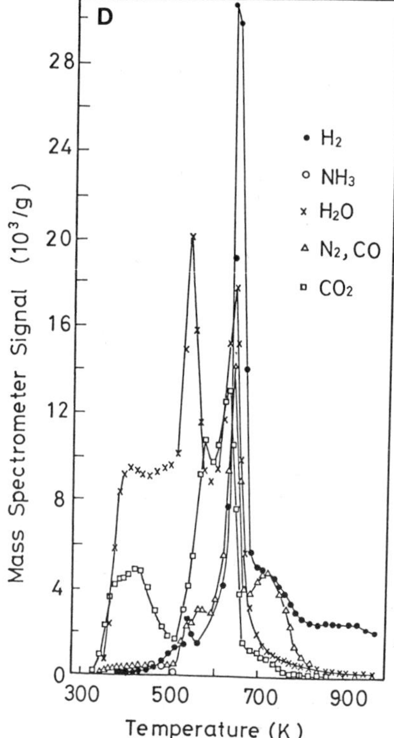

Fig. 9 Thermal desorption
    spectra of (A)NiUFPs,
    (B)TiN UFPs,(C) mixed UFPs
    and (D)composite UFPs.

mixed and the composite UFPs closely related to that of H_2O. Here, at the temperatures where NH_3 desorbed, the amounts of H_2O desorption decreased. NH_3 is considered to be produced by the surface reaction of TiN with H_2O. At temperatures above about 800K, N_2 desorption for the TiN and the mixed UFPs resulted from the decomposition of TiN.[14] Here, it is noted that N_2 desorption above 800 K was not observed for the composite UFPs and that (as was confirmed by CO adsorption-desorption experiments) the CO desorption at around 720 K was observed only for the composite UFPs.

The desorption behavior of mixed UFPs was summation of that of the Ni UFPs and the TiN UFPs. However, the desorption characteristics of the composite UFPs differed from those of the Ni or the TiN UFPs. These results imply that a strong nickel-titanium nitride interaction existed.

CONCLUSION

Mixed and composite Ni-TiN UFPs were successfully synthesized by an active plasma-metal reaction method. Ni UFPs were immobilized by mixing or combination with TiN UFPs. The desorption characteristics of the composite UFPs differed from other UFPs.

ACKNOWLEDGEMENTS

We wish to thank M. Uda, Y. Muramatsu, K. Halada and T. Uchikoshi for their helpful suggestions and discussions. We also wish to thank K. Honma for his TEM observation. This work was supported by the Special Coordination Funds for Promoting Science and Technology.

REFERENCES

1. S. Ohno and M. Uda, Generation Rate of Ultrafine Metal Particles in "Hydrogen Plasma-Metal" Reaction, <u>Nippon Kinzoku Gakkaishi</u>, **48**: 640 (1984).

2. M. Uda and S. Ohno, Preparation of Ultrafine Particles by Nitrogen-Molten Metal Reaction, <u>Nippon Kagaku Kaishi</u>, 862 (1984).

3. S. Ohno, Y. Sakka, H. Okuyama, K. Honma and M. Ozawa, Synthesis and Characterization of the Mixed Ultrafine Particles of Ni and TiN, <u>Proc.MRS International Meeting on Advanced Materials</u>, Vol.3 (1989) in press.

4. R. A. Andrievskii and S. E. Zeer, Specific Contact Phenomena During Sintering of Ultrafine Powders, <u>Science of Sintering</u>, **19**: 11 (1987).

5. See for example; J. S. Smith, P. A. Thrower and M. A. Vaaice, Characterization of Ni/TiO$_2$ Catalyst by TEM, X-ray Diffraction, and Chemisorption Techniques, J. Catalysis, **68**: 270 (1981).

6. S. Ohno and M. Uda, Room Temperature Oxidation of Ultrafine Iron Particles at Low Oxygen Partial Pressures, Nippon Kagaku Kaishi, 924 (1984).

7. K. Honma, H. Okuyama, S. Ohno and M. Uda, Preparation of Ultrafine Particles of TiN by Nitrogen-Hydrogen Gas Mixtures Plasma Method, J. High Temperature Soc., **13**: 199 (1987).

8. H. Mitani, H. Nagai and M. Fukuhara, On the Sintering of the TiN-Ni Binary Powder Compact, Nippon Kinzoku Gakkaishi, **42**: 582 (1978).

9. S. Ohno, H. Okuyama, K. Honma and M. Ozawa, Reprint of National Meeting of Japan Institute of Metals, 114 (1988. 3).

10. A. Hagimura, N. Tamari and A. Kato, Catalytic Effects of Various Materials on the Growth of Titanium Nitride Whiskers by Chemical Vapor Deposition, Nippon Kagaku Kaishi, 49 (1979).

11. A. Kato and N. Tamari, Some Common Aspects of the Growth of TiN,ZrN, TiC and ZrC Whiskers in Chemical Vapor Deposition, J. Cryst. Growth, **49**: 199 (1980).

12. G. A. Yasinskaya, The Wetting of Refractory Carbides, Borides, and Nitrides by Molten Metals, Soviet Powder Met. and Met. Ceram., No7(43), 557 (1966).

13. Y. Sakka, T. Uchikoshi and E. Ozawa, Low-Temperature Sintering and Gas Desorption of Gold Ultrafine Powders, J. Less-Common Metals, **147**: 89 (1989).

14. J. Hojo, O. Iwamoto, Y. Maruyama and A. Kato, Defect Structure Thermal and Electrical Properties of Ti Nitride and V Nitride Powders, ibid.,**53**: 265 (1977).

Part IV. SINTERING OF MULTIPHASE SYSTEMS

PHASE STABILITY AND SINTERING OF MULTIPHASE ALLOY SYSTEMS

G.S. Upadhyaya

Department of Metallurgical Engineering Indian Institute
of Technology, Kanpur - 208016, India

INTRODUCTION

A common objective of sintering studies is the achievement of an
understanding of the effect of sintering variables - time, temperature,
applied pressure (if used), powder size, and composition (including
additives and atmosphere control) - so that suitable control of
variables can lead to products with the required microstructure
(commonly small grain size) and high density.

In order to activate sintering, one can apply two methods -
physical and chemical. The detailed description are given elsewhere.[1]

The present paper shall be concerned more with the role of
additives in multiphase metal and ceramic systems from the viewpoint of
phase stability.

One major difficulty in understanding the exact means of operation
of an additive is that so many alternative mechanisms for these
processes exist. Thus, for example, the additive can work as a second
phase or as a solid solution.

As a Second Phase: (i) Providing a continuous high diffusivity
path way e.g. liquid phase at the boundaries, (ii) providing a
continuous low diffusivity pathway at the boundaries for diffusion
across the boundary which then acts to restrain grain boundary
movement.

As a Solid Solution: (i) Enhancing diffusion coefficients for the
controlling species in the lattice or parallel to the grain boundary by
effecting the point defect concentrations in the boundary or lattice,
(ii) slowing grain boundary movement by forming a segregated layer at
the boundary which must then be pulled along by the boundary, (iii)
altering the overall driving force for sintering by altering the ratio
of grain boundary energy to free surface energy, (iv) slowing intrinsic
grain boundary movement by reducing the diffusion coefficient for atom
movement across the grain boundary, again by affecting the defect
chemistry.

Either of the above situations brings forth the idea of phase

Science of Sintering
Edited by D. P. Uskoković *et al.*
Plenum Press, New York

stability which is based directly on the electronic stability of the
individual components. Though the theoretical Band Model satisfactorily
explains some of the physical properties, it is inadequate in fully
explaining the stability of alloy phases.

If we look at the electronic configuration of elements in the
periodic Table, we can divide them into blocks like

        s - elements
       sp - elements
 d - and f - elements

A multicomponent metal or ceramic system would revolve around
these elements and naturally, their overall stability during alloy
formation is vital. Various workers (like Engel and Brewer[2] and also
Samsonov[3]) did propose different models. According to the latter,
during the formation of a condensed solid state, there is an interplay
in the localised and non-localised valence electrons, so that electrons
form a fairly broad spectrum of configurations. The stable electronic
configurations fod d - or f - metals are $d^5 - d^{10} - d^0$ or $f^7 - f^{14} - f^0$,
according to descending energy stability. Similarly for sp elements, the
stable configurations are sp^3 and s^2p^6, while for s - elements, it is s^2
configuration. A detailed study of the types of phase diagrams on the
basis of electronic structure has been done in the past. With elements
of similar stability of their electronic configurations, there is more
and more possibility of the formation of solid solution.

The thermodynamic data also confirm the types of bonding (Fig. 1).

The stability of binary and ternary intermediate phases has also
been investigated. The intermediate phases are classified into three
categories

                Normal valency compounds
                Electron compounds
                Size factor phases

It is usually been found that several types of atomic interaction
are simultaneously involved in a metallic phase and of course the
structure and properties actually assumed depend on the resultant
effect. In extreme cases, the structure can be understood in terms of

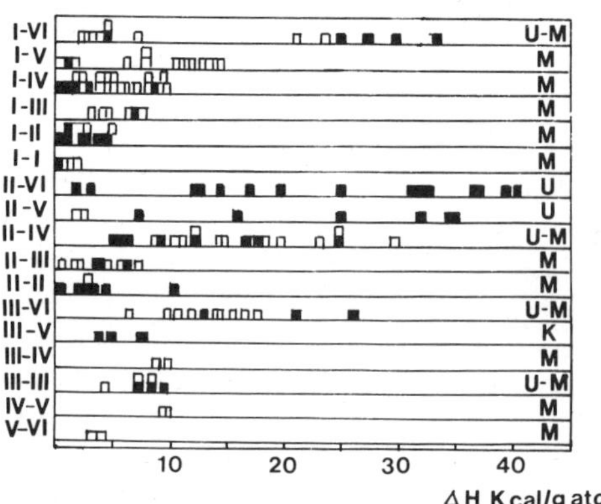

Fig. 1. Heat of formation of intermediate phases depending on the
        position of their components in Periodic Table: U-ionic bond,
        K-covalent bond, M-metallic bond.

one type of interaction only. The great majority of cases are
intermediate in character, and although the appearance of a given
structure and properties can frequently be correlated with some one
property of the atom concerned e.g. size or valency, it is generally
true that the other properties of the atom concerned must not be
unfavourable to the formation of the structure. This would imply, for
example that for the formation of electron phases the size factor
should not be unfavourable and vice-versa.

One of the intermediate phases the sigma phase, has received much
attention, chiefly because of the detrimental effects which the
formation of this phase has on the properties of industrial alloys. The
majority of sigma phases are found in systems involving the transition
elements of group V or VI with a transition metal of group VII or VIII.
Significant features of this phase are broad homogeneity ranges and the
composition and temperature range of stability are not the same in
different alloy systems.

There are more than 50 binary examples of the sigma phase; the
effect of a third element has been investigated in many important
systems and in a few ternary sistem,[4] the sigma phase is formed where
it is not found in the corresponding three binary systems (Fig. 2). For
example in Cr-Ni-Mo, Cr-Ni-W, Cr-Ni-Si and Cr-Ni-P systems, the sigma
phase exists in ternary alloys, while no such phase has been found in
the Cr-Ni binary system. Similarly in the binary systems V-Co, V-Ni,
V-Fe, Cr-Co, Cr-Ni and Cr-Mn, the sigma phase field is expanded by the
addition of Si. On the other hand, Al has been shown to have the
opposite effect reducing the composition range of formation in Fe-Cr,
Fe-V and Cr-Co alloys. This shows that relatively electronegative Si
(and P) increase the tolerance of the sigma phase for electropositive

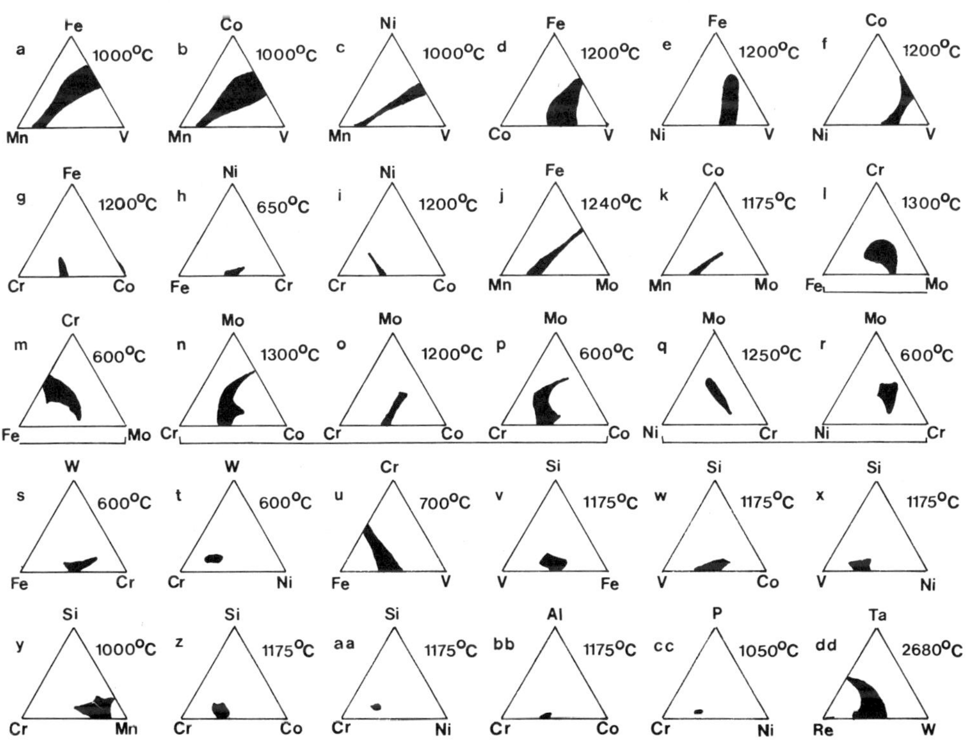

Fig. 2. Ternary sigma phase fields.

elements, such as V and Cr, by accepting the additional electrons made available by these elements, while the additional Al, a relatively electropositive element, increases the electron concentration and thus reduces the available number of V or Cr atoms in the structure. All these suggest that such phases are more near to electron compounds. It is significant that the phase does not occur in all cases at the same approximate composition, but in general, for phases having the same 'A' elements (V or VI group) the composition shifts towards A as the number of electrons outside filled shells increase in the 'B' elements (VII or VIII).

Coming to another intermediate phase, Laves phase (AB_2), it has been agreed that the main reason for their existence is of geometrical origin - that of filling of space in a convenient way. However, the electronic influences also appear to play an important role. For example, Laves and Witte[5] have shown that the electron concentration determines which of the three structures form in the pseudo binary systems of $MgCu_2$ and $MgZn_2$ with Al, Ag and Si. With increasing electron concentration, one or more of the Laves phases is found in the order $MgCu_2$-$MgNi_2$ and $MgZn_2$ respectively. There exist about 45 combinations of Sc group elements (including the lanthanide series) with Cu group elements whose radius ratios lie between 1.11 and 1.58, but none forms a Laves phase. Furthermore, the appearance of ternary Laves phases on addition of Si, where no such phases form in binary systems, again indicates a strong influence of the electronic effect on the stabilization of Laves phases. Elliott and Rostoker[6] have discussed this from the point of view of e/a ratios. The Laves phases in rare earth elements has been discussed in details by McMasters and Gschneider.[7] It is observed that the rare earth always behave as the 'A' element (larger size partner). The 'B' partner is Mn, Fe, Co, Ni and cogeners as well as Mg and Al. Such systems are of great importance in rare earth sintered permanent magnets.[8]

SINTERING AND PHASE DIAGRAMS

German[9] has postulated the predictability of enhanced sintering with the help of phase diagrams. The criteria are solubility, segregation and diffusion. Fig. 3 shows the characteristic features necessary to obtain enhanced sintering in either the solid or liquid states. The author[10] has investigated the liquid phase sintering response of Cu base alloys containing various additives such as Ag, Si, Sn and Pb, such that the liquid phase was 5 mass percent. The sintering temperatures selected were 1.1. and 1.2 times greater, respectively, than the corresponding isotherms in the binary phase diagrams. Results

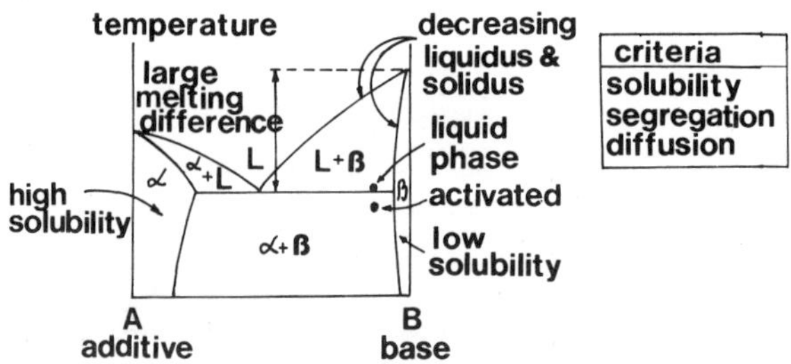

Fig. 3. A model binary phase diagram showing the characteristic features necessary to obtain enhanced sintering in either the solid or liquid states (after German).

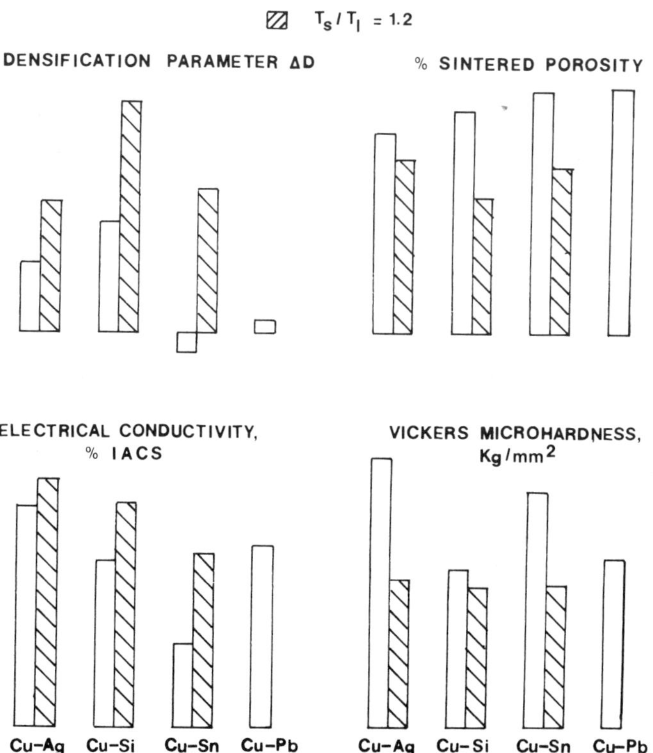

Fig. 4. Sintered properties of liquid phase sintered binary Cu-alloys.

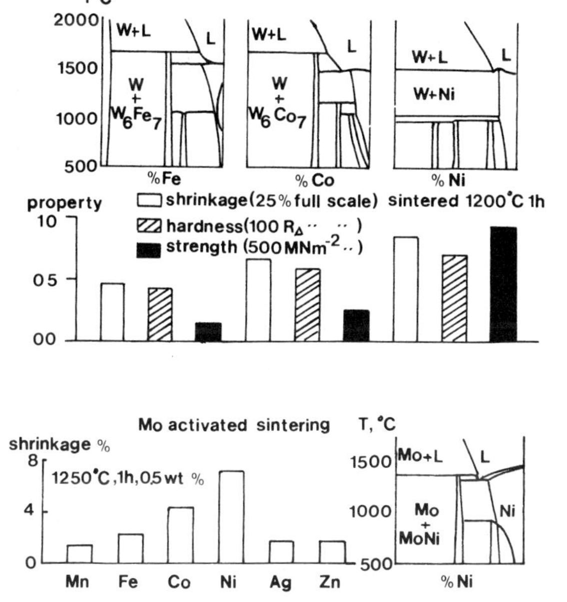

Fig. 5. Comparison of phase diagrams and sintered properties for
tungsten and molybdenum containing 0.5 mass % transition metal
additives (after German).

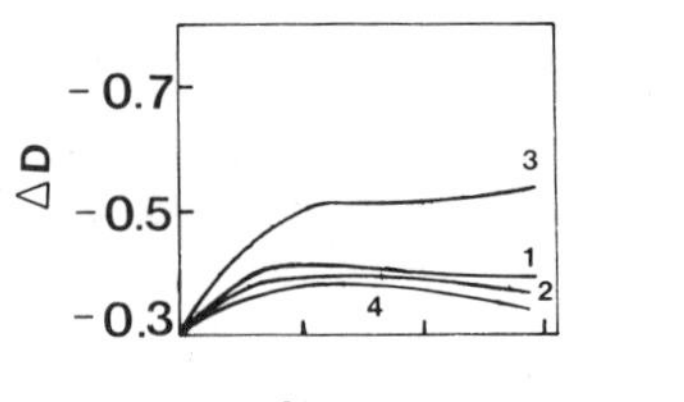

Fig. 6. Sintering behaviour of Cu-7 Al Alloy containing transition
metal additives.

(Fig. 4) showed that the highest densification parameter among all the
alloys considered was found in a Cu-Si alloy. It is worth noting that
Cu-Si forms a series of intermetallics whereas Cu-Ag does not form any
intermetallics. It is thus evident that in the former case the
sintering is more enhanced due to the better possibilities of a
segregation effect. As Si is a covalent bonding additive, there is more
directional bonding, thus accupying more space, which in turn would
demonstrate higher densification during sintering.

Another interesting example of the German model is in explaining
the enhanced sintering of tungsten or molybdenum containing nickel as
an additive (Fig. 5).

Al-BRONZE CONTAINING TRANSITION METALS

The sintering results by Upadhyaya and Singhal[11] and Cu-7Al premix
containing transition metals (0-3 mass percent) after 950°C sintering
revealed that such additions imparted growth to the compacts in the
descending order Co-Mo-Fe-Ni. The very promising influence of nickel
among all the transition metal additions in aluminium bronze has been
justified by Mitani and Yokota,[12] who showed a suppression of the
Kirkendall effect by means of a nickel addition, thus lowering growth.
For activated sintering to occur, one needs higher diffusivity, which
in other words means a higher contribution of nonlocalized valence
electrons or lower SWASC values. This is the reason why molybdenum and
cobalt, having higher SWASC d^5 and d^{10}, respectively, show relatively
poor sinterability. However a question still remains unsolved why
nickel with higher d^{10} SWASC than iron and cobalt still shows the
opposite behaviour. This could be explained on the basis of the Haworth
and Hume-Rothery[13] concept that with dissolution of nickel in copper or
Cu-Al, the normal tendency to fill the d-shells is counteracted by the
effect of the greater freedom of nickel atoms.

Diffusion data[14] suggest that the activation energy of
interdiffusion on the copper rich side of Cu-Ni alloys decreases with
increases in nickel concentration. Such a behaviour would naturally
enhance densification due to increased diffusivity. A question remains
unanswered why cobalt containing bronze also shows a similar pattern
with increase in cobalt content, although the activation energy pattern
is the opposite. This may be correlated with the possibility of an
increase in SWASC d^{10} in cobalt with a increase in temperature, a
feature more pronounced in cobalt, which has lower SWASC d^{10}, than
nickel at ordinary temperature.[3]

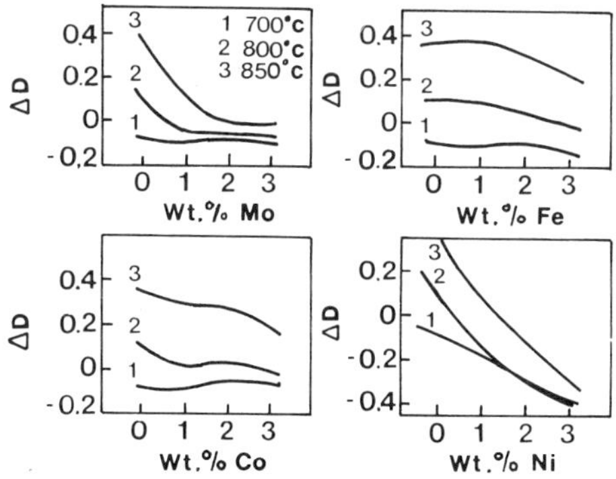

Fig. 7. Sintering behaviour of brass containing transition metal aditives.

SINTERING OF BRASS WITH TRANSITION METAL ADDITIONS

A detailed investigation on sintering of prealloyed Cu-35.6 Zn brass powder with transition metal additives like molybdenum, iron, cobalt and nickel (up to 3 mass percent) was carried out by Srivastava and Upadhyaya.[15] Sintering temperatures were 700, 800 and 850°C, while the sintering period was one hour; the atmosphere being dry hydrogen. The densification behaviours with iron or cobalt additions were similar, being superior to those with molybdenum or nickel addition (Fig. 7). This suggests that the metals with highest d^5 or d^{10} configurations exhibit indifferent behaviour, while those with intermediate range of d^5 configurations, i.e. iron and cobalt, have a better probability of interaction and thus impart densification.

Fe-P-X (X = Mo, Ni, Cu)

Hamiuddin and Upadhyaya[16-18] did extensive investigation on the liquid phase sintering of Fe-P-X systems, and found that Mo additions imparted maximum densification among all (Fig. 8). Next to this came nickel and finally copper, which have energetic stability in the descending order. It is note worthy that copper, having filled d^{10} electrons, tries to maintain its own environment during alloying, thus imparting growth. As far as nickel is concerned, which has two 3d-electron vacancies, one needs a greater proportion of its addition in iron for densification. In other words, in lower nickel concentration, the densification behaviour of iron is an extension of what is observed in Fe-Cu alloys i.e. growth.

HEAVY ALLOYS

Srikanth and Upadhyaya[19-22] studied the sintering of 90W-10 (X,Ni) heavy alloy system with varying proportions of X (X = Cu, Fe, Co or Cr) from 0-100 percent. The results established that after sintering at 1500°C for 1 h, densification increased in the sequence Cr-Cu-Fe-Co-Ni (Fig. 9). A cursory look at the solubility relationship of tungsten in the binder metals reveals that the maximum solubility falls in the sequence Ni (40 percent) – Co (35 percent) – Fe (33 percent) – Cu (0 percent). The W-Cr binary phase diagram is entirely different from the others in the sense that it forms isomorphous solid solutions at

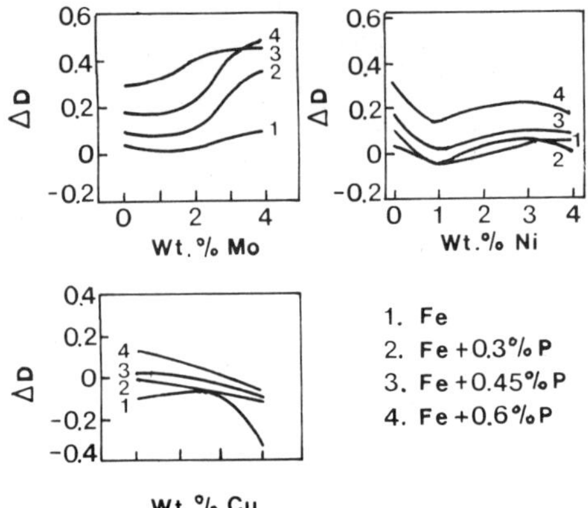

Fig. 8. Sintering behaviour of Fe-P premixes containing transition
metal additives.

elevated temperatures with some miscibility gap. This is the reason why
all the results on W-Cr-Ni heavy alloys presented by the author do not
follow a clear-cut correlation. As the difference in d^5 statistical
weight between tungsten and chromium is not very large, they form
isomorphous solid solutions. On the other hand, a very large difference
in statistical weight, as in the case of W-Cu (88 percent), results in

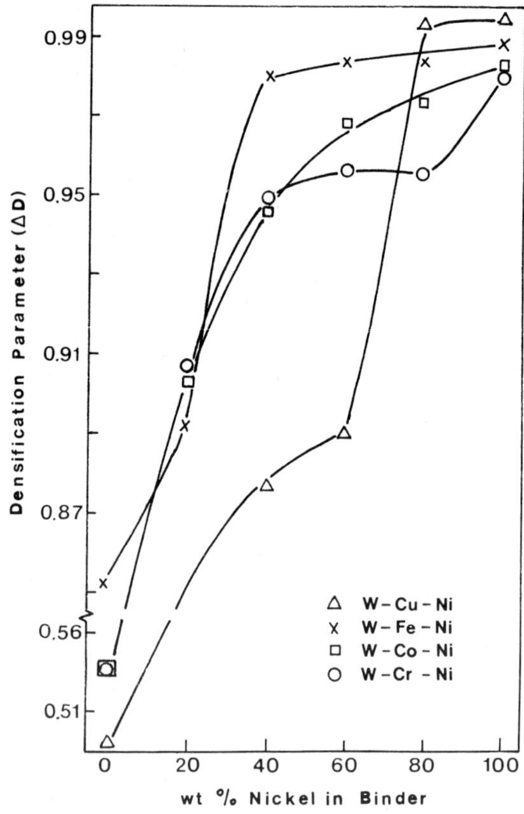

Fig. 9. Sintering behaviour of various tungsten based heavy alloys.

practically no interaction, whereas the situation for other transition
metal binaries are intermediate ones.

RARE EARTH - IRON - BORON MAGNETS

Coming to the rare-earth permanent magnets, these compounds must
combine the basic attributes of high saturation magnetisation, high
curie temperature and large magnetocrystalline anisotropy with a
magnetically unique crystallographic axis. Shortly after the
confirmation of $SmCo_5$ as a permanent magnet material, alloys containing
copper as well as rare-earths and cobalt emerged, which became known as
the precipitation-hardened family of R(CuCo) alloys. The addition of
copper was observed to increase the stability and homogeneity range of
RCo_5 phase and allow the incorporation of iron into the lattice
producing an increase in saturation magnetisation. Metallographic
examination has revealed that a fine precipitate structure of $R_2(CoFe)_{17}$
platelets in a matrix of $R(CoCu)_5$ occurs at the peak coercivity. In the
early 1970's, it was reported that cubic RFe_2 laves phases had
magnetocrystalline anisotropies similar to those of RCo_5 series. For
the rare earths La, Ce, Pr and Nd, few intermetallic compounds form and
those which do have unacceptably low curie temperatures. Hence adhering
to equilibrium RFe binary phases does not appear to be promising for a
study of potential magnet materials. However, two alternative
approaches exist:

(I)   extending the study to metastable/nonequilibrium phases instead
        of being limited to equilibrium phases.

(II)   extending the study to ternary/quaternary systems for stable
        phases. It was not until 1983 that Stadelmaier and coworkers
        pointed out the importance of earlier Russian studies of several
        R-T-B systems.

During 1983, Sumitomo Special Metals of Japan announced the
production of P/M Nd-B-Fe (12 at% Nd, 6 at% B, 82 at% Fe). The compound
had tetragonal structure, with a high uniaxial anisotropy and a curie
temperature of 585 K. A typical microexamination of such magnet reveals
the following:

(1)   A hard magnetic major tetragonal phase (27 wt% Nd)

(2)   A grain boundary Nd-rich phase (80 wt% Nd)

(3)   A B-rich phase (38 wt% Nd)

(4)   Numerous oxide inclusions

The Nd-rich grain boundary phase is an eutectic phase and contains a
large proportion of Nd. The melting point of the eutectic is thought to
be class to the NdFeB binary eutectic temperature of approximately
900K. Hence, during sintering, this phase is liquid and promotes
densification. Excess oxidation of the alloy composition during
processing results in a shift in the effective composition to the Fe-
rich side of the phase diagram.

AL-REFRACTORY CARBIDES COMPOSITES

Aluminium-refractory carbides are unique combinations in which
both aluminium and carbon belong to the sp electronic configuration.

Mishra and Upadhyaya[23] studied such composites with carbides such as TiC, ZrC, Cr_3C_2, NbC, Mo_2C and WC. TiC additions gave the highest densification, while Cr_3C_2 gave the least. As it is well known, TiC because of the highest degree of SWASC sp^3 has the highest hardness among all the carbides, which would promote residual stresses at the metal matrix interface and thus activate sintering process. The poor response by Cr_3C_2 having lower SWASC sp^3 is also due to its complex crystal structure (rhombohedral), which incorporates a higher degree of anisotropy in the system. While considering the stable configuration formation, the question arises as to which of the two sp type elements, C or Al, will have the higher energetic stability for sp^3 configuration. The answer is carbon, as it has lower principal quantum number in comparison to aluminium. A more conclusive picture can be drawn, when it is assumed that nonlocalised electrons strongly form hybrid states with both d^n and $s^x p^y$ states.

WC BASED CEMENTED CARBIDES

Upadhyaya and Basu[25] reported sintering results on WC-10 Co hard metals, where the cobalt binder was partially or fully substituted by other transition metals like iron and nickel. For efficient liquid phase sintering, wettability of the melt is an essential criterion. WC has excellent wettability with iron-group transition metals, as compared to the IVth and Vth group transition metal carbides. This has been explained elsewhere[25] on the basis of chemical bonding in the respective refractory carbides. The solubility of the carbide phase in the binder melt can also be explained on the basis of the above reasonings. Among all the binders studied by the author, WC-Fe and WC-Co form pseudobinary systems, whereas WC-Ni does not. The almost filled d configuration of nickel, brings forth the obvious conclusion why WC-Ni hard-metal is not an ideal one. Solubility data confirm that cobalt (M.P. $1495^\circ C$) dissolves 10-22 percent of WC at the eutectic temperature of $1320^\circ C$, whereas iron (M.P. $1535^\circ C$) dissolves a lower quantity (7.5 percent) of WC at the eutectic temperature ($1143^\circ C$). Our results show that iron may be an efficient binder candidate provided the WC solubility range in it could be increased. Althouth the solubility of WC in nickel is not small, the exploitation of the system could not be made as they do not fall in an equilibrium pseudobinary system.

COVALENT BONDED SOLIDS

The difficulties encountered in sintering of covalent compounds is mainly due to their higher degree of sp^3 electronic configurations, which have highly directional bonds. For silicon carbide, a breakthrough in sintering was achieved by using a small amount of sintering aids. The customarily utilized additives are B and C[26] and recently Al and C.[27,28] Although differing explanations have been offered, these development have surpassed complete understanding of the conventional sintering mechanisms so far. The detailed microstructural analyses of SiC doped with such additives have shown that clean grain boundaries were detected in (B,C) doped materials, while (Al,C) doped materials revealed the existence of thin Al-containing grain boundary.[29] It is worth noting that clean grain boundaries are very favorable e.g. with respect to high-temperature properties. From the electronic structure viewpoint it is obvious that all sintering aids must also belong to similar type of configuration i.e. sp. A detailed treatment of Snow[30] on the electronic structure of aluminium atoms in the condensed state has confirmed the s-p transition. It is but natural that a more covalent bonded element shall be more effective than a metal like aluminium, where the stock of sp^3 configuration even after the transition is fairly low.

Sintering aids in the sintering of silicon nitride are another example in which such additives are essential for sintering below $1700^{\circ}C$. The additive assists in a liquid phase of sufficiently low viscosity while combining with silica at the particle surface of the nitride. Such additives are MgO, Y_2O_3 and ZrO_2. The advantage in high-temperature mechanical properties of Y_2O_3 over alkaline earth oxide e.g. MgO is linked up with the appearance of a refractory grain boundary phase instead of glass as in the MgO case. This confirms that a transition metal oxide with a strong chemical bonding is more desirable as a sintering aid. In order to understand the roles of sintering aids more theoretically, one must take into consideration the reactions to form complex oxides, kinetics of reactions, surface and grain boundary energies, wettability of liquid phase, solid solution formation, diffusional motion of atoms and of ions and so on.

REFERENCES

1. D. P. Uskokovich, G. V. Samsonov, M. M. Ristić, Activated Sintering (in Russian), International Institute for the Science of Sintering, Belgrade, 1974.
2. L. Brewer, Science, 161(1968)115.
3. G. V. Samsonov, UKR. Kim. Zh., 31(1965)1233.
4. E. O. Hall and S. H. Algie, Metallurgical Reviews, 11(1966)61.
5. F. Laves and H. Witte, Metall wirt., 15(1936)840.
6. R. P. Elliott and W. Rostoker, Trans. ASM, 50(1958)617.
7. O. D. Mc Masters and K. A. Gshneidner Jr., In "Compounds of Interest in Nuclear Technology" (Ed. J. T. Waber, P. Chiotti and W. N. Miner), Met. Sc. of AIME, 1964, p. 93.
8. l. V. Mitchell (Ed.) Nd-Fe Permanent Magnets, Elsevier Applied Science Publishers, London, 1985.
9. R. M. German, Liquid Phase Sintering, Plenum Press, N.Y., 1985.
10. G. S. Upadhyaya, Unpublished work.
11. G. S. Upadhyaya and S. K. Singhal, In "Proceedings of Vth Powder Metallurgy Conference," Poznan (Poland), 3-5 October, 1979, Vol. III, p. 15.
12. H. Mitani and M. Yokota, Trans. J. I. M., 116(1975)502.
13. J. B. Howorth and W. Hume-Rothery, Phil. Mag., 43(1952)613.
14. D. B. Butrymowicz, J. R. Manning and M. E. Read, Diffusion Rate Data Mass Transport Phenomena for Copper Systems, National Bureau of Standards, Washington, 1977.
15. V. C. Srivastava and G. S. Upadhyaya, Trans.Powder Metallurgy Association of India, 12(1985)22.
16. M. Hamiuddin and G. S. Upadhyaya,Powder Metallurgy, No.3(1980)136.
17. Ibid, Int. J. of Powder Metallurgy and Powder Tech., 16(1980)57.
18. Ibid, Trans. PMAI, 6(1979)57.
19. V. Srikanth and G. S. Upadhyaya, Trans. PMAI, 12(1985)16.
20. Ibid, J. of the Less Common Metals, 120(1985)213.
21. G. S. Upadhyaya and V. Srikanth, in "Proceedings of 11th Plansee Seminar", 20-24 May 1985, Vol. 2, (Ed. H. Bildstein and H. M. Ortner), Metallwerk Plansee, Reutee, 1985, p. 203.
22. Ibid, in "Modern Developments in Powder Metallurgy", Vol. 15-17 (Ed. E. N. Aqua and C. I. Whitman), Metal Powder Industries Federation, Princeton, 1985, p. 51.
23. P. S. Misra and G. S. Upadhyaya, in "Sintered Metal-Ceramic Composites", (Ed. G. S. Upadhyaya), Elsevier, Amsterdam, 1984, p. 255.
24. G. S. Upadhyaya and D. Basu, in "Proceedings of 11th Int. Plansee

Seminar", 20-24 May 1985, Vol. 2, (Ed. H. Bildstein and H. M. Ortner), Metallwerk Plansee, Reutee, p. 203.

25. G. S. Upadhyaya, in "Sintered Metal-Ceramic Composites", (Ed. G. S. Upadhyaya), Elsevier, Amsterdam, 1984, p. 41.

26. S. Prochazka, in "Ceramics for High Performance Applications", (Ed. J. J. Burke et. al.), Brook Hill, Chestnut Hill, 1974, p. 239.

27. K. A. Schwertz and A. Lipp, Science of Ceramics, Vol. 10, 1980, 149.

28. H. Hausner, in "Energy and Ceramics", (Ed. P. Vincenzini), Elsevier, Amsterdam, 1980, p. 582.

29. R. Hamminger, G. Grathwohl and F. Thummler, in "Science of Ceramics", Vol. 12 (Ed. P. Vincenzini), Ceramurgia s.r.l., Faenze, 1984, p. 299.

30. E. C. Snow, Self Consistent Band Structure of Aluminium by an APW Method Report, Los Alamos Scientific Lab., 1967.

SOLID STATE SINTERING OF TWO COMPONENT SYSTEMS WITH SOLUBILITY

W. Schatt and Ch. Sauer

Dresden Technical University, GDR

INTRODUCTION

The driving force for "voluntary" (pressureless) sintering is the
degradation of the free energy of the disperse body. For solid phase
sintering of a one-component system, this is expressed in a reduction
of external and internal surface and in the annihilation of lattice
defects. In an heterogeneous system with components which are soluble
into one another in addition, there are efforts to decrease the
gradient of the chemical potential (homogenization) present in the
contact range of chemically different particles. For that reason,
during sintering of heterogeneous systems, a foreign diffusion caused
by concentration gradients superimposes on the self-diffusion caused by
capillary forces. That part of the sintering process where pore space
is eliminated as well as chemical reactions take place between the
initial components of the powder mixture (phase formations) is defined
as reaction sintering.

The way in which heterodiffusion influences the densification is
differently assessed. According to I. M. Fedorcenko and I. I. Ivanova,[1]
the volume shrinkage of Ni-Co-specimens made of mechanical powder
mixtures of different composition and sintered isothermally for two
hours is correlated with the partial Ni and Co diffusion coefficients
(Fig. 1). Therefore the shrinkage should be the greater, the higher the
diffusion coefficients are, but in any case lower than shrinkage of the
pure components, i.e. that of the Ni and Co sintered compacts. However,
dilatometer measurements performed by F. Thümmler and W. Thomma[2] using
the same system show at a sintering temperature of 900°C for the alloys
always greater shrinkage than for Ni and Co, whereas, at the final
sintering temperature of 1200°C, the Co specimens were maximally but
the Ni compacts minimally shrunk. The compositions in between alternate
and exhibit a flat minimum of shrinkage at Co60-Ni40 and a flat maximum
for Ni60-Co40.

In view of such differing results which, in addition, originate
from different investigations with regard to the method and
experimentation, it is difficult to state something fundamentally on
the correlation and interaction of densification and heterodiffusion in
reaction sintering. But starting points can be found if the increasing
degree of homogenization is measured simultaneously with the shrinkage ε

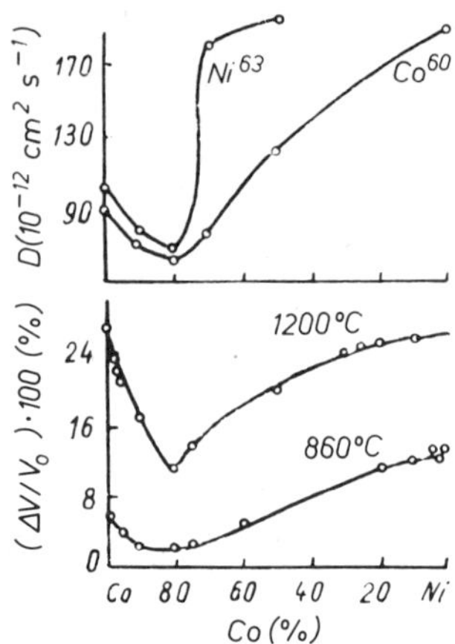

Fig. 1. Partial diffusion coefficients and volume shrinkage $\Delta V/V_o$ in the system Co-Ni (according to[1]).

and its rate $\dot{\varepsilon}$. Additionally, because of the temperature and time dependence of diffusion and homogenization, it is recommended to vary the velocity v_A of heating up to the isothermal sintering temperature. Furthermore, special attention should be paid to the processes in the (non isothermal) heating phase, in which, the main part of densification is realized.

SYSTEM OF UNLIMITED SOLUBILITY

The two-component system Ni-Co already mentioned was selected out of the systems of unlimited solubility of the components in the interesting temperature range (> 400°C). The Ni- and Co-powders were sieved to a definite average particle size L_p. For compacts of Ni- and Co-powders and for those made of mechanical powder mixtures of graded composition, the ε- and $\dot{\varepsilon}$-curves were measured during the heating period and the first 15 minutes of isothermal sintering (Fig. 2). In addition, the state of homogenization H was determined for two representative compositions following the method described in[5] (Fig. 3). H represents the ratio of the respective measured average mixed crystal concentration and the concentration of the saturated solid solution.

Solely by the occurrence of $\dot{\varepsilon}$-extreme values, it cannot be concluded that the densification in the phase of intense shrinkage is controlled by heterodiffusion. In the ε- and $\dot{\varepsilon}$-curves of the sintered alloys show, in conformity with their composition, the ε- and $\dot{\varepsilon}$-characteristics of the Ni and Co sintered specimens (Fig. 2). These are attributed to the superficial influence of the different deformability (creepability) of the disperse basis·components, i.e. of Ni and Co.

In Fig. 3, further information is given on the sintering behaviour. In addition to the ε- and $\dot{\varepsilon}$-curves, the corresponding H-curves are represented for the extremly different heating rates v_A. Whereas ε varies comparatively slightly depending on v_A, but mainly on concentration, H shows practically no concentration influence but only a distinct v_A- dependence. The H curves of the sintered alloys Ni80-Co20 and Co80-Ni20 nearly coincide. This cannot be brought into coincidence with thoughts of a correlation between densification and

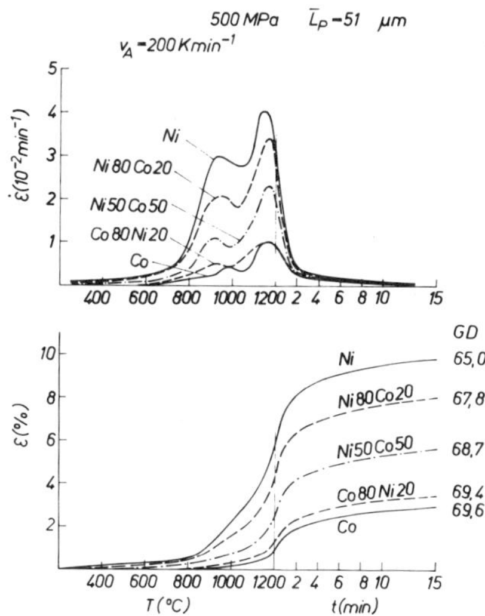

Fig. 2. ε- and ε̇-curves of sintered Ni and sintered Co and of
different NiCo sintered alloys (according to K. Brand, W.
Schatt); green density values GD are indicated in percent of
the theoretical density TD.

diffusion coefficients,[1] as otherwise under the same conditions, the
homogenization gradient of the sintered alloy Co80-Ni20, because of the
minimum of the partial diffusion coefficients (Fig. 1) present for this
composition, should be always essentially below the minimum of the Ni-
rich alloy.

On the other hand, according to the facts that the alloys rich in
Ni and Co show nearly the same homogenization behaviour, it can be
concluded that in the investigated range of intensive shrinkage,
heterodiffusion does not take place under equilibrium conditions and
therefore, with regard to a correlation with shrinkage, the equilibrium

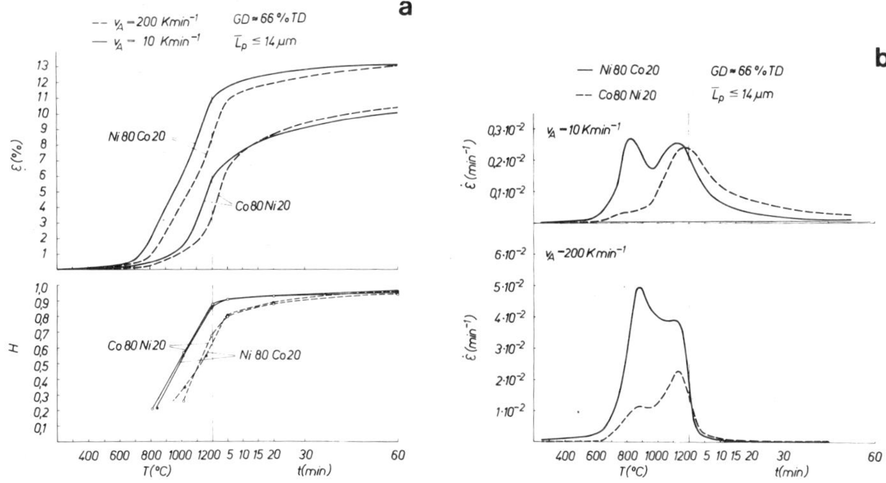

Fig. 3. ε- and H-curves (a) and ε̇-curves (b) of Ni80-Co20 and
Co80-Ni20 sintered alloys for different v_A; in all cases, GD
was ≈ 66% TD (according to K. Brand, W. Schatt).

diffusion coefficients should not be utilized. The partial diffusion
coefficients given in Fig. 1 and the volume shrinkage measured after
two hours of isothermal sintering extensively correspond to the
equilibrium state. The main part of shrinkage is realized in a T-, t-
range, where there is a condition characterized by a high lattice
defect density[3,4] which significantly influences diffusion as well as
shrinkage.

SYSTEM OF LIMITED SOLUBILITY

Phenomena linked with solid phase sintering of a heterogeneous
system, the components of which are soluble into one another to a
limited extent, are discussed using the sintered alloy Cu-Si4 as an
example. At an isothermal sintering temperature of $800^{\circ}C$, Cu solves up
to about 5.3 mass-% of Si.

Mechanical Cu-4 mass-% Si-powder mixtures were chosen as the
material pressed into compacts (Cu-Si4) of different green density (GD
= 69%, 76% TD) and of different integral Cu-Cu- and Cu-Si-contact
surface. Varying the heating rate (v_A = 10, 50, 200 K min^{-1}), for these
compacts the ε-, $\dot{\varepsilon}$- and H-curves for the heating phase and the
isothermal sintering (up to 120 minutes) were measured. In addition,
for comparison, the ε- and $\dot{\varepsilon}$-curves of copper powder compacts of the
same green density (Cu) as well as the ε-curve of compacts were
registered made by a pre-alloyed powder of the Cu Si4 co,position
(MC) (Fig. 4, 5).

As it can be seen in the Figs. 4 and 5, the ε- and $\dot{\varepsilon}$-curves of
the Cu-Si4 and Cu sintered specimens are of the same type. For both
materials, the influence of v_A and GD manifests itself nearly in the
same way.[3] However, there are principal differences in the curve level:
for reaction sintered Cu-Si4, the ε = f(T,t)-values and $\dot{\varepsilon}$ = f(T,t)-
values are lower than those of sintered copper. That means, that the
densification during sintering is obstructed by the homogenization.

During slow heating (v_A = 10 K min^{-1}) at the beginning of a
measurable shrinkage, homogenization advanced to such an extent and, as
a consequence, the solid solution hardening is strongly developed so
that the deformation processes (creep) being the basis for
densification are considerably impeded (Fig. 4). This becomes more
apparent for compacts of higher green density, as the integral contact
surface Cu-Si is greater and, hence, for the same T and v_A
homogenization is more developed than for heterogeneous sintered alloys
of lower green density GD (Fig. 5). In the same way higher partial
diffusion coefficients (or their difference) are influencing the
sintering behaviour, as they accelerate the homogenization at
comparable T and v_A ($800^{\circ}C$: $D_{Si \to Cu} \approx 27.10^{-10}, D_{Cu \to Si} \approx 5.10^{-5}$ cm .s^{-1}).[6]

The densification process can be optimized when the heating rate
is increased to such an extent (v_A = 50, 200 K min^{-1}) that, due to the
temporal dependence of the diffusion, the fast advancing
homogenization and the course of shrinkage producing process in that
part of the basis component Cu, which is not yet solid solution
hardened, take place in the same sintering interval. However, it should
also be taken into consideration that the deformation processes causing
shrinkage are diffusion controlled, and therefore $\dot{\varepsilon}$, but not ε can be
optionally increased via v_A. Among the cases investigated, an optimal
densification behaviour seems to be given using v_A = 50 K min^{-1}. When
the deformation processes are nearly excluded, as in the case of a
solid solution hardened (pre-alloyed) powder, the shrinkage of the

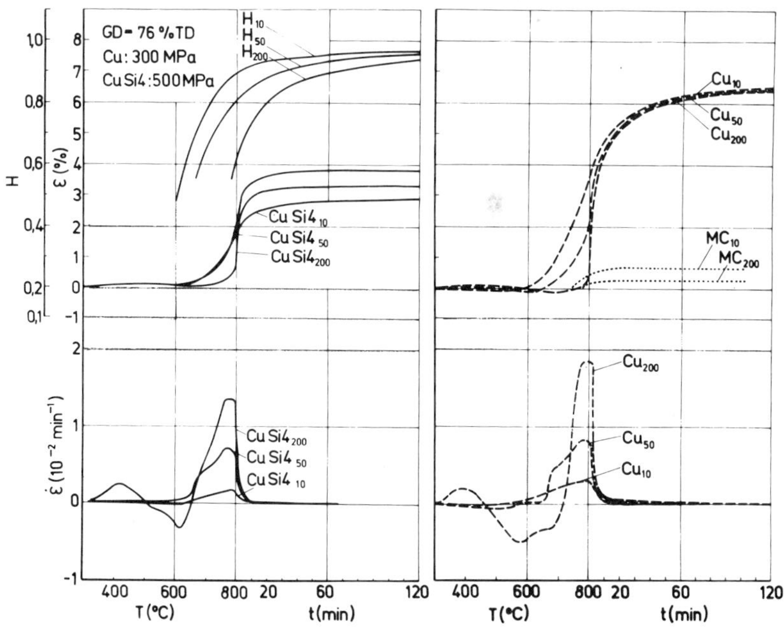

Fig. 4. ε – and ε̇– and H-curve of different Cu based sintered
materials of a green density GD = 76% TD (according to W.
Püsche); indices 10, 50 and 200 indicate the heating rate v_A.

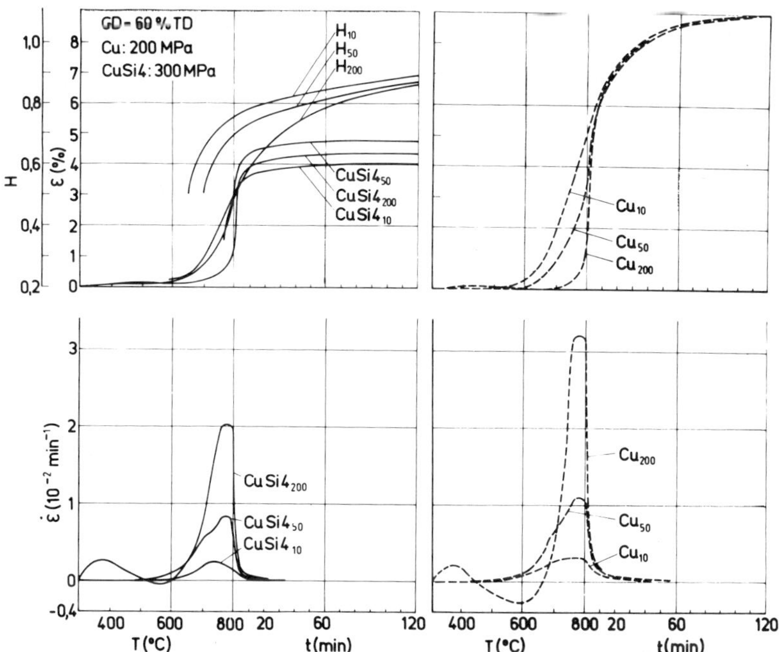

Fig. 5. ε-, ε̇- and H-curves of different Cu based sintered materials of
a green density of GD = 69% TD (according to W. Püsche);
indices 10,50 and 200 indicate the heating rate v_A.

sintered alloy is minimal (MC curves in Fig. 4). The v_A dependence of these curves leads to the assumption that, in this case, the shrinkage is mainly due to a diffusion material transport.

DISCUSSION AND CONCLUSIONS

The results of sintering experiments carried out at a system of complete solubility (Ni-Co) and a system of partial solubility (Cu-Si) point to the assessment that the sintering behaviour of the heterogeneous disperse body in the phase of intense densification is mainly determined by the processes in the basis component. The curves ε, $\dot{\varepsilon}$ = f(T,t) of the heterogeneous sintering compact show the same characteristics and properties which can be influenced in characteristics and properties which can be influenced in the same way by v_A, L_p and GD as the ε, $\dot{\varepsilon}$ = f(T,t) curves of the base component, sintered as a one-component system under the same conditions.

In a disperse one-component system during the non-isothermal sintering phase, vacancies are emitted by the contact, due to the conversion of powder particle contact boundaries into "normal" large angle grain boundaries (lower in energy). These vacancies agglomerate in its adjacent volume and form multiple vacancies. At appropriately high homologous temperatures (Cu: $700...750^{\circ}C$, Ni: $800...850^{\circ}C$), the vacancy configurations change into climbable dislocations. Such a change is promoted when the stacking fault energy exceeds the critical value ($\approx 5 \cdot 10^{-3}Gb$). This condition is fulfilled for Ni ($16,2 \cdot 10^{-3}Gb$) and Cu ($15,4 \cdot 10^{-3}Gb$), but not for Co,[7] hence, Ni and Cu during sintering show good but Co poorer deformation ability. The high dislocation densities obtained in the contact range allow rapid densification (deformation) via dislocation creep (climb, non conservative dislocation motion) in the late non-isothermal and early isothermal sintering phases. The actual processes of the required material transport are a movement of the particles as a whole at simultaneous shape-accommodated creep-deformation of their surface ranges as well as immediate dislocation creep of contact zone substance into the pores. In each case the dominance of one this mechanisms is indicated by discrete ε-maxima (Fig.2). Its relative size and the occurrence of only one $\dot{\varepsilon}$-maximum (Fig. 4, 5), respectively, depend on the technological parameters (GD, L_p, v_A) and on the "activity" of the starting powder.[3,4,8] The ε-maximum, which can be observed at lower temperature, is assigned to the particle movement, but the maximum occuring at more increased temperature and a single maximum, respectively, are assigned to dislocation creep of contact zone substance.

Such sintering behaviour is shown also by the basis component of a heterogeneous system. Adding a solid solution forming component, in general two types of phenomena occur. Heterodiffusion is intensified due to the temporally increased vacancy concentration in the contact ranges to such a extent, that H behaves nearly independent on concentration (Fig. 3). The proceeding homogenization is connected with a solid solution hardening, obstructing the deformation processes (creep mechanisms) which densify the porous sintering body (Fig. 2, 4).

According to Shen Yangyun and R.J. Brook,[9] for reaction sintering, three fundamental variants of the behaviour, schematically represented in Fig. 6, can be differentiated: (1) in the system homogenization goes on more rapidly than densification, (2) homogenization rate and densification rate are nearly the same, (3) densification proceeds

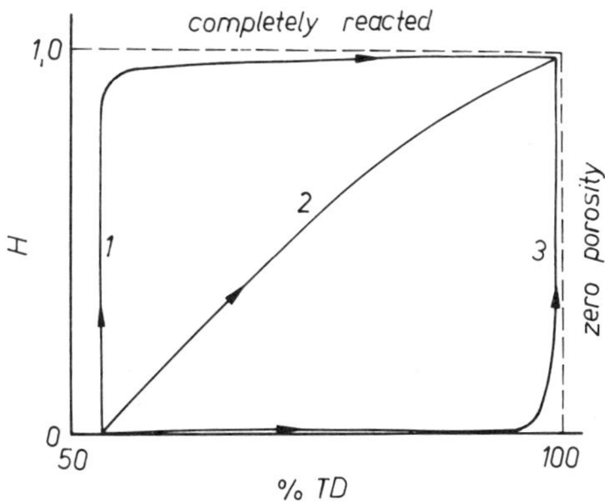

Fig. 6. Schematic representing of fundamental processing variants for
reaction sintering (according to[9])

considerably faster than homogenization. With respect to the properties
attainable of the sintered alloy, the last mentioned possibility is the
most favourable. According to the authors opinion it can be realized
when the activation energy for the reaction (heterodiffusion) is higher
than that for the densification processes. This is hardly the case for
metallic systems as they occur in practice. In addition is mentioned,
that with decreasing powder particle sizes, reaction rate and
densification rate increase, but the compacting rate increases to a
greater extent and, in this way, the sintering system could be
influenced in the sense of variant (3). In the sintering technology of
ceramics the powder particle sizes used are commonly lower than 1 μm,
but in the powder technology of metals, such small powders are rarely
utilized, i.e. this possibility to optimize reaction sintering of
technical relevant metallurgical systems will not be taken into
account. Hence, at present, for sintering heterogeneous compacts of
metal alloys, reaction sintering can be only optimized in the way of a
variant (2).

ACKNOWLEDGEMETS

The authors would like to thank Ms. K. Brand and Mr. W. Püsche for
assistance in experiments.

REFERENCES

1. I. M. Fedorcenko and I. I. Ivanova, in: Teorija i technologija
   spekanija, G. V. Samsonov (Ed.), Kiev, Naukova Dumka, 253
   (1974).
2. F. Thümmler and W. Thomma, in: Modern Development in Powder
   Metallurgy, Vol. 1: Fundamentals and Methods, New York and
   London, Plenum Press 361,(1966).
3. W. Schatt and M. Hinz, Powder Metallurgy International 20, 17
   (1988).
4. W. Schatt, Attempt of a uniform consideration of sintering
   processes, Conference reports of the GDR Academy of Sciences
   (1989).
5. A. N. Klein, Doctoral thesis, Fakultät für Maschinenbau der
   Univarsitet Karlsruhe (1983).
6. R. N. Hall and I. H. Racette, J. Appl. Phys. 35,379(1964).

7. S. Mader, in: Modern Problems of Metall Physics, 1st Vol., A. Seeger (Ed.) Berlin, Heidelberg, New York, Springer Verlag 218 (1965).

8. W. Schatt and E. Friedrich, in: Sintering '85, G. C. Kuczynski et al. (Ed.), New York and London, Plenum Press 133 (1987).

9. Shen Yangyun and R. Brook, Science of Sintering $\underline{17}$, 35(1985).

PROGNOSIS OF SINTERING OF SYSTEM W-Ni

IN THE PRESENCE OF LIQUID PHASE

Z.S. Nikolić, R.M. Spriggs, and M.M. Ristić

Faculty of Electronic Engineering, Nish, YU
Alfred University, Alfred, NY, USA
Serbian Academy of Sciences and Arts, Belgrade, YU

INTRODUCTION

The great number of experimental results obtained from sintering studies of W-Ni alloys certainly provides rich material to study the kinetics
of this process. From such information, one can presume that a solution-
reprecipitation process takes place in a liquid-solid system, manifesting
itself by dissolution of the material in the liquid phase, its transport
through a thin layer of liquid and its reprecipitation at an other location.

This paper describes a study of the sintering of the W-Ni system from
the point of view of both process modeling and the application of a simulation method. This is all the more appropriate because liquid phase sintering is a very complex process. Application of the simulation method enables the introduction of an arbitrary number of parameters and their observation during real time simulation of the development of the process [1-4].

For this purpose, a SIMTFS computer program for two-dimensional simulation of liquid phase sintering has been developed.

PROCESS MODELING

For modeling of the liquid phase sintering process, the investigations
of Yoon and Huppmann [5,6] are particularly interesting. They came to the
conclusion that under certain experimental conditions, the driving force
for the solution-reprecipitation process is governed by the difference in
chemical potential between pure W, which dissolves, and the W-0.45 at.%
Ni alloy formed as the reprecipitation product.

Let the spherical W particles of concentration C be in the liquid phase. The solid obtained by reprecipitation is the W-Ni alloy of equilibrium
concentration C_o; its chemical potential will be less than that of spherical W particles, i.e.

$$\mu_o = \mu - RT \ln \frac{C}{C_o}.$$

The difference in chemical potential $\Delta\mu = \mu - \mu_o$ causes the dissolution
of tungsten which is transported trough the liquid phase and precipitates
in the place where the W-Ni alloy is formed.

Science of Sintering
Edited by D. P. Uskoković *et al.*
Plenum Press, New York

Let the grain growth as a result of the solution-reprecipitation process at the boundaries of solid and liquid phase be the consequence of the diffusion in the system. Supposing that both processes develop at some rate, it may be assumed that the concentration of solid phase atoms, dissolved in the liquid phase, is continually in equilibrium with the solid phase, which is in the contact region of the solid-liquid phase, while in the space between the contact regions, there is a concentration gradient of liquid phase. Its concentration distributions depend directly on the diffusion rates in the system.

If C_L^{W-Ni} is the concentration of liquid phase in contact with the W-Ni alloy of concentration C^{W-Ni} and C^W is the concentration of pure W, then at constant temperature the concentration of liquid phase in contact with W particles which dissolves in it, will be

$$C_L^W = C_L^{W-Ni} \frac{C^W}{C^{W-Ni}} .$$

For defining the probable system state at which the precipitation process starts, we shall begin with the supposition that the solution of the solid phase atoms of one component in the liquid phase takes place. Reaching the equilibrium of liquid phase concentration does not mean, however, the formation of the two-component system. In the contrary, the solution process continues. The liquid phase concentration continues to increase up to the saturation point. Only then do the random liquid phase fluctuations cause the formation of a two-component solid phase at the boundary between the solid and the liquid phases.

Note that such an analysis requires the use of a two-component system in which the formation of the two-component alloy at sintering temperature takes place in the presence of a liquid phase.

Such a consideration of the sintering process enables both the application of the simulation method and the theoretical calculation of the grain growth and geometry changes during this process.

NUMERICAL SOLUTION OF THE DIFFUSION EQUATION

We discuss here some of the problems occuring in the numerical solution of the diffusion equation and the analysis of the numerical method. For the sake of simplicity in nomenclature, we shall consider only the two-dimensional case. However, all results given in the following can be generalized to three dimensions in a straightforward manner.

If D_L is a diffusion coefficient in the liquid then, for the two-dimenzional case, the diffusion through the liquid phase is defined by the partial differential equation

$$C_t = D_L (C_{xx} + C_{yy}), \tag{1}$$

where the subscripts t, xx and yy denote partial derivatives.

From the viewpoint of numerical analysis there are two types of numerical methods: finite element method and finite difference method. We have applied the second one for numerical solution of the partial differential equation (1).

Let the domain of a rectangular shape (experimental region), in which a solution of the equation (1), is sought, is partitioned into subregions by a mesh which is a set of meshlines parallel to the coordinate axes. The first and the last line coincide with the boundaries. We put n lines parallel to the y-axis and m lines parallel to the x-axis through the rectangular domain.

We shall use the following abbreviations:

$$h_x = x_{i+1} - x_i, \quad i=1,n-1$$

$$h_y = y_{j+1} - y_j, \quad j=1,m-1$$

$$h_t = t_{k+1} - t_k, \quad k=0,1,2,\ldots$$

where t is time variable, and

$$C_{i,j,k} = C(x_i,y_j,t_k), \quad i=1,n; \; j=1,m; \; k=0,1,\ldots$$

Assuming that C is continuously differentiable three times, we can replace the partial derivatives by:

$$C_{t,i,j,k} = \frac{C_{i,j,k+1} - C_{i,j,k}}{h_t} + 0(h_t)$$

$$C_{xx,i,j,k} = \frac{C_{i+1,j,k} - 2C_{i,j,k} + C_{i-1,j,k}}{h_x^2} + 0(h_x^2)$$

$$C_{yy,i,j,k} = \frac{C_{i,j+1,k} - 2C_{i,j,k} + C_{i,j-1,k}}{h_y^2} + 0(h_y^2).$$

If we restrict ourselves here to the classical five point approximation then, for an approximate solution of equation (1) and

$$h_t \leq \frac{(h_x h_y)^2}{2D_L(h_x^2+h_y^2)},$$

we have [7]

$$C_{i,j,k+1} = 0.25(C_{i+1,j,k} + C_{i-1,j,k} + C_{i,j+1,k} + C_{i,j-1,k})$$

$$i=2,n-1; \; j=2,m-1; \; k=0,1,2,\ldots, \tag{2}$$

where h_x and h_y are equal ($h_x = h_y = h$).

INITAL AND BOUNDARY CONDITIONS

Let

$$C_{i,j,0} = g_{i,j}, \quad i=2,n-1; \; j=2,m-1$$

be the initial conditions, and

$$\left. \begin{array}{l} C_{i,1,k} = f_1{}_{i,k} \\[2em] C_{i,m,k} = f_2{}_{i,k} \end{array} \right\} \quad i=1,n; \ k=0,1,2,\ldots$$

$$\left. \begin{array}{l} C_{1,j,k} = f_3{}_{j,k} \\[2em] C_{n,j,k} = f_4{}_{j,k} \end{array} \right\} \quad j=1,m; \ k=0,1,2,\ldots$$

be the boundary conditions. With a set of values $C_{i,j,k}$ at a given time t the concentration values $C_{i,j,k+1}$ at the time $t+h_t$ can be obtained by applying finite difference equation (2).

PROCESS SIMULATION

Grain growt in liquid-solid systems is a very complex process, since it is the result of the action of number of elementary processes of which the modern science of sintering knows very little. To this end, our investigations, which introduce models and define the possible states as elements of process modeling, are intended to give some new explanation and interpretations of the sintering process using the simulation method.

If we know the concentrations $C_{i,j,k}$ $(i=1,n;\ j=1,m)$ then the concentrations $C_{i,j,k+1}$ are determined by equation (2) and mass transfer is determined by equation

$$j = - D_L \left(\frac{C_{i,j,k+1} - C_{i-1,j,k+1}}{h_x} + \frac{C_{i,j,k+1} - C_{i+1,j,k+1}}{h_x} + \frac{C_{i,j,k+1} - C_{i,j-1,k+1}}{h_y} + \frac{C_{i,j,k+1} - C_{i,j+1,k+1}}{h_y} \right).$$

Then, a determination of a new geometry of the model system starts. This procedure is iterative and ends by reaching the required simulation time.

Keeping in mind such a consideration, a computer program SIMTFS for two-dimensional simulation of liquid phase sintering has been developed (Fig. 1).

PROGNOSIS OF SINTERING IN THE W-Ni SYSTEM

When we speak of the system state, it must be pointed out that a definition of the initial state is generally a problem that is characteristic of a numerical forecast which occurs due to the impossibility of always precisely knowing the initial state. That is why must introduce certain theoretical assumptions. However, we may discuss their validity and acceptability only after the forecast results are obtained, simultaneously noticing possible corrections of those assumptions. Since the future sys-

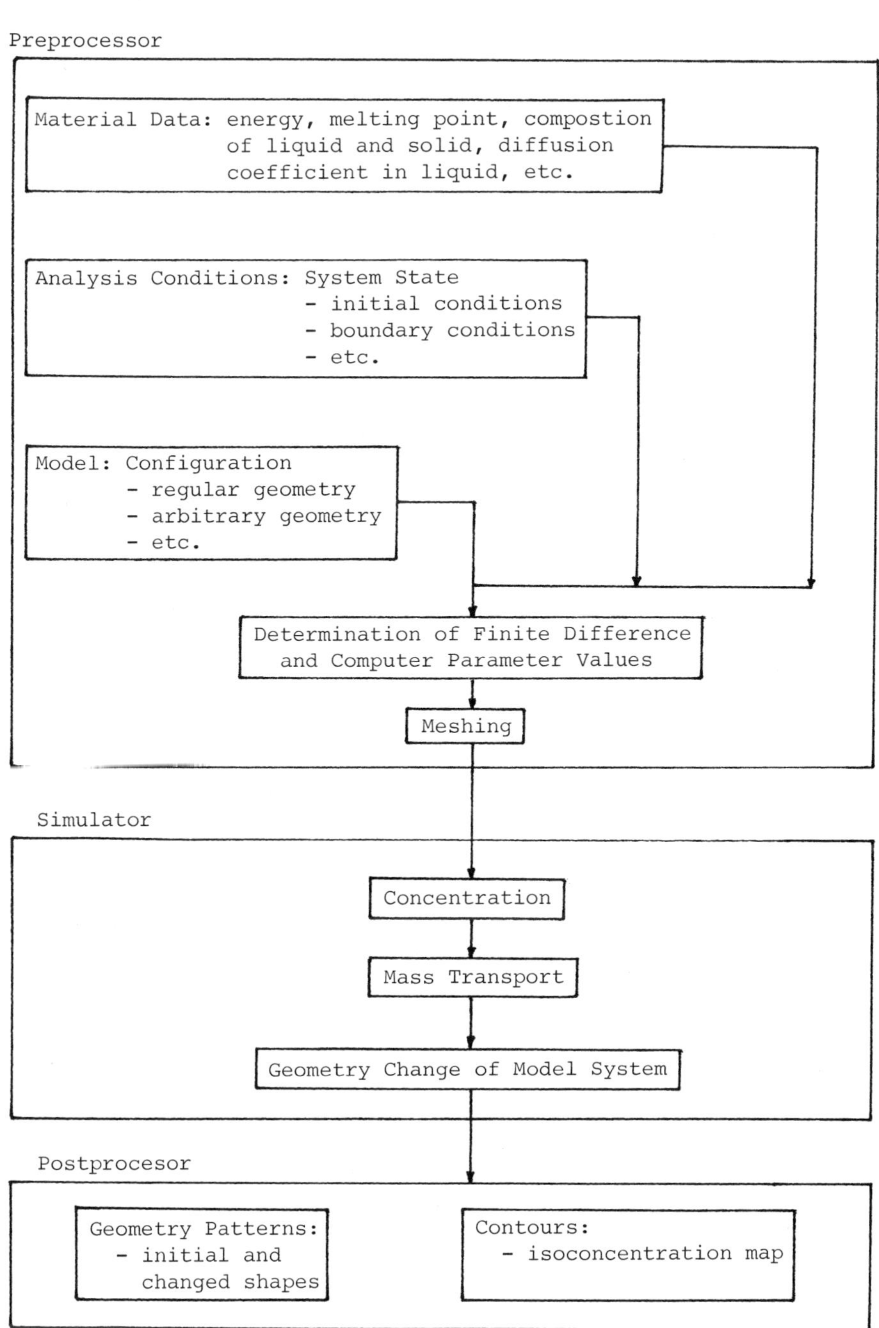

Fig. 1. Structure of computer program SIMTFS.

tem state depends on the previous one, the underterminatess present in the known initialstate may cause the wrong simulation results and rherefore should always be kept in mind in the interpretation of the result. The problem is even more difficult because, in additin to the rich experimental results, many phenomena accompanying the sintering process in the presence of liquid phase are still unknown or insufficieny explained.

In that sense, computer program SIMTFS used the following conditions:

Initial conditions

$$g_{i,j} = \begin{cases} c_L^W, & \text{solid phase} \\[2em] c_L, & \text{liquid phase} \end{cases}$$

$$i=2,n-1; \quad j=2,m-1$$

Boundary condition

At time t=0.

$$f_{1_{i,0}} = f_{2_{i,0}} = f_{3_{j,0}} = f_{4_{j,0}} = C_L, \quad i=1,n; \quad j=1,m$$

At time t>0.

$$C_{i,1,k} = C_{i,2,k} \quad \text{for} \quad C_L \leq C_{i,2,k} \leq C_L^W$$

$$C_{i,m,k} = C_{i,m-1,k} \quad \text{for} \quad C_L \leq C_{i,m-1,k} \leq C_L^W$$

$$i=1,n; \quad k=1,2,\ldots$$

$$C_{1,j,k} = C_{2,j,k} \quad \text{for} \quad C_L \leq C_{2,j,k} \leq C_L^W$$

$$C_{n,j,k} = C_{n-1,j,k} \quad \text{for} \quad C_L \leq C_{n-1,j,k} \leq C_L^W$$

$$j=1,m; \quad k=1,2,\ldots$$

Studing of the liquid phase with W-Ni system was accomplieted the parameter values corresponding to the experimental condition [6]: the composition of the recipitated alloy, $c^{W-Ni} = 99.55$ at.% W; the composition of the liquid in contact with this alloy; $c_L^{W-Ni} = 35$ at.% W; the composition of liquid in contact with pure W, $c_L^W = 35.16$ at.% W; and the diffusion coeficient in the liquid, $D_L = 10^{-5}$ cm^2/s. The mesh of points for approhimate solving of equation (1) was 61x61 points and the size of the experimental region was 600x600/um.

The results of the study of liquid phase sintering from the point of view of process modeling and the application of the simulation method are shown in Fig. 2 and Fig. 3.

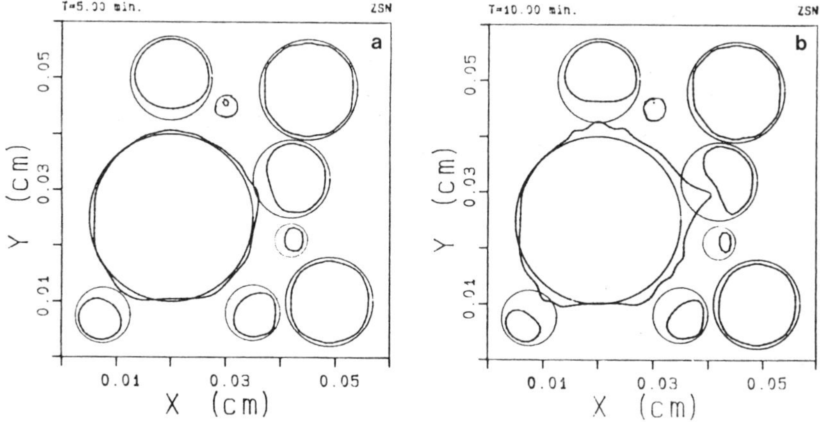

Fig. 2. Microstructural changes due to diffusion
controlled solution-reprecipitation. Nine
spheres model. (a) t=5 min. (b) t=10 min.

Figure 2 shows the results of simulation for the model system of nine spheres of radius 150, 90, 80, 75, 50, 50, 30 and 20 /um respectively. Of greater interest are the geometrical changes in the contact region between the bigest sphere and the other ones. These changes are in reasonable agreement with the characteristic microstructural changes observed in experiments with spherical particles [8].

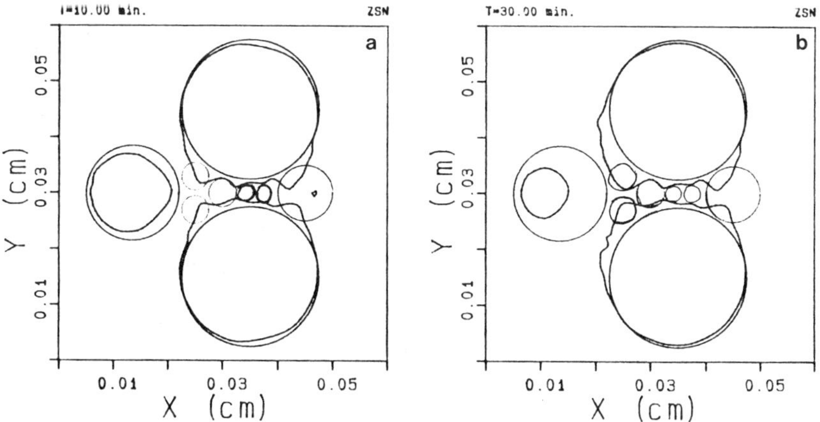

Fig. 3. Microstructural changes due to difusion
controlled solution-reprecipitation. Nine
spheres model. (a)t=10 min. (b) t=30 min.

Figure 3 shows the simulation results for nine spheres model of radius 125, 125, 85, 50, 25, 25,25, 15 and 15 /um, respectively. The calculation for this starting geometry yields results that are similar, to the results in Fig. 2.

These results support the assumption that directional growth during
liquid phase sintering of W-Ni system is controlled by diffusion through
the liquid phase.

REFERENCES

1. Z.S.Nikolić, Ph.D.Thesis, Faculty of Electronic Engineering, University
   of Nish (1980)

2. Z.S.Nikolić, W.J.Huppmann, Acta Met., $\underline{28}$, 475 (1980)

3. Z.S.Nikolić, M.M.Ristić, W.J.Huppmann,1980 International Powder Metal.
   Conf. and Eximition, Washinton, D.C. USA June 22-27, 1980, Modern development in powder metallurgy, 12, Principles and Processes, pp. 497-502

4. Z.S.Nikolić, M.M.Ristić, W.J.Huppmann, Science of Sintering, $\underline{12}$, 19
   (1980)

5. D.N.Yoon, W.J.Huppmann, Practical Metall., $\underline{15}$, 399 (1978)

6. D.N.Yoon, W.J.Huppmann, Acta Met., $\underline{27}$, 973 (1979)

7. G.E.Forsythe, W.R.Wasow, Finite Difference Methods for Partial Differential Equations, John Wiley, New York (1960)

8. D.N. de G.Allen S.C.Dennis, Q.J.Mech. appl. Math., $\underline{4}$, 199 (1951)

INFLUENCE OF SINTERING AND THERMOMECHANICAL TREATMENT ON

MICROSTRUCTURE AND PROPERTIES OF W–Ni–Fe ALLOYS

M. Mitkov and W.A. Kaysser*

Institute Boris Kidrič, Belgrade, Yugoslavia, POB 522
*Max–Planck–Institut für Metallforschung
Pulvermetallurgisches Laboratorium
Heisenbergstr. 5, 7000 Stuttgart 80

ABSTRACT

Densification and microstructural development during sintering of W–Ni and W–Fe–Ni alloys between 1100 and 1470°C were investigated. In solid state sintering densification and grain growth ceased at Ni or Ni/Fe contents above 0.15 wt.%. During liquid phase sintering, densification was enhanced by increasing the amount of lower melting additives. Dense liquid phase sintered samples were cold– or warm–worked to different degrees by rolling and rotation forging. The deformation of individual grains was measured. The distribution of the dislocation densities in individual grains was calculated from the shape and size of grains which formed during annealing of the worked samples in solid or partially liquid state. Dislocation densities of up to 10^{15} m^{-2} were deduced in the vicinity of contact areas of adjacent W grains after only a slight macroscopic deformation of the material by 5%. At higher degrees of deformation recrystallization of deformed initial W grains and their subsequent disintegration by the penetration of liquid phase lead to considerable grain refinement. The mechanisms of recrystallization, stress–induced boundary migration and melt penetration into the recrystallized fine grained materials are discussed.

INTRODUCTION

In spite of the improved understanding of the major mechanisms leading to microstructural changes during liquid phase sintering of heavy metal alloys (1–7), only minor attention has been paid to the influence of prior material deformation on the microstructural development. Since liquid phase sintering in heavy metal alloys occurs before recrystallization starts, the increased dislocation density in some regions of the material may modify the grain growth behaviour. This was indirectly suggested by previous work (8–14). Clear explanations of the mechanisms were impossible, however, because a broad variety of driving and retarding forces were effective during the experiments reported in the literature. Major causes for the broad variety of forces were the presence of pores and chemical non–equilibria. Pores hinder or retard the migration of grain boundaries or other interfaces.(15, 16). Chemical non–equilibria, in particular the non–equilibrium concentration of the W solid solution, have been shown to induce liquid film migration (6, 13, 17–20).

In the present work liquid phase sintered 90W–7Ni–3Fe alloys were cold rolled and rotation forged, yielding various degrees of strain in the W particles. The deformed material was subsequently resintered. Chemical driving forces due to concentration

Science of Sintering
Edited by D. P. Uskoković *et al.*
Plenum Press, New York

changes of the liquid and solid phase during resintering were excluded by performing the initial sintering and the resintering treatment at the same temperature. Due to the absence of pores the microstructural development during resintering was essentially determined by the presence of an increased dislocation density.

EXPERIMENTAL PROCEDURE

The W–Ni–Fe alloys were prepared from fine metal powders by liquid phase sintering at 1470°C in flowing dry hydrogen. The powders had mean particle sizes of 1.5 μm (W), 5 μm (Ni) and 3 μm (Fe). The final density of the sintered material reached the theoretical density. The sintered samples were either cold rolled to give height reductions of 5, 15 and 50 % or rotation forged with a sectional area reduction of the bars of 25 % (without intermediate anneals). Subsequently, the samples were liquid phase sintered at 1470°C. Another set of deformed samples was solid state sintered at 1420°C. All samples were polished and etched by the usual metallographic procedures.

RESULTS

After initial liquid phase sintering (prior to deformation) the alloy consisted of single crystal W(Fe,Ni) solid solution grains with bcc structure, (subsequently called original W grains) embedded in a Fe– and Ni–rich matrix with fcc structure. The W grains were connected at the contact regions by boundaries. Fig. 1 shows the microstructure of samples which were annealed in solid state at 1420°C for 3

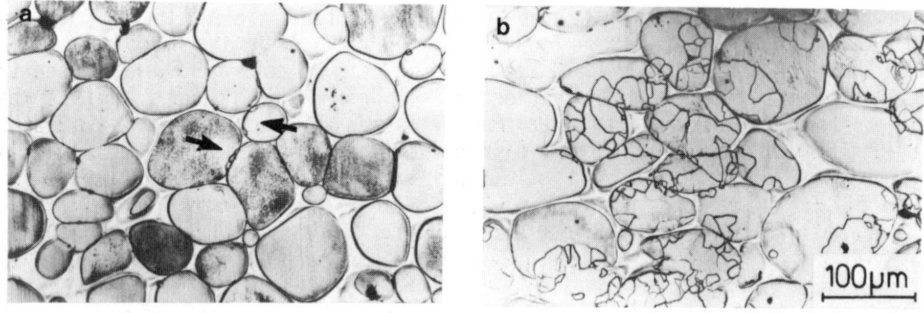

Fig. 1.   Microstructures of prestrained W–7Ni–3Fe heavy metal alloys after solid state sintering at 1420°C for 3 min: (a) 15% reduction, (b) 50% reduction.

min after cold rolling. In the sample reduced by 15 %, new grains with diameters of approximately 6 μm had formed in a few contact areas between the original W grains (arrows in Fig. 1a). In samples reduced 50 %, several new grains formed at most contact areas between the original W grains (Fig. 1b). The microstructures of the forged sample (25% reduction) which was subsequently solid state sintered at 1420°C for 3 and 40 min, are shown in Fig. 2. In the sample reduced by 25% and solid state sintered for 3 min (Fig. 2a) some new grains with diameters of a few microns recrystallized also at contact areas. During further sintering up to 40 min (Fig. 2b) some recrystallized grains grew into the interior of the original W particles forming new grain boundaries, some of them being wavy. The microstructure of prestrained samples after liquid phase sintering at 1470°C are shown in Figs. 3 to 6.

After short liquid phase sintering of the samples, six distinct microstructural features were found which had not been present in the microstructures of the material after initial liquid phase sintering.

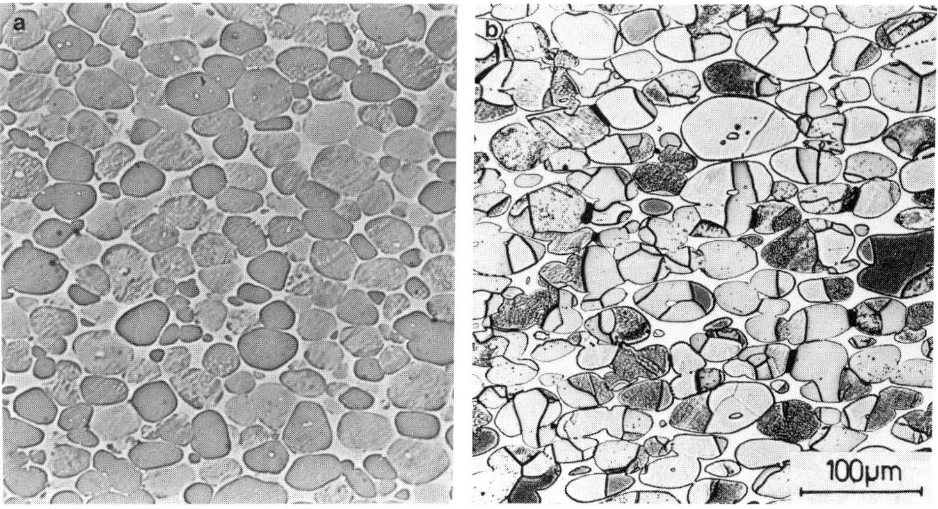

Fig. 2.   Microstructures of a prestrained alloy after solid state sintering at 1420°C for (a) 3 min and (b) 40 min (forged, 25% reduction).

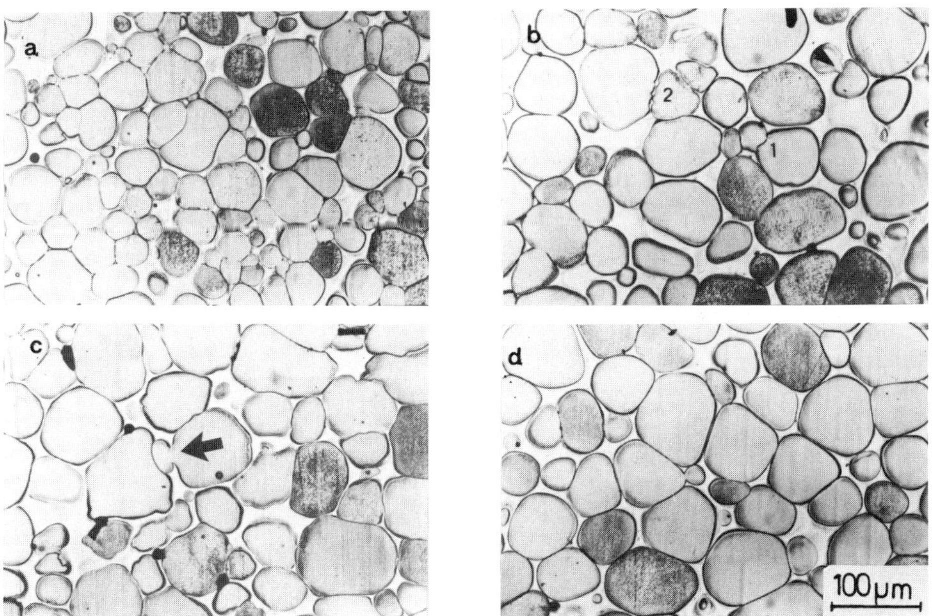

Fig. 3.   Microstructures of a 5 % prestrained W–7Ni–3Fe heavy metal alloy after liquid phase sintering at 1470°C for (a) 1 min, (b) 10 min, (c) 40 min and (d) 720 min.

{i}   New grains with diameters between 5 and 35 μm showed up at the necks and in the interior of original W grains (Fig. 5a, indicated by 3 in Fig. 3a)

{ii}   Original W grains grew at the expense of their immediate neighbour despite the simultaneous increase in interface area caused by this process (indicated by arrows in Figs. 3b, 3c and 6a).

{iii}   Thin liquid films penetrated along the grain boundaries and triple junctions of the newly formed grains, leading to detachment and to disintegration of the original W grains (Figs. 5b and 6c).

{iv}   Thin liquid films were present in many contact areas replacing the initial boundaries (Figs. 4a and 6b).

{v}   Some thin liquid layers between the original W grains developed a wavy shape (indicated by 2 in Figs. 3b, 3c and 6b).

{vi}   Grain boundaries in the necks between original W grains developed a wavy shape (indicated by 1 in Figs. 3b and 6b).

After a short liquid phase sintering (up to 10 min) of samples which were reduced by 5%, the formation of new grains was restricted to the immediate contact region. The diameters of the new grains were between 5 and 7 μm. In addition the features {iv}, {v} and {vi} were found to a large extent.

During prolonged sintering (40 min and more) the features {ii} became more characteristic (Fig. 3c). In 15 % reduced samples liquid phase sintering at 1470°C for 1 and 10 min led to the formation of new grains in the contact areas between the particles with average diameters of 5 and 10 μm (Fig. 4a and b). Some of the newly formed grains still appear to grow into the interior of the single crystal grains after prolonged liquid phase sintering (arrows in Fig. 4c). After liquid phase sintering for 720 min the boundaries between the particles were mostly straight again (Fig. 4d). In 25 % reduced forged samples some new grains appear in the neck regions. Some of them were already grown forming wavy shaped boundaries (Fig. 6a). The shape of the deformed initial grains is still present in the microstructure. Wavy shaped boundaries are also observed in the neck regions between the initial W grains. During prolonged sintering, liquid film penetration became more pronounced (Fig. 6b) and led to further particle disintegration (Fig. 6c). The microstructural analysis (Fig. 6c) yields evidence for miscellaneous grain growth processes.

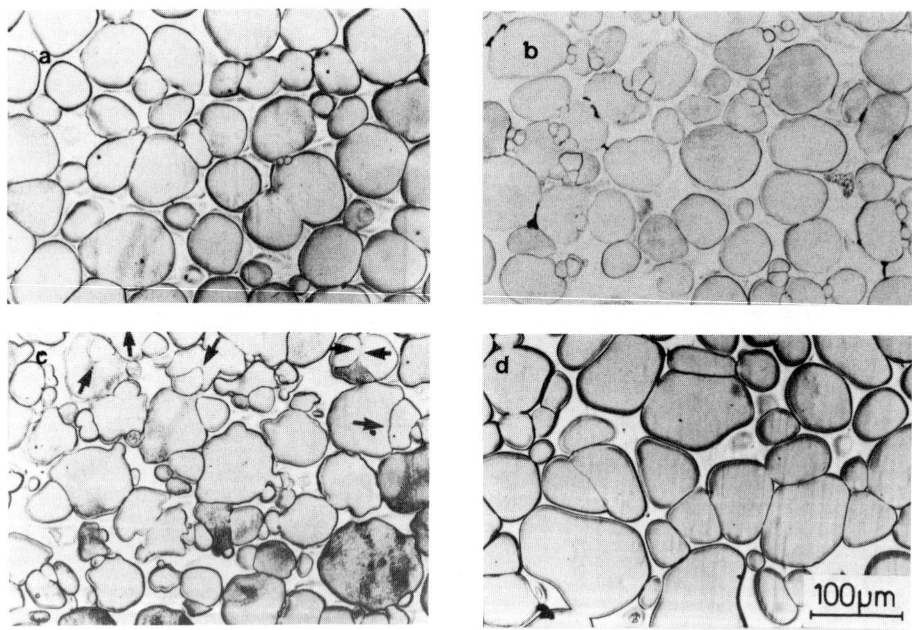

Fig. 4.   Microstructures after liquid phase sintering of a 15% prestrained W–7Ni–3Fe heavy metal alloy at 1470°C for (a) 1 min, (b) 10 min, (c) 40 min, (d) 720 min.

As mentioned above the majority of the initial boundaries in the neck areas were replaced by thin liquid layers during liquid phase sintering of the prestrained samples. The ratio R_1

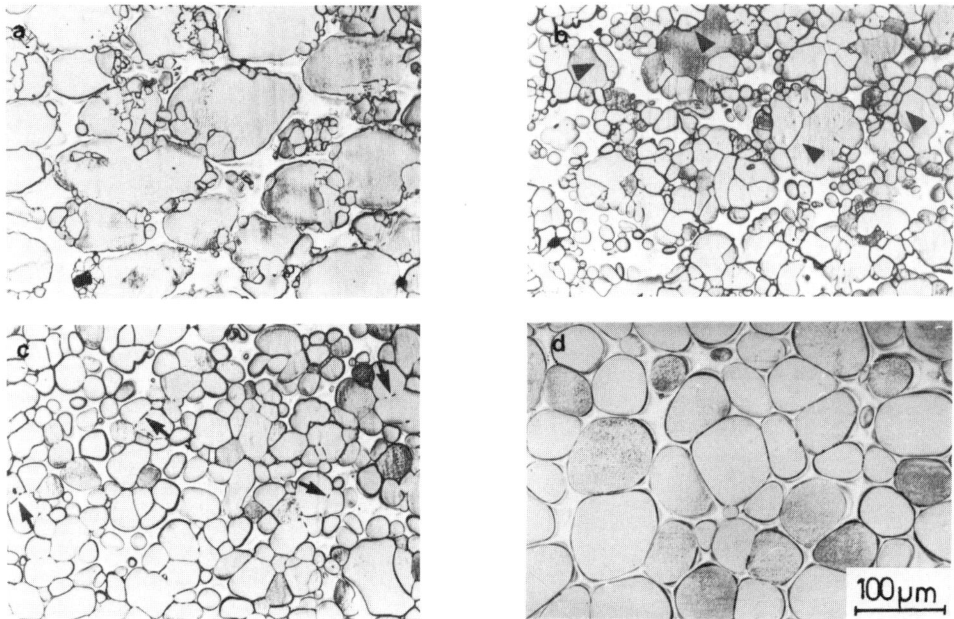

Fig. 5. Microstructures after liquid phase sintering of a 50% prestrained W–7Ni–3Fe heavy metal alloy at 1470°C for a) 1 min, b) 10 min, c) 40 min, d) 720 min.

is defined as the ratio of the contact area where two particles are separated by a thin liquid film to the particle contact area where the particles are separated by a boundary of the type present after initial liquid phase sintering. The changes in R_1 during sintering at 1470°C in samples deformed by 5 and 15% are shown in Fig. 7. The replacement of grain boundaries by liquid films is more pronounced in alloys reduced by 5% than in alloys which were reduced by 15% during cold rolling.

Fig. 5 shows the microstructures of the cold rolled (reduction 50 %) W–Ni–Fe heavy metal alloy after liquid phase sintering at 1470°C. After annealing 1 for min, many new grains with diameters smaller than 5 μm have formed at most contacts (Fig. 5a). Larger grains formed in the interior of the initial W grains (Fig. 5b). After liquid phase sintering for 1 min, only a few areas in the particles were left where no new grains had formed (triangles in Fig. 5b). The formation of new grains is followed by the penetration of melt along the newly formed grain boundaries. At the periphery of the original W grains, the new grains are detached and appear to move toward melt–rich pools. After 40 min of annealing, disintegration of the original W grains is pronounced (Fig. 5c). Small newly formed and detached grains, which were visible in the microstructure after sintering for 10 min, have partly disappeared or spheroidized in the annealing interval between 10 and 40 min (compare Fig. 5b and c). After prolonged liquid phase sintering (720 min, Fig. 5d) the microstructure shows the usual coarsened grains, very similar to those in the inital microstructure of the undeformed liquid phase sintered samples.

Figure 8 shows the average intercept length L_3 of the W(Ni,Fe)–solid solution particles and the detached grains, after sintering, in a sample which had been reduced 50% by cold rolling. The average intercept length after solid state sintering at 1420°C for 3 min is indicated by the asterisk. During liquid phase sintering, the formation of new grains and particle disintegration result in a drastic grain refinement which is still present after annealing for 40 min. During prolonged liquid phase sintering the grains coarsen again. The average intercept length values, L_3, measured after annealing more than 40 min, fit the relation

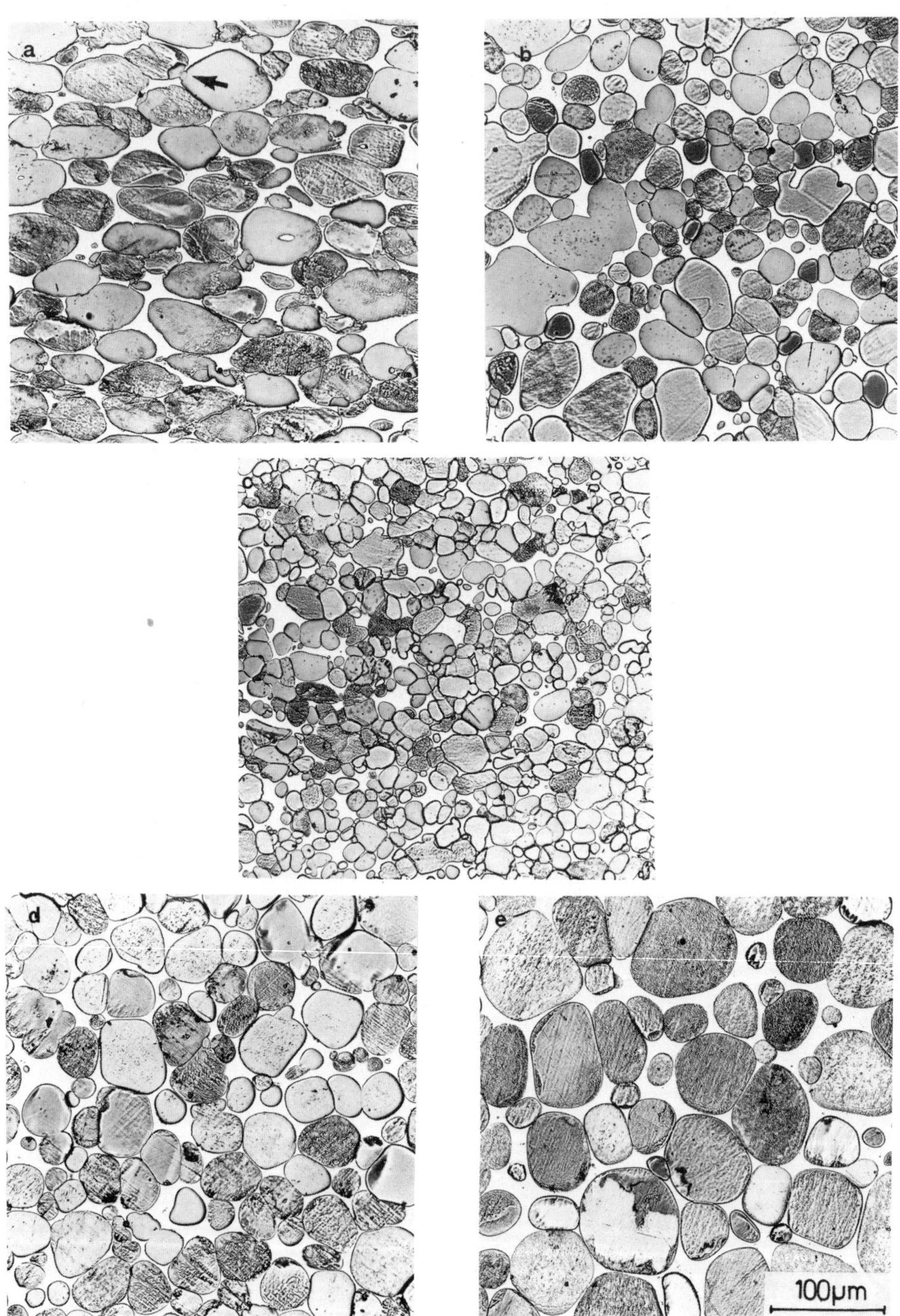

Fig. 6. Microstructures after liquid phase sintering of forged (25 % reduction) alloy at 1470°C for (a) 1 min, (b) 10 min, (c) 40 min, (d) 120 min, (e) 480 min.

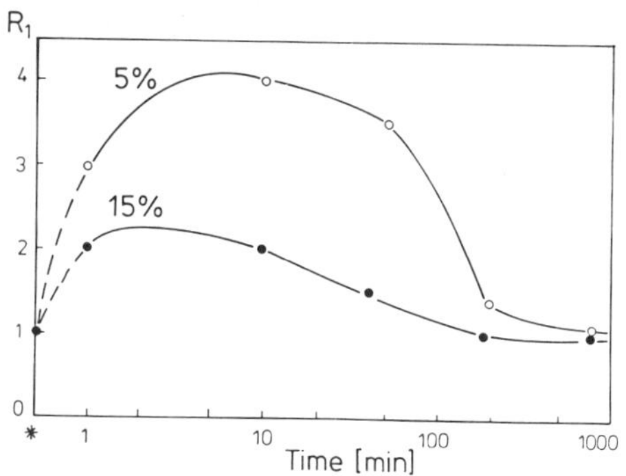

Fig. 7. Particle contact area ratio R_1 in cold rolled W–7Ni–3Fe after liquid phase sintering at 1470°C. * Specimens originally solid state sintered at 1420°C for 3 min. R_1 is defined as the ratio of contact area where two particles are separated by a thin liquid film to the area where the particles are separated by a boundary of the type present after initial liquid phase sintering.

$$L_3(t)^n = L_3(2400)^n + k \cdot t_s, \tag{1}$$

with $n = 1/3$, $k = 3.41 \cdot 10^{-18}$ m³s⁻¹ and $t_s = t - 2400$ s, where t is the isothermal annealing time. Figure 9 shows a schematic summary of the microstructural development observed during solid state and liquid phase sintering of the prestrained W–Ni–Fe alloy.

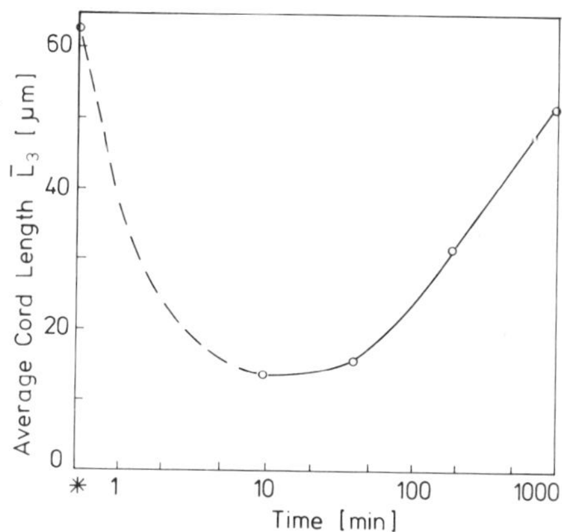

Fig. 8. Average intercept (chord) length of grains of a W–7Ni–3Fe heavy metal alloy, resintered at 1470°C in the presence of a liquid phase. (After initial liquid phase sintering the sample height was reduced 50 % by cold rolling; * indicates the state after solid state sintering of the prestrained sample at 1420°C for 20 min).

DISCUSSION

Recrystallization, Grain Growth and Dislocation Density

During solid state and liquid phase annealing of cold deformed W heavy metal alloy, new grains form in contact areas and in the interior of the original single crystal W(Ni, Fe) grains. The formation and growth of the new grains depends on energy reduction accomplished through the elimination or rearrangement of dislocations. The reduction in strain energy must overcome the energy increase which results from the formation of additional interface area when the new grains form and grow. The present

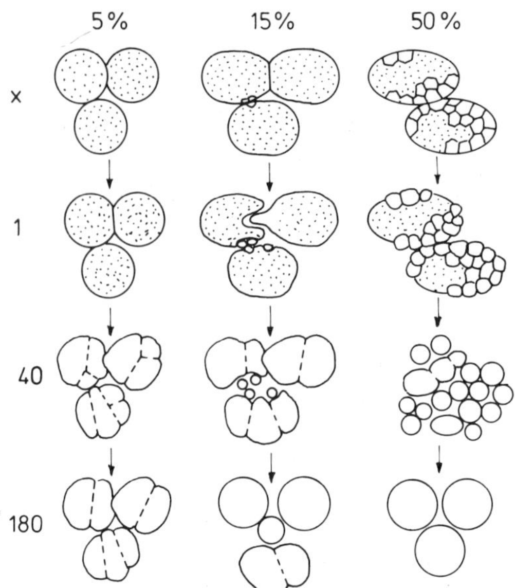

Fig. 9.   Schematic description of the microstructural development during solid state and liquid phase sintering of a prestrained W–7Ni–3Fe alloy. x after solid state sintering at 1420°C for 3 min; 1, 40, 180 after liquid phase sintering at 1470°C for 1, 40 and 180 min.

observations will be discussed first, with respect to the magnitude of both energies and second, with respect to the resulting driving forces. The maximum potential difference at a grain boundary due to the dislocation density difference at either side is obtained when the coherency of the adjacent grains is completely lost and the energy of the dislocations transforms directly without relaxation into an average increase of the chemical potential, $\Delta\mu_1$. For this case, the increase in the chemical potential is estimated by (21, 22)

$$\Delta\mu_1 = \Omega E_v \, \Delta\rho \tag{2}$$

where E_v is the average specific line energy of the dislocations, Ω is the molar volume of the host material and $\Delta\rho$ is the change in the dislocation density. The isolated approximately spherical grains in the originally single crystal W grains tend to shrink due to their curved interfaces. The unrelaxed potential difference at either side of the grain boundary due to the curvature is (22).

$$\Delta\mu_2 = -4\gamma_{gb} \, \Omega/G \tag{3}$$

where $4/G$ is the sum of the two principial curvatures. In approximation, G is the grain size, and γ_{gb} is the grain boundary energy. Calculations with $\Delta\mu_1 = \Delta\mu_2$ indicate that the reduction of the dislocation density by the newly formed grains has to be considerable to overcome the surface tension of the curved interface.

After liquid phase sintering of 5% reduced samples new grains were observed in the contact areas with sizes between 5 and 7 μm, which is equivalent to a reduction in dislocation densities on the order of 10^{15} m^{-2}. The dislocation density initiating recrystallization in deformed metals is usually considered to be approximately 10^{16} m^{-2} (26). However, recrystallization nuclei, in general, are not ideally perfect and can contain initial dislocations. For that reason, the difference in dislocation density which provides sufficient driving force for primary recrystallization is always less than 10^{16} m^{-2}.

Larger grains were not observed. This indicates that the deformation occuring outside the intimate contact areas is small and that the dislocation density decreases steeply to lower values within 3 μm from the contact plane.

After sintering, samples which were reduced 15 % at 1420°C for 3 min, showed diameters of the newly formed grains (Fig. 1a) which were less than 6 μm. The formation of the grains obviously required a minimum dislocation density greater than 10^{15} m^{-2}. The deformation of contact areas yields dislocation densities in this order of magnitude and higher, within 3 μm from the contact planes. During liquid phase sintering, the decrease of the dislocation density in the wake of moving boundaries caused thin liquid films to migrate from the contact areas against the directions of their centers of curvature into one of the adjacent grains. This indicates a dislocation density of 10^{14} m^{-2} at distances of 15 and 25 μm from the contact areas, respectively.

After liquid phase sintering of samples reduced by 50 %, grains smaller than 5 and 9 μm in size were present in contact areas and in other peripheral areas of the original W grains. This corresponds to equivalent dislocation densities above $2 \cdot 10^{15}$ and $6 \cdot 10^{14}$ m^{-2}, respectively. In the interior of the initial W grains, dislocation densities of $1.3 \cdot 10^{14}$ m^{-2} were indicated. The microstructures in Fig. 5a and 5b indicate that the recrystallization starts at the periphery of the particles and proceeds toward the interior of the particles. The delayed start of recrystallization and the larger size of the newly formed grains in the inner parts of the particles are likely to be due to the lower dislocation densities in these areas.

Fig. 10 shows schematically the typical minimum dislocation densities in various areas of the prestrained initial W grains obtained from the size of grains newly formed during solid state and liquid phase sintering of the prestrained material.

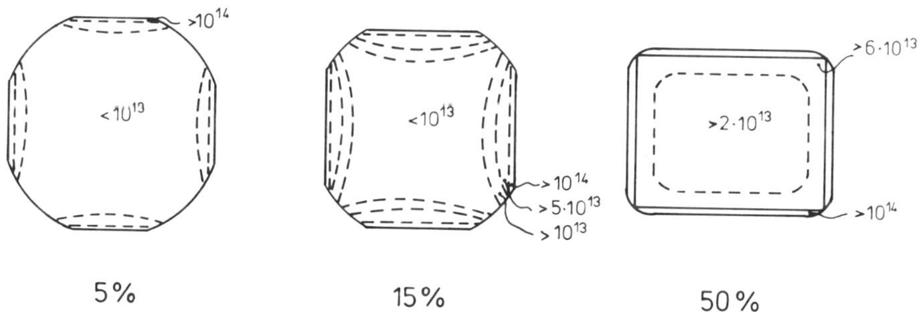

Fig. 10.   Dislocation density (in m^{-2}) in W grains of prestrained W–Ni–Fe. Reduction by cold rolling in %.

Liquid Film Penetration during Liquid Phase Sintering

The penetration mechanism of thin liquid films along the boundaries between original W grains and along the grain boundaries of newly formed grains is not yet clear. Nevertheless, important changes of the microstructure are caused by the penetration, e.g., a reduction of the grain size instead of the usual grain coarsening.

After liquid phase sintering of the slightly deformed alloys (5 % reduction), thin liquid films (thickness < 1 μm) became visible in many contact areas. This means that the initial boundaries were replaced by liquid phase layers (Figs 3a, 3b and 5). The replacement requires that the specific boundary energy, γ_{gb}, be larger than twice the interface energy liquid/solid, γ_{sl}. It is suggested that the excess free energy of the boundary in the presence of high dislocation densities located near the boundaries, i.e., after deformation is composed of (a) the increased energy of the atoms at the interface itself and (b) of the increased energy of thin adjacent bulk layers.

Fig. 11 shows a model where a liquid film of approx. 1 μm thickness penetrates along prior grain boundaries (23). At the top of the penetrating liquid film, material with high dislocation density dissolves in the melt, diffuses along the melt layer and precipitates at solid/liquid interfaces which are not part of the contact areas (24). With an initial thickness of the "liquid film" of $1.24 \cdot 10^{-10}$ m (roughly a monolayer of Ni being segregated at the boundary) the model yields 1016, 164 and 32 s for the complete penetration into a boundary at the contact area with the dislocation densities of 10^{13}, 10^{14} and 10^{15} m^{-2}. These values fit well to the experimentally observed times (compare Figs. 3, 4 and 5). In addition, some evidence of the described penetration mechanism is given by the microstructures (x in Fig.4).

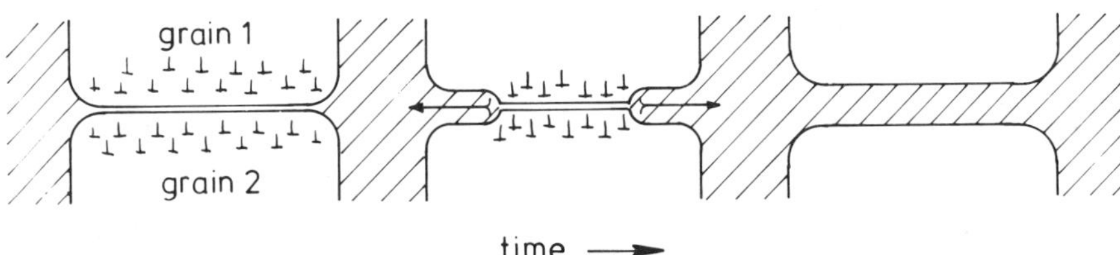

Fig. 11.   Schematic representation of the penetration of a melt layer along a deformed boundary between original W grains.

Disintegration of Original W Grains

During prolonged liquid phase sintering, grains which were newly formed in original W grains were detached from the original grains and moved toward adjacent liquid pools (Fig. 5b). The detachment of grains from the polycrystalline areas requires the penetration of melt along the grain boundaries. The detachment also requires sufficient cross sectional area for the melt to flow into the small separating areas during the grain movement. In addition, a certain spheroidization of the grains at triple junctions and at the nodes of four adjacent grains is necessary. The spheroidization is equivalent to the opening of triangular melt channels along the triple junctions of grain boundaries. The melt channels at triple junctions of grain boundaries tend to increase in size if γ_{gb} is larger than $2\gamma_{sl}$. Calculations on various models suggest that a fast opening of the melt channels along triple junctions depends on a prior melt penetration along the planar boundary areas (25).

CONCLUSIONS

In samples which are 5 % reduced the deformation is essentially restricted to the immediate contact areas. Due to thin dislocation—rich layers at the boundaries, most of the initial boundaries between the particles were replaced by thin liquid films up to 1 μm in thickness. The penetration occurs by a mechanism where material of increased dislocation density dissolves at the front of the penetrating liquid film and precipitates with lower dislocation density at other grains.

After 15 % reduction the resintered samples show liquid film migration driven by the difference in dislocation density of the dissolving and the reprecipitating material. The dislocation densities are 10^{15} m^{-2} in the contact areas with a depth of up to 10 μm and between $1 \cdot 10^{14}$ and $8 \cdot 10^{14}$ m^{-2} in the interior of the original W grains.

A 50 % reduction of the original W grains yields an excess dislocation density of $5 \cdot 10^{15}$ m^{-2} at the periphery of the grains which decreases continuously to $5 \cdot 10^{13}$ m^{-2} toward the interior of the original grains. The boundaries of grains forming in the original W grains during liquid phase sintering are subsequently penetrated by thin liquid films. Melt channels at triple junctions widen until detachment of the grains occurs, resulting in a complete disintegration of the recrystallized original grains. This mechanism reduces the average grain size by a factor of four.

ACKNOWLEDGEMENT

We thank the International Büro of the KFA–Jülich for the continuous support of this work.

REFERENCES

1. W.J. Huppmann et al., "The Elementary Mechanisms of Liquid Phase Sintering — I. Rearrangement", Z. Metallkde., **70** (1979) 707–713.
2. W.J. Huppmann et al., "The Elementary Mechanisms of Liquid Phase Sintering — II. Solution–Reprecipitation", Z. Metallkde., **70** (1979) 792–797.
3. G. Petzow and W.A. Kaysser, "Basic Mechanisms of Liquid Phase Sintering," Sintered Metal–Ceramic Composites, ed. G.S. Upadhyaya, Amsterdam, Elsevier Science Publishers, (1984), 51–70.
4. W.A. Kaysser and G. Petzow, "Present State of Liquid Phase Sintering", Powder Metallurgy, **28** (1985) 145–150.
5. D.N. Yoon and W.J. Huppmann, "Grain Growth and Densification during Liquid Phase Sintering of W–Ni", Acta Metall. **27** (1979) 693–698.
6. D.N. Yoon and W.J. Huppmann, "Chemically Driven Growth of Tungsten Grains during Sintering with Liquid Nickel", Acta Metall., **27** (1979) 973–977.
7. S.S. Kim and D.N. Yoon, "Coarsening Behaviour of Mo Grains Dispersed in Liquid Matrix", Acta Metall, **31** (1983) 1151–1157.
8. E.G. Zukas and D.T. Easch, "Possible Reinforcement of the W–Ni–Fe–Composite with W Fibers", J. Less–Comm. Met., **32** (1973) 345–353.
9. E.G. Zukas, P.S.Z. Rogers and R.S. Rogers, "Unusual Spheroid Behaviour During Liquid Phase Sintering", Int. J. of Powd. Met. and Powd. Tech., **13** (1977) 27–33.
10. G. Petzow and W.J. Huppmann, "Flüssigphasensintern", Z. Metallkde., **67** (1976) 575–582.
11. M. Hofmann–Amtenbrink et al., "Kosseluntersuchungen zur Korngrenzenwanderung in verformtem Wolfram in Anwesenheit von Nickel", Z. Metallkde., **77** (1986) 368–376.
12. W.A. Kaysser, F. Puckert and G. Petzow, "Recrystallization and Grain Boundary Migration During Sintering", Powder Metallurgy International, **12** (1980) 188–191.
13. W.A. Kaysser and S. Pejovnik, "Grain Boundary Migration During Sintering of Mo with Ni Aditions", Z. Metallkde., **71** (1980) 649–653.
14. W. Schatt et al., "Influence of Dislocations on Solid State and Liquid Phase Sintering", Powder Metallurgy International, **19** (1987) 14–19, and **19** (1987) 37–39.
15. R. Brook, "Pore–Grain Boundary Interactions and Grain Growth", J. Am. Ceram. Soc., **52** (1969) 56–57.
16. C.H. Hsueh, A.G. Evans, and R.L. Coble, "Microstructure Development during Final/Intermediate Stage Sintering — I. Pore/Grain Boundary Separation", Acta Metall., **30** (1982) 1269–1279.
17. C.A. Handwerker, J.W. Cahn, D.N. Yoon, and J.E. Blendell, "The Effect of Coherency Strains on Alloy Formation: Migration of Liquid Films", Atomic Transport in Alloys: Recent Developments", G.E. Murch and M.A. Dayananda (eds), (Dayton, OH, TMS/AIME Publications, 1985).

18. D.N. Yoon, J.W. Cahn, C.A. Handwerker, J.E. Blendell, and Y.J. Baik, "Coherency Strain Induced Migration of Liquid Films through Solids", Interface Migration and Control of Microstructure (Washington D.C., NBS; 1985).

19. W.–H. Rhee, Y.–D. Song, and D.N. Yoon, "A Criitical Test for the Coherency Strain Effect on Liquid Film and Grain Boundary Migration in Mo–Ni–(Co–Sn) Alloy", Acta Metall., **35** (1987) 57–60.

20. M. Hofmann–Amtenbrink, W.A. Kaysser, and G. Petzow, "Grain Boundary Migration in Recrystallized Mo Foils in the Presence of Ni", J. Physique, **C4** (1985) 545–552.

21. P.A. Beck, P.R. Sperry, and H. Hu, "The Orientation Dependence of the Rate of Grain Boundary Migration", J. Appl. Phys., **5** (1950) 420–425.

22. H.P Stuewe, "Driving and Dragging Forces in Recrystallization", Recrystallization of Metallic Materials, ed. F. Haessner, Stuttgart, Riederer Verlag, (1978), 11–21.

23. W.A. Kaysser, W.J. Huppmann, and G. Petzow, "Analysis of Dimensional Changes During Sintering of Fe–Cu", Powder Metallurgy, **23** (1980) 86–91.

24. S.–J.L. Kang et al., "Growth of Mo Grains around Al_2O_3 Particles during Liquid Phase Sintering", Acta Metall., **33** (1985) 1919–1926.

25. W.A. Kaysser, "Influence of Dislocations on the Basic Mechanisms of Liquid Phase Sintering in W–Ni–Fe", to be submitted to Int. J. Refractory and Hardmetals.

26. S.S. Gorelik, "Recrystallization in Metals and Alloys", Mir Publishers, Moscow, 1981.

REMOVAL OF Ni-ACTIVATOR FROM THE ACTIVATED SINTERED W-COMPACT BY HIGH
VACUUM TREATMENT

In-Hyung Moon, Yuong-Hwan Kim, and Myung-Jin Suk

Department of Materials Engineering, Han Yang University
Seoul 133-791, Korea

INTRODUCTION

A small amount of nickel added to W-powder as an activator can
increase greatly the sinterability of the W-powder compact, enabling
full densification to be achieved at the relatively low sintering
temperature of about $1200^{\circ}C$-$1400^{\circ}C$. However, the presence of Ni-
activator causes a negative effect on the mechanical properties of the
activatedly sintered W-parts, resulting in extreme brittleness
associated with a brittle Ni-rich phase. Therefore, Ni-activated
sintered W-compacts can not be subjected to further plastic working
process for a final shaping.[1] If after playing a positive role as an
activator in the first stage of sintering,the Ni-activator can be
removed fully or partly from the activated sintered W-compact, the
degree of detrimental effect might be substantially reduced.

Nickel added to W-powder as an activator we usually found to be
segregated mainly at W-grain boundaries as well as in the grain
boundary junctions, forming a Ni-rich interlayer phase or a Ni-rich
phase aggregate pool. Only the minor portion of nickel is soluble in W-
grain. The amount of Ni-solute atom in W-matrix cannot exceed the
maximum solubility of nickel in W at the pertinent sintering
temperature. It is about 0.06 wt.% at $1400^{\circ}C$. The equilibrium vapor
pressure of Ni is relatively high at $1400^{\circ}C$, the order of 10^{-3} torr.[3]
Therefore Ni can be effectively evaporated in high vacuum at this
temperature. If nickel existing as Ni-rich phase at a W-grain boundary
can be removed appropriately by evaporation, the residual nickel
content in the sintered W-compact will be reduced to 1/4 - 1/8 of the
initial content of nickel, 0.2 - 0.4 wt.% Ni. Furthermore, this
remainder after evaporation treatment will be existing in the W-matrix
as solute atoms. Therefore it may be no longer be so harmful to the
sintered properties, as in the abovementioned case, where it formed a
brittle Ni-rich second phase.

In the present study, the possibility of Ni-removal by evaporation
treatment was investigated in order to improve the microstructural
features and mechanical properties of activated sintered W-compacts for
use in engineering applications.

Science of Sintering
Edited by D. P. Uskoković *et al.*
Plenum Press, New York

SOME CONSIDERATION

Evaporation Rate

The net vaporization flux from the clean metallic surface is generally expressed in the form of the ideal equation of Hertz and Knudsen[4] as follows;

$$J = \frac{\alpha(p_e - p)}{(2\pi mkT)^{1/2}} \tag{1}$$

where p, the vapor pressure; p_e, the equilibrium vapor pressure; m, molecular mass; k, Boltzman's constant; T, the absolute temperature; and α is the vaporization coefficient, which is dependent on the surface properties of the metal and is nearly equal to unity for a polycrystalline surface free of any severe facetting. If the reaction chamber is maintained in high vacuum during evaporation treatment, then, $p_e > p$, and Eq.(1) can be expressed as

$$J = \frac{p_e}{(2\pi mkT)^{1/2}} \tag{2}$$

for the polycrystalline specimens.

Equilibrium Vapor Pressure of W-Ni System

The equilibrium vapor pressure of each constituent in the present system at 1400°C was calculated from the data given in literature as follows:

$$p_W = 1.2 \times 10^{-15} \text{torr}, \quad P_{WO_3} = 8.4 \text{ torr and } p_{Ni} = 1.14 \times 10^{-3} \text{ torr}.$$

The possibility of NiO formation is excluded at 1400°C for the present system, because of very low oxygen partial pressure in the high vacuum state.

However, nickel does exist in this case, occuring as a Ni-rich phase with maximum solubility of W at 17 at.% at 1400°C. The equilibrium vapor pressure of this Ni(W) rich phase at 1400°C can be calculated by using the excess free energy data[2] and further by assuming that it is a regular solution:

$$P_{Ni(w)} = 9.4 \times 10^{-4} \text{ torr at } 1400°C$$

In the present experimental condition (high vacuum, better than 1×10^{-4} torr), the evaporation of W-atoms is negligible, because of the very low equilibrium vapor pressure of tungsten at 1400°C. The evaporation rate of WO_3 will be controlled by the partial pressure of oxygen on the W-surface, because the formation of WO_3 should precede the evaporation of WO_3. But Ni-evaporation rate is dependent only on the temperature, as long as Ni is supplied continuously to the surface of the specimen from the inner part of the specimen. By putting the above value of $P_{Ni(w)}$ into Eq.(2), one can obtain Ni evaporation flux:

$$J_{Ni} = 6.14 \times 10^{-4} \text{ g/cm}^2 \cdot \text{min}$$

Evaporation Area of Nickel

The volume fraction of Ni-rich phase in W-Ni system with X amount of Ni can be obtained by following relation for the temperature of 1400°C

$$f_{Ni} = \frac{\rho_{th}(X - 0.06)}{8.9 \times 10^2}$$

where ρ_{th} is the theoretical density of the W-Ni system, 8.9 is the density of Ni in g/cm^3 and 0.06 is the maximum solubility of Ni in W in wt.% at 1400°C, and the molar volume of Ni-rich phase is assumed to be the same as that of nickel.

If Ni is distributed homogeneously in the specimen, the surface area fraction of Ni-rich phase in the total specimen surface area should be a function of the 2/3 power of the volume fraction value given above:

$$f_{Ni}^{s} = [\frac{\rho_{th}(X - 0.06)}{8.9 \ X \ 10^2}]^{2/3} \tag{3}$$

Values of f_{Ni}^{s} = 0.0378 and 0.0209 are obtained for specimens containing 0.4 wt.% and 0.2 wt.% Ni, respectively.

If A_t and A_g are the real surface area and the geometrical surface area of the specimen and its surface roughness factor is r, then the total evaporation area of nickel phase, A_{Nl}, can be expressed as

$$A_{Ni} = A_t \cdot f_{Ni}^{s} = A_g \cdot r \cdot f_{Ni}^{s} \tag{4}$$

The sintered specimen may contain some open pores. If it can be assumed that the surface of the open pores has about the same roughness factors as that of the specimen surface, then Eq.(4) can be rewritten as

$$A_{Ni} = (A_g + A_p) \cdot r \cdot f_{Ni}^{s} \tag{5}$$

where A_p is surface area of open pores.

Measurement of Ni-Removal Rate

The amount of Ni removed from the specimen during vacuum treatment can be determined directly by measuring the weight loss of the specimen, dM_{Ni}, during the Ni evaporation. This quantity is related to Ni-evaporating flux as follows:

$$J_{Ni} = \frac{1}{A} \frac{dM_{Ni}}{dt} \tag{6}$$

where M_{Ni} is the total amount of Ni removed during the evaporation treatment over time t and it represents the difference between the total weight loss and the weight loss due to WO_3-evaporation. For a specimen with open pore structure, one obtains from Eq.(5) and Eq.(6):

$$J_{Ni} = \frac{1}{(A_g + A_p)r \cdot f_{Ni}^{s}} \frac{dM_{Ni}}{dt} \tag{7}$$

Total Pore Surface Area

In principle the surface area of closed pores cannot contribute to the measured rate of Ni-evaporation in the present case. The pores in a sintered compact having a porosity of less than 6-8 % will consist exclusively of closed pores,[5] and those in a specimen having porosity higher than 10 % will be mainly of open pore structure. Therefore, the evaporating atoms from the surface of the open pores can be removed out of the specimen through open pore channels via specimen surface, while the evaporating atoms from the surface of the closed pores are confined within these pores at an equilibrium vapor pressure.

In a porous part, the specific surface area of pores, S_v, defined as the pore surface area per unit volume of the sintered compact (in

cm^2/cm^3), is related to sintered density D_s and the theoretical density D_{th} of the sintered material for the intermediate stage of sintering as follows:

$$D_s = -m\, S_v + D_{th} \qquad (8)$$

where m is constant. Therefore,

$$S_v = \frac{D_{th} - D_s}{m} = \frac{(D_{th} - D_s)}{D_{th}} \cdot \frac{D_{th}}{m} = P_o \cdot \frac{D_{th}}{m} \qquad (9)$$

where P_o is porosity of the specimen.

As given in Eq.(9), the surface area of pores is directly proportional to the porosity in some range that is dependent on the sintered density.

EXPERIMENTAL

Activated sintered W-specimens with 0.2 or 0.4 wt.% Ni additive and with various levels of porosity were prepared by a conventional method; 0.2 or 0.4 wt.% Ni was added to W-powder, which had an average particle size of 1.53 μm and 4.22 μm with a purity of 99.9 %, by a Ni-salt solution and reduction method.[6] Ni-added W-powder was compacted into a disc shape of 10.8 mm diameter and 8 mm height: various green densities were obtained by applying bidirectional compacting pressures over the range 200 MPa - 400 MPa. These green compacts were sintered at 1400°C in H_2-atmosphere for various times. The sintered density of the specimen was found to range from 70 to 98 % of theoretical density, depending on the sintering time.

The sintered W-specimens with 0.2 or 0.4 wt.% Ni content together with a pure W reference specimen of same porosity, were subjected to evaporation treatment in a vacuum better than 1×10^{-4} torr at 1400°C, achieved by using oil diffusion vacuum pump. In the heating up stage, a relatively lower heating rate was adopted in order to avoid the sudden change of vacuum due to the rapid heating. The amount of nickel removed was determined by means of weight loss measurements and by analyzing the residual content of Ni with the aid of an AA spectrometer. The weight loss was measured by subtracting the weight loss amount of the W reference specimen from that of the Ni doped W-specimen, and the residual Ni-content in the specimen was determined after each vacuum heat tratment cycle with an AA spectrometer (Hitachi model 180-30).

The Ni-evaporation rate was obtained by dividing the removed amount of Ni by the geometrical surface area of the specimen and its evaporation treatment time. Furthermore the microstructural changes due to the evaporation were examined and qualitatively analyzed with SEM and EDS.

EXPERIMENTAL RESULTS AND DISCUSSION

Fig. 1 shows the relationship betwen weight loss and evaporation treatment time for the 0.4 wt.% Ni-added W-sintered specimen with a porosity of 3.9 % and for the pure W-reference specimen at full density. The former specimen was prepared by an activated sintering at 1400°C for 1 hour in H_2-atmosphere and the latter was a part of commercial pure W-TIG electrode with a purity of 99.75 %W. As shown in this figure, the weight loss due to evaporation, either by Ni-atoms or by WO_3-compound, is to some extent proportional to the time of

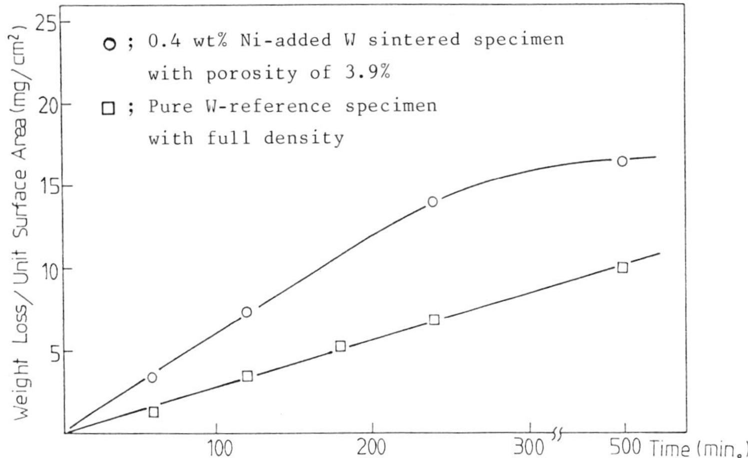

Fig. 1. Weight loss due to evaporation as a function of evaporation
time for 0.4 wt.% Ni added W-sintered specimens of 3.9 %
porosity and pure W-reference specimens of full density.

evaporation treatment for up to 4 hours for both specimens. The weight
loss _vs._ treatment time-curve decreased its slope slightly after a
prolonged treatment time for the Ni activated sintered W specimen,
while the curve for the reference specimen of the pure tungsten kept
its linearity constant. Such a decrease in evaporating rate after a
certain time might be attributed to a decreasing amount of total
residual nickel available in the specimen as well as to a decrease in
the fraction of open pores due to closing of pore channels associated
with further densification effects during heat treatment. However, the
latter case should be excluded in the present condition, because the
present specimen had an initial porosity of 3.9 %, and thus contained
only closed pores.

The measured weight loss rate of the pure W-specimen was about
2.6×10^{-5} g/cm^2 min. As discussed in a previous section during
evaporation treatment the W-atoms were removed from the surface of W-
specimen in the form of the oxide compound, WO_3. This evaporation rate
has a somewhat higher value than the one given in the literature.[7] The
weight loss rate of the Ni-activated sintered W-specimen was 5.5×10^{-5}
g/cm^2 min., which is higher than that of a pure W-reference specimen by
a factor of two. Such a high weight loss was derived, of course, from
the superposition effect of the Ni-evaporation flux and the WO_3-
evaporation flux. Therefore, the evaporation rate for nickel can be
obtained simply by subtracting the weight loss rate of pure W-specimens
from that of the Ni-activated sintered W-specimens, as far as it is
assumed that the small amount of Ni added to W does not influence the
reaction rate of WO_3 formation. It is about 2.9×10^{-5} g/cm^2 min. If
this measured value of Ni-evaporation rate is put into Eq.(6), one can
estimate the total evaporation area of nickel, A_{Ni}, which is related
with Ni-content and surface roughness factor as given in Eq.(4). The
value of A_{Ni} was 0.0472 cm^2 per geometrical unit area for the present
specimen. As f_{Ni} = 0.0376 for the specimen with 0.4 wt.% Ni, the
surface roughness factor of the test specimen should be 1.25 according
to Eq.(4). This result seems to be reasonable one, because the surface
of the test specimen was finished by polishing with 1000 mesh abrasive.[8]

Fig. 2 shows the dependence of the residual amount of nickel on
the evaporation treatment time in the 0.2 wt.% Ni added W-sintered

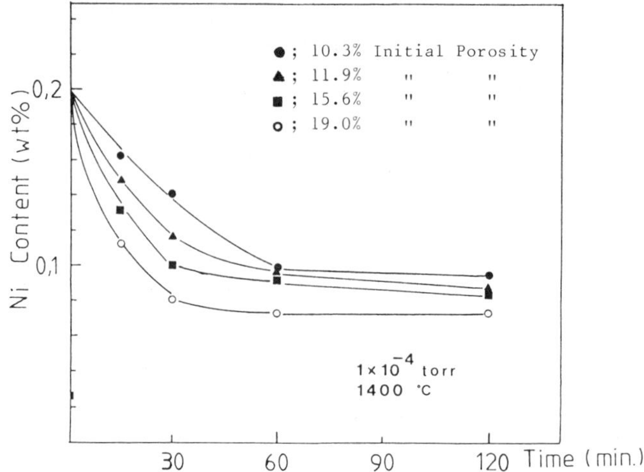

Fig. 2. Dependence of residual amount of nickel on the evaporation time
in 0.2 wt.% Ni added W-sintered specimens with initial porosity
between 10 and 20 %.

specimens with initial porosities between 10 and 20 %. The amount of
residual Ni was determined by AAS analysis. The residual Ni content
decreased with increasing time in an approximately exponential form.
Therefore, the residual Ni content in the specimen was kept nearly
constant after 60 min.; the Ni-removal rate decreased very rapidly to
negligible value within 120 min. The residual Ni content was less than
0.07 wt.% for a specimen of open pore structure with a porosity of 19%
after it was vacuum treated for 120 min. at 1400°C. It might be
therefore concluded that nickel existing originally as a Ni-rich phase
in W-specimen was fully removed, except for those remnant Ni atoms
dissolved in the W-matrix.

The Ni-removal rate, especially, in the initial stage of treatment
for up to 60 min. was also dependent on the porosity of the sintered
specimen, as shown in the same figure. The higher the porosity, the
higher the Ni-removal rate obtained. This result is expected, because
the surfaces of the open pores increase with increasing porosity. In
this figure one can find also a rapid decrease in residual amount of
nickel with evaporation treatment time. It should be attributed firstly
to the decrease of total Ni-content near the free surface area and
partly to the decrease of open pores by densification effects, as
described previously. If Ni located on the surface of the specimen as
well as on the surface of the pores is completely removed by
evaporation, the Ni-removal kinetics will be controlled by a
diffusional process capable of transporting nickel atoms from inner
part to the exposed surface area through either grain boundaries or
lattice volumes.

Fig. 3 shows the fractographs of one of the tested specimens, for
which the different positions are marked as A, B and C. This specimen
with 0.4 wt.% Ni additive had an initial porosity of 9.1 %, i.e. the
pore volume consisted of both open pores and closed pores. As shown in
Photo A of this figure, the relatively large open pores were found at
the grain boundary junctions, in addition to the small pores with in W-
grains. Such large pores were seldom observed in the center part, as
shown in Photo C. The large pore found at a grain boundary junction
near the specimen surface had a sharp angular shape just as did the Ni-

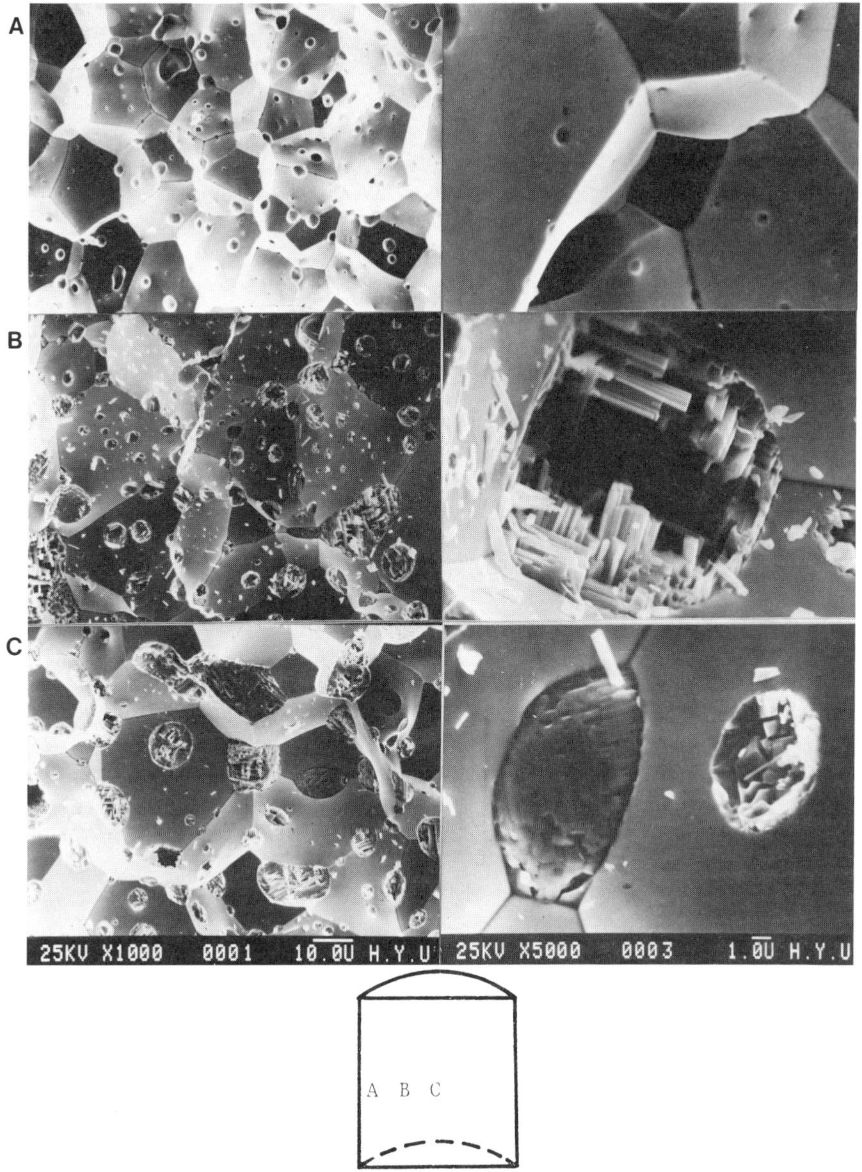

Fig. 3. SEM micrographs of fractured surfaces at different positions
marked as A, B, and C in one of the test specimens. This
specimens contained 0.4 wt.%. Ni and had an initial porosity of
9.1 %.

rich phase pools, which were usually formed at triple or quadrople
points of the grain boundaries in the Ni-activated sintered W-compact.[9]
Furthermore, there was no the measurable Ni-concentration around this
pore. Therefore this large pore seemed to be formed by depletion of the
Ni-rich phase pool by evaporation. Some rounded large pores were also
found in the center part of the specimen, and bar shaped crystalline
growths were frequently observed on the surface of these closed pores,
as shown in Photo C. This bar-shaped crystalline matter was revealed to
be tungsten crystals by EDS analyses. Such a columnar W crystalline
form was supposed to be created by a process similar to the formation

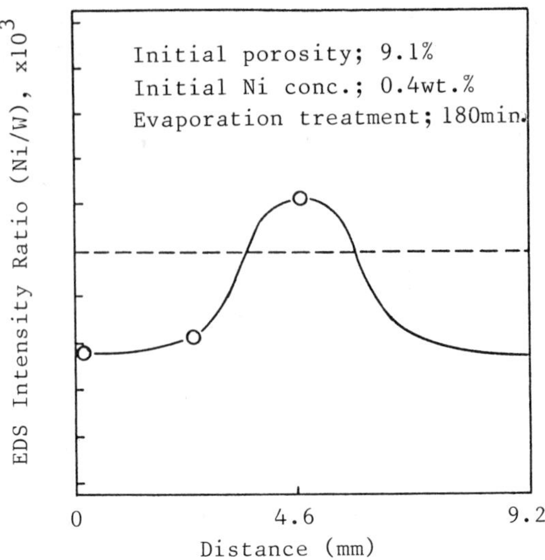

Fig. 4. Variation of relative Ni-concentration with the distance from specimen surface. The relative Ni-concentration was determined from the EDS intensity ratio of Ni to W.

of W-whiskers, which are often found in W-powder obtained by hydrogen reduction of WO_3 powder with Ni-impurity more than 0.1 wt.%.[10] WO_3 compound formed possibly by reaction of W-atoms with the residual impurity enclosed within the pore might be evaporated together with Ni-impurity and subsequent reduced to W in hydrogen atmosphere during sintering. It could have grown further to large bar shaped structure within the closed pores during the latter evaporation treatment cycle.

Fig. 4 shows the relative Ni-concentration profile along the diameter of the middle part of the same specimen given in Fig. 3. As shown on Fig. 4, the EDS intensity ratio of Ni to W was higher in the central part of the specimen than in the area near the surface. Such a Ni concentration profile agrees well with the fractographic features described above.

As discussed partly above, the Ni removal rate was strongly dependent on porosity of the specimen as well as on the type of pore. The level of porosity is the principal determinant of the type of pore. It is therefore very important to control the porosity of the Ni-added W-sintered specimen in order to remove the nickel successfully by evaporation.

Fig. 5 shows the dependence of weight loss on porosity for the 0.2 wt.% Ni doped W-sintered specimens of various porosities and for the pure W-sintered reference specimen in evaporation treatment after first 15 minutes. The test specimens were prepared with different sizes of W-powders for the convenience of the porosity control, using the fine powder for the specimen of low porosity and a coarser one for high porosity. As shown in this figure, the weight loss rate increased continously with increasing porosity. However it is remarkable that there was some change in slope of the curve for Ni-added W-specimen at the porosity of about 10% as well as at the porosity of higher than 20 %. The change of weight loss rate, in other word, the evaporation rate, near 10 % porosity is likely to be related to the transition of pore structure from an open to a closed configuration. The slight decrease in slope of curve at porosities higher than 20 % might be

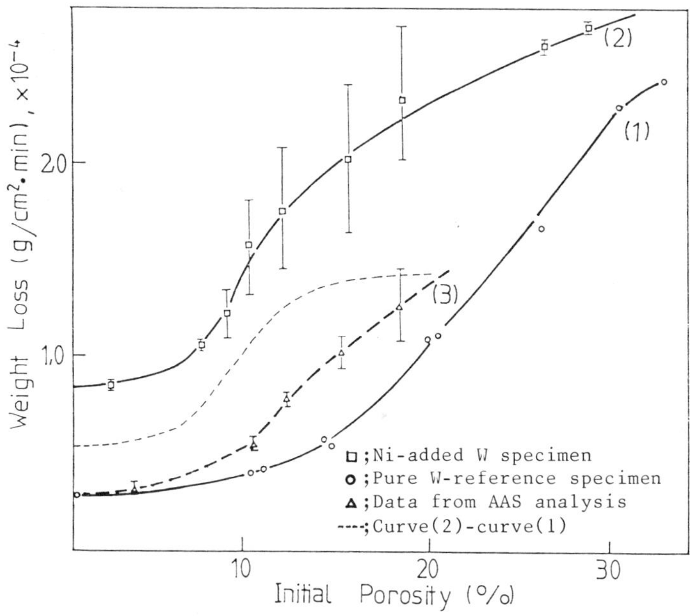

Fig. 5. Dependence of weight loss rate on porosity for the 0.2 wt.% Ni
added W-sintered specimens of various porosity and for the pure
W-reference specimen in evaporation treatment after first 15
minutes.

attributed to no more measurable difference in total surface area in
open structure with higher porosity as well as to the decreasing
contribution of Ni evaporation rate to the total weight loss rate due
to the early removal of the most of the available nickel.

The amount of nickel removed during evaporation treatment for 15
min is just the difference between the value of curve 2 and that of
curve 1 for a given porosity in Fig. 5. The evaporation rate of nickel
obtained by such a weight loss difference was somewhat higher than that
determined by measurement of residual Ni amount with the aid of AAS
analysis, as given in dotted line in the same figure. Such slight
descrepancies between the two measurements might be partly compensated
by considering the possible weight loss portion of W-atoms dissolved in
Ni-rich phase. Furthermore, such differences might also in part be
caused by differences in microstructure other than the porosity. The
grain size as well as the particle size of W-powder used can affect the
evaporation rate, because the vapor pressure on a curved surface is
different from the equilibrium vapor pressure on a flat surface. After
calculation according to the Kelvin equation, the ratio of vapor
pressure of 1 μm W-powder to that of 4 μm W-powder should be about 1.5
for the given experimental condition in the present study; i.e. the
vapor of 1 μm W-powder is higher than the equlibrium vapor pressure on
a flat surface by a factor of 1.74 at $1400^{\circ}C$.

As discussed above, nickel contained in the activated sintered
compact could mostly be removed by the evaporation treatment in vacuum,
especially when the sintered compact has a porosity level that assumed
on open pore structure.

ACKNOWLEDGEMENT

The authors acknowledge gratefully the financial support of Korea Science and Engineering Foundation.

REFERENCES

1. I. H. Moon and Y. D. Kim, Relationship Between Workability and Structural Features of Tungsten Produced by Activated Sintering, Modern Developments in Powder Met. 15:541 (1985).
2. A. Gabriel, H. L. Lukas, C. H. Allibert and I. Ansara, Experimental and Calculated Phase Diagrams of the Ni-W, Co-W and Co-Ni-W Systems, Z. Metallkde. 76:589 (1985).
3. O. Kubaschewski and C. B. Alcock, "Metallurgical Thermochemistry", 5 ed., Pergamon Press, Oxford (1979).
4. J. P. Hirth, Evaporation of Crystal, in: "Metal Surfaces: Structure, Energetics and Kinetics", ASM, Metals Park, Ohio (1962).
5. R. M. German, "Powder Metallurgy Science", MPIF, Princeton (1984).
6. I. H. Moon, J. S. Kim and Y. L. Kim, The Effect of the Doping Method on the Sinterability of Nickel-Doped Tungsten Compacts, J. Less-Common Metals 102:219 (1984).
7. J. C. Batty and R. E. Stickney, Oxidation of Metals 3:370 (1978).
8. Th. Heumann and I. H. Moon, Sauerstoffadsorption auf Kupfer und Bestimmung der Rauhigkeitsfaktor, Surface Science 24:370 (1971).
9. I. H. Moon and Y. S. Kwon, Some Observations on Sintering of the Nickel-Doped Tungsten Compacts, Scripta Met. 13:33 (1979).
10. R. Haubner, W. D. Schuber, E. Lassner and B. Lux, Einfluss von Eisen und Nickel auf die Reduktion von Wolframoxid zu Wolfram, Intern. J. Refractory & Hard Metals 7:47 (1988).
11. A. W. Adamson, "Physical Chemistry of Surfaces", 4 ed, John Wiley & Sons, New York (1982).

Part V. PRESSURE SINTERING

Part V. PRESSURE SINTERING

EXPLOSIVE COMPACTION OF POWDERS: PRINCIPLE AND PROSPECTS

R. Prümmer

Fraunhofer Company
München, W. Germany

INTRODUCTION

It is the wish of each powder metallurgist to posses presses with
great capacities developing high pressures. Hard powders are especially
difficult to compact. For this reason, the Hot Isostatic Pressing
procedure was developed. Explosive Compaction on the other hand has the
potential of developing very high pressures, dynamically applicable to
powders. Its achievements include not only relatively high densities
for green compacts (approximately 100% of theoretical density), but also
the possibility of creating new materials. The main features of the
method are explained and a survey of the latest developments is given.

THE DIRECT METHOD OF EXPLOSIVE COMPACTION

There exist several methods of explosive compaction. For instance,
a piston is accelerated explosively in a gun-barrel-like apparatus
hitting a powder at a dead end and compacting it. Another method
consists of a metal container submerged in water in a closed vessel.
When an explosive charge is detonated in water, the powder is
compacted. These indirect methods of explosive compaction today play
only a minor role in application. Gas guns, however, are important for
research purposes in order to investigate the physical parameters and
shock wave behaviour during dynamic compaction.

The arrangement for direct explosive compaction is very simple.[1-7]
A metal tube is filled with powder, closed with end plugs, and evenly
surrouded with a layer of explosive of proper detonation velocity. As
the detonation is initiated at the upper end of the container, a ring-
shaped detonation front travels in the axial direction over the
container compressing it in a manner similar to an extrusion press. The
magnitude of the pressure is controlled by the detonation velocity of
the applied explosive, which ranges from V_D = 1700 m/s to V_D = 8400 m/s
and gives a pressure range up to 30 GPa.

Higher pressures can be obtained using two coaxially arranged
tubes:[8,9] the inner tube is the container for the powder to be
compacted. The outer tube called a driver tube, arranged coaxially with
a space of about double its thickness is surrounded in the same manner

with an explosive as in the method previously described. The difference
from the first method is that during detonation of the outer layer of
an explosive, the driver tube is accelerated to a high velocity. When
colliding with the container tube very high pressures are generated.
The amount of pressure can be calculated from the velocity of the
driver tube and shock impedance data (product of shock wave velocity
and specific weight) of the materials of the tubes involved.

EXPLOSIVE COMPACTION - ISOSTATIC COMPACTION

If one attempts to compare the method of explosive compaction with
isostatic compaction methods, difficulties arise. When pressure is
applied to an encapsulated powder during isostatic compression, each
volume element of the powder undergoes the same densification with time
(if friction at the container wall is neglected). In explosive
compaction, however, a compaction wave travels through the powder with
the velocity of sound of the powder, leaving consolidated material
behind.

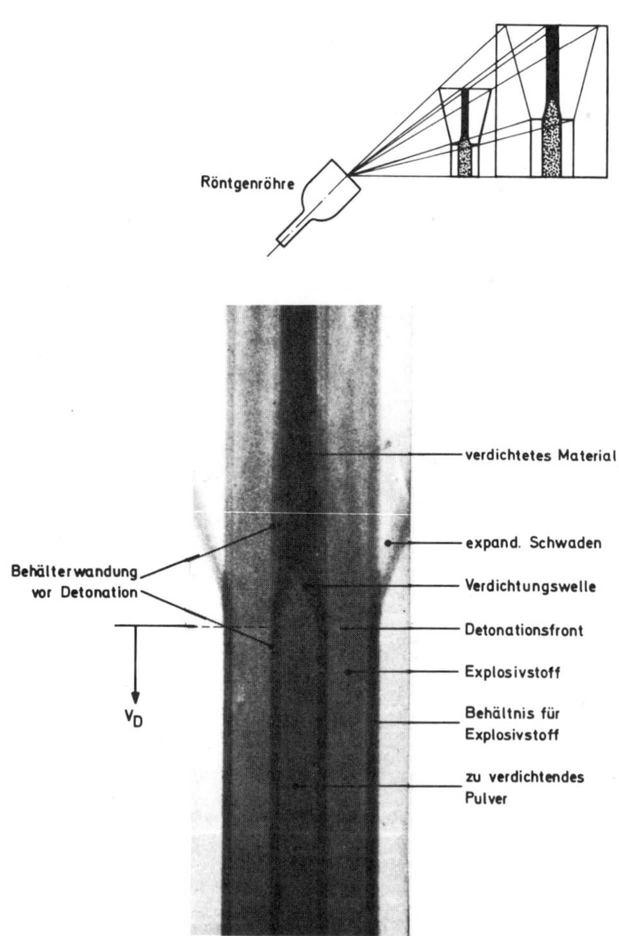

Fig. 1. Explosive compaction of aluminium powder. X-ray flash pattern,
revealing the shape of the shock front as a hollow cone.

SHOCK WAVE CONFIGURATION

As the shock wave proceeds from the container wall towards the center of the sample, two opposing effects determine the strenth of the shock wave:

- The shock wave is a converging one, so its pressure and velocity are increasing.

- Energy is consumed during compaction (plastic deformation of ductile or crushing of brittle particles), causing the shock wave intensity to attenuate while proceeding towards the center of the cylindrical sample.

It is the aim of practical experimental designs to allow both processes to compensate each other. Only in this case can a uniform densification over the cross-section of the cylindrical sample be obtained.[5-7] That means that a conical shock wave configuration has to be achieved. Many investigations therefore have been made to determine the shock wave front during compaction of cylindrical samples. The simplest way is to apply an X-Ray flash, shuch as shown in Fig. 1. The sample is photographed from a side position during explosive compaction. As the exposure time of the X-Ray flash is only 120 ns, the detonation front proceeding at velocity V_D = 3500 m/s, can be made visible.

The high pressure is obtained in accelerating the container tube in a radial direction. A shock wave is created this way in the powder. In the case shown (Fig. 1) of the compaction of an aluminium powder, indeed the shape of a hollow cone is obtained.

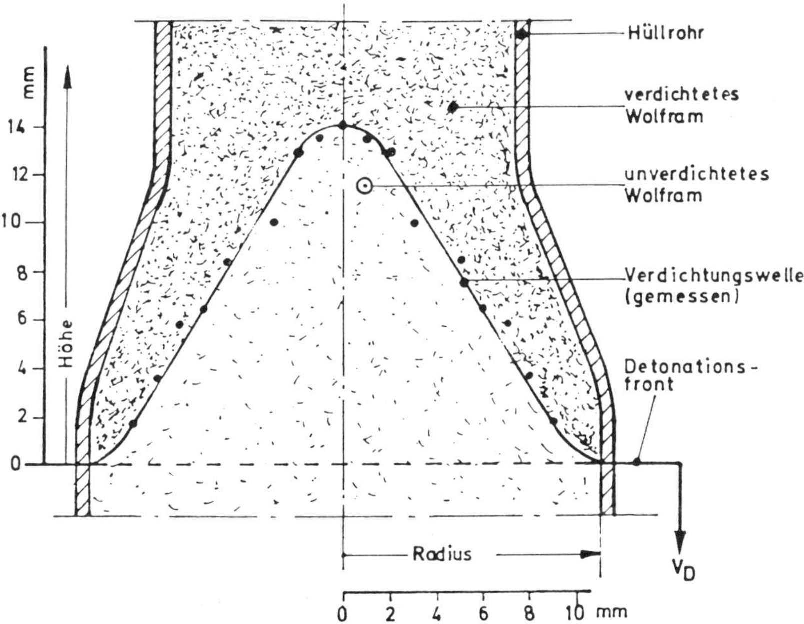

Fig. 2. Shock front in explosive compaction of tungsten powder, measured by the electric pin method.[10]

When powders with large specific weight or when large diameter
samples are compacted, another method for the determination of the
shock wave configuration is advantageous: elecrically isolated pins are
arranged at different radial distances in the powder. When hit by the
shock wave the pins give rise to a signal which can be recorded by
means of an oscilloscope or counter. For an explosive compaction of
tungsten powder with parameters leading to a uniform density over the
cross section of the sample, the shock configuration was measured[10]
this way and is shown in Fig. 2. It indeed is of the shape of a hollow
cylindrical cone, as expected.

Another means to control the shock wave configuration is computer
modeling. The HEMP-code[11] was developed and applied for modelling of
explosive compaction. Using Prümmer's[10] experimental data, Reaugh[12]
simulated the shock wave configuration during explosive compaction of
tungsten, as shown in Fig. 3. Also in this case, as expected a shape of
a hollow cylindrical cone for the shock wave front is obtained.

It is also interesting to note that the pressures necessary for
full compaction of tungsten, obtained experimentally and by computer
modeling, correspond well[12] Computer modeling is a good tool for
forecasting experiments and reducing the number of experiments
necessary for optimization of the process of explosive compaction.

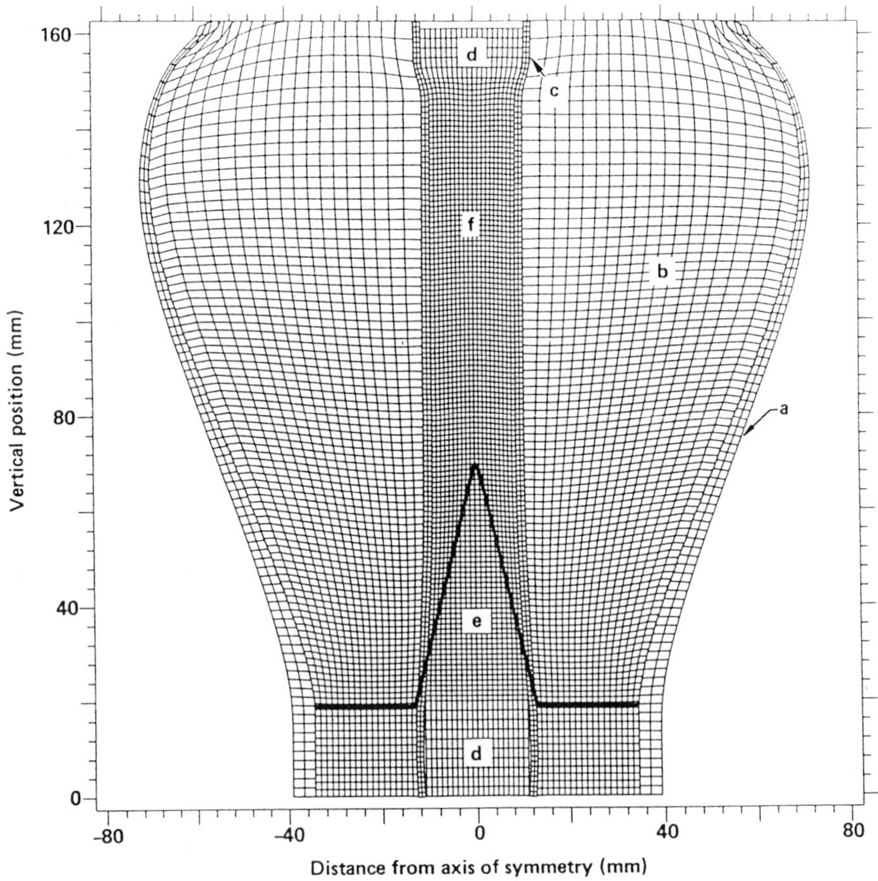

Fig. 3. Shock front in explosive compaction of tungsten powder[10],
simulated[12] by the HEMP-code.[11]

PARAMETERS FOR EXPLOSIVE COMPACTION

In order to obtain cylindrical samples in explosive compaction
with uniform density over the cross section, the cone shaped shock-wave
configuration has to be achieved. To solve this task by computer
modeling, the Hugoniot-data of the powder have to be known. They
describe the relation between volume and pressure under the action of
shock waves. Such data are available for many substances, also powders[13]
however, if new powders, like rapidly quenched or melt spun alloys have
to be compacted, such data are missing. Experimental investigations
therefore are of great importance.

Numerous experimental investigations[5,10] reveal a linear relation
between the pressure required to explosively compact a powder and the
hardness of the metal powder particles. Aluminum powder, for instance,
requires only moderate pressures to achive a densification of 100% of
theoretical density, whereas to compact, for instance, a highly alloyed
steel powder of hardness 500 HV, a pressure of as much as 6 GPa is
required for full densification. The linear relationship between the
pressure required for densification and the hardness of powder
particles shown in Fig. 4 is obtained. Data taken from other authors
and obtained with other methods than explosive compaction, especially
by using a ballistic gun or a flyer plate technique also are included.
Also the most recent results obtained with explosive compaction of
diamond powders[14] by means of a flyer plate technique are included.
Although different experimental methods were used in order to determine

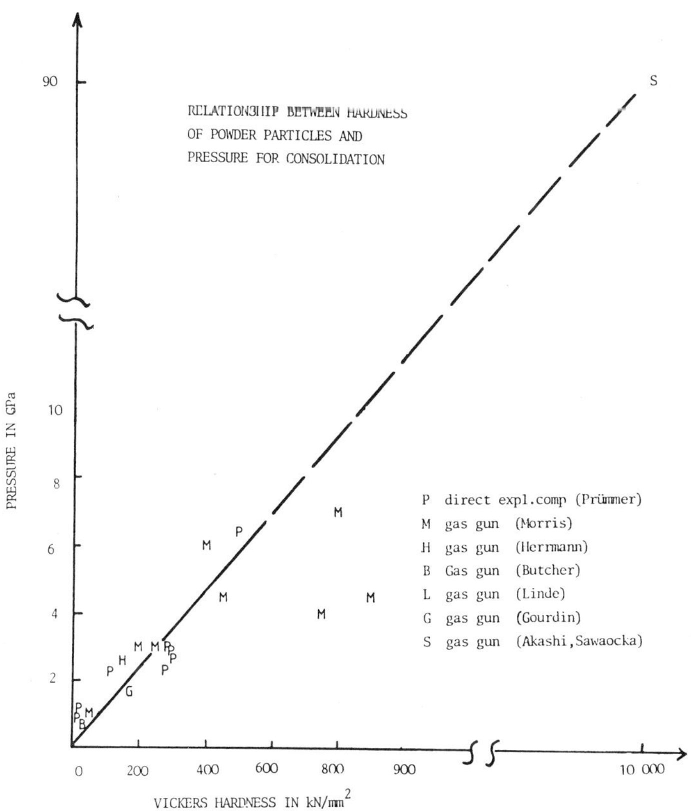

Fig. 4. Pressure required for densification of powders as a function of
hardness of powder particles.[10]

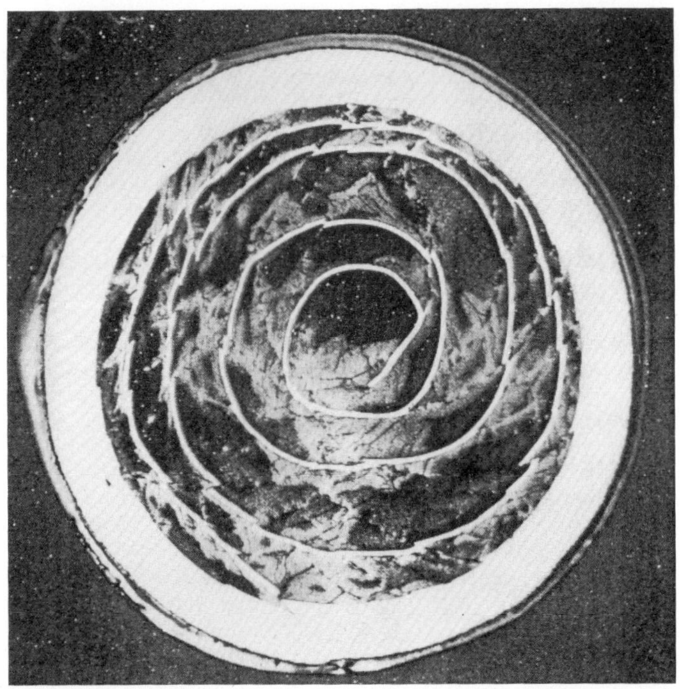

Fig. 5. Cross Section of a rod of 18 mm diameter made by explosive
compaction of Al_2O_3-powder (grain size = 5 μm) with a spirally
shaped metal foil (niobium) to demonstrate the occurence of
adiabatic shearing when compaction pressure is too high.[15]

the pressures, the linear relationship in Fig. 4 holds. This fact
allows us to determine one of the explosive's parameters: an explosive
with a minimum detonation velocity of

$$V_{Dmin} = 2\sqrt{\frac{1 \cdot 2 \cdot HV}{\rho_E}}$$

is required to densify a metal powder, where HV is the hardness of its
particles and ρ_E = density of the explosive.

For ceramic powders, however, difficulties arise due to the effect
of adiabatic shearing. During compaction, as the shock wave proceeds
towards the center of the cylindrical sample, an already consolidated
outer shell has to be crushed in order to allow the center part of the
sample to be compacted. Adiabatic shearing is the mechanism for this
purpose. Fig. 5 shows the cross section of an alumina rod, made by
explosive compaction of Al_2O_3-powder of 5 μm particle size.[15] Thin
metal foils (100 μm thick) were used to make the shearing visible. When
the density of the consolidated rod is limited to 95% of the
theoretical density, adiabatic shearing can be avoided.

TEMPERATURE EFFECTS

The compression of matter by means of shock waves is accompanied
by a sudden rise of temperature. There are distinctive differences in
the behaviour of solid and porous (powders) materials.

Energy Consideration

Fig. 6 gives the pressure-volume relationship (Hugoniot-Curves) for shock loading of solid and porous materials. When a sample starting with specific volume V_O is compressed by means of a dynamic pressure P to the smaller specific volume V_1, its internal energy increases by

$$E = \frac{1}{2} (P_O + P_1)(V_1 - V_O)$$

where P_O is the starting pressure. Most of the energy is dissipated by heat; a part of it, however, is stored in the material (lattice defects and -distortions). From Fig. 6, it can be seen that porous materials are accompanied by a higher increase of internal energy, since the starting specific volume V_O is larger.

Explosive Liquid Phase Sintering

Most of additional energy E in the explosive compaction of porous materials arises as heat at the powder-particle's surface. If this heat is supplied rapidly and if its dissipation occurs comparatively slowly, surface melting of the particles can take place.[16-19] After cooling, a structure of the consolidated materials is obtained existing of a shock-wave-hardened interior of the individual particles welded to neighbouring powder particles by a rapidly solidified interphase. Therefore this process is called "Explosive Liquid Phase Sintering". Fig. 7 shows the micrograph[10] of an explosively compacted superalloy IN 100. It can be seen that the dendritic structure in the welded interphase is much finer than in the original powder particles. It can be estimated from dimensions of the dendritic structure[17] that the colling rate in the welded interphase is very high (on the order of 10^{14} per sec). A compacted sample made by "Explosive Liquid Phase Sintering" therefore exists of a work hardened interior of the powder particle surrounded by a network of extremely rapidly quenched material.

It is self-evident that the explosive liquid phase sintering (ELPS)-process allows one to sinter amorphous materials: during compaction of amorphous powders, heating occurs and leads to a melting of the surface of the particles, also, rapid cooling does not allow re-crystallization to set in[10,15,20,21], provided that the compaction pressure was not too high.

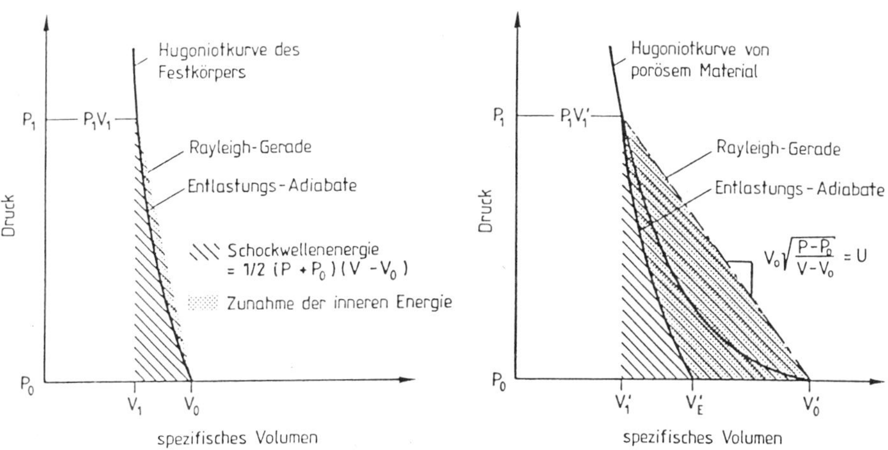

Fig. 6. Hugoniot-curves for solid and porous materials.

In more recent investigations, it has been shown that bulk samples made of superalloy powders by explosive compaction can have superior strength compared to conventionally prepared (hot isostatically pressed) samples. Fig. 8 shows the yield strength (Fig. 8a) and ultimate tensile strength (Fig. 8b) versus annealing temperature. Annealing was applied after dynamic compaction with a gun. The compacted sample has a yield strength of 1 GPa, which is about 10% higher than that of the hot isostatically pressed sample. After heat treatment at 620 $^{\circ}$C, the shock compacted sample has a 20% higher yield strength and a 40% higher UTS than the hot isostatically pressed sample. The increase in strength is atributed to interparticle melting (ELPS, as described above). In addition a high density of fine and uniformly dispersed γ''-precipitates is observed. In comparison to hot isostatically pressed material, the shocked material is activated, resulting in faster nucleation kinetics and therefore in finer (about half size) and larger number of γ''-precipitates in the shock compacted material.

RESIDUAL STRESS, STORED ENERGY

Part of the shock wave energy is absorbed in the crystal lattice. Lattice defects (point defects and dislocations) are created, leading to an increase of lattice distortion and a decrease of subgrain size. Both can be determined by X-ray diffraction methods. A summary of all existing data[10] reveals that after explosive compaction a dislocation density – derived from lattice distortion – and values of stored energy can be created in the material in magnitudes similar to such after heavy plastic deformation of metals. This rule applies both to compacted metal and ceramic materials. Stored energies of up to 9 Joule/g have been observed, e.g. in TiC.[23] Such high values of stored energy can lead to enhanced diffusion. An activation of the sintering process was observed.[24,25] There is also evidence for enhanced chemical[26,27] and catalytic[28] reactivity of ceramic materials after shock wave treatment.

SHOCK WAVE SYNTHESIS

The large amounts of energy that it is possible to supply to the particle's surfaces during explosive compaction of powder mixtures enables us to perform shock wave synthesis of materials. The first observed chemical synthesis of this kind was that of the formation of zinc ferrite with its constituents[29] and shortly after that of TiC.[30] It was found that the amount of explosive necessary to initiate the chemical reaction is related to the reaction temperature under ordinary conditions.[31,32] Also exothermic reactions are possible, under shock wave conditions. Superconductive materials have been synthetized by means of shock waves, like Nb_3Sn[33] and Nb_3Si[34] using stochiometric mixtures of the constituents.

Of special interest are present methods of fabricating high temperature suprconductors starting either from superconductive powders or its constituents.[35] The superconductive properties can be maintained in the high density green compact as long as the pressure during compaction is limited.[35,36] If superconductivity gets worse, a heat treatment in an oxygen atmosphere is a good means to restire the superconductivity. Explosive compaction was already been used to make superconductive coils starting from the powder.[37]

Also widely investigated is a process for explosive compaction of

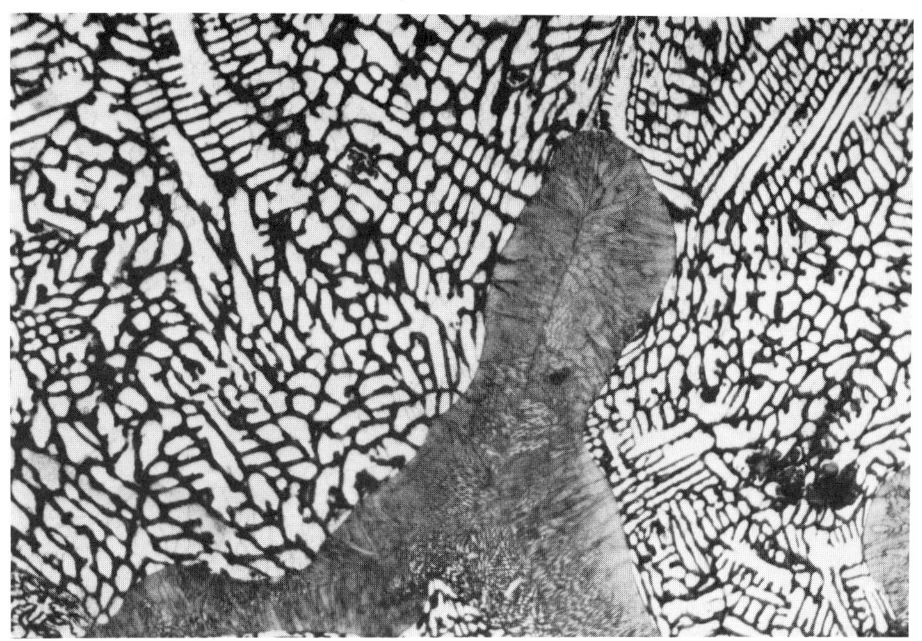

Fig. 7. Explosively compacted Superalloy IN 100: individual particles welded to one another.[10]

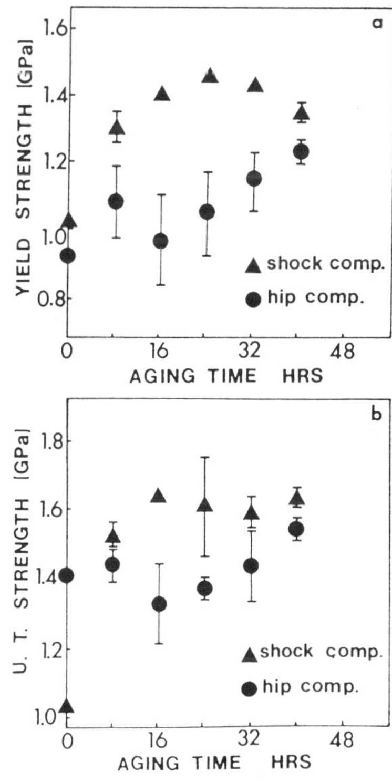

Fig. 8. Yield Strength and Ultimate Tensile Strength vs. aging temperature of shock consolidated and isostatically pressed material.

mixtures of powders which are associated with an exothermic reaction.
This process, called combustion synthesis, has been investigated in
order to form hard materials, like, e.g., titanium diboride TiB_2.[38]
Also parts of intermetallic phases, like e.g., Ti_3Al can be made[39] by
means of combustion synthesis. Further research is necessary to
demonstrate the superiority of such products compared to conventional
fabrication, however.

EXPLOSIVE TRANSFORMATION OF MATTER

Polymorphic transformations of materials are also observed under
conditions of explosive compaction. The most significant discovery was
that of the transformation of graphite to diamond. In massive
production of diamond powder, a mixture of graphite- and metal powders
is explosively compacted. The metal powder serves as a shock plate and
a heat sink. The diamond powder obtained with a particle size of 0,2 -
10 µm is used for grinding and surface finishing of metal and ceramic
materials and as a constituent in cutting tools. Also, the
transformation of graphite-like boron nitride to its cubic modification
can be performed in a similar manner.[43-45] Both, diamond powder and
cubic boron nitride powder produced by explosive transformation reveal
severe lattice distortions which are dependent upon the cooling rate after
passage of the shock wave.

SUMMARY

The explosive compaction method existing of a cylindrical
container surrounded by a proper type and amount of explosive is a zero
investment method to achieve high densities of almost as much as
theoretical density. The explosive's parameters have to be adjusted to
the type of the powder to be compacted: the required explosive's
pressure is linearly related to the Vickers hardness of the metal
powder particles. If higher pressures are applied, an "explosive liquid
phase sintering"-process can be achieved, allowing the welding of
individual particles. The residual properties of the material are
characterized by a high defect structure; revealing dislocation
densities and subgrain sizes comparable to those after heavy plastic
deformation of metals as well as in compacted ceramic powders. Enhanced
sintering reactivity, chemical and catalytic reactivity may be the
result of high values of stored energy observed in the ceramic
materials, shock wave treated under conditions of explosive compaction.
The properties of the materials produced by shock wave synthesis or by
shock wave transformation are also determined by a high density defect
structure.

REFERENCES

1. Ya. N. Riabinin: Certain Experiments on Dynamic Compression of
        Substances, Sovj. Phys.-Techn. J. 1 (1956) 2575.
2. S. W. Proembka: Compacting Metal Powders with Explosives,
        Powder Metallurgy 6 (1960) 125.
3. J. Pearson: The Explosive Compaction of Powders, in: Adv. in High
        Energy Rate Forming, ASTME, Detroit (1961) SP 60-158.
4. R. W. Leonard, D. Laber and V. D. Linse: Advances in Explosive
        Powder Compaction, Proc. 2nd. Int. Conf. HERF, Estes Park, Co.,
        U.S.A. (1969), 8-3-1.
5. R. A. Prümmer: Latest Results in the Explosive Compaction of

Metal and Ceramic Powders and their Mixtures, Proc. 4th Int.
Conf. HERF (1973) Vail, Co. U.S.A.

6. A. A. Deribas, A. M. Staver: Shock Compression of Porous
    Cylindrical Bodies, Fizika Gorenija i Vzryva 10 (1974), No. 4,
    568-578.

7. R. A. Prümmer, G. Ziegler: Structure and Annealing Behaviour of
    Explosively Compacted Alumina Powders, Powder Metallurgy Int. 1
    (1977), 374.

8. M. A. Meyers and S. L. Wang: An Improved Method for Shock
    Consolidation of Powders, 2nd Workshop on Industrial
    Application Feasibility of Dynamic Compaction Technology,
    Tokyo, (December 1988).

9. G. R. Cowan: Plug Closure in a Container for Subjecting Sample to
    Shock Wave, US Patent 3.568.248 (March 9, 1971).

10. R. Prümmer: Explosivverdichten pulvriger Substanzen, Springer
    Verlag, Berlin, Heidelbg., New York, London, Paris, Tokyo
    (1987), ISBN 3-540-17029-4, p.37.

11. M. L. Wilkins, in Methods of Computational Physics, Vol. 3
    (1964), B. Alder, S. Fernbach and M. Rotenberg (eds.), Academic
    Press, New York.

12. J. E. Reaugh: The Explosive Consolidation of Rods, J. Appl. Phys.
    61 (1987) No.3, 962-968.

13. M. V. Thiel, A. S. Kusubov et al., eds.: Compendium of shock wave
    data, UCRL-50108 (TID-45000).

14. T. Akashi and A. B. Sawaoka: Shock Consolidation of Diamond
    Powders, J. Mat. Science 22 (1987) 3276-86.

15. R. Prümmer: Dynamic Compaction of Powders, Proc. 19th Univ. Conf.
    Emergent Process Methods for High Technology Ceramics, R.F.
    Davis, H. Palmour III and R. L. Porter eds (1984), Plenum Press
    New York, London, 621-636.

16. D. Reybould: The Cold Welding of Powders by Dynamic Compaction,
    Int. J. Powder Met. and Techn. 16 (1980), 9-12.

17. D. G. Morris: The Compaction and Mechanical Properties of Metallic
    Glass, Metal Science J.15 (1981), 116-124.

18. R. B. Schwartz, P. Kasiraj, T. Vreeland Jr., and T.J. Ahrens: A
    Theory for the Shock Wave Consolidation of Powders, Acta Met.
    in press.

19. H. W. Gourdin: Energy Deposition and Microstructural Modification
    in Dynamically Consolidated Metal Powders, J. Appl. Phys. in
    press.

20. C. F. Cline and M. L. Wilkins: Dynamic Consolidation of a Rapidly
    Solidified Ni-Mo-B-Alloy, 8th Int. HERF Conf., San Antonio, Tx,
    U.S.A. (1984).

21. O. V. Roman, V. G. Gorobtsov, B. S. Mitin and V. A. Vasiljev:
    Structure and Properties of Iron-Base Amorphous Materials,
    Proc. 4th Int. Conf. RQM, (1981), Sendai, Japan.

22. N. N. Thadhani, A. H. Mutz and T. Vreeland J.: Structure/Property
    Evaluation and Comparison between Shock-Wave Consolidated and
    Hot-Isostatically Pressed Compacts of RSP Pyromet 718 Alloy
    Powders, Acta. Met. 37 (1989) No.3, 897-908.

23. H. Palmour III, et. al: Effect of Dynamic and Isostatic Compaction
    on the Microstructure and Mechanical Behavior of AlN, TiB_2 and
    TiC, APS Conf. Interaction of Shock Waves with Condensed
    Matter, Santa Fe, N.M., U.S.A. (1983).

24. K. Y. Kim, A. S. Batchelor, K. L. More and H. Palmour III: Rate
    Controlled Sintering of Explosively Shock Conditioned Alumina
    Powders, Proc. 19th Univ. Conf. Emergent Process Mesthods for
    High Technology Ceramics, Raleigh, N.C., U.S.A. (1982).

25. E. K. Beauchamp, R. A. Graham and M. J. Carr: Densification of
    Shock Wave treated Aluminum Nitride and Aluminum Oxide, Int.

Conf. Interaction of Shock Waves with Condensed Matter, Santa
Fe, N.M., U.S.A. (1983).

26. D. L. Hankey, R. A. Graham, W. F. Hammetters, and B. Morosin:
    Shock Induced Reactivity Enhancement of ZrO_2 - Powders, J. Mat.
    Sci. Letters 1(1982), 446-447.

27. S. S. Batsanov: Synthesis under Shock Wave Pressures, in:
    Preparative Methods in Solid State Chemistry, (1987), Academic
    Press Inc, New York and London, 133-146.

28. J. Golden, F. Williams, B. Morosin, E. L. Venturini and R. A.
    Graham: Catalytic Activity of Shock Loaded TiO_2 Powder, AIP.
    Conf. Proceedings 78 (ed H.C. Wolfe) Shock Waves in Condenses
    Matter - 1981 (Menlo Park) American Institute of Physics (1982)
    New York, 74-76.

29. Y. Horguchi and Y. Nomura: Formation of Zinc Ferrite by Explosive
    Compaction, Jap. J. of Appl. Phys. 2 (1963) 312.

30. Y. Horiguchi and Y. Nomura: Explosive Synthesis of TiC by Contact
    Technique, Bull. Chem. Soc. 36 (1963) 486-496.

31. S. S. Batsanov and E. S. Zolotova: Shock Synthesis of Chromium II
    Calcogenides, Dokl. Akad. Nauk SSSR 180 (1968), 93.

32. S. A. Batanov et al.: Impact Synthesis of TiN Chalcogenides, Dokl.
    Akad. Nauk SSSR 185 (1969), 33-331.

33. G. Otto, O. Y. Reece and U. Roy: Synthesis of Nb_3Sn by Shock
    Waves, Appl. Phys. Letters 18 (1971), 418.

34. D. D. Hughes and V. D. Linse: Formation of Superconducting Nb_3Si
    by Explosive Compression, J. Appl. Phys. 50 (1979), 3500.

35. L. E. Murr, A. W. Hare and N. G. Eror: Fabrication of Novel Bulk
    Superconductor Composites by Simultaneous Explosive
    Consolidation and Bonding, in; Shock Waves for Industrial
    Applications, E. Murr ed., Noyes Publ., Park Ridge, N.J. USA
    (1989), 473-527.

36. R. A. Prümmer, C. Politis, H. Keschtkar: Synthesis of High
    Temperature Superconductors by Explosive Compaction, X Int.
    HERF Conf. Ljubljana, Jugoslavia, Sept.89.

37. S. Hagino et al.: Microstructures and Superconducting Properties
    of YBaCu Oxide Coils Repared by the Explosive Compaction
    Technique, Proc. 1st Int. Conf. Superconductivity, 1988,
    Nagoya, Japan.

38. T. Kottke and A. Niiler: Effects of Thermal Conductivity on the
    SHS-Reaction Kinetics, Material Processing by SHS, MTL-SP-87-3
    (1987).

39. M.A. Meyers, N. N. Thadani and Li-Hsing Yu: Explosive Shock Wave
    Consolidation of Metal and Ceramic Powders, in Shock Waves for
    Industrial Application, L. Murr ed. Noyes Publications, Park
    Ridge, N.J. USA, (1989).

40. P. S. DeCarli and C. J. Jamieson: Formation of Diamond by
    Explosive Shock, Science 133 (1961), 1821.

41. P. S. DeCarli: Shock Wave Synthesis of High Pressure Phases, in:
    Science and Technology of Industrial Diamonds, ed. J. Burls,
    Industrial Diamond Inf. Bureau, London (1967) 49-64.

42. O. R. Bergman: Detaclad Explosion Bonded Metals and Shock
    Synthesized Polycrystalline Diamond, Proc. 7th Int. Conf. HERF,
    Leeds, US (1981), 142-151.

43. N. L. Coleburn and J. V. Forbes: Irreversible Transformation of
    Hexagonal Boron Nitride by Shock Compression, J. Chem. Phys.
    48(1968), 555.

44. S. S. Batsanov and L. R. Batsanova: Effect of Explosions on
    Matter: Formation of Dense Modifications of Boron Nitride, Zh.
    Strukt. Chim. 9 (1968), 1024.

45. G. H. Zhadanovich et. al.: Method of Obtaining Diamond and/or
    Diamond-like Modifications of Boron-Nitride, UK-Pat. 2090239
    (1980).

THEORETICAL ASPECTS OF HIGH PRESSURE SINTERING

P.S. Kisly

Institute for Superhard Materials
Ukrainian Academy of Sciences
Kiev, USSR

INTRODUCTION

Pressureless sintering and hot pressing are considered to have
been much studied theoretically and experimentally. The results of
these studies are, however, unusable to describe the mass transfer
processes at high pressure. Sintering under these conditions is due to
mechanisms of plastic deformation and diffusional creep and at high
hydrostatic pressure, these processes may be considered to be
competitive.

Initiation and development of the processes of plastic deformation
under hydrostatic pressure are possible, first of all, at the contacts
of the particles. Under high pressure, the activation energy of plastic
deformation, and hence its rate, change notably:[1]

$$\dot{\varepsilon} = A \cdot \exp\left[-\Delta F/RT\right]; \tag{1}$$

$$\dot{\varepsilon} = A \cdot \exp\left[-(\Delta F - \Delta VP)/RT\right], \tag{2}$$

where ΔV is the volume to be activated, P is the stress (pressure), ΔF
is the change in free energy during the transition of the complex from
the normal state into the activated one:

$$\Delta F = E_a - T\Delta S, \tag{3}$$

where E_a is the activation energy, ΔS is the change in entropy during
transition from the normal state into the activated one. Thus,
according to Eq.(2)., the rate of the processes related to plastic flow
increases steadily with pressure.

At the same time, with any creep mechanism, the rate of the
process is also proportional to the stress:

$$\dot{\varepsilon} = A \cdot P^n \cdot \exp[-E_a/RT], \tag{4}$$

where A and n are the constants. The creep rate increases up to
particular values because diffusional processes down slow with pressure
increases:[2]

Science of Sintering
Edited by D. P. Uskoković *et al.*
Plenum Press, New York

$$D = D_o \exp[-\Delta V P / RT] \qquad (5)$$

Therefore, an analysis of high pressure sintering by diffusional creep mechanisms is of particular interest.

DIFFUSIONAL CREEP THEORY FOR POROUS SOLIDS UNDER PRESSURE

According to Nabarro-Herring, diffusion under pressure is due to the transfer of vacancies into the normal pressure region and that of atoms in the opposite direction.[3] In a porous solid, grain boundaries are sinks of vacansies, and atoms transfer to free surfaces of pores. If a simple model of a porous solid is assumed, in which equal grains are contiguous to round pores, then under external hydrostatic pressure and the Laplace pressure, the stress in every grain on the surface of pores and boundaries will be defined by equations:

$$\sigma_p = \frac{2 \cdot \gamma}{\rho}; \qquad (6)$$

$$\sigma_g = -\frac{2\gamma g}{r} - P_g, \qquad (7)$$

where γ, γ_g are the free surface energy and the boundary energy, respectively, r, ρ are the radii of the grains and pores, respectively, P_g is the pressure at the grain boundary and is given by:

$$P_g = P \frac{r^2}{x^2}, \qquad (8)$$

where x is the radius of the neck (boundary). Due to these stresses, the chemical potential of the vacancies on the surface of pores and grain boundaries varies according to the Thompson-Gibbs equation and is given by:

$$\mu_p = \mu_o + \frac{2\gamma}{\rho} \Omega; \qquad (9)$$

$$\mu_g = \mu_o - \frac{2\gamma g}{r} \Omega - P \frac{r^2}{x^2}, \qquad (10)$$

where Ω is the vacancy volume. The formation of vacancies on the surface of pores and their annihilation at grain boundaries result in a deviation of their concentration from the equilibrium value. This is also related to changes in the chemical potential, i.e:

$$\mu_p = \mu_o + RT \ln(C_p/C_o); \qquad (11)$$

$$\mu_p = \mu_o - RT \ln(C_g/C_o), \qquad (12)$$

where C_o, C_p, C_g are the equilibrium concentration of vacancies, the concentration of vacancies on the pore and boundary surfaces, respectively.

Substitution of logarithms of the vacancy concentration ratio by relative variation in vacancy concentration, $\Delta C_p/C_o$ and $\Delta C_g/C_o$, gives the following expressions for changes of vacancy concentration on the

pore surface and at the boundary as compared with the equilibrium
concentration:

$$\Delta C_p = \frac{C_o \Omega}{RT} \frac{2\gamma}{\rho} \; ; \tag{13}$$

$$\Delta C_g = \frac{C_o \Omega}{RT} \left(\frac{2\gamma_g}{r} + P\frac{r^2}{x^2} \right), \tag{14}$$

Taking into consideration $\gamma_g/r \ll \gamma/\rho$, the value for vacancy
concentration supersaturation between the pore free surface and the
boundary surface may be written as:

$$\Delta C = \Delta C_p - \Delta C_g = \frac{C_o \Omega}{RT} \left(\frac{2\gamma}{\rho} + P\frac{r^2}{x^2} \right) \tag{15}$$

Because of this vacancy concentration supersaturation, atom
transport to the free surface occcurs, i.e. the pore size, and
consequently the volume of the solid, decrease. Further, according to
Refs. 4, 5, the macroscopic approach to the deformation process of a
porous solid can be reduced to the deformation rate evaluation by the
velocity of vacancy flow from the pore surface to the grain boundary.
The diffusion path which corresponds to the vacancy concentration
change, ΔC, may be considered to be equal to a grain radius, r;
therefore, the density of vacancy flow on this diffusion path may be
given as

$$J = \frac{D_v \cdot \Delta C}{r} = \frac{D}{r} \cdot \frac{\Delta C}{C_o}, \tag{16}$$

where D_v is the vacancy diffusion coefficient and is equal to D/C_o.

The velocity with which one grain approaches another is equal to
the deformation rate and is a function of the vacancy precipitation on
the boundary unit area. As the vacancy flow at the boundary proceeds
from the two neighboring pores, then the velocity of the boundary
displacement v equals the doubled flow value, $2J$, and the deformation
rate $\dot{\varepsilon}$ equals v multiplied by the number of boundaries per unit grain
length $1/r$. Thus, the deformation rate is defined as

$$\dot{\varepsilon} = \frac{v}{r} = \frac{2J}{r} \tag{17}$$

Taking into account Eqs.(15) and (16), the deformation rate for a
porous solid under hydrostatic pressure can be written as:

$$\dot{\varepsilon} = \frac{2D\Omega}{RT} \left(\frac{2\gamma}{\rho r^2} + \frac{P}{x^2} \right) \tag{18}$$

In Eq.(18), the diffusion coefficient depends on pressure according to
Eq.(5). Because of this, the velocity of the diffusional creep may be
given by:

$$\dot{\varepsilon} = \frac{2\Omega D_o \exp(-\Delta VP/RT)}{RT} \left(\frac{2\gamma}{\rho r^2} \frac{P}{x^2} \right) \tag{19}$$

To establish the pressure values at which the creep rate is at its
maximum or minimum, we define the first derivative of Eq.(19):

$$\frac{d\varepsilon}{dP} = \frac{2D\Omega}{RT} \left(-\frac{\Delta V}{RT}\frac{2\gamma}{\rho r^2} - \frac{\Delta V \cdot P}{RTx^2} + \frac{1}{x^2} \right) \tag{20}$$

Taking the first derivative to be zero, the critical pressure is

$$P^* = \frac{RT}{\Delta V} - \frac{2\gamma}{\rho}\frac{x^2}{r^2} \tag{21}$$

The second derivative of Eq.(19) is negative when $P = P^*$, therefore,
Eq.(19) is at its maximum at P^*, i.e. the deformation rate at first
increases with pressure and then at $P > P^*$ it decreases.

Practical Assessment of the Theory

It is worth evaluating the critical pressure values for
temperatures of substantial atomic mobility. The activation volume ΔV
is usually defined as a difference between volumes of the initial
compound structure and of the structure in activated (transient) state.
Proceeding from the rigid spheres model, the volume may be represented
as the vacancy volume Ω.[6] Due to relaxation processes occurring around
vacancies, however, it is considerably lower, as the activation energy
of self-diffusion, E_a, surpasses by far that of vacancy formation. Data
taken from different works and gathered in Ref.7 really show that $\Delta V \approx$
$(0.6-0.7)\,\Omega$ for metals with close-packed lattices and $\Delta V \approx (0.3-0.5)\Omega$
for substances with loose-packed lattices. The activation volume can be
calculated using the following relation:

$$\Delta V = \Omega E_v / E_a \tag{22}$$

The energy of vacancy formation in sublattices of metals and nonmetals
in compounds is the same, and as it is seen from Ref.8, is equal to the

Table I. Activation volumes and critical pressures for high-melting-
point compounds

Com-pound	Vacancy Volume, cm^3/mol	Vacancy Formation Energy, E_v, kJ/mol	Self-diffusion Activation Energy, E_a kJ/mol	Activation Volume V, cm^3/mol	Critical Pressures P^*, MPa	
					1000K	2000K
TiC	6.06	231.3	529	2.65	3137	6274
ZrC	7.73	199.5	570	2.70	3079	6158
HfC	7.51	266.8	624	2.75	3023	6046
VC	5.25	101.8	438	1.22	6814	13628
NbC	6.56	140.5	582	1.57	5295	10590
TaC	6.45	144.8	638	1.46	5694	11388
TiN	5.77	336.4	483	4.02	2068	4136
ZrN	7.03	365.3	487	5.27	1578	3156
HfN	6.86	369.2	490	5.17	1608	3216
VN	5.32	251.0	348	3.84	2165	4330
TaN	6.19	246.8	504	3.03	2744	5488

heat of compound formation. Activation energies of self-diffusion are
reliably defined only for some of the high-melting-point compounds.
Therefore, the known relations between the activation energy of self-
diffusion and the melting temperature are preferably used for
estimation. From tabulated data on melting temperatures and activation
energies of self-diffusion for f.c.c. metals this relation is:

$$E_a = (0.15 \pm 0.02) \, T_{melt}, \, kJ/mol \tag{23}$$

High-melting-point compounds with the NaCl-structure have f.c.c.
metal sublattices and, therefore, the above relation can be used and
the activation volumes and critical pressures can be estimated for
these compounds (see Table I).

As can be seen in Table I, critical pressures for high-melting-
point carbides and nitrides with the NaCl-structure are rather high,
such as used in up-to-date high pressure apparati. However, since even
higher pressures can be attained, experimental conditions are
experimentally attainable under which diffusional creep slows down and
even ceases.

REFERENCES

1. T. Yokobori, "An Interdisciplinary Approach to Fracture and
   Strength of Solids", Metallurgiya, Moscow (1971).
2. P. G. Shewmon, "Diffusion in Solids", McGraw-Hill Book Co., N.Y.,
   San Francisco, Toronto, London (1963).
3. C. Herring, Diffusional Viscosity of Polycrystalline Solid,
   J.Appl.Phys. 5:437 (1950).
4. P. S. Kisly and M. A. Kuzenkova, "Sintering of High-Melting-Point
   Compounds", Naukova Dumka, Kiev (1980).
5. J. Hornstr, "Dynamic Properties of Grain Boundaries", in: "Science
   of Ceramics", G. H. Stewart, ed., Metallurgiya, Moscow (1967).
6. C. Roomans, "Structural Investigation of Some Oxides and Other
   Chalcogenides at Normal and Very High Pressures", Mir, Moscow
   (1969).
7. P. C. McCormick, and A. L. Rouff, Creep under high pressures, in:
   "Mechanical Behaviour of Materials under Pressure", H. L. Pugh,
   ed., Mir, Moscow (1973).
8. L. Kaufman and E. Clougherty, in: "Metallurgy at High Pressures
   and Temperatures", AIME, N.Y. (1964).
9. Properties of Elements (Handbook), G. V. Samsonov, ed., Metallurgiya,
   Moscow (1976).

DYNAMIC COMPACTION OF AMORPHOUS $Ni_{78}P_{22}$

B. Mihelić, B. Šerbedžija, V. Petrović, M. V. Šušić*, and
D. P. Uskoković

Institute of Technical Sciences of the Serbian Academy of
Sciences and Arts, Belgrade, Yugoslavia
*Institute of Physical Chemistry, Faculty of Sciences
Belgrade, Yugoslavia

ABSTRACT

The changes ocuring in the structure of dynamically compacted
amorphous $Ni_{78}P_{22}$ powder in the pressure range 50–150 kbar were studied.
Powder was compacted by shock waves obtained during explosion of high
explosives. At 50 kbar $Ni_{78}P_{22}$ does not crystallize, but increases in
pressure cause both crystallization, which is completed at 150 kbar,
and the formation of stable crystalline Ni and Ni_3P phases. The minimum
porosity was obtained at 70 kbar. The mechanism of hydrogen sorption
seems to be the same on dynamically compacted samples as well as on the
initial amorphous powder. Any shift of crystallization temperature of
the amorphous phase remaining after dynamic compaction was not
detected, but the activation energy for crystallization was somewhat
lower ($\sim$ 150 kJ/mol) than the value corresponding to the
crystallization of unshocked amorphous powder (212 kJ/mol). These data
may indicate that shock waves cause a certain "destabilization" of
amorphous $Ni_{78}P_{22}$.

INTRODUCTION

Amorphous metal powders and processes for their consolidation have
been extensively studied over the last few years with the main goal
being to obtain compacted amorphous materials displaying extraordinary
properties.[1,2] Dynamic compaction is one of the most promising methods
for reaching this goal, because it enables powders to be consolidated
to near-theoretical density in a very short time through the action of
intensive shock waves.[3] Dynamic compaction is usually accompanied by
local overheating at particle-particle contacts and by complete
crystallization of the amorphous phase.

There are different experimental techniques for dynamic
consolidation[4] including direct contact high explosives, projectile
impacts, high explosives, powder propelled guns, electrical explosions,
gas-and electromagnetically-impelled guns, pulsed radiation (from laser
and electron beams), magnetic impulsions, and other special methods.
All these methods have found wide application in powder consolidation.

Science of Sintering
Edited by D. P. Uskoković *et al.*
Plenum Press, New York

The majority of papers devoted to the study of dynamic compaction have attempted to obtain non-porous amorphous compacts with properties similar to those of amorphous ribbons.[5-12] Different amorphous systems prepared either by rapid quenching or by mechanical alloying, have been studied. Only a few data on crystallization of amorphous Ni-P and its dynamic consolidation have been reported. Using differential scanning calorimetry and X-ray diffraction, Šušić and Uskoković[13] have shown that crystallization of $Ni_{78}P_{22}$ obtained by chemical methods occured in two steps (at $\sim 615K$ and $\sim 670K$, depending on heating rate) and that stable Ni and Ni_3P phases were formed. Karpusha et al.[14] obtained similar results from optical adsorption spectra in the case of $Ni_{81}P_{19}$ prepared either by rapid cooling of melt or by evaporation from the gas phase. Limited data concerning dynamic compaction of compounds in the Ni-P system, specifically, the compound 89% Ni, 11% P, can be found in Prümmer's review paper.[15] Dynamic compaction at 37 kbar was reported to result in a rather high porosity; however, it dissapeared at 44 kbar, and the sample remained in the amorphous state. Some local crystallization, detected in X-ray diffractograms, occured at 51 kbar. More detailed information about this cited study could not be found.

The purpose of this paper was to study such consolidation processes in $Ni_{78}P_{22}$, including associated amorphous-crystal transformation. A further goal was to estimate sorption kinetics, as well as the kinetics of crystallization of any amorphous phase remaining after dynamic compaction. The stability of amorphous systems under the action of shock waves was also a matter of some interest.

EXPERIMENT

The amorphous $Ni_{78}P_{22}$ was obtained by electroless reduction of divalent nickel ions by sodium hypophosphite in buffered aqueous solution at 380K[13] and conventionally pressed at 3 kbar ($\sim 50\%$ theoretical density). The obtained pellets (10 mm diameter and ~ 3 mm height) were put in copper ampoules mounted in a steel support. The dynamic compaction was accomplished by shock waves generated by explosion of high explosives. The intensity of shock waves was controlled by the thickness of polyamide used as attenuator between the explosive and the loaded ampoules. Dynamic pressures in the attenuator at the boundary of the copper ampoules were calculated as 50, 70, 90 and 150 kbar, respectively.

Changes in the amorphous $Ni_{78}P_{22}$ powders after chock compaction were determined with a Du Pont Thermal Analyser 1090 operating in the differential scanning calorimeter (DSC) mode, in flowing hydrogen. The structure analyses were based on X-ray diffraction data, obtained with $Cu_{K\alpha}$ radiation and a graphite monohromator.

RESULTS AND DISCUSSION

Upon heating of samples of 10-20 mg in DSC cell in a hydrogen atmosphere, a thermogram with three exothermal maxima was obtained at the corresponding temperatures (Fig. 1a). The first maximum on the thermogram corresponds to hydrogen adsorption on the amorphous sample. The other two maxima correspond to the first and second stages of crystallization, occuring at $\sim 615K$ and $\sim 670K$, respectively, and depending somewhat on the heating rate.[13] After reheating a cooled sample which had been exposed to air for 10-15 min., we obtained a thermogram with only one maximum, corresponding to hydrogen adsorption on the amorphous sample, similar to that previously analysed.[13]

Thermograms of dynamically compacted samples (Fig. 1b-d) show that the abovementioned hydrogen sorption still occurs, and is independent on the compaction pressure. The degree of crystallization does depend on the compaction pressure: an increase of compaction pressure causes a decrease in the amorphous phase content. At 50 kbar, no significant changes in the amorphous phase content were observed but at 150 kbar, complete crystalization of $Ni_{78}P_{22}$ was observed in X-ray diffractograms (Fig. 2). They were practically identical to similar diffractograms obtained from samples heated to temperature exceeding 673K. In both cases i.e. static and dynamic, stable Ni and Ni_3P crystal phases were obtained.

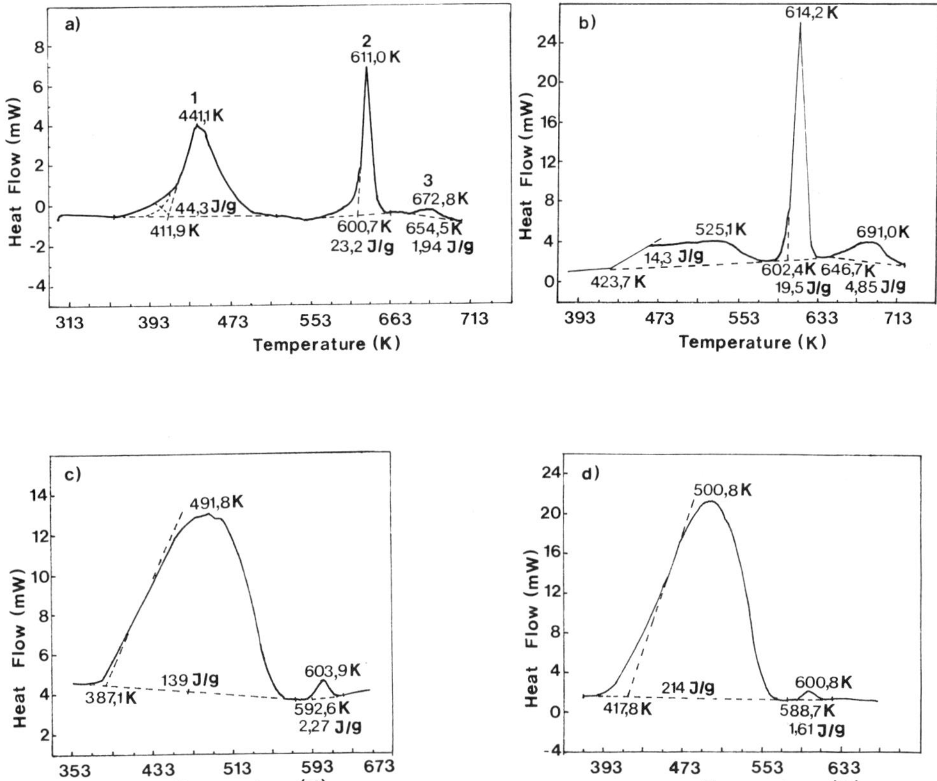

Fig. 1. DSC thermograms of amorphous and dynamically compacted $Ni_{78}P_{22}$: (a) amorphous powder; (b) 50 kbar; (c) 70 kbar; (d) 90 kbar.

Microstructures of dynamically compacted samples were investigated, with viewing planes chosen to be both parallel with and normal to shock waves. In all experiments where microstructural nonhomogeniety of such dynamically compacted samples was observed, it was also investigated and confirmed by DSC analyses. In Fig. 3, optical micrographs of polished surfaces of dynamically compacted $Ni_{78}P_{22}$ specimens are compand. The highest porosity and the largest pores were detected in samples compacted at only 50 kbar, but even there it did not exceed 5% porosity. The lowest porosity was obtained at 70 kbar

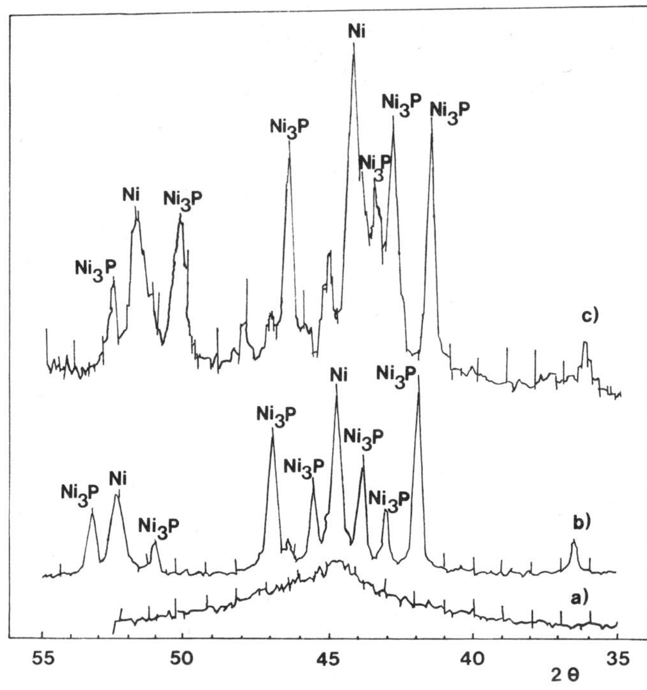

Fig. 2. X-Ray difractograms of the amorphous and crystalline $Ni_{78}P_{22}$
(a) amorphous powder; (b) after crystallization resulting from
thermal treatment; (c) after dynamic compaction at 150 kbar.

($\sim$2.5%). In the latter case, the pore sizes were $\sim$1 μm and their
distribution was essentially homogeneous. For samples compacted at
90 kbar, somewhat larger pores (1-2 μm), and higher overall porosity,
but still with homogenous pore distribution, was observed. In samples
compacted at 150 kbar, pores were distributed in microchains,
indicative shearing or slipping of bulk material along oblique planes
during shock compaction.

Hydrogen sorption in dynamically compacted samples was studied by
non-isothermal DSC techniques, developed on the basis of the non-
isothermal differential enthalpy method (DEC) used for chemical
reactions. It has already been shown[13] that a linear relation, log
k-1/T, is obtained for chemical reactions of the first order. The
activation energy for this process can be derived from the slope and
other kinetic parameters as well. The rate constant k is ($\Delta H/\Delta t$)/(A-a),
where ΔH is the enthalpy, A is the total area below the maximum in the
thermogram, a is the area below the maximum recorded in time t for the
reaction (i.e., up to the programmed temperature), $\Delta H/\Delta t$ is the slope
of a tangent to the thermogram at time t. According to the DSC method,
k=ΔmW (A-a), where ΔmW is the value measured on the ordinate, in
milliwats.

The maximum enthalpy of hydrogen sorption during the second
heating in hydrogen was 562 J/g and it was found to increase after
repeated heating, accompanied by more and more pulverization of the
sample (Fig. 4). The relation log k-1/T for hydrogen sorption on the
samples compacted at 50, 70 and 150 kbar, respectively, is shown in
Fig. 5. The sample compacted at 50 kbar was previously pulverized and
activated by heating in hydrogen up to 723K (Fig. 4, thermogram a), and

288

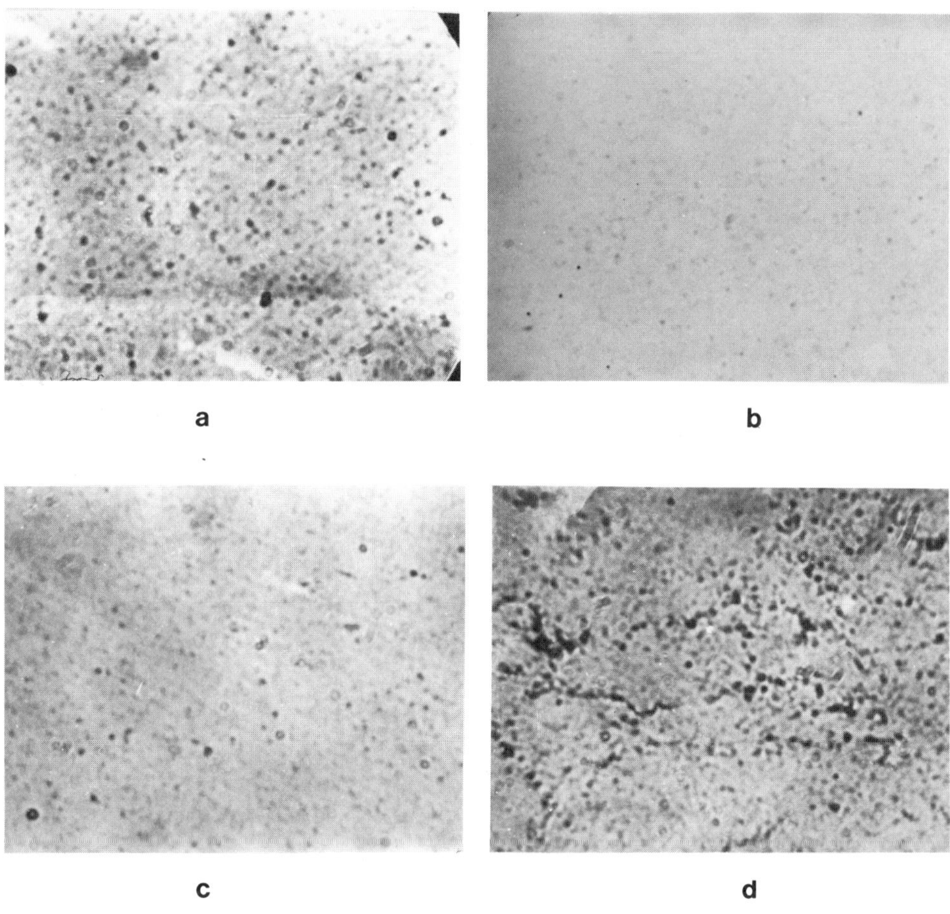

Fig. 3. Light microscopy of polished surfaces of dynamically compacted
$Ni_{78}P_{22}$: (a) 50 kbar; (b) 70 kbar; (c) 90 kbar; (d) 150 kbar.

after that, hydrogen adsorption was determined (Fig. 4, thermogram b).
It is obvious from Fig. 5 that log k-1/T is a linear relation with
three slopes, similar to hydrogen sorption on unshocked amorphous $Ni_{78}P_{22}$
powder. Thus we may conclude that the mechanism of hydrogen sorption is
identical in both cases, although the rate constants, the activation
energies, and the frequency factors are slightly different. This means
that the process of hydrogen sorption is an inherent characteristic of
the material, and thus it does not depend on compaction pressure.
Kinetic and thermodynamic parameters obtained by the analyses of
thermograms from Fig. 5 are shown in Table I.

However, it should be emphesised that the kinetic and
thermodynamic parameters, especially those for activation energy of the
first step, do depend on prior history, especially the level of that
samples previous activation. For example, in the case of the sample
compacted at 50 kbar, which has been pulverized and activated by
previous heating in hydrogen, the calculated activation energy was
significantly lower than for the other two similar cases, where
previous activation treatments had not been applied.

Crystallization of remnant amorphous phases in dynamically
compacted samples was investigated by Barker's method[13] for all samples

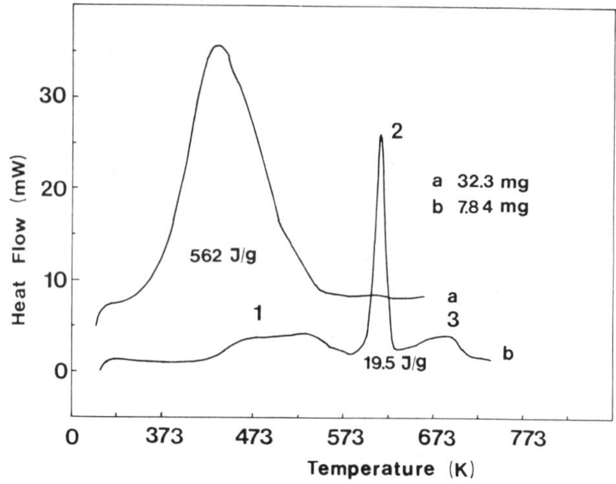

Fig. 4. DSC thermograms of dynamically compacted $Ni_{78}P_{22}$ at 50 kbar (a) first heating; (b) second heating in hydrogen (30 K/min).

which maintained amorphous structures after compaction (50,70 and 90 kbar, Fig. 6). For each such samples a linear relation, $(\log \beta /T^2)-1/T$, was obtained, where β is the heating rate of the sample and T is the peak temperature of the exothermic maximum. The activation energy can be derived from the slope. The frequency factor, Z, is calculated from the relation $Z =[\beta E \exp(E/RT)]/RT^2$, and the crystallization rate constant, k, is determined from the relation $k = Z \exp (-E/RT)$. The activation energies of the samples compacted at 50, 70 and 90 kbar were found to be 151.02; 148.28 and 149.44 kJ/mol., respectively. The mean value of Z is $6.83x10^{12}$ min^{-1} and the mean value of k at 602K is 0.98 min^{-1}. The rate constant may be calculated for each temperature, as well.

No significant change was observed in the crystallization temperatures of the remnant amourphous phase after shock compaction, as compared to the initial amorphous powder. Moreover, the activation

Table I. Kinetic and thermodynamic parameters of hydrogen sorption on dynamically compacted $Ni_{78}P_{22}$ in hydrogen flow

P,kbar	E,kJ/mol			Z,s⁻¹		
	E_1	E_2	E_3	Z_1	Z_2	Z_3
50	61.77	28.90	98.00	$7.10x10^6$	$5.10x10^2$	$6.96x10^9$
70	145.88	41.03	107.70	$4.88x10^{16}$	$1.13x10^3$	$7.00x10^8$
150	163.65	39.61	97.07	$1.48x10^{15}$	$3.65x10^3$	$7.57x10^8$

energies of crystallization were lower when compared to the initial
powder (212.36 kJ/mol)[13], indicating that after dynamic compaction, the
amorphous phase was less stable. Thus it crystallize more easily than
did amorphous phase in the original unshocked powder.

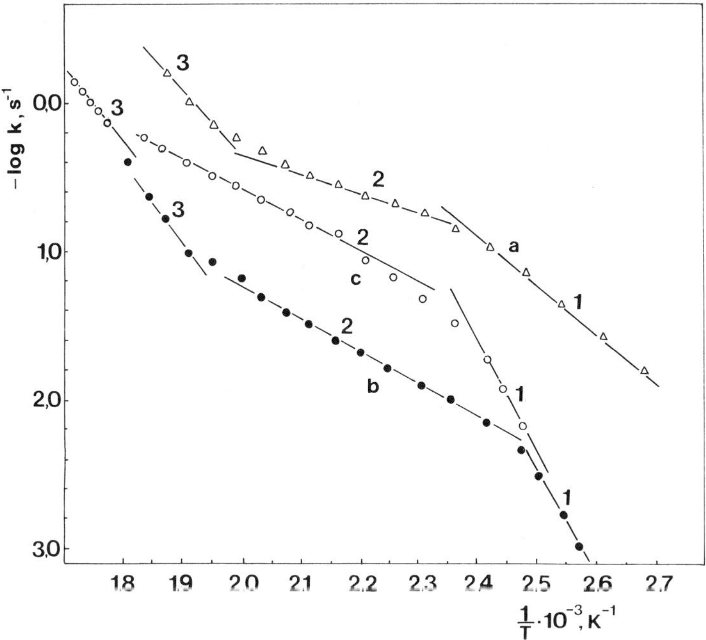

Fig. 5. Rate constant k vs 1/T for hydrogen sorption on dynamically
compacted $Ni_{78}P_{22}$ at (a) 50 kbar; (b) 70 kbar; (c) 150 kbar.

CONCLUSION

Processes occurring during dynamic compaction of amorphous $Ni_{78}P_{22}$
were studied over the pressure range 50-150 kbar. Particular attention
was given to the analyse of hydrogen sorption and to determinations of
the kinetics of crystallization of the remnant amorphous phase. Samples
were compacted by shock waves obtained by contact explosion of high
explosives. At 50 kbar no significant change in crystallization of the
initial amorphous phase was registered. Further increases in shock
pressure resulted in reductions of amorphous phase content. At 150
kbar, crystallization was complete, and stable Ni i Ni_3P phases were
formed. The resultant microstructures are rather inhomogeneous. Samples
obtained at 50 kbar had the highest porosity. At 70 kbar both the pore
volume and the pore sizes were minimal (~ 1 µm). Further increases in
shock pressure caused retention of more pore volume, and yielded larger
pore sizes. At 150 kbar, the pores were distributed like microchains,
indicating that plastic deformation had occurred during consolidation.

291

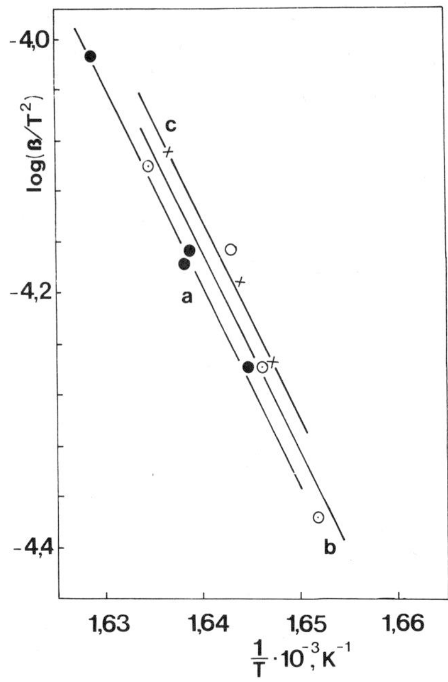

Fig. 6. Relationship between $\log(\beta/T^2)$ and 1/T for the crystallization of the remnant amorphous phase after dynamic compaction of amorphous $Ni_{78}P_{22}$ (a) 50 kbar; (b) 70 kbar; (c) 90 kbar.

Kinetic and thermodynamic analyses of hydrogen sorption on dynamically compacted $Ni_{78}P_{22}$ specimens, spanning the range 50-150 kbar, indicates that the sorption does not differ significantly from that observed on unshocked amorphous powder, and that it does not depend on compaction pressure. When compared to the initial unshocked powder no significant changes in the crystallization temperature of the amorphous phase remaining after shock compaction were observed. However, the activation energy for crystallization of the remnant amorphous phase after shock compaction was ∿150 kJ/mol, considerably lower than the value corresponding to the crystallization of the initial amorphous powder (212 kJ/mol). These finding indicate that the amorphous phase remaining in the samples after dynamic compaction was less stable, and it can be crystallized more easily than the initial unshocked $Ni_{78}P_{22}$ powder.

ACKNOWLEDGEMENTS

The investigations presented in this paper were financially supported by the Republic Committee for the Science of Serbia through the project "Physical Chemistry of Condensed Systems". The authors are grateful Dr. Lj. Karanović for the X-ray analyses.

REFERENCES

1. M. Zdujić, D. Uskoković, "Rapidly solidified crystalline and amorphous powders-problems and perspectives" in SINTERING '87, Ed. S. Somiya et al., Publ. Elsevier, 1988,1504.

2. S. A. Miller, "Amorphous metal powder: production and consolidation", in AMORPHOUS METALLIC ALLOYS, Ed. F.E. Luborsky, Butterworths, 1983, 506-521.

3. R. Prümmer, "Explosive compaction of powders-state of art", in HIGH ENERGY RATE FABRICATION, Proc. of Ninth Intern. Conf., Novosibirsk, Aug. 18-22, 1986, 169-178.

4. G. E. Duvall, R. A. Graham, "Phase transition under shock wave loading", Rev.Mod.Phys., $\underline{49}$ (1977), 523-579.

5. D. G. Morris, "Compaction and mechanical properties of metallic glass", Met. Sci. J., $\underline{14}$ (1980), 215-220.

6. O. V. Roman, V. G. Gorobcov, B. S. Mitin, V. A. Vasilev, "Struktura i svoistsva amorphnih materialov na zheleznoi osnove", Porosh. Metal., No.6, Minsk, Vischaya shkola, 1982, 8-13.

7. O. V. Roman, G. A. Adadurov, Yu. V. Baldohin, A. P. Bogdanov, Yu. N. Voloshin, V. G. Gorobcov, A. M. Kaplan, M. Yu. Messiner, I. M. Pikus, "Pressovanie poroshkov amorphnih splavov v dinamicheskih uslovyah", Porosh. Metal., No. 5 (1984), 17-23.

8. V. F. Nesterenko, A.V. Muzikantov, "Ocenka uslovya sohraneniya amorphnoi strukturi materialov pri kompaktirovanii vzrivom" Fiz. Gorenya i Vzryva No. 2 (1985), 120-126.

9. V. F. Nesterenko, "Vozmozhnosti udarno-volnovih metodov polucheniya i kompaktirovaniya bistrozakalenih materialov", ibid., No. 6 (1985), 85-98.

10. A. A. Deribas, "Obrabotka materialov energiei vzryva", ibid., No. 5 (1987), 148-158.

11. V. F. Nesterenko, S. A. Pershin, E. D. Tabachnikova, "Temperaturnaya zavisimost prochnosti na szhatie materilov, poluchenih impulsnim pressovaniem bistrozakalennih poroshkov", ibid., No. 4 (1988), 125-129.

12. N. N. Gorshkov, E. Y. Ivanov, A. V. Plastinin, V.V. Silvestrov, T. M. Sobolenko, T. S. Teslenko, "Vzrivnoe kompaktirovanie amorphnogo poroshka Cu-Sn polushenogo metodom mechanicheskogo splavleniya", ibid., No. 2 (1989), 125-133.

13. M. V. Šušić, D. P. Uskoković, "Crystallization kinetics of amorphous $Ni_{78}P_{22}$ powders and hydrogen adsorption on both amorphous and crystal alloy powders", J. Mater. Sci., $\underline{23}$ (1988), 4076-4080.

14. V. D. Karpusha, V. N. Korzhik, Yu. A. Kunickii, L. V. Poperenko, S. L. Revo, I. A. Shakevich, "Opticheskoe pogloschenie i elektronnaya struktura amorphnih splavov $Ni_{81}P_{19}$ i Ni_3B", Metallophisika $\underline{10}$ (1988), 61-66.

15. R. Prümmer, "Dynamic compaction of powders" in EMERGENT PROCESS METHODS FOR HIGH-TECHNOLOGY CERAMICS Mat. Sci. Res., $\underline{17}$ (1984), 505-517.

THE PREDICTION OF HIP PARAMETERS FOR INTERMETALLIC

PREALLOYED Ni - Al POWDER

R. Laag, W.A.Kaysser, R. Maurer and G. Petzow

Max-Planck-Institut für Metallforschung
Pulvermetallurgisches Laboratorium
Heisenbergstraße 5, 7000 Stuttgart 80, FRG

INTRODUCTION

Rapid solidification processes have received considerable attention as a suitable method to produce spherical powders with fine grain size, increased chemical homogeneity and extended solid solubility. Depending on the type of process the material's response to rapid heat extraction differs widely. Microstructures of high-grade homogeneity can be obtained if process control during atomization allows a high undercooling of the particles prior to solidification. Undercooling controls the solidification rate. The solidification front velocity must exceed the diffusivity of atoms in the melt at solidification temperature if segregation is to be avoided. High pressure Ar gas atomization is one of the preferred processes for producing polycrystalline, narrow size distribution spherical powder particles with inner grain sizes of only a few micrometers. These attributes are beneficial for HIP consolidation. Transferred to the production of fine grained intermetallic compounds these attributes offer promising properties for technological applications. In a previous paper experiments were described where $Ni_{50}Al_{50}$, $Ni_{75}Al_{25}$ and nonstoichiometric prealloyed powders were produced by Ar-gas atomization (1). The consolidation of these fine grained intermetallics with ordered lattice structure by plastic deformation was limited to the very beginning of the densification. Grain boundary and volume diffusion were also limited by the ordered structure. Full density material was achieved only if HIP cycles reach approx. 0.8 T_m. Grain growth decreased the ductility (2) and, hence, cycle times should be optimized to a minimum length to obtain full density and fine grained HIPed material. The purpose of this paper is to demonstrate the collection of data from an actually used material tuning the sensitive parameters in HIP diagram calculations (3).

EXPERIMENTS

Powders of the composition $Ni_{76}Al_{24}$, $Ni_{75}Al_{25}$, $Ni_{70}Al_{30}$ and $Ni_{50}Al_{50}$ were produced by high pressure Ar-gas atomization with constant gas pressures of 100 MPa. The as-atomized powders were prepared and characterized using procedures described elsewhere (1).

For dilatometry experiments, spherical powder particles about 150 μm in size were selected by the shadow cut method in a microscope. The plane parallel plates of two Alumina pushers were prepared as a specimen holder by polishing with 1 μm diamond paste and coating with a carnuba wax fixative to conserve the adjusted uniaxial load during mounting. The wax evaporates completely at approx. 150 °C and the specimen is fixed between the two pushers. Calibration of the dilatometer was performed with an Iridium sphere 250 μm in diameter to evaluate the creep of the alumina holders, the influence of thermal expansion and the drifting of the sensors. From each alloy nine dilatometer cycles were carried out at 1000, 1100 and 1200°C at a constant load of 0.2, 0.4 and 0.8 N. During isothermal heating, the length change, normalized to a reference of alumina, and the load were monitored. In order to ensure a representative sample, the results were compared.

Comparative hot isostatic pressing of powder compacts was carried out with three different sieve fractions of each Ni aluminide. After encapsulation in DURAN glass containers and vacuum degassing to a rest pressure of 10^{-7} MPa at room temperature the capsules were sealed. The charges were heated at 30 K/min up to 1100 °C and pressed with 50, 75, 100, 125 and 150 MPa for 60 min in an ASEA HIP.

CHARACTERIZATION OF THE POWDERS

All powders used in this investigation, were atomized under the same conditions with superheating of $1.25 \cdot T_m$, where T_m is the liquidus temperature of the alloys. The particles were spherical and free of satellites (i.e. $Ni_{50}Al_{50}$; Fig. 1). Cooling rates for $Ni_{50}Al_{50}$ were calculated on the basis of 300°C undercooling using thermodynamical data (Fig. 2) (4). $Ni_{50}Al_{50}$ shows a single phase microstructure. The powders with 24 and 30 at.-% Al show a dual phase microstructure with $(\gamma' + \gamma)$ in $Ni_{76}Al_{24}$ and $(\gamma' + \beta\text{-NiAl})$ in $Ni_{70}Al_{30}$ (Figs. 3,4). TEM observations showed in $Ni_{75}Al_{25}$ powder the β-NiAl phase with a martensitic twinned structure (Figs. 5,6) which transforms to Ni_3Al during HIPing. Grain size measurements were employed using the mean intercept length of a random test line by dividing the total length of test line by the number of grains traversed. Table I lists the results of the chemical analysis, the average grain size and the tap density of the powders, which was the initial density in the HIP cycle.

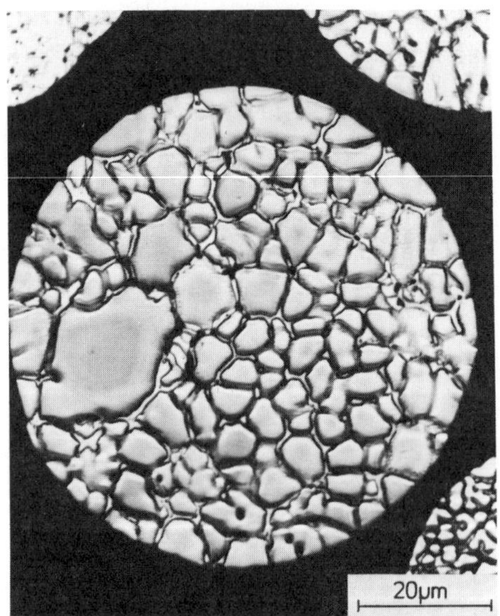

Fig. 1. $Ni_{50}Al_{50}$ powder particles, 63 μm in size, show an inner grain size of 8 to 10 μm.

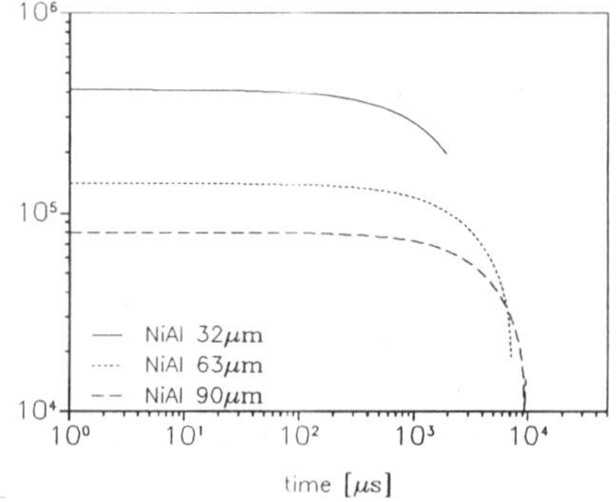

Fig. 2. The calculation of cooling rates gives an average solidification time of about 100 ms and a rate of $4 \cdot 10^5$ Ks^{-1} for 32 μm.

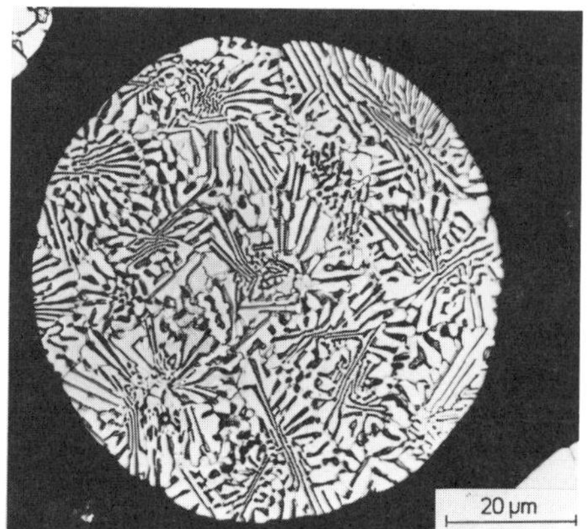

Fig. 3. Ni₇₆Al₂₄: Powder particle, 63 μm in diameter, shows a dual phase (Ni₃Al+γ) microstructure.

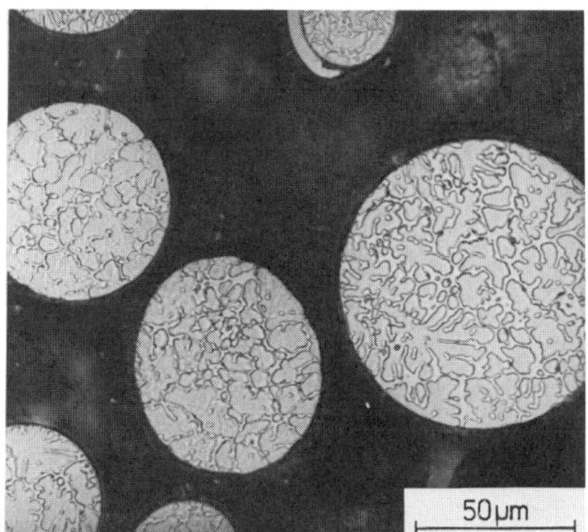

Fig. 4. Ni₇₀Al₃₀: Powder particle, 63 μm in diameter, shows a dual phase (Ni₃Al+β-NiAl) microstructure.

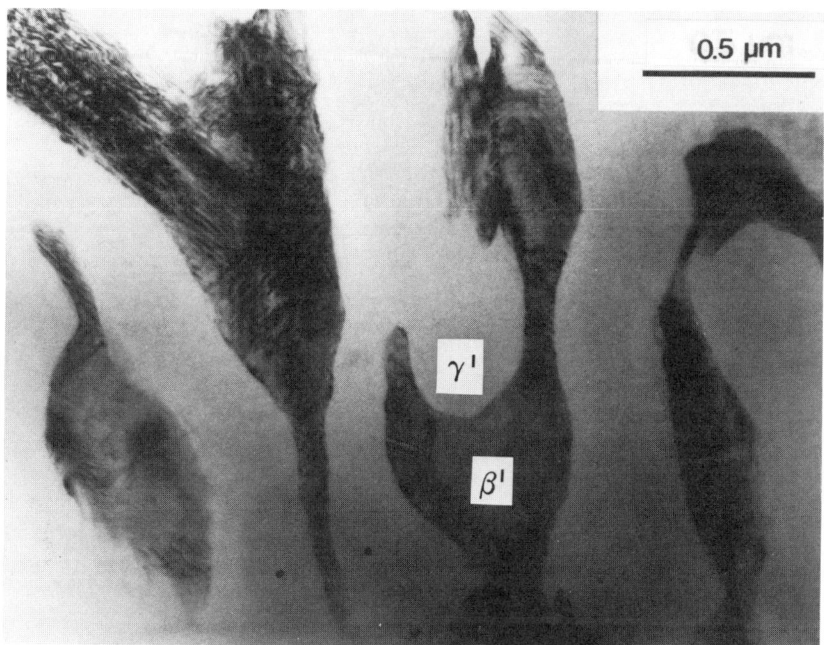

Fig. 5. TEM photographs of initial Ni₇₅Al₂₅ powder, as atomized. Powder particle consisting of the two phases γ' and β' (martensitic phase).

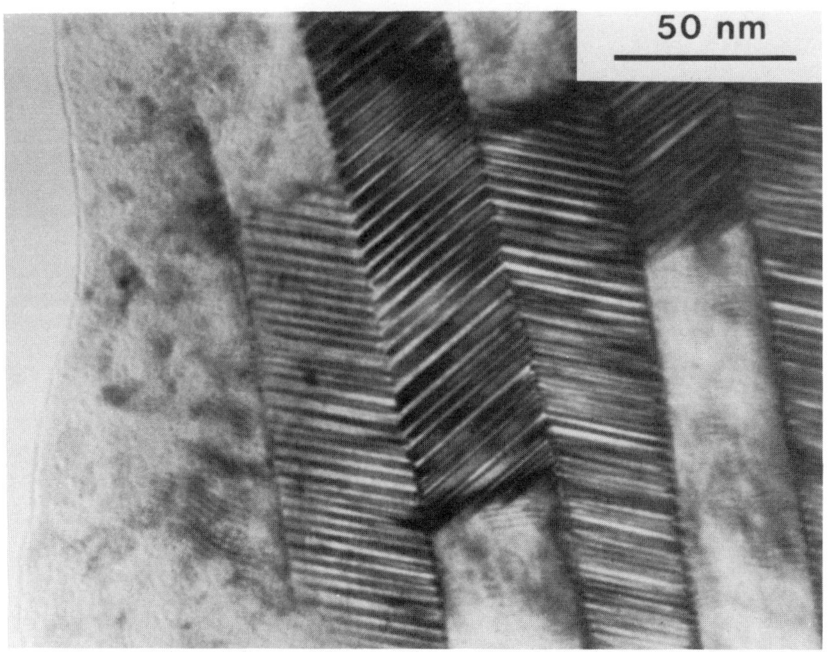

Fig. 6. Magnification of the internally twinned martensitic phase.

Table I. Analyzed chemical composition, inner grain size and tap density of powders with a diameter of 63μm.

Alloy (nom.)	Al Content	Grain Size	Tap Density
$Ni_{76}Al_{24}$	24.1	6.10	4.58
$Ni_{75}Al_{25}$	25.2	8.06	4.50
$Ni_{70}Al_{30}$	29.8	7.32	4.25
$Ni_{50}Al_{50}$	49.7	8.00	3.74
	[at.-%]	[μm]	[gcm^{-3}]

DEFORMATION BEHAVIOR

The spherical powders of all alloys tested in the single particle dilatometer responded to loads and isothermal temperatures with typical deformation, monitored as contact flattening. The measurement, which is described in detail by Aslan (5), can be reproduced with an accuracy of ± 5% of the total change for a second particle. Figure 7 shows the arrangement in the dilatometer and Fig. 8 shows a deformed $Ni_{70}Al_{30}$ powder particle at the end of a dilatometer run. The actual contact radius x(t) which leads to the new radius R(t) can be expressed as a function of the deformation h(t) (5). For a given t the contact radius is

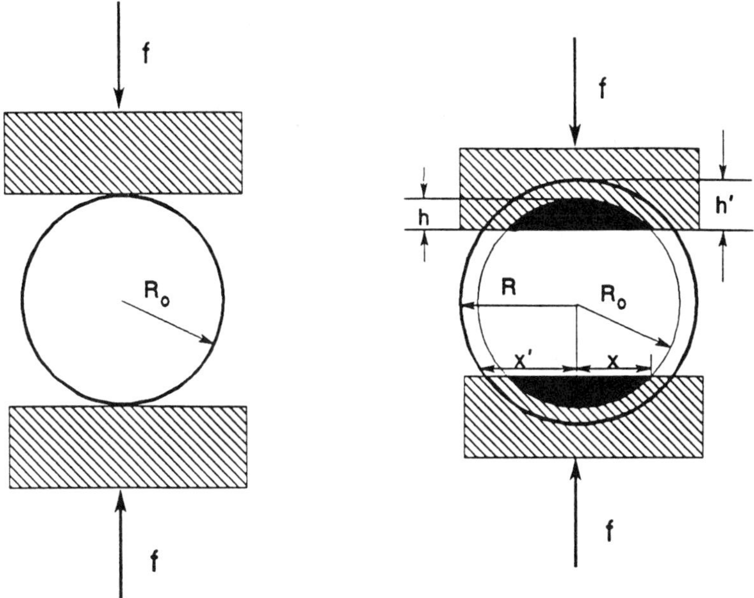

Fig. 7. Deformation of a sphere between two parallel plates (schematic).

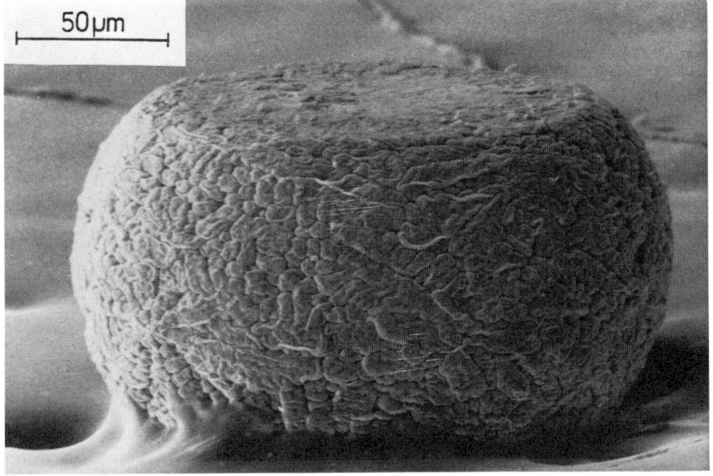

Fig. 8. A powder particle ($Ni_{70}Al_{30}$) after deformation in the dilatometer.

$$x = \left[2 \cdot h \left(R_0 + \frac{h^2}{3 \cdot (R_0 - h)} \right) \right]^{1/2} \tag{1}$$

and the new contact radius is

$$R = \left[\left[2 \cdot R_0^3 + (R_0 - h)^3 \right] \cdot \frac{1}{3 \cdot (R_0 - h)} \right]^{1/2} \tag{2}$$

with R_0 as the radius of the undeformed sphere. The pressure, p, at the contact area, generated by the load, f, can be calculated by

$$p = \frac{f}{\pi \cdot x^2} \tag{3}$$

In Fig. 9 the normalized deformation h/R_0 is plotted. The results show the expected temperature dependence of the creep rates. For $Ni_{76}Al_{24}$ and $Ni_{75}Al_{25}$ the creep rate and the deformation, h/r_0, are higher than for $Ni_{70}Al_{30}$ and $Ni_{50}Al_{50}$ during the isothermal testing. Between 1000 °C (Fig. 9a) and 1100 °C (Fig. 9b) the increase in creep rate is higher for the $Ni_{76}Al_{24}$ alloy than for $Ni_{75}Al_{25}$.

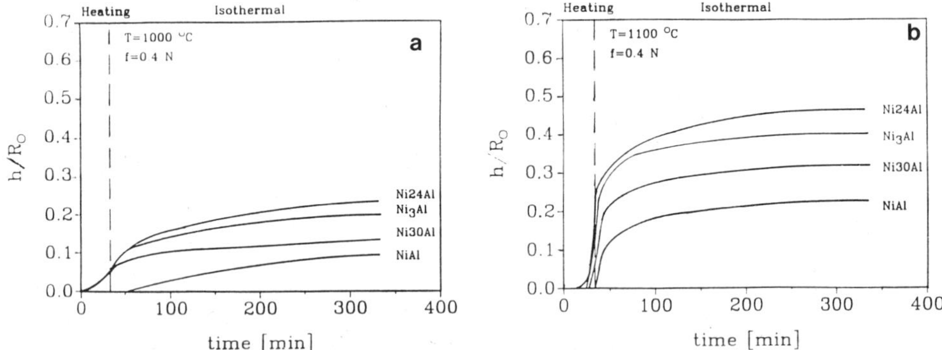

Fig. 9. Deformation curves normalized to R_0 for $Ni_{76}Al_{24}$, $Ni_{75}Al_{25}$, $Ni_{70}Al_{30}$ and $Ni_{50}Al_{50}$ at 1000 °C (a) and 1100 °C (b).

PREDICTION OF HIP-PARAMETERS FROM DILATOMETRY

The procedure to predict HIP-parameters from dilatometer experiments is described in (5), (6). The prediction is valid in a range where the densification is provided by plastic deformation and power law creep because the contribution of surface and grain boundary diffusion to the densification is not covered by the analysis. The observation of an increasing single contact between two particles gives a quite exact picture of the contact deformation during HIPing as long as the individual contact areas do not influence each other. At relative HIP densities between 85% and 95% the particle contacts touch each other and the porosity closes. This state is not covered by the calculations.

Different equations are proposed in the recent literature to correlate the deformation of a single powder particle and the HIP densities of powder compacts

(7). For gas atomized superalloy AP1, Cu- and mild steel powders, the empirical relation

$$D = D_0 + \frac{2h}{R_0} \cdot \left[1 + \frac{h}{R_0} \right] \qquad (4)$$

provide a good agreement between predicted and HIP densities. If the powder particle between the two dilatometer pusher rods is treated as a particle inside a cube deforming under triaxial pressures of equal value (isostatic condition) then equation

$$D = \frac{D_0}{1 - \frac{1}{9}\frac{1}{R_0^3}\left[\left[7.3 + 8.6\,\frac{x^2}{R_0^2} \right] h^2(3R_0 - h) \right]} \qquad (5)$$

can be obtained, using the random density function (RDF) of Scott (8) and its approximation given by Fischmeister and Arzt (9) for monosized rigid spheres. In HIP diagram calculation the approximation

$$D = D_0 \left[\frac{R}{R_0} \right]^3 \qquad (6)$$

(10) is used. In Fig. 10 the densities calculated with eqs. (4), (5) and (6) for a particle 100 μm in diameter are plotted as a function of the deformation, h. For a given h a wide range of densities can be obtained dependent on the equation used for the calculation. The external isostatic HIP pressure relates to the force on the particle in the dilatometer experiment and can be calculated using the equation given by (9)

$$p_{HIP} = \frac{Z \cdot D}{4\pi \cdot R_0^2} \cdot f^* \qquad (7)$$

where f^* is the average force on a particle. Setting $f^* = f$ from the dilatometer experiment results in an effective HIP pressure p_{HIP} which can be taken as $p_{HIP} =$ p from eq. (3). The contribution of the specific densification mechanism at a certain density is also dependent on the coordination number, Z, which is itself sensitive to the size distribution of the particles in the green body. There are sets of functions available to calculate this deviation from a monosized packing (11). With the approximation of Scott's RDF, Arzt (12) gives the most exact approximation with

$$Z = Z_0 + 9.5 \cdot (D - D_0) \qquad (8)$$

for $D \leq 0.85$ and

$$Z = Z_0 + 2 + 9.5 \cdot (D - 0.85) + 881 \cdot (D - 0.85)^3 \qquad (9)$$

for $D > 0.85$, setting $Z_0 = 7.3$ as the initial coordination number and $D_0 = 0.64$ as the initial density. Another approximation on the relation between Z and the density was made using the RDF for ideal elastic spheres directly (9). Up to full density, the densification was described with

$$Z = Z_0 + 15.5 \cdot \left[\left[\frac{D}{D_0} \right]^{1/3} - 1 \right] \qquad (10)$$

Eq. (10) neglects the material redistribution at all densities which leads to considerable differences from eq. (9), especially at higher densities. Calculation of

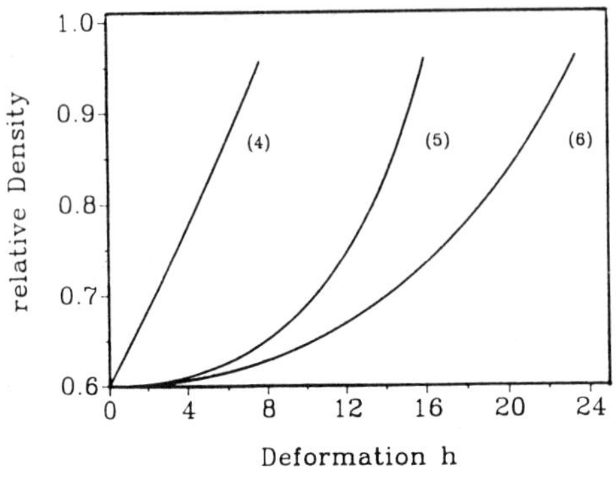

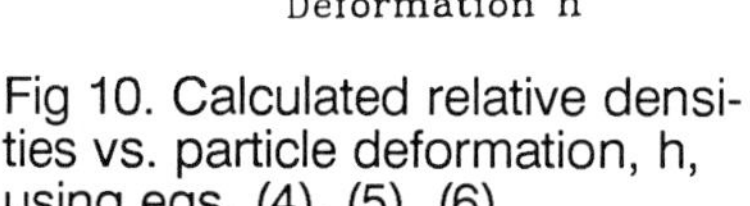

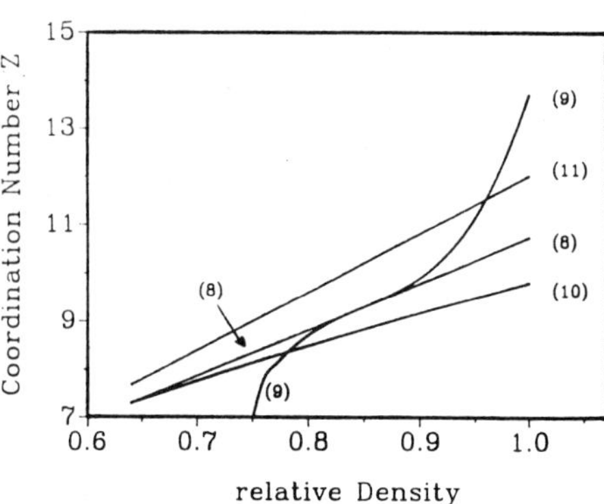

Fig 10. Calculated relative densities vs. particle deformation, h, using eqs. (4), (5), (6).

Fig 11. The coordination number plotted as a function of relative density using eqs. (8), (9), (10), (11).

HIP diagrams by the recent programs (3) is based on a linear approximation of the contact number as a function of density

$$Z = 12 \cdot D \tag{11}$$

which is a rough approximation to eq. (10), but it simplifies the calculation. Figure 11 shows the coordination number as a function of relative density for the different relations.

The conditions outlined at the beginning of this paragraph allow the prediction of HIP parameters from the deformation of a single sphere using eqs. (1), (2) and (5). This prediction is valid starting from green density, D_0, (provided the powder attains the theoretically assumed green density) to a density where diffusion driven mechanisms may start to dominate densification. Figure 12 shows the predicted relative density versus the HIP pressure calculated from dilatometer experiments with the loads, f, of 0.4, 0.6 and 0.8 N (open symbols). The measured HIP densities of $Ni_{50}Al_{50}$ and $Ni_{75}Al_{25}$ are included for comparison. The agreement between predicted and measured HIP densities is good up to densities of 90%. Multiple dilatometer runs with different loads result in a prediction of the minimum pressure which is necessary to reach a desired HIP density. HIP pressures above 50 MPa cannot be simulated realistically by the dilatometer at present.

TUNING HIP DIAGRAM CALCULATIONS WITH NEW MATERIAL DATA

In addition to the calculation of equivalent HIP pressures, the deformation of single spheres allows the extraction of data from the tested material which may improve the precision of the prediction of HIP conditions using Ashby's program for HIP diagram calculation /3/. For many metallic materials the yield stress, which forms the lower bound for consolidation by rate independent plastic flow, and the parameters for power law creep are most important to characterize the densification. The determination of temperature dependent yield stress data with the dilatometer is realized by stepwise rapid heating of the specimens. Figure 13 shows data for Ni$_{75}$Al$_{25}$, which were obtained from dilatometer experiments. For comparison, a imilar stoichiometric material of 10 μm grain size and a cast material with 100 μm grain size, which were tested under the same compression rate were also

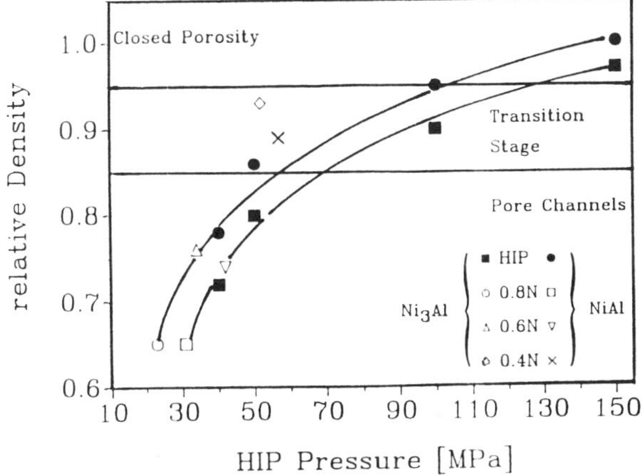

Fig. 12: Predicted relative densities vs. HIP pressure calculated from the load, f, with 0.4, 0.6, 0.8 N. Plotted data of HIPed densities for Ni$_{75}$Al$_{25}$ and Ni$_{50}$Al$_{50}$ correspond to the densities obtained after HIPing at 1100 ºC, 60 min for powder with a particle size range between 160 and 200 μm.

Table II. Creep constant and prefactor A for the tested alloys between 1000 and 1200 °C taken from single particle dilatometry. Load f=0.4 N.

Alloy (nom.)	Temp.	n	Log(A)
Ni$_{76}$Al$_{24}$	1000.0	4.31	-34.0
Ni$_{76}$Al$_{24}$	1100.0	5.33	-39.3
Ni$_{76}$Al$_{24}$	1200.0	8.15	-59.2
Ni$_{75}$Al$_{25}$	1000.0	4.10	-40.0
Ni$_{75}$Al$_{25}$	1100.0	4.90	-49.3
Ni$_{75}$Al$_{25}$	1200.0	7.05	-59.2
Ni$_{70}$Al$_{30}$	1000.0	3.57	-27.9
Ni$_{70}$Al$_{30}$	1100.0	14.86	-33.1
Ni$_{70}$Al$_{30}$	1200.0	17.53	-36.9
Ni$_{50}$Al$_{50}$	1000.0	1.28	-13.0
Ni$_{50}$Al$_{50}$	1100.0	2.54	-19.1
Ni$_{50}$Al$_{50}$	1200.0	3.51	-26.5
[at.-%]	[°C]		

shown /13/. The data from usual specimen sizes and single powder particles are in satisfying agreement. Hence the dilatometer measurements seem to provide a valid prediction of HIP parameters. Dilatometry also has the advantage of providing the data directly at HIP-temperatures with compression rates similar to those under HIP conditions.

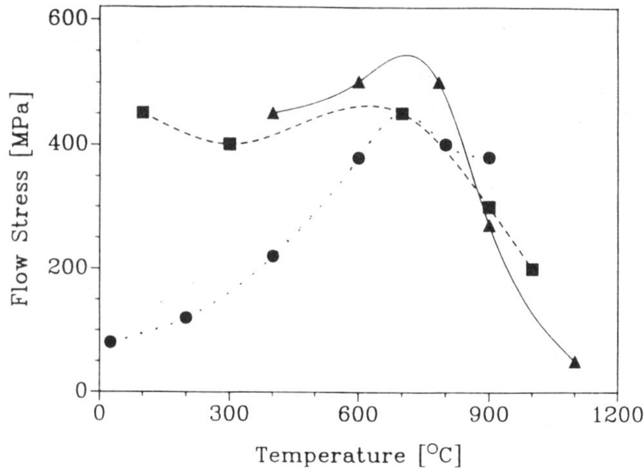

Fig. 13. Flow stress measured by single powder particle dilatometry and of cast Ni$_{75}$Al$_{25}$ (13).

The prefactor, A, and the power law creep exponent can also be extracted from the ratio of deformation rate to effective pressure in the neck at isothermal temperatures. The deformation rate by power law creep is given by

$$\frac{\dot{h}}{x} = \left[\frac{9 \cdot \pi}{16} \dot{\epsilon}_0 \, \sigma_0^{-n}\right] \cdot \left[\frac{p_{eff}}{3}\right]^n = A \cdot \left[\frac{p_{eff}}{3}\right]^n \tag{12}$$

Plots of $\log(\dot{h}/x)$ versus $\log(p_{eff}/3)$ give the creep exponent, n, whereby $n \leq 3$ indicates creep mostly controlled by bulk diffusion and $3 \leq n \leq 5$ indicates creep strongly depending on dislocation movement inside the crystal (14). For $\log(p_{eff}/3)$ = 0 the prefactor, A, can be obtained. Prefactors and creep exponents for a number of Ni aluminides tested at a load of 0.4 N at temperatures between 1000 and 1200 °C are listed in Table II.

Ni$_{76}$Al$_{24}$ and Ni$_{75}$Al$_{25}$ powders in the range of 1000 to 1100° C have creep exponents of about 5 which indicates deformation by power law creep. At higher temperature (T > 1000 °C) under the same load, which is equivalent to a HIP pressure of approximately 40 MPa, it is expected that the densification occurs by plastic deformation due to the rapid decrease in flow stress above 85% of the melting temperature. The extremely high creep exponents of Ni$_{70}$Al$_{30}$ reflect the presence of second phase particles and indicate the influence of "swelling stresses". A creep exponent of 1.28 measured for Ni$_{50}$Al$_{50}$ at 1000° C under a load of 0.4 N indicates primary creep in this stage. This is thought to be due to the insufficient number of slip systems in the B2 structure.

HIP-DIAGRAM CALCULATION

HIP-diagram calculation requires numerous material data. The foregoing paragraphs described a technique to collect some of these data from single powder particle dilatometry. However independent of the source of the data, either from literature or from dilatometry, we often found that the diagram calculations predict fully dense materials for HIP cycles that were obviously either too low in pressure, too low in temperature or too short in time to achieve full density in real HIP experiments. Minor modifications of the program's source code were made to give more flexibility in calculating flow stress variations with temperature such as for example with the abnormal increase in flow stress of $Ni_{75}Al_{25}$ and $Ni_{76}Al_{24}$ as a function of temperature. Furthermore the linear approximation between coordination number and relative density of eq. (11) was replaced by the exact geometrical solution given in eq. (8) and (9), because the coordination number is the controlling variable for the densification mechanism in HIP diagram calculation. The difference between both approximations is that eq. (11) predicts a higher contact number in the first stage of densification (open porosity) yielding a lower effective stress at the contacts. In this stage plastic deformation can be suppressed if the pressure on the particle contact is only slightly lower than the yield criteria requires. The suppression of plastic flow can promote power law creep with relatively high deformation rates. This would reduce the densification rate. At densities where closed porosity exists (second HIP stage), the contact number in the linear approximation (eq. (11)) is too low. This results in an overestimation of the pressure on the individual contacts and predicts power law creep as the dominant densification mechanism up to full density. The overestimation of contact pressure leads to concomitant increase in predicted density.

HIP diagrams calculated with the unmodified HIP-487 program and with data from dilatometer experiments are shown in Fig.14 for $Ni_{50}Al_{50}$ and $Ni_{75}Al_{25}$ (Fig. 15). The predicted densification is too fast because power law creep is the dominant densification mechanism in these calculations.

After modifications of the coordination number calculation and of the abnormal temperature dependence of flow stress in $Ni_{75}Al_{25}$, the diagrams changed predicting clearly longer densification times, higher temperatures and / or higher HIP pressure to achieve higher densities. In the case of $Ni_{50}Al_{50}$ (Fig. 16) the slope of the isochrone lines decreases with increasing density. Figure 17 shows a TEM

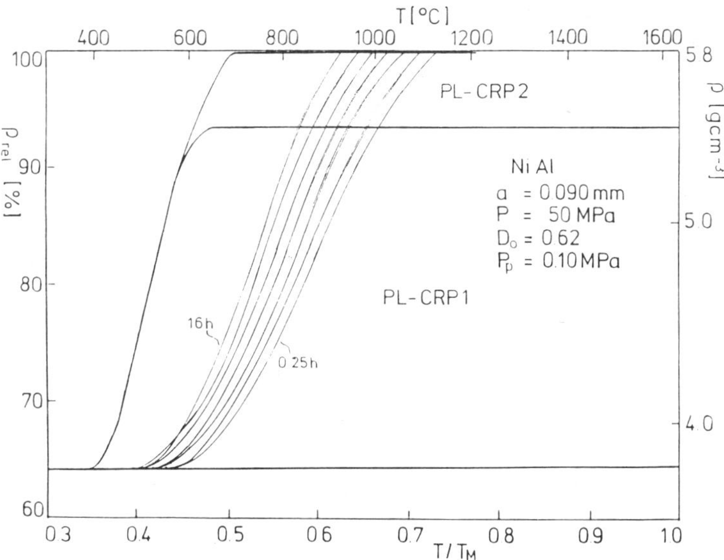

Fig. 14. HIP diagram calculated with the unmodified HIP-487 program and material data from dilatometer experiments of $Ni_{50}Al_{50}$.

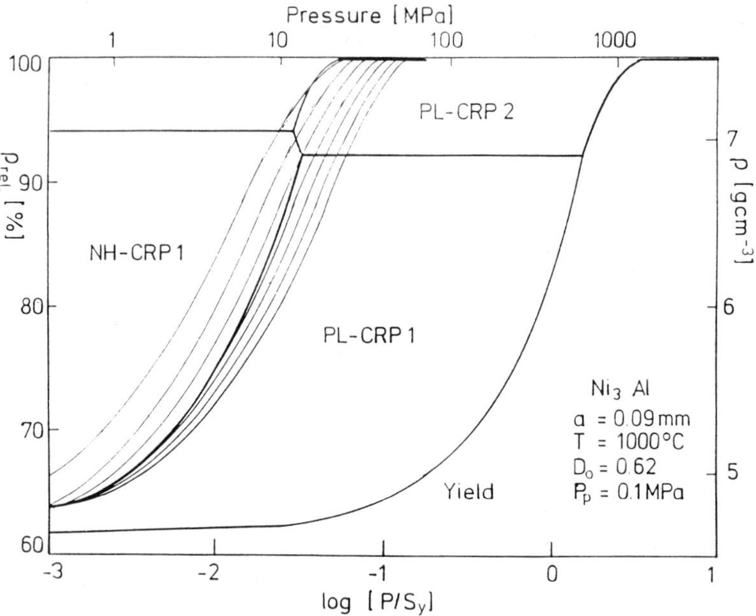

Fig. 15. HIP diagram calculated with the unmodified HIP-487 program and material data from dilatometer experiments of Ni$_{75}$Al$_{25}$.

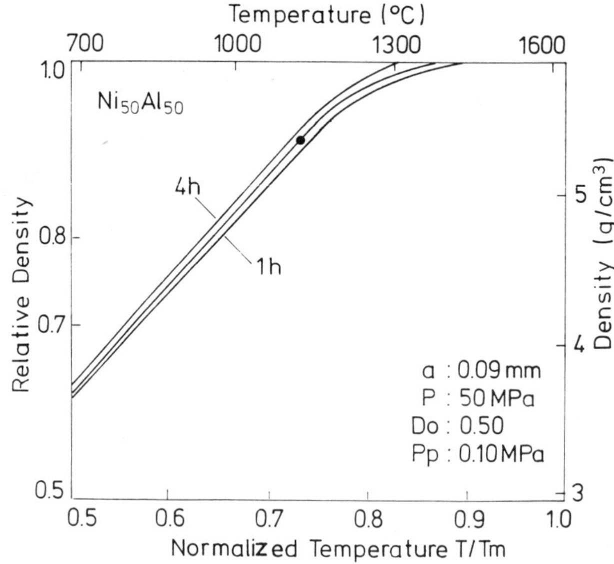

Fig. 16. HIP diagram calculated with the modified HIP-487 program and material data from dilatometer experiments of Ni$_{50}$Al$_{50}$.

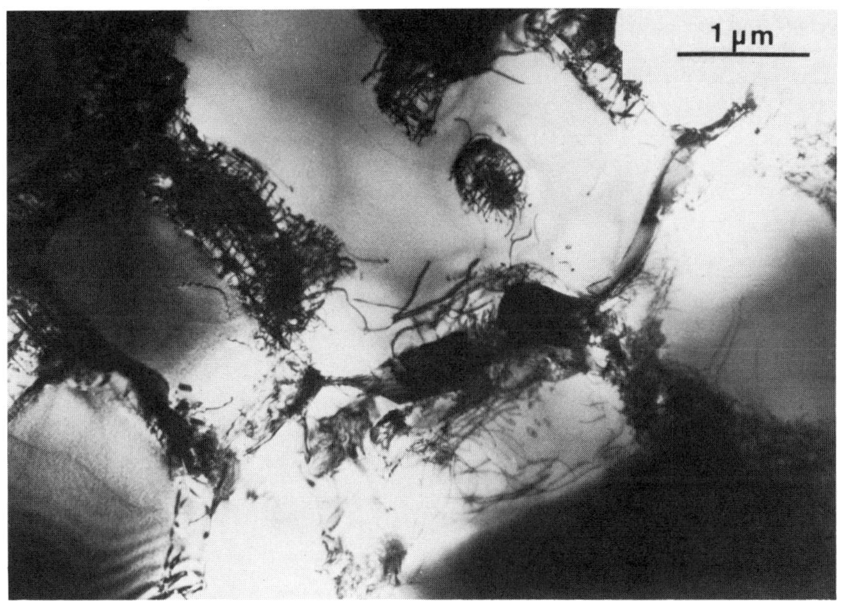

Fig. 17. TEM photograph of a specimen with 93% theoretical density HIPed at 1100 °C, 50 MPa for 2 h.

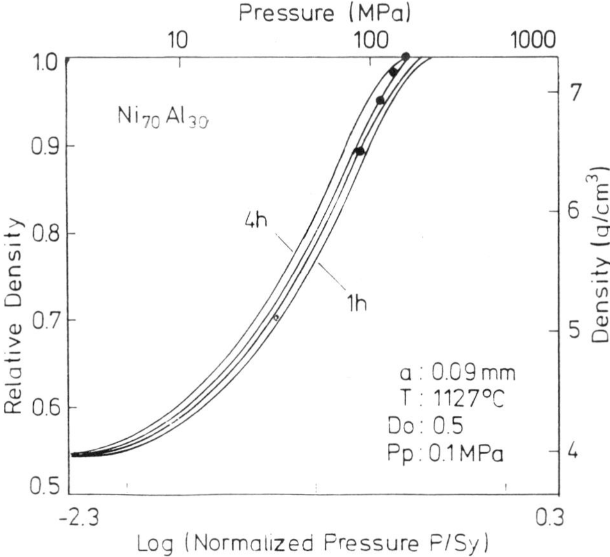

Fig. 18. HIP diagram calculated with the modified HIP-487 program and material data from dilatometer experiments of Ni$_{70}$Al$_{30}$.

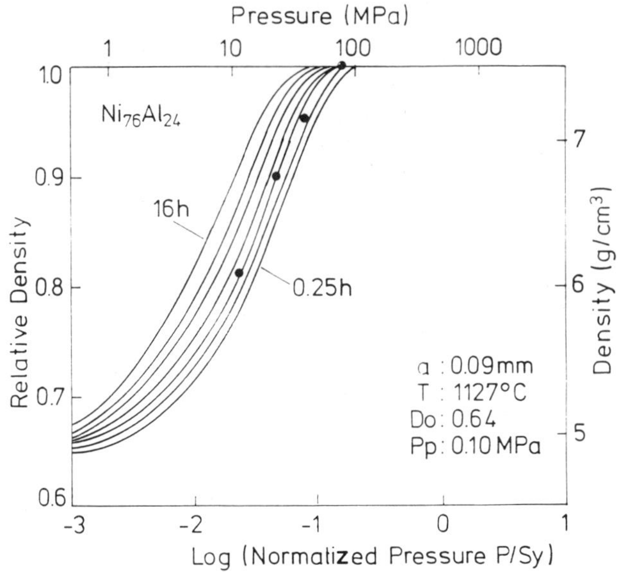

Fig. 19. HIP diagram calculated with the modified HIP-487 program and material data from dilatometer experiments of $Ni_{76}Al_{24}$.

photograph of a specimen with 93% relative density, HIPed at 1100 °C at 50 MPa for 2 h. The dislocation networks limited to the grain boundary regions indicate that, from this HIP stage to full density, grain boundary diffusion may be more important than power law creep. This is also indicated by a decreasing slope of densification curves at higher densities. $Ni_{70}Al_{30}$ and $Ni_{76}Al_{24}$ show good agreement between predicted densities and measured HIP densities (Fig. 18, 19). The HIP conditions to achieve equal densities were similar for $Ni_{76}Al_{24}$ and $Ni_{75}Al_{25}$.

CONCLUSION

The program HIP-487 for HIP diagram calculation is a powerful tool to predict HIP parameters. It is applicable for a wide variety of materials. This paper shows its ability to predict valid HIP parameters for complex materials such as intermetallic nickel aluminides, provided the special properties of these materials are taken into account. Single powder particle dilatometry allows further refinement of the material data for HIP diagram calculation.

REFERENCES

1.   R. Maurer, G. Galinski, R. Laag, W.A. Kaysser, submitted to: *High Temperature Ordered Intermetallic Alloys III*, edited by C.C. Koch, C.T. Liu, N.S. Stoloff, A.I. Taub; (Mater.Res.Soc. Pittsburgh, PA, 1989), 133, in press.

2.   E.M. Schulson, D.R. Barker, *Scripta Metall.* 17, 519-522 (1983).

3.   M.F. Ashby, *HIP 487: A Program for Constructing Hot Isostatic Pressing Diagrams*, University of Cambridge, UK (1987).

4.   K. Rzesnitzek; private communication.

5.   M. Aslan, Ph.-D. Thesis, University of Berlin (FRG), 1987.

6.   W. Kaysser, M. Aslan, G. Petzow, Proceedings *International Conference on Hot Isostatic Pressing of Materials: Applications and Developments* (The Royal Flemish Society of Engineers, The Metallurgical Section, Antwerpen 1988), pp. 1.11-1.16.

7.   W.A. Kaysser, M. Aslan, E. Arzt, M. Mitkov, G. Petzow, *Powder Metall.* **31**, 63-69.

8.   G.D. Scott, *Nature* **194**, 956-957 (1962).

9.   H.F. Fischmeister, E. Arzt, *Powder Metallurgy* **26**, 82-88 (1983).

10. A.S. Helle, K.E. Easterling, M.F. Ashby, Technical Report *Hot-Isostatic Pressing Diagrams: New Developments* (Luleå University of Technology, ISSN 0349-3571, 1985).

11. R.M. German, *Powder Particle Packing Characteristics*, Metal Powder Industries Federation, Princeton NJ. (1989).

12. E. Arzt, *Acta Metall.* **30**, 1883-1890 (1982).

13. J.K. Tien, S. Eng, J.M. Sanchez, *High Temperature Ordered Intermetallic Alloys II*, edited by N.S. Stoloff, C.C. Koch, C.T. Liu, O. Izumi; (Mater.Res.Soc. Pittsburgh, PA, 1987) **81**, pp. 183-193.

14. W.R. Cannon, O.D. Sherby, Met. Trans. 1 1030-1036 (1970).

ACKNOWLEDGEMENT

The authors would like to thank Mrs. Ute Bäder for the preparation of TEM specimen. They wish to acknowledge the financial support of the DFG (Deutsche Forschungsgemeinschaft).

DEFECT HEALING MECHANISMS DURING SINTER/HIP

OF POLYPHASE MATERIALS

A. Frisch, W.A. Kaysser, and G. Petzow

Max-Planck-Institut für Metallforschung
Institut für Werkstoffwissenschaft
PML, Stuttgart, FRG

ABSTRACT

The Sinter/HIP-Process is a technology for sintering and post-densification in one cycle with low pressures. It is a beneficial and economical method for the production of dense and homogeneous materials. Sinter/HIP may be a favored technology for polyphase materials, such as cemented carbides, heavy metal alloys or ceramics, which consist at the sintering temperature of hard grains and a highly mobile or viscous phase. For industrial applications, however, it is necessary to investigate the densification behavior and the defect healing mechanisms to establish the conditions under which a dense and homogeneous microstructure is obtained. In tungsten with nickel and in alumina with anorthite-glass the densification behavior and the elimination of large defects were investigated during Sinter/HIP. The defects were similar to macropores which may result from differential sintering of agglomerated powders or after debinding of injection molded greens. To verify the densification behavior, creep parameters and the viscosity of the systems were measured in a load dilatometer by single sphere deformation. The predictions made from the deformation experiments are in good agreement with the measured densities of HIPed samples. Macropore elimination by a simultaneous flow of grains and the highly mobile intergranular phase is obtained by changes of the amount of intergranular phase and the application of different Sinter/HIP-conditions. The conditions necessary for a mixed flow during postdensification with low pressures by HIP are quantified and discussed as a function of temperature, time, pressure and initial grain size.

1. INTRODUCTION

Besides the pressure assisted powder consolidation and diffusion bonding of composite materials by Hot Isostatic Pressing (HIP), a major purpose of the HIP-process is the post-densification of sintered materials to eliminate micro- and macroporosity to obtain a dense and homogeneous microstructure. Macropores which may result from differential sintering of agglomerated powders or from debinding of injection molded parts have to be eliminated to improve the mechanical properties of the material. A special challenge in this respect is posed to the recently developed Sinter/HIP-technology. During Sinter/HIP, the material is presintered until only closed porosity remains, followed by a post-densification treatment with an isostatic pressure to cause complete densification. Both steps are performed in one cycle. The

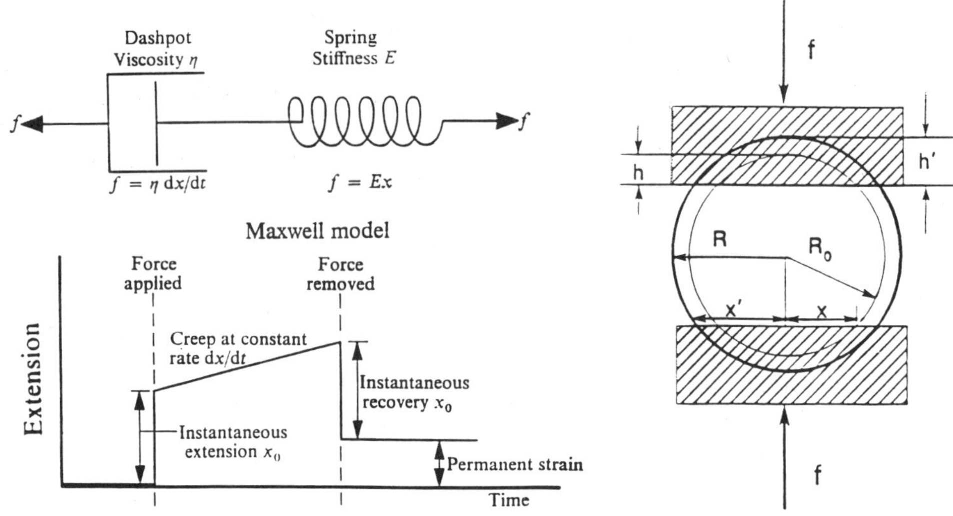

Fig. 1. A spring – dashpot model
simulating the creep process

Fig. 2. Schematic single
sphere deformation (2)

main advantages in comparison to "conventional HIPing" are the reduction to one
heat treatment and the application of relatively low pressures ($\leq$ 50 MPa) which are
both important benefits from energetic aspects (1). So, Sinter/HIP may be a favoured
technology as a densification process after new forming methods, such as injection
molding or slip casting.

To enhance pore elimination in a material a highly mobile or viscous phase is
often introduced intentionally, e.g. in heavy metals, cermets or high–tech–ceramics.
To predict a complete and homogeneous densification of these polyphase materials,
two questions have to be answered. The first question adresses the initial
densification behavior during Sinter/HIP. The prediction of the HIP–parameters is
often difficult because required material data are normally not available, especially
those of complex materials. In the first part of the paper, material data, such as creep
rate and viscosity of polyphase materials consisting of hard grains and a mobile grain
boundary phase are presented. The materials used were tungsten with an
intergranular Ni–phase and Al_2O_3 with additions of anorthite glass
($CaO \cdot Al_2O_3 \cdot 2SiO_2$). The data were measured by deformation of single spheres (2)
under equivalent HIP pressures. The HIP parameters can be predicted by a simple
relationship including the viscosity (temperature), pressure and time. The second
part of the paper concerns the microstructural development during Sinter/HIP. In
polyphase materials, residual macropores are eliminated during Final Stage HIPing
by a mixed flow of both the mobile phase and the grains or by a flow of the mobile
phase alone. Pore elimination by the second mechanism leads to soft spots and an
inhomogeneous microstructure. The transition to a mixed flow depends strongly on
the volume fraction of the mobile phase and the grain size of the solid constituents. In
W–Ni and Al_2O_3–anorthite, the elimination of spherical artificial macropores has
been investigated during Final Stage HIPing. A model for a mixed flow into the
macropores is presented including microstructural and HIP parameters.

2. PREDICTION OF THE HIP BEHAVIOR DURING SINTER/HIP

The application of low pressures during Sinter/HIP requires densification
maps which predict the essential HIP parameters to adjust theoretical density. Some
years ago HIP diagrams were developed (3,4) to simplify the engineer's choice of the
required HIP parameters for a complete densification. For the calculation of these

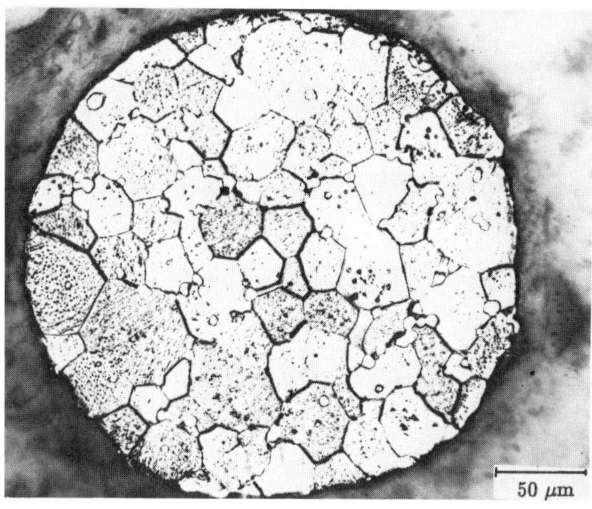

Fig. 3. Microstructure of a W – 5 vol. – % Ni sphere

diagrams, a considerable amount of material data are needed which are often not available from the literature.

In polyphase materials with a mobile grain boundary phase, such as heavy metals, cermets or ceramics, diffusion controlled creep (5) and grain boundary sliding (6) may be the dominating mechanisms for shrinkage and neck growth. So, the bulk viscosity of these materials is the main parameter controlling the densification behavior. The response of these systems to applied stresses may be described by the viscoelastic behaviour. The viscoelastic properties are assumed to be linear (7). That means creep deformation consists of an elastic and viscoelastic part, according to a simple rheological Maxwell – model shown in Fig. 1, and is related by

$$\epsilon = \frac{P}{E} + \frac{P}{\eta} \cdot t \qquad \langle 1 \rangle$$

with

P : pressure
E : Young modulus
η : viscosity
t : deformation time

In our polyphase materials the elastic deformation is small compared to the viscoelastic part and so the elastic part is negligible. The first derivation of equation $\langle 1 \rangle$ leads to the creep rate $\dot{\epsilon}$ as a linear function of pressure :

$$\dot{\epsilon} = \frac{1}{\eta} \cdot P \qquad \langle 2 \rangle$$

For measurements of the viscosity, single spheres were uniaxially deformed under a controlled load, as schematically shown in Fig. 2, by the method developed by Aslan and Kaysser (2). Tungsten powder with different amounts of nickel were mixed and spherical agglomerates were prepared by tumbling of the powder in a cylinder tilted at an angle of 30^0 to the horizontal axis of rotation by the method developed of Claussen and Petzow (8). After tumbling, the spheres were sintered to full density in a H_2 atmosphere. The particle size was in the range of 200–300 μm with an average internal grain size of 25 μm (Fig. 3).

313

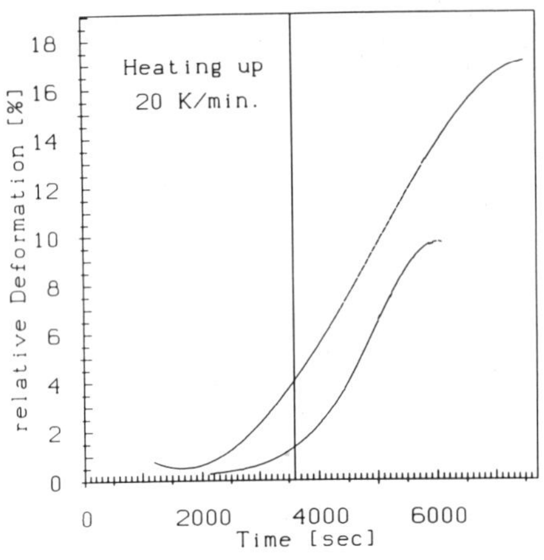
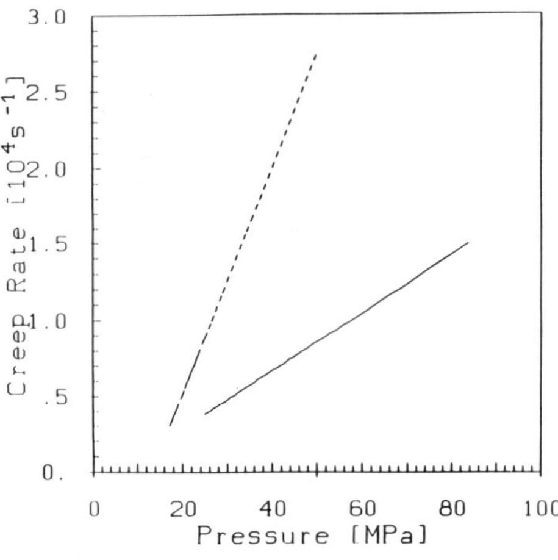

Fig. 4a. Deformation vs. time   Fig. 4b. Creep rate vs. pressure

Fig. 4a and 4b. Deformation of W–Ni spheres in a load dilatometer
(——— : 1400 °C, – – – : 1440 °C, d = 300 μm)

The spheres were deformed in a load dilatometer under forces between 0.3 and 0.6 N at different temperatures for up to 60 min. The heating rate was 20 K/min. Figure 4a shows the deformation in the dilatometer during heating up and subsequent isothermal annealing at 1400 °C and 1440 °C. Figure 4b shows the corresponding creep rates; a linear function of creep rate versus deformation pressure is obtained . Following equation ⟨2⟩, the viscosity is calculated as the reciprocal from the straight line's slope. In Table 1, viscosities of tungsten with 2 and 5 vol.–% Ni at different temperatures are presented. For comparison, the viscosity of a borosilicate glass at 600 °C is 10^{12} Pas .

Knowledge of the viscosity permits the prediction of HIP parameters during Sinter/HIP. After Evans and Hsueh (9), the creep rate is related to the pore shrinkage rate by

$$\dot{\epsilon} = \frac{4 \cdot \pi \cdot n \cdot (1-f) \cdot r^2}{3 \cdot V_S} \cdot \dot{r} \qquad \langle 3 \rangle$$

with $\dot{r}$: pore shrinkage rate
n : number of pores
f : volume part of the pores
V_S: volume of solid, with

$$V_S = \frac{4}{3} \cdot \pi \cdot r^3 \cdot \frac{n \cdot (1-f)}{f} \qquad \langle 4 \rangle$$

Combining equations ⟨2⟩, ⟨3⟩ and ⟨4⟩ and integrating with the boundary conditions of $r(t=0) = r_0$ and $r(t=t_c) = 0$ leads to

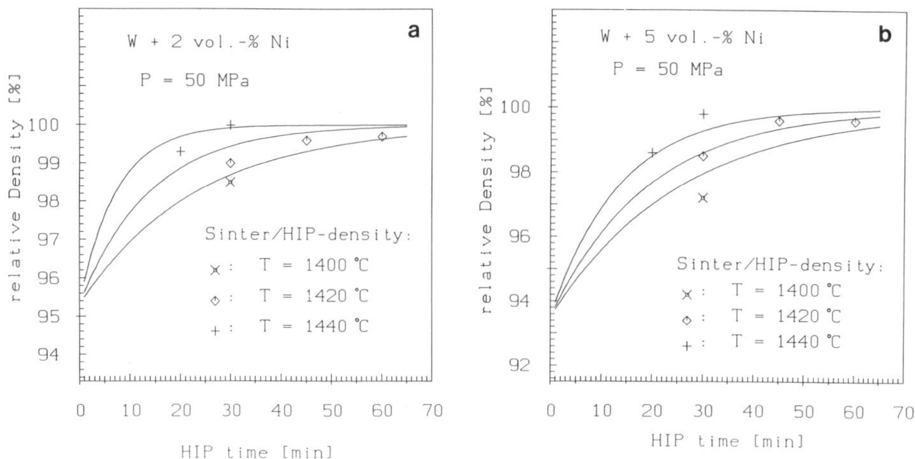

Fig. 5a and 5b. Density vs. time at different temperatures and Ni contents

$$\frac{r}{r_0} = e^{-t \cdot p / \eta} \qquad \langle 5 \rangle$$

Expressing equation $\langle 5 \rangle$ by the relative final density results in the prediction of the density as a function of time, pressure and viscosity (temperature) during Sinter/HIP:

$$\rho = 1 - f \cdot e^{-t \cdot p /(f \cdot \eta)} \qquad \langle 6 \rangle$$

with

f : volume fraction of pores
p : HIP pressure
t : HIP time

The parameter f represents the volume fraction of the pores after sintering during the Sinter/HIP-process.

To verify the prediction of the HIP parameters, corresponding HIP experiments were carried out. The HIP equipment used was a microprocessor controlled Sinter/HIP apparatus from ASEA. Samples of tungsten with different amounts of nickel were presintered under vacuum at 1200 ⁰C to closed porosity and subsequently hot isostatically pressed with gas pressures up to 50 MPa at different temperatures. The HIP time varied between 10 and 60 minutes. The comparison between calculations and experiments is shown in Figures 5 and 6. Figures 5a and b show the densification for W with 2 and 5 vol.-% Ni at different temperatures at a HIP pressure of 50 MPa. Figure 6 shows the densification at different pressures. The diagrams indicate a good agreement between the predictions based on viscosity measurements by single sphere deformation and the densities measured from Sinter/HIP experiments.

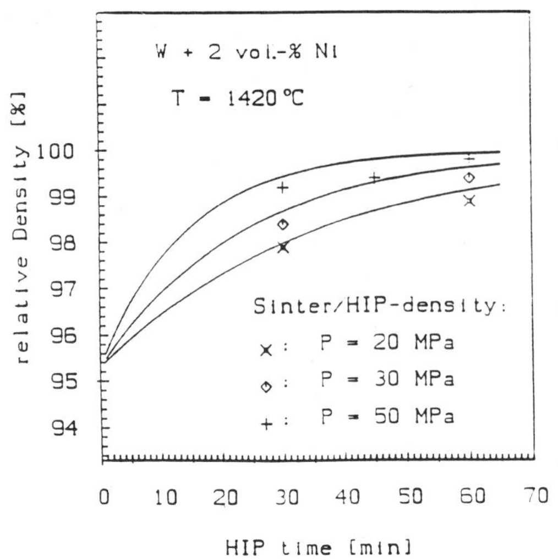

Fig. 6. Density vs. time at different HIP pressures

Table 1. Viscosities of W–Ni measured by single sphere deformation

vol.–% Ni	Temperature [°C]	Viscosity [Pas]
2,0	1380	$2,2 \cdot 10^{13}$
	1400	$3,8 \cdot 10^{12}$
	1420	$3,0 \cdot 10^{12}$
	1440	$2,2 \cdot 10^{12}$
5,0	1380	$6,3 \cdot 10^{12}$
	1400	$1,2 \cdot 10^{12}$
	1420	$9,0 \cdot 10^{11}$
	1440	$6,3 \cdot 10^{11}$

3. MACROPORE ELIMINATION DURING FINAL STAGE HIPING

During pressureless densification, residual macropores can result from irregularities in the initial particle arrangement or from void formation due to differential sintering in the vicinity of undestroyed agglomerates of fine powders (10). To improve the mechanical properties of the material, these pores have to be eliminated by pressure assisted post–densification, such as Final Stage HIPing. In polyphase materials, consisting of immobile grains and a highly mobile grain boundary phase, macropores are eliminated by a mixed flow of grains and mobile phase or by a flow of the mobile phase alone. The filling of the pores by the mobile phase alone results in weak spots and an inhomogeneous microstructure. To avoid this unfavourable microstructural development with certainty, it is necessary to establish conditions under which the pore elimination occurs by a mixed flow of grains and mobile phase.

The elimination of artificial spherical macropores with an average diameter of 60 μm was investigated during Final Stage HIPing in the tungsten–nickel system and the Al_2O_3–anorthite glass system ($CaO \cdot Al_2O_3 \cdot 2SiO_2$). The pores were produced by mixing polymeric spheres in the powder mixture. After cold isostatic compaction, the polymeric spheres burnt out during annealing at 500 ^{0}C. The Sinter/HIP programs used were the same as described above. Figure 7 shows the elimination of macropores in W–Ni samples with different Ni–contents. In samples with small Ni–contents, macropores were mainly filled by a flow of the mobile Ni–phase (Fig. 7b). At higher Ni–contents, the large pores were eliminated by a mixed flow of grains and the Ni–rich grain boundary phase (Fig. 7c).

During Sinter/HIPing of Al_2O_3–anorthite, the macropores were eliminated by a mixed flow of grains and intergranular glass phase (Fig. 8a to c). With increasing bulk content of anorthite, the volume fraction of anorthite also increased in the filled prior pores (13). At larger initial grain sizes, which were obtained after long presintering times, the amount of glass increased in the filled pores considerably.

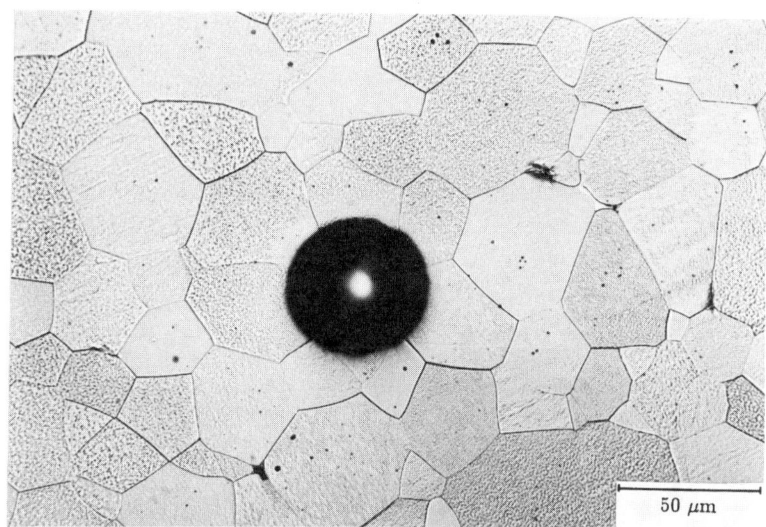

Fig. 7a. Residual artificial macropore in presintered W – 0.35 vol. – % Ni after HIPing for 30 min at 50 MPa.

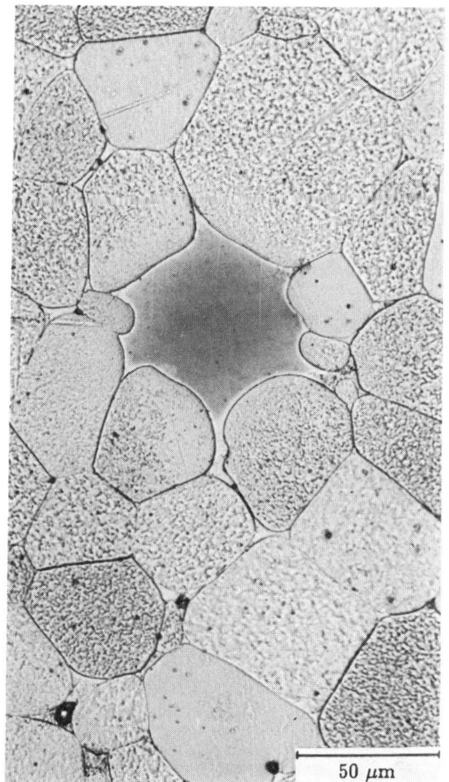

Fig. 7b. Pore closing by Ni–phase in W + 2 % Ni, T: 1420 ^{0}C, P: 50 MPa

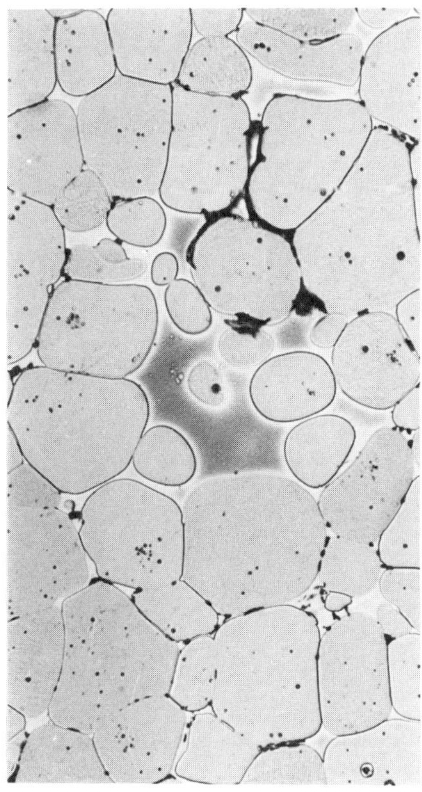

Fig. 7c. Pore closing by a mixed flow in W + 5 % Ni T: 1420 ^{0}C, P: 50 MPa

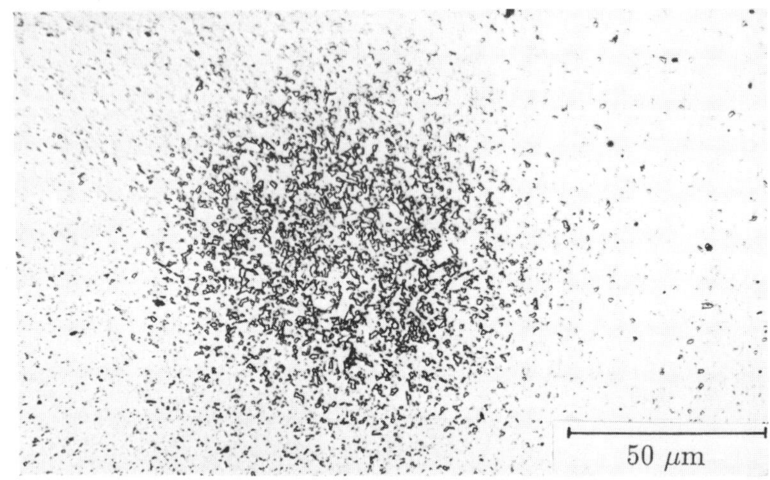

Fig. 8a. Elimination of macropores in presintered Al$_2$O$_3$−5.0% anorthite after HIPing for 30 min at 1600 ^{0}C with 20 MPa

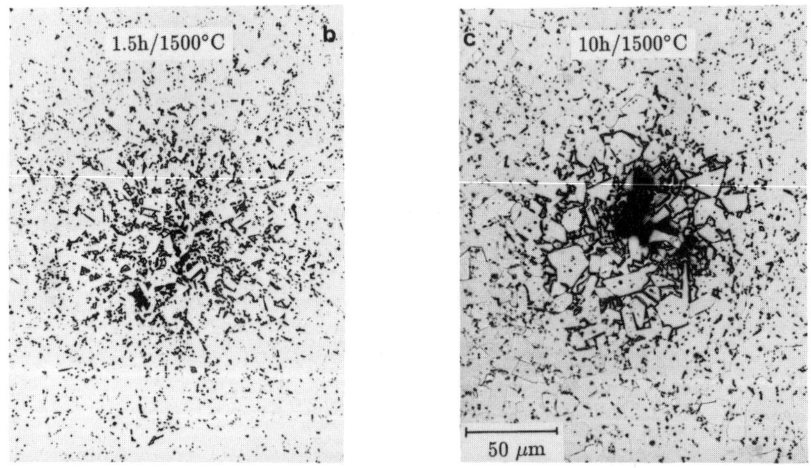

Fig 8b and 8c. Pore elimination in Al$_2$O$_3$−2.5% anorthite at different presintering conditions, HIP : T =1750 ^{0}C, P =50 MPa, t =30 min

4. DISCUSSION

In polyphase materials, which contain small amounts of additives, the polycrystal can be regarded as a continuum. Macropore elimination can be described by viscoelastic deformation or diffusional creep, which means that the mobile grain boundary phase provides a high diffusivity path. This allows a diffusive transport of the solid phase through the mobile phase from regions of higher compression to those of lower, due to the existing pressure gradient around the macropores. With the assumption that the lattice diffusion is negligible the viscosity of these polyphase materials is given by (9)

$$\eta = \frac{k \cdot T \cdot d^3}{14 \cdot \Omega \cdot \pi \cdot \delta_b D_b} \qquad \langle 7 \rangle$$

with

d : grain size
$\delta_b D_b$: grain boundary diffusion coefficient
Ω : atomic volume

After (10), the pore shrinkage rate dr/dt is given by

$$dr/dt = - \frac{p \cdot r + 2 \cdot \gamma}{4 \cdot \eta} \qquad \langle 8 \rangle$$

with

γ : specific surface energy of the pores

Combining equation $\langle 7 \rangle$ and $\langle 8 \rangle$ and integrating with the boundary conditions of $r(t=0)=0$ and $r(t)=r_p$ yields

$$r_p = \frac{2 \cdot \gamma}{P} \cdot \left(\exp\left(\frac{7 \cdot P \cdot t \cdot \Omega \cdot \pi \cdot \delta_b \cdot D_b}{2 \cdot k \cdot T \cdot d^3}\right) - 1 \right) \qquad \langle 9 \rangle$$

Equation $\langle 9 \rangle$ relates pore removal by diffusional creep to the required elimination time, HIP pressure and grain size.

In polyphase materials with higher contents of viscous or liquid phase, the flow of the mobile phase through melt channels at triple junctions additionally contributes to pore elimination. The flow of the mobile phase through melt channels is approximated with the Hagen–Poiseuille equation to

$$\frac{dV}{dt} = \frac{\pi \cdot P \cdot r_c^4}{8 \cdot \eta_L \cdot d} \qquad \langle 10 \rangle$$

with

η_L : viscosity of the mobile phase
r_c : radius of the channel

The channel radius r_c is obtained as an equivalent radius after (11) by

$$r_c^4 = \frac{\phi_0 \cdot d^4}{64} \qquad \langle 11 \rangle$$

with ϕ_0 : volume content of the mobile phase in the channels.

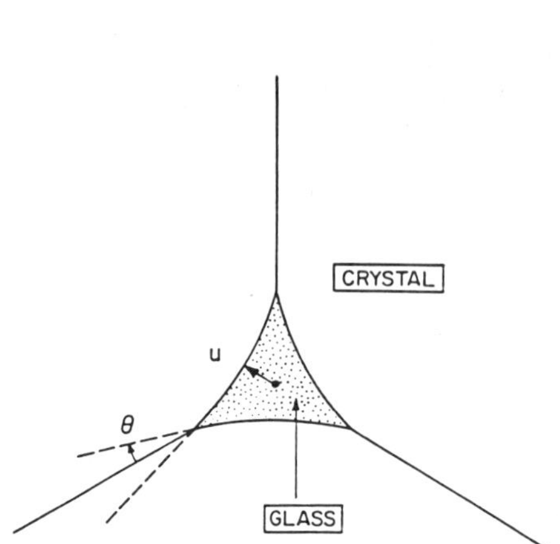
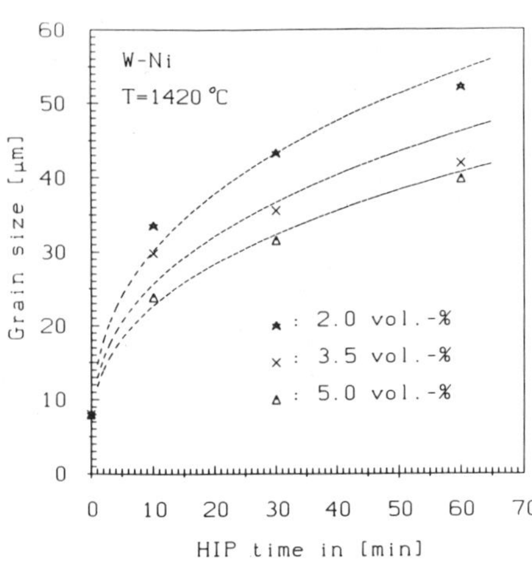

Fig. 9. Aggregation of mobile phase segregated at triple junctions

Fig. 10. Grain growth in W–Ni

The viscosity η_L of the mobile phase is expressed by the Stokes–Einstein equation,

$$\eta_L = \frac{k \cdot T}{8 \cdot \Omega^{1/3}} \cdot \frac{1}{D_L} \qquad \langle 12 \rangle$$

D_L : diffusion coefficient of solid in mobile phase

Combining equations $\langle 10 \rangle$, $\langle 11 \rangle$ and $\langle 12 \rangle$ and integrating with the boundary condition of $V(t=0) = 0$, we obtain the volume of the mobile phase V which has flowed at certain HIP conditions

$$V = \frac{\pi \cdot P \cdot \phi_0{}^2 \cdot D_L \cdot \Omega^{1/3} \cdot d^3}{64 \cdot k \cdot T} \cdot t \qquad \langle 13 \rangle$$

The volume was replaced by an equivalent sphere radius r_V, defined as

$$r_V = \left(\frac{3}{4 \cdot \pi} \cdot V\right)^{1/3} \qquad \langle 14 \rangle.$$

Equation $\langle 14 \rangle$ for the flow through melt channels can be directly compared with the flow caused by diffusion controlled creep, according to equation $\langle 9 \rangle$. A comparison of equations $\langle 9 \rangle$ and $\langle 14 \rangle$ shows that at constant HIP conditions, the grain size, and hence grain growth, will control the volume fracture of the mobile phase in the prior macropores.

In the following, pore removal by diffusional creep and channel flow are compared and the influence of grain size and grain growth is discussed. During HIPing of W–Ni grain growth was observed which could be formally described after (12) by

$$d = (d_0{}^3 + c \cdot t)^{1/3} \qquad \langle 15 \rangle$$

with

d_0 : initial grain size
c : growth constant

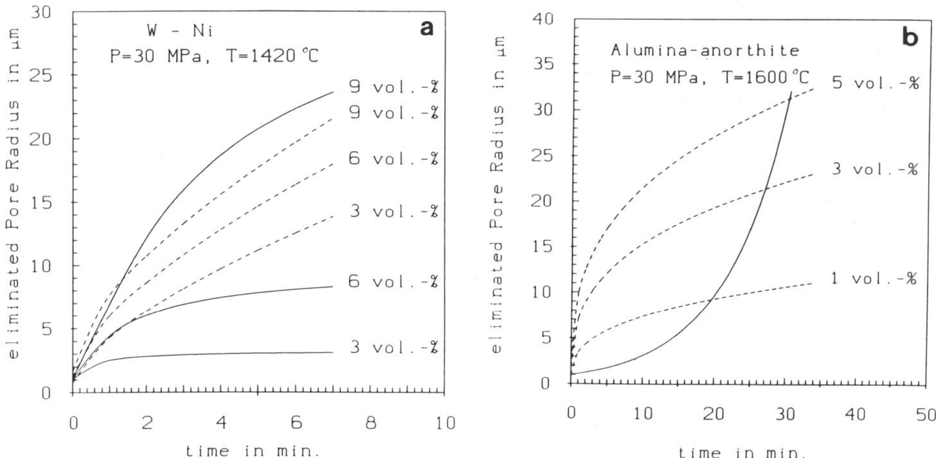

Fig 11. Pore elimination in W–Ni (a) and in Al$_2$O$_3$–anorthite (b)
Comparison between diffusional creep (——) and channel flow (– – –)

The growth constant c depends on the volume fraction of the mobile grain boundary phase and the temperature. With increasing Ni–content, the grain growth rate decreased, so a lower growth constant, c, is obtained as shown in Figure 10. The grain size, d, in equations ⟨9⟩ and ⟨14⟩ was then replaced by relation ⟨15⟩ for grain growth to model the microstructural development during HIPing.

In Figure 11a the pore elimination by diffusion controlled creep and flow through melt channels is compared. Due to an extended grain growth, the eliminated pore radius caused by diffusion controlled creep yields a threshold value. This value increases with lower grain growth constants (see Fig. 10). The residual pore volume will be filled by the flow of the mobile phase through the melt channels. Increasing HIP pressures leads to an increase of diffusional creep, whereas at low pressures, pore removal by the mobile phase alone is dominant. Figure 11b shows the results for Al$_2$O$_3$–anorthite. In the MgO doped Al$_2$O$_3$, no measurable grain growth was observed during HIPing. There was also no difference in grain size for different anorthite contents. The predicted flow through channels increases with increasing anorthite content, whereas the creep rate does not change. Hence the volume fraction of the glass in prior pores increases with increasing bulk contents of glass, as shown at the intersection points. HIP pressure in the usual range up to 200 MPa does not influence the volume fraction of glass in the macropore at constant initial bulk contents of glass. Higher HIP pressures shorten the required time for complete pore elimination.

Equivalent to the HIP diagrams (3,4), defect healing diagrams for polyphase materials may be constructed which indicate the healing mechanism for different defect sizes under special HIP conditions. In Figure 12, defect healing diagrams are shown for W+5 vol.–% Ni (Fig. 12a) and Al$_2$O$_3$+5 vol.–% anorthite (Fig. 12b) at T=1420 °C and T=1600 °C, respectively. The region for a mixed flow of grains and mobile phase was defined for the volume fraction of the mobile phase in the pores in the range of 40 to 60% (grey fields). With extended grain growth (Fig. 12a), the dominating healing mechansim is the flow of the mobile phase through melt channels. At low HIP pressures, macropores are eliminated by the mobile phase alone, which yields an inhomogeneous microstructure. For a homogeneus microstructural development, higher HIP pressures (here: 40 to 80 MPa) have to be used to increase creep rate and to allow pore removal by a mixed flow or diffusional creep.

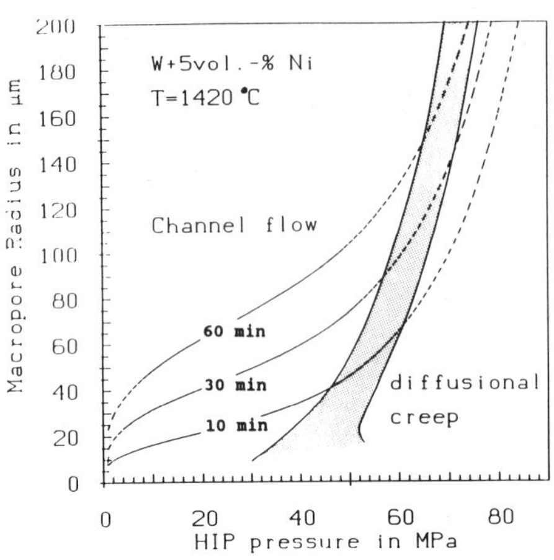
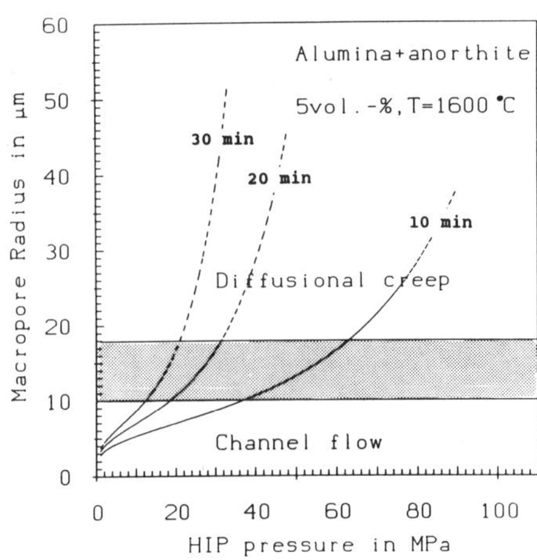

<table>
<tr><td>

Fig. 12a. Defect Healing Diagram for W+5 vol.-%Ni at 1420 ⁰C

</td><td>

Fig. 12b Defect Healing Diagram for Al$_2$O$_3$–anorthite at T = 1600 ⁰C

</td></tr>
</table>

In the absence of grain growth during HIPing (Fig. 12b), small pores will also be filled by the mobile phase alone. Macropores will be eliminated by a mixed flow or diffusional creep, but with increasing initial grain sizes, the region for mixed flow will be shifted to higher defect sizes. The volume fraction of glass in the macropore will also increase with increasing initial glass content.

5. SUMMARY

The Sinter/HIP-technology is a beneficial and economical method for the production of polyphase materials consisting of grains and an intergranular highly mobile phase such as cemented carbides, heavy metal alloys or ceramics. To predict the HIP parameters of such materials, the viscosity and creep rate in W–Ni were measured by deformation of single spheres, due to the method proposed by Aslan and Kaysser. The predictions made from the deformation experiments are in excellent agreements with the measured densities of HIPed samples.

Residual macropores may result after sintering due to differential sintering of fine powders or after using forming methods, such as injection molding or slip casting. A model was developed which describes the elimination of these large pores in polyphase materials during Final Stage HIPing. In materials with small amounts of additives, pore removal will occur by diffusional creep, whereas in materials with higher additive contents, the flow through melt channels at triple junctions additionally contributes to pore elimination. Defect healing diagrams which indicate the healing mechansim at different defect sizes under special HIP conditions were developed. At small grain sizes or with a smooth grain growth during HIPing, diffusion controlled creep will dominate pore elimination even at low HIP pressures and a mixed flow of grains and intergranular phase into the macropores occurs. If initial grain sizes are large or if an extended grain growth occurs during HIPing, then high HIP pressures have to be used to avoid pore filling by the mobile phase alone which would lead to inhomogeneities in the microstructure.

REFERENCES

1. A.G.Nyce, Containerless HIPing of PM Parts: Technology, Economics and Equipment Productivity, <u>Metal Powder Report</u>, 38, 387–392 (1983).
2. M.Aslan, W.A.Kaysser, and G.Petzow, Deformation of Spheres under Non-Constrained Conditions and during HIPing, <u>in</u> Proc. of "International Conference on Hot Isostatic Pressing–Theory and Applications", Lulea, 51–55 (1987).
3. E.Arzt, M.F.Ashby, and K.E.Easterling, Practical Applications of Hot Isostatic Pressing Diagrams: Four Case Studies, <u>Metall. Trans.</u>, 14A, 211–221 (1983).
4. A.S.Helle, K.E.Easterling, and M.F.Ashby, Hot Isostatic Presing Diagrams: New Developments, <u>Acta Met.</u>, 33, 2163–2174 (1985).
5. R.L.Coble, Diffusion Models for Hot Pressing with Surface Energy and Pressure Effects as Driving Forces, <u>J. Appl. Phys.</u>, 41, 4798–4807 (1970).
6. R.Raj and M.F.Ashby, On Grain Boundary Sliding and Diffusional Creep, <u>Metall. Trans.</u>, 2A, 1113–1127 (1971).
7. C.H.Hsueh, A.G.Evans, R.M.Cannon, and R.J.Brook, Viscoelastic Stresses and Sintering Damage in Heterogeneous Powder Compacts, <u>Acta Met.</u>,34, 927–936 (1986).
8. N.Claussen and G.Petzow, Wachstum und Festigkeit kugeliger Agglomerate aus Pulvern hochschmelzender Werkstoffe, <u>High Temperatures–High Pressures</u>, 3, 467–485 (1971).
9. A.G.Evans and C.H.Hsueh, Behavior of Large Pores During Sintering and Hot Isostatic Pressing, <u>J. Am. Ceram. Soc.</u>, 69, 444–448 (1986).
10. W.A.Kaysser, A.Frisch, M.Aslan, and G.Petzow, Microstructural Development during Final Stage HIPing, <u>in</u> Proc. of "International Conference on Hot Isostatic Pressing of Materials: Applications and Developments", Antwerp, 2.1–2.9 (1988).
11. R.Raj, Morphology and Stability of the Glass Phase in Glass–Ceramic Systems, <u>J. Am. Ceram. Soc.</u>, 64, 245–248 (1981).
12. W.A.Kaysser and G.Petzow, Ostwald–Ripening and Shrinkage during Liquid Phase Sintering, <u>Zeitschrift für Metallkunde</u>, 76, 687–692 (1985).
13. J.Kunesch, V.Carle, R.Monteiro, and W.A.Kaysser, Porenfüllung beim HIPen von Al_2O_3–Glas–Systemen, <u>Praktische Metallographie</u>, 20, 537–544 (1989).

RESIDUAL STRESS CHARACTERISTICS OF CERAMIC COATINGS

AND THEIR CRACKING BEHAVIOR

Y. Ishiwata, Y.Z. Itoh, and H. Kashiwaya

Heavy Apparatus Engineering Laboratory
Toshiba Corporation
Tsurumi-ku, Yokohama, 230, Japan

ABSTRACT

The residual stress of metal-to-ceramic laminates (stabilized ZrO_2/Nb and Al_2O_3/Nb) produced by solid state bonding was studied using finite element method (FEM) analysis. The results indicated that the dimensionless residual stress at the ceramic coatings decreased with increasing coating thickness ratio, (hc/hs), and elastic modulus ratio, $(1-\mu s)Ec/[(1-\mu c)Es]$.

Further, using the results of residual stress analysis and the bending strength of sintered stabilized ZrO_2 and Al_2O_3, the critical temperature for coating to prevent cracking was estimated. The analytical results were in good agreement with the cracking behavior of HIP treated ceramic coatings, but in the case of Al_2O_3 coated specimens, the dependence of cracking on HIP temperature was not in agreement with the analytical results, because of the phase transformation of the Al_2O_3 coating.

INTRODUCTION

Plasma-spraying of ceramic coating is a useful technique to improve properties, such as heat, corrosion and wear resistance of metal components. This technique has been widely applied to gas turbine and fusion reactor components. It is well known that there are a large number of pores existing in the coatings. Such a porous structure in coatings is good for thermal barrier coatings. However, the metal substrate which has been surface coated with ceramics by plasma-spraying is sometimes attacked by corrosive gases and solutions have penetrated into the coating layer. As a useful method to improve such properties of plasma-sprayed ceramic coatings, hot isostatic pressing (HIP) treatments have been investigated by some authors.[1,2] However, cracks transverse to the coatings are often observed in some HIP treated components, as shown in Fig. 1.

In this paper, the residual stress characteristics of ceramic coatings after HIP treatment were analysed using a finite element method (FEM). Furthermore, the relationships between residual stress distribution and cracking behavior which had been actually observed in

Science of Sintering
Edited by D. P. Uskoković *et al.*
Plenum Press, New York

some HIP treated coatings, were investigated and the optimum conditions of HIP treatment were considered for several combinations of ceramic coating and metal substrate.

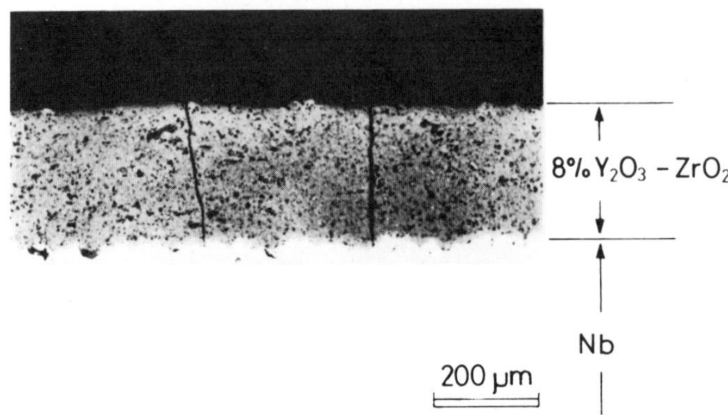

Fig. 1   Micrograph of cracks observed in plasma-sprayed stabilized ZrO_2 coating after HIP treatment (1300°C, 98.1 MPa. 1hr.)

RESIDUAL STRESS CHARACTERISTICS OF CERAMIC COATINGS

Condition of Residual Stress Analysis

As the cracking observed in some HIP treated ceramic coatings would be caused by the tensile residual stress which was developed from the mismatch in the thermal expansion coefficient between ceramic coating and metal substrate, the following residual stress analyses were carried out to investigate the residual stress characteristics of coatings.

The residual stress of metal-to-ceramic laminates produced by solid state bonding was studied. As the analytical model, a metal disk coated on one side with ceramic was used, as shown in Fig. 2. The residual stresses were analysed by finite element method. In this analysis, the

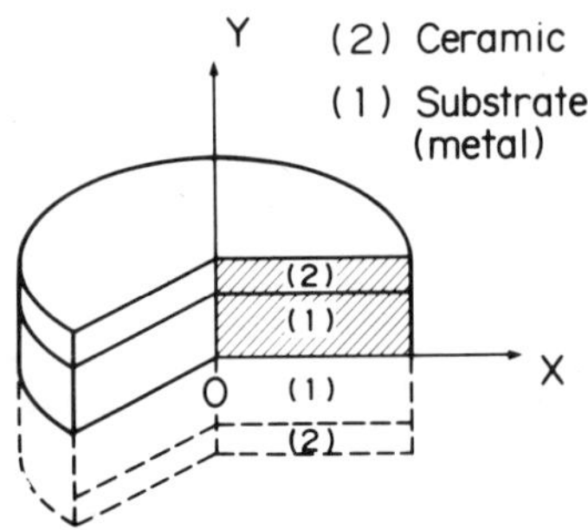

Fig. 2   Analytical model of ceramic coating

residual stresses were defined as the thermal stresses induced by the difference between room temperature and bonding temperature at which the residual stresses were zero, on the assumption that the ceramic and the metal behaved elastically.[3]

In order to clarify the residual stress distribution at the bonding interface, analyses were performed for the case of Nb substrate, 25 mm in diameter and 5 mm in thickness, coated with 1 mm thickness of stabilized ZrO_2. Material constants used for FEM analysis are shown in Table 1. Also, the effects of elastic modulus and coating thickness on the residual stress of ceramic coatings were investigated for the one side coated disk and the both sides coated disk.

Table 1 Material constants used for FEM analysis

Materials	Young's modulus (GPa)	Poisson's ratio	Thermal expansion (/°C)
$8\%Y_2O_3\text{-}ZrO_2$	200	0.3	11×10^{-6}
Nb	100	0.3	8×10^{-6}

Computational Results of Residual Stress

Through thickness residual stress distributions within stabilized ZrO_2/Nb after cooling from bonding temperature to room temperature ($\Delta T \fallingdotseq 1000°C$) are shown in Fig. 3. The remarkable stress concentrations of σ_y and τ_{xy} are observed near the edge of the bonding interface, but σ_y and

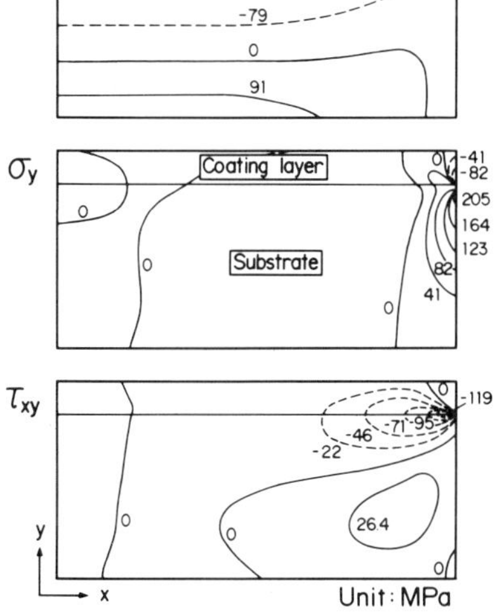

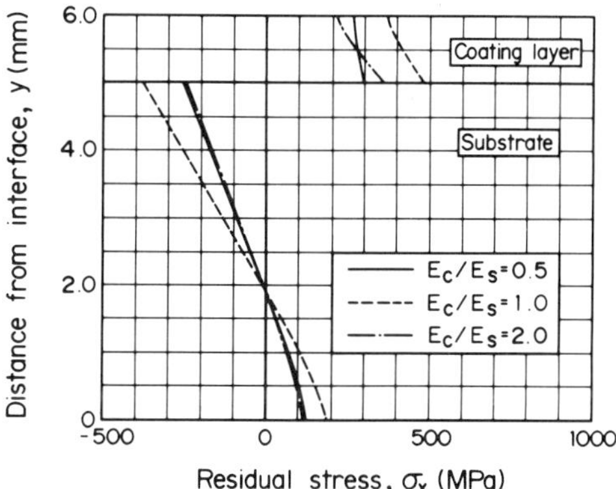

Fig. 3 Stress distributions on the cross-section of ceramic coated disk

Fig. 4 Effect of Ec/Es on through thickness stress distribution (Ec; elastic modulus of coating, Es; elastic modulus of substrate)

τ_{xy} become radically lower toward the inside of the laminate. Further, σ_y has a negative value within the ceramic coating, the stress concentrations of σ_y and τ_{xy} would not have a large effect on the cracking of the coating. On the other hand, σ_x has a high positive value within the ceramic coating and it becomes constant inside of the laminate. This tendency of σ_x is in good agreement with the cracking tendency of HIP treated stabilized ZrO_2 coatings, as shown in Fig. 1. Fig. 4 shows the σ_x stress distribution across the center of the laminate. For each elastic modulus ratio (Ec/Es = 1.0, 1.5, 2.0), a high tensile stress develops near the bonding interface in the ceramic coating, and on the contrary, a high compressive stress develops near the bonding interface in the substrate. In these calculations, the tensile stress in the coating shows a higher value when the elastic modulus ratio, (Ec/Es), equals 1.0. From the above analytical results, σ_c, which is the maximum value of σ_x, is considered the main stress which causes the cracking observed in the HIP treated ceramic coatings.

It has been suggested that the dimensionless parameter[3] $\sigma_c(1-\mu c)/[Ec(\alpha c-\alpha s)\Delta T]$ (μc;poisson's ratio of coating, Ec;elastic modulus of coating, ($\alpha c-\alpha s$); difference of thermal expansion coefficient between coating and substrate, ΔT;difference of temperature) is useful for presenting the stress in coatings. Concerning σ_c for a plate coated on one side with different material, the following solution has been proposed.[4]

$$\frac{\sigma c(1-\mu c)}{Ec(\alpha c-\alpha s)\Delta T} = \frac{(1-\mu c)[4m^3+3m^2)n+1]}{(m^4n^2+mn)(1-\mu s)+3(1+m)^2mn+(m^3n+1)(1-\mu c)}$$

$$hc/hs=m, \quad Ec/Es=n \qquad \dots\dots\dots(1)$$

The results of σ_c analysed by FEM and calculated results by using equation (1) are shown in Fig. 5. Both results show a similar tendency but there is a little difference of the σ_c value between FEM results and the calculated results by equation (1), because equation (1) is the solution obtained for plane stress condition and it is not considered the three-dimensional bending deformation caused by the residual stress. From

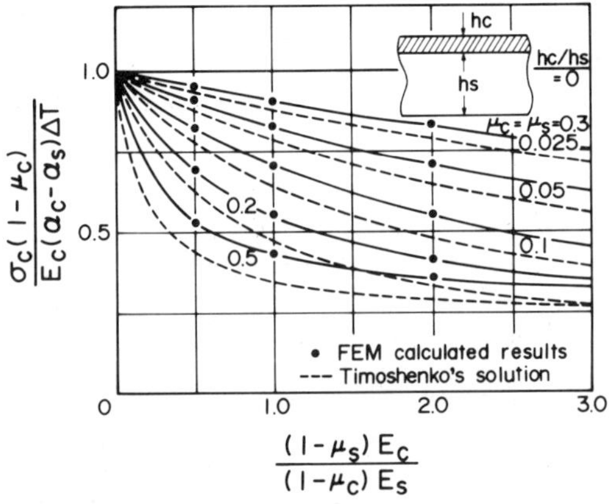

Fig. 5   Residual stress at ceramic coating

the results shown in Fig. 5, the dimensionless stress become higher with
decrease in coating thickness and the elastic modulus ratio $(1-\mu_s)E_c/[(1-\mu_s)E_s]$.

Using the results shown in Fig. 5, if the material constants, the
thickness of both coating and substrate and the bonding temperature are
all known, the residual stress developed in the coating can be estimated.
Further, if the strength of the coating is known, the critical temperature
for coating to prevent the cracking can be determined. For example, the
critical temperatures for coating of stabilized ZrO_2/Nb and Al_2O_3/Nb were
calculated.

It was assumed that the crack would occur when the maximum stress σ_c
was greater than the bending strength of ceramic coating. For the present
study, the bending strengths obtained for sintered stabilized ZrO_2 and
Al_2O_3 were used as the strengths of the HIP treated coating. The critical
temperatures for coating are shown in Fig. 6. The mean bending strength
of stabilized ZrO_2, which is 432 MPa, is much higher than that of Al_2O_3
which is 269 MPa, but the critical temperature for coating of stabilized
ZrO_2 coated on Nb substrate tends to be lower than that in the case of
Al_2O_3 coated on Nb substrate. This means that the developing residual
tensile stress in the stabilized ZrO_2 coating is higher than that of the
Al_2O_3 coating.

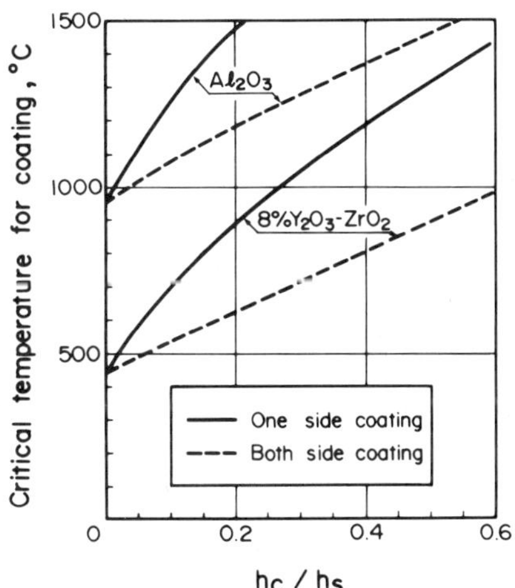

Fig. 6   Calculated results of critical
heating temperature

Fig. 6 shows the critical temperature for coating of a disk coated on
both sides. As compared with a one side coated disk, the critical
temperature for coating is lower because the bending deformation induced
by thermal stress should not occur.

EXPERIMENTAL PROCEDURES

In order to examine the results of the residual stress analysis by
FEM and to make clear the effect of HIP treatment on improving the
properties of plasma-sprayed ceramic coatings, the following experiments
were conducted.

For experiments, the following two cases were considered. One is the
case where the thermal expansion of the ceramic coatings is greater than

that of the metal substrate (stabilized ZrO_2/Nb and Al_2O_3/Nb) and the tensile residual stress developed in coatings after HIP treatment. Another is the case where the thermal expansion of the coating is smaller than that of the substrate (stabilized ZrO_2/Type 304 and Al_2O_3/Type 304) and the compressive residual stress developed in coatings. Experimental procedures are described below.

Disk shape substrates, 25 mm in diameter and 5 mm in thickness were made from Nb and Type 304 stainless steel and the disk surface was grit-blasted using alumina. Plasma spraying was carried out using a Metro 9MB gun; these specimens were coated with a 0.2 mm thickness of stabilized ZrO_2 (8 wt.% Y_2O_3) and Al_2O_3. The thermal expansion coefficients at 1000°C of the materials used are shown in Table 2. After cleaning, the ceramic coated specimens were enclosed in a vacuum can, which was made from mild steel, and the HIP treatments using argon gas were carried out at various temperatures from 1100°C to 1300°C at a fixed pressure of 98.1 MPa for 1 hour.

After HIP treatment, the specimens were taken out of the can and the cracking behavior of the ceramic coating layer was observed by optical microscope and secondary electron microscope (SEM). In order to confirm the effect of HIP treatment on improving the properties of plasma sprayed ceramic coatings, the bonding strength was measured in accordance with the ASTM standard. The porosity in the coatings was also measured.

Table 2   Thermal expansion coefficient
of used materials at 1000°C

Materials	Thermal expansion[1] $(/^{\circ}C)$
Nb	8.1×10^{-6}
Type 304 steel	19.0×10^{-6}
$8\%Y_2O_3\text{-}ZrO_2$	10.7×10^{-6}
Al_2O_3	8.8×10^{-6}

1. at 1000°C

EXPERIMENTAL RESULTS AND DISCUSSION

Cracking Behavior of HIP Treated Coatings

Micrographic cross-sections of as-sprayed, and 1100°C 1300°C HIP treated specimens are shown in Fig. 7 and their cracking behaviors are summarized in Table 3. As shown in Fig. 7 and Table 3, no cracking was observed in all as-sprayed coatings but some cracks, which were through the coatings perpendicular to the surface, were observed in the HIP treated specimens of Nb, on which tensile residual stress developed in the ceramic coating after cooling from HIP temperature. Especially, in the case of stabilized ZrO_2, the HIP temperature was much higher than the critical temperature for coating calculated by FEM as shown in Fig. 6, (thickness ratio, (hc/hs), equals 0.05). There was a small number of broad cracks in the stabilized ZrO_2 coating layer in comparison with the Al_2O_3 coating layer. While, according to the residual stress analysis, it

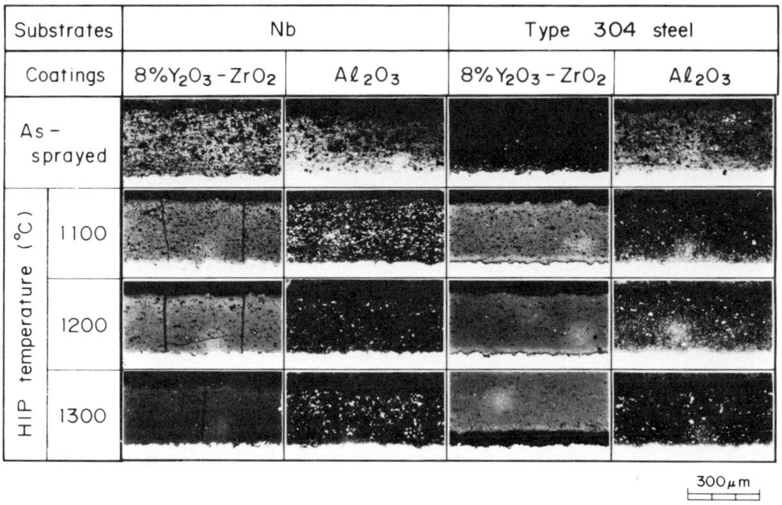

Fig. 7 is a grid of micrograph cross-sections with the following column and row labels:

Substrates	Nb		Type 304 steel	
Coatings	$8\%Y_2O_3-ZrO_2$	Al_2O_3	$8\%Y_2O_3-ZrO_2$	Al_2O_3
As-sprayed				
HIP temperature (°C) 1100				
1200				
1300				

300 μm

Fig. 7 Micrograph cross-sections of as-sprayed and HIP treated specimens

might be thought that the Al_2O_3/Nb specimens treated by 1100°C HIP had no cracking, by experiments, the micro-cracks could be observed in the Al_2O_3/Nb specimens. Thus, the cracking behavior of Al_2O_3/Nb specimens was not in agreement with the analytical results.

Besides the crack was not observed in the specimens of Type 304 stainless steel, on which compressive residual stress developed after HIP treatment. However, only the stabilized ZrO_2 coating treated at 1300°C HIP was detached from the substrate.

Table 3 Cracking of coated specimens

Substrates		Nb		Type 304 steel	
Coatings		$8\%Y_2O_3-ZrO_2$	Al_2O_3	$8\%Y_2O_3-ZrO_2$	Al_2O_3
As-sprayed		◯	◯	◯	◯
HIP temp. (°C)	1100	✕	△	◯	◯
	1200	✕	△	◯	◯
	1300	✕	✕	—	◯

◯ ; No cracking, △ ; Micro cracking
✕ ; Cracking

In order to prevent cracking in the coatings, HIP treatment should be carried out below the critical temperature for the coatings analysed by FEM. However, the densification and the improvement of bonding strength would not be achieved sufficiently at such a low temperature. Accordingly, the effects of HIP temperature on improving the properties of plasma-sprayed ceramic coatings were investigated.

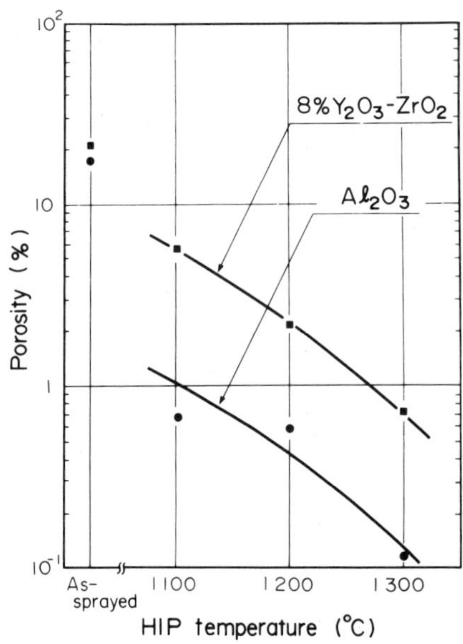

Fig. 8   Porosity measured in as-sprayed and HIP treated ceramics coating

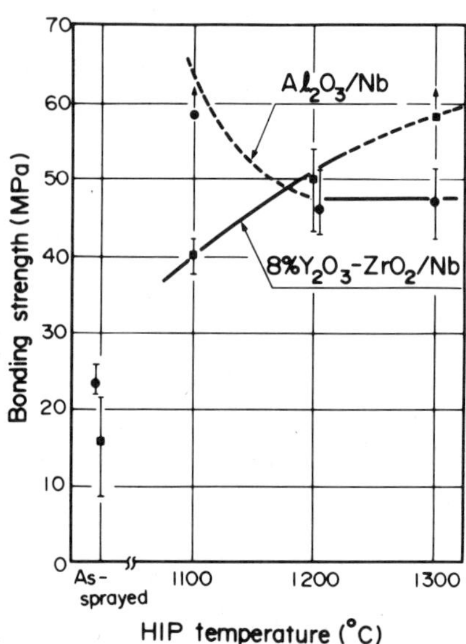

Fig. 9   Bonding strength measured in as-sprayed and HIP treated ceramics coating using pull-off test with epoxy cement

Fig. 8 shows the porosities measured within as-sprayed and HIP treated stabilized ZrO$_2$ and Al$_2$O$_3$ coatings. The porosities in both as-sprayed coatings are more than 10%, but the porosities of HIP treated coatings are decreased remarkably with increasing HIP temperature. Especially after 1300°C HIP treatment, the porosities became less than 1% in both coatings. As compared with the stabilized ZrO$_2$ and the Al$_2$O$_3$ coatings treated by HIP, the porosities in Al$_2$O$_3$ coating tend to be lower than that in stabilized ZrO$_2$ coating. With regard to the Al$_2$O$_3$ and the stabilized ZrO$_2$, it is reported that the temperature required to achieve full density using HIP treatment is more than 1200°C and 1400°C, respectively. Such a difference in porosity between the Al$_2$O$_3$ coating and the stabilized ZrO$_2$ coating treated by HIP would be due to the difference between HIP temperature and the required temperature to achieve full density.

Measurement of Bonding Strength

The bonding strengths among the stabilized ZrO_2 and the Al_2O_3 coatings and Nb substrate are shown in Fig. 9. It is obvious that the bonding strength is remarkably improved by the HIP treatment for both coatings. For the stabilized ZrO_2 coated specimens, the bonding strength increases when increasing HIP temperature. After $1300^\circ C$ HIP treatment the bonding strength becomes more than 60 MPa. Most of the as-sprayed coatings were perfectly detached from the substrate, but the HIP treated specimens were fractured within the coating layer. It has been considered that the improved bonding strength by HIP treatment is due to the mechanical interlocking effect, which is caused by the plastic deformation of metal substrate, and the improvement of coating strength by densification.

On the other hand, for Al_2O_3 coated specimens, the maximum bonding strength was observed when HIP treatment was carried out at $1100^\circ C$, while and the bonding strength was extremely decreased in the specimens treated at $1200^\circ C$ and $1300^\circ C$ HIP. The experimental results showed that the porosity was decreased with increasing HIP temperature as well as stabilized ZrO_2 (Fig. 8). Also, a chemical interaction between Al_2O_3 coating and Nb substrate was not obviously observed, but the bonding strength of Al_2O_3 coated specimens treated by $1200^\circ C$ and $1300^\circ C$ HIP was much lower than that of Al_2O_3 coated specimen treated by $1100^\circ C$ HIP. Consequently, it is anticipated that another factor to reduce the bonding strength would exist.

Table 4 shows the phase constitutions which were determined by an X-ray powder diffractometer for the starting powders, the as-sprayed specimens and the HIP treated specimens of both stabilized ZrO_2 and Al_2O_3. The stabilized ZrO_2 starting powder was the mixture of cubic, tetragonal and a small amount of monoclinic, almost the same phase constitution was identified for the as-sprayed specimen and all HIP treated specimens with little change in the composite ratio. Concerning Al_2O_3, several phases were observed. Namely, the starting powder consisted of $\alpha\text{-}Al_2O_3$ while the as-sprayed Al_2O_3 exhibited a diffraction pattern of strong lines which were indexed as $\eta\text{-}Al_2O_3$, as well as considerably weaker lines which indexed as $\alpha\text{-}Al_2O_3$. $1100^\circ C$ HIP treated Al_2O_3 exhibited a diffraction pattern of considerably strong lines which were indexed as $\alpha\text{-}Al_2O_3$, as well as some weaker lines which were indexed as $\delta\text{-}Al_2O_3$. HIP treatment at $1200^\circ C$ resulted in full reversion to $\alpha\text{-}Al_2O_3$. The diffraction patterns of $\gamma\text{-}Al_2O_3$ and $\eta\text{-}Al_2O_3$, $\theta\text{-}Al_2O_3$ and $\delta\text{-}Al_2O_3$ are very similar, they can be distinguished from the shape of the diffraction pattern. Such a transformation of plasma-sprayed Al_2O_3 has been reported by some authors.

Table 4 Phase constitutions determined by X-ray powder diffractometer

	Starting powder	As-sprayed	HIP temperature ($^\circ C$)		
			1100	1200	1300
$8\%Y_2O_3$ $-ZrO_2$	cubic tetragonal (monoclinic)	cubic tetragonal	cubic tetragonal	cubic tetragonal	cubic tetragonal
Al_2O_3	$\alpha\text{-}Al_2O_3$	$\eta\text{-}Al_2O_3$ ($\alpha\text{-}Al_2O_3$)	$\alpha\text{-}Al_2O_3$ $\delta\text{-}Al_2O_3$	$\alpha\text{-}Al_2O_3$	$\alpha\text{-}Al_2O_3$

The density of δ-Al_2O_3 is about the same as that of η-Al_2O_3, but the density of α-Al_2O_3 is relatively greater than that of η-Al_2O_3 and δ-Al_2O_3. Therefore, in accordance with $\eta \rightarrow \alpha$ and $\delta \rightarrow \alpha$ transformations, a volume shrinkage of Al_2O_3 is induced and a tensile stress will be developed. It might be that an extreme decrease of bonding strength, which was observed in Al_2O_3 coated specimens treated by $1200^{\circ}C$ and $1300^{\circ}C$ HIP, would be due to the effect of internal stress caused by the volume shrinkage with the phase transformation.

CONCLUSIONS

1. The residual stress characteristics of the one side coated disk and the both sides coated disk (stabilized ZrO_2/Nb and Al_2O_3/Nb) were analysed by the finite element method, and the critical temperatures for coating was calculated with the bending strength of sintered ceramics. The critical heating temperature to prevent cracking of stabilized ZrO_2/Nb, which has a large difference in the thermal expansion, tend to be lower than that of Al_2O_3/Nb, which has a small difference in the thermal expansion

2. In the case of stabilized ZrO_2/Nb, the dependence of cracking on HIP temperature was in good agreement with the analytical results, while the cracking behavior of Al_2O_3/Nb was not in agreement with the analytical results because of phase transformations in the Al_2O_3 coating.

REFERENCES

1. H. Kuribayashi, K. Suganuma, Y. Miyamoto and M. Koizumi, Effect of HIP Treatment on Plasma-Sprayed Ceramic Coating onto Stainless Steel, J. Am. Ceram. Soc., 65: 1306 (1986).
2. A.W. Mullendore, J.B. Whitley and D.M. Mattox, Thermal Fatigue Properties of Coated Materials for Fusion Device Applications, J. Nuc. Mat'l., 103 & 104, 251 (1981).
3. M. Toyoda, T. Komatsu, K. Satoh and M. Nayama, Elastic Thermal stresses of Bonded Joints of Dissimilar Materials, IIW, Doc., X-1161-88 (1988).
4. T. Timoshenko and J.N. Goodier, "Theory of Elasticity", McGraw-Hill 433 (1970).
5. A.S. Helle, K.E. Easterling and M.F. Ashby, Hot-Isostatic Pressing Diagrams: New Development, Acta Metall. 33, 2163 (1985).
6. V. Gross and M.V. Swain, Mechanical Properties and Microstructure of sintered and Hot Isostatically Pressed Yttria-Partially-Stabilized Zirconia (Y-PSZ), J. Australian Ceram. Soc., 22, 1 (1986).
7. G.C. Lippens and J.H. DE Boer, Study of Phase Transformation During Calcination of Aluminum Hydroxides by Selected Area Electron Diffraction, Acta Cryst., 17, 1312 (1964).
8. G.F. Hurley and F.D. Gac, Structure and Thermal Diffusibility of Plasma-Sprayed Al_2O_3, Ceramic Bulletin, 58, 509 (1979).

Part VI. RATE CONTROLLED SINTERING

RATE CONTROLLED SINTERING FOR CERAMICS

AND SELECTED POWDER METALS

Hayne Palmour III

Department of Materials Science and Engineering
North Carolina State University
Raleigh, NC 27695-7907
USA

Abstract

After almost a quarter-century of research and development at NCSU, the
process optimization method for densification of ceramics now known as Rate
Controlled Sintering (RCS) has more-or-less 'come of age'. For a wide range
of ceramic materials, it has been found to provide an efficient and
effective means for determining near-optimum (though typically non-linear)
temperature-time pathways for densification. In doing so, it appears to
beneficially influence and regulate the 'path of morphological change,'
thereby facilitating refinement and control of final sintered microstructures. More recently, it has also been found useful for the sintering
of a special class of high-porosity metal compacts (e.g., those that have
been formed by injection molding).

In this paper, the background and evolution of RCS concepts and experimental procedures are reviewed, and some recent examples of its applications
for both ceramics and powdered metals are introduced to illustrate its
versatility. Consideration is then given to certain unifying factors that
are thought to enable RCS methods to operate effectively across such a
diverse range of materials and conditions. Finally, possible extensions of
RCS methodologies into other classes of materials, higher temperature
regimes and/or hybrid process routes are evaluated and discussed.

INTRODUCTION

Novel rate controlled sintering concepts for ceramic materials were first
developed at NCSU more than two decades ago. Over the intervening years,
those ideas have been refined, and supporting, specialized experimental
facilities that permit <u>rate controlled sintering</u> (RCS) to be accomplished
effectively in the laboratory - and beyond - have also been developed. It
has been demonstrated that these RCS procedures really do work, and that
beneficial refinements of final microstructures, as well as better thermal
efficiencies during firing operations, can result from their proper
applications. The approach taken for RCS has been largely experimental, so
it has almost inevitably led to the development of sophisticated computer-
based research methodologies, e.g., <u>precision digital dilatometry</u> (PDD) and
<u>computer aided design of optimal path(s) for sintering</u> (CADOPS), as well as
its own working nomenclature (See separate Glossary of Terms). From a
relatively small number of such well-instrumented experiments, one reliably
arrives at near-optimal sintering pathways that typically yield fine,
controlled microstructures. Principles and practices involved in RCS methods
have been summarized elsewhere. [1-10]

Science of Sintering
Edited by D. P. Uskoković *et al.*
Plenum Press, New York

Table I. Examples of Ceramic Systems Found Responsive to Rate Controlled Sintering.

Sintering in the Solid State	Sintering in the Presence of Liquid Phase(s)
Al_2O_3 *undoped; doped*	Clay-based Bodies for Brick, quarry tile, etc.

$MgAl_2O_4$
    undoped; toughened
	Whiteware Porcelain
	    all the way to prototypes!

MgO
    undoped
	$BaTiO_3$
	    modified; multi-layer

SYNROC
    35 cation species!
	$BaFe_6O_{19}$
	    modified; pre-oriented

UO_2
    initial stages only
	Cordierite-Zirconia
	    crystal- and glass-based
	    composites

ZrO_2	$YBa_2Cu_3O_{7-\delta}$

Until very recently, the RCS approach has been thought of mostly as an exclusively 'ceramic' processing methodology. Though high purity oxide ceramics (aluminas, spinels, magnesias, etc.) certainly continue to be of interest as 'model materials' {and as important articles of commerce for some industrial sponsors}, it is important to note that RCS practice is not limited to such 'ideal' materials alone. Systems that sinter in the solid state, as well as those that sinter in the presence of liquid phases, appear to benefit - morphologically and/or kinetically - from proper shaping of the densification rate profile. Over the years, many different kinds of ceramics - including some that have been exceedingly complex - have been successfully investigated and optimized (See Table I).

RECENT BACKGROUND

During the four year interval since RCS topics were last discussed before this sintering community at the VIth WRTCS[10], an already-established pattern of working with rather complex materials has continued in NCSU's laboratories. For such 'difficult' materials, it has been found that precision digital dilatometry, with its selective control of atmospheres, can be quite usefully employed - along with DTA, TGA, etc. - in a purely diagnostic sense (e.g., to aid in scientific understanding), as well as for engineering optimization of RCS profiles.

Among the complex ceramics that have been studied in recent time have been whiteware porcelains [11] (See Fig. 1) and '1-2-3' Y-B-C-O oxide superconductors [12-15]. More recently, this laboratory's long-time focus on oxide ceramics has been deliberately broadened to include metal injection-molded (MIM) powders; they display initial fractional densities, and hence sintering shrinkages, that are not much different from those of most ceramics.[16] Non-oxide {nitrogen-based} ceramics are also currently under investigation.

During this four year interval (and to some degree, during the preceding one as well), concepts, procedures, and results that are characteristic of the RCS approach to sintering have gradually become more widely understood, better accepted, and/or usefully practiced elsewhere. For example, RCS

Fig. 1. Rate controlled sintering of prototype porcelain ware entirely under computer control; the decorative underglaze characters can be translated from the Chinese as "rate control sintered porcelain" (After Palmour 11).
   The ware was thrown, decorated and dip-glazed entirely by hand, but its sintering schedule was developed entirely within the computer environment, i.e., from initial gathering of densification data through design of the final optimized RCS profile. Firings were then done one piece at a time in a small box kiln, but still under precise computer control.

concepts have been discussed in major reviews of broad issues relating to sinterability, microstructural development and/or end-use properties of ceramics.[17-19]

The basic principle that undergirds RCS[a] has also been successfully adapted to other ceramic processing issues, including nitridation [20] and calcining and drying [21]. A growing number of investigators have been applying RCS concepts and procedures {or functional equivalents} to a broadening range of ceramic materials, including barium titanates[22-25]; piezoelectrics[26]; toughened, strengthened spinels[27]; cordierite/zirconia composites[28-29] (See Fig. 2); aluminum nitrides[30], and silicon carbides[31].

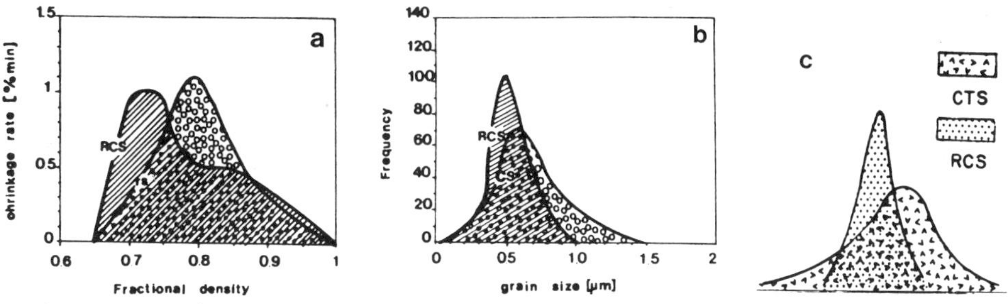

Fig. 2. Rate controlled vs. conventional sintering of cordierite/zirconia composites: (a) densification rate profiles and (b) resultant grain size distributions, 1986 (After Nieszery, *et al.*[28]); (c) postulated grain size distributions, RCS vs. CTS, 1977 (After Palmour, *et al.*[7]).

---

[a] i.e., establishing well-instrumented control of the process through reliance upon some form of feedback from a properly chosen independent variable.

APPLICATIONS FOR INJECTION MOLDED POWDER METALS

Because ceramics and injection molded metal powders have similar fractional green densities ($D_0 \sim 0.60$-0.65), the 'paths of morphological change' that necessarily accompany densification processes must also be fairly similar. Furthermore, the transport processes involved in each case are thermally activated, so for each, sintering (densification) is expected to respond to the main variables (P, T, t) that affect the kinetics of such processes. Thus it is not unreasonable to expect that proper applications of existing RCS concepts and methodologies to metallic (MIM) materials could be similarly effective, and could, therefore, also result in refined microstructures.

Credit for recognizing, nurturing, and promoting the idea of commonality in sintering behavior between these two rather different materials fields (ceramics, MIM metals) can best be attributed to German (e.g., see refs. 32a-d). Substantial reinforcement has come from the early experimental work of Beuers and Poniatowski [33], who first applied RCS concepts to magnetic metallic materials, and from Palmour's initial RCS demonstrations with injection-molded carbonyl iron compacts [16,34] (See Fig. 3).

When compared to ceramics, the rate controlled sintering of MIM-processed metal powders now appears to be in some respects a little easier, but in other respects, probably somewhat more difficult. On the 'easy' side, in early experiments conducted under RCS conditions with carbonyl iron, it was observed that better removal of porosity and related refinement of microstructure was not very difficult to accomplish; that the thermal processing conditions employed for RCS tended to be more moderate than normal powder metallurgy practice would dictate [b]; finally, that the microstructure of as-sintered RCS specimens could be further refined through standard metallurgical heat treatment techniques.[16,34]

On the 'difficult' side, the high mobility and/or volatility of interstitials [c] (e.g., carbon) in metals can introduce some 'nuisance' variables (e.g., composition gradients) that tend to make RCS methodologies less than ideally precise. Therefore, rigorous control of the interstitial-stabilizing ambient atmosphere composition is strongly indicated. Very close control of the binder removal step that precedes the sintering of MIM materials is also

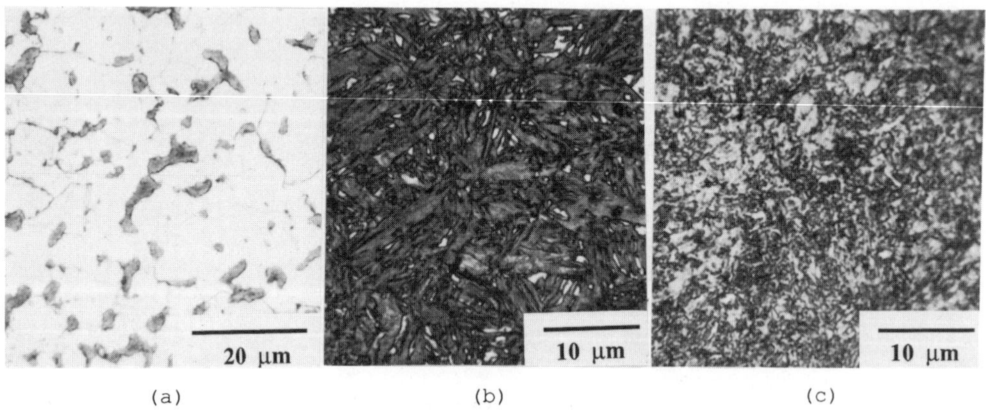

Fig.3. Microstructures of injection molded, rate control sintered carbonyl iron powders: (a) as-sintered, (b) solution heat treated and quenched; (c) same as (b), but subsequently tempered (After Palmour[16]).

---

[b] e.g., RCS profiles will typically require lower temperatures, presumably with some attendant benefits in terms of demands on kilns or furnaces.

[c] i.e., relative to the more-or-less fixed (valence-stabilized) anions commonly found in ceramics.

indicated. Finally, it can be anticipated that the adaptation of current
generation PM production furnaces and their instrumental controls to operate
successfully on RCS profiles for MIM-processed metals d may present design-
ers and operators with some interesting (though not intractable) engineering
challenges.

PROBABLE UNIFYING FACTORS

In some sectors of the sintering community, rate controlled sintering is
apparently still regarded mostly as a 'thing apart', a kind of 'engineering
anomaly' that admittedly works well, but can't really be understood in terms
of sintering science. In this section, we will hope to dispel that view by
(1) considering certain unifying factors that permit RCS methodologies to
work with demonstrable effectiveness over a disparate group of ceramic and
metallic materials, and (2) where appropriate, relating those factors to
existing theories of sintering and to other pertinent scientific considerations.

Rate controlled sintering is, in effect, primarily an engineering optimization methodology. In its purest form, it represents a deliberate 'fine
tuning' of the sintering process for a particular material. Strictly speaking, such an RCS optimization is valid only for its own specific material,
pre-processed to its own specific 'green' condition. As with other forms of
engineering, one seldom is called upon to treat just 'ideal' cases, but
rather, to address generally more complex situations. In doing so, it is not
unexpected for one to find that various trade-offs have to be accommodated,
and some deliberate risks occasionally taken.

In providing such services to a largely industrial clientele, one learns
that potential sponsors cannot usually afford, and thus do not often seek,
outside help on process-related matters such as this until most of the key
concerns with science issues (e.g., those that might affect the inherent
sinterability of their product) have already been dealt with, at least to
some degree. Note, however, that as used here, the term 'dealt with' needs
to be broad enough to cover many different points of view!

Therefore, as one proceeds toward a given RCS optimization, it is necessary
to exercise constant vigilance over *all* the basic issues that can affect the
sinterability of a given material. Among the more obvious ones are: (a)
existing impurity levels; (b) deliberate doping, if any; (c) existing
particulate properties and their distributions; (d) powder processing and
compaction behavior, (e) overall thermodynamic behavior of the system,
including known phase transformations; (f) atmosphere effects, if any; and
finally, and most usefully, (g) one's own experimentally determined
parameters, including kinetics field responses, carefully measured fractional green and fired densities, resultant microstructural features, etc.

In attempting to link such engineering optimization practices with science,
many of the detailed answers one might seek are likely to be quite complex,
even very eclectic (e.g., strongly material-dependent and/or environment-
specific, or perhaps highly proprietary). For these reasons, they are often
very difficult to obtain. However, on the basis of considerable experience
in this sphere of activity, it is the author's opinion that the RCS approach
is likely to have worked well with 'real' industrial materials for one or
more of these several reasons: (a) it is not limited to simple materials nor
is it confined to easily-modeled sintering situations; (b) it depends mostly
upon straightforward laboratory methods e; and (c) it also relies on one's
cumulative engineering design experience. Note that in using CADOPS on a
literal-minded microcomputer, that kind of cumulative experience is rather
quickly developed. One has to make - and input to the computer - successions

---

d   typically non-linear in T vs. t; at times, even calling for negative
    slopes.
e   well established, experimentally based, thoroughly documented.

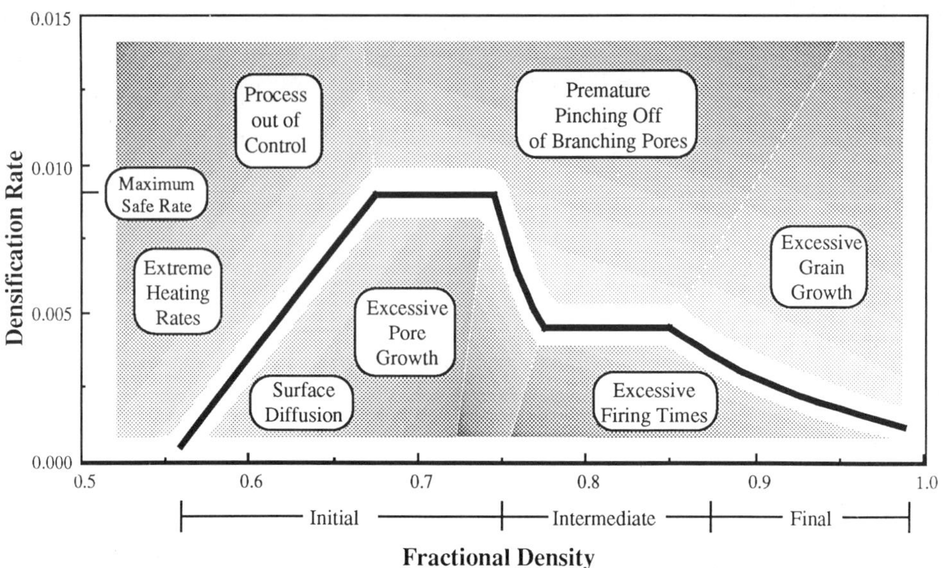

Fig. 4. Typical RCS profile, expressed as dD/dt vs. D, annotated to indicate some practical constraints as well as some underlying science issues that can - and should - influence path selection.

of rational decisions ('judgement calls') throughout the firing cycle. They are typically expressed in engineering quantities, but in terms of the kinds of understanding needed, those decision-making steps do in fact embrace the whole set of thermally activated processes that we simply call 'sintering'.

RCS pathways tend to be inherently conservative with respect to energy inputs 3 and and thus they minimize the risk of overfiring 5,6: they are carried out at the lowest possible temperature(s) consistent with maintenance of the densification rate(s) being imposed in a particular stage of sintering 3. Three-staged RCS pathways are programmed in an initially fast manner to suppress adverse surface diffusion effects 3,7 and associated rearrangement processes 35,36. Thereafter, they become progressively slower as densification becomes more and more difficult due to increasing morpho-logical constraints and concomitant diminutions of surface-related driving forces 1-3,10 (See Fig. 4).

Under RCS profiling, the pore channels tend to remain open for longer times than in other modes of sintering, thereby promoting proper removal (e.g., by permeation) of occluded and/or chemisorbed gases, burnout of other anion impurities, etc.16 Since most of that deliberate 'slowing down' occurs quite early in the densification process, e.g., over the range D = 0.75 - 0.85, the grain boundaries are effectively pinned by the pore network. Even though sintering times do become longer in this regime as a consequence of RCS profiling, for the case of well-made compacts (i.e., with high D_0), it is apparent that there is not an inherent imposition of a significant grain growth penalty.6,37

The multi-staged rate profiles characteristic of RCS pathways also appear to assist in establishing near-equilibrium 'paths of morphological change' for both grains and pores.10,16 With such RCS pathways, at a critical point in fractional density (i.e., at or near the onset of final stage sintering, D ~ 0.93-0.95), that evolutionary morphological process will typically have yielded only fairly narrow distributions of rather closely spaced pores, located mostly at the triple points.16 Thereafter, those very stable pore locations are thought to facilitate further *pore control* of grain growth, a process that can continue far out into the final stage of sintering.16,37

There are certain inherent process-related limitations that can adversely affect practical experimentation in the RCS mode. Among them are: (a) poor dispersion of powders,[5] (b) strong agglomeration of particles,[38] (c) inadequate initial compaction,[5,6] and (d) deleterious materials interactions, especially those occurring at the higher temperatures. Some of these limitations are specimen-related. Since RCS optimizations tend to focus primarily on control of morphological evolution during sintering, the achievement of good particle-particle coordination in the compact is regarded as essential in assuring that desired final microstructures can result [5,6,16,37,38].

From practical experience, it has been learned that, for best results, the initial fractional density, D_o, should be ≥ 0.60 (binder-free basis); variations in D_o need to be minimized, even on a very local scale [5,6,10,16,37,38]. These very practical and frequently confirmed experimental observations also have very significant links with the findings of many other investigators. Though this seemingly innocuous experimental variable, D_o, has often been treated rather casually by some industrialists (and scientists as well f) it is in fact highly germane to important issues (See Fig. 5): they include particle-particle coordination during compaction [37]; morphology of subsequent rearrangement and sintering processes [36,37]; shrinkage-related stress generation and resultant sinter-driven creep processes[39]; the role of pore-grain coordinations in determining whether pores will shrink or grow [40]; premature formation of deleterious micro-dense regions, which then trigger uncontrolled late-stage grain growth [41]; and the whole issue of grain growth that can occur during sintering [6,10,16,38,42-49].

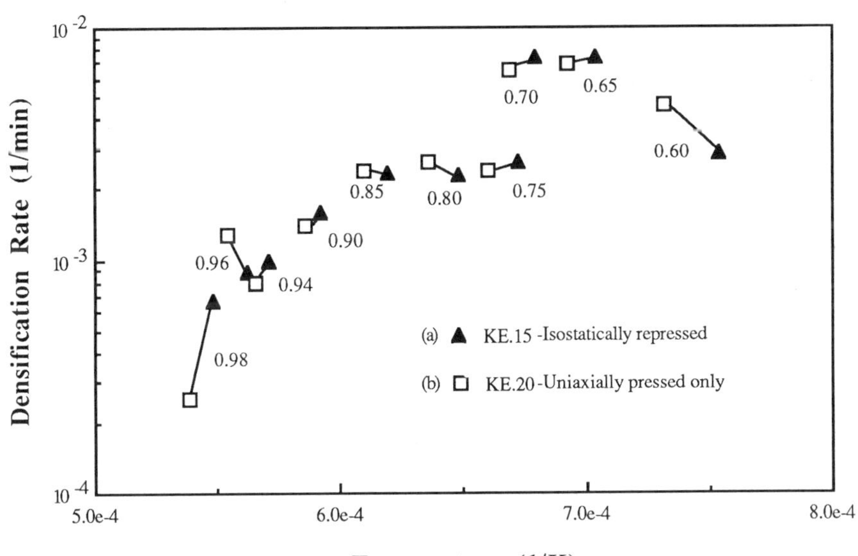

Fig. 5. Effect of uniformity of compaction (and thus of particle-particle coordination) on kinetics field response(s) of undoped MgAl2O4 spinel when sintered under RCS conditions: (a) isostatically repressed to $D_0 = 0.585$, (b) same as (a), but only pressed uniaxially to $D_0 = 0.571$.
 Short lines connect points of equal density, as indicated. The denser compact followed the designed RCS path much more accurately; it also reached successive iso-density points at consistently <u>lower</u> temperatures.

---

f e.g., how often have we all been told, or have we read, that a D_o near 0.50, or even some lesser value, was considered to be 'good enough'?

There are also some important apparatus-oriented limitations for practical
RCS experimentation. Because PDD instrumentation depends on a 'touch-based'
LVDT system that works through sapphire extension rods 9, adverse reactions
between the specimen and the sapphire components (support and probe rods,
hot stage, isolation shims) must be avoided. As much as possible, materials
containing large amounts of highly volatile constituents must also be avoided, because they tend to contaminate the system with undesirable cation
and/or anion impurities (e.g., Li, Pb, Bi, F).

The design-limited temperature capabilities of the existing equipment must
not be exceeded. The present 'safe' limit is considered to be ~ 1750°C. When
equipped with the newest, best and most expensive furnace expendables
(including elements, refractories and thermocouples, as well as sapphire
stage components), the maximum service limit for this particular PDD design
is only expected to equal or slightly exceed 1800°C.

Under normal conditions, in accordance with the University's safety standards, neither highly explosive nor very toxic gas atmospheres are utilized
in these dilatometers. However, if a real need were to exist for utilizing
such hazardous gas atmospheres, and if the added costs could be afforded,
then the possibility of installing the required additional protective equipment, as well as providing hazardous gas monitors, related safety gear and
other special precautions, could presumably be considered.

THE ROLE OF RATE CONTROLLED SINTERING IN HYBRID PROCESSING

For a number of special high-value-added applications involving both ceramic
and powder metals, it has become increasingly common to utilize various
forms of hybrid powder processing, particularly those that are completed at
high temperature and pressure by <u>hot isostatic pressing</u> (HIP). Advantages of
such HIP-based hybrid cycles are considered to include healing of flaws,
higher final fractional densities, better microstructures, and greater
reliabilities of end-use properties. The disadvantages would include limitations on sizes and/or shapes of ware, some loss of overall process flexibility, particularly if 'canning' or 'glassing' of the product should be
required, as well as unavoidably higher capitalization and operating costs.

Consider, then, the potential technological and economic advantages of rate
controlled sintering as the critical first step in such a hybrid densification cycle (i.e., RCS-plus-HIP). Under well-optimized RCS conditions, it is
known, for example, that most of the observed grain growth occurs *very* late
in the overall cycle, i.e., only after the fractional density has moved well
out into the final stage of sintering (typically at $D \geq 0.97$). By deliberate
truncation of that RCS pathway, sintering can be stopped at a fractional
density just high enough to assure the attainment of technical density
(i.e., containing only closed pores).

In this way, one one can rather easily obtain technically dense specimens
that are still quite fine-grained and also display the needed uniformity of
pore-grain distributions within their microstructures to facilitate further
densification. RCS-sintered specimens can then be straightforwardly HIPed to
full density without having to add containment, usually under conditions
that promote the retention of quite fine grain sizes. Very sophisticated and
quite expensive integrated (sinter/HIP) devices may therefore not be needed.

In such a hybrid process, the RCS sintering step can be efficiently accomplished in more-or-less standard sintering furnaces, where the needed large
volume shrinkages can be better and more economically accommodated. Important quality control and quality assurance inspections can then be effectively performed, and further sizing and finishing operations readily carried
out. Finally, the separate, higher-cost HIP step can be economically completed (e.g., with very effective volumetric loadings) in an ordinary HIP
apparatus, thereby bringing the ware to its full density. Depending on the
lattice structure of the starting material (anisotropic; isotropic), ceramics densified by these RCS-plus-HIP procedures can become highly trans-

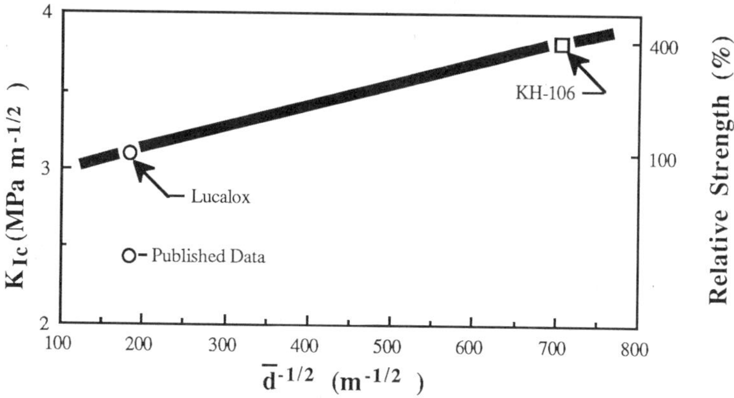

Fig. 6. Comparisons of fracture toughness and relative strength for two
highly translucent aluminas: coarse-grained lighting-grade alumina
(Lucalox®); fine-grained alumina (KH-106) from NCSU-developed hybrid cycle.
K_{Ic} values were determined from indentation-induced crack lengths
(~ 50 indentations per data point), using the method of Anstis, *et al.*50
Values for relative strength were obtained from a comparative - but
proprietary - industrial test method involving complex states of stress
and strain.

lucent, even transparent. Since they are typically still quite fine-grained,
it is not surprising that they can also display unusual levels of fracture
toughness and mechanical strength (See Fig. 6).

EXTENSION OF RCS TO OTHER MATERIALS, OTHER TEMPERATURE RANGES

Considerable curiosity has been expressed about prospects for applying RCS
methods to other, more difficult classes of material, including advanced
structural ceramics. For many investigators, the pertinent questions are:

> Would RCS help?
> > Can RCS be done under the high temperature conditions
> > > that prevail for the sintering of nitrides or carbides?

At the conceptual level at least, a positive answer must be given:

> Yes, RCS probably would help, because it deals mostly with the safe
> > removal of pores during densification.

Pores are just as common to advanced structural ceramics as they are to most
other classes of green ceramic ware. Therefore, one would have to expect RCS
methods to be effective, *if* the data needed can be acquired experimentally
(e.g., see results recently obtained for SiC by Lihrmann, et al. 31).

Relative to conventional sintering practice, pores could probably be removed
with greater reliability under RCS conditions, and perhaps with enhanced
thermal efficiency as well. Improved control of final grain sizes and size
distributions would also be fairly likely, and thus RCS might also be
expected to enhance the predictability of the resultant useful mechanical
properties. The typically non-linear *T vs. t* profiles that are associated
with RCS pathways might also be expected to cause some beneficial alterations
in composition(s), volume(s) and disposition(s) of the added liquid phase(s)
that are needed to promote sinterability in most such non-oxide ceramic
systems. Ultimately, possible RCS-related reductions in the needed volumes,
compositions, and/or fluidities of those liquid phases might also be expected
to result in improved high temperature creep properties.

345

At the experimental level, the proper response would still be positive, but more along the lines of:

Yes - maybe.

Many of the needed experimental resources are already available, including proven basic instrumentation concepts and related control and data reduction software, proven profile design concepts and software, extensive engineering research and equipment design experience, etc. However, for sintering to become effective, most Si-based structural ceramic materials need temperatures higher than the present capabilities of existing PDD units. The very refractory nature of the non-oxide ceramic systems of interest (as well as known chemical affinities, e.g., of Si) can be expected to present other experimental difficulties, including some potentially destructive hazards that could result from excess reactivity or volatility of the various chemical species involved. In treating such interrelated apparatus design and materials selection issues, one is likely to find that the most vexing problems are materials-dependent, or temperature-sensitive, or both.

DISCUSSION

In this section, a few specific research examples are used to illustrate some other interesting aspects of rate controlled sintering, and to point up their various relationships with both engineering concerns and the science of sintering. The three examples selected are (1) artware porcelains, (2) cordierite/ZrO2 composites, and (3) highly translucent, fine grained alumina ceramics produced by NCSU's hybrid RCS-plus-HIP cycle.

Artware Porcelains

It is interesting that the set of small hand-thrown pots shown in Fig. 1 turned out, to the best of our knowledge, to be the first such finished ware ever to be sintered optimally entirely under computer control, via RCS, PDD, and CADOPS.11 It is also interesting to note that convergence upon that RCS

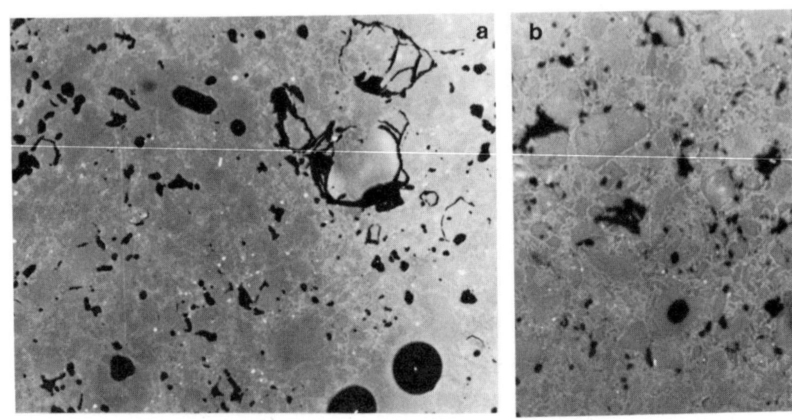

250 μm

Fig.7. Microstructures for artware porcelain specimens subjected to different densification paths: (a) conventionally sintered, (b) rate control sintered (See also Fig. 1; after Palmour 11).
   Note overall microstructural refinement in (b); note also that the typical ring-cracks around the large α-quartz grains in (a) have been avoided in (b). Polished, unetched. (Reflected light microscopy by Viktoria Carle, MPI/PML, Stuttgart)

346

profile design was achieved - and confirmed - with notable engineering
efficiency; it required only *six* experimental runs to create the data base.

However, from the standpoint of sintering science, an even more interesting
aspect of this small study resides in its microstructures (See Fig. 7),
wherein it is shown that, under RCS conditions, there is little or no
evidence of normally-expected internal cracking of the matrix surrounding
the larger quartz grains. The densification path followed does exert a marked influence on the temperature-time history, and thereby also can influence
the localized composition, quantity, and viscosity of the liquid phase(s)
generated, as well as the crystallization habit of the strengthening mullite
phase. These observed path-dependent influences on microstructural development also affect the ability of the matrix to withstand localized mismatch
stresses and strains associated with the α to β quartz transition at
573°C.[11] As mentioned earlier, it is not unreasonable to think that somewhat similar microstructural benefits might accrue from RCS profiling of
other liquid-phase sintering systems, including those that are classed as
structural ceramics.

<u>Cordierite/Zirconia Composites</u>

The data for cordierite/zirconia composites illustrated in Figs. 2a and 2b,
from the research of Nieszery, *et al.*[28], were probably the first to provide
clear and independent confirmations of earlier postulates about the role of
RCS profiling in bringing about a narrowing of the grain size distribution
relative to that attainable by conventional sintering (See Fig. 2c, after
Palmour, *et al.*[7]). Thus these results made an interesting and useful
contribution to sintering science.

They also served to document significant rates of progress in the development of experimental and computational capabilities for RCS. While on
sabbatical leave in mid-1984, this author spent four months in the Powder
Metallurgical Laboratory at the Max Planck Institut in Stuttgart; some key
concepts for the present CADOPS design capability were then under development, but they had progressed only to a fairly low level of mathematical
sophistication. Among the several data sets from ceramic studies there that
were available for such plotting and calculation was one drawn from
Nieszery's cordierite/zirconia study, then ongoing. From contemporaneous
notes, it would appear that only three such dilatometer runs were available,
from which no more than 16 useful data points could be extracted. They were
then transformed on a pocket calculator, hand plotted, and subsequently
employed to establish, graphically, the locations of just four iso-density
lines, spanning the range 0.75 - 0.90. An RCS design was similarly calculated and transposed graphically, so that it could be scaled to run on an
available eight-step programmer. Though it had hardly been better than a
crude first approximation g, that first RCS design did its job rather well,
as indicated in Nieszery's data (Fig. 2). In comparison to earlier efforts,
it resulted in higher densities (D ~ 0.95) and finer grain sizes, but with
some minor problems (e.g., reaction of constituents to form zircon, specimen
cracking) still being noted.

Three years later, in mid-1987, this author again became involved with that
same composite material, but this time, the research was to be carried out
on PDD units at NCSU, which were by then fully computerized and equipped
with CADOPS. The resolution of the PDD system was about an order of magnitude better than the dilatometer used by Nieszery; an *abundance* of data
spanned the whole range of experimental interest; the computer-based instrumentation provided *many* segments for control, etc. As reported by Semar, *et
al.*[29], the resultant RCS profiling via CADOPS made it possible to attain a

---

g beyond design of the RCS profile, this author had no direct involvement in
   the research itself.

very high sintered density (≥ 0.99) and a fine microstructure (both phases <1μm in dia.) for this composite material, while keeping it free of any zircon reaction product. From these comparisons of the rate controlled sintering of essentially the same ceramic composite material, it seems evident that if microstructural control is to be the goal, then it not only matters *how* the sintering is done, but also, how *precisely* the RCS profiling can be designed and executed!

Highly Translucent Alumina from NCSU's hybrid RCS-plus-HIP cycle

As indicated in an earlier section, alumina that has been hybrid processed properly in the RCS-plus-HIP mode is both highly translucent and quite fine-grained, as well as being demonstrably tough and strong (See Fig. 6). In the particular example cited, the average final grain size was ≤ 2μm, at a final density experimentally indistinguishable from that of sapphire.

Remnant pores in this highly translucent material were characterized with a confocal scanning laser microscope (CSLM) [51], yielding results that are summarized in Fig. 8. By direct measurement (see Fig. 8a), 119 very fine (<< 1μm dia.) light reflecting pores were shown to be present in the volume studied, yielding a number concentration of 7.28×10^{-3} μm-3. They were also found to be statistically distributed in a three-dimensional pattern in-dicative of pore clustering at spacings about equivalent to the average grain diameter (Fig. 8b).[51] SEM fractographs of this material (See Fig.9), as well as polished and thermally etched plane sections.(not shown), con-firmed the presence of very fine pores in all alumina specimens from that particular RCS sintering profile; they were found to be separated from, and trapped just within, the grain boundaries. At applied pressures up to 30 kpsi, variations in HIP parameters were found to alter the average final grain sizes somewhat, but *not* to remove the entrapped fine pores (e.g., see arrows, Fig. 9).

When the sintered specimens from that particular RCS profile (similar in profile shape to that diagrammed in Fig. 4, and giving rise to data of Figs. 6, 8 and 9) were examined prior to final HIPing, i.e., at D ~ 0.97, it was found that the grain sizes were discernibly bimodal in character (See Fig. 10a); one can infer, therefore, that those specimens may have been vulner-able to the kinds of micro-density problems discussed by Bennison and Harmer [41]. However, when that same RCS profile was run again, but over a longer time base, thereby reducing all its densification rates in proportion,[5]

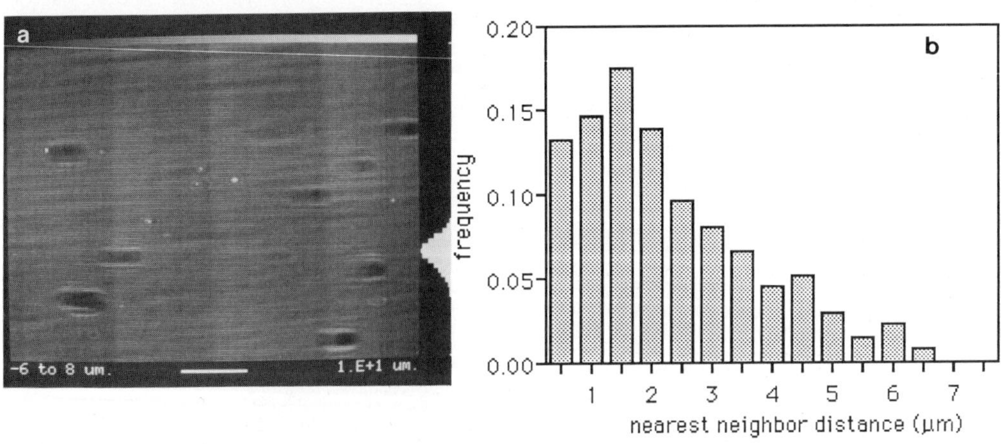

Fig. 8. Direct imaging of very fine pores in a dense, fine-grained trans-lucent alumina: (a) CSLM image, showing light scattering by a few very small pores (bright spots only); (b) histogram constructed from a set of such 'serial section' images, indicative of clustering of pores at slant range spacings of ~ 1.5μm (After Russ, *et al.*[51]).

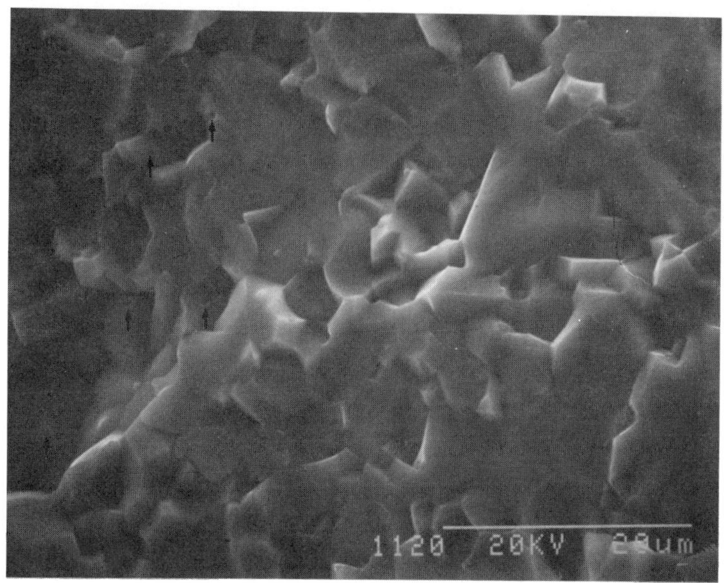

Fig.9. Microstructure of RCS-plus-HIP-processed dense, fine grained alumina, showing very fine pores entrapped during hybrid processing.
The observed fracture mode is entirely transgranular, in keeping with measured high values for K_{Ic} and relative strength (See Fig.6).

it was found that the resultant as-sintered microstructure was very uniformly distributed (See Fig. 10b). For specimens sintered by this slower version of that same RCS profile, it was found after HIPing that the entrapment of very fine pores within grains had been effectively eliminated (not shown), and that the resultant optical translucence had been correspondingly enhanced.

In this author's opinion, there are several important aspects of sintering science to be extracted and/or elucidated from these different observations.

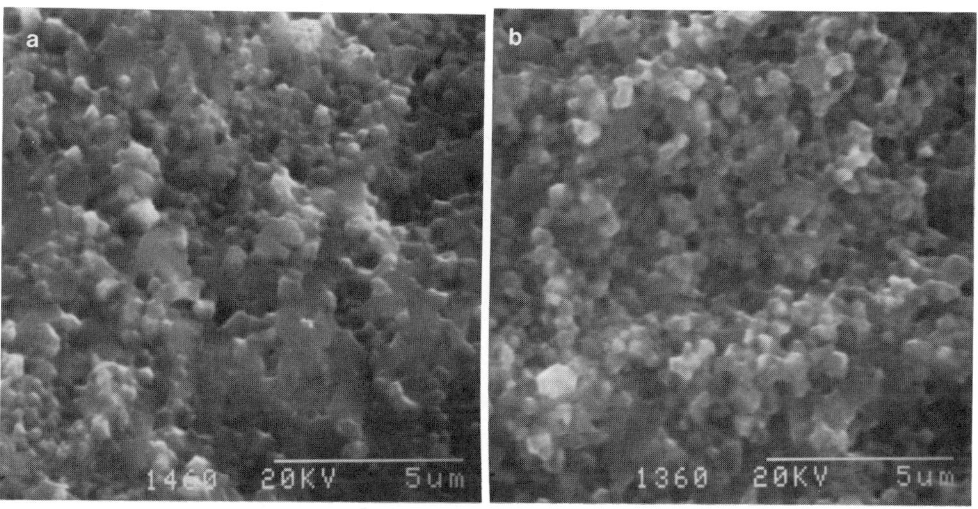

Fig.10. Microstructures of RCS-sintered alumina at D ~ 0.97, prior to final HIPing: (a) sintered on RCS profile over an originally designed time base, (b) sintered on RCS profile of the same shape, but with the time base having been deliberately lengthened.

The first deals with the importance of allowing sufficient time within the open pore regime of the sintering schedule for the pores and grains to evolve naturally, under near-equilibrium conditions [16]. For best results, it will be insufficient just to have access to the right shape of the RCS profile; clearly, it must also have been scaled properly with respect to time *per se* [5].

The important concept of a 'maximum safe rate' in rate controlled sintering has again been revalidated, e.g., for Al2O3 in this research, as well as for ZrO2, reported in recent studies by Oberacker, *et al.* [52] It would appear that the region of greatest risk, e.g., for setting up rate-dependent morphological anomalies that can later cause pore entrapment and thus lead to the micro-dense grain growth mechanism, is to be found in the *intermediate*, open pore, range of fractional densities, between ~ 0.70 and 0.85. It is there that the intrinsic sinterability of the material is typically at a maximum, and also where densification rates *must* be limited, even sharply reduced, if one is to gain the full morphological benefits of an RCS pathway.

If one is to gain control over microstructural development, then the interactions existing between how a compact has been *prepared*, and how it is to be *fired*, take on special significance. Though it is still in some circles taken much too casually, the dominant role of proper prep-aration of the starting compact as a key determinant of microstructural evolution simply *must* be recognized. The pore-grain distributions thus established will assuredly determine all the FIRST ORDER effects that occur during subsequent sintering (See Fig. 11).

Yet, even when one works with very well prepared compacts, it has been repeatedly demonstrated that in comparison with other sintering modes, properly optimized RCS pathways do yield better final microstructures. Furthermore, the beneficial effects of RCS profiling have been observed across a diverse range of sinterable materials. Thus it has become

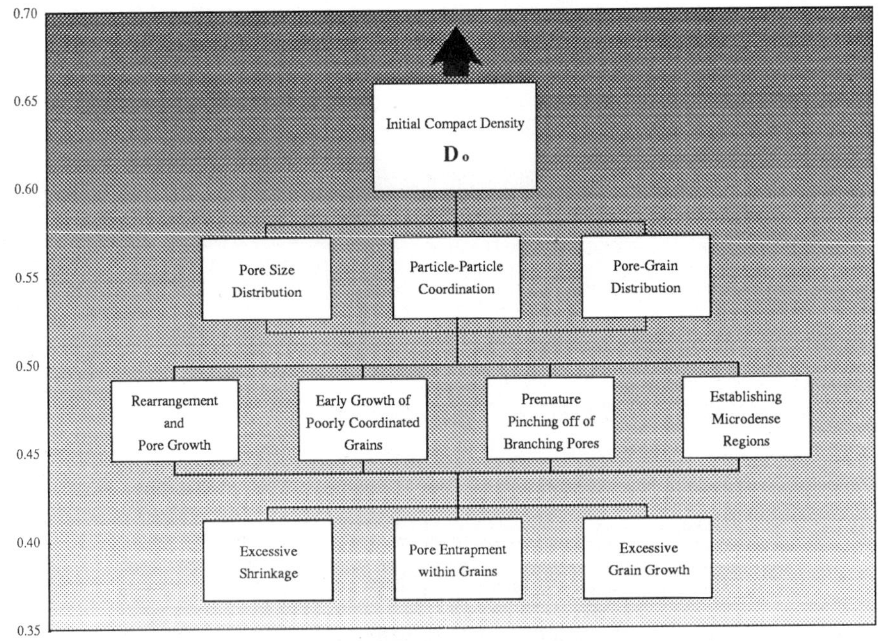

Fig. 11. For a well dispersed, reasonably monosized starting powder, the initial compact density, D_o, is the most important determinant of all *first order* influences on morphological evolution during subsequent sintering: the higher D_o, the better the final microstructure!

increasingly clear that even for the best of compacts, the sintering
pathway chosen - rightly or wrongly - will also influence the microstructural outcome, by means of significant SECOND ORDER effects.

In the future, it is hoped that - among scientists and engineers, at
least - it will be considered insufficient just to make very good green
compacts and thereafter be content to fire them in any old way.
Instead, the obverse should be true: the better the quality of the
starting compact and the higher the expectation for its close microstructural control, the more critical will become the need for access
to a well planned, carefully controlled sintering profile.

In principle, if only a single transport mechanism is involved, the
deliberate alteration of densification rates in successive stages of
sintering should have *no bearing* on sintering behavior or microstructural evolution (e.g., see Kaysser and Ahn 53). Therefore, the consistent microstructural successes and the wide scope of applicabilities
that can now be ascribed to the use of RCS methodologies by a number of
investigators would appear to indicate that, in fact, very few 'real'
materials can be fully densified by means of any *single* transport
mechanism.

The final points to be made in this section relate to the modeling of
sintering and grain growth as influenced by such pore entrapments, etc.
For more complete examinations of the important but subtle kinds of
interplay that appear to exist between sintering mechanisms, morphological parameters and densification rate histories, the interested
reader is referred to the elegantly structured *stereological theory of
sintering* by DeHoff 54, as well as to other previously cited references, including, e.g., 2, 7, 10, 24, 36, 42-45, 47 and 49.

In extending a morphologically based statistical model for sintering
42-45, Fang and Palmour 37,47 recently introduced a new term, Z, to
represent the fraction of *active* pores; it was also postulated that Z
would be process dependent, and that its value (ranging from 0 to 1)
would be subject to change as sintering proceeded through the different
stages. For example, in well-made compacts having pores on the grain
boundaries, the value of Z would approach 1. Along with the usual terms
for expression of thermal activation and pre-exponential dependences,
in this model the normalized densification rate is treated as a function of two terms, $F_i(y,Z) \cdot (p/p_o)^{mi}$, where y is the width of the pore
size distribution, p the porosity, p_o the initial porosity, and Z the
the fraction of *active* pores (i.e., those causing either shrinkage or
bloating). Therefore, at any point in the sintering history, the
determinative *morphological* parameters are considered to be y, Z, and
m_i.

Very fine pores entrapped in dense aluminas resulting from an RCS-
plus-HIP hybrid cycle - which were carefully observed, and thereafter
eliminated by means a simple time-elongation of the rate controlled
sintering pathway (see Figs. 8-10) - clearly seem to meet all the
criteria needed to establish, at least experimentally, the essential
validity of that new term, Z. Those fine entrapped pores were (a) found
to be strongly process-dependent; (b) found responsible, once so
entrapped, for sharp alterations of the observed densification
kinetics; (c) in fact, found to be resistant - even at applied HIP
pressures of 30 kpsi - to *any* further densification, thereby establishing a clear case where Z = 0. But when that same material was sintered
in a slightly different manner, few if any entrapments occurred, and
densification was able to continue, i.e., again the value of Z was ~ 1.

Another relevant example comes from recent studies of rate control
sintered ZrO_2 by Oberacker, et al. 52, where it was found that for the
fine, reactive material investigated, the 'maximum safe rate' was $\leq$ 1
vol%/min, occurring at about D = 0.65. Specimens sintered at or below
the 'maximum safe rate' densified fully (i.e., Z ~ 1), thereby yielding

dense, fine and uniform final microstructures. However, those sintered at higher rates developed pores larger than the grain size; thereafter, those coarser pores could *not* be removed by sintering (i.e., Z ~ 0).

SUMMARY AND CONCLUSIONS

It is probably sufficient here to reiterate just three summary thoughts - in slightly rearranged form - that were first presented at the International Conference on Densification and Sintering held at Hakone, Japan in 1978 [6]:

- Do not underestimate the importance of good processing and compaction prior to firing!

- Do not underestimate the importance of densification rate dependencies among competing transport processes (e.g., as they may affect sintering and grain growth theories!)

- Do not underestimate the advantages that rate controlled sintering may have for your material!

ACKNOWLEDGEMENTS

Based in part on hitherto unpublished material from three invited lectures presented at Alfred University (Apr.'88); Rennselaer Polytechnic Institute (Sept.'88); New England Section of the American Ceramic Society (Oct. '88). Sponsorship of research and development activities by various industrial organizations [h] is gratefully acknowledged. Many interested persons from industrial and academic institutions and research institutes at home and abroad (regrettably, far too numerous to be mentioned individually here) have also contributed to a variety of challenging problems, stimulating questions and enlightening discussions.

At NCSU, T. M. Hare, A.D. Batchelor, M. J. Paisley and R. L. Russell have made indispensable contributions to the development of apparatus and software, and to the various research efforts being undertaken. S. Srinivasan made the very detailed measurements needed to establish K_{Ic} values. Special assistance in the preparation of this paper was graciously provided by R. L. Russell (data acquisition; computer graphics; literature searches) and J. D. Mahaffee (photography). Finally, the manuscript was critically reviewed by three experienced ceramists and fellow educators: T. M. Hare, R. L. Porter and R. M. Spriggs.

Travel and subsistence support - provided by the International Division, U.S. Department of Energy; by host institutions in Yugoslavia, including Institut 'Jozef Stefan' (Ljubljana), Institut of Chemistry 'Boris Kidric' (Ljubljana), and the International Institute for the Science of Sintering (ETAN, Beograd); as well as by North Carolina State University - is also gratefully acknowledged.

A GLOSSARY OF TERMS (After Palmour[16])

<u>Rate Controlled Sintering</u> (RCS) - A sintering (firing) profile that has been rationally designed with respect to an optimal, near-equilibrium 'path of morphological change.' Because pore-grain morphologies *must* undergo change as sintering proceeds, RCS profiles are commonly multi-staged; they can best be understood when plotted in coordinates of *dD/dt vs. D* or *D vs. t*. In the latter, they display their most characteristic feature: a smooth and

-----

[h] since 1985: Baikowski Int'l Corp.; Coors Ceramic Company; Eastman Kodak Co.; Kenny Associates, Inc.; PXA Technologies, Inc.; Raytheon; Schott Glaswerke (Mainz).

closely controlled transition from (a) fast to (b) slower to (c) progressively slower densification rates. The resultant {dependent}T vs. t curves
used to control actual firings are typically non-linear, with the temperatures required to attain given density levels being mostly lower than for
conventional sintering. As a result of the deliberate profiling, the final
microstructures of ceramics sintered under RCS contitions are found to be
finer and more uniform than those of their conventionally sintered counter-
parts.

<u>Maximum Safe Rate</u> - In an RCS profile, the *fastest* rate allowed, usually
occurring only in the last part of the initial stage of sintering ($D \sim 0.70$
- 0.75). For most ceramic materials, the maximum safe rate is found to be on
the order of 1 to 2 vol. %/min; in some instances, < 0.5 vol. %/min.

<u>Roll-Over Point(s)</u> - In an RCS profile, the density points at which changes
in densification rate are deliberately imposed. Typically, such roll-over
points are found at or near fractional densities of 0.7, 0.75, 0.775 and
0.85, respectively.

<u>Turn-Down Ratio(s)</u> - In an RCS profile, the ratio(s) of densification rate
reductions imposed. The rate allowed in the intermediate stage is typically
no more than 1:2 with respect to that permitted in the initial stage; beyond
$D \sim 0.85$, rates become progressively slower, resulting in a further overall
reduction of 1:3 to 1:4.

<u>Precision Digital Dilatometry</u> (PDD) - A computer-based instrumental tech-
nique used in studies of the sintering behavior of ceramics and related
particulate materials. As developed at NCSU, precision digital dilatometry
now provides for closely-controlled sintering of a wide range of materials
at temperatures up to $\sim$ 1750°C. Runs can be carried out at heating rates up
to $\sim$ 30°C/min in a variety of different gas atmospheres. Limiting precisions
for the digitally-acquired and computer-logged raw variables (temperature -
displacement - time) are on the order of 0.1°C, 0.25 μm, and 0.001 sec,
respectively. A typical sintering run, when completed, thus yields *hundreds*
of fully corrected, digitally-precise, suitably transformed data points.

<u>CADOPS</u> - An acronym for "Computer Aided Design of Optimal Path(s) for
Sintering." As developed at NCSU, the CADOPS software now provides for (a) a
modeling capability, for systematic numerical analysis of whole sets of
dilatometer-generated data, (b) a *design* capability, for calculating resultant temperature-time relationships for any given sintering path which can be
expressed in *dD/dt vs. D* coordinates, (c) an *auto-editing* and *file-
structuring* capability, to obtain practical input commands for computer-
based control systems, thus enabling them to successfully execute the
designed response, and (d) a supporting *graphics* and *numerical display*
capability, to facilitate the designer's selection among the available
design choices - and permitting him, *in advance*, to understand their
consequences - throughout the entire design process.

<u>Kinetics Field Response</u> - An expanded graphical representation of the
numerically modeled sintering response, expressed in Ahrrenius coordinates
- *ln (dD/dt) vs. 1/T*. It covers the whole sintering range of interest, as
experimentally determined by precision digital dilatometric means. As
plotted, the kinetics field response displays a 'family' of isodensity
lines; collectively, they represent all the densification rates, all the
temperatures, and all of the density levels investigated.

REFERENCES

1. H. Palmour III, D. R. Johnson, Phenomenological model for rate controlled
     sintering, p 779-791 in G. C. Kuczynski et al, eds., Sintering and
     Related Phenomena. Gordon & Breach, NY, 1967.
2. H. Palmour III, R. A. Bradley, D. R. Johnson, A reconsideration of stress
     and other factors in the kinetics of densification, pp 392-407 in
     T.J.Gray, V.D.Frechette, eds., Kinetics of Reaction in Ionic Systems.
     Mater. Sci. Res., Vol. 4, Plenum Press, NY, 1969.
3. H. Palmour III, M. L. Huckabee, U.S.Patent 3,900,542, Process for
     Sintering Finely Divided Particulates and Resulting Ceramic Products,
     1975.

4. T. M. Hare, H. Palmour III, Process optimization and its effect on properties of alumina sintered under rate control, pp 307-320 in G.Y.Onoda Jr., L.L.Hench, eds., Ceramic Processing Before Firing. John Wiley & Sons, NY, 1978.

5. M. L. Huckabee, T. M. Hare, H. Palmour III, Rate controlled sintering as a processing method, pp 205-215 in H. Palmour III, R. F. Davis, T. M. Hare, eds., Processing of Crystalline Ceramics. Mater. Sci. Res., Vol. 11, Plenum Press, NY, 1978.

6. H. Palmour III, M. L. Huckabee, Rate controlled sintering, pp 278-297 in S. Somiya, ed., Proc Intl Symp Factors in Densification and Sintering of Oxide and Non-Oxide Ceramics. Tokyo Inst. Tech, 1978.

7. H. Palmour III, M. L. Huckabee, T. M. Hare, Rate controlled sintering: principles and practice, p 46-56 in M.M.Ristic´, ed., Sintering - New Developments. Elsevier Science Publ, Amsterdam, 1979.

8. H. Palmour III, T. M. Hare, Sintering of SYNROC: case history for phase formation and densification of complex oxide systems, pp. 185-192 in D. Kolar, S. Pejovnik and M. M. Ristic', eds., Sintering - Theory and Practice. Mat'ls Sci. Monographs 14, Elsevier Scientific Publishing Company, Amsterdam, 1982.

9. A. D. Batchelor, M. J. Paisley, T. M. Hare, H. Palmour III, Precision digital dilatometry: a microcomputer-based approach to sintering studies, pp. 233-251 in R. F. Davis, H. Palmour III, R. L. Porter, eds., Emergent Process Methods for High Technology Ceramics. Mat. Sci. Res., Vol 17, Plenum Press, New York, 1984.

10. H. Palmour III, T. M. Hare, Rate controlled sintering revisited, pp 17-34 in G. C. Kuczynski, D. P. Uskokovic´, H. Palmour III and M. M. Ristic´, eds., Sintering '85. Plenum Press, New York, 1987.

11. H. Palmour III, Rate controlled sintering of a whiteware porcelain, Ceramic Engineering & Science Proceedings (Amer. Ceram. Soc.), 7 (11-12) 1203-1212 (1986).

12. A. Kingon, S. Chevacharoenkul, S. Pejovnik, R. Velasquez, R. Porter, T. Hare, H. Palmour III, Processing and microstructures of $YBa_2Cu_3O_{7-\delta}$, pp. 335-348 in W. E. Hatfield and J. E. Miller, eds., High Temperature Superconducting Materials: Preparation, Properties and Processing. Marcel Dekker, Inc. New York - Basel, 1988.

13. S. Pejovnik, T. Hare, A. Kingon, R. Porter, H. Palmour III, Sintering and microstructure development in superconducting $YBa_2Cu_3O_{7-\delta}$, pp. 1472-1477 in S. Somiya, M. Shimada. M. Yoshimura and R. Watanabe, eds., Sintering '87.Elsevier Applied Science, London, New York, Tokyo, 1989.

14. S. Pejovnik, R. L. Porter, A. I. Kingon, T. M. Hare, H. Palmour III, Calcination, milling and pressing of sinterable $YBa_2Cu_3O_{7-\delta}$ powders from mixed oxides and carbonates, *ibid.*, pp. 1478-1483.

15. S. Pejovnik, Z. Lengar, V. Hudnik, T. Meden, A. I. Kingon, H. Palmour III, Control of synthesis and ionic states of copper atoms in Y-Ba-Cu-O system, Journal of the Slovenian Chemical Society [*Vestn. Slov. Kem. Drus.*] 35 (4) 379-385 (1988).

16. H. Palmour III, Rate controlled sintering technology for PM and composites, Powder Metal Report, 572-579, September, 1988.

17. M. F. Yan, Microstructural control in the processing of electronic ceramics", Materials Science and Engineering 48 53-72 (1981).

18. R. J. Brook, Additives and the sintering of ceramics, Science of Sintering 20 (2/3) 115-118 (1988).

19. P. Greil, Opportunities and limits in engineering ceramics, Powder Metal. Intl., 21 (2) 40-46(1989)

20. P. Wong, D. R. Messier, Procedure for fabrication of Si_3N_4 by rate controlled reaction sintering, Amer. Ceram. Soc. Bull. 57 (5) 525-526 (1978).

21. W. J. Lackey, P. Angelini, A. J. Caputo, C. E. Devore, J. C. McLaughlin, D. P. Stinton, R. E. Hutchens, Rate controlled technique for calcining and drying, Commun. Am. Ceram. Soc.(7) C-102 - C-104 (1984).

22. L. Schaepelynck, J. M Haussonne, Application de la nouvelle
    dilatométrie: étude théorique du frittage et frittage a vitesse de
    retrait controlée, Science of Ceramics, 12 313 (1984).
23. J. M. Haussonne, L. Schaepelynck, A new concept of dilatometry:
    technology and applications, pp 19-38 in Advances in Ceramics, Vol. 11,
    Amer. Ceram.Soc.(1984).
24. C. Genuist and J. M. Haussonnne, Sintering of BaTiO3: dilatometric
    analysis of diffusion models and microstructural control, Ceramics
    International 14 169-179 (1988).
25. B.-S. Chiou, C.-M. Koh and J.-G. Duh, The influence of firing profile
    and additives on the PTCR effect and microstructure of $BaTiO_3$, J. Mater.
    Sci. 22 (11) 3893-3900 (1987).
26. W. Wersing, H. Wahl and M. Schnöller, PZT-Based multilayer piezoelectric
    ceramics with AgPd-internal electrodes, Ferroelectrics 87 271-294
    (1988).
27. K.-L. Weisskopf, H. Palmour III, N. Claussen, Strengthening of ZrO_2-
    containing Al_2O_3-rich spinel ceramics, presented at Fall Meeting, The
    American Ceramic Society, San Fransisco, CA, October 29, 1984.
28. K. Nieszery, K.-L.Weisskopf, G. Petzow, W. Pannhorst, Sintering and
    strengthening of cordierite with different amounts of zirconia, Science
    of Sintering 20 (2/3) 149-154 (1988).
29. W. Semar, W. Pannhorst, T. M. Hare, H. Palmour III, On the sintering of
    a crystalline cordierite/ZrO_2 - composite," Glastechnische Berichte 62
    (2) 74-78 (1989).
30. A. Kranzmann, P. Greil, G. Petzow, Pressureless sintering of aluminum
    nitride, Science of Sintering 20 (2/3) 135-139 (1988).
31. J.-M. Lihrmann, E. Kostic´, H. Schubert, G. Petzow, Rate-controlled
    sintering of silicon carbide with additives (a) B and C, (b) Y_2O_3 and
    Al_2O_3, this volume.
32. (a) R. M. German and K. W. Lay, eds., Proceedings of Symposium on
    Processing of Metal and Ceramic Powders. The Metallurgical Society of
    AIME, 1982.
    (b) R. M. German, Powder Metallurgy Science. Metal Powder Industries
    Federation, 1984.
    (c) R. M. German, Liquid Phase Sintering, Plenum Press, New York-London,
    1985.
    (d) R. M. German, Particle Packing Characteristics. Metal Powder
    Industries Federation, Princeton, N. J., 1989.
33. J. Beuers, M. Poniatowski, Metal injection molding - a progressive
    shaping process requires sophisticated sintering techniques, pp 230-236
    in S. Somiya, M. Shimada. M. Yoshimura and R. Watanabe, eds., Sintering
    '87. Elsevier Applied Science, London, New York, Tokyo, 1989.
34. H. Palmour III, I. K. Simonsen, H. H. Stadelmaier, The Eutectoids in
    iron - carbon alloys prepared by powder metallurgy, Z. Metallkde. 80 (8)
    563-565 (1989).
35. H. E. Exner, G. Petzow, P. Wellner, pp 352-362 in G. C. Kuczynski, ed.,
    Sintering and Related Phenomena, Plenum Press, New York-London, 1973.
36. T. M. Hare, Statistics of early sintering and rearrangement by computer
    simulation, pp 77-93 in G. C. Kuczynski, ed., Sintering Processes.
    Mater.Sci. Res., Vol. 13, Plenum Press, New York - London, 1980.
37. T. T. Fang, H. Palmour III, Evolution of pore morphology in the
    sintering of powder compacts.[In Press, Ceramics International.]
38. T. M. Hare, K. L. More, A. D. Batchelor, H. Palmour III, Sintering
    behavior of overcompacted shock-conditioned alumina powder, pp 265-280
    in G. C. Kuczynski, A. E. Miller and G. A.Sargent, eds., Sintering and
    Heterogeneous Catalysis. Plenum Press, New York - London, 1984.
39. H. J. Leu, T. M. Hare, and R. O. Scattergood, A computer simulation
    method for particle sintering, Acta Metall. 36 (8) 1977-1987 (1988).
40. F. F. Lange, Sinterability of agglomerated powders, J. Am. Ceram. Soc.
    67 (2) 83 (1984).
41. S. J. Bennison, M.P.Harmer, Grain growth and cavity formation in MgO
    doped Al_2O_3, Advances in Ceramics, VI 171-183 (1983).
42. G. C. Kuczynski, Statistical approach to the science of sintering, pp
    325-337 in Mater. Sci. Res., Vol. 10, Plenum Press, New York-London,
    1975.

43. G. C. Kuczynski, Statistical theory of sintering, Z. Metallkunde, 67 (9)
    606-610 (1976).
44. G. C. Kuczynski, Statistical theory of pore shrinkage and grain growth
    during powder compact densification, pp 233-245 in R. M. Fulrath and J.
    A. Pask, eds., Ceramic Microstructure-'76. Westview Press, Boulder,
    Colorado, 1977.
45. G. C. Kuczynski, Statistical Theory of sintering and microstructure
    evolution, pp 37-44 in D. Kolar, S. Pejovnik and M. M. Ristic´, eds.,
    Sintering - Theory and Practice. Mater. Sci. Monographs, Vol. 14,
    Elsevier, Amsterdam, 1982.
46. H. Palmour III, A. D. Batchelor, K. L. More, T. T. Fang, M. J. Paisley,
    G.T. Goudey, T. M. Hare, The later Kuczynski: an appreciation of his
    'statistical theory' from the experimentalists' viewpoint, Science of
    Sintering (5) 77- 89 (1984).
47. T. T. Fang, H. Palmour III, Useful extensions of the statistical theory
    of sintering.[In Press, Ceramics International.]
48. T. T. Fang, H. Palmour III, Non-destructive characterization of morpho-
    logical development in sintered powder compacts.[In Press, Ceramics
    International.]
49. H. E. Exner, Principles of single phase sintering, in F. V. Lenel, ed.,
    Reviews on Powder Metallurgy and Physical Ceramics (Fruend Publishing
    House, Tel Aviv) (1) 1 (1979).
50. G. R. Anstis, P. Chantikul, B. R. Lawn and D. B. Marshall, A critical
    evaluation of indentation techniques for measureing fracture toughness:
    I. direct crack measurements, J. Am Ceram. Sc. 64 (9) 533 (1981).
51. J. C. Russ, H. Palmour III, T. M. Hare, Direct 3-D pore location
    measurement in alumina [In press, Journal of Microscopy, London].
52. R. Oberacker, K. Dorfschmidt, T. Liu and F. Thümmler, Application of
    rate controlled sintering in the production of ZrO2-based ceramic
    materials, this volume.
53. W. Kaysser and I. S. Ahn, Rate controlled sintering of Ni doped W, in
    Modern Developments [Proceedings of PM '88, Orlando; In Press].
54. R. T. DeHoff, Stereological theory of sintering, this volume.

APPLICATION OF RATE CONTROLLED SINTERING IN THE PRODUCTION OF ZrO_2-BASED CERAMIC MATERIALS

R. Oberacker, K. Dorfschmidt, T. Liu, and F. Thümmler

Institut für Keramik im Maschinenbau
Universität Karlsruhe
Haid-und-Neu-Str. 7, D-7500 Karlsruhe

ABSTRACT

The principles of rate controlled sintering (RCS) were applied to compacts from a $ZrO_2(Y)$-TZP powder. The time-temperature profiles for RCS of this material can be approximately predicted from a few dilatometer experiments carried out at constant rate of heating, if the densification- and heating rates remain within some certain limits. In this way optimum sintering cycles could be designed, leading to advantages concerning density and grain size, compared to conventional isothermal sintering. Limitations for using the full potential of RCS for the investigated practical system arise from processing defects, which cannot be healed during sintering.

INTRODUCTION

Previous investigations on sintering behaviour of Y-TZP-materials have shown that the densification in a conventional sintering cycle (constant rate heating and isothermal sintering at 1100 to 1500 °C) is extremely sensitive to the chosen heating rate dT/dt [1]. Increased heating rates resulted in a decrease in sintered density, indicating that microstructural damages could have developed in the early and intermediate parts of the sintering cycle, hindering densification in the final stage (Fig. 1).

Besides the application of low heating rates during the total heating period, rate controlled sintering (RCS) could possibly solve this problem. Rate controlled sintering was shown in literature to result in enhanced microstructures for a variety of ceramic materials, especially when three stage RCS-profiles are used, beginning on a high level of densification rate which is reduced in the intermediate and later stages of the process [2]. Such RCS cycles can be realized optimally by a computer controlled closed loop system [3]. In the case of so called "well behaving materials", the RCS-temperature-time profile can also be calculated on the basis of a few dilatometer experiments carried out at constant rate of heating (CRH). This is most interesting for practical application of RCS, as it allows to establish RCS-sintering schedules for industrial furnaces without dilatometer control. Therefore the applicability and the consequences of this procedure were investigated for the Y-TZP material mentioned above.

Science of Sintering
Edited by D. P. Uskoković *et al.*
Plenum Press, New York

EXPERIMENTAL PROCEDURE

The experiments were carried out with a commerical press grade powder of 3 mole-% Y_2O_3-stabilized ZrO_2 (TOSOH, Japan). The powder was compacted into plates of 65 x 45 x 8mm, which were sealed into PVC and isostatically recompacted at 600 MPa. The green density of these specimens was 3.16 g/cm^3, corresponding to 51.9% of the theoretical density.

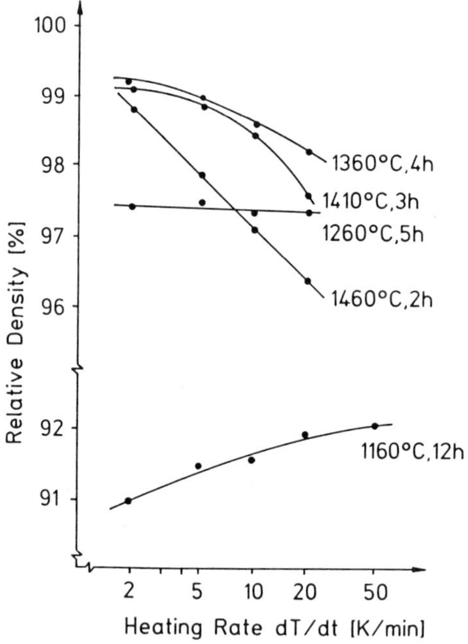

Fig. 1. Influence of heating rate in conventionally sintered Y-TZP [1].

For the dilatometer experiments specimens of 8 x 8 x 15mm were cut from the green plates. The dilatometer measurements were carried out in a commerical dilatometer system (NETZSCH 402) with Al_2O_3 sample holder. Flowing air was used as sintering atmosphere. The change in length of the specimens was monitored by a PC and converted into the rel ative shrinkage $(\Delta L/L_0)$ of the specimen, taking into account the thermal expansion of the specimen holder and of the specimen itself. The holder was calibrated against a sapphire standard. The expansion of the specimen was measured on dense (post-HIPed) specimens of the same composition. The shrinkage rate $d(\Delta L/L_0)/dt$ was calculated by differentiation of $(\Delta L/L_0)$. The relative density D and the densification rate dD/dt can then be calculated from the initial (green) density D_0 and $(\Delta L/L_0)$ by equations (1) and (2):

$$D = \frac{D_0}{[1 - (\Delta L/L_0)]^3} \qquad (1)$$

$$\frac{dD}{dt} = \frac{3 \cdot D_0}{[1 - (\Delta L/L_0)]^4} \cdot \frac{d(\Delta L/L_0)}{dt} \qquad (2)$$

This procedure requires an isotropic shrinkage behaviour and constant specific volumes. Both conditions were fairly fulfilled in our experiments. The relative density was calculated by assuming a theoretical density of 6.09 g/cm^3.

The final density of the materials was measured by the Archimedian method. Microstructural investigations were carried out by SEM on polished and thermally etched (1320 °C/1 h) specimens. The average grain diameter was estimated from SEM-micrographs by counting about 500 grains. For the evaluation of mechanical properties the complete plates were sintered and cut into specimens of 3.5 x 4.5 x 45 mm. The strength was measured by four point bending with 40 and 20 mm outer and inner span at a loading rate of 2 N/s.

EXPERIMENTAL RESULTS

Densification in Constant Rate Heating (CRH) Experiments

The heating rate was varied between 1 and 40 K/min. Fig. 2 shows the densification rate of the materials in form of an Arrhenius plot, the so called "kinetics field response" [4]. The numbers on the curves are representing the actual density D. The isodensity lines can be fitted in a good approximation by straight lines, which holds for a wide range of

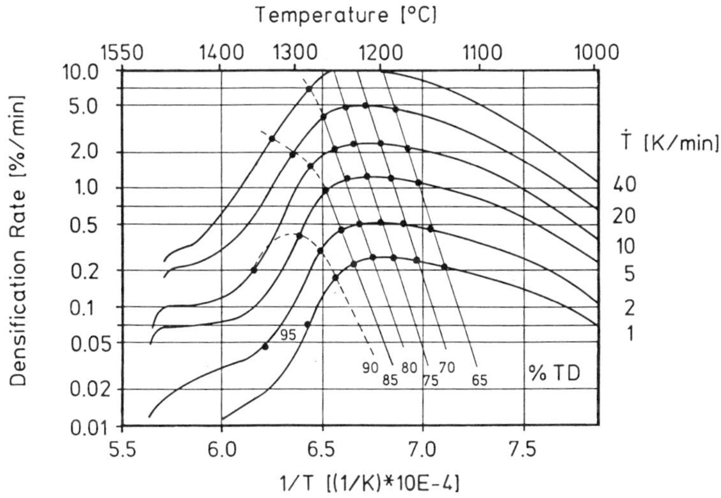

Fig. 2. "Kinetics field response" of Y-TZP.

relative densities. The activation energies, calculated from the slopes of the isodensity lines are in the range of 300 to 400 kJ/mole at the density level of 65 to 85 %. Only at higher densities, the isodensity curves deviate from linearity, which becomes more pronounced at the higher heating rates. The densification rate shows its maximum at densities of 70 to 80 % TD, depending on the heating rate. This maximum exceeds 10%/min at the heating rate of 40 K/min. A densification rate of 1%/min, which is often used in a first approximation as the "maximum safe rate" demands for a heating rate of less than 5 K/min during the critical density interval.

The microstructure at the end of the heating stage, which was finished at 1500 °C is shown in Fig. 3. The only significant difference is the size and number of pores, which increases with increased heating rate. But even at dT/dt = 40 k/min, the pore size remains in the order of the grain size. No microstructural damages able to hinder densification in a subsequent isothermal stage become visible from these micrographs.

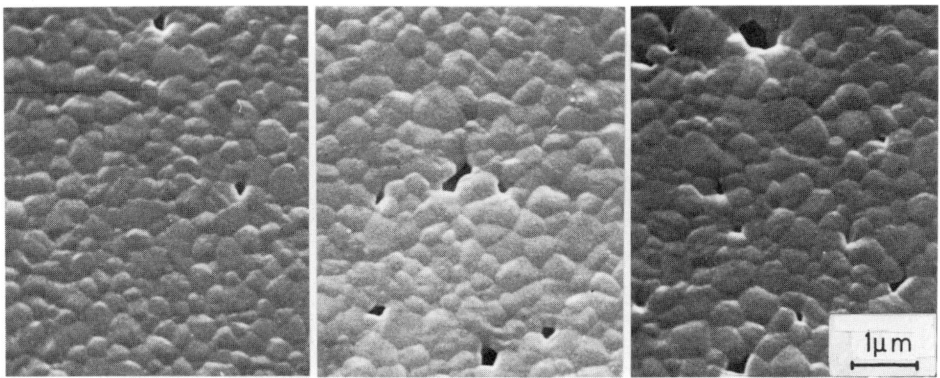

Fig. 3. Microstructure of CRH-sintered Y-TZP (1; 10; 40; K/min left to right).

Rate Controlled Sintering (RCS) Experiments

This type of kinetics field response should make it possible to calculate temperature-time schedules for any given densification regime, as long as the densification rate and the temperature remain within the limits where dD/dt is a one-valued function of the heating rate at a given density. Usually the desired densification behaviour is formulated as the dependence between densification rate and actual density. For every point of this function, a corresponding value of 1/T can be taken from the kinetics field response, resulting in a function T vs. D. The temperature-time function itself follows then by integration of the corresponding densification rate, which can be approximated by the summation of constant average densification rates within narrow temperature intervals. The temperature-time schedule can then be realized by a usual programmable temperature controller.

In this way, different rate controlled experiments were carried out, schematically plotted in Table 1. The first cycle is a three stage process as proposed by H. Palmour et al with a density-time profile with a) linear, b) slower linear, and c) logarithmic decreasing shape [2]. For comparison, cycles with constant, as well as increasing or decreasing densification rates were chosen. The densities reached in these experiments are included in Table 1. They are very close to each other. The highest value is reached with a constant densification rate at a relatively low level of 0.3 %/min. Increasing dD/dt leads to a reduction in sintered densities. The density reached with the three stage profile at dD/dt = 1 %/min in the early and 0.5 %/min in the intermediate stage falls between the values reached at the corresponding constant rates. Cycles with increasing or decreasing densification rates at a maximium dD/dt of 0.45 %/min give a slightly enhanced density compared to the three stage cycle. Taking into account the results of conventional sintering cycles with constant rate heating and 3 h of isothermal sintering at 1410 °C (Fig. 1), it becomes visible, that the maximum densification rate seems to be the main parameter which determines the final density.

Table 1. Density of Y-TZP after rate controlled sintering.

Densification Mode	$\dot{D}_{max}$ [%/min]	Sintered Density [% TD]
	1.0	99.4
	0.3	99.9
	0.6	99.6
	1.0	99.2
	0.45	99.8
	0.45	99.7

The average grain size, which should be a sensitive parameter with regard to changes in the sintering cycles is 0.35 μm without significant differences within the rate controlled sintered materials. The conventionally sintered materials on the other hand exibit a significantly increased grain size of about 0.5 to 0.6 μm over the whole range of heating rate variation. This is presumably a consequece of the relatively long soaking time, as a part of the rate controlled experiments end up with a much higher temperature of about 1500 °C, but only for a short time interval.

Strength Properties of Rate Controlled Sintered Materials

Two rate controlled cycles were realized with the complete plates of 65 x 45 x 8 mm in a laboratory chamber furnace with a programmable temperature controller, according to the cycles calculated from the kinetics field response in Fig. 2. The first cycle is constant

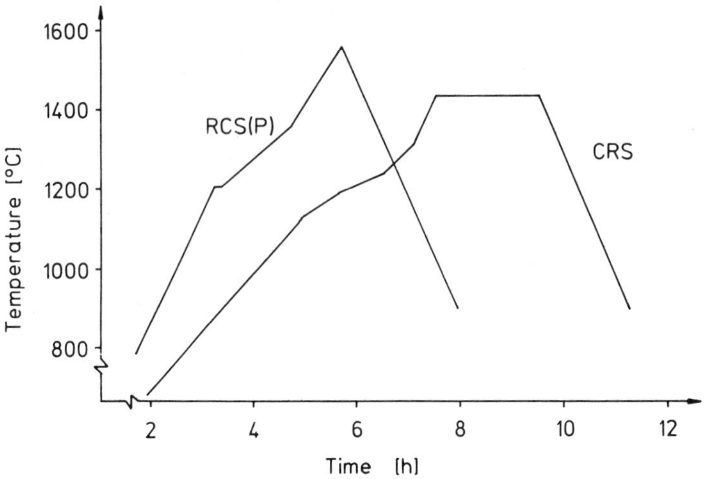

Fig. 4. Temperature-time schedule calculated for constant rate sintering (CRS) and 3-stage rate controlled cycle (RCS(P)).

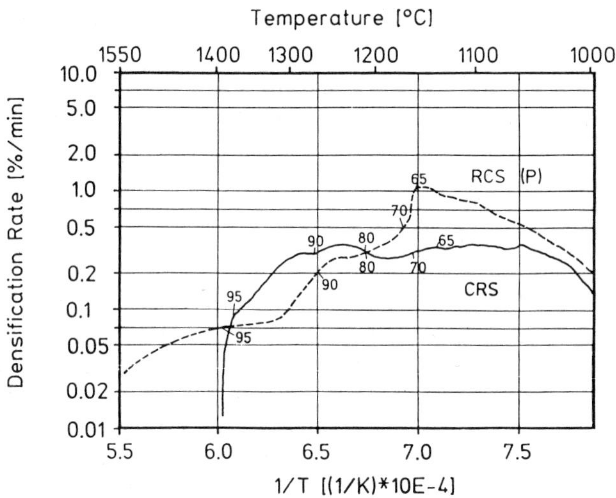

Fig. 5. Densification rate vs. 1/T according to schedule from Fig. 4.

densification rate sintering (CRS) with dD/dt = 0.3%/min. The second cycle (RCS(P)) is a three stage process with a densification rate of about 1% up to a density of 65%, followed by dD/dt = 0.3%/min up to 85% and a decreasing rate for the final densification. The temperature-time profiles are included in Fig. 4. The corresponding densification rate is

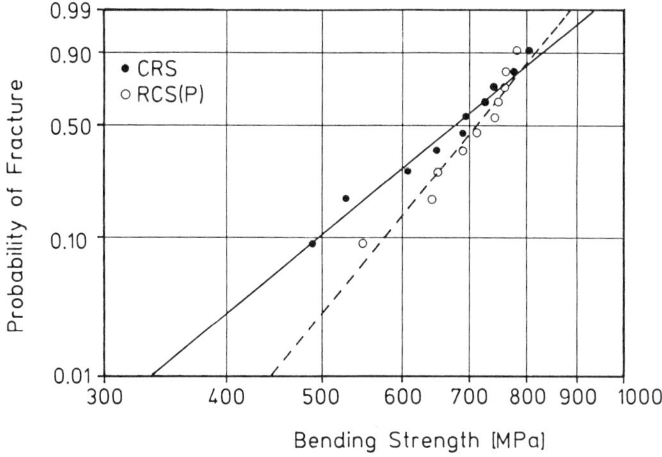

Fig. 6. Strength distribution of CRS and RCS(P) specimens.

given in the Arrhenius plot in Fig. 5. The intended constant rate cycle is met fairly well, while the three stage cycle shows some deviations from the intended shape, which could be avoided only by use of a higher number of time segments, which were limited to 10 in the chamber furnace used for sintering of the plates. As these deviations are not expected to be essential, the plates were sintered under the same conditions as used in the dilatometer experiments in Fig. 5.

The density of the sintered plates was on a significantly lower level as that of the small dilatometer specimens and reached only 98.4% for the CRS- and 98.3% for the RCS-material. The Young's modulus was at 210 and 211 GPa, respectively. The strength is plotted in Fig. 6 in form of a Weibull distribution. The distributions of the two materials deviate at lower strength levels, while they are very close to each other at higher strength values. From this figure, the three stage RCS-cycle seems to be the superior one, despite of the slightly lower density reached with both types of specimens -the small dilatometer specimens and the bigger plates- in this cycle.

DISCUSSION

RCS, either in a three stage cycle as proposed in literature, or at constant densification rates below 1%/min was shown to result in improved densities of the investigated Y-TZP compared to conventional sintering, even when very low heating rates are used. Nevertheless, a lot of questions arise from the results of the experiments presented above.

Apart from the very early stages where no densification takes place, the heating rates in the RCS-experiments did not exceed 7 K/min. Therefore, it becomes not really clear, whether the densification rate or some other consequence of high heating rates are responsible for the behaviour observed in Fig. 1. Such other consequences are, e. g., stresses arising from thermal gradients at high heating rates. Another question is, whether a temperature-time profile for a given RCS-schedule really can be predicted on the basis of the limited number of experiments carried out to establish the kinetics field response in Fig. 2. This proceeding assumes that it is possible to switch over from one densification curve in this figure to every other at any desired temperature, and follow the new curve by simply keeping the corresponding heating rate. In last consequence, this means, that the history of the specimen prior to a regarded moment has no influence on its further behaviour. And this is in contradiction with the observation in Fig. 1.

For this reason, some experiments were carried out where the specimens were cooled down at different density levels from a constant heating rate of 10, respectively 40 K/min to the densification curve reached with 1 K/min, and subsequently were heated with 1 K/min. The results are presented in Fig. 7 for the experiments started with dT/dt = 10 K/min. Only the specimen switched over at 60% TD behaves approximately like expected. The densification curve after reduction of the densification rate follows the corresponding curve reached without an enhanced initial densification rate. But already here, a slight difference can be observed at densities above 97%. If dD/dt is reduced in a later stage, the densification rate remains below that of the CRH-cycle at the same density levels. The 90% and 95% density level is therefore reached at much higher temperatures than without an initially increased rate. If the initial heating rate is chosen to 40 K/min, it is impossible to meet the 1 K/min curve again, even by switching over at 60% of TD. Although the experiments were finished at the same final temperature of 1530 °C, the residual porosity of the materials increases with the density level at which the rates are reduced. The size of the remaining pores is comparably small, as visible from Fig. 8, and no signs for differential sintering or cracks induced by thermal stresses become evident. Therefore, the explanation of the negative influence of too high initial heating rates may rather be

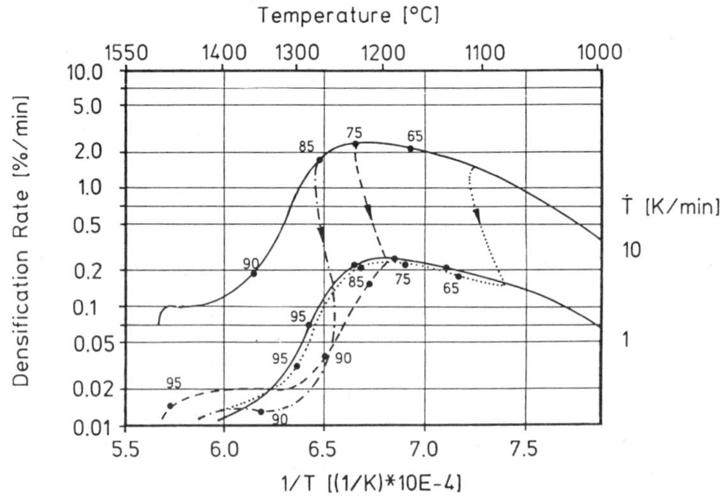

Fig. 7. Consequences of change in heating (and densification) rate.

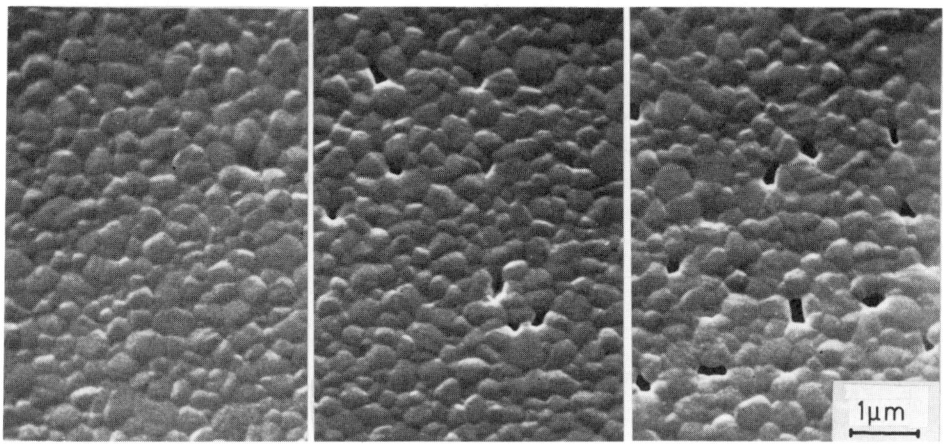

Fig. 8. Microstructure after sintering with changed heating rates (change at 60; 75; 85; %TD, left to right).

entrapped gases or an incomplete rearrangement of the particles than the effects mentioned above.

Nevertheless, it becomes clear that a densification rate of 1 %/min is indeed a first approximation for a maximum safe rate as described in literature. Whether enhanced densification rates during the early stages are advantageous compared to constant densification rates remains in doubt for the regarded material, although the strength distribution in Fig. 6 indicates such advantages. The differences in strength at the lower strength level clearly result from processing faults prior to sintering as can be seen from the fractographs in Fig. 9. These imperfections and the small grain size make it problematic to draw a proper conclusion on this question.

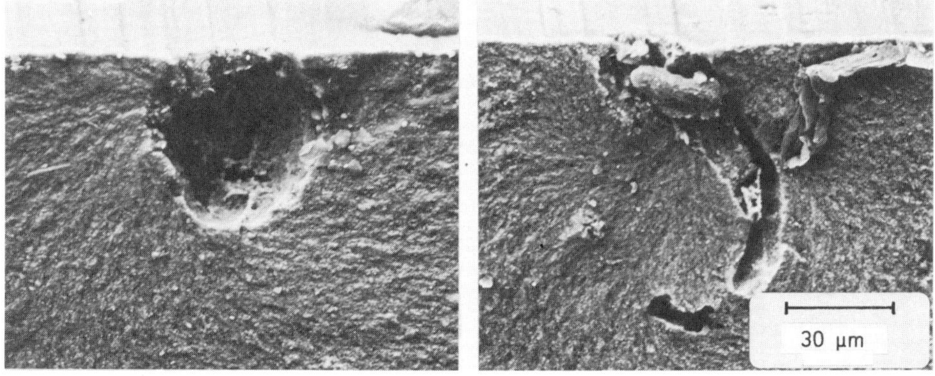

Fig. 9. Processing flaws at the lower end of strength level (CRS left; RCS(P) right).

The different final densities of the dilatometer specimens and the larger plates make another problem of RCS visible. Enhanced densification rates, which require enhanced heating rates lead to temperature gradients in larger parts. This is especially a problem with materials of low thermal conductivity like ZrO_2. For the dimension of our plates, the temperature difference between surface and center can be roughly estimated to be 10 to 50 K at heating rates between 2 to 10 K/min and a surface temperature of 1000 °C. This increases possibly the problem of internal stresses and of gas entrapment, due to a faster densifying surface.

CONCLUSIONS

The concept for the estimation of RCS-cycles from the so called kinetics field response, established by a few CRH-dilatometer experiments in principle is applicable for the investigated Y-TZP material. Limitations are due to the "memory" of the specimens to their densification history, if they are densified at rates above 1 %/min to a higher density level as 60% TD.

The main parameter for the final density of the materials is the maximum densification rate. This parameter has to be kept below 1 %/min. A variation within these limits leads only to small differences in properties, which cannot be correlated to the densification cycle, as they are overcompensated by processing flaws from production steps prior to sintering. Possible benefits of optimized sintering cycles can be only evaluated when defects in microstructure caused by powder packing or shaping are thoroughly avoided.

The low thermal conductivity of ZrO_2 leads to limits for RCS, as high temperature gradients occur in bigger parts at enhanced heating rates.

LITERATURE

[1]   T. Liu and K. Dorfschmidt; Sinteroptimierung und -verhalten von teilstabilisiertem ZrO_2, Annual Meeting DKG, München 18.-20. Okt. 1988.

[2]   H. Palmour III, M. L. Huckabee and T. M. Hare; Microstructural Development during Optimized Rate Controlled Sintering, in Ceramic Microstructures '76, R. M. Fulrath, J. A. Pask, ed., Westview Press, Boulder CO, 1977.

[3]   A. D. Batchelor, M. J. Paisley, T. M. Hare and H. Palmour III; Precision Digital Dilatometry: A Microcomputer-Based Approach to Sintering Studies, Mat. Sci. Res., 17: 233 (1984).

[4]   H. Palmour III; Rate Controlled Sintering Technology for Powder Metals and Composites, Seminar on Advanced Sintering, P/M '88, Orlando, June 5-10, 1988.

RATE CONTROLLED SINTERING OF SiC WITH ADDITIONS OF (a) Al_2O_3 $+Y_2O_3$ (b) B + C ; (c) B_4C + C

J.-M. Lihrmann[*], P. Halary, E. Kostic[**] and H. Schubert

Max-Planck Institut für Metallforschung, Heisenbergstrasse
5, D - 7000 STUTTGART 80 (West Germany)
[*]Current Address : Université Paris XIII, CNRS-LIMHP
Avenue J.B. Clément, F 93430 - VILLETANEUSE (France)
[**]Current Address : Boris Kidric Institute of Nuclear
Sciences, Materials Department, 11001 BELGRADE (Yugoslavia)

ABSTRACT

Densification of α-silicon carbide with different types of additives
has been performed in view of the rate-controlled sintering (R.C.S) concept
developed by Professor Palmour. Kinetic results suggest that this concept
should apply to high temperature materials such as SiC, whether
densification predominantly occurs by liquid state or by solid state
diffusion. However, a minimum amount of additives seems necessary in order
to observe a controllable and reproducible sintering behavior, although
high final densities may be achieved with lower amounts of additives. The
homogeneous distribution of additives throughout the sample is thus thought
to be a key issue in the rate-controlled sintering of silicon carbide.

INTRODUCTION

After early recognition that powders of silicon carbide (either α-Sic
or β-Sic) could be densified under pressure with additives (1),
pressureless sintering of β-Sic was first reported in 1973 by Prochazka
(2). Simultaneous small additions of boron (introduced as elemental B, B_4C,
or $LiBH_4$) and carbon (as polymethyl phenylene or carbon black) ensured a

density of 3.08 (96 % th.d.) after heating for 15 minutes in Argon at 2040°C ± 20°C. Later the pressureless sintering of α-SiC with B and C was reported by Carborundum Company (3). Densification of α-SiC requires temperatures in excess of 2050°C (4) but final microstructures are in general more homogeneous because they are unaffected by the $\beta \rightarrow \alpha$ acicular transformation (5). Successful attempts to lower the densifying temperature of silicon carbide to the range 1950°C - 2050°C have been made by using moderate pressures and Al_2O_3 as additive (6), or more recently, by using normal pressure and additives in form of Al_2O_3 - Y_2O_3 or Al_2O_3 - Dy_2O_3 powder mixtures (7).

Our experiments were attempts to apply the concept of rate - controlled sintering to α-silicon carbide doped with three types of additives:(a) Al_2O_3 + Y_2O_3, (b) B + C , (c) B_4C + C. Developed by H.Palmour in the last 25 years (8), rate controlled sintering (RCS), in contrast to conventional temperature sintering (CTS) which is characterized by a constant rate of heating, refers to a three-stage densification profile, thought to optimize the structural changes occuring during sintering, such as gas entrapment, exaggerated grain growth, etc ... Established for Al_2O_3 and a variety of ceramic oxides (10) with a potential extension to other materials, this profile consists of: a. a fast densification rate (1%/min) up to D = 75% theoretical density ; b. a slower densification rate (0.3-0.4 %.min^{-1}) to about D = 85% t.d.; c. log-decreasing densification rates up to D_f in a time $\geq$ 50% of total sintering time. In the case of high-purity, 0.1% MgO-doped alumina (10), such a RCS profile has been shown to result in substantially finer and more uniform grain sizes than those obtained by CTS. Our purpose was to examine this possibility in the case of variously doped α-silicon carbide.

EXPERIMENTAL

The starting materials used in this study were commercial powders of α-SiC (A-10, Hermann C.Starck), B (cristalline Boron,Aldrich), Carbon black (FW 18 and FW 200, Degussa), B_4C (F 1500, H.C.Starck), Al_2O_3 (A-16, Alcoa) and Y_2O_3 (russian origin). FW 18 was used in mixture 2 (98 wt% SiC + 0.5 wt% B + 1.5 wt% C), whereas the variety FW 200 was used in mixture 3 (96.86 wt% SiC+0.64 wt% B_4C + 2.5 wt% C). The α-SiC had as major impurity 0.85 wt% oxygen. The specific surface areas (m^2/g) measured after outgassing for 3 h at 400°C were as follows:α-SiC, 16.1 ± O.3; B, 5.79 ± 0.05; FW 18, 274 ± 1; FW 200, 460 ± 2.

Additive 1 was 10 wt% of the eutectic composition 60 wt% Al_2O_3- 40 wt% Y_2O_3 (11), since it has recently been found to densify α-SiC at 1950°C with an optimum concentration of second phase(7).Mixture 2 was selected (5) from results indicating that it yielded the highest final density when heated at 2060°C for 30 minutes. Eventually, mixture 3 had the same boron content as mixture 2 whereas the carbon content was purposely increased.

Mixture 1(90 wt% α-SiC + additive 1) was vibratory-milled in water for 2 h in proportions 50 g/50 ml, whereas mixture 2 was dispersed in isopropanol and attritor-milled for 1 h at 625 rpm using SiC media. Mixture 3 was dispersed in a solution of ethanol containing 5wt% diluted NH_3 (25%) and stirred for 15 min at 10000 rpm using a high velocity stirrer. The solvent was eliminated with a rotating evaporator and the powder, dried at 60°C.

After sieving at 120 μm, powder mixtures packed in rubber molds were cold-isostatically pressed (200 MPa, 3 min) into cylinders (14 mm diameter X 12 mm height) of approximately 62% green density. Heating was done in an Astro furnace equipped with a high temperature dilatometer, and changes in length of samples were automatically recorded. Heating rates within the relevant ranges of temperature varied between 2 and 20 K/min; maximum temperature achieved was 2240°C. The temperature inside the graphite furnace was measured with a boron- graphite thermoelement BCT-2 (Astro Industries, USA). Samples were heated in a graphite crucible and covered with a graphite lid connected to the dilatometer head with a graphite rod. Those of mixture 3 were exposed to static Argon, all others to flowing Argon (flow rate 1 cc/s). After sintering all samples were characterized by density measurements and microstructural observations.

RESULTS

Applying the RCS concept requires a series of dilatometric curves recorded at different heating rates. These curves are then exploited as detailed in Refs 8-10. The runs performed are characterized in Table I in terms of final density and weight loss.

Mixture 1

Exploiting dilatometric curves yielded a set of points of coordinates $(\frac{dD}{dt}, \frac{1}{T})$ which could be satisfactorily joined by a family of straight iso density lines as shown in Fig.1, referred to as the kinetic field response or kinetic plot. As yet no attempt has been made to convert Fig.1 into RCS profiles.

An additional experiment consisted of heating a sample to 1980°C at 3.33 K/min followed by a 20 min hold at 1980°C. The microstructure of this sample is shown in Fig.2a in comparison with sample 1.2 (Fig.2b).

Table 1 . Characterization of typical runs performed

	MIXTURE 1 (calculated t.d. 3.28)				MIXTURE 2 (t.d. assumed 3.21)			MIXTURE 3 (t.d. assumed 3.21)			
Sample	1.1	1.2	1.3	1.4	2.1	2.2	2.3	3.1	3.2	3.3	3.4
Heating Rate (K/min)	2.0	3.3	5.0	10.0	10.0	10.0	10.0	2.5	5.0	10.0	20.0
Maximum Temperature ($^{\circ}$C)	2000	2000	2000	2000	2240	2240	2240	2200	2200	2200	2200
Hold at T_{max} (min)	20	20	20	20	20	20	20	30	30	30	30
Density (% t.d.)	88.2	95.0	96.6	97.4	95.0	92.9	93.4	96.6	95.9	95.6	92.3
Weight Loss (%)	10.64	9.95	9.47	8.59	2.13	2.26	2.17	3.10	3.08	3.07	2.96

Mixture 2

Fig.3 illustrates an important scattering in the kinetic field response of three identical runs. All three samples were prepared from the same batch of attritted powder, identically packed, simultaneously pressed, heated according to the same heating schedule and exposed to the same environment.

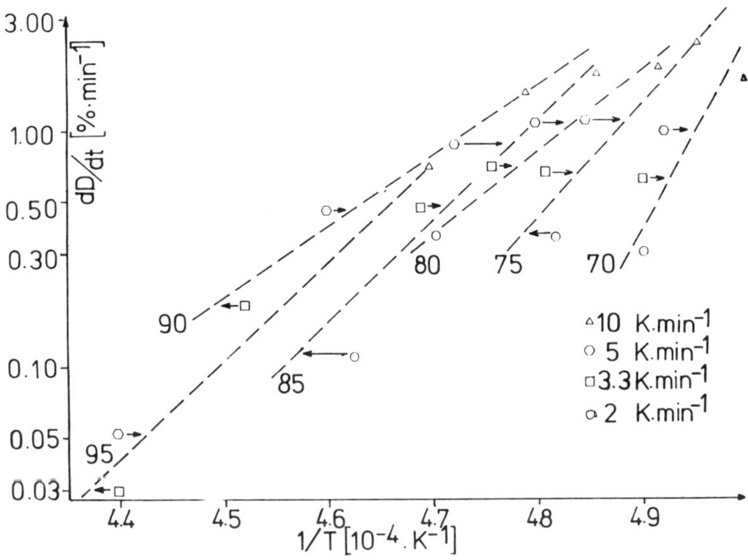

Fig. 1. Kinetic field response obtained for samples of mixture 1.

Mixture 3

Here as in case 1 the points $(\dfrac{dD}{dT},\dfrac{1}{T})$ could reasonably well be joined by a set of straight isodensity lines (Fig.4). Fig.4 permitted to define an optimized temperature-vs-time heating profile, whose experimental responses are given in Fig.5 (dotted curve) and Fig.6. The RCS schedule has resulted in a final density of 3.06 (95.2% t.d.) and a microstructure depicted on Fig. 7a. For comparison Fig. 7b shows the microstructure of sample 3.3, characterized by a similar density (95.6% t.d.) and corresponding to the C.T.S. profile of Fig.5 .

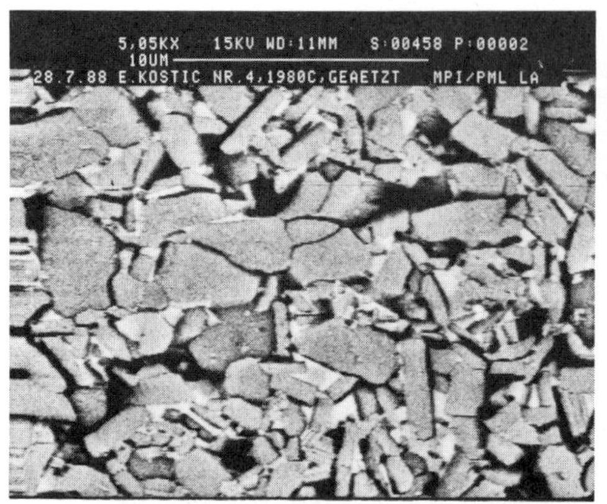

Fig.2a. SEM micrograph of a sample
(mixture 1) heated to 1980°C

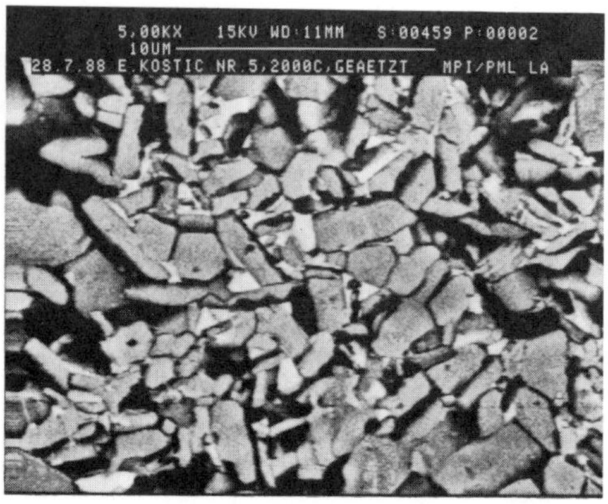

Fig.2b. SEM micrograph of a sample
(mixture 1) heated to 2000°C

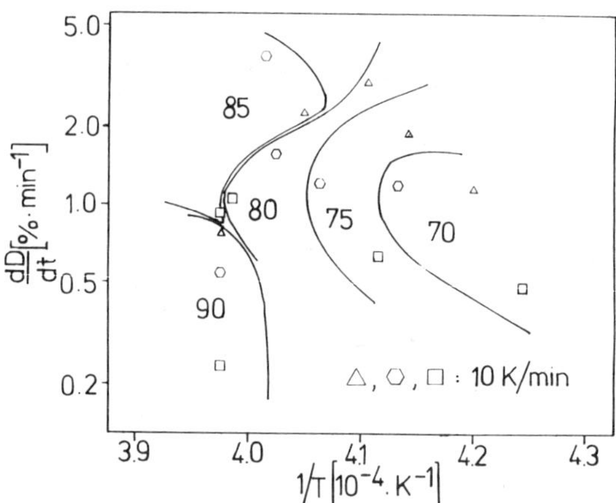

Fig. 3 . Illustration of the scattering observed in the kinetic field
response of samples of mixture 2

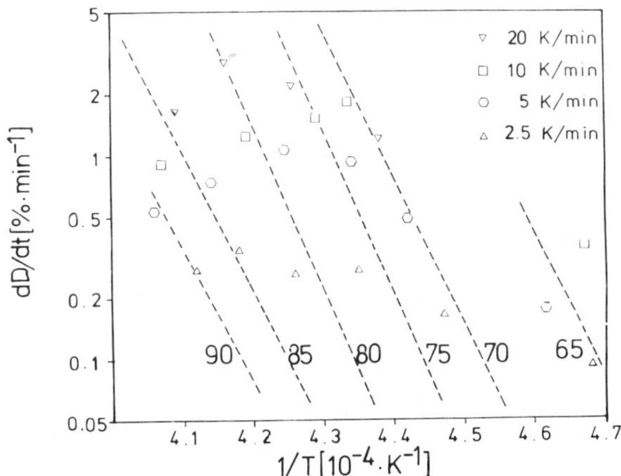

Fig. 4. Kinetic field response of samples of mixture 3

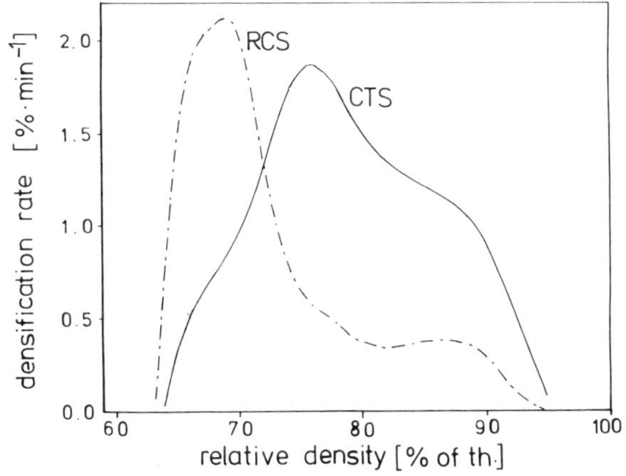

Fig.5. Densification rates vs Density curves in the case of rate-
controlled sintering and conventional sintering

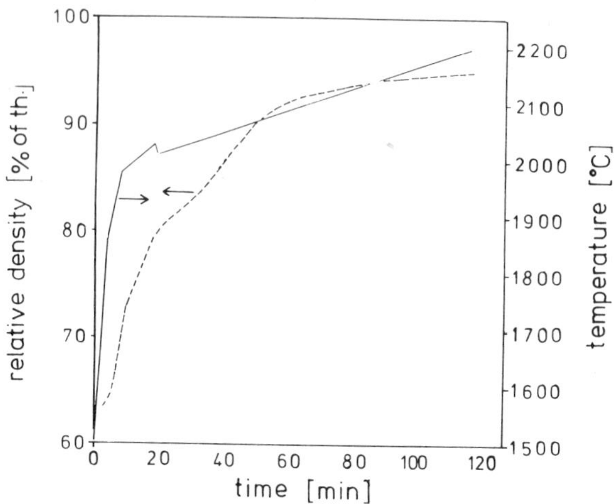

Fig. 6. Relative density vs time and Temperature vs time profiles in the
case of rate-controlled sintering

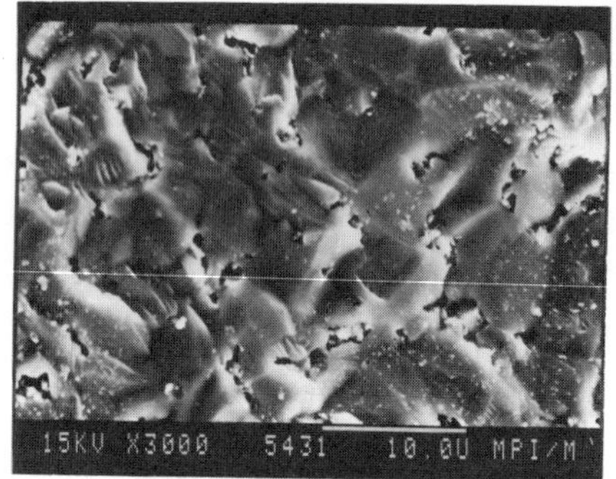

Fig.7a. SEM micrograph of a sample
(mixture3) heated after the
RCS profile of Fig. 5

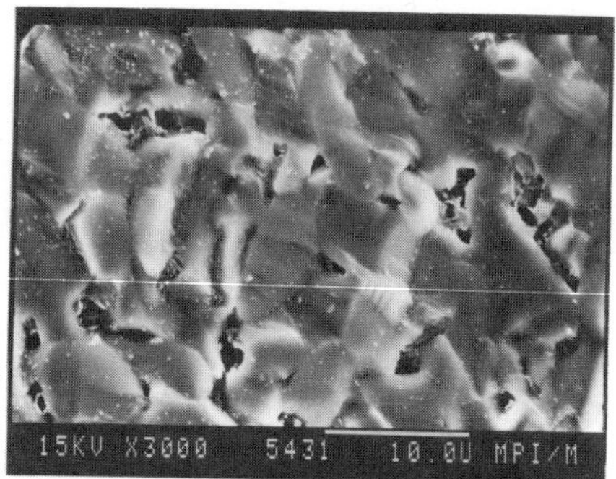

Fig.7b. SEM micrograph of a sample
(mixture3) heated after the
CTS profile of Fig. 5

DISCUSSION

Mixture 1

The acceptable alignment of isodensity points in Fig.1 indicates that this kinetic field response should be exploitable into RCS schedules. However such attempts have not been done as yet due to the short amount of time available.

Importantly, the slopes of all lines are such that densification rates decrease with increasing temperatures. This strongly suggests that the predominant mechanism in the densification of α-SiC with additive 1 is a particular case of liquid phase diffusion, whereby the amount of liquid phase diminishes as the temperature increases. As shown in the next paragraph this is indeed what is observed for additive 1, which corresponds to the eutectic of the system $Al_2O_3 - 3\ Y_2O_3 . 5\ Al_2O_3$ (the latter compound known as Yttrium-Aluminum Garnet or YAG).

Eutectic liquid forms at 1800°C. Above 1800°C the liquid phase is thought to separate into constituent units of Al_2O_3, which increasingly evaporate as the temperature increases, and of YAG, which increasingly crystallize as the temperature increases. As a net result, the amount of liquid phase importantly decreases with increasing temperature and the liquid phase diffusion is consequently reduced.

Both the vaporization of Al_2O_3 and the crystallization of the garnet phase have been reported in the literature (7,12). In our study, evidence for the vaporization of Al_2O_3 can be seen in the weight losses of samples 1.1 through 1.4 (Table I), which decrease as the time of exposure to high temperature increases. The overall decrease in the amount of liquid phase as the temperature increases can be appreciated by comparing Fig.2a and Fig.2b. It is thought that the microstructure shown in Fig.2a contains a higher amount of secondary phase than the one shown in Fig.2b, which we explain by the 20°C difference in the heat treatments of the samples (1980°C and 2000°C, respectively). As for the nature of the secondary phase, it has been satisfactorily identified as Y.A.G. by semi-quantitative elemental mapping.

Mixture 2

It has been firmly established (2,13) that densification of silicon carbide is preceded by surface reactions of carbon and selective segregation of boron. Differential sintering rates are introduced (14) as soon as the concentration of boron and/or carbon is unequal around the particles of SiC.

In this context agglomerates of boron can be frequently observed: Fig.8 illustrates a boron defect, the size of which is 18 times the average B particle size. Boron inclusions of size 60μm have been reported elsewhere (15). Nonuniform particle size distribution of silicon carbide further enhances differences in sintering rates. It is believed that the amount of additive 2 (0.5 wt% B + 1.5 wt% C) is enough to ensure densification of α-SiC to high final densities, as is indeed observed, but insufficient to ensure a reproducible sintering rate from one sample to another, as is illustrated by the scattered data points of Fig.3, due to the inhomogeneous distribution of B and C particles. The amounts of B and C in mixture 2 seem therefore inappropriate to rate-controlled sintering.

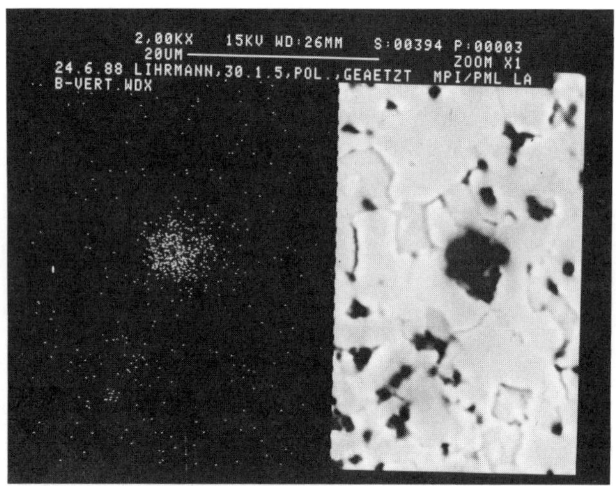

Fig. 8. SEM micrograph of a boron agglomerate in a sample of mixture 2 and
characterization by WDX .

Mixture 3

The kinetic field response of Fig.4 is satisfactory indicating that the amount of additives in mixture 3 is appropriate to rate-controlled sintering.

Firstly we notice from the isodensity lines of Fig.4 that in contrast to mixture 1, the densification rates increase with increasing temperatures. In this case the predominant mechanism in the densification of α-SiC is therefore solid state diffusion, which does not preclude minor contribution of a transient liquid phase such as a B_4C-SiC eutectic melt (16).

From Fig.4 a tentative, non-linear temperature-vs-time heating

schedule was derived, which was corrected by successive trial-and-error
experiments. It can be seen (Fig.5) that the RCS curve which resulted from
these experiments contained the 3 stages sought, but was not optimal in
that the first stage occurred at too high a densification rate whereas the
last one was too short. The non-ideal profile combined with the slight
scattering of the isodensity points (Fig.4) may at least partially explain
the similarity of microstructures 7a and 7b, although porosity appears more
finely distributed in the case of RCS. Eventually, it is worthwhile
noticing the temperature decrease near 80% t.d. for a short amount of time
(Fig.6).

SUMMARY AND CONCLUSIONS

α-silicon carbide doped with the appropriate amounts of Al_2O_3 and Y_2O_3
can be reproducibly densified by liquid phase diffusion at temperatures not
exceeding 2000°C (mixture 1). Due to the decrease in the amount of liquid
phase the total time of high temperature exposure must be limited.

Sintering with B and C (mixture 2), or with B_4C and C (mixture 3) is
controllable provided that a minimum amount of additives, near 3 wt%, be
used to ensure a complete and homogeneous distribution throughout the
sample. Higher temperatures are needed than with mixture 1. When these
conditions are fulfilled, the concept of rate-controlled sintering applies.
Encouraging results have been observed with mixture 3, but both for
mixtures 1 and 3, more experiments need to be done in order to determine
the optimal RCS profiles. In mixture 2, the amount of carbon is most likely
insufficient and should be increased to 2.5 wt%.

ACKNOWLEDGEMENTS

Authors wish to thank Professor Petzow for the opportunity to conduct
this research at PML. They are indebted to Mr. Passing and Mr. Bazin for
assistance in powder preparation and dilatometer runs, and to Mr. Labitzke
for SEM observations. Fruitful discussions with Professor Brook are
gratefully acknowledged.

REFERENCES

1. R. A. Alliegro, L. B. Coffin and J. R. Tinklepaugh, "Pressure-Sintered
   Silicon Carbide", J.Am.Ceram.Soc., 39:386 (1956)
2. S. Prochazka, "Sintering of Silicon Carbide", in "Proceedings of the
   Conference on Ceramics for High Performance Applications", Eds. J.J.
   Burke, A.E. Gorum and R.N. Katz, Brook Hill Publ. Co., Hyannis,
   Massachusetts (1974)

3a. J. A. Coppola and C. H. Mc Murtry, "Substitution of Ceramics for Ductile Materials in Design", in "Proceedings of the National Symposium on Ceramics in the Service of Man", Carnegie Institution, Washington (1976)

3b. J. A. Coppola, H.A. Lawler and I.H. Mc Murtry, U.S.Patent 4123286, Oct.1978

4. S. Dutta, "Densification and Properties of α-silicon carbide", J.Am.Ceram.Soc., 68(10):C269 (1985)

5. W. Bocker and H. Hausner, "The Influence of Boron and Carbon Additions on the Microstructure of Sintered Alpha Silicon Carbide", Powder Met. Int., 10(2):87 (1978)

6. F. F.Lange, "Hot-Pressing Behavior of Silicon Carbide with Additions of Aluminum Oxide", J.Mat.Sci., 10:314 (1975)

7. E. Kostic, "Sintering of Silicon Carbide in the Presence of Oxide Additives", Powder Met. Int., 20(6):28 (1988)

8. H. Palmour III, "Rate-Controlled Sintering for Ceramics and Selected Powder Metals", this Proceedings

9. H. Palmour III, "Practical Applications of Dilatometric Data", Internal Report, MPI-PML, Stuttgart (July 1984)

10. H. Palmour III, M.L. Huckabee and T.M. Hare, in "Proceedings of the Conference on Ceramic Microstructures", Eds. R. M. Fulrath and J.A. Pask, Westview Press, Boulder, Colorado (1977)

11. E. M. Levin and H.F. Mc Murdie, in "Phase Diagrams for Ceramists", 1975 Supplement, Ed. and Publ. the American Ceramic Society, Westerville, Ohio (1975)

12. K. Suzuki, "Pressureless Sintering of Silicon Carbide with Addition of Aluminum Oxide", Reports Res. Lab. Asahi Glass Co., 36(1):25 (1986)

13. G. Greskovitch and J.H. Rosolovski, "Sintering of Covalent Solids", J.Am.Ceram.Soc., 59(7-8):336 (1976)

14. M. N. Rahaman, L.C. de Jonghe and R.J. Brook, "Effect of Shear Stress on Sintering", J.Am.Ceram.Soc., 69(1):53 (1986)

15. R. Hamminger, G. Grathwohl and F. Thummler, "Microanalytical Investigation of Sintered SiC", J. Mat. Sci., 18:353 (1983)

16. S. R. Billington, J. Chown and A. E. S. White, in "Special Ceramics 2", Ed. P. Popper, Academic Press Inc., London and New York (1965)

Part VII. MICROSTRUCTURE CONTROL

GRAIN BOUNDARIES IN SINTERING

J.A. Pask

College of Engineering, and Lawrence Berkeley Laboratory
University of California, Berkeley, USA

INTRODUCTION

The objective of this presentation is to contribute to the
understanding of the role that grain boundaries play in sintering. It
is a phenomenological overview of how they behave and why. Specific
functions to be covered are the formation of grain boundaries, their
nature and movement. This discussion is based on the foundation of our
knowledge on the sintering of ceramics established by an excellent
tutorial review by Coble and Burke in 1963.[1] Their chapter lists many
references on specific aspects of sintering.

In all cases definitions are important since they provide a basis
tor intelligent communication. The most general term is an "interface"
which refers to a boundary for a condensed phase, either a solid or a
liquid. When such a boundary is in contact with an atmosphere, it is
specifically a "surface". When a crystalline particle is in contact
with another particle of the same composition, the interface is a
"Grain Boundary". If the grains are of different compositions, then the
interface is generally referred to as an "Interphase Boundary".

An interface for a given grain has an excess free energy over an
equivalent amount of material within its bulk. When the interface is in
contact with a vapor or atmosphere, then the excess energy is defined
as "Surface Energy" (γ_{sv}, kg/m^2). When it is in contact with a liquid,
it is also referred to as surface energy in contact with liquid and
designated as γ_{sl}. When it is in contact with another grain, the
interfacial energy is referred to as "Grain Boundary Energy" (γ_{ss} or
γ_{GB}). As the terminology suggests, normally γ_{GB} is used for the total
interfacial energy. In actual fact, the total energy should be
specified as the sum of the surface energy of one grain relative to the
other grain and the surface energy of the second grain relative to the
first ($\gamma_{GB} = \gamma_{ss} = \gamma_{s1s2} + \gamma_{s2s1}$). Also, the same question can be raised
in regard to γ_{sl} However, the surface energy of the liquid in contact
with the solid (γ_{ls}) is considered to be small enough to be ignored.
The excess free energy arises from the fact that the structure of the
surface is distorted and modified such as less screening for surface
atoms, so that it is at a higher free energy level than an equivalent
unit within the bulk. Any adsorbed species, chemical or physical, will
affect the screening of interface atoms and thus the interfacial
energy.

Science of Sintering
Edited by D. P. Uskoković *et al.*
Plenum Press, New York

There is, however, another complication that is generally
overlooked. Crystalline particles with clean surfaces are expected to
have anisotropic surface energies, i.e., different crystallographic
faces have different surface energies because the energies are
basically dependent on the respective surface structures which are
different. It is thus also important to realize that distortion of a
surface structure, undoubtedly enhanced by segregation or adsorption of
impurities, to the extent that long range atomic order at the surface
is destroyed for all crystallographic orientations leads to an
amorphous surface structure and an isotropic surface energy. A perfect
crystalline particle under favorable circumstances will tend to assume
the lowest total surface free energy structure which will be a
polyhedron bonded by crystallographic faces with the lowest surface
energy, whereas a crystalline particle with sufficient surface
imperfections for the surfaces to become amorphous will thus assume
isotropicity and the configuration of a sphere. On the other hand,
liquids do not have long range atomic order throughout their structure
and thus inherently are amorphous and have isotropic surface energies.

SINTERING MODELS

In order to adequately cover the role of grain boundaries in
determining the stages of sintering it is necessary to develop
sintering models utilizing the interfacial energy approach. Sintering
is a complex process. It is thus highly desirable to first evolve
models with the least complicating conditions. For our purpose then the
surface and interface energies for the crystalline particles will be
assumed to be isotropic throughout the sintering process so that the
particle could retain perfect sphericity without faceting on the
surface. It will also be assumed that grain boundaries can form and be
retained at every contact between spheres throughout the sintering
process.

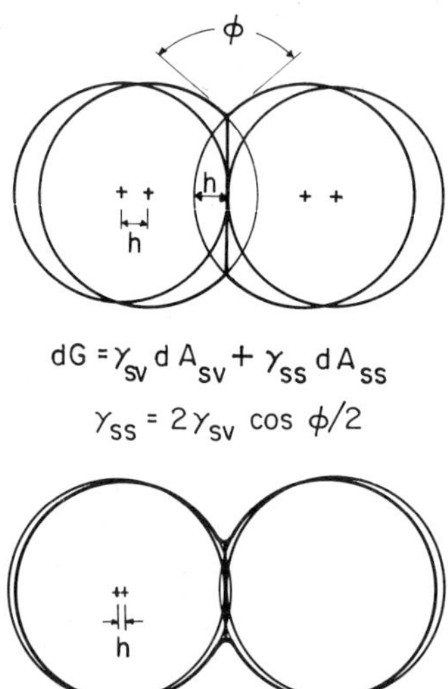

$$dG = \gamma_{sv}\, dA_{sv} + \gamma_{ss}\, dA_{ss}$$

$$\gamma_{ss} = 2\gamma_{sv} \cos \phi/2$$

Fig. 1. Schematic of two-sphere model showing shrinkage,(top) when mass
transport from developing grain boundary to neck region, Step
1, is the slow step, and (bottom) when mass transport from the
neck region to free surfaces, Step 2, is the slow step.

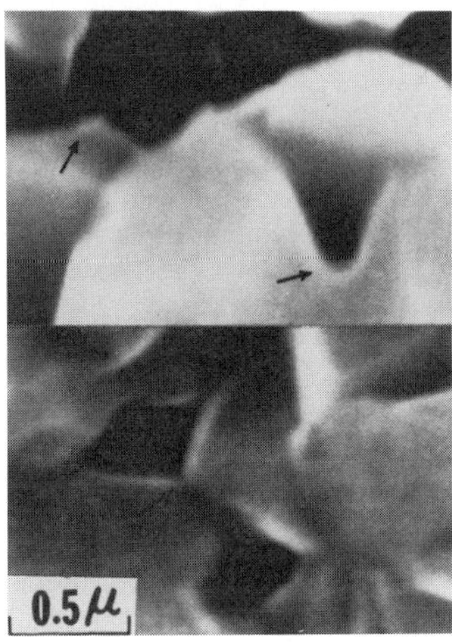

Fig. 2. Scanning electron micrographs of fracture surfaces of MgO
        powder compacts sintered to ≈ 81% of theoretical density in
        (top) static air and in (bottom) flowing water vapor.

Two-Sphere Model

The first model to be discussed will be the two-sphere model that
was first introduced by Kuczynski[2] and expanded by Kingery and Berg[3]
mathematically to enable determination of the kinetics and mass
transport mechanisms that take place during sintering. In our
discussion we will consider geometric and thermodynamic conditions
under which the centers of two spherical particles move toward each
other to realize shrinkage and the growth of a grain boundary by mass
transport of material from the contact regions to the periphery of that
grain boundary (Fig. 1). The usual observation is that consequently a
neck forms at the periphery of the grain boundary as the material
accumulates. However, the necks sometimes do not form. Figure 2 shows
SEM photographs of MgO powder sintered to the same density: with necks
at grain/grain contacts when sintered in air and without a neck when
sintered in water vapor (Wong and Pask).[4] The kinetics in the two
atmospheres were also significantly different as shown in Fig. 3. In
the presence of water vapor the rates of densification were
significantly faster. The associated absence of the neck suggests that
material moves away from the neck region much faster. This data further
suggests that presence of sorbed water on the free surfaces results in
surface diffusion being much faster.

The question then arises as to the implications of these
observations. They suggest that the mass transport is essentially a
two-step process. Step 1 is movement of material from the grain
boundary to the neck region, and Step 2 is movement of the material
from the neck region to the spherical free surfaces. Concurrently, if
Step 1 is much faster than Step 2, then a neck forms which is commonly
the case as experimentally observed. If step 2 is much faster than Step
1, no neck forms in the neck region and essentially spherical curvature
of the spheres is maintained (common terminology that has evolved in
this case refers to this condition as a sharp neck). Presumably,
intermediate cases also occur.

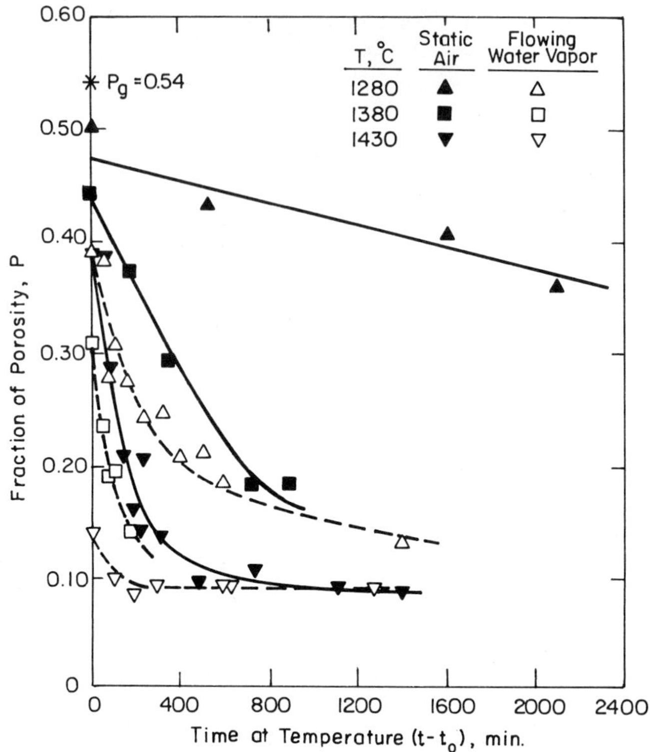

Fig. 3. Porosity versus time for sintering of MgO powder compacts in static air and flowing vater vapor at 1280, 1380 and 1430 $^\circ$C.

Schematically, the top sketch of Fig. 1 represents a condition when a neck does not form (Step 2 >> Step 1). A dihedral angle $\emptyset$ forms at the triple point between the spheres s_1 and s_2 and the vapor phase. At the beginning of sintering, the dihedral angle (as well as the area of the grain boundary) is zero and increases with sintering as material moves from the grain boundary to the neck region until the equilibrium dihedral angle is reached. Thermodynamically, equilibrium occurs when dG reaches zero according to

$$dG = \gamma_{sv}dA_{sv} + \gamma_{GB}dA_{GB} \tag{1}$$

A balance of interfacial energies occurs at this point as given by

$$\gamma_{GB} = 2\gamma_{sv}\cos\emptyset/2 \tag{2}$$

when it is assumed that the spheres are of uniform size and have the same isotropic surface energy. If each sphere had a different surface energy, then Eq 2 would be given as

$$\gamma_{GB} = \gamma_{ss} = \gamma_{s1s2} + \gamma_{s2s1} = \gamma_{s1v}\cos\emptyset_1 + \gamma_{s2v}\cos\emptyset_2 \tag{3}$$

where $\emptyset_1$ plus $\emptyset_2 = \emptyset$.

This steady state configuration is metastable. The stable configuration would be a single sphere of the same total volume. It should be recognized, however, that the stable configuration can not be reached because the grain boundary can not move. Any movement away from the neck region in either direction would require an increase in area that would result in a positive dG. In the presence of a liquid film at the grain boundary, however, such movement could occur.

The bottom sketch of Fig. 1 represents a condition when a neck does form (Step 1 ≫ Step 2). Mass transport from the neck region, however, goes on continuously, presumably because of favorable curvature of the neck surface. When a final equilibrium state is reached, a configuration similar to that represented in the top sketch will result.

Multi-Sphere Ideal Models - fcc Packing

In order to determine the nature of the sintering stages multi-sphere models must be developed. The hypothetical ideal materials specifications indicated at the beginning of this section must be used to simplify the first model, i.e., uniform spherical crystalline particles with amorphous surface structure, and formation and growth of grain boundaries at every particle/particle contact. The simplest assembly is 3 spheres with a pore that has a C.N. of 3 relative to particles. Figure 4 shows grain boundaries forming at the sphere/sphere contacts. Shrinkage is accompanied by growth of grain boundaries at the three contact points and continuous closure of the open pore until closed. Necks have not been shown in the top sketch because one of the conditions in this case was that Step 2 >> Step 1 in sintering. Furthermore, the three dihedral angles increase from zero degrees, and as the pores close, the angles approach 60° which corresponds to a $\gamma_{s1s2}/\gamma_{s1v}$ or $\gamma_{GB}/2\gamma_{sv} = 0.866$.[5] This corresponds to a statement that the surface energy of the grain in contact with another grain (γ_{s1s2}) must be 0.866 of the surface energy of the grain (γ_{s1v}) or less.

This is the smallest pore which occurs in the (111) plane of fcc packed spheres which have a 0.26 fractional void volume. Figure 5 shows steps in the densification of representative clusters on three crystallographic faces in the fcc packing of uniform spheres. On sintering, interpenetration and grain boundary formation continues uniformly according to the model until the pores on all of the (111)

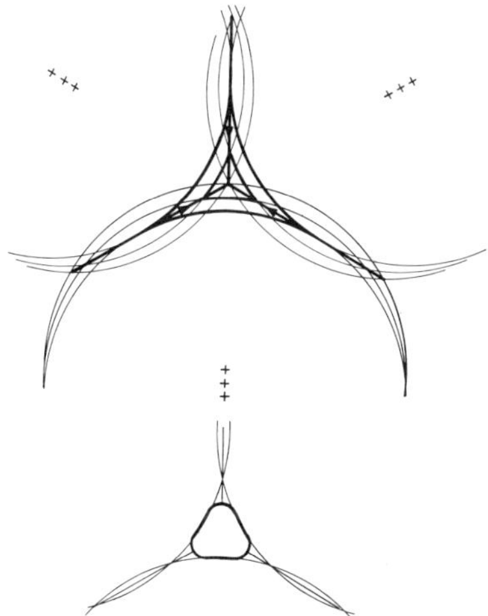

Fig. 4. Sintering of the three-sphere pore, (top) when Step 1 is the slow step, and (bottom) when Step 2 is the slow step. In top sketch, dihedral angles increase to 60° as pore closes and sphere centers move toward each other.

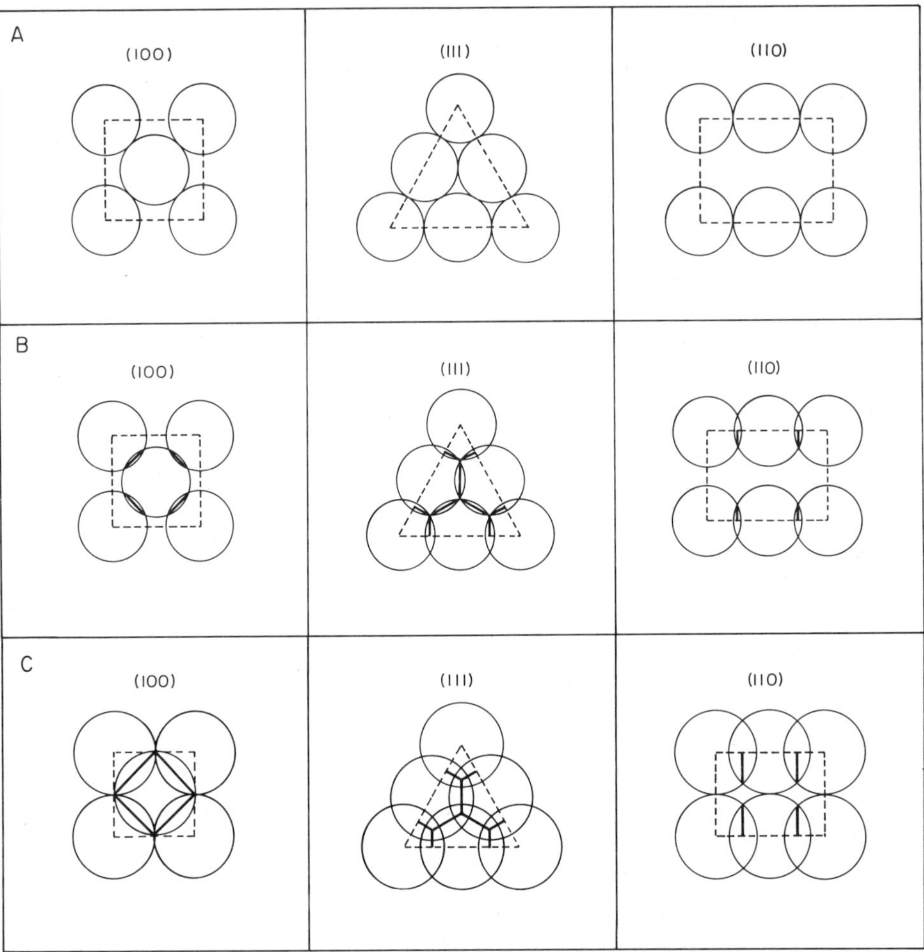

Fig. 5. Shrinkage along (100), (111) and (110) crystallographic planes in fcc packing of spheres.

planes (Fig. 5) close as pictured in Fig. 4. At this point closed pores are formed within fcc packing with an overall fractional void volume of 0.035, a theoretical density of 0.965, and a fractional linear shrinkage of 0.084 for the sintering compact. It should be noted that during this period the peripheries of all of the forming grain boundaries are terminated by open pore channels and thus can not move as described for the 2-sphere model (Fig. 1). Continued sintering results in complete densification of all of the closed pores with a total franctional linear shrinkage for the compact of 0.095 and $\emptyset$ of 109° ($\gamma_{s1s2}/\gamma_{s1v} = 0.581$).[5]

In this specified model system the sintering process consists of two stages. An Initial Stage with all open pores continues until the pores are all closed with no grain growth, a 8.4% linear shrinkage and a 96.5% theoretical density. The Final Stage starts with the densification of all closed pores until theoretical density is reached with subsequent potential grain growth. It should be noted that if this

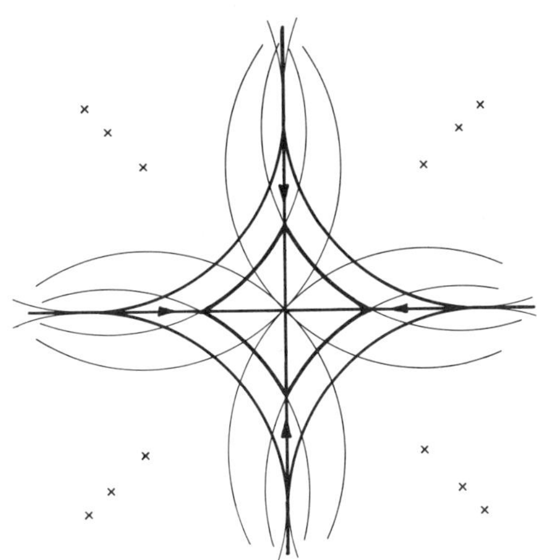

Fig. 6. Sintering of a planar pore formed by four spheres. Dihedral angles increase to 90° as pore closes and sphere center move toward each other.

material had a ratio of $\gamma_{s1s2}/\gamma_{slv}$ greater than 0.866 an equilibrium dihedral angle would be reached before any pores closed. A ratio slightly less than 0.866 would result in all closed pores and an end point density. Finally, a ratio of 0.581 or less would be required to obtain theoretical density geometrically. It is evident that complete densification beyound 96.5% theoretical density is a more complex process.

Multi-Sphere Ideal Models – sc Packing

Another simple assembly for analysis is the closing of a planar pore formed by four spheres (Fig. 6) which is a building unit in the (100)plane of sc packed spheres with an 0.48 fractional void volume (Fig. 7). As grain boundaries grow at contact points (Fig. 6), and assuming that step 2 of mass transport is faster than step 1, the overall dihedral angle $\emptyset$ increases and approaches 90° when the pore closes which corresponds to a $\gamma_{s1s2}/\gamma_{slv}$ ratio of 0.708. At this point closed pores are formed with an overall fractional void volume of 0.036, a theoretical density of 0.964, and a fractional linear shrinkage of 0.184 for the sintering compact. In the Final Stage continued sintering results in complete densification of all of the closed pores with a total fractional linear shrinkage for the compact of 0.196 and $\emptyset$ of 109°($\gamma_{s1s2}/\gamma_{slv}$ = 0.581). As in the fcc system, the sintering process would consist only of Initial and Final Stages.

It should be noted that the maximum $\gamma_{s1s2}/\gamma_{slv}$ ratio that will permit the formation of closed pores decreases with less dense packings of spheres, i.e., the surface energy must be larger relative to the grain boundary energy. This point is further illustrated by the densification of a model system of uniform spheres with diamond cubic (dc) packing with an initial 0.68 fractional void volume. On sintering, closed pores form with $\emptyset$ at 105° which is equivalent to a $\gamma_{s1s2}/\gamma_{slv}$ ratio of 0.613. At this point the fractional void volume is 0.101, a theoretical density of 0.899, and a fractional linear shrinkage of 0.277 for the sintering compact. Complete densification requires a $\gamma_{s1s2}/\gamma_{slv}$ ratio of 0.537 and a linear shrinkage of 0.316.

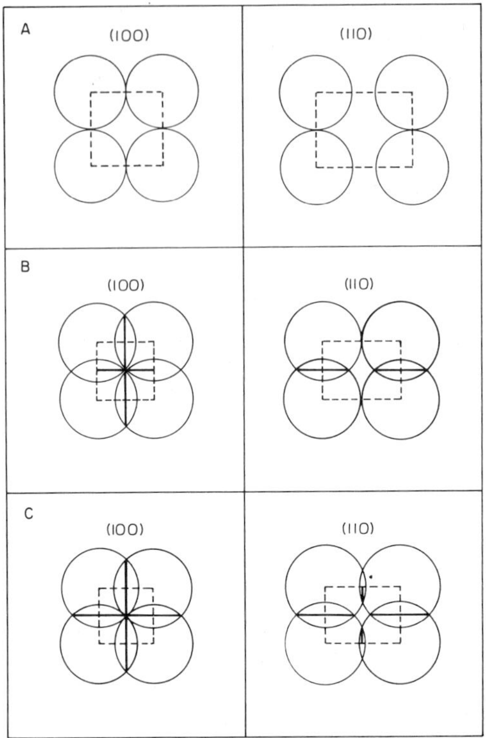

Fig. 7. Shrinkage along (100) and (110) crystallographic planes in sc packing of spheres.

Multi-Sphere Ideal Models-Summary and Discussion

It can be seen that in ideal systems, defined as consisting of uniform size crystalline particles with regular uniform packings, with amorphous surface structures (isotropic surface energies) and with grain boundaries formed at all sphere/sphere contacts, only two sintering stages exist - Initial and Final (Fig. 8). No grain boundary movement and thus no grain growth can occur in the Initial Stage, but in the final stage both phenomena can occur. However, if the same size particles are subjected to non-uniform or irregular packings, then regions of varying porosities could exist that would be dependent on different packings. With uniform size particles the initial shrinkage rate will be the same regardless of the type of packing. At a linear shrinkage of 8.4% the smallest pores formed by the closing of 3 spheres will close as indicated in Fig. 4 and Table I. This early stage is the Initial Stage of sintering. Continued sintering causes the less dense regions to densify progressively until all of the pores become closed. This stage is the Intermediate Stage which is not present in uniformly packed ideal systems. Continued heating results in the densification of the closed pores and constitutes the Final Stage.

Another modification of the hypothetical ideal system consists of the presence of a number of regions of uniform size particles with each region having a different size particle; however, all of the regions have the same type of packing (Fig. 9). The shrinkage rate of each region will depend on the size of its particles, the rate being inversely proportional to the size. The initial shrinkage for the compact will occur at some average rate until the smallest particle

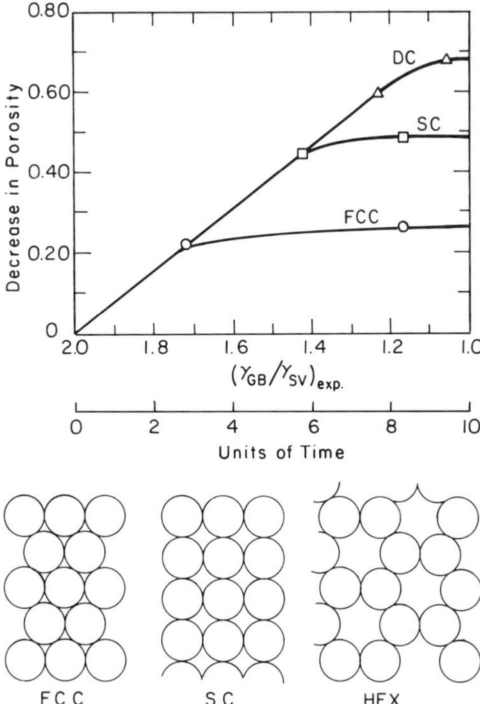

Fig. 8. Effect of fcc, sc and dc packings of uniform size isotropic spheres on decrease of porosity vs time curves.

regions form closed pores; this early stage is the Initial Stage of sintering. Continued heating causes the increasing particle size regions to densify progressively until all of the pores become closed; this stage is the Intermediate Stage of sintering. Continued heating from this point constitutes the Final Stage when no open pores are present and all the closed pores become completely densified. Undoubtedly, some of the earliest formed closed pores will densify completely before the end of the Intermediate Stage is reached, which increases the complexity of the Intermediate Stage.

Table I. Parameters for Solid Phase Sintering Models

	FCC	SC	DC
Fractional initial void volume:	0.26	0.48	0,68
Starting theoretical density	74%	52%	32%
At formation of closed pores:			
Linear Shrinkage, h/R	0.084	0.184	0.277
Fractional void volume	0.035	0.036	0.101
$\emptyset$ (dihedral angle)	60	90	105
$\gamma_{s1s2}/\gamma_{slv}$	0.866	0.708	0.613
At theoretical density:			
Linear shrinkage of unit cube	0.095	0.196	0.316
$\gamma_{s1s2}/\gamma_{slv}$	0.581	0.581	0.537

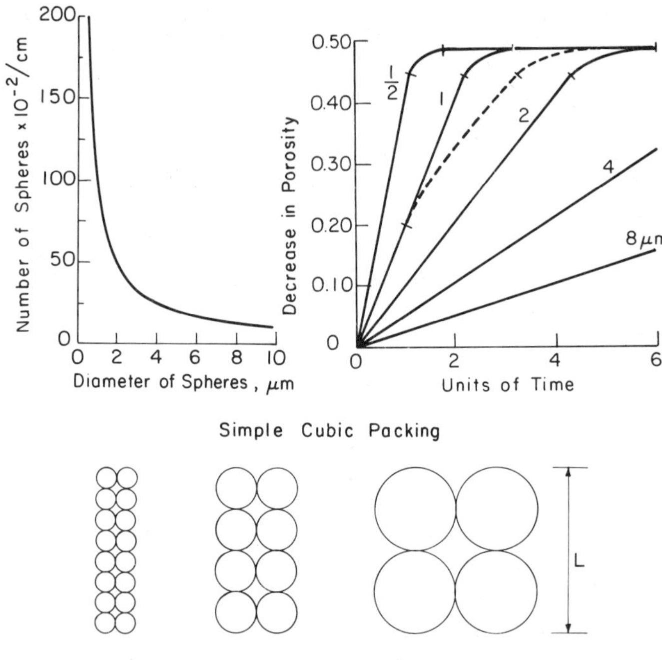

Fig. 9. Effect of isotropic sphere size in compacts with sc packing on
decrease of porosity vs time curves.

The final modification of the ideal system consists of an overall
mixture of particles with different packings and of different sizes.
Agglomerates and aggregates that are present in practically all powders
can be included into our sequence of complexity at this point. Similar
behavior to the previous modifications exists. As before, the Initial
Stage will have an average shrinkage rate until closed pores begin to
form in the regions of smallest particles with densest packing. The
Intermediate Stage starts at this point and continues until all of the
open pores become closed. The Final Stage constitutes the balance of
the firing process when the system becomes completely densified. Grain
boundaries within the closed pore regions are not as rigidly held and
presumably can move under certain conditions.

Multi-Sphere Real Systems

The final modification that can be introduced into this sequence
of increasing complexity is the removal of the boundary condition of
isotropic interfacial energies. The normal condition of anisotropic
interfacial energies introduces an additional factor that plays a role
in grain boundary movement and consequently grain growth. This
contribution is significant in the Intermediate and Final Stages of
sintering because of the formation of closed pores and releasement of
grain boundaries trapped by pores. The introduced anisotropy, however,
should have no effect in the Initial Stage of sintering, and also in
the Intermediate Stage in regions with open pores that have not yet
closed.

It seems logical that the kinetics of the three stages of
sintering should be different. The least complex of the three stages

appears to be the Initial Stage. The mass transport mechanism would be
expected to be similar to that for the two-sphere model. The
densification rate, however, would not be uniform throughout the
compact since the times for individual transports will not be the same
because the distances of transport will be dependent on the particle
sizes. A related complication is that variability in packing density
will be reflected in variability of localized shrinkage. This situation
undoubtedly leads to adjustments in arrangement of particles if rigid
grain boundaries did not form at all apparent particle/particle
contacts. The next simplest analysis of kinetic mechanisms is that for
the Final Stage. However, in this case there is an additional
complication. In order to realize the additional shrinkage of the
compact to compensate for the decrease in bulk volume due to the
densification of closed pores, a removal of some atomic or molecular
units of material has to occur from some of the crystallographic planes
that had been densified by closure of pores. An example, is the
necessary reduction in area of the dense (111) plane in fcc packing of
particles formed by closure of trigonal pores (Fig. 4 and 5).

Likewise, it appears logical that the kinetics of densification
during the Intermediate Stage would be the most difficult to quantify
since the structure throughout the specimen changes significantly. It
is a transition period from the Initial Stage of no closed pores to the
Final Stage of all closed pores; with time it will range from a
structure of open pores and some closed pores to one of closed pores
and some open pores. A difficulty in kinetic studies is that the degree
of densification in the Initial and Intermediate Stages (which are the
ones that receive the greatest experimental attention) is not fixed and
is strongly dependent on the range of particle sizes, variability in
packing of particles, and uniformity. Another difficulty is the
interpretations of isothermal experiments. If shrinkage is measured
from the time of reaching the selected isothermal temperature, a
certain amount of prior sintering has taken place. An example of this
point is illustrated by Fig. 3. In many cases if the experimental
temperature is high enough and the character of the powder is such that
it has a short Initial Stage, the specimen may have actually reached
the Intermediate Stage on reaching the test temperature. Such factors
are frequently overlooked or unrecognized.

GRAIN BOUNDARIES AND THEIR MOVEMENTS

Sintering occurs because the grain boundary energy is less than
the surface energies of the surfaces that were replaced. A parallel
definition is that the surface energy of the solid in contact with
another solid is less than the surface energy of the solid in contact
with vapor. This differential provides the driving force for sintering.
Consequently, as the ratio $\gamma_{s1s2}/\gamma_{s1v}$ decreases, the driving force for
sintering increases. The same situation holds for the other half of the
grain boundary with the ratio $\gamma_{s1s2}/\gamma_{s2v}$; because of anisotropy they
normally are not equal. If the grain boundaries and surface energies
are isotropic and thus equal, then $\gamma_{s1s2}= \gamma_{s1v}\cos\emptyset/2$ and $\gamma_{GB}= \gamma_{ss}= 2\gamma_{sv}$
$\cos\emptyset/2$.

As already emphasized, grain growth is dependent on grain boundary
motion. Such motion can only occur in regions where closed pores have
formed during the Intermediate and Final Stages of sintering. Such
closures eliminate thermodynamic barriers permitting potential movement
of grain boundaries to positions at a lower free energy state. A result
of such movement can be the entrapment of pores within growing grains
that are subsequently difficult, if not impossible, to eliminate. It is

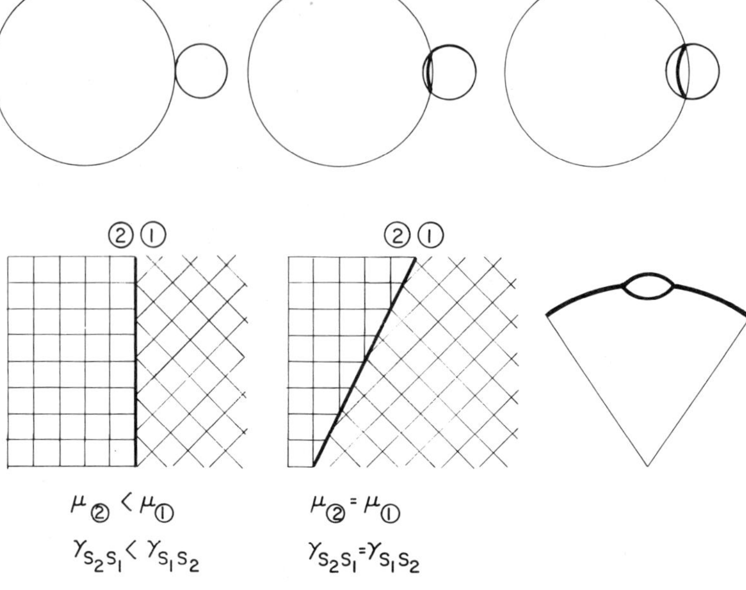

Fig. 10. Schematic of two adjoining misoriented crystalline particles. (Left) Anisotropic grain boundary, and (Right) after movement of grain boundary to isotropic position.

postulated that this movement past a pore is not favorable with isotropic grain boundaries but favorable with anisotropic grain boundaries. Isotropic grain boundaries apparently can move only if they are curved. This behavior has been extensively discussed in the literature on the basis that movement toward their center of curvature would result in reduction of the grain boundary area and free energy which becomes the driving force for passing a pore. However, it is postulated here that this driving force is not sufficient for bypassing a pore and entrapping it. On the other hand, an anisotropic grain boundary has an additional driving force for movement due to the fact that the interfacial energies of the adjoining grains at the grain boundary interface are not equal providing a driving force for the boundary to move to a balancing free energy position.

A misorientation of adjoining grains are shown in Fig. 10. The sketch on the left indicates an unbalance of interfacial energies with $\gamma_{s2s1} < \gamma_{s1s2}$ resulting in movement even though the boundary is not curved. As a result of this unbalance of energies, the chemical potential μ and the vacancy concentration in S_2 grain at the grain boundary is greater than in grain S_1. The grain boundary could then move into grain S_2 and also reorient itself as shown in the sketch on the right so that γ_{s1s2} becomes equal to γ_{s2s1}. In this position no further movement occurs because of loss of anisotropy and the associated thermodynamic driving force.

With this hypothesis it is evident that isotropic grain boundaries are necessary to prevent movement of boundaries past closed pores. It is postulated that an amorphous structure, or lack of long range, order, is necessary in the interfacial layers for an isotropic grain boundary. It is presumed that certain additives to a sintering powder, e.g. MgO additions to Al_2O_3 powder, preform this function. In general, it should be theorized that any imposed contamination or distortion of surfaces and interfaces that would eliminate the anisotropic behaviour due to perfect and clean crystallographic faces would possibly achieve the desirable isotropic characteristics.

CONCLUDING REMARKS

The principal objective of this tutorial report has been an effort to emphasize the importance of the grain boundary and the controlling features that it possesses throughout the sintering of a powder to its theoretical densification. Particular attention is directed to the fact that the grain boundary energy is the sum of the interfacial energies of both grains constituting the grain boundary. Such a boundary normally has an anisotropic boundary energy. This structure provides a driving force for movement of the grain boundary irrespective of any curvature. It is proposed that specific additives to the sintering powder tend to make the grain boundaries amorphous in structure and thus the interface energies isotropic. As a result a major component of the driving force for grain boundary movement is eliminated. Isotropic grain boundaries, however, because of curvatures, experience localized movements and adjustments in order to realize favorable boundary angles at triple and quadruple points leading to the formation of equiaxed grain microstructures.

Sintering ideal models have been developed to show that the Initial Stage does not have any closed pores forming and that no grain growth has occurred. The Intermediate Stage starts with the formation of some closed pores and continues until all of the pores become closed. The Final Stage starts at this point. With the closure of pores grain boundaries that were associated with the pores before closure become free to move.

REFERENCES

1. R. L. Coble and J. E. Burke, "Sintering in Ceramics", in Progress in Ceramic Science, v. 3, J. E. Burke, Ed., 197-251 (1963).
2. G. C. Kuczynski, "Self-Diffusion in Sintering of Metallic Particles", Trans. AIME, 185, 169-177 (1949).
3. W. D. Kingery and M. Berg, "Study of the Initial Stages of Sintering Solids by Viscous flow, Evaporation-Condensation, and Self-Diffusion", J. Appl. Phys. 26, 1205-1212(1955).
4. B. Wong and J. A. Pask, "Experimental Analysis of Sintering of MgO Compacts", J. Am. Ceram. Soc., 62, 141-146 (1979).
5. C. E. Hoge and J. A. Pask, "Thermodynamic and Geometric Considerations of Solid State Sintering", Ceramurgia Int., 3, 95-99 (1977).

COMPUTER SIMULATION AND EXPERIMENTAL ANALYSIS OF ABNORMAL GRAIN GROWTH IN

BaTiO3 CERAMICS

U. Kunaver and D. Kolar

"J. Stefan" Institute
University of Ljubljana
Ljubljana

INTRODUCTION

Abnormal or discontinuous grain growth (sometimes called secondary recrystallization or exaggerated grain growth) is characterised by rapid growth of a limited amount of grains to sizes much larger than those of the average grain population. It is particularly likely to occur when normal grain growth is inhibited by the presence of impurities and pores (1). It may originate from many possible causes, such as a wide initial particle size distribution, a small amount of liquid phase, chemical heterogeneity, heterogeneous density distribution, intrinsic boundary properties (preforential grain boundary mobility) and others.

A characteristic feature of discountinous grain growth is rapid coarsening of the microstructure. Diameter of discountinuosly growing grains increases linearly with time, until impingement on an other large grain occurs. When this happens, grain growth practically ceases or strongly reduces. Discontinuous growth can result in grains larger than those ever produced by normal grain growth. Microstructures with average grain size of several tens of micrometers may evolve in only few minutes.

Early studies of discontinuous growth in Al2O3 ceramics revealed similarity with recrystallisation in metals (2). Although the mechanism for nucleation proposed for metals would not apply to sintered ceramics, discontinuous evolution of microstructure in ceramics may be treated as in metallic systems, as composed of two processes, i.e. nucleation and nuclei growth.

The same nomenclature as used in metals is used also in ceramics, with the understanding that the term "nuclei" indicates those grains which are capable of much faster growth than the surrounding grains.

In metals, two limiting cases of recrystallization are distinguished on the basis of the relative kinetics of nucleation. In one extreme, studied in detail by Johnson and Mehl (3), the preferred nucleation sites are regarded as equivalent and inexhaustible. Nuclei form randomly at a constant net rate within the untransformed material. In the second limiting case, analysed by Avrami (4), the nuclei are assumed to be activated simultaneously at the beginning of the transformation.

Science of Sintering
Edited by D. P. Uskoković *et al.*
Plenum Press, New York

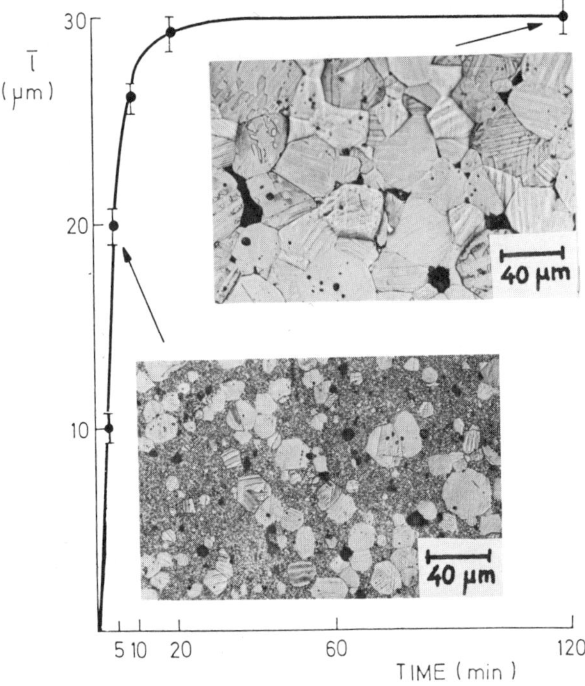

Fig.1. Average intercept length vs. sintering time and corresponding microstructures for BaTiO3-2 mol % Ba6Ti17O40 ceramic, sintered at 1350°C.

Quantitative experimental data for abnormal grain growth in ceramics are limited. Recently, Hennings et al.(5) reported data on the temperature dependence of nucleation rate and grain growth rate in BaTiO3-TiO2 ceramics. It was observed that the number of recrystallized grains (nuclei) increased exponentially with time. Our experiments in the BaTiO3-SrTiO3-TiO2 system (6) gave comparable results.. (Fig.1), reported in (6) illustrates some characteristics of discontinuously developed microstructure. Growth kinetics is linear and abruptly stops after recrystallization. After short sintering time, microstructure is bimodal consisting of large dis- continuously growing grains embedded in fine grained matrix. The final microstructure is quite homogeneous. It was pointed out earlier that a uniform microstructure can eventualy evolve from a discontinuously growing duplex structure, when all the small grains dissappear in favor of the large ones (7). However, in most cases such microstructures preserve a broad grain size distribution because all grains do not start to grow simultaneously.

In a process composed of nucleation and growth, the final grain size depends on the relative values of the rate of nucleation $\dot{N}$ and rate of nuclei growth $\dot{G}$. If the ratio of $\dot{N}$ to $\dot{G}$ is large, the final grain size will be small. Conversely, if the ratio of $\dot{N}$ to $\dot{G}$ is small, the final grain size will be larger. Both rates strongly depend on processing parameters, such as composition, initial grain size, grain size distribution and temperature.

Nucleation in ceramics is not well understood. It is generally accepted that large grains in the initial powder assembly may act as nuclei. This view is supported by experimental observation that discontinuous growth is more likely to occur in powder mixtures with broad size distribution. This fact was used to produce coarse grain structure in varistor ceramics by the addition of large "seed" grains (8). On the other hand, it was also demon- strated that by addition of larger amount of seeds of moderate size, it is possible to produce uniform, controlled grain size structure (5).

Several authors applied computer technique to simulate grain growth in a two dimensional matrix (9-12). Interesting recent observation (10) was that the inhibition of normal growth and the introduction of abnormally large grains is not a sufficient condition for the occurrence of abnormal grain growth, In present work, the simple computer model was employed to simulate the abnormal grain growth and to analyse the influence of the initial grain size distribution and seed grain addition on microstructure characteristics. The results were compared with experimental data on BaTiO3 ceramics reported in (5).

COMPUTER SIMULATION PROCEDURE

Computer simulation is performed on square lattice of 348 x 348 lattice sites. Each lattice site can be assigned to nonrecrystallized area or to one of the nuclei. Nuclei are initially generated and presented as circles on the lattice. Simulation is iterative. Each cycle is considered to be one simulation time unit and the distance between lattice sites is one simulation length unit.

Simulation is driven by two main rules, which were observed experimentally: growth rate is time independent and nucleation rate per unit of nonrecrystallized area is exponentially dependent on time.

$$\frac{dR}{dt} = \text{const.} \tag{eq.1}$$

$$\frac{dN}{dt} = A \cdot e^{B \cdot t} \tag{eq.2}$$

In the simulation growth rate is one length unit per time unit; i.e., each nucleus in one simulation cycle grows for circular layer which is one length unit thick. Simulation length unit was fixed to be 0.5 μm. Experimental data reported in (5) were used to transform simulation time unit into seconds (Table 2). All other quantitites are calculated according to those two units. Grain sizes (R) are expressed as equivalent circular radii.

Initial grain size distribution was assumed to be log-normal expressed by equation:

$$f(R) = \frac{1}{\sigma_g \cdot (2\pi)^{1/2}} \cdot e^{-\frac{(\log R - \log \bar{R}_g)^2}{2 \cdot \sigma_g^2}} \tag{eq.3}$$

where σ_g is geometric standard deviation, which is a measure for dispersion of distribution, and $\bar{R}_g$ is geometric mean size. $\bar{R}_g$ was corrected so that all used distributions have the same arithmetic mean size of 1 μm. In case of simulation with the seed grains, the size of seed grains was 4 μm. Initial size distribution is normalized to area of the lattice. It means that total surface of all grains defined with distribution is equal to lattice area.

At the beginning of simulation a starting situation is generated. It is assumed that all grains, larger than the twice mean grain size, are nuclei already from the beginning of simulation (Hillert criterion). Individual sizes of initial nuclei are calculated from initial size distribution, their positions being chosen randomly. The amount of initial nuclei varies with

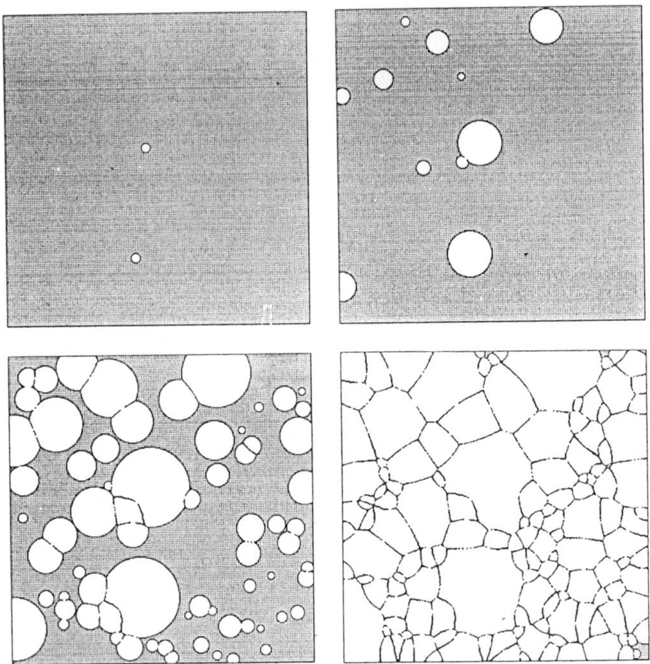

Fig.2. Time-lapse sequence of computer simulated development of
microstructure.

the broadness of initial size distribution and amount of seed grains.

After generating starting situation, simulation continues with
iteration. In each step nuclei grow, so that their radii increase for one
lattice site. Where two nuclei impinge into each other, growth stops and a
boundary between them is formed. Such nuclei proceed to grow only into
nonrecrystallized area. Boundaries between nuclei do not move.

In each step of simulation process a certain amount of nuclei are
generated according to defined nucleation rate. Nucleation rate is
determined by two constants A and B (eq.2), which are fixed on the beginning
of the program. Constants, which strongly depend on temperature, were
derived from experimental data. Positions of new nuclei are choosen randomly
from nonrecrystallized area. Their sizes are determined according to matrix
grain size distribution: it is assumed that distribution and size of matrix
grains do not change during simulation. Nonrecrystallized area is shrinking
due to nucleation and growth of nuclei, so the number of matrix grains
decreases with time. In each step of simulation the matrix grain
distribution is normalized to nonrecrystallized area. New nuclei are
generated from the biggest grains left in matrix. By that means the number
of the biggest matrix grains decreases. Consequently nuclei generated in
later phases of nucleation are smaller then the proceeding ones.

When all area is recrystallized, nucleation stops because there is no
place to generate new nuclei. Simulation stops as well. Fig.2 shows time-
lapse sequence of simulated microstructure.

The simulation was performed on two sets of initial size distributions.
In the first case, different broad log-normal distributions with geometric
standard deviations (σ_g) 0.15, 0.20, 0.25, 0.30, 0.35 were assumed.
Corresponding initial nuclei area (which is equal to weight %) are presented

in table 1. Second case simulates log-normal distribution with geometric standard deviation 0.15 with added 0.1 wt %, 0.5 wt %, 1 wt %, 5 wt %, 10 wt %, 20 wt %, 4 μm seed grains. Arithmetic mean size (equivalent circular radius) was in both cases 1 μm.

TABLE 1. Initial nuclei area for different initial distributions.

σ_g	0.15	0.20	0.25	0.30	0.35
area	0%	1.4%	6.3%	14.7%	24.9%

Both sets of initial distributions were subjected to simulated growth at 3 temperatures. Data were extracted from (Ref.5).

TABLE 2. Input data for simulation.

T ($^{\circ}$C)	1315°C	1320°C	1330°C
EXPERIMENTAL DATA (Ref.5)			
dR / dt (μm/s)	0.0292	0.0434	0.0473
$d \ln N / dt$ (s^{-1}mm^{-2})	$2.53 \cdot 10^{-3}$	$9.01 \cdot 10^{-3}$	$11.48 \cdot 10^{-3}$
N_0 (mm^{-2})	58.8	42.13	165.8
$d \ln N / dR$ (mm^{-3})	86.6	207.6	242.7
CALCULATED SIMULATION DATA			
1 simul. length unit		0.5 μm	
1 simul. time unit	19.84 s	11.52 s	10.55 s
observed area		0.03027 mm^2	
dR / dt		1 length u./time u.	
initial mean size		1 μm = 2 length units	
preexp.fakt. A	0.089	0.13	0.606
exp.fakt. B	0.0501	0.103	0.121

$B = d \ln N / dt$ (in simulation units)
$A = B \cdot N_0$ (in simulation units)

RESULTS AND DISCUSSION

Simulated microstructures demonstrate large influence of the initial nuclei area (as consequence of broad initial size distribution or added seed grains) on shape of final grain size distibution (Fig.3,4). Mean size of final grains is more sensitive to nucleation rate. Large initial nuclei area reduces nucleation (Fig.5a,5d), so in that case, nucleation rate (defined by temperature) will have smaller effect on grain size. There was not much difference in influence of broad initial size distributions and added seed grains with the same initial nuclei area.

At low initial nuclei area (narrow distribution or small amount of seed grains) final size distributions show broad log-normal shape. In contrast, high initial area results in narrow normal distributions. In between is a region where bimodal distibution occurs. It can be described by two competitive subpopulations of nuclei. In first subpopulation are nuclei, which are present at the beginning or have nucleated in early phases of

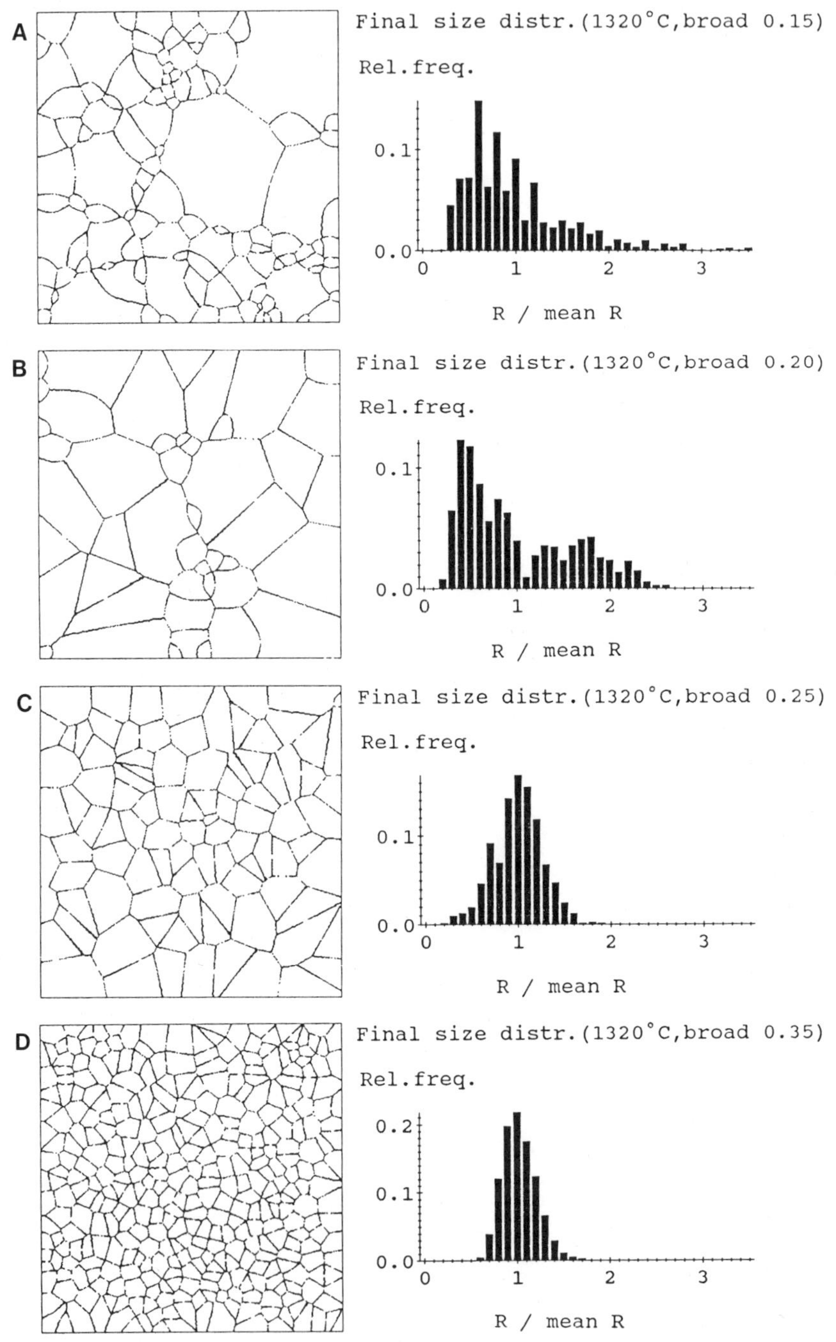

Fig.3. Simulated final microstructures and corresponding size distributons at 1320°C for various initial size distributions. Labels represent σ_g.

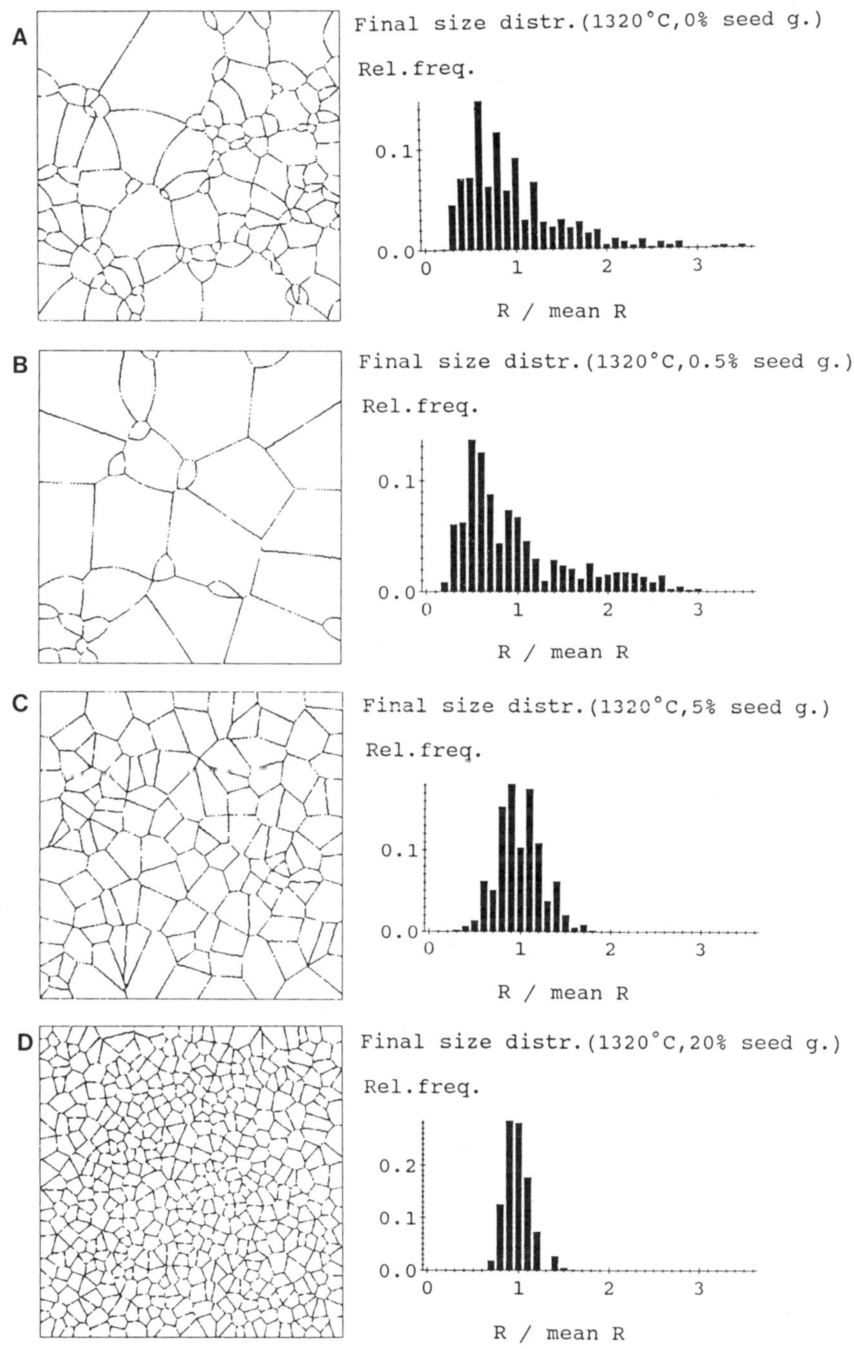

Fig.4. Simulated final microstructures and corresponding size distributions at 1320°C for various amount of seed grains added.

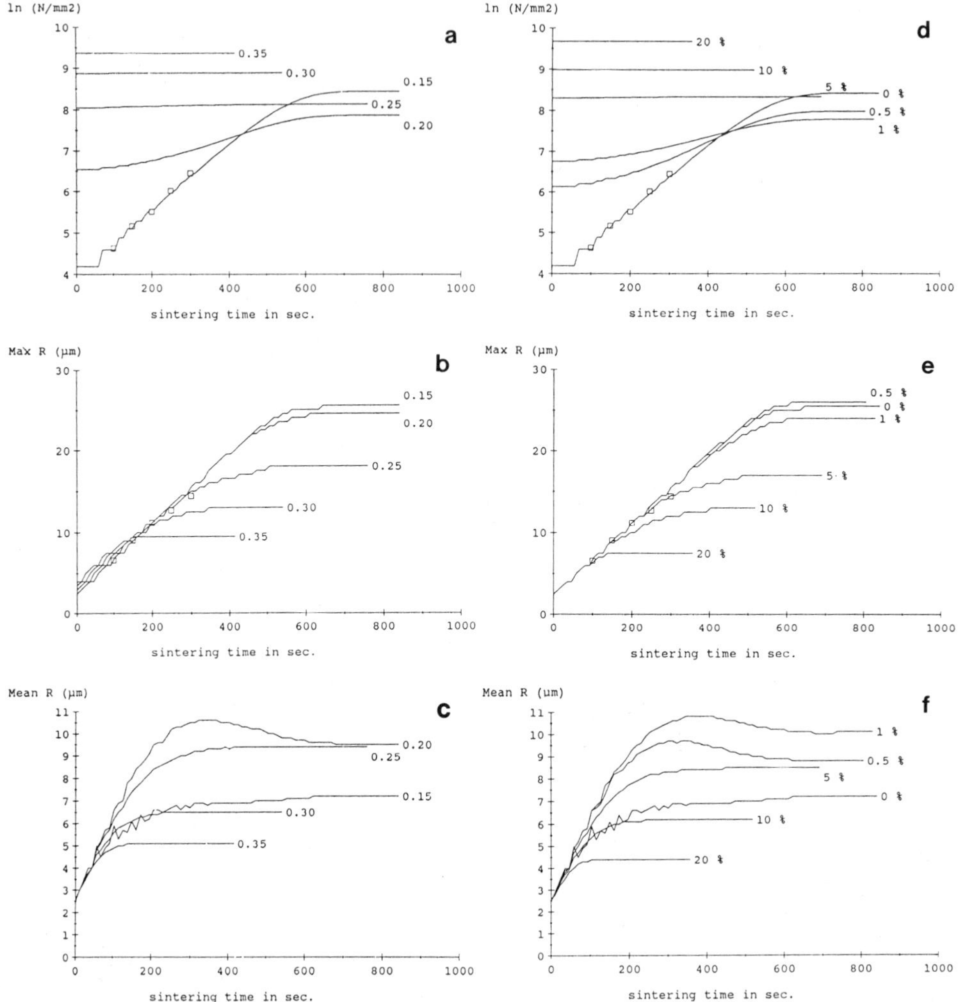

Fig.5. Number of nuclei (a,d), maximum nuclei size (b,e) and mean nuclei size (c,f) vs. time for differennt initial size distributions (a,b,c) and different amount of seed grains (d,e,f). Labels represent σ_g and amount of seed grains added, respectively. Squares represent data from ref. 5 for 1320°C.

simulation. They have narrower distribution and grow simultaneously forming normal final distribution. (Fig.3d,4d) This type of microstructure development takes place when initial nuclei area is large and decrease of nonrecrystallized area stops nucleation.

On the other hand nuclei which were nucleated later in simulating process, are forming a second subpopulation. It has log-normal form (Fig.3a, 4a). The number of new nuclei per time unit is increasing exponentially. Only few "oldest" nuclei can grow very large, forming typical "tail", and therefore broadening the distribution. Majority are "younger" nuclei, which remain small. This type occurs when there are only few initial nuclei. In the case of mixing of these two subpopulations, bimodal size distribution results.(Fig.3b,4b) Mixing takes place when subpopulation of initial nuclei is present, but is not large enough to prevent nucleation. In such a case, nucleation occuring in late phases of simulation still produces small nuclei

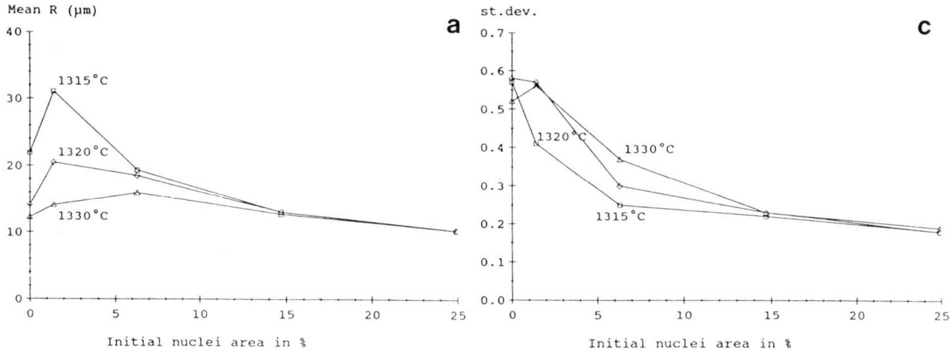

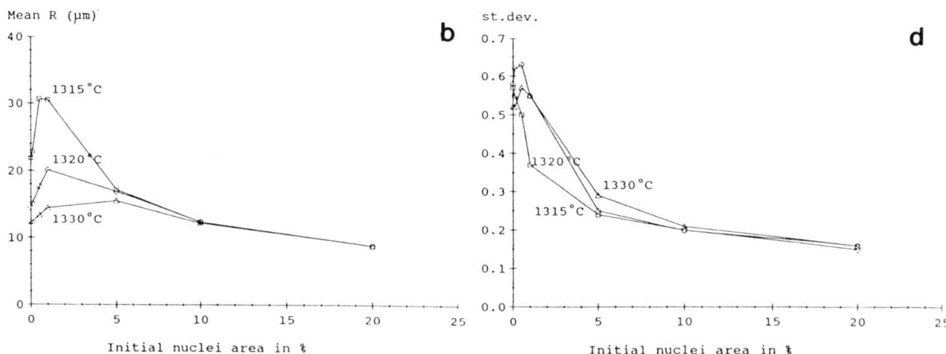

Fig.6. Mean final grain size vs. initial nuclei area for different initial
size distributions (a) and for different amount of seed grains added (b);
standard deviation of the final grain size distribution vs. initial nuclei
area for different initial size distributions (c) and for different amount
of seed grains added (d). Compare table 1.

of second population, as can be seen on time plot of mean nuclei size for
distribution of σ_g = 0.20 (Fig.5c) and for 0.5 and 1 wt% seed grains
(Fig.5f). New small nuclei forming in late phases of simulation cause drop
of final mean size of nuclei. Temperature dependance of final distribution
shape is weak.

Sizes of largest grains decrease with broadening of initial
distribution or with adding seed grains (Fig.6a,6b). Growth is also stopped
by higher nucleation rate (higher temperature). Large amount of initial or
later nucleated nuclei causes earlier impingement of the largest nuclei.

Final arithmetic mean grain size is increasing by small increase of
initial nuclei area. With larger initial nuclei area, decrease of mean size
is observed, forming a maximum between 0.5 and 5 % of initial nuclei area.
Position of maximum corresponds to appearance of bimodal size distribution
and is temperature dependent: at higher temperatures, the mean size maximum
occurs at a larger amount of initial nuclei (Fig.6a,b). At very narrow
initial distributions there are no initial nuclei, so all nuclei are formed
during simulation and final grain mean size is small. By small increase in
broadness of the initial distribution only a few initial nuclei appear,
which have good opportunity to grow very large, inhibiting nucleation and
therby resulting in large mean grain size. Further increase of amount of
initial nuclei imposes limit in growth due to available space restriction.
The effect is a decrease of the mean grain size. Maximum appears at a point
where influence of initial nuclei subpopulation exceeds influence of

nucleation process. At lower temperatures, this occurs at smaller amount of initial nuclei; this is so because nucleation is slower than at higher temperatures.

Temperature dependence of mean size plays major role at low initial nuclei area. At higher temperatures $\dot{N}$ / $\dot{G}$ increases, causing smaller final mean grain size. Large amount of initial nuclei prevents nucleation, so in that case, the effect of temperature is diminished.

Broadness of final distributions decrease with increasing initial nuclei area because of nucleation inhibition. Nucleation plays important role in broadening of grain size distribution. Broadness is slightly increasing with higher temperature due to higher nucleation rate (Fig.6c,d).

CONCLUSION

Microstructures simulated by presented simple computer model of abnormal grain growth correspond well to experimental data and qualitative observations. Initial grain size distribution influences the shape of the final grain size distribution curve as well as the final mean grain size. Different nuclation rates at different temperaures affect mostly final mean grain size at lower amounts of initial nuclei. Behaviors of systems with broad initial size distribution and added seed grains with the same initial nuclei area were found to be very similar.

ACKNOWLEDGEMENTS

We thank prof. J. Zupan from the Boris Kidric Institute for advices and valuable discussions.

REFERENCES

1. W. D. Kingery, H. K. Bowen and D. R. Uhlmann, "Intoduction to ceramics", 2.ed. pp 461-62 Wiley, New York, 1976
2. I. B. Cutler, "Nucleation and nuclei growth in sintered alumina", pp 120-127 in Kinetics of high temperature processes, (ed by W. D. Kingery), MIT Press Cambridge MA 1959
3. W. A. Johnson and R. F. Mehl, Trans. AIME $\underline{135}$, 416 (1939)
4. M. Avrami, J. Chem. Phys. $\underline{7}$, 1103 (1939); $\underline{8}$, 212 (1940)
5. D. F. K. Hennings, R. Jansen and P. J. L. Reynen, "Control of liquid-phase-enhanced discontinous grain growth in barium titanate", J. Am. Cer. Soc. $\underline{70}$ (1) pp 23-27 (1987)
6. D. Kolar and L. Marsel, "Microstructure evolution in (Ba,Sr)TiO3 ceramics, Science of Ceramics $\underline{14}$, pp 921-926 (1988)
7. F. M. A. Carpay and A. L. Stuijts, "Characterisation of grain growth phenomena during sintering of single phase ceramics", Sci. Ceramics $\underline{8}$ pp 23-38 (1975)
8. K. Eda, M. Enada and M. Matsuoka, "Grain growth control in ZnO varistors using seed grains", J. Appl. Phys. $\underline{54}$ (2) pp 1095-1099 (1983)
9. K. W. Mahin, K. Hanson, J. W. Morris, Jr., "Comparative analysis of the cellular and Johnson-Mehl microstructures through computer simulation", Acta Metall. $\underline{28}$ pp 443-453 (1980)
10. D. J. Srolovitz, G. S Grest and M. P. Anderson, "Computer simulation of grain growth - V. Abnormal grain growth", Acta Metall. $\underline{33}$ (12) pp 2233-2247 (1985)
11. T. O. Saetre, O. Hunderi and E. Nes, "Computer simulation of primary recrystallization microstructures: The effects of nucleation and growth kinetics", Acta Metall. $\underline{34}$ (6) pp 981-987 (1986)
12. K. Marthinsen, O. Lohne and E. Nes, "The development of recrystallization microstructures studied experimantally and by computer simulation", Acta metall. $\underline{37}$ (1) pp 135-145 (1989)

PORE REMOVAL DURING FINAL STAGE SINTERING OF MODIFIED YTTRIA

W. Rossner

Siemens AG, Corporate Research and Development
Munich, FRG

ABSTRACT

Starting with a relative density of about 95%, the final stage
sintering of Y_2O_3 doped with 30 mol% Gd_2O_3 is investigated in a
temperature range from 1700°C to 1900°C. The microstructural evolution
is characterized by grain size, bulk porosity, interpore spacing, and
pore size distribution. The pore removal is found to be grain boundary
controlled by enhanced, but normal grain growth. Due to pore entrapment
within grains, the densification stagnates before complete pore
elimination is achieved. In an intermediate stage of this process,
grain growth driven pore coalescence increases the size of pores
located at the grain boundaries.

INTRODUCTION

In the field of ceramic firing, it is desirable to reach sintering
densities close to the theoretical density of the base material. In the
particular case of ceramics for optical applications, this is the most
fundamental demand. To achieve such fully densified ceramic materials,
an advanced control of the formation of the microstructure is of
special interest. This necessitates a detailed understanding of all the
sintering processes and mechanisms involved, which are often
competitive, e.g. pore shrinkage and pore coalescence. The
technological interest is principally centered on the final stage of
sintering, where the pore system consists of individual pores and
ultimate densification occurs.

Focusing on the foregoing background, the final stage sintering of
a gadolinia-modified yttria was investigated in this study. Several
authors have previously reported on the sintering of yttria and related
compounds with various dopants.[1-7] However, most of the major work has
been concerned with the technological sintering aspects necessary to
produce a fully dense, highly transparent yttria ceramic. For that
reason, the purpose of this paper is to study the basic process of pore
removal during final stage sintering of a modified yttria in a more
analytical way.

EXPERIMENTAL

The Gd_2O_3 – modified Y_2O_3 powders containing 30 mol% Gd_2O_3 (both 99.999 grade) were prepared by an extended coprecipitation technique using the oxalate process with a final calcination at 650°C for 2 hours. By this process, a single-phase solid-solution powder was obtained with an average particle size of about 3 μm with 90% in the 0.5-6 micron range. Uniaxially cold-pressed powder compacts with green densities near 50% were fired in a tungsten shield resistance furnace in high vacuum. Isothermal sintering experiments were carried out in a temperature range from 1700°C to 1900°C with sintering times up to 8 hours and constant heating and cooling rates of 10K/min.

The microstructures were analyzed on polished and chemically etched (boiling 20% HCl) cross sections of the sintered compacts. The mean grain size was determined by the intercept length method, where the average distance between the intercepts, based on up to 200 counts, is multiplied by 1.5 due to a stereological correction. The porosity was measured statistically by means of an automatic image analyzing system (Quantimet 970, Cambridge Instr.) involving the bulk porosity, the interpore spacing, and the pore size distribution. The last one is represented by a frequency distribution of the circle diameter of the individual pore areas of the cross section. To get a sufficiently reliable picture of the pore size distribution, about 3000 pores were evaluated, even in cases where bulk porosities were less than 1%.

RESULTS AND DISCUSSION

Starting with a density of about 95% of theoretical, the pore removal and the formation of the microstructure were investigated in the final stage of sintering (1700°C – 1900°C) of Y_2O_3 doped with 30 mol% Gd_2O_3.

In this temperature range, solid state sintering is detected. The Y_2O_3 – Gd_2O_3 composition used in this investigation shows solid solution with a cubic structure up to high temperatures, so that isotropic sintering behaviour can be expected. However, after sintering at 1900°C for long times, a small but increasing amount of a secondary phase was observed, which was segregated at the grain boundaries. The

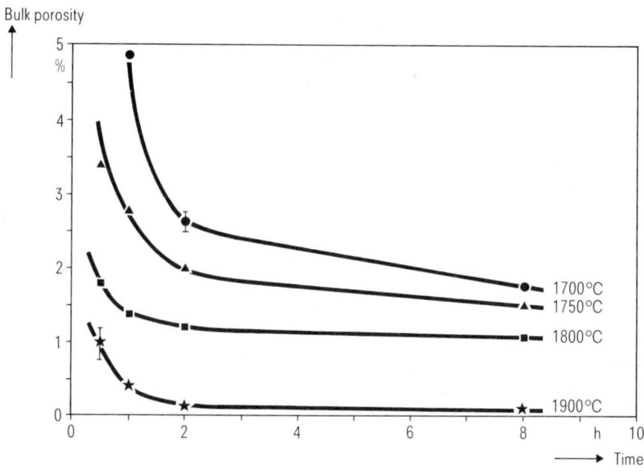

Fig. 1. Time dependence of the porosity for various sintering temperatures.

406

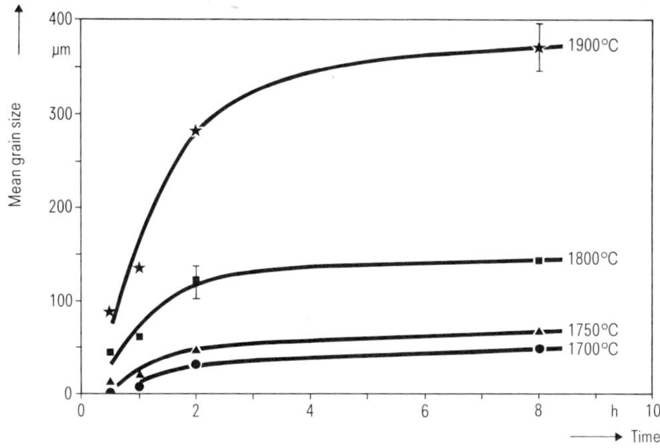

Fig. 2. Time dependence of the grain size for various sintering
temperatures.

nature of this phase is not investigated here, but it is assumed to be
similar to that reported for La_2O_3-doped Y_2O_3[7,8] where sintering in a
two-phase area results in secondary solid phase formation which retards
grain growth and allows complete densification.[1,5]

The progressive densification during isothermal sintering of this
Gd_2O_3-modified Y_2O_3 at various temperatures is plotted in Fig. 1 as
bulk porosity versus time. This illustrates that efficient pore
elimination is limited to short sintering times even at the higher
temperatures. Only for sintering at 1900°C is a nearly complete
densification achieved with a bulk porosity of less than 0.1%.

Referring to the grain size data in Fig. 2, it becomes obvious
that pore removal in this sintering stage may be governed by a
mechanism where pores are trapped within the grains and the
densification rate is reduced. Detailed analyses of the microstructures
verified that enhanced but, normal, continuous grain growth occurred.
The exaggerated, discontinuous growth mechanism mentioned previously,
which should result in a bimodal grain size distribution in the
intermediate growth period, could not be detected. The observed grain
growth roughly follows a kinetic relation described in the usual way by
$G^n \alpha t$ where n ranges from 1.5 to 3, with grain size, G, and time, t.
Although this varying time exponent is not sufficient in itself to
demonstrate a particular growth mechanism, it argues for its complex
nature. In particular, the interaction between migrating grain
boundaries and residual pores is expected to be most important for pore
removal.

The interpore spacing is a suitable parameter for studying this
process in detail. During sintering, the interpore spacing is
increased by pore coalescence or pore shrinkage up to the point of pore
elimination. Fig. 3 shows the evolution of the interpore spacing versus
grain size for specimens sintered at various temperatures and times.
The relationship between these two quantities shows a linear
characteristic when the grain growth has the same time dependence as
the increase of the interpore spacing. For a statistical consideration
assuming a homogeneous distribution of the pores initially located at
the grain boundaries, a slope equal to one suggests that pores stay at
the grain boundaries during sintering. For slopes greater than one, the
same behaviour would be obtained with suggest pore controlled grain
growth. Slopes smaller than one suggest grain boundary controlled pore

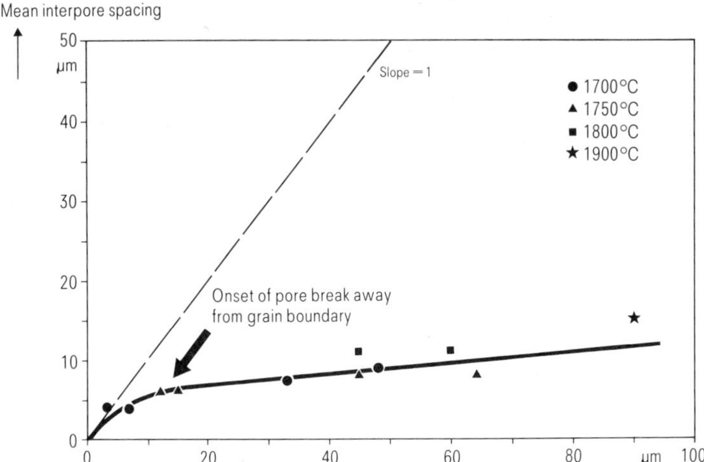

Fig. 3. Interpore spacing vs grain size for various sintering
temperatures and times.

removal where pore entrapment occurs within the grains. Most of the
data obtained in this study for sintering at 1700°C to 1900°C follow a
straight line with a slope less than one (Fig. 3). This emphasizes that
pores break away from grain boundaries at the beginning of final stage
sintering.

The process of pore entrapment is confirmed by the micrographs in
Fig. 4a-c. The fine-grained microstructure obtained after sintering at
1750°C for 1 hr (Fig. 4a) represents the initial situation of this pore
entrapment, where most of the pores are located at the grain boundaries
with a monomodal pore size distribution and an average pore size of
about 2 µm. Sintering at 1800°C for 2 hrs illustrates an intermediate
stage of this process, demonstrating the enhanced increase of grain
size (Fig. 4b). Now, a broad and bimodal pore size distribution exists
peaking at about 2 µm and 9 µm. The small pores ranging about 2 µm are
already trapped within grains, while the larger pores ($\approx$ 9 µm) are
principally located at the grain boundaries. Referring to the initial
situation shown in Fig. 4a, these larger pores are assumed to be formed
by a typically grain growth driven pore coalescence, which increases
the pore size.[10] At a later stage, shown by the microstructure of a
specimen sintered at 1900°C for 2 hrs (Fig. 4c), only pores trapped
within the enormous grown grains are left as a residue. The pore size
distribution becomes nearly monomodal again at about 2 µm, which is
close to the initial situation. This clearly emphasizes that only pores
in the grain boundary region were able to shrink to elimination due to
the higher effective diffusivity therein compared to the bulk region.

As a result of this study, the evolution of the microstructure by
grain growth, pore coalescence and pore removal reveals that
densificastion stagnates if early pore entrapment occurs (Fig. 1). The
fact that nearly complete pore removal is achieved for sintering at
1900°C (porosity $\leqslant$ 0.1%), where grain growth is most enhanced (Fig. 2),
is believed to be caused by a retardation of the grain growth. This may
be effected by pore interactions and/or the second phase formation
mentioned above, so that pores may stay at the grain boundaries long
enough to shrink to elimination.

408

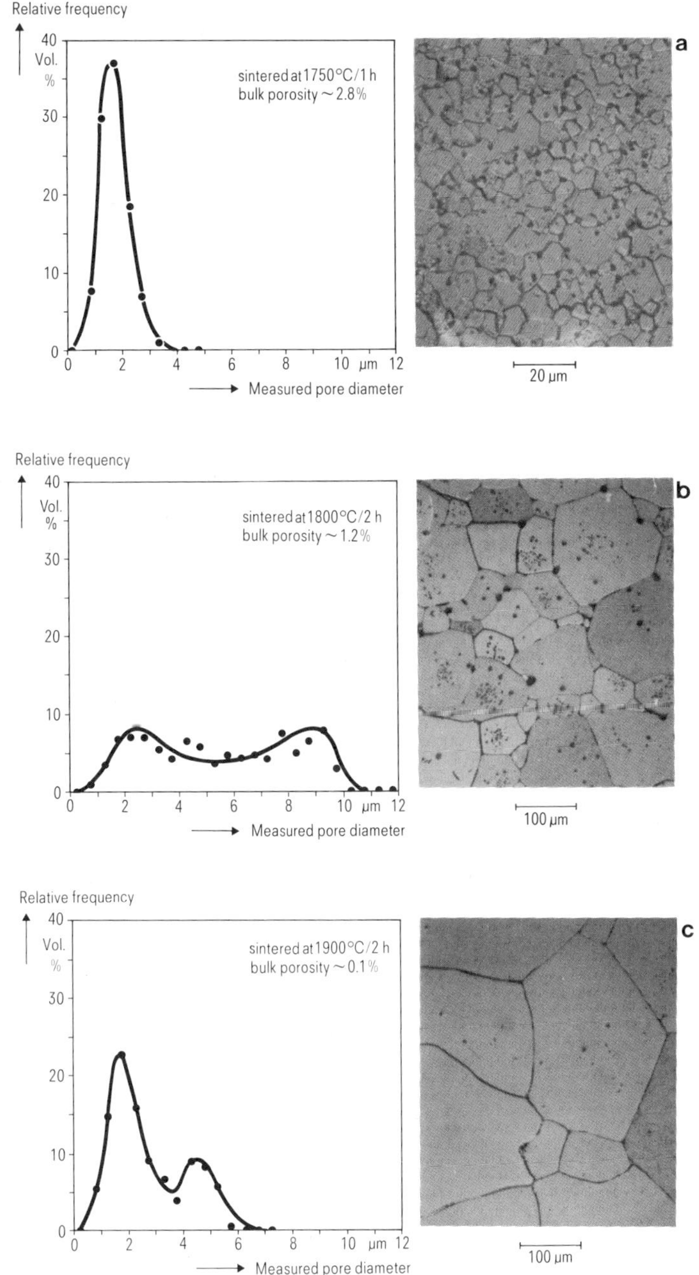

Fig. 4a-c: Microstructures and pore size distributions of Y_2O_3 doped with 30 mol% Gd_2O_3 sintered at various temperatures and times.

CONCLUSION

The mechanism of pore removal in the final stage of sintering of Y_2O_3 doped with 30 mol% Gd_2O_3 is shown to be principally grain boundary controlled by enhanced, but normal, grain growth. The relationship between interpore spacing and grain size indicates that pores break away from grain boundaries during sintering. Therefore densification stagnates if pore entrapment occurs at an early stage before complete pore elimination is achieved. In an intermediate stage of the pore removal, those pores located at the grain boundaries are increased by grain growth driven pore coalescence. Only these pores are removed during progressive sintering. For complete pore elimination, a retardation or inhibition of the grain growth is a favorable effect allowing pores to stay at the grain boundaries.

ACKNOWLEDGEMENTS

The author thanks J. Leppert for laboratory assistance, S. Kampermann for assistance with image analyses, and R. C. Harman for critical review.

REFERENCES

1. P. J. Jorgensen, R. C. Anderson, J. Am. Ceram. Soc. 50(11), 553 -558(1967).
2. R. A. Lefever, J. Matsko, Mat. Res. Bull. 2, 865-869 (1967).
3. E. Carnall, D. Pearlman, Mat. Res. Bull. 7, 647-654 (1972).
4. C. Greskovich, K. N. Woods, Ceram. Bull. 52(5), 473-478 (1973).
5. W. H. Rhodes, J. Am. Ceram. Soc. 64(1), 13-19 (1981).
6. K. Richardson, M. Akinc, Ceram. International 14, 101-108 (1988).
7. S. F. Horvath, M. P. Harmer, D. B. Willians, M. R. Notis, J. Mat. Sci. 24, 863-872 (1989).
8. G. C. Wei, T. Emma, W. H. Rhodes, J. Am. Ceram. Soc. 71(10), 820 -825 (1988).
9. M. K. -F. Lin, Ph. D. - Thesis, Lawrence Berkeley Lab., University California, (1988).
10. W. D. Kingery, B. Francois, J. Am. Ceram. Soc. 48(10), 546-547 (1965).

THERMOELECTRIC PROPERTIES OF LEAD TIN TELLURIDE COMPACTS

D.M. Rowe and M. Clee

University of Wales, College of Cardiff
School of Electrical, Electronic and Systems Engineering
POB 904, Cardiff, CF1 3YH, U.K.

ABSTRACT

In this paper the effects of grain size on the thermoelectric
properties of compacts of PbSnTe are investigated. The preparation of
compacts with a grain size L in the range 25<L<60, 10<L<25, 5<L<10 and
<5 m is described and the results of measurements of the thermal
conductivity, Seebeck coefficient and electrical resistivity reported.
At room temperature the thermal conductivity of compacts with a grain
size <5 μm is about 70% that of material with a grain size 25<L<60 μm
and this reduction increases further with an increase in temperature.
As pressed compacts possess degraded electrical properties compared to
"single crystal" material and heat treatment procedures are employed to
improve the electrical properties.

It is concluded that the thermal conductivity of PbSnTe alloys can
be reduced by the use of very small grain size compacted material. The
behaviour of the electrical properties both as a function of grain size
and temperature is problematic and attests the complicated relationship
between the physical and transport properties of compacted materials.
All three parameters which occur in the thermoelectric figure of merit
are sensitive to high temperature heat treatment and in judiciously
heat treated material the reduction in the thermal conductivity of the
smallest grain size material compared to that of the largest grain size
more than compensated for the degradation in electrical properties.
This resulted in an improvement in the thermoelectric figure of merit
of the specimen investigated.

INTRODUCTION

Alloys based upon Lead Telluride are established thermoelectric
materials and thermocouples fabricated using lead telluride technology
have been used in a number of space and military applications to
convert heat into electrical energy. Noteable examples of their
application are in the thermoelectric generators employed on the U. S.
Navy's TRIAD navigational satellite[1] (heat provided by a radioactive
isotope) and the range of multi-fueled thermoelectric generators
deployed by the United States Army.[2] Evidently in military and space
applications the fuel inventory is an important consideration and a

considerable effort has been made to improve the efficiency of the conversion system.

Having established the temperature conditions under which a thermoelectric generator operates, the conversion efficiency can conveniently be expressed in terms of a so-called figure of merit Z of the thermocouple material. $Z = \alpha^2\sigma/\lambda$ where α is the Seebeck coefficient, σ the electrical conductivity and λ the thermal conductivity. The thermal conductivity consists mainly of two components; λ_L due to the lattice vibrations and λ_e an electronic (hole) contribution. The parameters α, σ and λ_e are functions of the carrier concentration (n) and for lead telluride materials Z is optimized at carrier concentrations around 10^{24}-10^{25} m^{-3}. At these n values the lattice thermal conductivity still accounts for the major proportion of the thermal conductivity.[3] In recent years attempts to increase Z and hence improve the conversion efficiency have focussed on reducing the lattice thermal conductivity hopefully without adversely affecting the electrical properties. Some success has been achieved in reducing the lattice thermal conductivity of high temperature thermoelectric materials based on silicon germanium alloy by employing very small grain size compacted material (an effect attributed to phonon-grain boundary scattering).[4] This technique has also been extended to 2N-lead telluride material; in samples with mean grain size of 0.5 μm the thermal conductivity is about 9% less than equivalent "single crystal" material which is substantially greater than the 5% reduction predicted by theory.[5] However, as-compacted materials posses degraded electrical properties compared to "single crystal" material and heat treatment procedures have been employed to improve the electrical properties and in particular the electrical conductivity. Phonon-grain boundary scattering is enhanced in alloys and a theoretical investigation employing a realistic model for lead telluride alloys indicated that alloying PbTe with SnTe would result in a substantial reduction in thermal conductivity. In undoped PbTe-SnTe material with a grain size of <0.5 μm, the reduction in λ_L was estimated to be 15% which is reduced to around 11% in optimally doped material.[6] In this paper the results of an experimental investigation into the effect of grain size on the thermoelectric properties of compacts of lead tin telluride are reported.

MATERIAL PREPARATION

The starting material used in this investigation has a designation 3P* and was supplied by Global Thermoelectrics (Basano, Canada) either in the form of a pulled large grain size ingot or as coarse powder (stored under argon). The powder was seived through a set of British Standard microplate sieves using methanol as a vehicle and assisted by ultrasonic vibrations. Sieved fractions with size ranges 25<L<60, 10<L<25, 5<L<10 and L<5 μm were collected and used as the charge material. The presence of absorbed oxygen in the powdered material is thought to contribute to the fracture of compacts during pressing, consequently attempts were made to decrease the amount of oxygen

---

* Materials originally developed by the 3M Company (USA); they are identified by designations such as 2N. This signifies a specific semiconductor composed essentially of a compound of lead and tellurium with small additions of an electrically active impurity (0.03 mol percent PbI_2) to provide the required concentration of n-type electrical carriers, 3P signifies a lead tin tellurium combination doped with sodium and manganese – the actual composition is Pb-19.69%, Te-49.49%, Mn-3.46%, Na-0.47%.

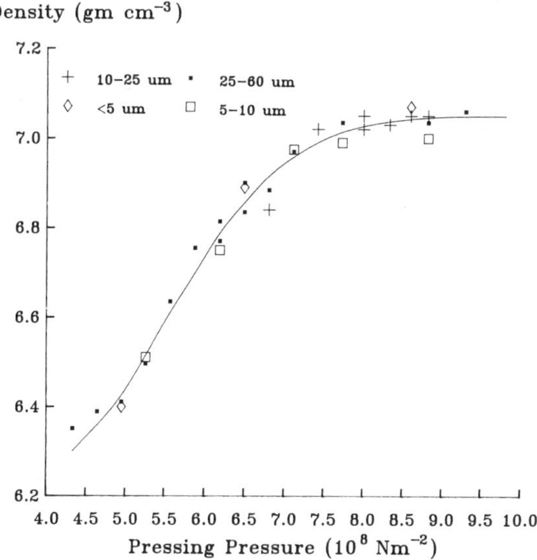

Fig. 1. Compact density vs compacting pressure.

present by heating the charge in hydrogen at 623K under an overpressure of 35 psi for 30 minutes.

Disc shaped compacts, having dimensions compatible with thermal diffusity measurements using laser flash technique (6.5 mm diameter) were pressed following a procedure reported at the previous meeting.[5] The die was loaded with an appropriate amount of powder which was sufficient to produce a compact between 1 and 2 mm thick. The high diameter to density ratio of the compacts assists in ensuring a uniform distribution of the pressing pressure and consequently of the density. Density determinations were made using Archimede's method. Single crystal density is 7.14 gm cm^{-3} and the density of the compacts increases with pressing pressure from about 6.4 gm cm^{-3} at 5 x 10^8NM^{-2} to 6.99 gm cm^{-3} at 9 x 10^8NM^{-2}, i.e. better than 98% of the density of single crystal material, as shown in Fig. 1.

The thermoelectric transport properties and in particular the electrical conductivity are very sensitive to the method of preparation. Following their removal from the die the compacts were annealed at 973K in an atmosphere of 45 psi argon plus 10 psi hydrogen for 2 hours.

TRANSPORT PROPERTIES MEASUREMENT

All three parameters which occur in the thermoelectric figure of merit were measured as a function of temparature. The Seebeck coefficient values were obtained by establishing a small temperature gradient (about 5K) along an ingot shaped specimen (5x2.5x2mm) and measuring the resulting Seebeck voltage. The accuracy of measurement was limited by the error in determining the slope of the Seebeck voltage vs temperature plot and was estimated to be about 3%. Electrical resistivity measurements were made on bridge shaped samples (4.4x1.2x2 mm) cut from the compacts using an ultrasonic drill. Unwanted electrical noise mainly associated with temperature fluctuations in the specimen were minimised by employing an a.c.

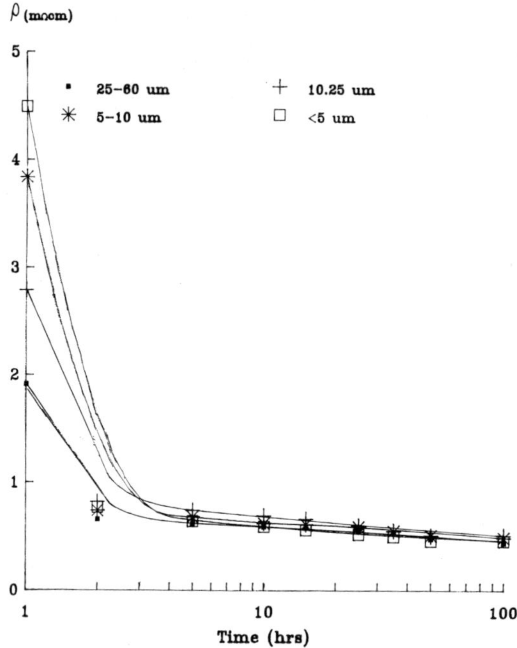

Fig. 2. Room temperature electrical resistivity ρ vs annealing time.

technique. Accuracy of measurements was about 3%. Thermal diffusivity
measurements were made on disc shaped specimens 6 mm in diameter using
a laser flash system.

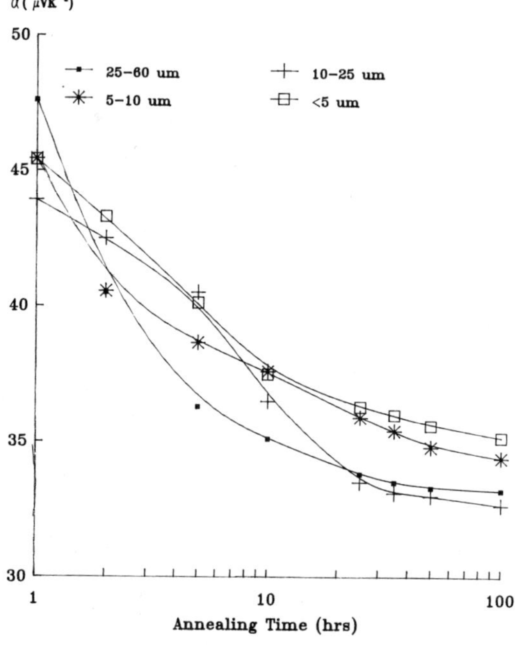

Fig. 3. Room temperature Seebeck coefficient α vs annealing time.

414

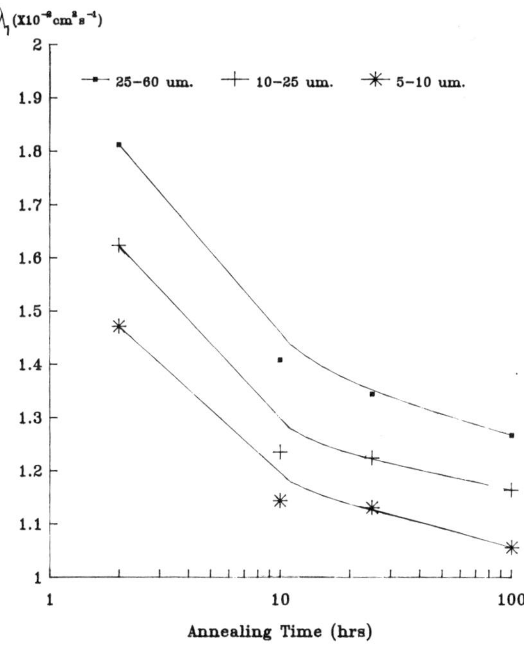

Fig. 4. Room temperature thermal diffusivity λ_1 vs annealing time.

RESULTS AND DISCUSION

The thermoelectric transport properties and in particular the electrical conductivity are sensitive to the preparation procedure. As-

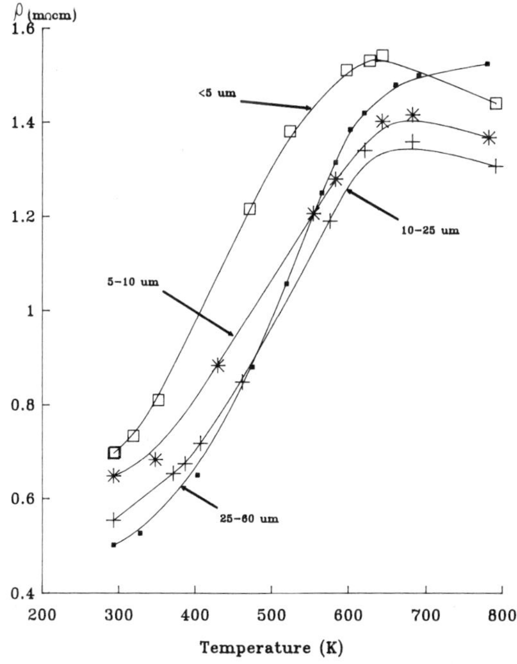

Fig. 5. Electrical resistivity ρ vs temperature.

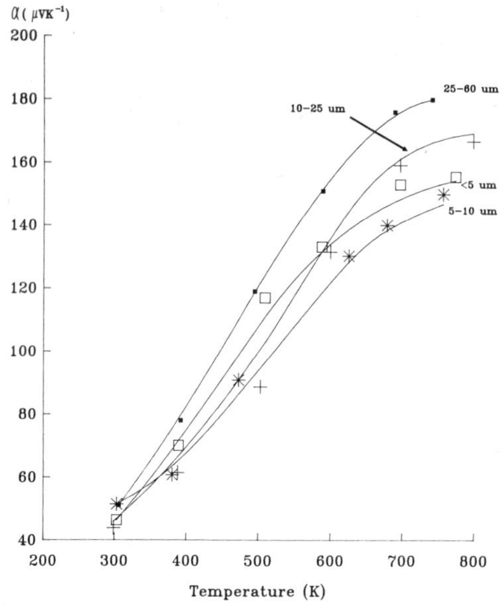

Fig. 6. Seebeck coefficient α vs temperature.

pressed materials possess electrical resistivity values many times higher than single crystal material. A decrease in the electrical resistivity can be achieved by high temperature heat treatment of the specimens although this improvement is accompanied by an unwanted reduction in the Seebeck coefficient. The effect of heat treatment on the electrical resistivity, Seebeck coefficient and thermal diffusivity are presented in Fig. 2, 3 and 4 for different grain size material. Of concern is the irregular behaviour of the Seebeck coefficient. However, both the electrical resistivity and the thermal diffusivity decrease

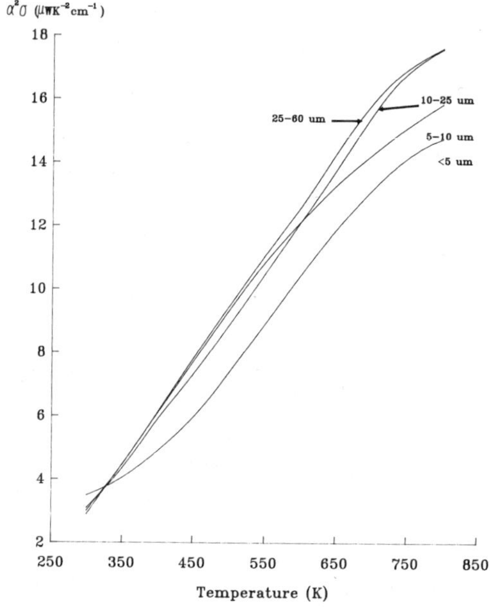

Fig. 7. Electrical power factor $\alpha^2\sigma$ vs temperature.

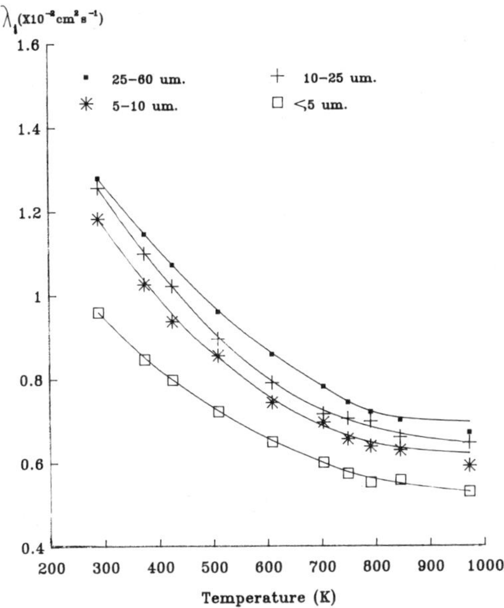

Fig. 8. Thermal diffusivity λ_1 vs temperature.

rapidly with time which is beneficial to increasing the figure of merit
while the reduction in the Seebeck coefficient, which is detrimental,
is less dramatic. As indicated previously, a heat treatment period of
about two hours at 973K appears to be about optimum and resulted in the
material attaining electrical properties approaching single crystal
values. Subsequent measurements were carried out on samples which had
undergone heat treatment for this period of time. The temperature
dependence of the electrical resistivity Seebeck coefficient and
electrical power factor for different grain sizes are presented if Figs.

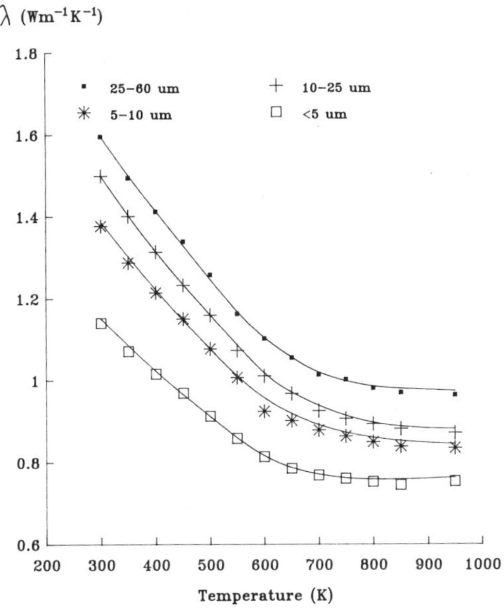

Fig. 9. Thermal conductivity vs temperature.

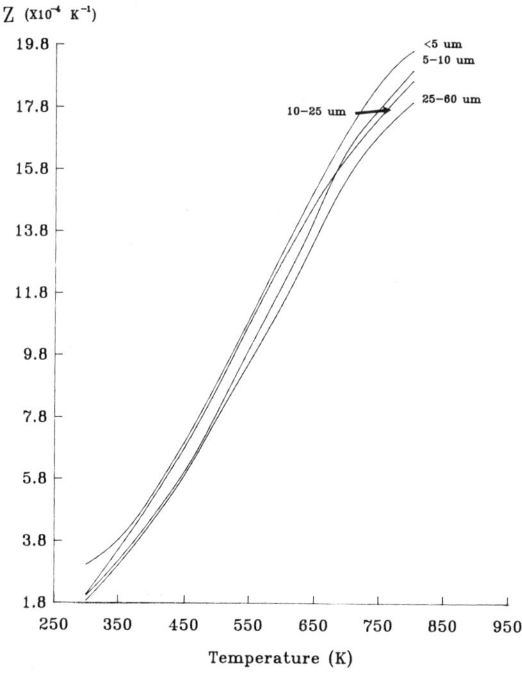

Fig. 10. Figure of merit z vs temperature.

5, 6 and 7. Although there are considerable inconsistencies in the electrical behaviour of these materials it is evident that over the temperature range from just above room temperature to the upper temperature limit of about 800K the electrical power factor suffers a degradation with a decrease in material grain size. The lattice thermal conductivity displayed in Fig. 8 decreases with reduction in grain size although the magnitude of the reduction is greater than predicted by theory. The thermal conductivity shown in Fig. 9 is derived from the diffusivity data using the relationship $\lambda = Cd\lambda_1$, where C is the specific heat capacity and d the density. The value of C was assumed to be independent of grain size and values provided by Global Thermoelectrics were used. Finally, the thermoelectric figure of merit is presented in Fig. 10 as a function of temperature and for different grain sizes. Up to a temperature of around 650K the figure of merit increases with a decrease in material grain size; the anomalous behaviour of the figure of merit of the intermediate 5-25 µm grain size material is considered to be a consequence of the behaviour of the electrical properties.

CONCLUSIONS

It is concluded that the thermal conductivity of PbSnTe alloys can be decreased by the use of very small grain size compacted material. The reduction in thermal conductivity is accompanied by a degradation in the electrical properties. The behaviour of the electrical properties both as a function of grain size and temperature is problematic and attests the complicated relationship between the physical and transport properties of compacted materials, and requires further investigation. All three parameters which occur in the thermoelectric figure of merit are sensitive to high temperature heat treatment, and in judiciously heat treated material the reduction in the thermal conductivity of the

smallest grain size material compared to that of the largest grain size
more than compensated for the degradation in electrical properties.
This resulted in an improvement in the thermoelectric figure of merit.

ACKNOWLEDGEMENT

The United States Army in Europe is thanked for providing support
during the initial stages of the project and Global Thermoelectrics for
kindly supplying the lead telluride materials. The University of Wales
is thanked for providing a research studentship for M. Clee.

REFERENCES

1. Kelly, C. E., 1975, Proc. 10th IECEC, Newark Delaware, p.880.
2. Guazzoni, G., 1987, Proc. 1st European Conf. on Thermoelectrics,
        Cardiff, U.K., Ed. D.M. Rowe, p. 302.
3. Rowe, D. M., 1986, Proc. 6th ICTEC., University of Arlington,
        Texas, p. 43.
4. Bhandari, C. M. and Rowe, D. M., 1988, Thermal Conduction in
        Semiconductors, Wiley Eastern.
5. Rowe, D. M., 1985, Proc. VI World Round Table Conference on
        Sintering, Herceg Novi, Yugoslavia, Eds. G.C. Kuczynski,
        D.P.Uskoković, Hayne Palmour III, M.M. Ristić, Plenum Press,
        New York, 1987, p. 215.
6. Rowe, D. M. and Bhandari, C. M., 1985, Appl. Phys. Letts., 47(3),
        p. 255.

MICROSTRUCTURAL DEVELOPMENT IN DENSE Si_3N_4 CERAMICS

M. Herrmann, S. Hess, H. Kessler, J. Pabst, and W. Hermel

Central Institute of Solid-State Physics and Materials
Research Academy of Sciences of the DDR, Dresden, DDR

INTRODUCTION

During the last several years, silicon nitride ceramics have
gained increasing importance. This has been brought about by both the
potential application possibilities and progress in the field of
materials development technology and microstructural design.[1]

Previous work[2,3] indicated that the relatively high fracture
toughness of Si_3N_4-materials is due to its needle-like grain
morphology. The needle-like grain growth is, according to[1,2], connected
with the α- to β-phase transformation in silicon nitride.

The crystal structures of the α- and β-Si_3N_4 phases are quite
similar. Both forms are hexagonal, only with a different packing
sequence of the Si_3N_4 tetrahedra in the c-direction. The first order
coordination in these two forms is the same.[3,4] The α- to β-phase
transformation is a reconstructive one of secondary coordination. It is
possible only in the presence of a liquid phase or at high partial
pressures of silicon and nitrogen near the sublimation point. In
powders with a high initial ratio of α- to β-Si_3N_4, the first stage of
the transformation of α- to β-Si_3N_4 is the nucleation period. Taking
into account the similar crystal structures of the α- and β-phases, we
have to consider beside homogeneous also heterogeneous nucleation.

However, the detailed mechanism of the α to β-phase
transformation, as well as its effect on microstructural development
during liquid phase sintering, is actually not fully understood. The
aim of this paper is to investigate the conditions of needle-like grain
growth of β-Si_3N_4 during sintering.

For this purpose, the microstructural development during sintering
of MgO doped reaction bonded silicon nitride (RBSN) and sintered
silicon nitride (SSN) and the influence of additional β-nuclei are
analysed.

EXPERIMENTAL PROCEDURE

One starting material was RBSN prepared by direct nitridation of

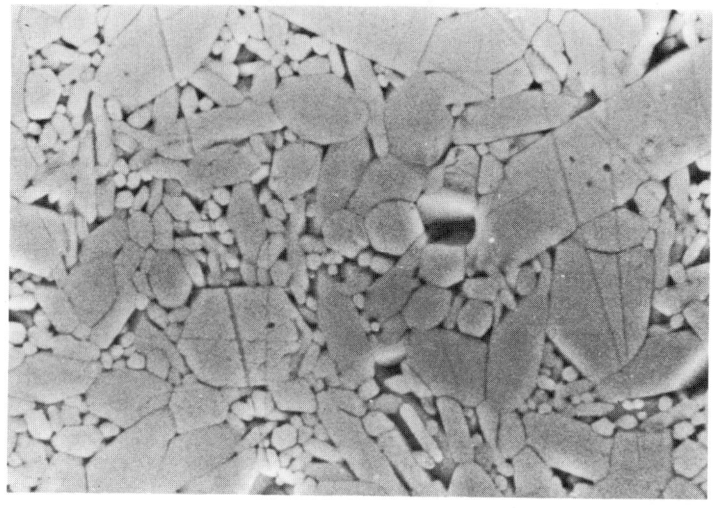

Fig. 1. SEM photomicrograph of an polished etched section of the
material RBSN-2.

silicon mixed with 8% MgO. Some characteristics of the RBSN materials
are given in Table I. The details of the preparation of these materials
were described in previous work.[5]

The starting mixtures of the materials RBSN-2 and -3 were prepared
by the addition of 4.5% and 9% β-Si_3N_4 powder, respectively.

The β-Si_3N_4 powder was obtained by heat treatment at 1850°C for 90
min at a nitrogen pressure of 30 atm. After heat treatment, it was
milled in a ball mill for 4 h. The resulting grain morphology was
nearly spherical with a mean grain size of 0.7 µm. The oxygen and iron
contents were 0.25% and 0.1%, respectively.

As starting powder for the SSN-material, a commercial silicon
nitride powder with the following characteristics was used: α/β-Si_3N_4
96:4, oxygen content 2.3%, mean grain size 0.83 µm, surface area
9,8 m^2/g. The sintering process was carried out in a gas pressure
furnace with graphite heating elements.

The microstructure of the sintered materials was characterised by
analysing SEM photomicrographs of etched polished samples (Fig. 1).

For all grains, the length and thickness were measured and the
apparent aspect ratio was calculated. The distribution of aspect ratios
determined in this way differs from the real distribution of aspect
ratios in the material, since the measured aspect ratio depends on the
orientation of the grain in relation to the polished section. In order
to determine the distribution of the real aspect ratio of the grains in
the volume of the material, a mathematical model was developed.[6] The
results of the analyses are shown in Table II.

In order to investigate the α-β-transformation, we used, in
addition to the pressed samples, Si_3N_4-powders, as follows:

1. α-Si_3N_4-powder, produced by nitridation of a mixture of Si and 8%
   MgO (α-Si_3N_4 content 98%, MgO content after nitridation: 2%),

2. the commercial powder mixed with 5% MgO.

Table I. Properties of the samples (RBSN-state)

Number	Composition	Phase content α/β Si_3N_4 %	Porosity %	Strength σ_{3B}, MPa
1	RBSN + 5 % MgO	78 : 22	32	200
2	RBSN + 5 % MgO + 4.5 % β-Si_3N_4	74 : 26	28.4	200
3	RBSN + 5 % MgO + 4.5 % β-Si_3N_4	69 : 31	29.2	2

Table II. Microstructure and properties of materials sintered at 1850 °C (A_m – maximal measured aspect ratio)

Material	Sintering time (min)	Thickness of grains (μm)	A_m	Portion of grains with aspect ratio larger than 3.5 (%)	5.5 (%)	Content of Mg (%)	0 (%)
RBSN 1	30	0.34	11	85	70	1.8	0.4
	90	0.50	14	–	–	1.8	0.6
	150	0.54	16	73	59	1.2	0.2
RBSN 2	30	0.78	8	33	18	1.3	0.9
	90	1.1	7	–	–	1.3	0.8
	150	1.3	8	30	21	0.9	0.2
RBSN 3	90	0.96	10	44	31	1.5	1.0
SSN	30	0.47	10			3.3	3.2
	150	0.65	10	71	62	3.1	3.3

These powders were heat treated with a heating-up rate of 70 K/min. The dependence of the grain morphology on the soaking time for both starting powders was characterised by scanning electron microscopy (SEM) and is shown in Fig. 2.

X-ray diffraction was used to determine the α- and β-Si_3N_4 contents.

RESULTS AND DISCUSSION

The investigation of the α/β-transformation in the RBSN and the commercial powders shows that after 10 min of isothermal sintering, the α/β-conversion was nearly complete in both powders (Fig. 2, Table III). However, the morphology of the two powders is quite different. The RBSN-powder shows a needle - like structure, while the commercial powder consists mainly of equiaxed grains. Only after 30 min soaking time does the shape of the grains changes to a needle - like one, but these needles are bigger than those of the RBSN-powder. These experiments show that the α/β-transformation and the needle - like grain

Table III. Dependence of the α- to β-phase transformation in RBSN and SSN powders doped with MgO on sintering times of 1750 $^{\circ}$C

Sintering time (min)	α/β -phase content	
	RBSN	SSN
initial state	98 : 2	96 : 4
0	86 : 14	98 : 11
10	1 : 99	15 : 95
30	0 : 100	0 : 100

growth depend on the initial state. We assume that their differences are due to the different state of homogenization of MgO in these two powder mixtures. During the preparation of the RBSN-powder, an homogenization of the MgO takes place[5] and the α/β -phase transformation is faster in this case (Tab. II). The experiments show, also, that the needle - like grain growth is possible after the complete α- to β-phase transformation. Needle-like grain growth after complete α- to β-phase transformation was also found during isothermal sintering of RBSN-1 at 1850°C and a nitrogen pressure of 30 atm (Tab. III). The data in Table II show that the maximum measured aspect ratio during sintering increases, but the quantity of grains with high aspect ratio decreases. In the materials RBSN-2 and -3 containing additional β-Si$_3$N$_4$, the resulting grain size after sintering was two times larger than in the material RBSN-1. Soaking times of more than 30 min do not lead to increasing aspect ratio. After sintering, in the material RBSN-3, only about 30% of grains have an aspect ratio greater than 3.5, whereas the amount of such grains in the sintered RBSN-1 materials is higher than 73%. Nearly the same situation with some differences was found after sintering the commercial powder. The mean aspect ratio was smaller and did not increase with soaking time. But the quantity of grains with an aspect ratio higher than 3.5 was nearly the same as after sintering the RBSN-1 material.

The growth in the thickness of grains can be described by the law which was found for Ostwald-ripening ($d^n-d_o^n$ = Kt; d-thickness of grains, t-time; with n $\approx$ 3). Since this law is based on the assumption of a dilute solution and equilibrium shape of the crystals, the exponent n cannot be used to distinguish between interface and diffusion controlled grain growth. The distribution of grain thickness d is a steady state one and can be described in the same way for all sintering times as a function of d/d_{50} (d_{50} - mean diameter). Taking into account that in the process of Ostwald - ripening, only a small portion of the large grains remains stable, it becomes clear why the addition of a small amount of β-Si$_3$N$_4$ to the starting powders RBSN-2 and -3 could so drastically change the microstructure and properties. These β-grains, acting as relatively large and therefore stably growing nuclei, determine the resulting microstructure.

As shown in Table III, there is a wide distribution of aspect ratios in all materials, which depends on the initial microstructure, the shape and size of the largest nuclei in the beginning of the sintering, and the sintering conditions. There is no evidence that only one aspect ratio exists in these materials for all grains, as is assumed in the literature.[1]

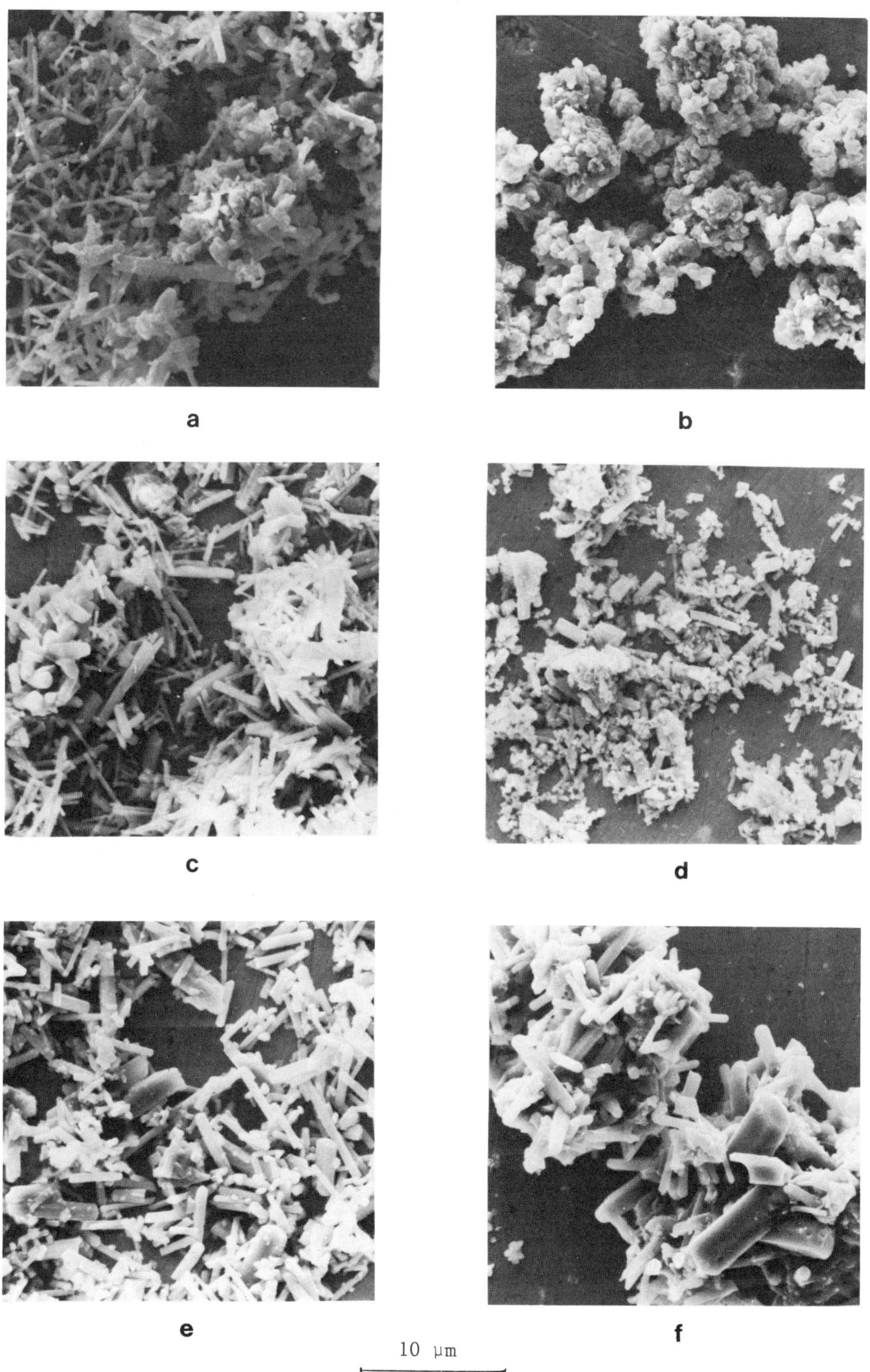

Fig. 2. SEM photomicrographs showing the morphology of the RBSN (a,c,e) an SSN powder (b,d,f) after different sintering times (a,b-0 min.; c,d-10 min; e,f-30 min.) at 1750 °C.

The question, what causes the development of a needle like microstructure in Si_3N_4-ceramics remains open. However, some arguments can be given that such anisotropic grain growth could be stimulated by an interface controlled growth mechanism during and after the α/β-phase transformation. A different roughness of solid-liquid interfaces found in reference 8 may be a reference to different growth mechanisms for the (210) and (001) planes. Another argument for such interface controlled growth is investigations found for hard metals that point out that the tracht of the resulting crystals consists only of a small number of several crystal planes. Analogously, the needle-like shape of the β-Si_3N_4 crystals may be the evidence for such a growth mechanism. Anisotropic grain growth can take place only when the supersaturation of the melt is not too low. If it is too low, the higher surface energy of the plane (001)[10] leads to a solution of this plane and to a growth of the thickness. This was found by G. Wötting and G. Ziegler[1] for hot-pressed silicon nitride with coarser microstructures.

SUMMARY

The investigations of microstructure of sintered reaction bonded silicon nitride and sintered silicon nitride conducted in this work can be summarized in the following items:

- The kinetics of the α- to β-phase transformation is different in the RBSN and SSN investigated.

- The needle – like grain growth can take place after completion of the α- to β-phase transformation.

- The microstructure of the resulting material depends on the size and shape of the biggest β-nuclei at the beginning of the sintering process. The addition of 4-10% of big β-Si_3N_4 nuclei leads to intensive grain growth and a decreasing amount of needle-like grains.

- In the investigated materials, a wide distribution of aspect ratios of grains was found.

REFERENCES

1. G. Ziegler, J. Heinrich, G. Wötting, "Review Relationships between Processing, Microstructure and Properties of Dense and Reaction-Bonded Silicon-Nitride", J. Mat. Sci. 22: 304 (1987).
2. F. F. Lange, "Fracture Toughness of Si_3N_4 as a Function of the Initial Phase Content", J. Am. Ceram. Soc. 62: 428 (1979).
3. K. Kato, Z. Inoue, K. Kijima, I. Kawada, H. Tanaka, "The Crystal Structure of α-Si_3N_4", J. Am. Ceram. Soc. 58: 90 (1974).
4. R. Grün, "The Crystal Structure of β-Si_3N_4", Acta Cryst. B. 35: 800 (1979).
5. J. Pabst, M. Herrmann, "The Kinetic of Postsintering of Reaction-Bonded Silicon Nitride with Different α/β Phase Content", to be published in Sci. of Sintering.
6. P. Obenaus, Ch. Schubert, H. Kessler, S. Hess, "Zur quantitativen Bewertung des Zusammenhanges zwischen Kornmorphologie des β-Si_3N_4 und der Bruchzähigkeit von dichtem Si_3N_4", in: IX. Pulvermetallurgische Tagung, Dresden (1989).
7. C. Waagner, "Theorie der Alterung von Niederschlagen durch Umlosung", Z. Elektrochemie 65: 581 (1961).

8. C. M. Hwang, T. Y. Tien, J-Wei Chen, "Anisotropic Grain Growth in the Final Stage Sintering of Silicon Nitride Ceramics", in Sintering '87, Tokyo: 566 (1987).

9. R. Warren, M. B. Waldron, "Microstructural Development during the Liquid-Phase Sintering of Cemented Carbides", Powder Metal. 15: 166 (1972).

10. P. Hartman, "Structure and Morphology", in: "Crystal Growth. An Introduction", P. Hartmann ed., North-Holland Publishing Company, Amsterdam (1973).

EFFECT OF SINTERING PARAMETERS ON MICROSTRUCTURE

AND PROPERTIES OF SIALON MATERIALS

P. Arató, E. Besenyei, A. Kele, and F. Weber

Research Institute for Technical Physics of the
Hungarian Academy of Sciences
Budapest, POB 76, H-1325

ABSTRACT

The influence of some parameters of pressureless sintering on mechanical
properties, microstructure and phase composition has been studied. By using
different types of silicon nitride as embedding powder it has been shown that,
besides weight change, density and bend strength of the samples also depend
on the type of powder bed. The composition of the material was varied by
changing the ratio of alumina/aluminum nitride while keeping the amount of
sintering aid constant. By increasing the oxygen content we found deterio-
rating mechanical properties and increasing weight loss. The observed variation of microstructure and phase composition with sintering time, temperature
and initial composition can be attributed to the changing composition of the
liquid phase.

INTRODUCTION

One of the most promising engineering materials is silicon nitride. It
has attracted great attention for years due to its unique properties of high
strength, thermal shock resistance, high toughness and high thermal resistance.
Silicon nitride cannot be densified without a sufficient amount of sintering
aid because its self diffusivity is very low at low temperature: at high
temperature, however, decomposition occurs if considerably high nitrogen overpressure is not used. Sintering aids such as Al_2O_3, AlN, and Y_2O_3 are used
commonly in pressureless sintering[1-4] and embedding powders are applied to hinder
decomposition processes.

Despite of the great amount of research on this material, the sintering
mechanisms are not fully understood. A generally accepted description of sintering involves the rearrengement of particles in the liquid formed by the
sintering aids, followed by the dissolution of α Si_3N_4 grains and the precipitation of β sialon. β sialon with the formula $Si_{(6-z)}Al_zO_zN_{(8-z)}$ incorporates Al and O in the β Si_3N_4 structure. Although β sialon is believed to
be the stable form, α sialon with the formula $Me_x(SiAl)_{12}(ON)_{16}$ which incorporates metallic cations such as Y^{3+} besides Al and O in the α Si_3N_4 structure, was observed in several cases. The appearance of α Sialon is not
explained by the above mechanism and only a few experiments have been reported[5,6]
on its effect on the microstructure and mechanical properties.

Science of Sintering
Edited by D. P. Uskoković *et al.*
Plenum Press, New York

In the present study, we have investigated the influence of some sintering parameters on the density, hardness, bend strength, phase composition
and microstructure of sialon materials. As in pressureless sintering, the
embedding powder has an important influence on the densification process,
thus different types of powders were applied. In order to study the influence of the oxygen content; a series of sample were prepared keeping
the amount of sintering aid constant but variing the ratio of alumina/alumi-
num nitride. We used a relatively large amount of sintering aid (31w%) to
enlarge the range of oxygen content without going above the β sialon region
in the phase diagram.[7]

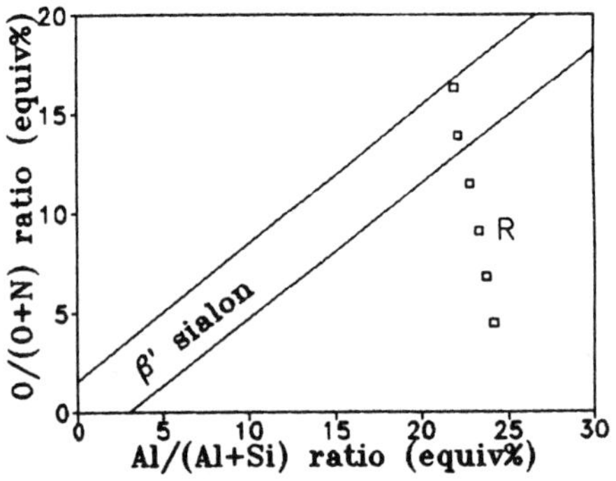

Fig.1.   Compositions of the samples

EXPERIMENTAL PROCEDURE

The compositions of the samples are shown in Fig.1. All samples contained 4% Y_2O_3 which was not included in the calculation of equivalent aluminum and oxygen content.[5,7]   Composition R of intermediate oxygen content
was used in the experiments of the investigation of the sintering parameters.
The mixed powders were milled in ethanol in a planetary type alumina ball
mill for 3 hours. Samples were dry pressed and pressureless sintered in a
nitrogen atmosphere in a graphite resistance furnace. The types of embedding
powders are shown in Table 1. We measured weight change and density by the
Archimedes method to characterize the sintering process. Hardness and indentation fracture toughness were determined using a Vickers diamond indenter
on a polished surface. The room temperature modulus of rupture (MOR) and the
modulus of elasticity were measured in four point bending with a span of
40 by 20 mm. In some cases three point bending was used on the as-sintered
surface. Phase analysis by the XRD method was carried out to determine the
phases present on the as-sintered samples. Microstructure of the fractured
surface was investigated by scanning electron microscope.

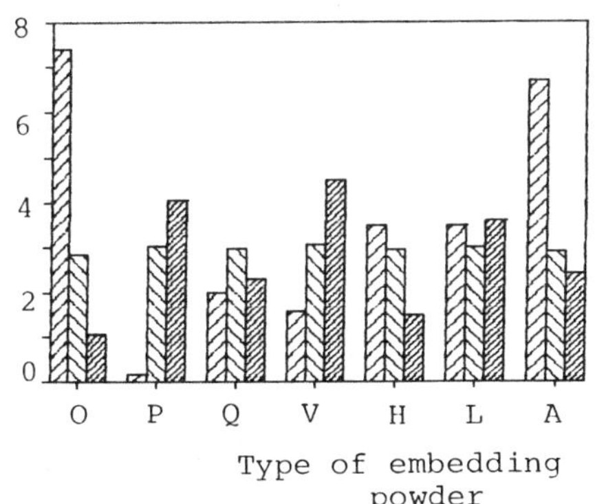

Table 1.   Type of embedding powders

Reference letter	Powder type	BET surface area m^2/g
O	no powder	-
P	β-Si_3N_4	0.4
Q	used P	-
$\tilde{V}$	α-Si_3N_4	3.1
H	used V	-
L	α-Si_3N_4	16,9
A	Al_2O_3	2.4

Fig.2   The effect of embedding powders on weight loss (▨ %),
density (◪ g/cm^3) and 3-point bend strength (▨ 100 MPa)
for samples sintered at 1730 $^{\circ}$C for 3 hours. (0 sintered
at 1700 $^{\circ}$C)

RESULTS AND DISCUSSION

The effect of powder bed on weight loss, density and 3-point bend
strength using different type of powders is shown on Fig.2. Without a powder
bed (0) great weight loss, porous surface and low bend strength were obtained
sintering even at 1700 $^{\circ}$C. With an inert powder bed (A) the weight loss was
less but the sample did not densify properly. Three different types of Si_3N_4
were tested. Powder L with a BET surface area of 16.9 m^2/g did not prove to
be suitable for embedding. The other two powders with low BET surface areas
(P,V) could be applied as embedding powder independently of their structure.
The second use of the powders (Q,H) resulted in higher weight loss, lower
density and bend strength. From XRD patterns we could observe the appearance
of SiC after the first run, but no considerable change in oxide content.

In pressureless sintering, a powder bed is applied to provide high enough
SiO pressure in order to hinder decomposition processes.[8] Either pure silicon
nitride or a mixture of similar composition to that of the compact were reported as packing powders[1,2]. In some cases, SiO_2 was added to silicon
nitride.[9]

Table 2.   The effect of sintering temperature

Sintering Temperature $^{\circ}$C	Sintering Time h	Weight loss %	Density g/cm^3	3-point bend strength MPa	$\alpha/(\alpha+\beta)$ %
1700	3	0	3.00	349	47
1730	3	0,2	3.05	406	44
1840	3	3.0	3.03	224	72
1700	6	0.5	3.06	394	37
1730	6	1.4	3.07	412	32
1750	6	1.3	3.06	352	59

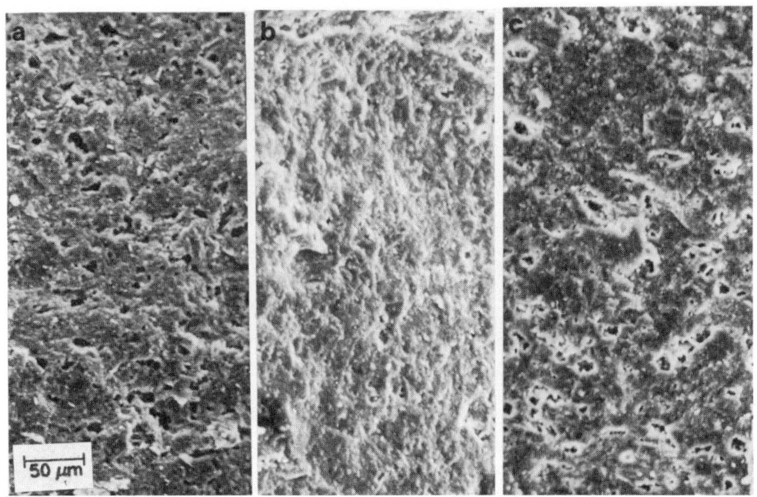

Fig.3. Microstructure of samples sintered at a. 1700 $^{\circ}$C 3 h
b. 1730 $^{\circ}$C 6 h, c. 1750 $^{\circ}$C 6 h.

If the SiO pressure is higher than that of the compact, SiO will enter the open pores and condense there, resulting in weight gain. If it is lower, decomposition of the material will occur and weight loss is observed, so the relative SiO pressure is indicated by the direction of weight change. In our experiments V powder proved to be the best embedding powder regarding density and bend strength while the weight loss was not the lowest one. It shows that it is not enough to seek zero weight change but density and bend strength also have to be considered. The contradiction regarding the reported us of pure silicon nitride and its mixture with silicon dioxide as embedding powders[1,9] can be explained by taking into account the dependence of the relative SiO pressure on the type of silicon nitride that we have shown here.

The effect of the sintering temperature on the properties of composition R is shown in Table 3. Samples were sintered for three or six hours at different temperatures. The P type of embedding powder was used. The best results were achieved at 1730 $^{\circ}$C for both times. In Fig. 3, the fractured surfaces of samples sintered at different temperatures are shown. After sintering at 1730 $^{\circ}$C for 6 hours, the least amount of pores can be observed.

Table 3. The effect of sintering time at 1730 $^{\circ}$C for composition R

Sintering time h	Weight loss %	Density g/cm^3	4-point bend strength MPa	Hardness HV GPa	$\alpha/(\alpha+\beta)$ %
0.1	-	2.88	-	-	22
0.5	0.4	2.93	319	11.7	3
1	0.7	2.97	367	12.8	5
3	1.3	3.05	333	14.1	53
6	2.3	3.11	323	14.5	32

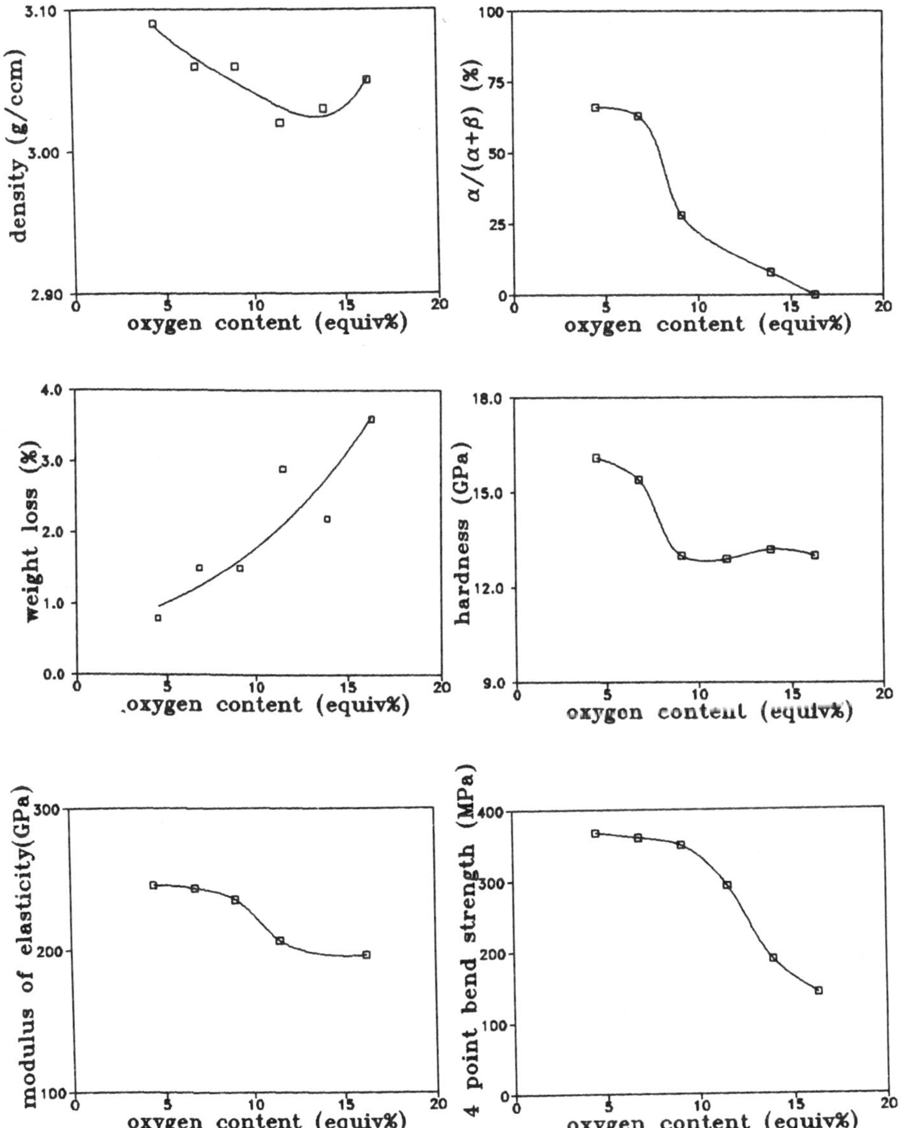

Fig. 4. The effect of oxygen content on weight loss, density, 4-point bend strength, hardness and modulus of elasticity for samples sintered at 1730°C 3 h.

At lower temperature, not all the pores have disappeared yet, while in the sample sintered at 1750 ^{O}C they seem to increase both in number and size probably due to decomposition. The decreased bend strength for the samples sintered at higher or lower temperatures may be attributed to the pores present in the material. Phase analysis of the as-sintered surface shows an increasing relative α sialon content at high temperature.

The effect of sintering time at 1730 ^{O}C is shown in Table 3. In these experiments, V powder was used for embedding. After the first hour of sintering, no major change in bend strength was obtained, while density and hardness increased with time and an increasing weight loss was observed. Analysing the phase composition of the samples, an initial decrease was followed by an increase in α content. The former is probably due to the dissolution of α Si_3N_4 grains while the latter can be attributed to the appearance of α sialon.

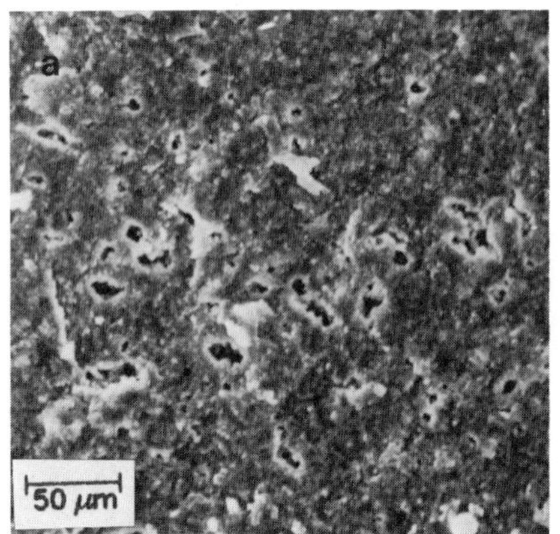
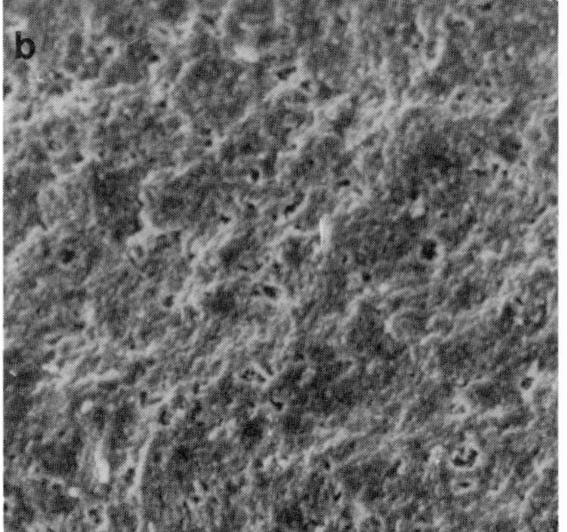

Fig. 5. Microstructure of samples of different oxygen content
a.13.9 equiv 0%, b. 4.5 equiv 0%

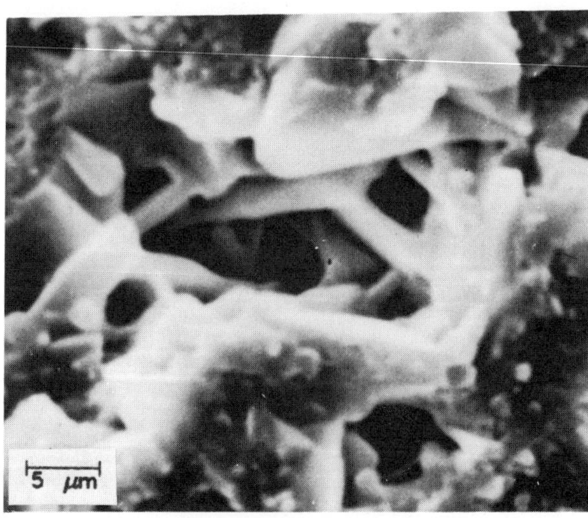

Fig.6.   Microstructure of a large pore into which elongated grains
grew

434

Samples with different oxygen content were used to determine the effect
of composition. The results are shown in Fig.4. By increasing the oxygen
content, increasing weight loss was observed. There was a slight minium in
density for the sample with 11.5 equiv 0%. 4-point bend strength, the modulus
of elasticity as well as the relative amount of α sialon considerably increased
when the oxygen content was decreased. Hardness increased with decreasing
oxygen content under 9 equiv 0%. Fracture toughness of the samples was
5 ± 1 MPam$^{1/2}$ and no systematic variation was found with oxygen content. In
Fig.5 the fractured surface of sample with 13.9 and 4.5 equiv 0% are shown.
Relatively smooth surface almost without large pores can be seen on the
samples with small oxygen content while a significant number of pores can be
observed if the oxygen content was high. At higher magnification, the pores
seem to be voids into which elongated grains grew (Fig.6.)

The experiments revealed that all the properties used to characterize
the sintering process are dependant on the composition of the liquid phase.
In the beginning of sintering of a material with a composition under the β
line in the phase diagram, the development of β sialon seems to be easier
due to the higher oxygen content of the first formed liquid. In the course
of sintering, the nitrogen content of the glass will grow because of the dissolution of Si_3N_4 and AlN and the precipitation of β sialon. The
increasing nitrogen content may lower the rate of precipitation of β and give
rise to the development of α sialon.

The sintering of the materials with different compositions can also be
explained by the oxygen content of the liquid phase. When its composition is
close to that of the β sialon the precipitation rate is probably high and it
will grow into the voids making it difficult for the material to density
further. If the oxygen content is lowered, the material can density while α
and β sialon are forming slowly. The increase in strength and modulus of
elasticity may be attributed to the absence of large pores.

CONCLUSION

It has been shown that by applying different types of Si_3N_4 as an
embedding powder in pressureless sintering, weight change, density and bend
strength of the sample all depend on the type of powder bed. It was also
shown that weight change depends not only on the powder bed but on the composition of the compact, too.

Sintering time and temperature, as well as the composition, were shown
to influence the mechanical properties and phase composition of the sintered
sialon materials. Bend strength, hardness and modulus of elasticity were
increased by decreasing the oxygen content. The higher the oxygen content
in the initial composition, the lower the amount of α sialon phase that
was found. The rate of precipitation of β sialon probably depends on the
composition of the liquid phase: either we deliberately change the initial
composition or it will change during the course of sintering, the process of
sintering will depend on it. At high oxygen contents, because of the fast
precipitation of β sialon, large pores remain in the structure resulting
in poor mechanical properties. At low oxygen content the development of α and
β sialon is not as fast, which results in a more favourable microstructure.

This work was financially supported by the Government Committee of
Research and Development.

REFERENCES

1. J.Briggs, Pressureless sintering of sialon ceramics, Mat.Res.Bull.
   12:1047 (1977)

2. S. Boskovic, L.J.Gauckler, G.Petzow and T.Y.Tien, Reaction sintering forming β -Si_3N_4 solid solutions in the system Si/Al/N/O: sintering of Si_3N_4-AlN-Al_2O_3 mixtures, <u>Powder Metall.Int.</u> 11(4):169 (1979)

3. Xu Jie and Sheng Xumin, The influence of protective powder bed on the properties of pressureless sintered Si_3N_4 based materials, <u>in:</u> "Proc.Conf. on Microstructure and properties of ceramic materials", T.S.Yen et al,ed., N.Holland, Amsterdam, 518 (1983)

4. M.Mimoto, N.Kuramoto and Y.Inomata, Fabrication of high strength sialon by reaction sintering, <u>J.Mat.Sci.</u> 14:2309 (1979)

5. T.Ekström and N.Ingelström, The relation between phase composition and hardness of different yttrium sialon materials, <u>in:</u> "Proc.Int.Conf. on Hot Isostatic Pressing" Lulea 15-17 June 367 (1987)

6. T.Ekström, N.Ingelström, R.Brage, M.Hatcher and T.Johansson, $\alpha-\beta$ sialons made from different silicon nitride powders, <u>J.Am.Ceram.Soc.</u> 71 (12):1164 (1988)

7. K.H.Jack, Review: Sialons and related nitrogen ceramics, <u>J.Mater.Sci.</u> 11:1135 (1976)

8. C.Greskovich and S.Prochazka,Stability of Si_3N_4 and liquid phases(s) during sintering, <u>J.Amer.Ceram.Soc.</u> 64 (7):C96 (1981)

9. S.Bandyopadhyay and J.Mukerji, Sintering and properties of sialons without externally added liquid phase, <u>J.Amer.Ceram.Soc.</u> 70(10):C273 (1987)

Part VIII. METALS AND COMPOSITE PROCESSING

POWDER PROCESSING OF HIGH TEMPERATURE ALUMINIDE-MATRIX COMPOSITES

R.M. German

Materials Engineering Department
Rensselaer Polytechnic Institute
Troy, New York 12180-3590 USA

ABSTRACT

Aluminide intermetallic compounds are the basis for high performance,
high temperature materials of the future. This presentation covers the
use of reactive powder processing techniques for the formation of
monolithic aluminide intermetallics and intermetallic-matrix composites.
A notable development has been the fabrication of homogeneous, high
density compacts from elemental powders by reactive sintering. A variant
process involving simultaneous pressurization in a hot isostatic press,
termed reactive hot isostatic pressing, is applicable to those compounds
that prove difficult to consolidate by reactive sintering. This paper
describes the effects of various processing factors on fabrication of
dense aluminides with primary emphasis on Ni_3Al. Results on the
fabrication of several other aluminide compounds will be discussed,
including NiAl, TiAl, $TaAl_3$, and $NbAl_3$. Current research is on the use
of these aluminides as the matrix for high temperature composites. A key
concern is with processing effects on microstructure, selection of
compatible ceramic reinforcing phases, and fiber alignment through
injection molding.

INTRODUCTION

Intermetallic compounds have emerged as the needed next generation
of high temperature, oxidation resistant materials for aerospace and
turbine applications. A recent surge of research on intermetallics has
taken place as the ceramics have failed to live up to their promise and
superalloys have apparently been exhausted [1,2]. The aluminide
intermetallic compounds have the attractive characteristics of low
density, high strength, good corrosion and oxidation resistance,
nonstrategic elements, and relatively low cost. In some cases these
intermetallics exhibit the unique characteristic of improved strength
with increasing temperature. Coupled with relatively high melting
temperatures, these attributes make for ideal high temperature materials.
Most importantly, recent research has demonstrated ductility in some
intermetallic systems. Thus, fabricability and reliability have been
improved, leading to new interest.

Sintering is ideal for the fabrication of complex shaped, high
performance intermetallic compound alloys [3,4]. The research covered in
this paper considers reactive sintering and reactive hot isostatic
compaction as fabrication routes for the production of aluminide
compounds and composites from mixed elemental powders.

BACKGROUND

Initial research focused on the processing of full density Ni₃Al. Subsequent efforts shifted to incorporation of fibers in a nickel aluminide matrix [5-9]. Because processing difficulties are expected from thermal expansion mismatches and interfacial interactions, every effort is being made to keep processing temperatures low in this early research. Further, injection molding has been applied to attain the desired fiber dispersion and orientation in the matrix while forming complex shapes. Final densification for these composites is by a new process termed reactive hot isostatic compaction.

Within this framework, the research has emphasized reactive sintering as one process for forming several aluminides. In general, reactive sintering involves a transient liquid phase [10]. The initial compact is composed of mixed powders which are heated to a temperature where they react to form a compound product. Often the reaction occurs on the formation of a first liquid, typically a eutectic liquid at the interface between contacting particles. Figure 1 shows a schematic binary phase diagram for a reactive sintering system, where a stoichiometric mixture of A and B powders is used to form an intermediate compound product AB. At the lowest eutectic temperature a transient liquid forms and spreads through the compact during heating. Heat is liberated because of the thermodynamic stability of the high melting temperature compound. Consequently, reactive sintering is nearly spontaneous once the liquid forms. By appropriate selection of temperature, particle size, green density and composition, the liquid becomes self-propagating throughout the compact and persists for only a few seconds. Like other transient liquid phase sintering treatments, the liquid provides a capillary force on the structure which leads to densification [10-13]. However, if the solubilities are unbalanced, swelling can occur due to the formation of Kirkendall porosity.

During slow heating solid state interdiffusion can generate intermetallic phases at the interfaces. Such compounds inhibit the subsequent reaction when the liquid forms; thus, reactive sintering is sensitive to processing parameters such as heating rates, interfacial quality, green density, and particle size. Because of the rapid spreading and reaction of the liquid, pore formation at prior particle sites is common, especially in systems with large aluminum particle sizes and large exotherms. Furthermore, dimensional control proves difficult if an excess of liquid is formed. Because of such problems, the application of reactive sintering has been limited. However, as demonstrated here, the potential for reactive sintering is large and it is well suited to forming dense intermetallic compound structures.

Figure 2 shows the aluminum-nickel binary phase diagram. The system is characterized by five intermetallic compounds, with initial interest in this study on Ni₃Al. For this system, reactive sintering treatments near the lowest eutectic temperature would be most appropriate (approximately 640°C). Figure 3 is a schematic diagram of the reactive sintering process. Nickel and aluminum powders are randomly mixed in a stoichiometric ratio. The powders have small particle sizes to aid intermixing and are compressed to create good particle-particle contact. During heating to the first eutectic temperature, solid state interdiffusion generates intermetallic compound phases and some self-heating. At the first eutectic temperature, liquid forms and rapidly spreads throughout the structure. The liquid consumes the elemental powders and forms a precipitated Ni₃Al solid behind the advancing liquid interface. Interdiffusion of nickel and aluminum is quite rapid in the liquid phase and the compound generates heat which further accelerates the reaction. If the reaction is controlled, then the compound will be nearly fully densified and suitable for containerless hot isostatic compaction to full density.

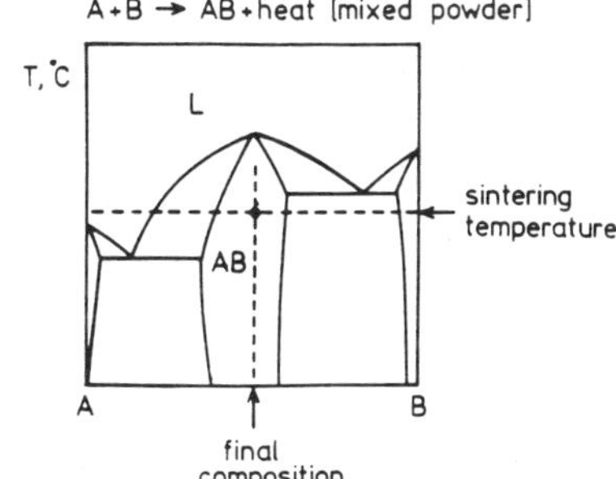

Fig 1. A schematic binary phase diagram showing the characteristics necessary for formation of the AB intermetallic compound from mixed A and B constituent powders.

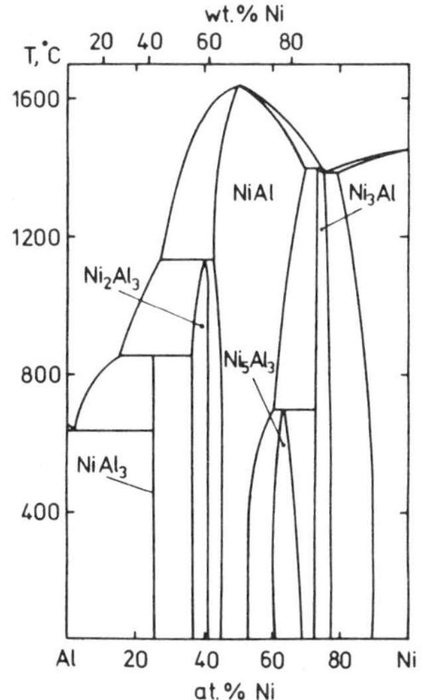

Fig 2. The aluminum-nickel binary phase diagram.

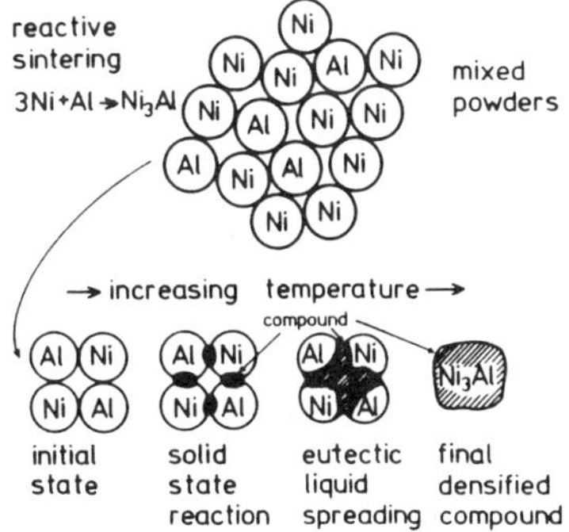

Fig 3. A sketch of the reactive sintering process for forming Ni₃Al from mixed elemental Ni and Al powders. As temperature increases, first a solid state reaction occurs, with subsequently a rapid reaction when the eutectic liquid forms. The final product is a densified compound.

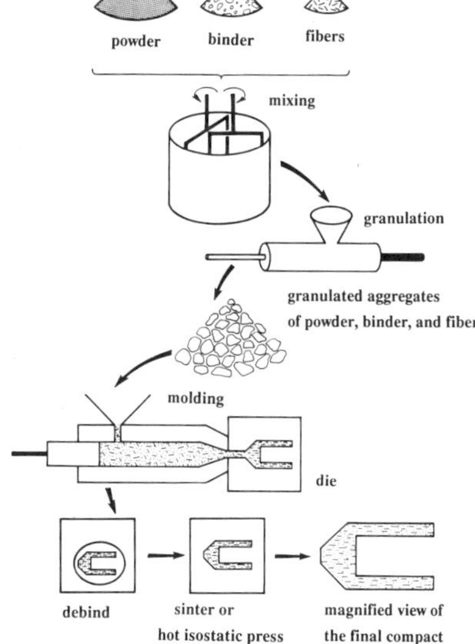

Fig 4. A schematic of the powder injection molding process for forming aluminide matrix composites.

The concept of reactive sintering has been applied to several
compounds in the past [14-19] and has been adapted to form NiAl, Ni$_3$Al,
TaAl$_3$, NbAl$_3$, and TiAl in this research. In one variant, compound
formation and densification are achieved in separate steps by mixing
elemental powders, reacting, pulverization (grinding), compaction, and
subsequent sintering. Variations on this basic scheme involve hot
pressing and pressure assisted sintering. In each case stoichiometry
control is important and is often achieved using an excess of the more
volatile ingredient or intermediate chemical leaching to remove unreacted
constituents. The current reactive sintering approach circumvents these
problems by using commercial elemental powders, low processing
temperatures, short process cycles with a classic press and sinter
technology.

EXPERIMENTAL

The sintering research presented here will emphasize formation and
densification of Ni$_3$Al where there is a large body of knowledge [5-10].
Parallel efforts on the other compounds will be summarized to illustrate
the possible adaptations of the Ni$_3$Al techniques to other aluminides.
The main process parameters were particle sizes of nickel and aluminum,
stoichiometry, alloying, milling time, green density, maximum sintering
temperature, heating rate, atmosphere, and duration of the sintering
treatment.

The characteristics of the nickel, niobium, titanium, and aluminum
powders used for reactive sintering are given in Table 1. The aluminum
has a minimum surface oxide due to the helium atomization process. Other
aluminum particle sizes (3, 10, 30 and 95 micrometers) and powder types
were examined, but the combination proved most successful in forming the
Ni$_3$Al compound. For the other aluminides, small particle sizes proved
generally most useful.

Table 1

Experimental Powder Characteristics

property	nickel	aluminum	niobium	titanium
vendor	INCO	Valimet	NRC	Micron Metal
designation	123	H-15	- - -	-325 mesh
powder type	carbonyl	gas atomized	hydride	sponge
purity, %	99.99	99.7	99.4	99.0%
FSSS size, micrometers	2.8	15.0	8.8	20
major impurities, ppm	Ca=10	Fe=1200	Ta=1100	O=1500
	Fe=30	water=200	O=3800	Cl=2500

As part of this research the Ni:Al stoichiometry was varied from 84
to 90 wt.% Ni; the Ni$_3$Al stoichiometry corresponds to 86.7 wt.% Ni. The
powders were mixed for 30 minutes in a turbula mixer. Various milling
times were also applied to the powder using a high intensity vibratory
mill to attain mechanical alloying. The carbonyl nickel powder is spiky
and agglomerated, giving clusters over 20 micrometers is size. The mixed
powders were compacted using a pressure between 200 to 330 MPa either by
uniaxial die pressing or cold isostatic compaction, giving typical green
densities near 70% of theoretical. Compact geometries included
cylinders, transverse rupture bars, and flat tensile specimens.

Sintering was performed in a horizontal laboratory tube furnace
capable of 1500°C temperatures (and several atmospheres, including

vacuum, dry argon, and dry hydrogen). Typically the specimens were loaded into an alumina crucible and inserted into a cold furnace. For vacuum sintering a pressure of 7×10^{-3} Pa was typical. Variations in degassing, heating rate, maximum temperature, and hold time were explored using either manual or automatic controls. The actual sample temperature was not measured, although parallel differential thermal analysis indicates considerable self-heating during sintering with temperatures exceeding 1300°C. From the reaction enthalpy and heat capacity the maximum self-heating is estimated at 1500 K. Through several experiments it was determined that temperatures from 550 to 750°C for times from 10 to 15 minutes gave nearly full density. Indeed, higher temperatures sometimes gave lower densities, due swelling of entrapped gas. After sintering, the samples were furnace cooled. Some material was additionally heat treated at 1350°C for one hour in dry argon to further homogenize the microstructure.

Measurements consisted of shrinkage, densification, density, temperature rise, hardness, bend strength, tensile strength, and tensile elongation. Additionally, fracture surfaces were examined using scanning electron microscopy. X-ray diffraction and transmission electron microscopy were applied to the samples for phase identification and to determine ordering, and electron microprobe analysis was conducted to identify the phases and stoichiometry. Optical metallography provided insight to the phases and pores present during and after reactive sintering. Dilatometry and differential thermal analysis were employed to identify reaction temperatures and assess the speed of the reaction. In all cases, these analyses were performed using standard procedures typically with computer interfaced data acquisition.

For the composites, various ceramic particles and fibers were incorporated into the powder mixture, including Y_2O_3, Al_2O_3, TiB_2, and SiC. The composites included up to 30 vol.% ceramic phase. The particles were directly mixed with the elemental particles. The fibers were aligned in the green microstructure using powder injection molding techniques as schematically outlined in Figure 4 with various wax-polymer binders. Debinding was by wicking and thermal degradation at temperatures below 500°C [9,20].

RESULTS

Mechanical alloying resulted in agglomeration of the powders. As a consequence there was more difficulty in sustaining the reaction and attaining densification. Consequently, unmilled powders were used in the balance of the research.

The atmosphere effect on sintered density is shown in Figure 5, again using the 15 micrometers aluminum powder. With the 3 K/min heating rate the samples sintered in argon and hydrogen swelled, resulting in low sintered densities. At a heating rate of 30 K/min densification occurred in all atmospheres, giving theoretical densities of 97.5% in vacuum, 96.4% in hydrogen and 93.1% in argon.

In light of the self-heating during reaction, experiments were performed to determine the maximum temperature needed for densification. Figure 6 shows example results for a 30 K/min heating rate to various maximum temperatures with a subsequent 15 minute hold time using two aluminum particle sizes. Temperatures below 550°C give higher porosities, most likely because no liquid is formed. At temperatures in the 550 to 600°C range there is good densification. With higher temperatures there is a gradual swelling phenomenon. Thus, the optimal reaction initiation temperature is relatively low. Note that generally the 3 micrometers aluminum powder gives less densification than the 30 micrometers powder. Indeed, the 15 micrometers aluminum powder proved

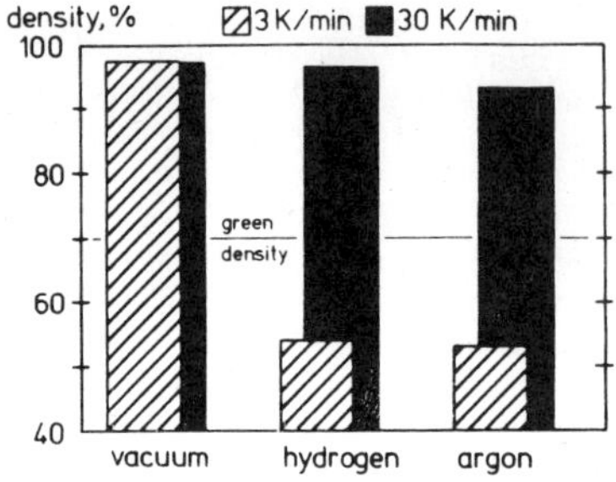

Fig 5. Sintered density shown for two heating rates and three atmospheres for unmilled 15 micrometers aluminum.

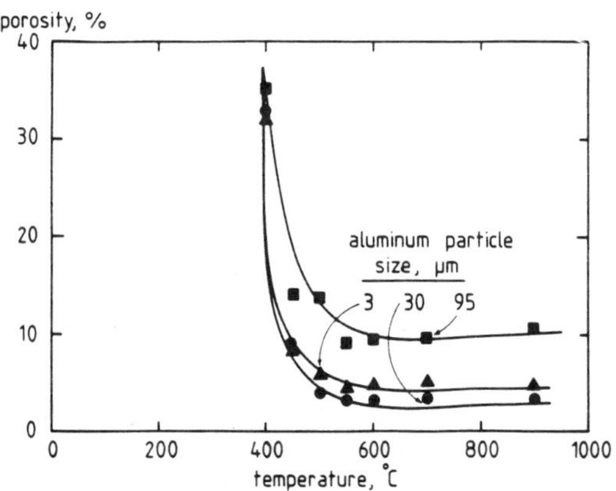

Fig 6. The effect of the maximum sintering temperature on porosity for three aluminum particle sizes.

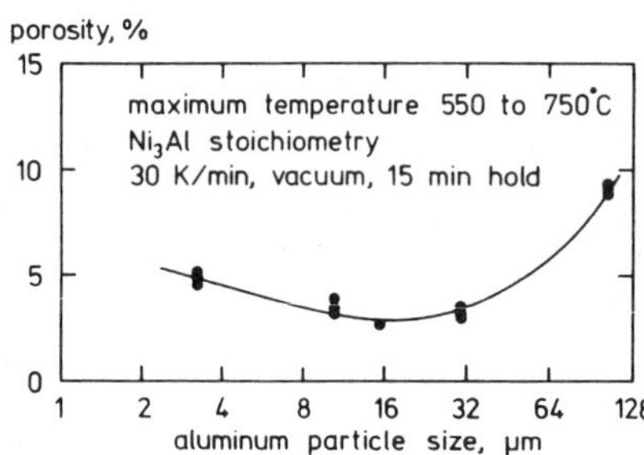

Fig 7. Porosity as a function of the aluminum particle size for various maximum furnace temperatures between 550 and 750°C.

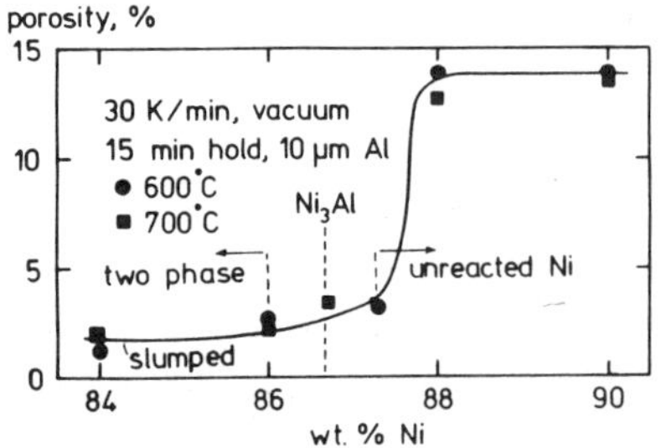

Fig 8. The stoichiometry effect on porosity for reaction sintered compositions near Ni₃Al using two maximum temperatures.

optimal as illustrated in Figure 7. This figure further demonstrates the
aluminum particle size effect by showing the final porosity versus
aluminum particle size for compacts sintered at temperatures ranging from
550 to 750°C. A particle size near 15 micrometers is best, giving a
final porosity less than 3%.

All of the above results are for a 3:1 atomic ratio of Ni to Al.
Experiments were conducted to determine the stoichiometry of the product
using 10 micrometers aluminum powder. Maximum temperatures of 600 and
700°C were employed along with a 15 minute hold and 30 K/min heating rate
in vacuum. Figure 8 shows the final porosity versus nickel content. A
dramatic change in behavior exists near the intermetallic compound
stoichiometry. The compacts slumped for the highest aluminum content
when sintered in vacuum; thus, the reactive sintering process appears
best suited to compositions close to the Ni_3Al stoichiometry.

The microstructure of samples with varying aluminum contents
exhibited two phase microstructures with minimal porosity. Densification
proved easy using unmilled 15 micrometers aluminum powder with a heating
rate of 30 K/min. The product structure exhibited a small amount of
porosity and two distinct phases Ni_3Al and $NiAl$. For stoichiometric
Ni_3Al the grain size after sintering is approximately 30 micrometers.
The bulk hardness was 52 HRA and the microhardness was measured as 264
Knoop (100 g load), which agrees favorably with a value of 240 measured
on a hot isostatically compacted and extruded prealloyed powder compact.

Chemical analysis after sintering gave the composition as 12.2% Al,
87.6% Ni (76.8 at.% Ni), with 0.02% Fe, 0.01% Si, 482 ppm O and 420 ppm
C. Electron microprobe analysis was used to identify the two phases,
giving Ni_3Al as the major phase with an aluminum level of 24.3 at.%. The
minor phase had an aluminum content of 34.8 at.%, approximately
corresponding to the Ni_5Al_3 compound. Since the Ni_5Al_3 compound is
unstable at temperatures over approximately 740°C, the reactive sintered
material was annealed at 1350°C for one hour to attain homogenization.
Microprobe scans confirmed the composition was uniform throughout the
sample after annealing. Transmission electron microscopy substantiated
that the product was ordered Ni_3Al.

Differential thermal analysis and dilatometry isolated the character
of the reactive sintering process. Figure 9 shows dilatometer and
differential thermal analysis scans performed on stoichiometric green
powder compacts. In the unreacted powder a large exotherm was evident at
approximately 600°C, demonstrating the onset of reactive sintering. This
is slightly higher than the temperature of 550°C which gave good
sintering. In the reacted sample, only an endotherm is evident when the
sample melts, indicating total consumption of the ingredients in the
reactive sintering process. The first eutectic temperature in the
aluminum-nickel system is at 640°C and aluminum melts at 660°C; thus, the
exotherm occurs prior to liquid formation and the compact undergoes self-
heating, leading to rapid liquid formation. The dilatometry results
correlated with the differential thermal analysis, indicating the
reaction begins below approximately 600°C. Furthermore, under optimal
conditions the duration of the reaction appears to be approximately two
seconds. Consequently, studies involving time at the maximum temperature
were not useful.

As a variant, elemental powders were mixed, cold isostatically
compacted, evacuated, and heated under pressure using a hot isostatic
press. This process is termed reactive hot isostatic pressing (RHIP).
Using the same mixed elemental powders and a maximum temperature of
1100°C, pressure of 172 MPa, and hold time of 1 h resulted in full
compact density. With the successful development of reactive sintering
and reactive hot isostatic pressing for Ni_3Al, some mechanical property
assessments were performed. A summary of results is provided in Table 2.

Improved properties were gained with boron doping, low temperature
degassing, and post-sintering heat treatments. With boron doping at
approximately the 0.1% level, and refined processing, the reactive
sintered product proved to be superior to that possible using fusion
metallurgy. Furthermore, preliminary oxidation tests indicate good
resistance up to 900°C.

Table 2

Mechanical Properties of Reactive Sintered Ni_3Al

process	yield strength, MPa	ultimate strength, MPa	elongation %
no boron	270	270	1
boron doped	353	682	12
boron doped, heat treated	591	827	5
RHIP, 1100°C, 1 h, 170 MPa	494	677	2

These results on Ni_3Al provide a basis for investigation of reactive
sintering and reactive hot isostatic compaction of several other
aluminides and intermetallic matrix composites. The many results on
these systems can not be detailed here; however, in most instances direct
variants on the above sintering processes have proven successful [21].
For example, composites with a Ni_3Al matrix reinforced with Al_2O_3 fibers
have been fabricated using RHIP at 800°C, 180 MPa, for 0.5 h. After RHIP
the compact is fully dense, but the microstructure is inhomogeneous as
shown in Figure 10. This indicates that pressure inhibits capillary
induced liquid flow that normally occurs in reactive sintering. As
indicated in Table 3, the composite showed increased strength, but
decreases ductility as the alumina content increased. Scanning electron
microscopy of the fracture surface shows that failure occurred by
preferential separation at fiber interfaces. Likewise, with $NbAl_3$ best
sintering densification occurs with a 9 micrometer niobium powder formed
by hydride-dehydride and a 30 micrometer aluminum (helium atomized)
powder. The mixed powders are compacted at 200 MPa giving a 77% green
density, heated to 500°C for degassing, then heated at 15 K/min to 1200°C
and held at that temperature for 1 h. The resulting product is 95% dense
and essentially pure $NbAl_3$. Reactive hot isostatic compaction follows a
similar processing route with a 170 MPa pressure giving 98% density. The
incorporation of 30 vol.% Al_2O_3 fibers gives a composite with a Rockwell
hardness of 87 HRA in contrast with HRA 72 for the monolithic $NbAl_3$. The
densified composite shows agglomerated ceramic regions, but minimal
residual porosity. Likewise, similar processing has taken place to
optimize the reactive consolidation of other aluminides.

DISCUSSION

The reactivity of the nickel and aluminum powders results in
relatively high sintered densities with a low apparent sintering
temperature and short sintering time. The high final densities occur
because a transient liquid is present during the sintering cycle.
Processing conditions which influence the reaction between the
constituent powders alter the amount, distribution, and duration of the
liquid. In transient liquid phase sintering these factors are critical
to the final sintered density and mechanical properties [11-13].
However, unlike other sintering studies, time at temperature is not a
significant factor since the process occurs rapidly once the liquid
forms. The role of the various process parameters can be explained in
terms of their effects on the liquid phase.

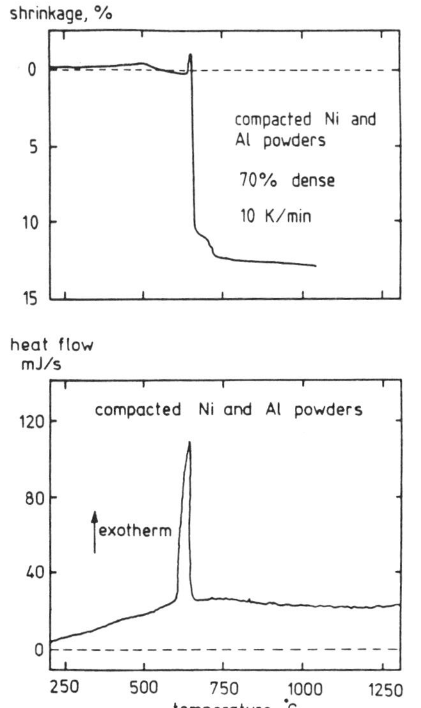

Fig 9.   The upper curve is the dilatometer scan and the lower curve is the differential thermal analysis scan for mixed nickel and aluminum powders showing the reactive sintering exotherm at approximately 600°C.

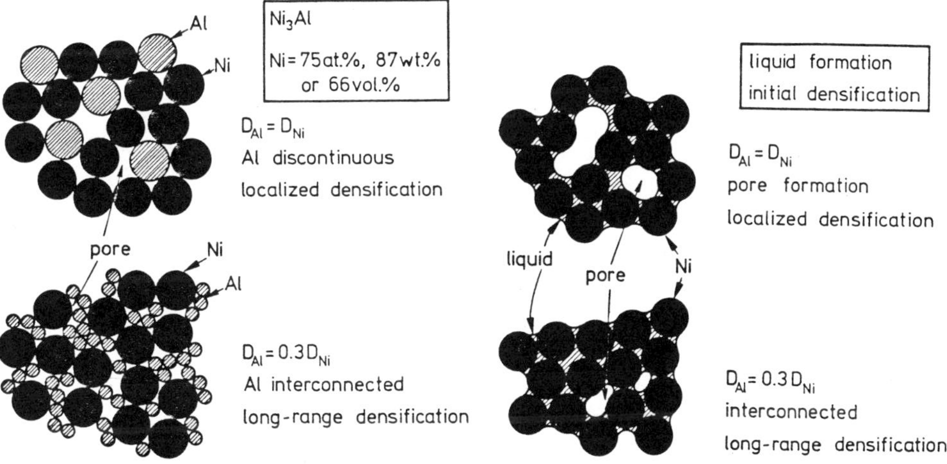

Fig 10.   An optical micrograph of Ni₃Al after reactive hot isostatic compaction.

Fig 11.   A comparison of the green microstructure effect on local liquid formation and densification.   Densification occurs when the Ni and Al are connected in the green microstructure prior to reactive sintering.

Table 3

Reactive Processing Results for Aluminides and Aluminide-Matrix Composites

compound	additive	process details	fractional density	strength, MPa
Ni_3Al	3% Al_2O_3	RHIP 800°C, 1 h, 170 MPa	1.00	548
Ni_3Al	20% Y_2O_3	RHIP 800°C, 1 h, 170 MPa	1.00	464
NiAl	-	RS 10% prealloyed 800°C	0.98	890 (c)
NiAl	15% TiB_2	RHIP 15% prealloyed 1200°C, 1 h, 170 MPa	1.00	1060 (c)
NiAl	20% TiB_2	RHIP 15% prealloyed 1200°C, 1 h, 170 MPa	1.00	1350 (c)
$NbAl_3$	-	RS 1200°C, 1 h	0.95	-
$NbAl_3$	30% Al_2O_3	RHIP 1200°C, 4 h, 170 MPa	0.98	-
$TaAl_3$	-	RHIP 1200°C	0.98	531
$TaAl_3$	8 at.% Fe	RHIP 1200°C	0.98	372
TiAl	-	RHIP 1370°C, 4 h, 170 MPa	1.00	533
TiAl	30% Al_2O_3	RHIP 1370°C, 3 h, 186 MPa	1.00	-

(c = indicates compressive test, all others are tensile)

Milling the powders decreased the sintered density because the liquid formed discontinuously in the microstructure. Also, an increase in the aluminum-nickel interfacial area due to milling increased the solid-state interdiffusion during heating, thereby reducing the amount of liquid during the reaction. With milling the liquid is consumed faster during sintering, and may be diminished by solid state interdiffusion prior to liquid formation. This concept is substantiated by the particle size experiments shown in Figure 7. A small aluminum particle size gives more rapid reaction (more interfacial area) and less densification. A coarse aluminum powder gives a poor distribution of liquid in the microstructure.

At the stoichiometric composition, aluminum constitutes 34 vol.% of the solid structure. This is insufficient to form an interconnected network of aluminum unless the aluminum particle size is less than the nickel particle size [10]. In the green compact, the nickel agglomerates are approximately 30 micrometers in intercept length, corresponding to a 45 micrometers diameter. According to Biggs [22], a particle size ratio of at least 2.4:1 (major phase diameter to minor phase diameter) is required at 34 vol.% to form a connected network in the minor phase. Thus, for 45 micrometers nickel agglomerates, an aluminum particle size of approximately 19 micrometers, or smaller is required. The optimal particle size for the aluminum (15 micrometers) was in agreement with this value. Thus, a connected aluminum matrix with minimal interfacial area is required for densification. Milling decreases the nickel particle size, thereby disrupting the aluminum connectivity.

The sintering atmosphere role in determining the sintered density is explained by heat conduction and entrapped gas effects. Heat is carried away from the compact during reaction by the higher thermal conductivity of a gas versus vacuum. Furthermore, because of the speed of the reaction, there is no time for adsorbed vapors and atmosphere captured in the pores to escape. Hydrogen has a higher solubility in Ni_3Al than argon. Thus, with hydrogen trapped in the pores there is some opportunity for gas escape even after the pores have sealed during densification. An examination of Figure 5 shows that hydrogen was better than argon with both heating rates; however, vacuum is superior, especially for TiAl, $NbAl_3$, and $TaAl_3$.

Heating rate effects on transient liquid phase sintering have been

explained based on solid state interdiffusion prior to liquid formation
[11-13]. With a slow heating rate there is more solid state reaction
with subsequently less liquid. Indeed, the intermetallic products from
solid state interdiffusion may actually inhibit liquid formation at the
reaction temperature. Typically, higher heating rates are beneficial
when sintering involves a transient liquid. However, there is a limit to
the benefit of rapid heating rates. Too rapid a heating rate gives a
loss of process control, and with massive samples proves difficult to
sustain with any uniformity. Additionally, as the heating rate increases
the reactivity of the liquid likewise increases, thereby decreasing
process control. Hence, intermediate heating rates prove most
reasonable.

A temperature over 550°C is required to optimally react nickel and
aluminum powders. Higher temperatures are not of benefit since the
reaction is fairly complete in short times as soon as a temperature near
600°C is attained. The calculated maximum 1500 K heating from the
exothermic reaction is more than sufficient to attain the 640°C eutectic
from a 550°C initiation temperature. Thus, the reaction is probably
spontaneous during heating, independent of the final temperature and hold
time. The heating rate is important only in its effect on outgassing and
interdiffusion prior to the reaction. Likewise, particle size is of
importance in determining the distribution of liquid in the
microstructure. As sketched in Figure 11, if the liquid forms in
isolated pools, then no long range capillary action is possible and
swelling is expected [10-13,23]. Alternatively, a connected aluminum
matrix will lead to rapid densification because of the long range
capillary action. This concept also helps explain the stoichiometry
results. With an excess of aluminum there is more liquid and an excess
over that required to form Ni_3Al. The final porosity is fairly constant
with an excess of aluminum, but slumping and shape loss are observed at
84% Ni. Alternatively, an excess of nickel reduces the amount of liquid
and decreases the aluminum connectivity, thereby separating the liquid
pools within the compact. For the experiments with a 10 micrometers
aluminum powder, the calculated loss of connectivity for the aluminum
would occur between 87 and 90 wt.% Ni, in agreement with the experimental
observation. As a consequence, the high nickel content compositions fail
to densify because of less liquid and a decreased connectivity of
aluminum in the microstructure.

The composition of the second phase in the as-sintered compacts
corresponds to Ni_5Al_3. This compound is stable over the approximate
composition range of 32 to 37 at.% aluminum, which agrees well with our
determination of 34.8 at.% Al. The one hour heat treatment at 1350°C
effectively dissolves this phase, leaving the equilibrium compound Ni_3Al.
The formation of the Ni_5Al_3 phase indicates that the reaction is not
directly from nickel and aluminum to Ni_3Al, but involves formation of
intermediate compounds. Removal of the residual porosity can be
facilitated by hot isostatic compaction. With sintered densities over
92% of theoretical, the pore structure consists of closed pores and
responds to containerless hot isostatic compaction to full density.

Variants on this reactive sintering process has been extended to
several other intermetallic compounds. The elemental powders are
available and provide experimental flexibility for the production of
novel compositions and composites. Table 3 provides a summary of some of
the other aluminides and aluminide-matrix composites that have been
fabricated using similar processes.

SUMMARY

Much progress has been made in applying reactive sintering to the
fabrication of aluminide intermetallics and composites using mixed

elemental powders. Densities in excess of 97% of theoretical are
possible through appropriate selection of particle sizes, composition,
green density, heating rate, atmosphere, maximum temperature, and hold
time. Although not all of these factors have been detailed here, the
concept behind this exciting development has been outlined. It is clear
that the sintered density depends on the amount of liquid formed at the
first eutectic temperature and the connectivity of this liquid. In this
sense, reactive sintering is analogous to transient liquid phase
sintering. Because the liquid persists for only a short time, it is
important that the several process parameters be carefully controlled to
optimize the sintered density. Subsequent processing (heat treatment and
containerless hot isostatic compaction) can then be used to remove the
residual porosity and homogenize the compact. A key to success is the
formation of a fully interconnected liquid phase. This dictates the
amount and particle size distribution of the constituents needed for
optimal densification. The fabrication process is appropriate for the
fabrication of intermetallic compounds and intermetallic-matrix
composites.

ACKNOWLEDGMENTS

This research was performed in the Powder Metallurgy Laboratories at
Rensselaer Polytechnic Institute under sponsorship of the Defense Advance
Research Project Agency with the help of several coworkers, especially
A. Bose, J. Murray, D. Alman, R. Oddone, N. Stoloff, and D. Duquette.

REFERENCES

1. C. C. Koch, C. T. Liu and N. S. Stoloff (eds.), High-Temperature
    Ordered Intermetallic Alloys, Materials Research Society, vol.39,
    Materials Research Society, Pittsburgh, PA (1985).
2. N. S. Stoloff, Inter. Metal Rev., 29:123-135 (1984).
3. W. M. Schulson, Inter. J. Powder Met., 23:25-32 (1987).
4. K. Vedula and J. R. Stephens, Powder Metallurgy 1986 State of the
    Art, W. J. Huppmann, W. A. Kaysser and G. Petzow (eds.), Verlag
    Schmid, Freiburg, West Germany, pp.205-214 (1986).
5. A. Bose, B. Moore, R. M. German and N. S. Stoloff, J. Metals, 40(9):
    14-17 (1988).
6. D. M. Sims, A. Bose, and R. M. German, Prog. Powder Met., 43:575-596
    (1987).
7. B. H. Rabin, A. Bose and R. M. German, Modern Developments in Powder
    Metallurgy, vol.21, P. U. Gummeson and D. A. Gustafson (eds.), Metal
    Powder Industries Federation, Princeton, NJ, pp.511-529 (1988).
8. A. Bose, B. H. Rabin and R. M. German, Powder Met. Inter., 20(3):25-
    30 (1988).
9. R. M. German and A. Bose, Mater. Sci. Eng., A107:107-116 (1989).
10. R. M. German, Liquid Phase Sintering, Plenum, New York, NY (1985).
11. W. H. Baek and R. M. German, Inter. J. Powder Met., 22:235-244
    (1986).
12. W. H. Baek and R. M. German, Powder Met. Inter., 17:273-279 (1986).
13. R. M. German and J. W. Dunlap, Metall. Trans. A, 17A:205-213 (1986).
14. P. P. Turner and L. J. Jones, U.S. Patent 2,755,184, 17 July 1956.
15. E. Fitzer, U.S. Patent 2,877,113, 10 March 1959.
16. S. T. Zegler and J. B. Darby, U.S. Patent 3,084,041, 2 April 1963.
17. R. Maier, U.S. Patent 3,260,595, 12 July 1966.
18. W. J. Werner, M. C. McIlwain and J. P. Hammond, U.S. Patent
    3,288,571, 29 November 1966.
19. L. S. Williams, U.S. Patent 3,353,954, 21 November 1967.
20. R. M. German, Powder Injection Molding, Metal Powder Industries
    Federation, Princeton, NJ (1990).

21. R. M. German, A. Bose and N. S. Stoloff, "Powder Processing of High
     Temperature Aluminides," presented at the 1988 Materials Research
     Society Meeting, Boston, December 1988, in press.
22. D. M. Biggs, <u>Metal-Filled Polymers</u>, S. K. Bhattacharya (ed.), Marcel
     Dekker, New York, NY, pp.165-226 (1986).
23. D. J. Lee and R. M. German, <u>Inter. J. Powder Met. Powder Tech.</u>,
     20:9-21 (1984).

DIFFERENTIAL SINTERING

Claudia P. Ostertag

National Institute of Standards and Technology
Ceramics Division
Gaithersburg, MD 20899

INTRODUCTION

Most sintered products are inhomogeneous when formed into green
compacts. Inhomogeneities in a green compact, such as areas of different
density or particle size, typically sinter at different rates than the
host powder and therefore, influence the sintering body in several ways.
Most importantly, stresses develop in association with the differential
shrinkage characteristics of the inhomogeneity relative to its
surroundings. In some instances, the stresses are of sufficient
magnitude and duration that crack-like damage forms during the sintering
cycle [1]. The paper discusses an extreme example of differential
sintering, namely, the sintering of fiber-reinforced composites.

Sintering of ceramics reinforced by fibers is impeded by stresses
generated by the differential shrinkage between the porous ceramic matrix
and the dense reinforcing agent. During the sintering of fiber-
reinforced composites, the ceramic matrix densifies approximately thirty
to forty percent by volume, while the fibers, which are already fully
dense do not densify at all. Fibers cause stresses to develop during
sintering, because of the constraints that they impose on the contracting
matrix. During the densification process, the fiber is put under
hydrostatic compression. Stresses in the matrix normal to the fiber axis
are tensile in the circumferential direction and compressive in the
radial direction. Parallel to the fiber axis, stresses in the matrix are
tensile because of shrinkage constraints (Fig. 1). Stresses that develop
due to the presence of fibers or dense inclusions are known as
heterogeneity stresses, to differentiate them from the applied stress and
from the sintering stress. Heterogeneity stresses occur primarily close
to the reinforcing agents and cause microcrack formation in the matrix.
Heterogeneity stresses also inhibit the densification of the matrix.
Several attempts have been made to predict theoretically these stresses
for spherical[2,3] and cylindrical[4] inclusions and thereby to assess the
potential for sintering damage. The models are valuable in evaluating
the effects of dense inclusions on sintering damage. However, they are
limited in the information that they can provide about the initiation of
stress (i.e.: Does the stress initiate at the sintering temperature or
already during the heating cycle? What is the critical temperature range
for stress development?).

Science of Sintering
Edited by D. P. Uskoković *et al.*
Plenum Press, New York

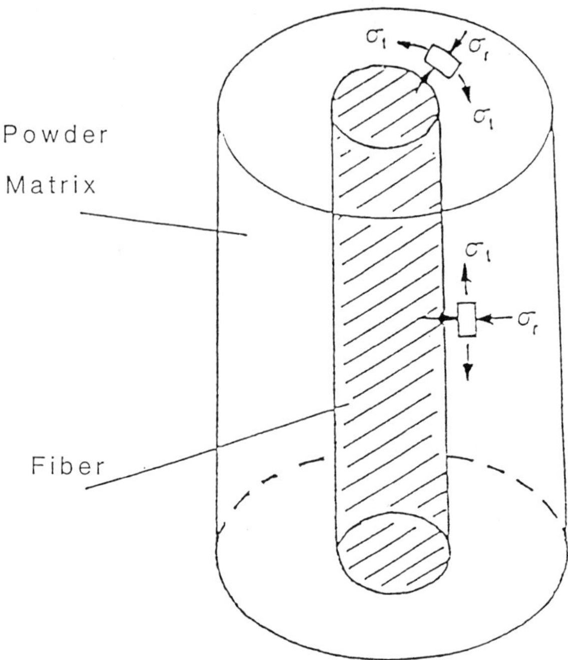

Figure 1.   Stress State in Matrix during the Sintering
of Fiber-Reinforced Composites

To avoid sintering damage, it is important either to reduce the
generation of heterogeneity stresses or to modify the response of the
matrix to these stresses such that no microcracks form. To employ
effective methods for reducing or avoiding these stresses, it is
necessary to know when they initiate and in which stage of the sintering
process they are responsible for crack formation. A model composite
system has been developed by Ostertag,[5] which can be used to assess
visually the stresses in unidirectionally oriented fiber-reinforced
composites. The stress is measured by means of a sandwich compact, which
consists of a layer, or plane, of SiC fibers sandwiched between two
layers of ceramic powder of **different thicknesses.** Upon sintering, this
configuration produces an asymmetric stress field across the thickness of
the specimen, which results in the bending of the compact. The stress
field resulting from differential sintering can be characterized by the
curvature of the fibers. To reveal the time and temperature at which the
stress initiates, the sintering experiment was observed in situ. A video
camera was mounted in front of an open tube furnace to monitor the
bending of the sample bars. The sintering experiments, including both
the heating and cooling cycles, were recorded on videotape. Bending of
the fibers and, hence, the development of the stress in the matrix began
during the heating cycle, before the actual sintering temperature was
reached. The resulting stress development depended strongly on the fiber
content and fiber spacing. For 2 fibers in the fiber plane, the bending
and hence the stress initiated at 1410°C, for 4 fibers at 1350°C and for
7 fibers in the fiber plane the bending occurred already at 1230°C.
Therefore, the higher the fiber content, the larger the bending and the
lower the stress initiation temperature.

In-situ observations of the sintering process revealed that
heterogeneity stresses initiate during the heating cycle. Based on

454

experiments such as those discussed above, it was hypothesized that
heterogeneity stresses which develop during the early stage of sintering
(at low matrix density, hence, at low matrix strength at the point of
stress initiation), are responsible for the sintering damage observed in
sintered fiber-reinforced composites. Therefore, to produce damage-free
composites, stresses have to be suppressed in the early stage of the
sintering process.

To prove the hypothesis that stresses which develop during the early
stage of sintering are responsible for the sintering damage observed in
sintered fiber-reinforced composites, the composite densification was
influenced only during the initial stage of sintering (i.e. heating
cycle) by a) changing the heating rate and b) by applying uniaxial
compressive stresses to the composite sample to avoid shrinkage along the
fiber axis and c) by retarding the densification of the matrix adjacent
to the fiber by coating the fiber with alumina of a different particle
size than the matrix.

EXPERIMENTAL PROCEDURE

Model ceramic composites were fabricated using spray-dried Al_2O_3
powder and continuous SiC fibers of 140 μm diameter as described in
detail in reference 5. A plane of continuous SiC fibers was placed off-
center into the Al_2O_3 powder matrix and uniaxially cold-pressed in a
rectangular steel die into bars of 25 mm length, 8 mm width, and 3 mm
height. The samples were heated at a heating rate of 50°C/min to 1450°C
and were sintered at 1450°C for 1 h. A sintering temperature of 1450°C
was chosen because this was the maximum temperature before significant
degradation of the fibers occurred.

The heating rate was lowered from 50°C/min to 16°C/min and the
sintering experiment was observed in situ and recorded on videotapes.
The shrinkage of the matrix along the length of the fibers was inhibited
during the heating cycle by applying a uniaxial compressive load
perpendicular to the fiber axis in a universal testing machine (Fig. 2)[6].
The experiment corresponds to a sinter-forging experiment, in that the
compact was not constrained by die walls. Therefore, the radial strain
of the sample was a variable. A critical uniaxial compressive stress was
determined for which no dimensional change was observed in the sample
parallel to the fiber plane (x direction of fig. 2). To determine the
critical stress, stresses in the range of 0.5 to 2.3 MPa were applied
perpendicular to the plane of the fibers. The compressive stress was
applied only during heat-up (from 1100°C to 1450°C and from 1100°C to
1350°C) and removed during the sintering temperature. The sample was
cooled to room temperature and the sample dimensions measured. Applying
the load during the heating cycle inhibits shrinkage of the matrix along
the length of the fibers and hence eliminates stress development in the
early stages of sintering. A colloidal processing route for coating
fibers with alumina was developed[7] and the effects of fiber coatings and
coating thicknesses on the stress initiation during the heating cycle was
studied.

RESULTS

The hypothesis was proven right based upon experiments where the
composite densification was influenced only during the initial stage of
sintering (i.e. heating cycle) by changing the heating rate, and by
applying uniaxial compressive stresses to the composite sample to avoid
shrinkage along the fiber axis and by large particle size coatings.

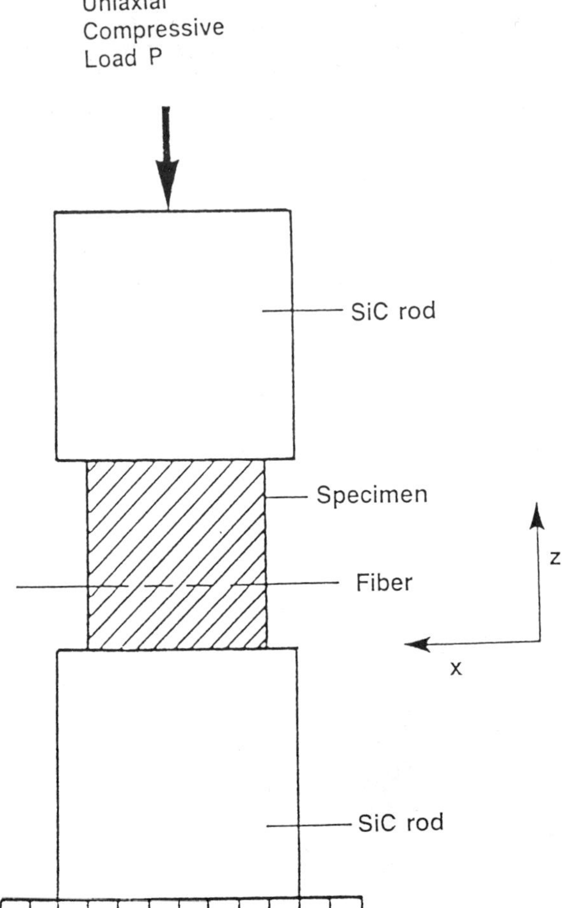

Figure 2. Experimental set-up for uniaxial compressive loading of composite

Lowering the heating rate from 50°C/min to 16°C/min reduced the stress development and, hence, the bending of the composite approximately 30%. Applying the load during the heating cycle inhibits shrinkage of the matrix along the length of the fibers and, hence, eliminates stress development in the early stages of sintering. A compressive stress of 1.5 MPa delays stress development until the density of the matrix is high enough to withstand heterogeneity stresses. A compressive stress of 1.5 MPa during the temperature range of 1100°C to 1350°C and 1.3 MPa and 1.6 MPa during the temperature range of 1100°C to 1450°C completely suppressed the sintering damage formation. The formation of cracks was avoided once the matrix density reached 65% of theoretical, even though stresses built up later as a consequence of sintering. Stresses which develop during the early stage of sintering are the most detrimental in achieving a damage free composite due to the low matrix density and hence the low matrix strength at the point of stress initiation. Hence, the stress development needs to be delayed until the matrix density is high enough to withstand the stresses associated with the reinforcing fibers.

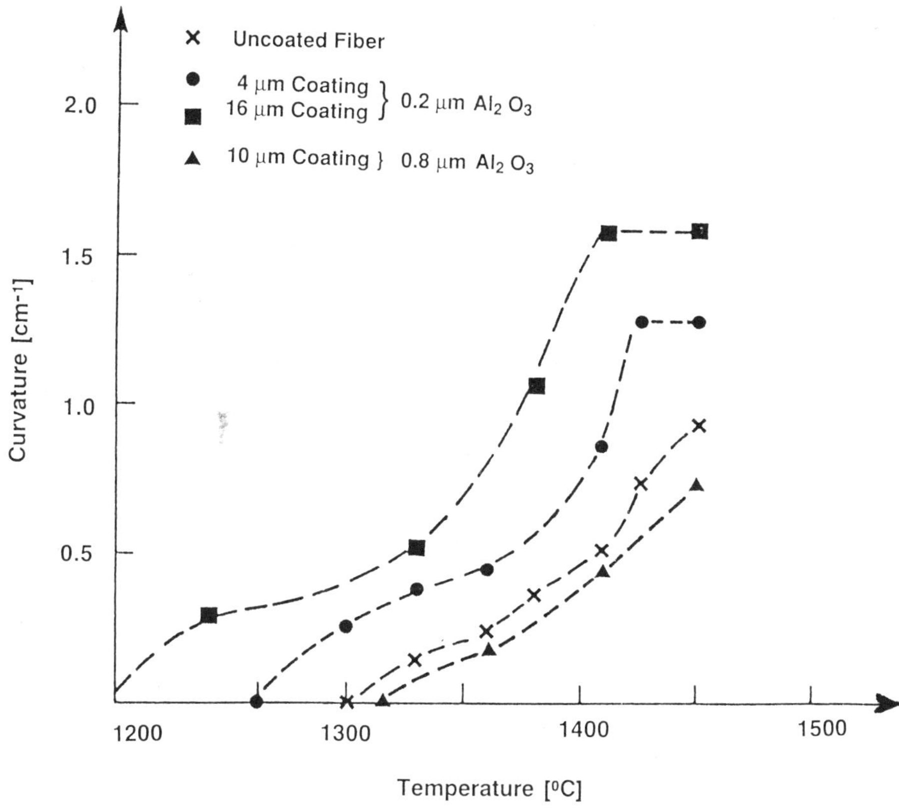

Figure 3. Curvature of model composite during heating cycle as function of coating thickness and coating particle size.

The curvature of the composite during the heating cycle for uncoated and coated fibers with small (d=0.2 μm) and large (d=0.8-1.0 μm) particle-size alumina powders is illustrated in Fig. 3. For the uncoated fibers in the fiber plane, the bending of the composite starts at 1300°C and increases with increasing temperature. However, the coatings of fine particle size alumina enhance the bending and, hence, the stress development in the composites during the heating cycle. During sintering of the model composite containing the fibers with a coating thickness of 4 μm, the bending and hence the stress initiates at 1265°C and increases steadily, exceeding the curvature of the composite of the uncoated fibers during the entire heating cycle. The strongest bending was observed for the coating thickness of 16 μm of the fine particle-size powder. Furthermore, the bending initiates at the earliest stage during the heating cycle, at 1200°C. Therefore, the densification rate is increased more than the creep rate in early stage sintering of alumina and, hence, the stress was enhanced by the coatings of the fine particles. The coarse particle coating achieved the desired results of delaying the stress initiation to higher temperature and producing high matrix strength. For the coarse size alumina coating to a thickness of 10 μm, the bending of the composite was considerably reduced and most importantly the bending was retarded, which resulted in delayed stress initiation to 1325°C. The coarse particle-size coatings appear to reduce the stresses by acting as a buffer layer between the matrix and the fiber.

CONCLUSION

Sintering of fiber-reinforced composites results in residual
stresses that develop in the early stages of sintering, usually during
the heating cycle. These stresses lead to crack formation, which is more
likely at lower matrix densities, despite the lower stress level.
Furthermore, cavities or cracks that form during the early stages in the
densification process do not close, but grow during subsequent
densification, since the shrinkage differential (and, hence, the stress
buildup) between the fiber and the matrix increases with increasing
temperature and annealing time. The stresses which occur in the early
stage of sintering are believed to be responsible for sintering damage
observed in fiber-reinforced composites. The reduction in curvature of
the composites by lowering the heating rate and the fabrication of damage
free composites by applying a uniaxial stress during the heat-up
confirmed this assumption. It is important to find processing routes
that delay the stress generation during the early stages of sintering,
specifically during the heating cycle. The magnitude of the stresses
that develop due to the differential shrinkage between fiber and matrix
are influenced by the shrinkage rate and the creep behavior of the matrix
material. The shrinkage rate tends to induce stresses, whereas creep or
viscous flow encourages stress relaxation. Coating the fibers with fine
particle-size alumina powder was assumed to enhance the creep and hence
the stress relaxation. However, as Fig. 3 illustrates, densification
rate dominates over creep rate and the stress was enhanced by the
coatings of the fine particles. The coarse particle coating achieved the
desired result of delaying the stress initiation to higher temperature
and hence higher matrix strength. To produce a stronger effect on the
stress retardation, experiments with coatings thicker than 10 μm have to
be performed. The coarse particle-size coatings appear to reduce the
stresses by acting as a buffer layer between the matrix and the fiber.
For matrices with high densification rates, thick, coarse particle-size
coatings are required.

These experiments reflect the importance of the fiber/matrix
interface with respect to stress enhancement (small particle-size powder)
and stress retardation (coarse particle-size powder) during the sintering
of fiber- reinforced composites. Even small interfacial layers of a few
microns considerably influence the stress state during sintering.

ACKNOWLEDGMENTS

The author is grateful to Dr. S. Malghan and Mr. Dennis Minor at
NIST for developing the fiber coating technique.

REFERENCES

1. C. P. Ostertag, P. G. Charalambides and A. G. Evans, "Observations
   and Analysis of Sintering Damage", Acta Metall., 37 [7]
   2077-2084 (1989)
2. C. H. Hsueh, A. G. Evans, R. M. Cannon, and R. J. Brook,
   "Viscoelastic Stresses and Sintering Damage in Heterogeneous
   Powder Compacts", Acta Metall., 34 [5] 927-936 (1986)
3. R. Raj, R. K. Bordia, "Sintering Behavior of Bi-Model Powder
   Compacts", Acta Metall., 32 [7] 1003-1019 (1984)
4. C. H. Hsueh, "Effects of Heterogeneity Shape on Sintering induced
   Stresses", Scripta Metall., 19 [8] 977-982 (1985)

5. C. P. Ostertag, "Technique for Measuring Stresses which occur during Sintering of a Fiber-Reinforced Ceramic Composite", J.Am. Cer. Soc., 70 (12) C 355-357 (1987)
6. C. P. Ostertag, "Reduction in Sintering Damage of Fiber-Reinforced Composites", Adv. Ceram., Ceram. Trans., ed. by C. N. Handwerker, J. E. Blendell, W. A. Kaysser, Am. Ceram. Soc. 1989, in press
7. C. P. Ostertag and Sughas Malghan, "Stress Relaxation in Sintering of Fiber-Reinforced Composites through Fiber Coating", Ceram. Eng. and Sci. Proc., 13th Ann. Conf. on Comp. and Adv. Cer. Mat., July 1989, in press

BORON/MAGNESIUM-ALUMINUM MACHINABLE COMPOSITE

T. Sato, E. Horikoshi, T. Iikawa, and K. Hashimoto

Fujitsu Laboratories Ltd., 10-1 Morinosato-Wakamiya
Atsugi, Japan

INTRODUCTION

Downsizing is the most outstanding trend in component design
today.

Typical of computer peripherals illustrating this trend is the
disk drive, in which components, such as head actuator arms and
carriages, require light-weight, high-strength materials to meet needs
for larger capacity, higher speed data access.

Magnesium-based materials, having the smallest mass of all
applicable metals, show the most promise in this regard.

We have previously reported a sintered boron/Mg-9wt%Al material
having a high modulus of elasticity and a controllable thermal
expansion coefficient.[1] Figure 1 shows the thermal expansion
coefficient of the composite which is close to that of aluminum, in
addition to having an improved modulus of elasticity, as the boron
content increases from 4 to 6 vol%.

This composite appears to be an excellent choice for high-
precision moving parts such as disk drive components.

Despite progress in powder metallurgy, the inability to produce
certain geometrical features, such as transverse holes and undercuts,
frequently necessitates machining, particularly drilling.

In this paper, we describe the effect of boron particle size on
the mechanical properties and machinability of the boron/Mg-Al sintered
composite.

EXPERIMENT

Table 1 shows the specifications of the starting powders for the
boron/Mg-9wt%Al system. A ground Mg powder (average particle size
50 μm), a reduced boron powder, and a gas-atomized Al powder (average
particle size 30 μm) were used. To examine the effect of boron particle
size on the mechanical properties and machinability of the composite,

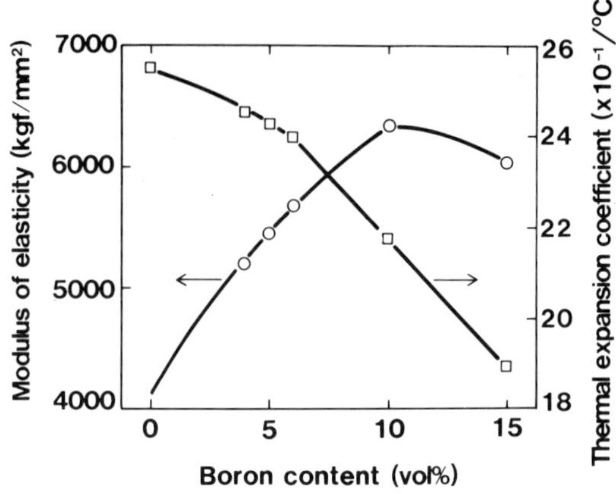

Fig. 1. Dependence of modulus of elasticity and thermal expansion
coefficient on boron content in boron/Mg-Al sintered composite

we varied the average particle size of the boron powder from 1 to
15 μm.

In mixing the different sizes of powder, we paid special attention
to the large difference between fine boron and coarse Mg. Powders were
mixed for 2 h in a V-shell mixer or a grinding mixer with ethyl
alcohol. Mixture were compacted into 97 mm x 4 mm x 6 mm trensile test
pieces under a pressure of 4 t/cm^2.

Samples were sintered at 600°C for 1 h in an argon atmosphere.

Table 1. Starting powders

Powder	Purity (%)	Particle size (μm)	Fabrication
Mg	99.9	50	Ground
Al	99.9	30	Gas-atomized
B	99.8	15	Crushed
B	99.7	10	Crushed
B	99.5	5	Crushed
B	99.7	1	Reduced

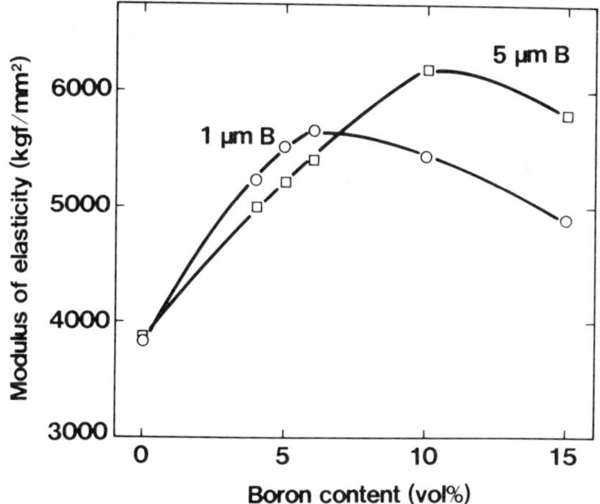

Fig. 2. Dependence of modulus of elasticity on boron content.

We measured the relative densities of the green compact and the sintered body by the Archimedes' method.

The mechanical properties, modulus of elasticity and tensile strength, were measured on an Instron Universal Testing Machine at a speed of 1 mm/min.

Machinability indices were determined by measuring tool wear, cutting resistance, and cut chips that developed during machining. Tool wear was determined by wear on the outer corner of a cemented-carbide drilling tool, after three holes were drilled in a test piece 6 mm thick. Cutting resistance was measured by an oblique-cutting test with

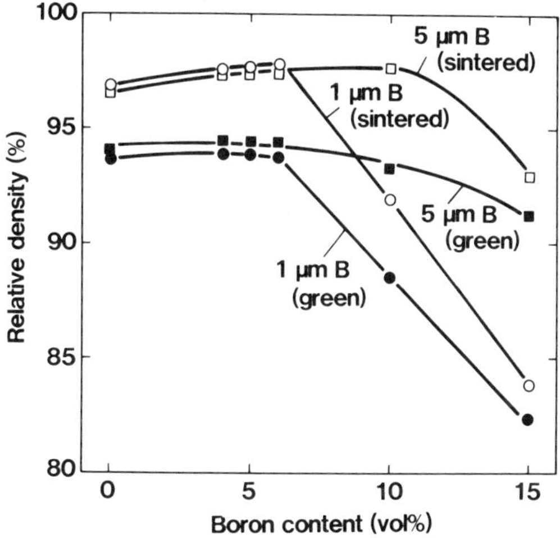

Fig. 3. Relation between relative density and boron content.

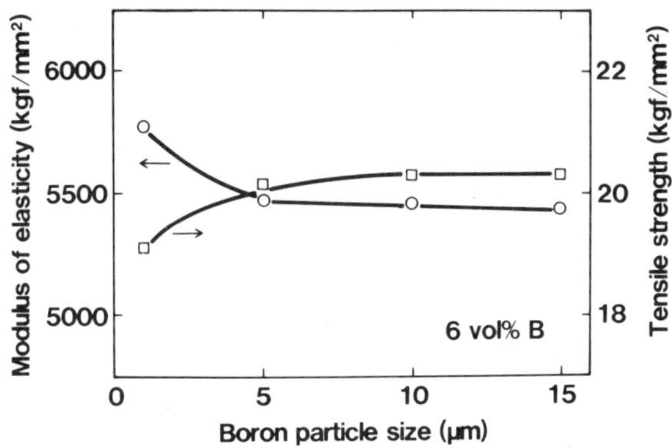

Fig. 4. Dependence of modulus of elasticity and tensile strength on boron particle size.

sintered diamond cutting tools. Features of cut chips were observed by an optical microscope.

RESULTS AND DISCUSSION

Mechanical Properties and Microstructures

Figure 2 shows the relation between the modulus of elasticity and the boron content in sintered Mg-9wt%Al alloys when the boron particles are   1 μm and 5 μm. The relationship for the sample with 1 μm boron deviates from the rule of mixtures having a smaller boron content than that for the sample with 5 μm boron. The results are discussed based on relative densities. The green density and the sintered densities for samples with boron particle of 1 μm and 5 μm are shown as a function of boron content in Fig. 3. The green densities decrease as boron content increases, and decrease abruptly when the 1 μm boron content exceeds over 6 vol%. Namely, changes correspond to changes in the moduli of

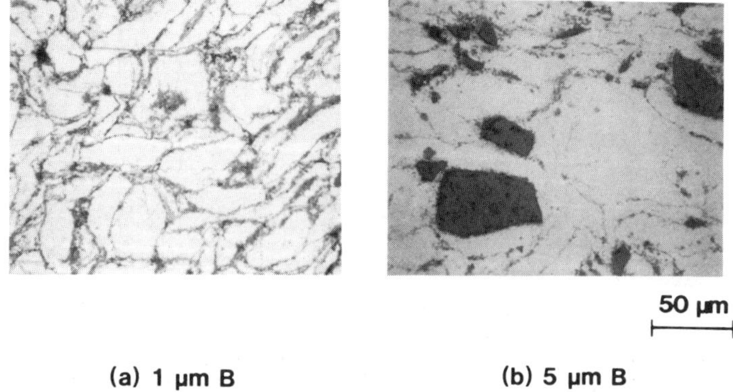

Fig. 5. Optical microstructures of the boron/Mg-Al sintered composite.

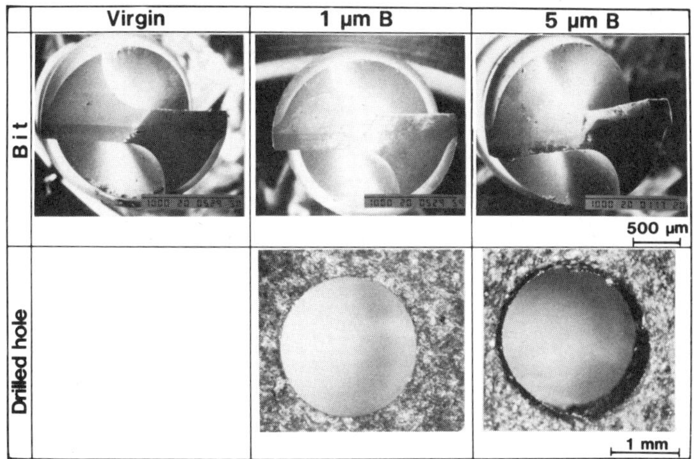

Fig. 6. SEM photographs of the worn bits and optical photographs of the holes.

elasticity in Fig. 2. Hard boron particles decrease the compactibility of the green sample. Also, the finer the particle size, the more difficult uniform blending with the other larger powders becomes. Finer boron powder particles agglomerate more easily, the higher the boron content, especially when boron content exceeds 6 vol%. This leads to poor compactibility and sinterability in a certain composition range. This is a problem in using 1 µm boron powder.

In the range with the rule of mixtures, particle-dispersed strengthening is slightly more effective when the boron particles are 1 m. The sample with 1 µm boron has a modulus of elasticity a few percent higher than that with 5 µm boron. Figure 4 shows the dependence of the modulus of elasticity and tensile strength on boron particle size when boron content is 6 vol%. The dependence is not the same, however, as tensile strength. Results showing the effect of boron particle size first appear meaningfully at a particle size of 1 µm. Regarding the modulus of elasticity, this may be because of slightly larger relative density. The microstructure observation may give one more reason for this. Figure 5 shows typical microstructures of the sintered composites. In sample (a), the fine boron particles collect together to form a network-like pattern, but individual particles are microscopically isolated in the sample with 6 vol% boron. This may be because Mg powder makes boron particles adhere to its surface during grinding. This network may suppress the transformation of the inner alloy matrix, resulting in an increase of modulus of elasticity.

The decreased tensile strength of the 1 µm boron sample may also be due to the network, which weakens the grain boundaries.

Machinability

Tool wear. Perhaps the most meaningful measure of machinability is tool wear.

Figure 6 shows bits before and after three holes were drilled in the test piece and the holes through which bits exited. The cutting

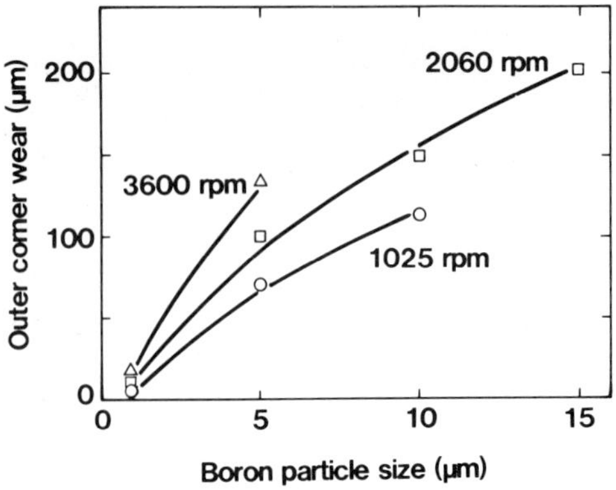

Fig. 7. Relation between boron particle size and outer corner wear.

edge of the bit used for the 1 μm boron sample had little damage, but that for the 5 μm boron sample was heavily damaged, especially on the outer corners. The holes reflect these facts. The test piece with 5 μm boron particles shows chips around the hole.

Figure 7 summarizes the outer corner wear as a function of boron particle size at different drilling speeds. Fine boron particles improve machinability.

Tool life is increased over 30 times by using 1 μm boron particles instead of 5 μm particles.

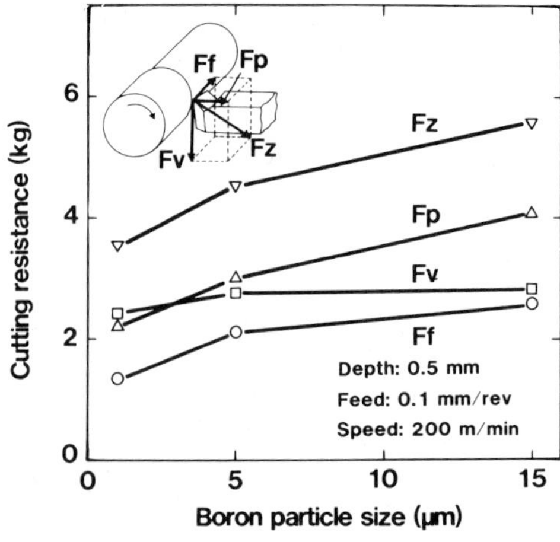

Fig. 8. Relation between cutting resistance and boron particle size.

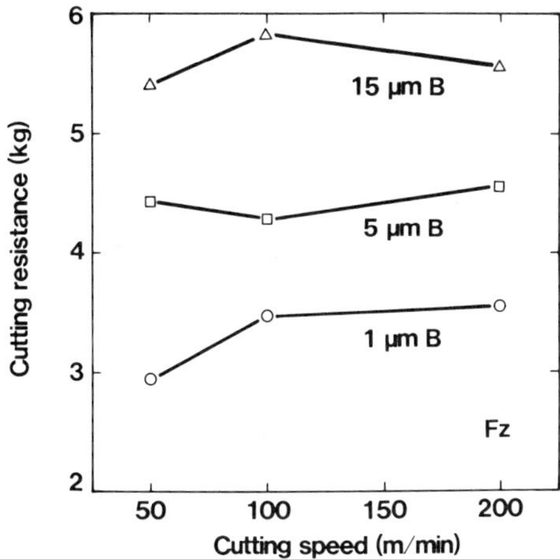

Fig. 9. Dependence of cutting resistance on cutting speed.

<u>Cutting resistance</u>. Cutting resistance during machining also can
be used to predict machinability. Low cutting resistance generally
means low tool wear and good machinability.[2]

Figure 8 shows the cutting resistances in an oblique-cutting test
for samples with different boron particle sizes. The Fv curve crossing
the Fp curve may have to do with the abrasion caused by boron particles
falling off cut chips, but further study is needed. The cutting
resistance, Fz, decreases with boron particle size, which agrees with
the machinability obtained from tool wear. This result is not changed
by cutting speed (Fig. 9).

<u>Cut chips</u>. The sample with 1 µm boron gives long curled chips,
which are a sign of good machinability (Fig.10). Chips of the sample
with 5 µm boron are cracked and broken. As was shown in sample (b) of
Fig. 5, the larger boron particles are dispersed at random in the alloy
matrix. In (b) of Fig. 11, note how even one large boron particle can
occupy a large part of the chip. The weak interface between the boron
and the alloy matrix may cause cracks under the bending stress caused
by the cutting tool. The chip is then easily broken. This tendency

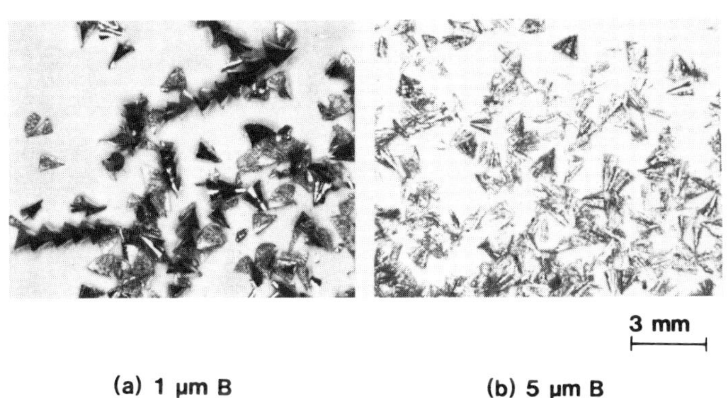

Fig. 10. Cut chips.

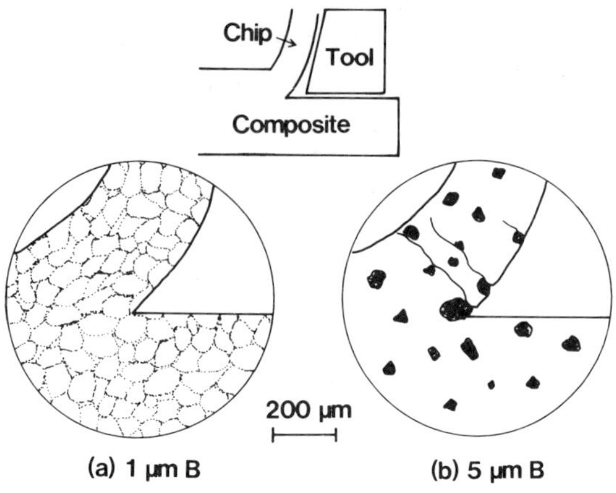

Fig. 11. Cutting chips.

increases as the ratio of cutting depth to boron particle size decreases. A small ratio causes boron particles to fall off and increases tool wear. Fine boron particles such as those in (a), do not cause this problem.

SUMMARY

We have discussed the effect of the particle size of boron powder on the mechanical properties and machinability of the boron/Mg-Al sintered composite.

(1) In the boron/Mg-Al sintered composite, boron particle size greatly affects machinability but only slightly affects mechanical properties.
(2) The larger the boron particles, the greater the tool wear.
(3) Tool life is increased over 30 times by using 1 μm boron particles instead of 5 μm particles.
(4) Fine boron gathers around large Mg powder particles in a network -like pattern. This suppresses the formation of the alloy matrix and improves the modulus of elasticity by a few percent.
(5) Unfortunately, smaller boron powders particles can result in poor blendability, compactibility, and sinterability as the boron content increases.

ACKNOWLEDGEMENTS

We thank Dr. K. Niwa and Mr. T. Yoshizawa for their encouragement. We would also like to thank Messrs. S. Terashima and K. Oishi for their work in the machining test.

REFERENCES

1. E. Horikoshi, T. Iikawa, and T. Sato: Modern Development in Powder Metallurgy, 19 (1988) 579.
2. J. S. Agapiou and M. F. DeVries, The International Journal of Powder Metallurgy, 24, (1988), 47.

EFFECT OF "INHERITANCE" IN POWDER METALLURGY

V.A. Ivensen and T.A. Rakotch

Institute for Refractory Metals and Hard Alloys
Moscow, USSR

The investigation of the kinetics of non-isothermal sintering have
facilitated the use of important new data which could not be explained
by conventional concepts of sintering processes.

The important characteristic feature of pore volume reduction on a
rising temperature is attributed to the fact that for a given powder
the rate of pore volume reduction depends not only on the instant
temperature value (T) but also on the rate of temperature increase ($\dot{T}$).
For active (rapidly sintered) powders the effect of $\dot{T}$ may be
predominent. This prevailing influence of $\dot{T}$ becomes most pronounced
when a powder compact is sintered such that its temperature is rapidly
increased between two isothermal periods. To describe the relationship
between the time-dependent rate of pore volume reduction and the time-
dependent $T(\tau)$, we use a relative value, i.e., the rate of pore volume
reduction referred to unit of pore volume, $W = -\dot{v}/v$ hour^{-1}. Such an
approach allows us to disregard the influence of "geometric factor",
i.e., the changes in pore geometry,[1] when describing the variation of
the rate of pore volume reduction with time.

The relation of $W(\tau)$ with time within the limits of temperature
rise between two successive isothermal periods is shown in Fig. 1 a, b,
c for $T(\tau)$ curves of different types. The peak value of W in Fig. 1(a)
almost coincides in time with the maximum rate of temperature increase.
The comparison of relations $T(\tau)$ and $W(\tau)$ shown in Fig. 1(b) confirms
the strong sensitivity of the rate of pore volume reduction (the value
of W) to the change in the rate of temperature increase. Any decrease
in the rate of temperature increase is accompanied by a sharp decrease
of W. If the rate $\dot{T}$ is constant up to the end of the increasing
temperature period, then the peak of W is shifted to the beginning of
the second isothermal period, as shown in Fig. 1(c). In this case the
decrease of W to the quasi-equilibrium level (inherent to isothermal
sintering,[1] occurs under isothermal conditions. The observed behaviour
may be explained by the fact that the accelerated flow of the heated
imperfect crystal is due to the "active" short-lived" defects generated
by the annihilation of the original crystal defects, rather than to the
original defects (of a dislocation type) themselves. This phenomenon is
described in more details in ref. 1.

Under the conditions of isothermal sintering, due to substantial

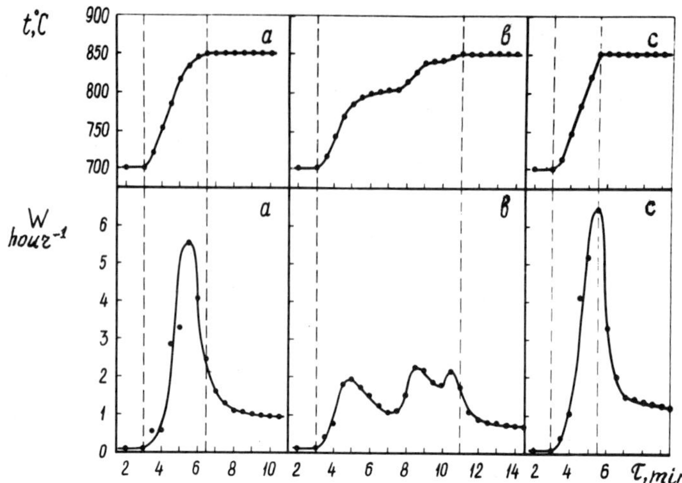

Fig. 1. Variation of the rate of pore volume reduction with time, $W(\tau)$, during temperature increase in the range of 700 to 850°C. On the top – $T(\tau)$ curves: (a) temperature-time curve with smooth approach to the isothermal holding; (b) the same with periodic speeding-up, slowing-down of temperature increase; (c) the same with an abrupt approach to the isothermal holding.

difference between the rate of the original defects annihilation (and, hence, the "active" defects generation) and much higher rate of original defects annihilation, dynamic equilibrium is established when the concentration of active defects and, therefore, the rate of crystal flow (and the rate of pore volume reduction) changes proportionally to the concentration of original defects. In this case the influence of the intermediate annihilation stage, i.e. the development of active imperfections, is not revealed. The direct relationship between the rate of pore volume reduction and the concentration of original imperfections was a basis of mathematical description of isothermal sintering with the use of quasi-chemical kinetics .[1]

On a rising temperature, the quasi-equilibrium condition is disturbed due to a substantial difference between activation energies of original and active defects annihilation (the former being much higher than the latter). Due to this fact the rate of generation of active defects increases far more intensively than that of their annihilation, which leads to a temporary accumulation of active defects. When the rate of temperature increase is lowered and, particularly, after the temperature of the next isothermal period has been reached, the concentration of active defects drastically decreases to the level that corresponds to dynamic equilibrium on a new temperature step, but that equilibration process takes a few minutes. The rate of pore volume reduction changes in accordance with the change in the concentration of active defects (Fig. 1c). Thus the influence of active defects could not be attributed immediately to the formation of excessive vacancies.

As shown earlier, the behaviour of powders of various origin, which differed by the conditions of crystal particle formation, is characteristically different within the period when temperature is rapidly rising (this period will hereinafter be referred to as a temperature jump).[1,2] For example, it was frequently observed in nickel

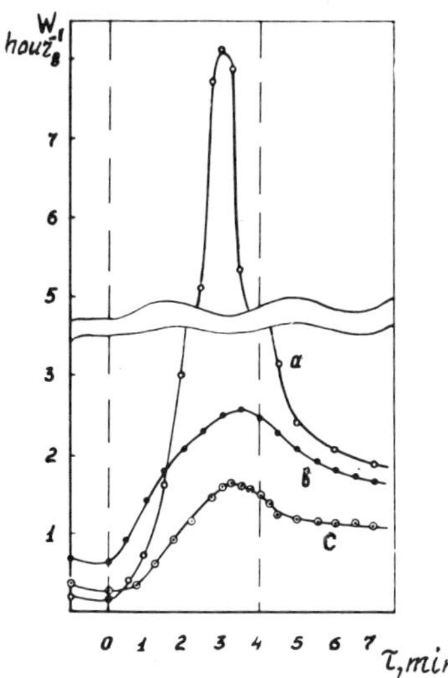

Fig. 2. Variation of the rate of pore volume reduction with time, $W(\tau)$, within the <u>temperature jump</u> (700 to 850°C in 4 min.) for nickel powders: (a) initial oxide nickel powder; (b) initial carbonyl nickel powder; (c) carbonyl nickel powder after preheating at 500°C.

powders obtained by hydrogen reduction of oxide and dissociation of carbonyl (in vapour-gaseous phase). In the sintering of nickel powders, time dependences of pore volume reduction are substantially distinguished by the extent of W increase within the period of rapid heating to sintering temperature (see Fig. 2). From the comparison of curves 2(b) and 2(c), it is evident that after preheating (calcining) carbonyl nickel powder (curve 2(c)) both the peak rate of W and the rate at the beginning of temperature rise period decrease while their ratio and the general form of the relation $W(\tau)$ remain almost unchanged. A similar effect is produced by variation in pre-sintering time before a temperature jump. Fig. 3 shows the relation $W(\tau)$ for powder compacts subjected to pre-sintering at 700°C for different periods of time (from 2 min. to 12 hours) followed by rapid heating, for 4 min. over the range of 700 to 850°C. Rapid heating was followed by isothermal holding at 850°C (sintering of oxide nickel powder). Under equal time and temperature conditions of the temperature jump, the characteristic form of all the curves $W(\tau)$ as well as the relative increase of W remained almost unchanged.

Similar experiments were made with carbonyl nickel powder, see Fig. 4. Since the experiments did not coincide in time, the relationships between the rate of pore volume reduction and that of temperature increase within a single jump were slightly different. Moreover, two methods were used to change the concentration of original defects, i.e., pre-heating (calcining) of powders before compaction and variation of pre-sintering time (at 700°). Nevertheless, in all the cases the shape of curves $W(\tau)$ peculiar to this powder remained unchanged, the increase of the rate of W within the temperature jump being much less than that for oxide nickel powder.

Thus, in spite of substantial reduction of the concentration of original defects after pre-heating of a powder or compact, the peculiarities of the "behaviour" of the powder of a given origin were completely retained within a temperature jump. It can be assumed from

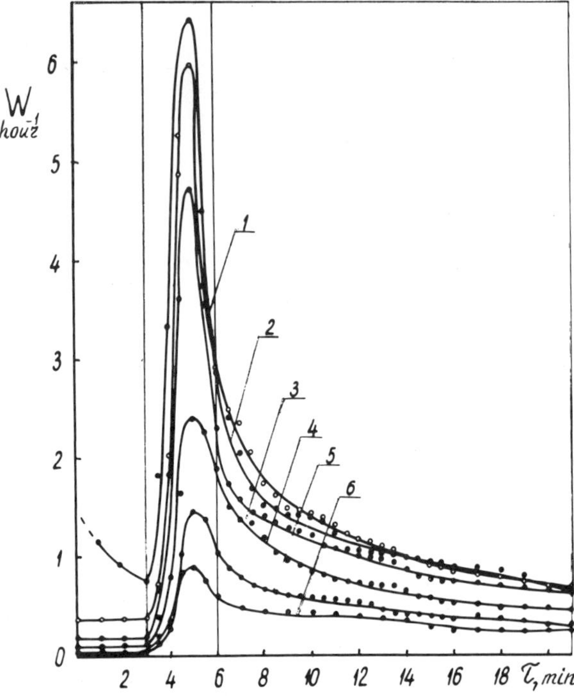

Fig. 3. Variation of the rate of pore volume reduction with time, $W(\tau)$, within the temperature jump (800 to 900°C in 3 min.) for oxide nickel powder after pre-sintering at 800°C during different periods of time: 1- 2 min.; 2 - 10 min.; 3 - 30 min.; 4 - 2 hours; 5 - 6 hours; 6 - 12 hours.

this fact that the "behaviour" of powder compacts within a temperature jump is determined only by the particular nature of lattice defects occuring during primary crystal growth. Calcining (or pre-sintering) reduces the concentration of original defects but does not change their "nature". Various dislocation structures developing under different conditions of crystal growth are supposed to have different abilities to generate active imperfections. The "nature" of original defects is retained although its concentration has been reduced after heat treatment.

It follows from this fact that the behaviour of a powder compact within a temperature jump is an important property of the powder of a given origin that reflects the peculiarities of the crystal substructure. This property may be conventionally expressed by a numerical value of W_M/W_B, where W_M is the maximum rate of pore volume reduction while W_B is the rate before the temperature jump. In the experimental data shown in Fig. 2 and 3 the value of W_M/W_B for carbonyl nickel was within the range of 4 to 6 while that for oxide powder was in the range of 38 to 56 (excluding curve 1, where the increase in temperature started before W reached quasi-equilibrium level after initial heating).

Further investigations have shown that the value of W_M/W_B for a given powder (under constant temperature-time conditions of a

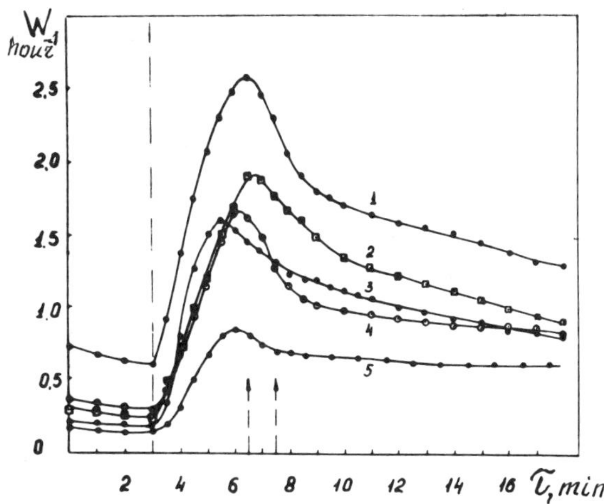

Fig. 4. Variation of the rate of pore volume reduction with time $W(\tau)$
within the temperature jump (700 to 850°C) for carbonyl nickel
powders treated under different conditions:
1, 4 - initial powder; 2, 3, 5 - after pre-heating at 500°C;
sintering time before the temperature jump: 1, 2 - 30 min.; 3 -
60 min.; 4, 5 - 120 min.

temperature jump) is also retained when a powder (or a compact) is
subjected to plastic deformation. Before considering experimental data,
let us remember that the principal conclusions result from the previous
investigations, in which it was shown that the "original" defects
occuring during primary crystal growth and "strain-induced" defects
occuring under mechanical loading that resulted in plastic deformation
have a substantially different influence on the rate of pore volume
reduction. It was found that the flow of crystalline substance is
accelerated only by "original" defects. Plastic deformation of a powder
or a compact results in a considerable increase in integral
defectiveness (according to X-ray diffraction data) and subsequently a
decrease in the rate of pore volume reduction which depends on the
degree of plastic deformation. [1,3] The defects induced by plastic
deformation are supposed to inhibit densification, thereby affecting
the "behaviour" of original defects. Defects of this type may
discourage the movement of structure components which constitute the
original defects, resulting in the reduction of their annihilation rate
and, subsequently, decreasing the rate of generation of active
imperfections. It is also possible that the original defects may be
"destroyed" or even change their "nature" by which the intensity of
active imperfection generation is determined. Possible mechanisms of
interaction between original defects and those induced by deformation
are considered. [3]

The influence of plastic deformation on the "behaviour" of a
powder compact being sintered was investigated by means of the
temperature jump for nickel powders obtained from oxide and carbonyl.
Powders of high-ductility metals (iron, nickel, copper etc.) may be
deformed by "rolling" them in a rotary drum filled with small steel
balls. This process makes it possible to deform the powder particles
almost without their being crushed. After such a treatment the powder
being sintered exhibits a considerably lower densification compared
with an untreated powder of the same dispersity. However, it was
interesting to consider not just the variation in pore volume

Table I. Comparison of W_M/W_B values for oxide and carbonyl nickel powders prior to and after strain hardening

Technique of obtaining powder	Mode of treatment	W_M/W_B	Metanol adsorbed, A mg/g	Relative spec.surf. A/A_{oxide}	Width of X-ray diffr. line, $\times 10^3$, rad.
Reduction of oxide	Untreated	26	0.166	1.00	8.2
	Rolling 24 hours	32	0.136	0.82	30.3
Dissociation of carbonyl	Untreated	5.3	0.065	0.39	7.4
	Rolling 24 hours	8.3	0.090	0.54	31.6

reduction after the powder deformation but especially the influence of deformation on the increase in the rate of pore volume reduction in a temperature jump, i.e. on the value of W_M/W_B, since this value is just concerned with the "nature" of original defects. Oxide and carbonyl nickel compacts made of deformed and initial (untreated) powders were sintered within the temperature jump of 700 to 850°C. Deformation of the powders was performed by "rolling" them with balls in the mode of "dry milling" during 24 hours. The change in dispersity was characterized by means of methanol vapour adsorption (under "standard" conditions) while the degree of strain hardening was determined by the change in the width of X-ray diffraction line (311). Comparison of these values with the increase in the rate of pore volume reduction within the temperature jump of 700-850°C in 4 min. is given in Table I.

It follows from Table I that an intensive strain hardening, resulting in about 4 times increase in the width of X-ray diffraction line, does not lead to substantial change in the value of W_M/W_B. The latter ratio remains in the limits typical for a given powder. Some variability in the values of W_M/W_B can be attributed to high sensitivity to the rate of temperature increase, and to the fact that in repeated experiments the temperature schedule cannot be exactly reproduced.

The treatment of a powder with balls in a liquid medium ("wet milling") instead of "rolling" (in the mode of "dry milling") makes it possible to achieve an intensive crushing of powder particles. Wet milling (in ethanol) was used to achieve dispersity of carbonyl nickel particles similar to that of oxide nickel powder (methanol adsorbtion increased from 0.065 to 0.183 mg/g). In this case the "behaviour" of crushed powder in a temperature jump did not change; W_M/W_B was 6.3, i.e., that is corresponding to a normal value of this ratio for carbonyl nickel powder (under the same temperature-time conditions of temperature jump). The absence of any influence of powder dispersity on the processes which are reflected by the value of W_M/W_B is confirmed by the above experimental data, along with the results of other investigations where powder dispersity varied in annealing.

The accelerated growth of the rate of pore volume reduction with increase of temperature is defined by a certain internal substructural characteristic which is retained after heat-or mechanical-treatment of powder. Both types of powder treatment result only in variation in the concentration of defects which, along with the powder dispersity, determines the value of W, while the value of W_M/W_B remains within the limits typical for a given powder. This observation confirms the "nature" of original defects to be retained even after substantial change in their concentration.

The fact that the "nature" of original defects is retained after deep plastic deformation is surprising. However, it is even more difficult to explain the fact that the "nature" of imperfections is retained after annealing of hardly deformed powder. In this case the intensive recrystallization process may seem to change the structure and substructure of particles so drastically that the retention of any original substructural elements is out of question. However, experimental results lead to the opposite conclusion. The result of one among many experiments investigating the phenomenon under consideration is given below. Oxide and carbonyl nickel powders were treated by "rolling" with balls for 24 hours followed by annealing at 550°C for one hour. Due to sharp decrease in activity of the powders after the treatment, sintering temperature was increased. The temperature jump was from 850 to 950°C in 3 min. Change in W (the rate of pore volume reduction) was compared with the "behaviour" of initial powders in a similar temperature jump, within the range of 100°C (700 to 800°C in 3 min.). Variation of $W(\tau)$ for initial and treated powders is shown in Fig. 5 (a,b). Relative arrangement and shape of the curves remain unchanged after strain hardening and annealing. The difference in the extent of W increase in oxide and carbonyl nickel powders is retained and is evidently as large as in the initial powders.

Thus, it was found that neither deformation, nor annealing, nor both of them performed in any sequence, could eliminate a certain particular nature of powders that defines the growth of W in temperature jump. This "nature" (which is probably due to some substructural elements of superstable origin) arises in the process of initial crystal growth. The difference in this "nature" becomes very distinct under substantially different conditions of crystal growth. As mentioned earlier, two types of powder are compared in the present experiments, i.e. powder obtained by reduction of oxides in hydrogen (topochemical reaction involving the shift of the boundary between the initial and new phases) and powders obtained by dissociation of compounds in vapour-gaseous phase (nucleation and growth of crystals). Depending on the method of synthesis of the powder, two substantially different types of dislocation structures seem to develop, exhibiting different abilities to generate active imperfections during crystal heating (an attempt to propose a plausible interpretation of this process is described earlier).[1]

Stable retention of the original properties of powders governing the kinetic mechanisms of densification has been described many times. In particular, it was shown that variation in the "behaviour" of powders is not attributed to small differences in the content of impurities. It was manifested by the experiments in which oxide nickel was obtained by oxidation of carbonyl nickel. After reduction in hydrogen "nickel from oxide" was obtained having similar content of impurities. In such a case the difference in the values of W_M/W_B remained unchanged.[4] It was further established that there is quite similar difference in the values of W_M/W_B in the sintering of iron powders prepared from oxide and carbonyl. In sintering tungsten carbide powders prepared under substantially different conditions the values of W_M/W_B varied about 2 times (within the temperature jump from 1300 to 1450°C in 4 min.). In sintering molybdenum and copper powders, a

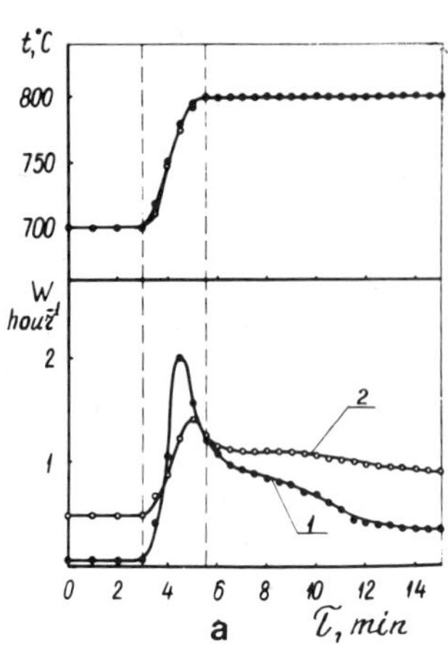
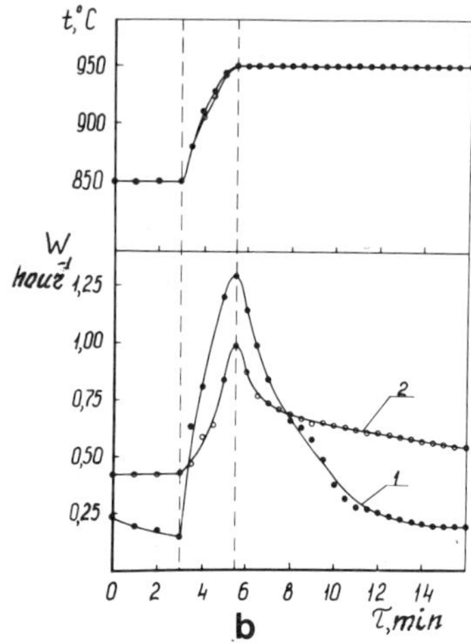

Fig. 5. Variation of the rate of pore volume reduction with time, $W(\tau)$, for oxide (1) and carbonyl (2) nickel powders: (a) before treatment within the temperature jump of 700-800°C in 3 min.; (b) for the same powders after strain hardening by "rolling" with steel balls and heat treatment at 550°C within the temperature jump of 850-950°C in 3 min.

typical relation, $W(\tau)$ with a prominent peak rate within the temperature jump, was obtained.

Attempts were made to find more adequate interpretations of the phenomena observed in experiments but they met with considerable difficulties. X-ray diffraction analyses aimed at measuring the degree of microdistortions (microstresses) and size of blocks (coherent scattering zones) in initial oxide and carbonyl nickel powders, as well as in powders after deformation and annealing, did not reveal any relation between the values of W_M/W_B and the substructural characteristics. On the other hand, the X-ray analyses data indicated that the degree of accumulation of defects induced by deformation under equal mechanical action differs depending on the "nature" of original defects. The recovery of the above-mentioned substructural characteristics during annealing also proceeds in a different manner if there are defects of a different origin in the powder being annealed. Thus, it can be noted that the existance of complex dislocation structures developing in topochemical reaction (with shifting phase boundary) affects to some extent the accumulation of defects induced by deformation (under mechanical action) and discourages the process of recovery (the elimination of microstresses) in annealing.[6] The higher apparent activation energy of densification of powders obtained by topochemical reaction observed in many investigations is probably attributed to higher stability of "topochemical" defects.

Far greater difficulties arise in the attempts to propose physical interpretations of the nature of active imperfections. The analyses of the phenomena referred to this problem made it possible to propose (as an alternative) the hypothesis of their development in the process of clustering of excessive vacancies. It is assumed that the product of clustering – "quasiliquid defect" – is much more "long-lived" than an excessive vacancy.[1] Although this hypothesis not only is consistent with the principal kinetic laws of isothermal and non-isothermal sintering but also explains them to some extent, it can be considered

476

only as a preliminary conception needed further confirmation and more
deep analysis.

Despite the absence of rigorous physical interpretation, the fact
that some properties depending on the conditions of crystal particle
formation are retained after mechanical and heat treatment of powders
or compacts can be considered established. This phenomenon may be of
interest not only for further research developments, but also for
sintering practice, since it has been found that the conditions in
which powders are obtained influence both the kinetics of densification
and the properties of finished products (in manufacturing hard-metals).
Such an influence is retained after crushing carbide particles to any
extent (within the limits of milling conditions used in practice).

The relationship between the final stage in powder metallurgy,
i.e., sintering, and the conditions of powder synthesis being retained
practically with any combination of sequence and conditions of
intermediate processing is proposed to be referred to as the effect of
inheritance in powder metallurgy. At present the theoretical basis of
this effect has just started to be developed. Its further development
will possibly require some essential revision to be made in the present
knowledge about physical processes occuring upon heating of an actual
("defective") crystal.

REFERENCES

1. V. A. Ivensen, Phenomenology of Sintering. "Metallurgiya", Moskva,
   1985, 244.
2. V. A. Ivensen, Some Characteristics of Powder Compact
   Densification in Non-Isothermal Sintering, Science of
   Sintering, 10(1978) 175-184
3. V. A. Ivensen, T. A. Rakotch, Influence of Original and Strain-
   -Induced Defects of Crystal Lattice on Densification in Non-
   -Isothermal Sintering, Poroshkovaya Metallurgia, No.1 (1984) 25
   -30.
4. V. A. Ivensen, T. A. Rakotch, Different Effects of Method of
   Powder Synthesis and Impurity Contents on the "Behaviour" of
   Powder Compact in Temperature Jump., Poroshkovaya Metallurgia,
   No 1, (1986) 24-28.
5. V. A. Ivensen, T. A. Rakotch, Relation of $\dot{V}(\tau)$ within Temperature
   Jump in Sintering Molybdenum and Tungsten Carbide Porous
   Bodies, Poroshkovaya Metallurgia, No 4(1986) 30-34.
6. V. A. Ivensen, T. A. Rakotch, T. A. Gorbatcheva, Influence of
   Annealing of Deformed Powder Particles on the "Behaviour" of
   Powder Compact within Temperature Jump and Its Substructural
   Characteristics, Poroshkovaya Metallurgia, No 2 (1986) 18-22.

HYDROSTATIC COMPACTION OF FINE COBALT POWDER

D. Dužević and M. Buchberger

Rade Končar - Electrotechnical Institute
Baštijanova bb, 41000 Zagreb, Yugoslavia

INTRODUCTION

During the past few decades, a large number of analytical
equations have attempted to describe the densification of powders under
pressure. A pertinent review of the subject is available in the recent
paper published by Chaklader and Bhattacharya.[1] The majority of the
empirical relations between the applied pressure and the corresponding
green-compact volume (density) has been designed to fit the data
obtained by the uniaxial technique of die-pressing, which suffers from
serious inconsistencies. The most striking one is the die-wall friction
with all of the accompanying anisotropies and inhomogeneities. A great
deal of these incongruities, as well as their masking effects, can be
avoided by applying the isostatic pressing technique. Basically, the
same pressing equations should fit the data obtained from both
isostatic and die-pressing experiments. Actually, on the basis of
Morgan and Sand's experiments, Lenel[2] points out that Heckel's (or
Konopicky's or Athy's) equation is the most suitable to describe the
phenomenon of isostatic powder densification. According to Mitin,[3] a
kind of inverse power law equation named after Zhdanowitsch, fairly
describes the process. With all the well known limitations of the
simple potentional, logarithmic or exponential equations, such a theory
does not appear tenable. The goal of this paper is, therefore, to test
the interpolating suitability of the empirical pressing equations by
treating the experimental data of our own.

EXPERIMENTAL

A 99.9 $^+$% pure cobalt powder with a mean particle diameter of
1.42 μm (as measured by the Fisher Sub-Sieve-Sizer) was pressed both
hydrostatically and uniaxially, without any additives. In the first
case, a cylindrical steel chamber, 30 mm in diameter and 50 mm in
height, with two rubber O-rings was used. The immersion fluid was a
light component mineral oil. Preforms were prepared by tapping the
powder into paper cubicles with 15 mm side length. An average mass of
the preformed samples was 4.4 g with deviations not exceeding 0.5 g,
giving a corresponding relative tap density of approximately 0.13 (fill
density 0.07). Pressing was accomplished on a 25 t precision Avery
laboratory press with the pressing speed increasing from 0.1 MPa/s at 5

MPa pressure to 3.0 MPa/s at 350 MPa. The pressure was relieved
instantaneously when the nominal applied pressure value was reached.
Density was determined from the mass and the volume of samples ground
to regular parallelopipeds, as well as by a water buoyancy method as
applied to the samples covered with an acetone solution of plastics.
Experimental error was calculated to be below 3%.

Uniaxial, top-stroke, pressing was performed in a 10 mm diameter
hardmetal pressing-tool die with hardened chromium steel puncheons in
the pressure range of up to 1900 MPa. Sample mass was kept in the
narrow range between 1.9-2.1 g, their height decreased with rising
pressure from 10 to 3 mm. With rising pressure, the pressing rates
increased from 0.4 to 10.6 MPa/s. Pressing speed was low enough to
exclude any measurable influence on the density obtained, as shown by
separate experiments.

It has to be stressed that there was no measurable density
difference between the samples pressed from the loose or tapped
powders. As regards the influence of the influence of the trapped air
on the hydrostatically pressed samples, Buren and Hirsch[4] have already
shown that it not does manifest itself systematically either on the
density or on the strenth of the compacts. Actually, starting with a
relative tap density of $\nu_0 = 0.13$ at zero applied pressure and finishing
with a maximum relative density of $\nu = 0.63$ at 350 MPa, the internal
air pressure of the non de-aired compacts grows from normal atmospheric
pressure to a pressure of $(1-\nu_0)/(1-\nu) \times 0.1 = 0.3$ MPa. Compared with
the single-measurement experimental error in density determination,
this is negligible.

RESULTS AND DISCUSSION

In Figs. 1-6, the densification data for both die- and
hydrostatically pressed powder samples are represented in diagrams
using coordinates in which the experimental points should lie on a
straight line if the corresponding pressing laws are obeyed. This
refers to the Balsin's, Smith's, Ballhausen's, Konopicky's, Jone's and
Nutting's equations. The same is attempted for Kawakita's equation in
Fig. 7 by taking into account hydrostatically pressed samples alone.
The result of fitting these data to the Cooper-Eaton probability
function is shown in Fig.8.

From the pressing diagrams given, it may be concluded that, except
for a systematic density surplus (5%) of hydrostatically pressed
samples, there is no essential difference in densification behaviour of
hydrostatically and die-pressed samples. Obviously, the surplus issues
from the die-wall friction by uniaxial die-pressing. As regards the
interpolating ability, except for the Cooper-Eaton's, none of the
pressing equations examined can assure correct fitting to experimental
data in the entire pressure range investigated. This also applies to
the equations of Athy, Tanimoto and Terzaghi, listed by Chaklader and
Bhattacharya in addition to those formerly mentioned. Actually, Athy's
equations are just another way to express mathematically the relation
implied by Konopicky's equation, whereas Tanimoto's equation represents
a slight modification of Kawakita's equation (Fig.6), with a linear
term added, which cannot eliminat its inherent curvature. The same
applies to Terzaghi's equation, which is only a simply modified
Balshin's one by adding linear members. As regards Murray's equation,
there is no possibility to test it by a "rectifying diagram". Laborious
trial-and-error work conducted in an effort to fit it to experimental
data resulted in four numerically very different expressions even in

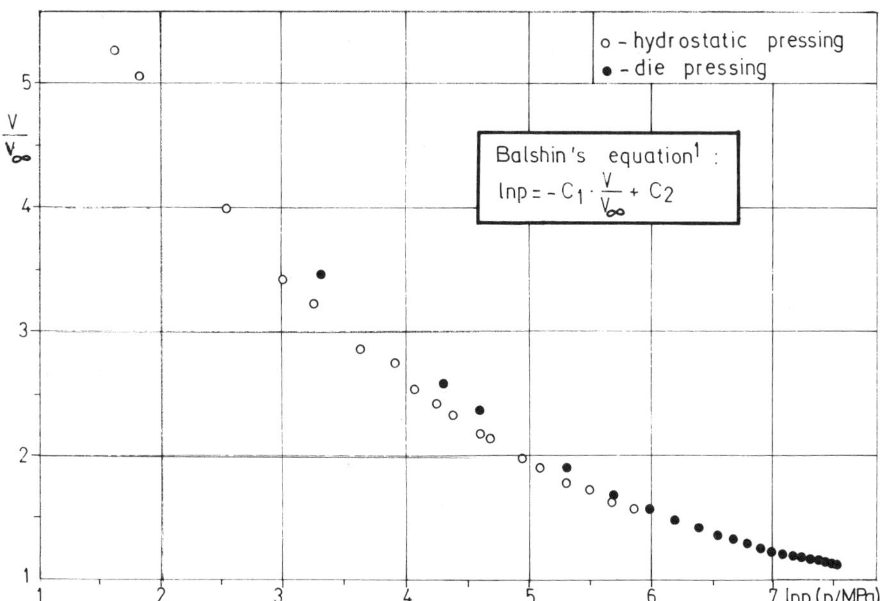

Fig.1. Experimental points in Balshin's rectifying
coordinates

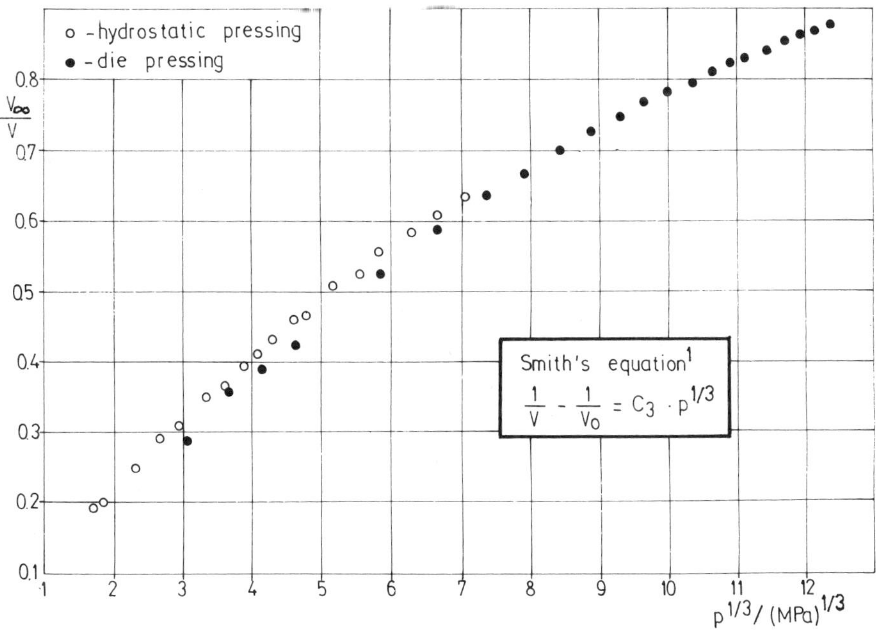

Fig.2. Experimental points in Smith's rectifying
coordinates

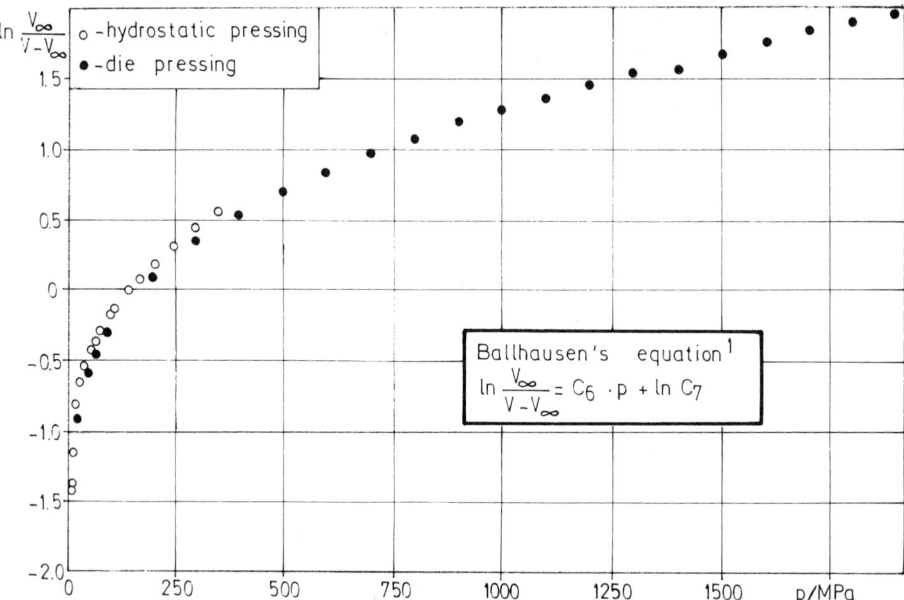

Fig.3. Experimental points in Ballhausen´s rectifying
coordinates

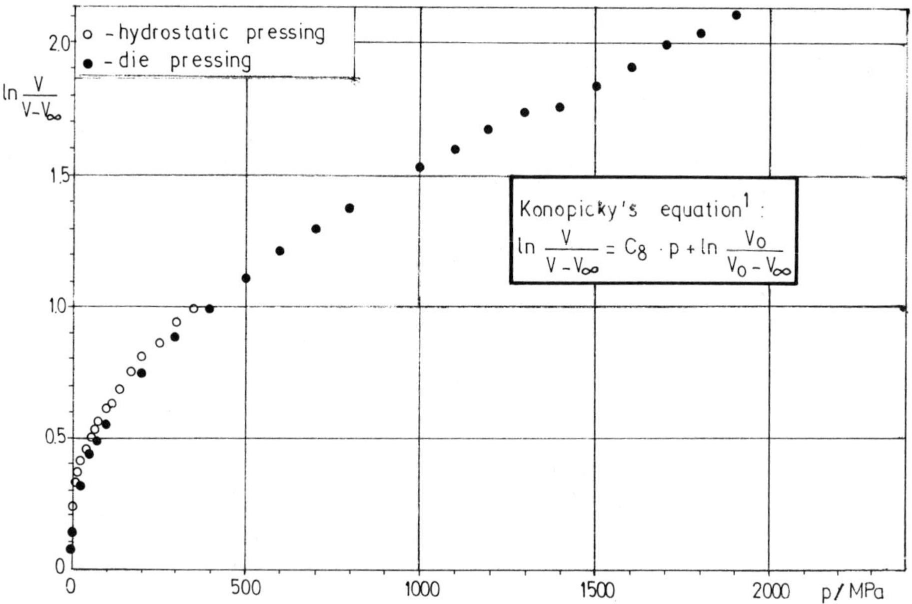

Fig.4. Experimental points in Konopicky's rectifying
coordinates

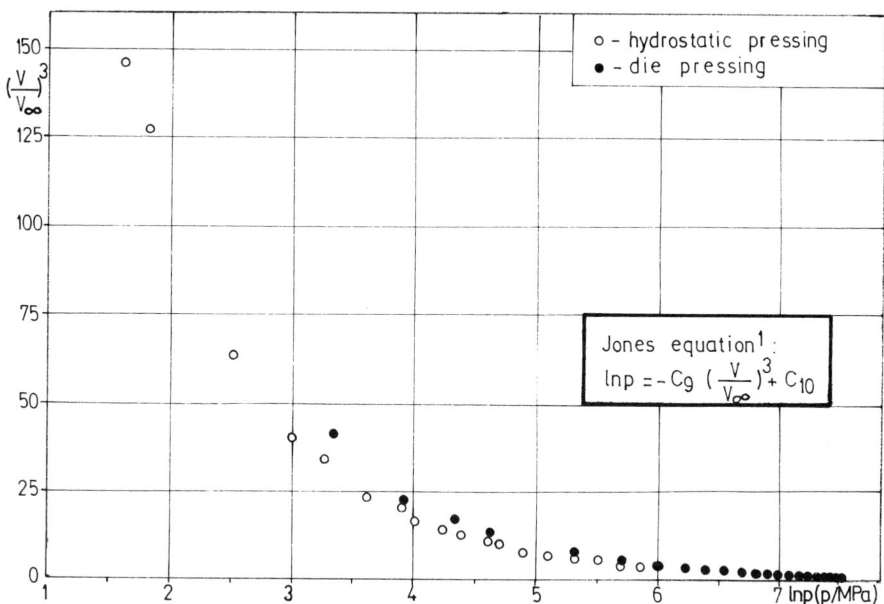

Fig.5. Experimental points in Jones´ rectifying coordinates

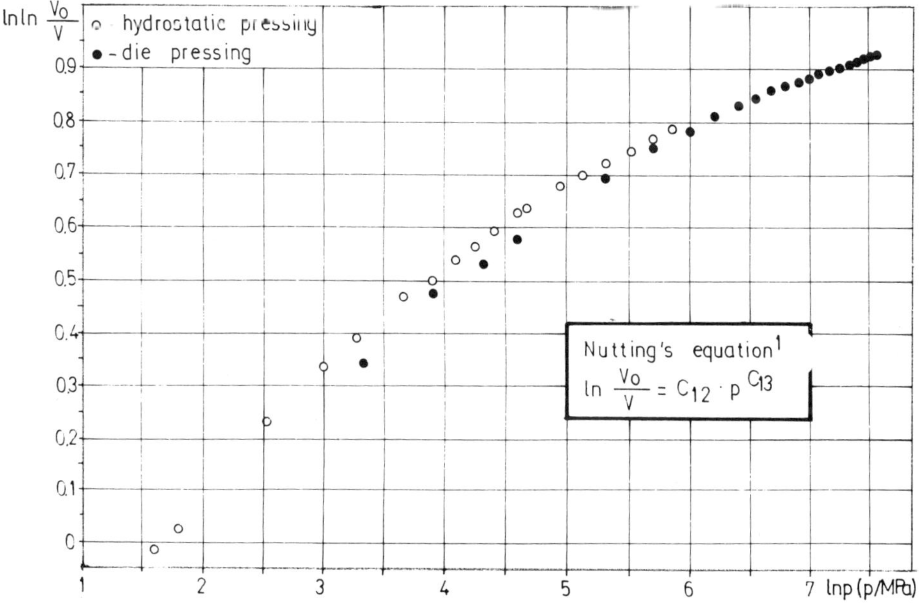

Fig.6. Experimental points in Nutting's rectifying coordinates

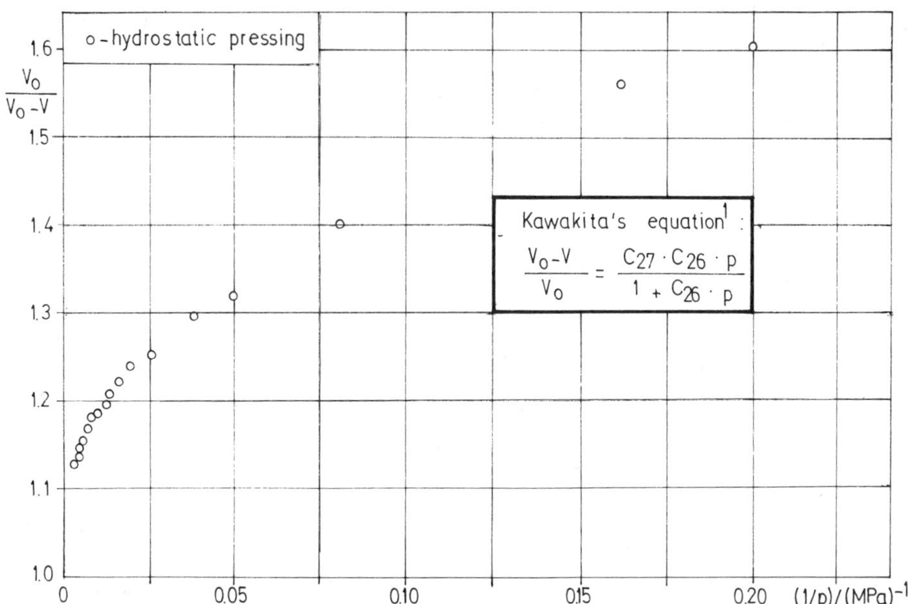

Fig.7. Experimental points in Kawakita´s rectifying
coordinates

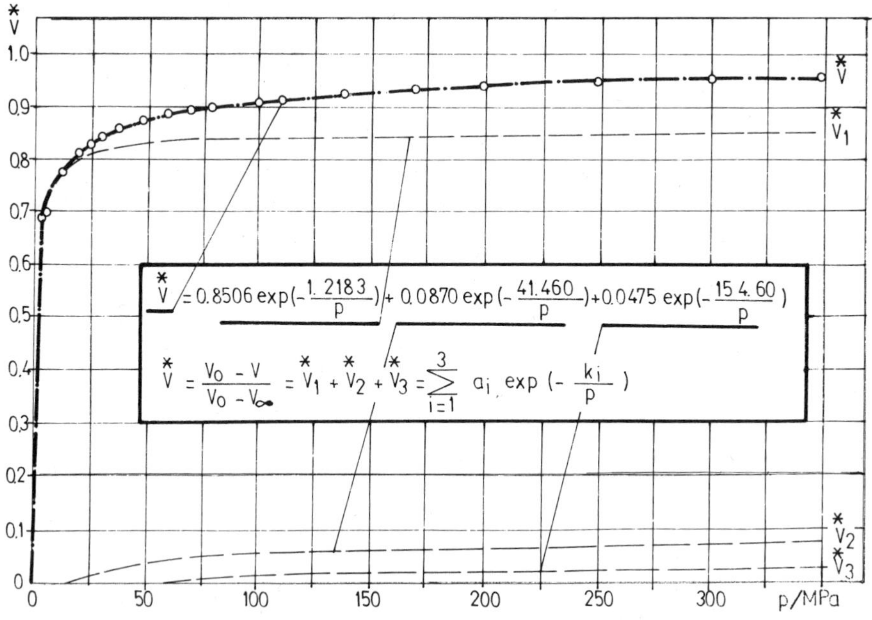

Fig.8. Fitting the experimental points to the
Cooper-Eaton three-distribution function

the case of relatively narrow pressure range of hydrostatically pressed
samples. As a matter of fact, the best (but not complete)
"rectification" is obtained through an extensive optimization procedure
in p $^{1/5}$ -ν coordinates. This corroborates the viewpoint about the high
fitting ability of power law equation as stressed by Mitin.

It is readily observed that the experimental points of die-pressed
samples in the high-pressure range conform well to a straight line. By
extrapolating these lines to theoretical relative density ($\nu=1$, $V=V_\infty$)
one obtains the value of limiting pressure p_1 needed to achieve
theoretical density at about 4000 MPa. This is true for Balshin's,
Smith's, Jone's, Nutting's and Kawakita's equations which
mathematically express the physical assumption the finite p_1 values. In
contrast, Ballhausen's and Konopicky's equations do not agree with this
viewpoint, expressing instead the obsolete view that the theoretical
density may be achieved only at infinite pressure.

A qualitatively higher level of fitting is accomplished by
applying the Cooper-Eaton interpolating function. The deviations of
experimental points from the positions predicted by this equation lie
within 1% for hydrostatically pressed green-compacts and even within
0.5% for die-pressed ones. The deviations prevail in the low-pressure
domain and vanish towards higher pressures. In both the hydrostatic and
die-pressing cases, the optimization procedure infers a necessity to
introduce three distribution terms instead of the original two, as
proposed by Cooper and Eaton.[5] By the way, they dealt with a number of
ceramic powders. The first distribution term of Cooper and Eaton refers
to filling holes of the same order of size as the original particles by
rearrangement motions. The pressure at which it exibits the highest
intensity is determined by halving the corresponding exponential
coefficient. In the case of Cooper-Eaton's ceramic powders, this
pressure is around 10 MPa. The second distribution of Cooper and Eaton
is associated with the process of filling voids that are substantially
smaller than the original particles and the operative mechanism
indicated is plastic flow in combination with fragmentation. One finds
the most intense effect of this process at about 150 MPa.

By halving the exponential coefficients of cobalt (given in Fig.
8) one finds the peak-value pressures of the three distributed
processes at 0.6, 21, and 77 MPa, respectively. In the case of die-
pressed compacts, the optimization procedure gives the expression:

$$\overset{*}{V} = \frac{V_0-V}{V_0-V_\infty} = \overset{*}{V}_1 + \overset{*}{V}_2 + \overset{*}{V}_3 = 0.9199 \exp\left(-\frac{3.4658}{p}\right) +$$

$$+ 0.0735 \exp\left(-\frac{203.34}{p}\right) + 0.0129 \exp\left(-\frac{1785.7}{p}\right) \tag{1}$$

in which V_0, V and V_∞ represent the powder volume at zero, applied, and
at infinite pressure, $\overset{*}{V}$ the overall compaction and $\overset{*}{V}_i$ the fractional
compaction associated with single mechanisms. According to (1), the
corresponding peak-value pressures obtained by die-pressing are 1.7,
102 and 900 MPa. In both hydrostatic and die pressing, the peak-value
pressure of the first and the most efficient distribution ($\overset{*}{V}_1$) lies
around 1 MPa. This signifficantly differs from the value of 10 MPa of
Cooper and Eaton's. Hence, the first cobalt distribution is hardly to
be identified with the one-particle-size pore filling process. It
appears reasonable to relate it to the well known arching or bridging
phenomenon of powder metallurgy, i.e. with arch break-down at low

pressures. The second distribution of cobalt densification ($\overset{*}{V}_2$) is
tentatively associated with the rearrangement process of filling large
pores, whereas the third one (V_3) is readily atributed to the most
inefficient process of filling small voids by plastic rearrangement and
fragmentation.

Starting from the obsolete standpoint that complete densification
to theoretical density takes place at infinite pressure, Cooper and
Eaton derived that the theoretical value of the sum of dimensionaless
distribution coefficients should attain the unity value ($\Sigma\, a_i = 1$). In
the case of die-pressed cobalt green compacts, this sum attains a value
of 1.0063, as seen from the formula (1). From this, one can compute
that the total densification ($\overset{*}{V}=1$) takes place at a finite pressure of
p_1=6000 MPa which is rather consistent with 4000 MPa indicated
previously. By substituting the value of 6000 MPa into (1), the percent
fractional compacts of the three mechanisms involved are calculated to
be 92,7 and 1%, respectively. Again, it seems proper to ascribe the
huge contribution of 92% to the arch breaking-down mechanism, while the
contribution of 7% of the second distribution is probably related to
the particle rearrangements. The rest of 1% coupled with the third
distribution is compatible with the process of filling small voids.

CONCLUSION

The Cooper-Eaton pressing equation appears to be the only one
which correctly describes the powder densification phenomenon in a wide
pressure range, both for hydrostatically and die-pressed green
compacts. In the case of cobalt, this equation comprises three
distribution members in contrast to the two originally proposed. This
indicates a densification process distributed three ways. The mechanism
of filling one-particle-size pores by particle rearrangements and
essentially smaller pores by plastic deformation and fragmentation are
tentatively recognized as the second and the third distribution terms.
The first and the most efficient distribution term is attributed to the
arch breaking-down process.

REFERENCES

1. A. C. D. Chaklader and S. K. Bhattacharya, Effect of Additives on
       the Cold Compaction Behaviour of SiC powder, in "Sintering
       '85", G. C. Kuczynski et al., eds., Plenum Press, New York and
       London (1987).
2. F. V. Lenel, "Powder Mretallurgy", MPIF, Princeton (1980).
3. B. S. Mitin, "Poroškovaya metallurgya i napilennie pokritiya",
       Metallurgiya, Moskva (1987).
4. C. E. Van Buren and H. H. Hirsch, Hydrostatic Pressing of Powders,
       in H. H. Hausner et al., eds., "New Methods of Consolidation of
       Metal Powders", Plenum Press, New York (1967).
5. A. R. Cooper and L. E. Eaton, Compaction Behaviour of Several
       Ceramic Powders, J. Am. Ceram. Soc.45:97 (1962).

MECHANICAL PROPERTIES OF Cu-P SINTERED ALLOYED STEELS: STUDY OF THE
COPPER INFLUENCE ON DIMENSIONAL CHANGES

L.E.G. Cambronero, J.M. Torralba, and J.M. Ruiz Prieto

Materials Engineering Department, Polytechnical University
of Madrid, c/Rios Rosas, 21, 28003, Madrid, Spain

INTRODUCTION

Phosphorus is usually considered a contaminating element in the
conventional steel industry, mainly because of its high segregation, the
fragileness it gives iron and the low rate of diffusion that it has in
iron. Nevertheless, phosphorus is one of the elements most often used as
an alloy of sintered steels, manufactured with methods based on powder
metallurgy.

Phosphorus modifies, in a positive way, the mechanical properties
of sintered steels, but its main disadvantage is that a high degree
of shrinkage is shown after sintering. The object of this study is to try
to control this excessive shrinkage that takes place in the sintering of
steel sintered with phosphorus, by adding Cu, as well as to study the
mechanical properties (tensile strength, hardness and elongation) and
microstructure (throuhg conventional optical microscopy and SEM) of this
type sintered alloyed steels.

EXPERIMENTAL PROCESS

Raw materials

The alloys have been obtained through the elemental mixture of three
types of powder: ferrophosphorus with a composition of Fe-O, 45%P known
commercially as PASC 45 (Höganäs, Sweden), highly pure electrolytic copper
(Norddeutsche Affinerie, West Germany) and natural graphite. Alloys of
Fe-O, 45P-C-Cu were obtained by adding carbon between O and 1.2%C and Cu
between O and 8% Cu.

The PASC 45 powder used is shown in the photograph of figure 1.

Compacting

The alloys obtained were compacted in a uniaxial matrix at a pressure
of 700 MPa, obtaining in all the samples a green density of 7.11 g/cm^3
which changed to 7.16 g/cm^3, after sintering. The samples that were made
were of tensile type according to ASTM Standard E8-69.

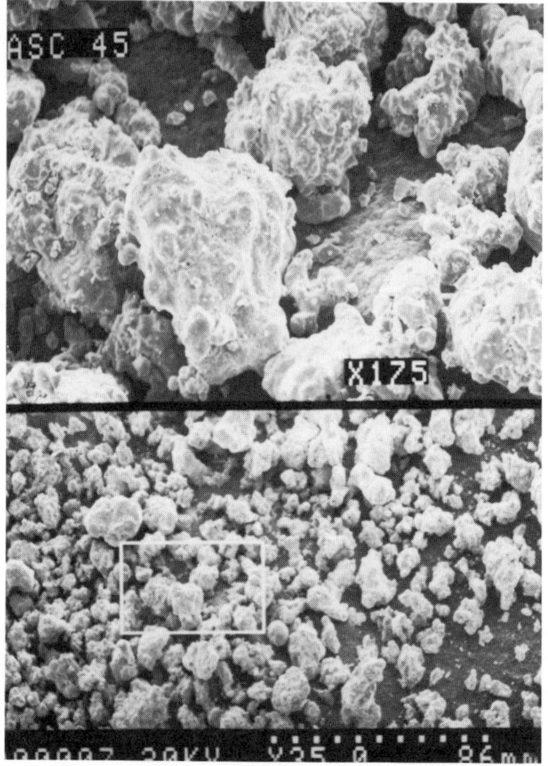

Fig. 1. PASC 45 powders (Höganäs, Sweden).

Sintering

The sintering was carried out in a tubular laboratory furnace at
1120°C/1 hr., with a previously purified Ar protecting atmosphere. The
furnace was heated at a rate of 15°C/min. and cooled at 5°C/min. to a
temperature of 450°C.

Physical and mechanical tests

From the tension test the tensile strength and elongation was
determined, for all the alloys studied. The hardness was determined
through a Rockwell test, scale A. The dimensional change was determined
by measuring the percentage difference in the length of the samples before
and after sintering.

Microstructural Study

A microstructural study was done on all the alloys prepared through
conventional optical microscopy and scanning electron microscopy.

RESULTS

The results are shown in figures 2 to 5 for the tensile strength,
hardness, elongation and dimensional change. As can be observed in figures
2, 3 and 4, for equal copper contents, the values for tensile strength and
hardness increase with carbon content, while elongation shows the opposite
tendency. As for the dimensional change and according to what is seen in

488

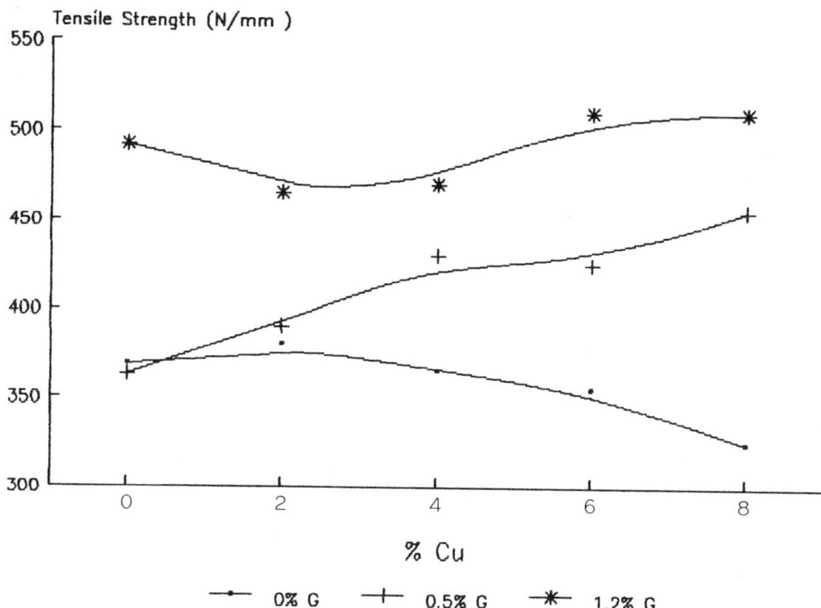

Fig. 2. Tensile strength vs. copper content for Fe-0.45P-C-Cu
alloys sintered at 1.120 °C/1 h.

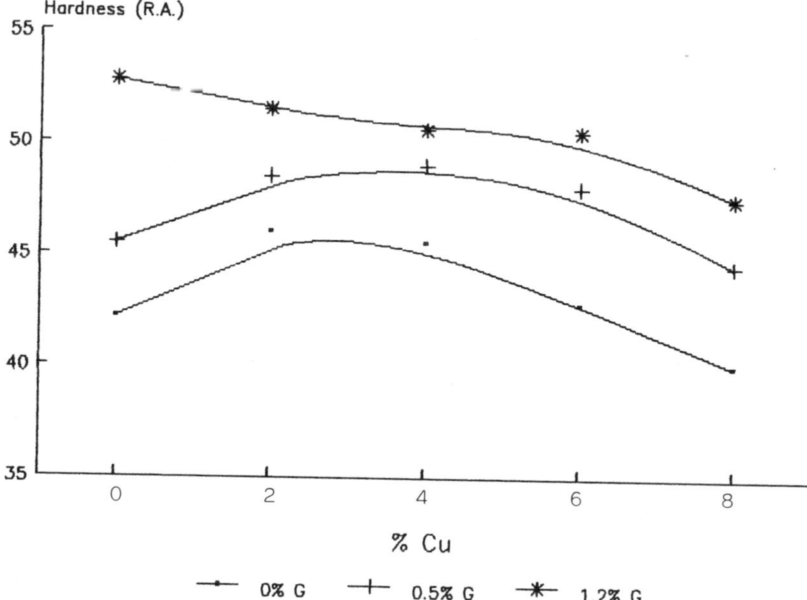

Fig. 3. Hardness vs. copper content for Fe-0.45P-C-Cu alloys
sintered at 1.120 °C/1 h.

figure 5, we can note that among all the ranges of compositions used we
can achieve a zero dimensional change for several different copper and
carbon contents.

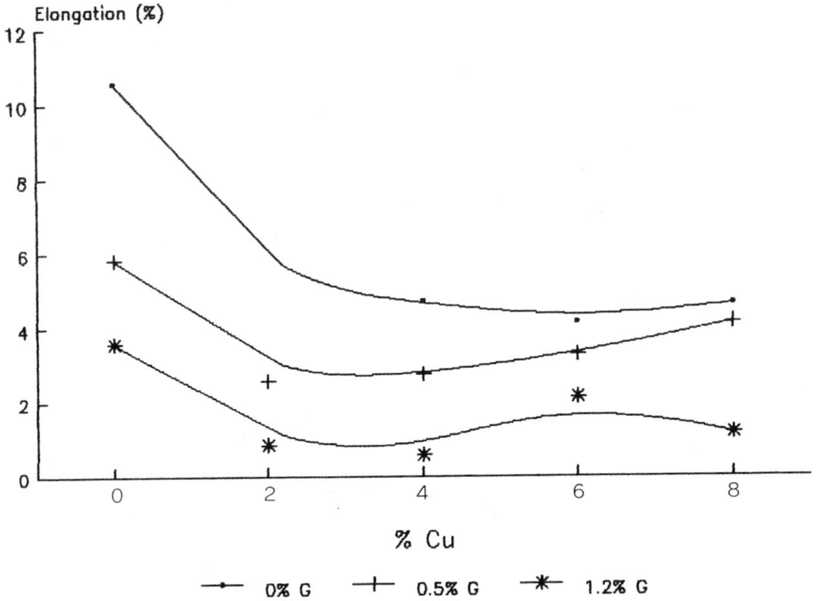

Fig. 4. Elongation vs. copper content for Fe-0.45P-C-Cu alloys sintered at 1.120 °C/1 h.

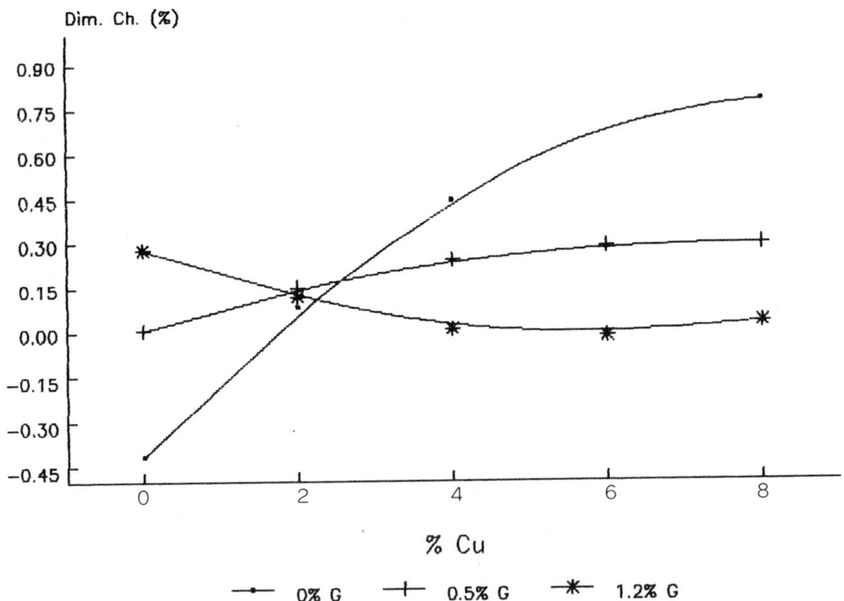

Fig. 5. Dimensional changes vs. copper content for Fe-0.45P--C-Cu alloys sintered at 1.120 °C/1 h.

From the microstructural study we can note that the microstructures
obtained, in the case of the Fe-P-C alloys, correspond with the usual ones
for this type of alloys, that is, ferrite-pearlitic, with retained
austenite, in hypoeutectoid compositions and pearlitic structures with
some signs of free cementite in hypereutectoid compositions. In the case
of the Fe-P-Cu-C alloys, for each carbon content, there appears a higher
proportion of ferrite than normal in the hypoeutectoid compositions, and
a network of cementite bordering the grains of pearlite clearly appears
in hypereutectoid compositions. All this is reflected in some of the
microstructures obtained (figures 6 to 10). Also, for copper contents above
5%, free copper grains have been observed, this being confirmed by scanning
electron microscopy. No abnormal distribution of eutectic Fe-Fe3P, through
scanning electronic microscopy, has been observed. Lastly, we should note
that from the microstructural study, a decrease in porosity was shown with
the increase in copper content.

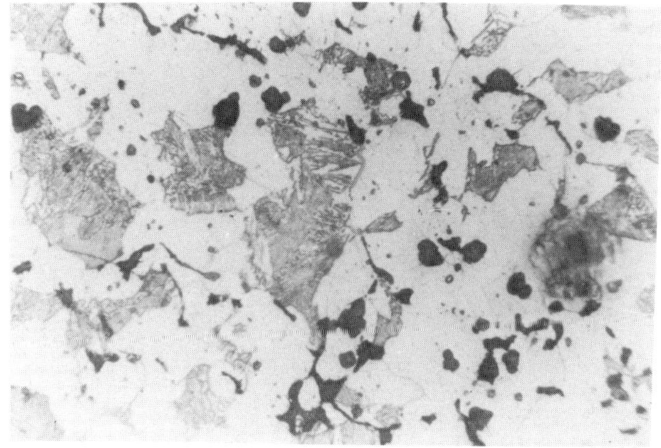

Fig. 6. Fe-0.45P-0.5G sintered at 1.120 °C/1h. 250X.

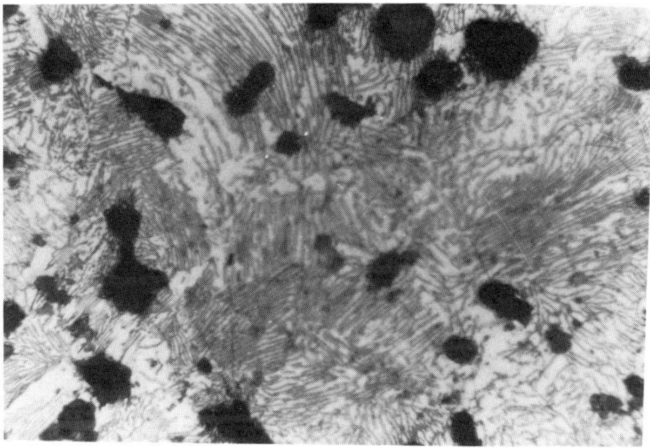

Fig. 7. Fe-0.45P-1.2G sintered at 1.120 °C/1h. 250X.

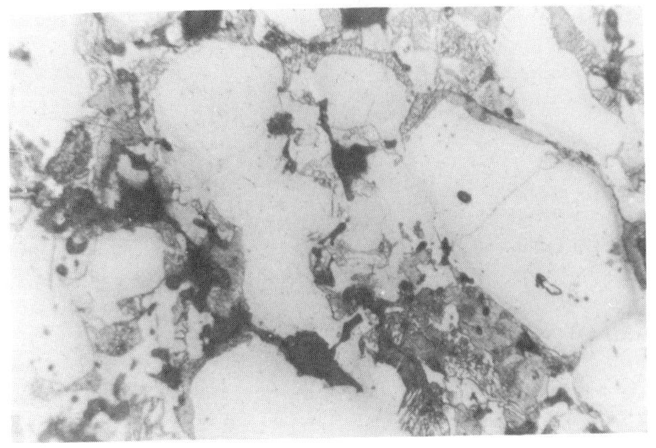

Fig. 8. Fe-0.45P-0.5G-2Cu sintered at 1.120 °C/1h. 250X.

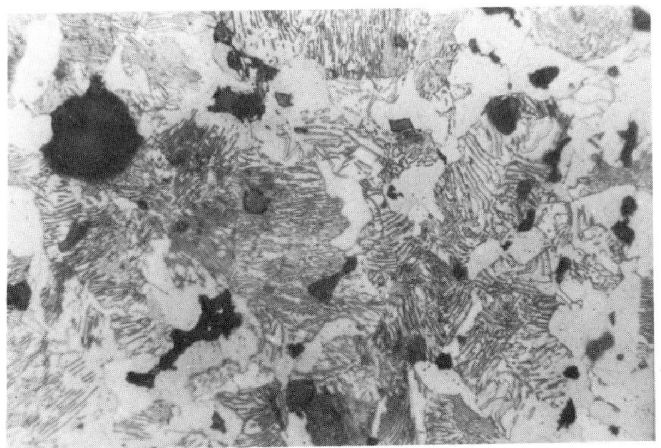

Fig. 9. Fe-0.45P-1.2G-0.4Cu sintered at 1.120 °C/1h. 100X.

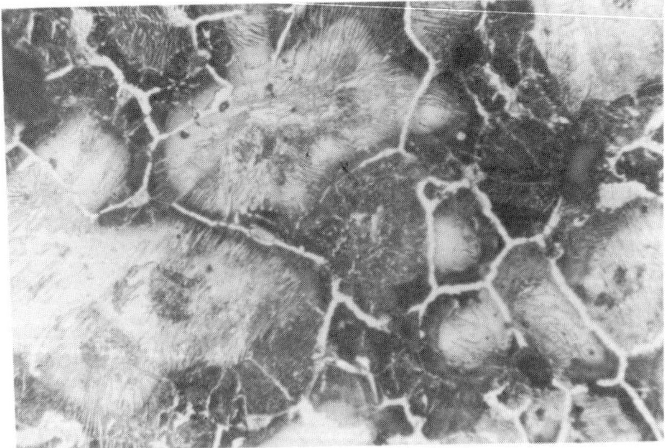

Fig. 10. Fe-0.45P-1.2G-8Cu sintered at 1.120 °C/1h. 250X.

In general, we can say that the mirostructures obtained for the Fe-O, 45P-Cu-C alloys do not differ much from those that are obtained for the alloys of Fe-Cu-C[1].

DISCUSSION

The behaviour of the alloys studied, compared to the mechanical properites of hardness and tensile strength, could be considered normal, given that these properties improve with carbon contents, a result that contrasts with those of the other authors[2,3], as well as the drop in hardness for high copper content, this being due, mainly, to the appearance of free copper in the microstructure. As for the improvement in tensile strength for high copper contents, despite the appearance of an extended cementite network, and in the case of high carbon content as well, this could be explained by the greater densification of the material produced by sintering with a transient liquid phase (in the case of eutectic Fe-Fe3P)[4] and a permenent one, in the case of copper. The decrease in porosity, with the increase in copper content, can be contrasted in figures 8, 9, and 10.

The dimensional change for the Fe-P-Cu alloys depends on the swelling effect ot the carbon and the strongly densifying effect of the phosphorus, which to a certain degree offset one another, but not sufficiently.

The dimensional change for the Fe-P-Cu alloys also depend on the strong shrinkage caused by the phosphorus and the strong growth caused by the copper[3]. It is well-known that the strong growth of the copper is primarily due to the penetration of the liquid copper between the dihedral angles that are null or near zero[5]. In the Fe-P-C-Cu alloys, while the phosphorus maintains its role, the carbon, nevertheless, helps to palliate the expanding effect of the copper because it opens the dihedral angloo and decreases the penetration of the liquid copper through them[5]. Therefore, in figure 5, we can observe, for increasing values of the added graphite, that we get smaller dimensional changes. For copper contents of greater than 6%, nevertheless, we observe, in all cases, a decrease in the growth, due to the phenomena of rearrangement of the sintering with a liquid phase of the copper.

CONCLUSIONS

The main conclusions of this study are the following:

-Adding Cu to the sintered Fe-P-C alloys brings about better dimensional control, allowing them to reach, in some cases, dimensional changes of zero.

The mechanical properties of the Fe-O,45P, such as hardness and tensile strength are improved with small additions of Cu and C, although at the cost of a strong decrease in elongation.

ACKNOWLEDGEMENTS

We thank the companies of Höganäs A.B (Sweden) and Norddeutsche Affinerie (West Germany) for generously providing the powders used, as well as SINTERMETAL S.A. (Spain) for its help in carrying out this study.

REFERENCES

1. Ruiz, J.M., Torralba, J.M., Sánchez, J.L. and Ferri, J.M. Estudio de las propiedades y microestructura de los aceros sinterizados aleados con níquel y cobre, in "VIII Congreso Mundial de Mineraªia y Metalurgia", Principado de Asturias, Ed., Oviedo, Spain (1988).

2. Siddiqui, M.A. and Hamiuddin, M., Mechanical properties of sintered atomized iron alloy powder premises containing phosphorus, Powder Met., 29: 217 (1986).

3. Lindskog, P. Tengzelius, J. and Kuist, S.A., Phosphorus as an alloying element in ferrous P/M.

4. Lindskog, P. The effect of phosphorus additions on the tensile, fatigue and impact strength of sintered steels based on sponge iron powder, Powder Met., 16:32:36 (1973).

5. N. Dautzemberg, H.J. Dorweiler, Dimensional Behaviour of copper-carbon sintered steels, Powder Metall. Int., 17:6:279-282 (1985).

Part IX. SINTERING OF OXIDE CERAMICS

ULTRA-RAPID SINTERING OF CERAMICS

D. Lynn Johnson

Department of Materials Science and Engineering
Northwestern University
Evanston, IL 60208-3108

INTRODUCTION

While most ceramics are sintered slowly in batch or pass-through
procedures with processing times measured in hours, simple shapes can be
fired in minutes or even seconds. Early work by Morgan et al.[1-3] and
Vergnon et al.[4-5] involved rapid heating of small low density, high
surface area powder compacts, using a low thermal mass furnace or rapid
insertion into a preheated furnace, respectively. Maximum heating rates
of 30 K/S and 100 K/s were reported. High densities were not attained in
either work. Wynn-Jones and Miles[6] sintered β-alumina tubes by passing
them through an induction heated alumina muffle tube at 12.5 mm/min with
a maximum heating rate of about 3 K/s, resulting in good densification.
Harmer et al.[7] used a similar scheme to sinter MgO-doped α-alumina at
heating rates up to 40 K/s, obtaining high densities and grain sizes that
were significantly smaller than those attainable by conventional
sintering schedules. They attributed the high densification rates and
fine fired grain size to the suppression of surface diffusion, which
dominates sintering at the lowest temperatures.

Rapid heating has been carried to an extreme using zone sintering in
gas plasmas and a hollow cathode discharge in the author's laboratory,
where thin wall tubes and small diameter rods have been translated
through high temperature plasmas, resulting in maximum heating rates well
in excess of 100 K/s.[8-12] Sintered densities for MgO-doped α-alumina have
exceeded 99% of theoretical density with a final grain size of 2-5 μm,
depending upon conditions. The time interval from the onset to
completion of densification at any given point on the specimen was as
short as 10 s. Although translation rates of 40 to 60 mm/min were usual,
3 mm diameter rods have been sintered to 99% density at rates as high as
100 mm/min.[13] Maximum linear shrinkage rates were on the order of
1-3%/s, but exceeded 4%/s in the hollow cathode discharge.[10]

The success of ultra-rapid sintering was observed to be critically
dependent upon the characteristics of the starting powder. Thin wall
tubes of appropriate powder could shrink from a green diameter of 12 mm
to a fired diameter of 8 mm over a distance of <10 mm at a translation
rate of 60 mm/min, while compacts of coarser powders cracked and were
destroyed under all translation rates attempted.

Science of Sintering
Edited by D. P. Uskoković *et al.*
Plenum Press, New York

Several aspects of ultra-rapid sintering can be considered in light of relevant sintering theory.

SINTERING THEORY

Although none of the sintering kinetics models has been refined sufficiently to quantitatively predict sintering behavior, some useful generalizations can be made. It is well known that densification can occur by motion of atoms from regions of compressive stress on the grain boundary to the solid-vapor interface at the surface of necks between individual particles in the initial stage of sintering, or the surface of the pores in the intermediate and final stages of sintering, under the action of chemical potential gradients generated by capillarity forces. Additionally, atoms will move along or beneath surfaces down the chemical potential gradients caused by curvature gradients at the solid-vapor interface within a compact. These latter fluxes cause coarsening without densification and, in fact, inhibit densification because of the resulting reduction in driving force and increase in diffusion distance.

Since the source and sink of the atom fluxes for grain boundary and volume diffusion from the grain boundary are the same, it can be assumed to first approximation that their chemical potential gradients are identical, resulting in the following generalized equation for the linear shrinkage rate from combined grain boundary and volume diffusion from the grain boundary to the neck or pore surface:

$$- \frac{dL}{L dt} = \left(\frac{\gamma \Omega}{kT} \right) \left[\frac{D_v F_v(\rho)}{G^3} + \frac{\delta D_b F_b(\rho)}{G^4} \right] \tag{1}$$

where L is the instantaneous length of the specimen, t is time, γ is the surface energy, Ω is the molecular volume, D_v is the volume diffusion coefficient, D_b is the grain boundary diffusion coefficient, δ is the thickness of the region of enhanced diffusion at the grain boundary, G is the grain or particle diameter, ρ is the fractional density, $F(\rho)$ is a function of density, k is Boltzmann's constant, and T is the absolute temperature. The $F(\rho)$ parameters depend not only on the density but on all aspects of the microstructure except the magnitude of the grain size. Expressions for these for various sintering models are shown in Table I. Actual values of $F(\rho)$ are unknown and may not be predictable to a high degree of precision. Of course, they could be obtained from Eq. (1) if the other parameters were known.

It can be shown from Eq. (1), using the $F(\rho)$ values in Table I, that grain boundary diffusion should be the predominating densification mechanism for all particle and grain sizes encountered in ultra-rapid sintering. Therefore the volume diffusion term of Eq. (1) will be dropped in the further discussion.

Precise values of the surface, grain boundary, and volume diffusion coefficients in oxides are rare or nonexistent. However, some generalizations can be made. It is expected that the activation energy for surface diffusion, at least in the low temperature regime, should be less than that for grain boundary diffusion which, in turn, would be less than that for volume diffusion. The initial stage sintering models also predict that surface diffusion should predominate at the onset of neck growth between particles and also is favored at smaller particle sizes. This is in agreement with observations that surface diffusion causes significant coarsening of microstructures in the earliest stages of sintering, even before measurable densification occurs.[14] Rapid heating should minimize the deleterious effects of surface diffusion by carrying

Table I. Density functions for Equation (1)

Model	Stage	$F_v(\rho)$	$F_b(\rho)$	Reference
A	Initial $(\Delta L/L_o < 0.05)$	$\dfrac{20}{\Delta L/L_o}$	$\dfrac{12.5}{(\Delta L/L_o)^2}$	16
B	Initial $(\Delta L/L < 0.05)$	$\dfrac{16A\ (X+R)}{\pi X^4 R}$	$\dfrac{64\ (X+R)}{X^3 R}$	16
C	Intermediate	250	$\dfrac{330}{(1-\rho)^{1/2}}$	17
D	Intermediate	$\dfrac{240}{\rho}$	$\dfrac{27}{\rho(1-\rho)^{1/2}}$	18
E	Intermediate	$\dfrac{16\ HS_v}{G}$	$\dfrac{16\ HL_v}{G}$	17
F	Final	$\dfrac{288f(\rho)}{\rho}$	$\dfrac{1047f(\rho)}{\rho(1-\rho)^{1/3}}$	19

$$f(\rho) = \frac{1 - (1-\rho)^{2/3}}{3\ (1-\rho)^{2/3} - [\ 1 + (1-\rho)^{2/3}]\ \ln\ (1-\rho)\ -3}$$

Note: A = normalized neck surface area; X = normalized neck radius; R = normalized minimum radius of curvature of the neck surface; H = mean surface curvature; S_v = pore surface area per unit volume; L_v = pore-grain boundary perimeter length per unit volume.

the compact, essentially with its green microstructure, into the temperature range where the higher activation energy of grain boundary diffusion increases its importance relative to surface diffusion to the extent that densification can proceed with minimal coarsening.

Although $F(\rho)$ is not known precisely, further insights into ultra-rapid sintering can be obtained by computing sintering curves using approximate values. Also needed are the dependence of the grain size on density and the temperature as a function of time. For the purposes of the following calculations, the grain size was assumed to be proportional to the square of density, for instance starting at 250 nm at a green density of 50% and reaching 3 μm at a final density of 99%. These starting and ending grain sizes are typical of the powder used in the plasma sintering of MgO-doped alumina[*]. Although the dependence of grain size on density during ultra-rapid sintering is unknown at the present time, the assumed dependence is probably not greatly in error since grain growth will be low initially and increase as higher densities are achieved. If anything, the dependence on density may be higher than second order.

The values of $F(\rho)$ for the simple models shown in Table I are incompatible between the initial, intermediate, and final stages. The

---

[*]CR30, deagglomerated, Baikowski International Corp., Charlotte, NC 28210

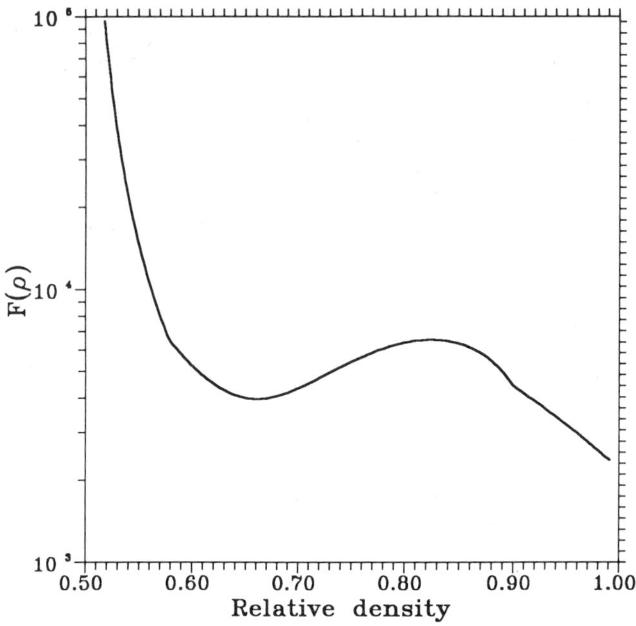

Fig. 1. Assumed dependence of F(ρ) on relative density.

initial stage models predict a rapid decrease of F(ρ) with increasing density, while the intermediate stage models show an increase and the final stage predicts a decrease. In fact, there must be a smooth transition from initial through intermediate to final stage. For the purposes of the calculations, the initial and final stage values for the simplified geometries shown in Table I (Models A and F, respectively) were used and matched through the intermediate stage by the arbitrary curve shown in Fig. 1, where the initial stage extends from 50 to 58% and the intermediate stage extends from 58 to 90% density.

The temperature-time profile for passage through a plasma was assumed to be of the form shown in Fig. 2, in which a maximum heating rate of 100 K/s was assumed at 1850 K. Although the exact dependencies of F(ρ), grain size and temperature on time will differ quantitatively from those assumed, it is felt that the differences will not be sufficiently great to negate the general discussion that follows.

The instantaneous linear shrinkage, dL/Ldt, was calculated using the grain boundary diffusion coefficient obtained from creep data by Cannon and Coble[20], (δD_b=8.6 x 10^{-10} exp (-418 kJ/mol/RT) m^3/s) as well as other parameters for alumina ($\gamma\Omega/k$=3.08 x 10^{-6}m). Starting particle sizes of 25, 250, and 1000 nm were selected. The assumed grain growth rate curves were offset to the starting particle size.

The density is shown as a function of time from the onset of heating in Fig. 3 and as a function of temperature in Fig. 4. Note that the 1000 nm powder failed to densify fully under this heating schedule. The 25 and 250 nm powder reached 99% density at temperatures of 2175 and 2195 K, respectively.

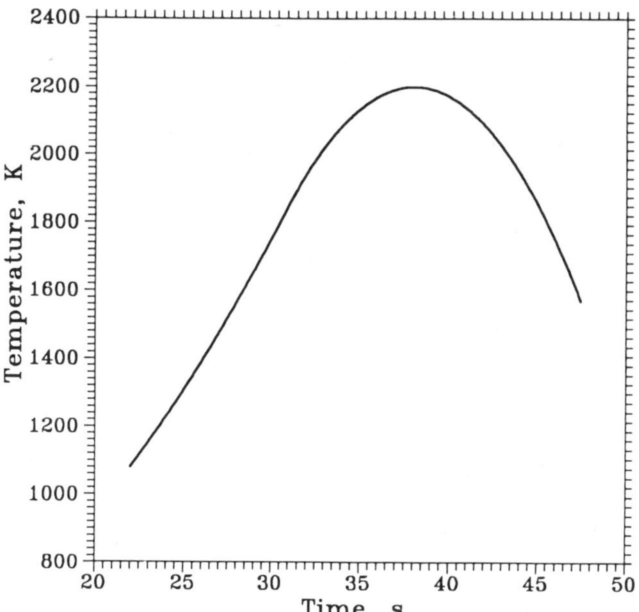

Fig. 2.   Assumed temperature-time profile.

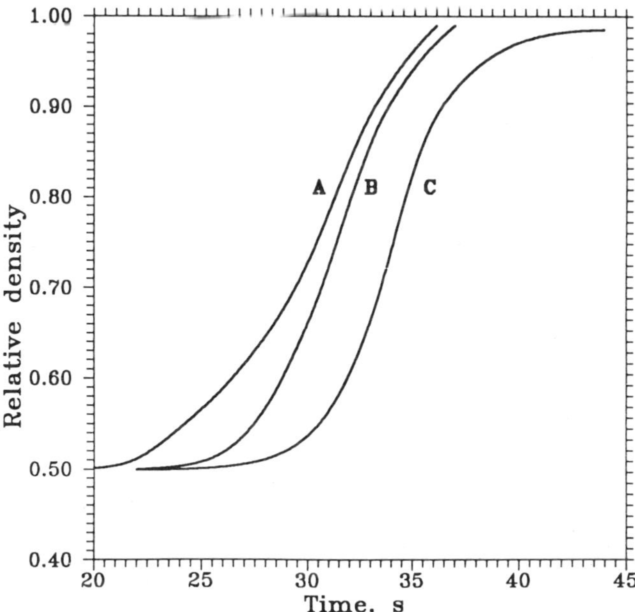

Fig. 3.   Relative density for initial particle diameters of
(A) 25 nm, (B) 250 nm, and (C) 1 μm.

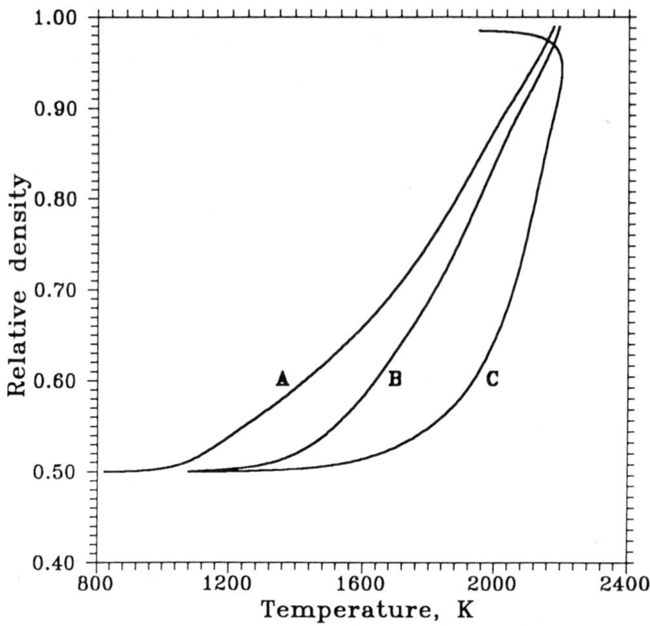

Fig. 4. Relative density for initial particle diameters of
(A) 25 nm, (B) 250 nm, and (C) 1 μm.

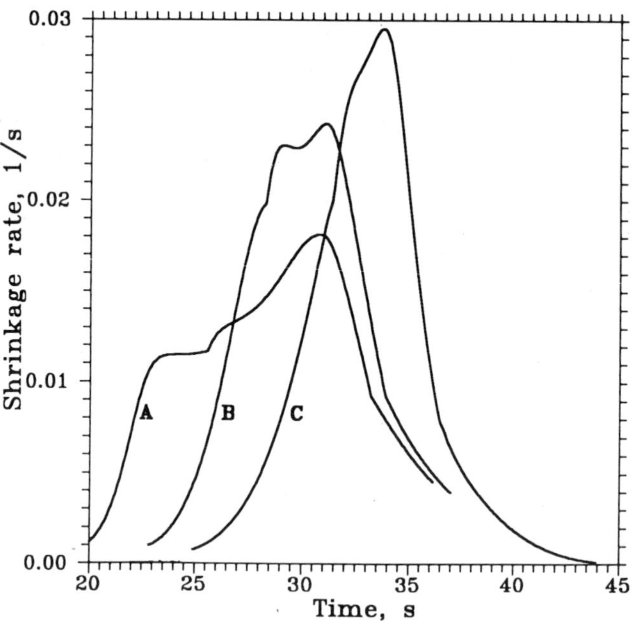

Fig. 5. Shrinkage rates for initial particle diameters of
(A) 25 nm, (B) 250 nm, and (C) 1 μm.

The fractional shrinkage rate is plotted as a function of time in Fig. 5. Here the inflections and discontinuities in the slopes of the curves are artifacts of the assumed form of F(ρ), but the general trends are apparent. The calculated shrinkage rates of the 250 nm powder are of the same magnitude as those observed experimentally for powder of this size[12].

The grain diameter during sintering is shown in Fig. 6. Here the curves terminate at the 99% density point; no attempt was made to simulate grain growth beyond that density.

The shapes of all of these curves will be functions of the heating rate through its influence on surface diffusion. Figure 7 shows the variation of F(ρ) on relative density obtained from previous simulations of initial stage sintering of compacts of identical spheres[15], using Model B, Table I, which takes into account the neck dimensions explicitly. The curve labeled "no surface diffusion" represents the maximum value. The curve labeled "isothermal" uses the same grain boundary diffusion parameters but includes a significant surface diffusion contribution. Here the ratio of surface diffusion to grain boundary diffusion was 3 at 1% linear shrinkage and fell to unity at 5% shrinkage. The third curve, labeled "Ag,CRH", was simulated for 1 μm diameter silver spheres heated at 100 K/s. At lesser heating rates the curve would be depressed even further. This depression of F(ρ) is thought to be an important factor in the inhibition of densification during slow heating.

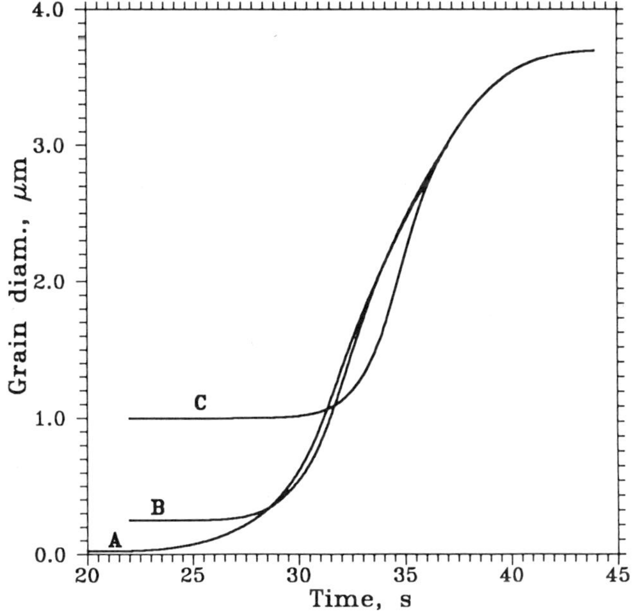

Fig. 6. Grain diameters for initial particle diameters of (A) 25 nm, (B) 250 nm, and (C) 1 μm.

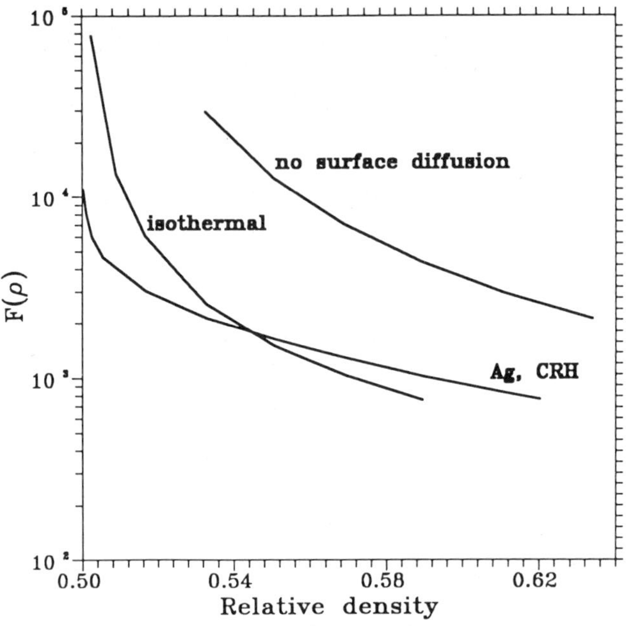

Fig. 7. Computed values of $F(\rho)$ for no surface diffusion, isothermal sintering with surface diffusion, and simulated sintering of 1 μm diameter silver spheres heated at a constant rate of 100 K/s, using Model B, Table I.

DISCUSSION

The above calculation of shrinkage rates probably overestimates the shrinkage rate at the lower densities. In fact, however, the value of $F(\rho)$ probably will not be as much as an order of magnitude less than the values shown for the final stage sintering model. It thus appears that the observed shrinkage rates are compatible with expectations based on the published diffusion coefficient for alumina. It should be pointed out that the temperature gradients that must exist will affect the sintering behavior, but it is not clear at this point whether these gradients would enhance or inhibit densification of the over-all compact at any given point.

The expected strong role of particle size is readily apparent. Thus the smaller particle size materials begin to sinter at much lower temperatures, although the maximum shrinkage rate does not vary strongly with particle size. A compact of particles on the order of 1000 nm or larger would not sinter to full density under the temperature time profile used in the present calculations, even if it survived the rapid temperature rise.

Final grain sizes for the 25 and 250 nm compacts would be larger than the values shown in Fig. 6 because grain growth would continue during the balance of the temperature excursion. Experimentally, compacts of CR30 alumina doped with MgO have a final grain size on the order of 4 to 5 μm at a density of 99.7%.

There are two additional observations that bear further discussion. First is the remarkable deformation that was described above, namely

504

shrinkage of the compact over a relatively short distance. A further
result, which has not been discussed previously, is the fact that
although the green densities of the CR30 powder were relatively low,
typically less than 50%, the sintered densities of the MgO-doped powder
were uniformly above 99% and frequently above 99.5%. Fracture surfaces
revealed only occasional voids smaller than the grain size. The compacts
were prepared after the addition of magnesium nitrate and binder to the
as-received powder without milling or other processing. One would have
to assume that the green microstructure was far from ideal, and would not
be expected to result in the high densities observed.

These two results are related to each other, and probably are the
result of the rapid excursion of the fine grain green microstructure into
a temperature range at which extremely rapid diffusion takes place. The
powder would become essentially superplastic and be capable of rapid
deformation in response to local and global stresses imposed by the
temperature gradient, the shrinkage gradient, and powder packing
inhomogeneities. It appears as though ultra-rapid sintering of powders
for which it is feasible overcomes at least some of the problems
associated with local nonuniformities in the compaction of agglomerated
high surface area small particle size ceramic powders.

Further work is required to further refine the dependencies of $F(\rho)$
and grain size on densification under the rapid heating schedules typical
of plasma sintering, and to address the issues of temperature gradients
and any transient phenomena that may occur.

CONCLUSIONS

The high shrinkage rates observed for ultra-rapid sintering by
plasma heating can be rationalized on the basis of published diffusion
coefficients in aluminum oxide.

The starting particle size is of critical importance in the success
of ultra-rapid plasma sintering. Particle sizes above about 1000 nm
result in failure to achieve high density and cracking in the strong
temperature gradients that are imposed on the specimen.

Compacts prepared of suitable powders undergo rapid deformation
during sintering which preserves specimen integrity, and results in high
sintered densities.

REFERENCES

1. C. S. Morgan and C. S. Yust, Material Transport during
    Sintering of Materials with the Fluorite Structure, J.
    Nucl. Mater., 10:182 (1963).
2. C. S. Morgan, C. J. McHargue, and C. S. Yust, Material
    Transport in Sintering, Proc. Brit. Ceram. Soc., 3:177
    (1965).
3. C. S. Morgan, Densification Kinetics During Nonisothermal
    Sintering of Oxides, in: "Kinetics of Reactions in Ionic
    Systems, Materials Science Research Vol. 4", T. J. Gray and
    V. D. Frechette, eds., Plenum Press, New York (1969).
4. P. Vergnon, F. Juillet, and S. J. Teichner, Effect of
    Increasing Rate of Temperature on Sintering of Pure Alumina
    Homodispersed Particles, Rev. Int. Hautes Temp. Refract.
    3:409 (1966).

5. P. Vergnon, M. Astier, D. Beruto, G. Brula, and S. J. Teichner, Sintering of Very Finely Divided Particles, II. Flash Heating Technique and Kinetic Study of Shrinkage in Titanium Oxide and Aluminum Oxide Compacts, <u>Rev. Int. Hautes Temp. Refract.</u> 9:271 (1972).

6. I. Wynn-Jones and L. J. Miles, Production of β-Al$_2$O$_3$ Electrolyte, <u>in</u>: "Proc. Brit. Ceram. Soc. Vol. 19", Stoke-on-Trent, England (1971).

7. M. Harmer, E. W. Roberts, and R. J. Brook, Rapid Sintering of Pure and Doped α-Al$_2$O$_3$, <u>Trans. J. Br. Ceram. Soc.</u> 78:22 (1979).

8. D. L. Johnson and R. R. Rizzo, Plasma Sintering of β''-Alumina, <u>Am. Ceram. Soc. Bull.</u> 59:467 (1980).

9. Joung Soo Kim and D. Lynn Johnson, Plasma Sintering of Alumina, <u>Ceram. Bull.</u> 62:620 (1983).

10. D. Lynn Johnson, Wayne B. Sanderson, Jennifer M. Knowlton, and Eric L. Kemer, Sintering of α-Al$_2$O$_3$ in Gas Plasmas, <u>in</u>: "Advances in Ceramics Vol. 10", W. D. Kingery, ed., The American Ceramic Society, Columbis, OH (1984).

11. Wayne B. Sanderson and D. Lynn Johnson, Sintering of Al$_2$O$_3$ in a Hollow Cathode Discharge, <u>in</u>: "High Temperature Materials Chemistry III", Z. A. Munir and D. Cubicciotti, eds., Electrochemical Society, Pennington, New Jersey (1986).

12. Eric L. Kemer and D. Lynn Johnson, Microwave Plasma Sintering of Alumina, <u>Am. Ceram. Soc. Bull.</u> 64:1132 (1985).

13. Donald C. Lynch, Microwave Plasma Sintering of Magnesia and Alumina, <u>M.S. Thesis, Northwestern University, June 1987</u>.

14. Svante Prochazka and Robert L. Coble, Surface Diffusion in the Initial Sintering of Alumina Part III, Kinetic Study, <u>Phys. Sint.</u> 2:15 (1970).

15. D. Lynn Johnson, Ultra Rapid Sintering, <u>in</u>: "Sintering and Heterogeneous Catalysis", G. C. Kuczynski, Albert E. Miller, and Gordon A. Sargent, eds., Plenum Publ. Corp., New York (1984).

16. D. Lynn Johnson, New Method of Obtaining Volume, Grain Boundary and Surface Diffusion Coefficients from Sintering Data, <u>J. Appl. Phys.</u> 40:192 (1969).

17. D. Lynn Johnson, A General Model for the Intermediate Stage of Sintering, <u>J. Am. Ceram. Soc.</u> 53:574 (1970).

18a. R. L. Coble, Sintering Crystalline Solids. I. Intermediate and Final State Diffusion Models, <u>J. Appl. Phys.</u> 32:787 (1961).

  b. R. L. Coble, Sintering Crystalline Solids. II. Experimental Test of Diffusion Models in Powder Compacts, <u>J. Appl. Phys.</u> 32:793 (1961).

  c. R. L. Coble, Intermediate-Stage Sintering: Modification and Correction of a Lattice-Diffusion Model, <u>J. Appl. Phys.</u> 36:2327 (1965).

19. E. Arzt, M. F. Ashby, and K. E. Easterline, Practical Applications of Hot-Isostatic Pressing Diagrams: Four Case Studies, <u>Metall. Trans. A</u> 14a:211 (1983).

20. R. M. Cannon and R. L. Coble, Review of Diffusional Creep of Al$_2$O$_3$, <u>in</u>: "Deformation of Ceramic Materials", R. C. Bradt and R. E. Tressler, eds., Plenum Press, N.Y. (1975).

ACKNOWLEDGEMENT

This material is based upon work supported by the National Science Foundation under Grant No. 8719077.

HYDROTHERMAL ZrO_2 POWDER AND ITS SINTERING BEHAVIOUR

S. Somiya, M. Yoshimura, Y. Suwa, T. Akiba, Z. Nakai,
K. Hishinuma, and T. Kumaki

The Nishi Tokyo University, Tokyo Institute of Technology
and Chichibu Cement Co. Ltd., Japan

INTRODUCTION

Recently strong attention has been directed to hydrothermally
processed powders, especially ZrO_2 powders. There are reports
concerning hydrothermal synthesis of ZrO_2 and metal-doped ZrO_2 powders.

V.A.Kuznetzov[1] reported growth of ZrO_2 single crystals $1\sim2.5X$
$1\sim2.5X$ $0.3\sim0.8$ mm, from NH_4F solution; the conditions included 550-
590°C dissolution temperature, 560-610°C growth temperature, 0.5
occupation factor, 7 days cooking time. His aim for this study was to
make single crystals rather than to make powders.

Powders for advanced ceramics need to meet these criteria: (1)
mean particle size must be <1 μm (2) must be homogeneous (3)
composition and purity must be controllable (4) particle size
distribution must be narrow (5) little or no macroscopic agllomeration
permitted (6) particles must be as spherical as possible etc.

For meetings these criteria, one of the possible approaches is
hydrothermal synthesis. The characteristics of powders made by
hydrothermal processing include (1) relatively low cost, (2)
particulate properties attained without milling (grinding) in many
cases, (3) ultra fine, (4) free flowing, (5) controllable particle
size, (6) controllable particle shape, (7) high degree of crystalline
perfection, (8) homogeneity, (9) little or no chemically bound water,
(10) little or no agglomeration, (11) pollution minimized, (12) etc.

Many original papers and/or reviews related to hydrothermal
studies have appeared, e.g., G.W.Morey,[2] R.Roy,[3] R.A.Laudise and
J.W.Nielsen,[4] A.A.Ballman and R.A.Laudise,[5] W.Eitel,[6] R.A.Laudise,[7]
A.N.Lobachev,[8] R.A.Laudise,[9] A.Rabenau,[10-12] and W.J.Dawson.[13]

PREVIOUS STUDIES

In my laboratory, studies relating to ZrO_2 powder have been
underway since 1970, and papers were presented at a high pressure
conference in 1981. After this time, many other papers appeared.

Science of Sintering
Edited by D. P. Uskoković *et al.*
Plenum Press, New York

P.Reynen et al.[14] reported his results in 1981. His starting
material was zircon. K. Haberko and W. Pyda[15] mentioned preparation of
powder under hydrothermal conditions and its sintering behaviour. Their
starting materials were zirconyl chloride and calcium chloride and
sodium hydroxide; they were coprecipitated at 35°C and 95°C. Then
hydrothermal treatments were at 220°C for 4 h and at 260°C for 8 h. The
resultant crystallite size was about 12nm, and firing shrinkages were
22% for the 220°C, 4h powder and 18% for the 260°C, 8h powder. After
sintering at 1300°C under a high forming pressure of 198 MPa, the
relative density of the sintered body was 89%.

E.S.Stambaugh[16] and J.H.Adair and E.P.Stambaugh[17] reported the
production of partially stabilized zirconia. The starting material was
zirconyl solution as the oxychloride or oxynitrate at a concentration
of 0.18 to 1.5 moles per liter. The pH was adjusted by additions of
ammonium hydroxide, sodium hydroxide, or tetraethylammonium hydroxide.
The autoclave was teflon lined and the hydrothermal temperature was
$170-190^{\circ}$C. The washing agents were acetic acid and water.

For successful use of a stabilizing agent, pH is one of important
factor. If the pH is not above pH 4, the powders contain very little of
the stabilizing agent. The hydrothermally derived micro-crystallized
powders are highly reactive with respect sintering. With a heating rate
of 300°C/h, the optimum sintering temperature is between 1350°C and
1450°C.

S.Komarneni et al.[18] reported the synthesis of ZrO_2 gel from Zr
isoproxide solution. Zirconia was precipitated (1) by slow hydrolysis
in air from Zr propoxide in propanol and (2) from $ZrOCl_2$ in 1N NaOH.
Hydrothermal conditions were from 120°C to 700°C under 100 MPa for
times of 3h to 20h. By TEM, the sizes of particles were 6-10 nm,
ranging upward to 20-60 nm for plates.

G.Gunnarsson et al[19] reported supercritical drying; they used
$ZrOCl_2$ and YCl_3 as starting materials and precipitated by NH_3. The pH
was kept at $9.3^{\pm}0.1$. After filtration, the filter cakes were washed
with methanol. Then the mixture was placed in the "201 autoclave" and
heated to $240-260^{\circ}$C under 120-140 bars. Then a valve was opened and the
pressure controlled to 120-140 bars, and heated was continued at that
pressure until the temperature reached $320-340^{\circ}$C. The pressure was
reduced to 1 atm while the temperature was kept above 300°C. Then the
autoclave was connected to a vacuum pump to remove the last remaining
liquid. The solvent mixture coming from the autoclave was condensed and
collected. The supercritical drying is one of characteristic features
of this method. The powder was calcined at 1000°C and sintered at
1300°C and 1400°C. When sintered at 1400°C, the maximum relative
density was attained on 100%.

G.W.Kriechbaum et al[20] synthesed ZrO_2 from $ZrOCl_2$, YCl_3, $MgCl_2$,
NaOH, NH_4OH solution. The gels were prepared by coprecipitation at a pH
of 10 at 25°C, and were then treated hydrothermally at 200°C or 300°C
for 2 h. The less agglomerated powder sintered to a density higher than
98%. The most important factor influencing the size of particles
obtained was the pH of the solution at the this of co-precipitate.

J.H.Adair et al[21] explained solubility and ionic relations for Zr-
containing aqueous solutions as a function of the pH of the solution,
and its temperature. ZrO_2 powder obtained from acidic solution at high
initial supersaturations have high specific surface area and yield
monoclinic crystals. Alkaline solutions produce ZrO_2 powders with lower
specific surface areas that initially have the cubic structure. But

$$
\begin{array}{c}
\text{(a) Hydrothermal homogeneous precipitation method} \\
\text{1M/l } ZrOCl_2 \text{ solution} \\
Co(NH_2)_2 \\
200\,^{\circ}C, \ 6.3 \text{ MPa, 24h}
\end{array}
$$

$$
\begin{array}{c}
\text{(b) Hydrothermal hydrolysis method} \\
\text{0.5M/l } ZrOCl_2 \text{ solution} \\
215\,^{\circ}C, \ 2 \text{ MPa, 24h}
\end{array}
$$

$$
\begin{array}{c}
\text{(c) Hydrothermal crystallization method} \\
Zr(OH)_4 \\
300\,^{\circ}C, \ 100 \text{ MPa, 24h}
\end{array}
$$

Fig. 1. Preparation of ZrO_2 Powders by Hydrothermal Method

this is subsequently transformed to the monoclinic form of ZrO_2 with increasing reaction time, and increasing temperature. They studied values of pH ranging from 1.2 to 13.8 for temperatures of $150\,^{\circ}C$, $170\,^{\circ}C$ and $200\,^{\circ}C$.

OUR WORKS

My laboratory has published many papers related to fine ZrO_2 powders. They are listed separately, in the Appendix.

Pure ZrO_2 Powder Preparation

There are three methods to prepare pure ZrO_2 (Fig.1). The resultant X-Ray diffraction patterns are shown in Fig. 2. Depending upon hydrothermal processing and starting materials, the ZrO_2 products were found to differ with respect to morphology of particles and degree of agglomeration etc.

Y_2O_3-doped ZrO_2 Powder Production

A mixed solution of $ZrOCl_2 \cdot 8H_2O$, $YCl_3 \cdot 6H_2O$, and urea of 120% of the equivalent quantity was treated under hydrothermal conditions between $150\,^{\circ}$ and $220\,^{\circ}C$ (3 to 7 MPa) in a zirconium-lined autoclave (1000 ml to 20 l capacity). During the hydrothermal treatment, this mixed solution was stirred at 500 rpm. Under the hydrothermal conditions, urea acted as a precipitation agent by decomposing into NH_3 and CO_2 according to Eq. (1).

$$(NH_2)_2CO + H_2O \rightarrow 2NH_3 + CO_2 \tag{1}$$

with homogeneous precipitation occuring at a pH value of 7 to 8. The precipitate formed was hydrous ZrO_2 with Y_2O_3, and this crystallized into yttria-doped ZrO_2 under hydrothermal conditions. The products were washed in distilled water and/or ethanol by decantation and centrifugation to remove the Cl^- and NH_4^+ ions. These washed powders were dried at $120\,^{\circ}C$ for 10 hours, then calcined at different temperatures from $500\,^{\circ}$ to $1100\,^{\circ}C$ in air for one hour, in order to investigate the phase transformation and sinterability. After calcination, in some cases, the products were wet-milled with water in an alumina-lined mill by TZP balls for 12 hours and then dried at $120\,^{\circ}C$ for 10 hours.

The procedure for production of ZrO_2 is shown in Fig. 3. This

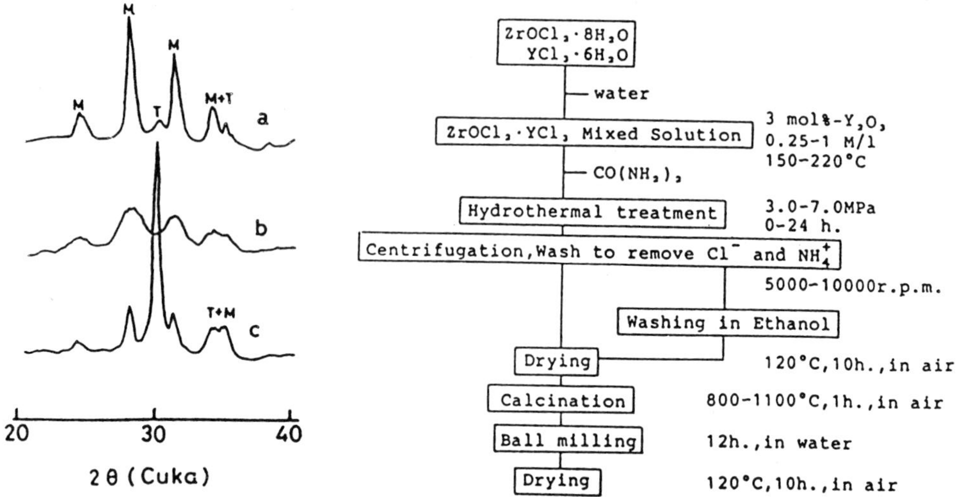

Fig. 2. X-ray diffraction patterns of pure ZrO_2 prepared by various hydrothermal methods.

Fig.3.Processing flow sheet of hydrothermal homogeneous precipitation method.

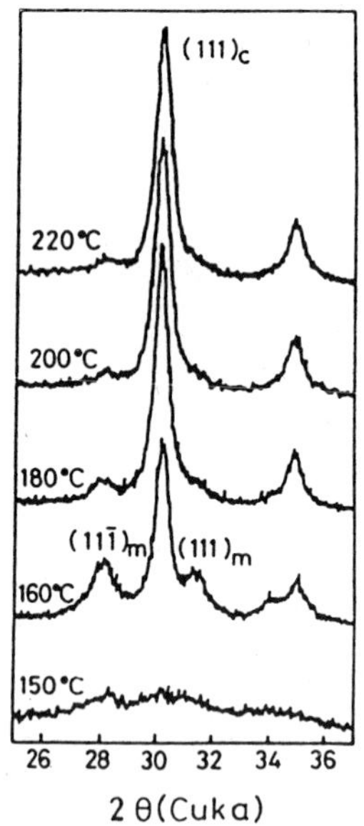

Fig. 4. X-ray diffraction patterns of PSZ (3 mol%-Y_2O_3) powders prepared by hydrothermal homogeneous precipitation methods at various temperatures (1M/1, 5 h)

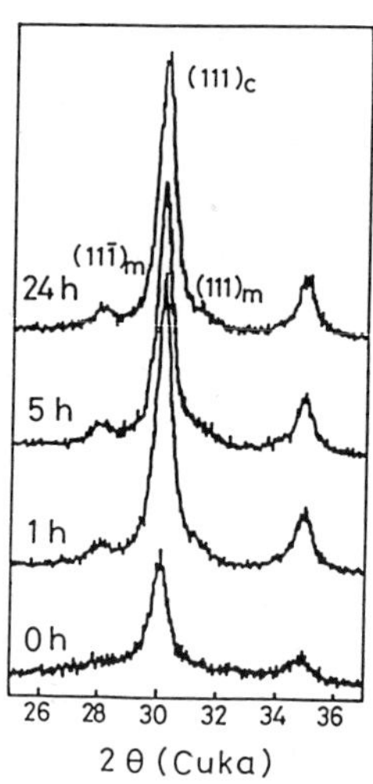

Fig. 5. X-ray diffraction patterns of PSZ (3 mol%-Y_2O_3) powders prepared by hydrothermal homogeneous precipitation method for various treatment times (1M/1, 180°C)

method is called the hydrothermal homogeneous precipition or the hydrothermal homogeneous urea precipition.

RESULTS AND DISCUSSION

<u>Characteristics of the Powders</u>

Above 160 °C, fine crystalline powders were obtained after treating for five hours. Figure 4 shows the X-ray diffraction (XRD) patterns after drying of 3 mol% Y_2O_3-ZrO_2(3Y-PSZ) powders prepared under hydrothermal conditions. The quantity of urea used was 120% of the equivalent, but it was noted that the pH of the suspension was almost constant, even if the quantity of urea was varied between 110% and 140% of the equivalent. It seems that CO_2 acted as a buffer.

The milled powders were cold isostatically pressed at a pressure of 196 MPa and sintered at different temperatures for two hours. The density of each sintered body was determined by the Archimedes method or by weight/dimension measurements, the microstructure of the sintered bodies being observed by scanning electron microscopy (SEM) of their polished and thermally etched surfaces. The effect of cooking time and temperature for hydrothermal homogeneous precipitation, concentration of Y_2O_3 in solution on X-ray powder diffraction patterns are shown in Fig. 5 and 6.

Well-crystallized 3Y-PSZ powder of 11.6 nm crystallite size was obtained by hydrothermal treatment at 220°C under 7 MPa for five hours. The crystallite size was determined by using the Scherrer-Warrens equation on the half-width of the (111) line of tetragonal/cubic ZrO_2. The crystallite size decreased from 15.0 to 11.6 nm by increasing the temperature from 160° to 220°C, as shown in Fig. 7. However, the surface area (BET) of the products remained constant at 100 m^2/g regardless of the temperature. Above 800°C, the products were transformed to the tetragonal form, as shown in Fig. 8.

<u>Sintering Behaviours</u>

Relations between calcination and green density and sintered density and also the effect of washing agents on green and sintered density are shown in Fig. 9.

The green density increased with increasing calcination temperature. The sintered density, however, was at its maximum at 1000°C calcination. The author obtained over 99% dense TZP ceramics with a homogeneous microstructure by sintering powder 1001 E at 1400°C for two hours. This powder had been washed in ethanol and calcined at 1000°C for one hour (Fig.10). When sintered at low temperatures, the sinterability of the powders washed in ethanol was better than that of the powders washed in water, this difference seeming to have stemmed mainly from the agglomeration of particles. The agglomeration of the former was soft, but that of the latter was relatively hard. Intraagglomerate pores were observed in compacts containing soft agglomerates, the radii of which were measured by Hg porosimetry to be about 15 nm. On the other hand, the intraagglomerate pore size in the compacts containing hard agglomerates was 10 to 40 nm. During the drying process, capillary forces caused a significant agglomeration of powder due to the small capillary dimensions and the high surface tension of water. These capillary forces could be reduced by washing the powder in a liquid of lower surface tension, for example, ethanol. The sintered density of 1001 E decreased above 1500°C, as shown in Fig.10.

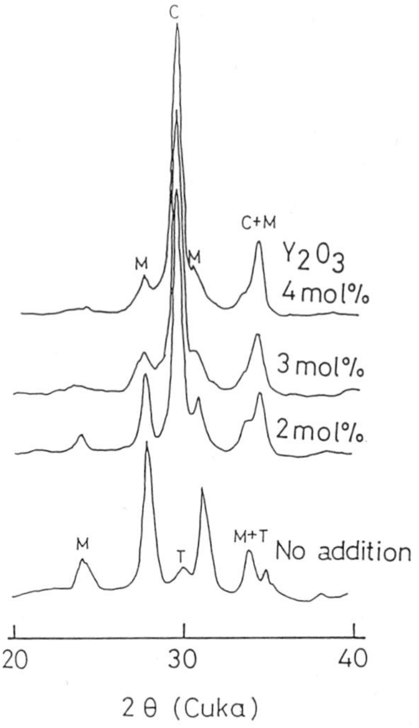

Fig. 6. X-ray diffraction patterns
of pure ZrO$_2$ and PSZ powders prepared
by hydrothermal homogeneous
precipitation method at 200°C under
a pressure of 6.3MPa for 24h (Y$_2$O$_3$
amount in solutions).

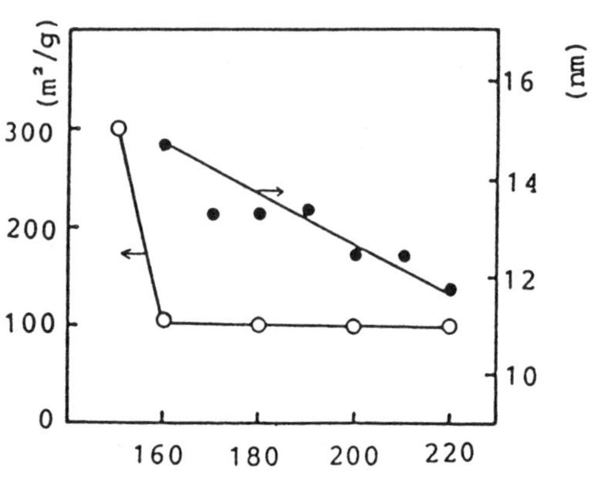

Hydrothermal temperature (°C)

Fig. 7. Variation of surface area
and crystallite size of PSZ
(3 mol% Y$_2$O$_3$) powders with
temperature under hydrothermal
conditions.

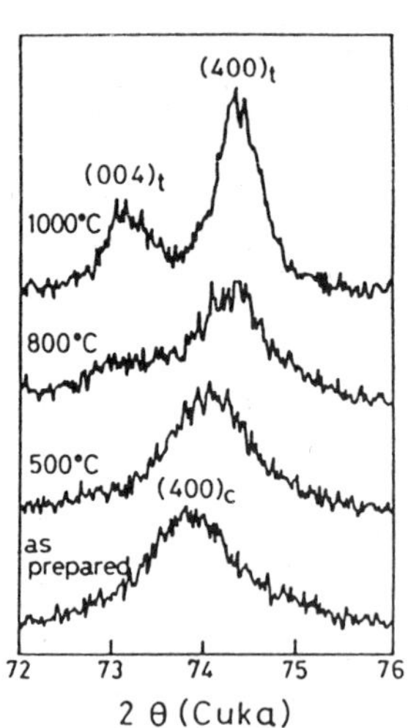

Fig. 8. X-ray diffraction patterns
of PSZ (3 mol% Y$_2$O$_3$) powders calcined
at various temperatures.

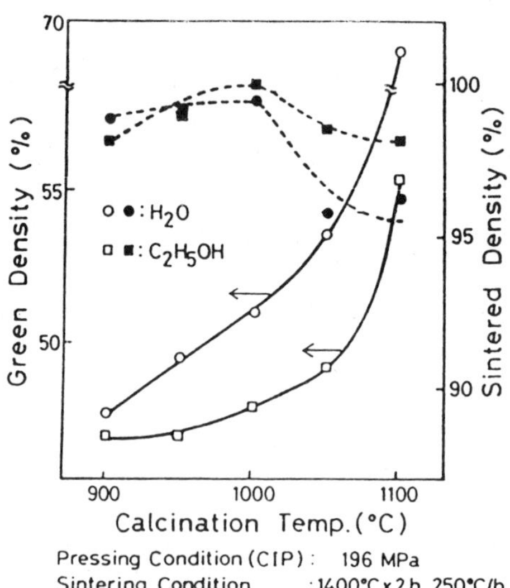

Pressing Condition (CIP): 196 MPa
Sintering Condition : 1400°C x 2 h, 250°C/h

Fig. 9. Effect of calcination
temperature on the green and
sintered densities of TZP
(3 mol% Y$_2$O$_3$).

512

Table I. Typical Properties of 3Y-TZP Ceramics Prepared from PSZ
Powders by Hydrothermal Homogeneous Precipitation Method

Sintered density ($1400\,^{\circ}$C, 2h)	600 (g/cm^3)
Bending strength (Three-point bending test, 30 mm span)	1012 MPa
Compressive strength	>2.9 GPa
Elastic modulus (Ultrasonic pulse method)	200 GPa
Coefficient of linear expansion ($200\,^{\circ}$– $1100\,^{\circ}$C)	11 x 10^{-6} (1/K)
Fracture toughness (M.I Method)	$6 \approx 8$ MPa·m$^{-1/2}$
Content of tetragonal ZrO	$\approx 100\%$

The 3Y-TZP ceramics were composed of uniformly fine grains. The
average grain size of the sintered body depended on the temperature,
and it was observed that the average grain size increased from 0.4 to
1.3 µm when sintering was between $1300\,^{\circ}$ and $1600\,^{\circ}$C for two hours. In
the TZP ceramics sintered at $1300\,^{\circ}$C small intergranular pores (<0.1 µm)
existed, these small pores disappearing above $1400\,^{\circ}$C. A scanning
electron micrograph of the TZP ceramic 1001E that was sintered at
$1400\,^{\circ}$C for two hours is shown in Fig.11, the average grain size being
0.53 µm. After sintering sample 1001E at $1600\,^{\circ}$C for two hours, larger
intergranular pores of the order of 0.5 µm were observed. These pores
resulted from the intraaglomerate pores of the powder remaining and
aggregating or bloating during the sintering process. It seems that
dragging of these large intergranular pores was a slower process than
densification at or above $1500\,^{\circ}$C. On the other hand, these pores were
not observed in sample 1001 sintered at $1600\,^{\circ}$C for two hours, the
sintered density of 1001 being higher than that of 1001E (Fig.10). This
difference seems to have stemmed from the agglomerate conditions and
densification rate.

3Y-TZP ceramics sintered below $1400\,^{\circ}$C consisted of only the
tetragonal phase. With increasing sintering temperature up to $1500\,^{\circ}$C, a
small amount of the cubic phase was present with the tetragonal phase.
With a further increase in temperature, a gradual increase of the cubic
phase was observed.

Table II. Quantitative EDS Analysis of 30 wt% $Eu_2O_3 \cdot HfO_2$ Cubic Solid
Solution Particles

Particle No.	Composition (wt%)	
	HfO_2	Eu_2O_3
1	68.1	31.9
2	69.7	30.3
3	69.7	30.3
4	71.3	28.7
5	70.6	29.4
6	71.1	28.9
Average	70.1	29.9
Standard deviation	1.1	

Crystallized hydrothermally at $300\,^{\circ}$C, 10 MPa, for 3 h from the
coprecipitate.

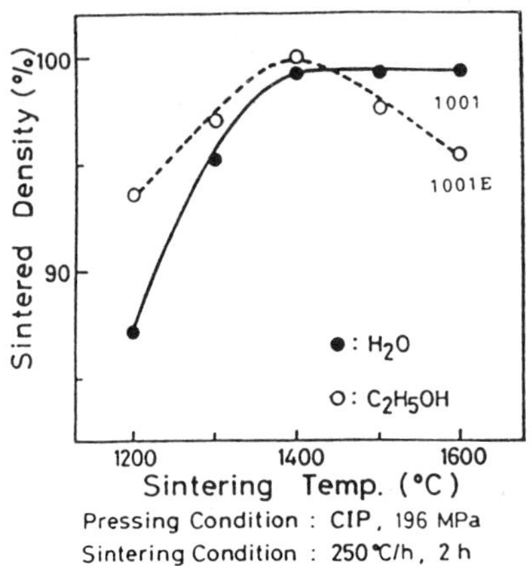

Fig. 10. Variation of the sintered density of TZP (3 mol% Y_2O_3) with sintering temperature.

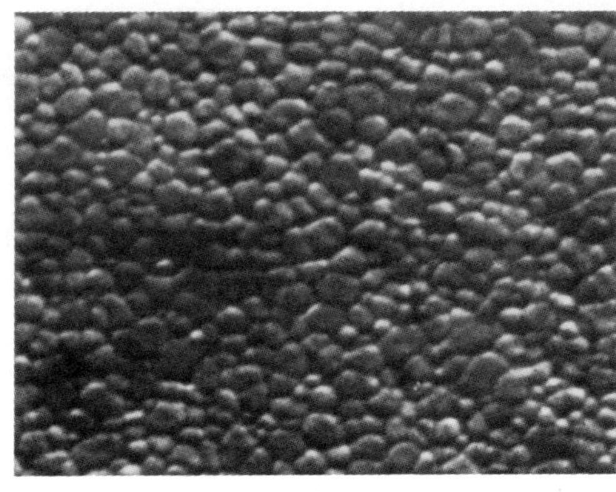

Fig. 11. Microstructure of 3Y-TZP sintered at $1400^\circ C$ for 2h (1001E). (Bar = 1 μm.).

In the present study, the transformation from the tetragonal to the monoclinic phase was not observed even after an annealing test at $200^\circ C$ for 1000 hours. The small grain size (<1.0 μm) of these ceramics leads to improvements in the stability at low termperatures (200° to $300^\circ C$) and in the mechanical behaviour. Table I summarizes the typical properties of the 3Y-TZP ceramics prepared from these powders.

Schematic illustrations of sintering of the hydrothermally produced $YZrO_2$ powders is shown in Fig. 12. This picture shows the effect of washing agent (water or ethanol), sintering temperature, etc., on microstructure, relative density, etc. As can be seen, there is an optimum sintering temperature for each of these hydrothermal powders.

Distribution of Dopant

Distributions of dopant (such as Y_2O_3 and/or other materials) were checked by EDS. Table II was one of the examples to shown distribution of dopants. The measured values are good agreement with each stoichiometric value.

SUMMARY

Ultrafine zirconia powders were produced by the hydrothermal homogeneous precipitation method. The crystallite size and sinterability were influenced by temperature, concentration of solution and precipitant, time, washing medium, and calcination. Hydrothermally processed Y_2O_3 doped ZrO_2 powder was found to be:

(1) Fine grained
(2) Free of agglomeration or soft agglomeration
(3) Narrow in its particle distribution
(4) Controllable in the chemical composition
(5) Good with respect to distribution of dopants
(6) Good with respect to homogeneity
(7) Controllable in terms of particle shape
(8) Anhydrous

514

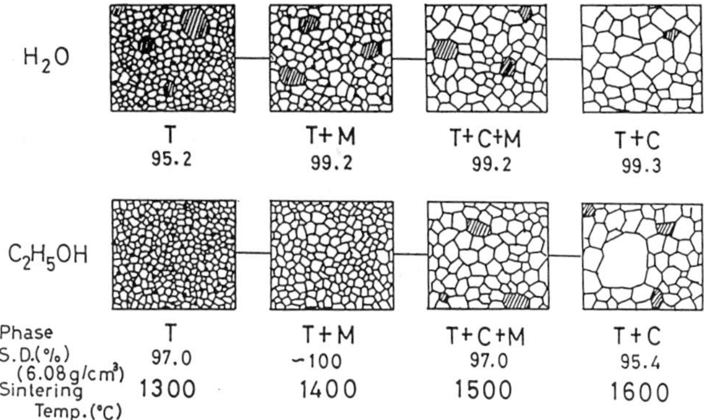

Fig. 12. Schematic representation of TZP (3 mol%-Y$_2$O$_3$) microstructures sintered at various temperatures.

(9) Available as single crystals and/or amorphous
(10) Stress free
(11) Relative freedom of defects in crystals
(12) etc.

The resulting TZP ceramics were composed of uniformly fine grains and had good mechanical properties and stability for annealing at low temperatures. Hydrothermal ZrO$_2$ powder is one of the best ZrO$_2$ powders for the production of advanced ZrO$_2$ ceramics. Good powder is one of the keys to make the excellent advanced ceramics.

REFERENCES

1. V. A. Kuznetzov, Crystallization of the oxides of Titanium subgroup metals, pp. 43-55 in Crystallization Processes under Hydrothermal Condition, A. N. Lobachev, editor, Consultant Bureau, N.Y.,1971.
2. G. W. Morey, Hydrothermal synthesis, J. Am. Ceram. Soc. 36(9) 279-285 (1953).
3. R. Roy, Hydrothermal experimentation II, method of making mixture for both "dry" and "wet" phase equilibrium studies, J.Am.Ceram.Soc.39, (4), 145-146 (1956).
4. R. A. Laudese and J. W. Nielsen, Hydrothermal crystal growth pp.149-222, in F. Seitz and D. Turnbull editors, Solid State Physics 12,1961
5. A. A. Ballman and R. A. Laudise, Hydrothermal growth, The Art and Science of Growing Crystals, pp.231-251 in J.J. Gilman editor, John Willey and Sons, Inc. N.Y., 1963.
6. W. Eitel, Silicate Science, Vol. IV, Hydrothermal Silicate System pp.615, Academic Press,1966.
7. R. A. Laudise, Hydrothermal growth, pp. 275-293 in The Growth of Single Crystals, Prentice Hall, Inc.1970.
8. A. N. Lobachev, Hydrothermal synthesis of crystals, pp 153 and 255 in Crystallization Processes Under Hydrothermal Conditions, A.N.Lobachev, editor, Consultant Bureau, N.Y.,1973.
9. R. A. Laudise, Hydrothermal growth, pp. 162-197, in P. Hartman, editor, North-Holland Publishing Co. 1973.

10. A. Rabenau, Hydrothermal synthesis in acid solutions, pp.198-209
    in P.Hartman, editor, Crystal Growth: An Introduction, North
    -Holland Pub.Co., 1973.

11. A. Rabenau, The Role of hydrothermal synthesis in preparation
    chemistry, Angew Chem. Int. Ed. Engl. $\underline{24}$, 1026-1040 (1985).

12. A. Rabenau, The role of hydrothermal synthesis in materials
    science, J. Mat. Education. $\underline{10}$(5) 543-592 (1988).

13. W. J. Dawson, Hydrothermal synthesis of advanced ceramic powders,
    Am. Ceram. Soc. Bull. $\underline{67}$(10) 1673-1678 (1988).

14. P. Reynen, H. Bastius, B. Pavlovski, D. von Mallenckrodt,
    Production of high purity zircon ($ZrSiO_4$), Advances in Ceramics
    3, Science and Technology of Zirconia, A. H. Heuer, L.W.Hobbs,
    editors, Am.Ceram.Soc., 1981.

15. K. Haberko and W. Pyda, Preparation of Ca-stabilized ZrO_2
    micropowders by a hydrothermal method, pp.774-783 in N.
    Claussen, H.Ruhle A. H. Heuer, editors, Proc. the Second
    Int.Conf. on the Science and Tech. or Zirconia,
    Amer.Ceram.Soc.,1984.

16. E. P. Stambaugh and J. F. Miller, Hydrothermal precipitation of
    high quality inorganic oxides, pp. 859-872, The 1st
    International Symposium on Hydrothermal Reactions,1983.

17. J. H. Adair, E. P. Stambaugh, Hydrothermal preparation of
    partially stabilized Zirconia powders, Presented at the 2nd
    International Symposium on Hydrothermal Reaction,1985.

18. S. Komarneni, R. Roy, E. Breval, M. Ollinen, Y. Suwa, Hydrothermal
    route to ultrafine powders utilizing single and diphasic gels,
    Advanced Ceramic Materials $\underline{1}$ (1)87-94(1986).

19. G. Gunnarsson, E. L. Sveinsdottr, A. K. Westman, T. Lepisto, T.
    Mantyla, O. T. Sorensen, Production of Zirconia powders by
    hydrothermal methods and supercritical drying and their
    properties, Presented 2nd International Conference on Ceramic
    Powder Processing Science and Technology, Oct. 12-14, 1988.

20. G. W. Kriechbaum, P. Kleinschmit, D. Peukert, Wet chemical
    synthesis of Zirconia powders, pp.146-153 in G. L. Messing, E.
    R. Fuller, H. Hausner, editors, Proceedings of the First
    International Conference on Ceramic Powder Processing Science,
    Ceramic Transactions: Ceramic Powder Science, Am Ceram. Soc.,
    1988.

21. J. H. Adair, R. P. Denkewicz, F. J. Arriagada, K. Osseo-Asare,
    Precipitation and in-situ transformation in the hydrothermal
    synthesis of crystalline Zirconium dioxide pp.135-145 in G. L.
    Messing, E. R. Fuller, H. Hausner, editors, Ceramic
    Transactions. Ceramic Powders Science, Am. Ceram. Soc. 1988.

APPENDIX

  Our papers written in English, related to Hydrothermal ZrO_2

1. K. Nakamura, S. Hirano and S. Somiya,Hydrothermal growth of
    Y O -stabilized cubic ZrO_2 crystals, Am.Ceram. Soc. Bull.,
    $\underline{56}$(5), (1977), 513.

2. M. Yoshimura and S. Somiya, Fabrication of dense, nonstabilized
    ZrO_2 ceramics by hydrothermal reaction sintering, Am. Ceram.
    Soc. Bull., $\underline{63}$, (2),246(1980)

3. M. Yoshimura and S. Somiya, Hydrothermal reaction sintering of
    monoclinic Zirconia, pp.455-463, in A. H. Heuer and L. W.
    Hobbs, editors Advances in Ceramics 3, Am. Ceram. Soc. (1981).

4. E. Tani, M. Yoshimura and S. Somiya, Hydrothermal preparation of ultrafine monoclinic ZrO_2 powder, J. Am. Ceram. Soc., 64 (12), 181(1981)

5. H. Toraya, M. Yoshimura and S. Somiya, Preparation of fine monoclinic Hafnia powder by hydrothermal oxidation, J. Am. Ceram. Soc., 65 (5) May, C-72, (1982).

6. M. Yoshimura, S. Kikugawa and S. Somiya, Preparation of Zirconia fine powders by reactions between Zirconium metal and high temperature-high pressure solutions, pp.793-796 in C. M. Backman, T. Johannsson and L. Tegner, editors, High Pressure Research and Industry, Vol.2. (1982).

7. M. Yoshimura and S. Somiya, Synthesis and sintering of Zirconia fine powders by hydrothermal reactions from Zirconium metal and high temperature high pressure solutions, pp.417-422, in D.Kolar, S. Pejovnik and M.M.Ristić, editors, Sintering-Theory and Practice, Proceedings of the 5th International Round Table Conference, Elsevier Pub. Co., Amsterdam, (1982).

8. H. Toraya, M. Yoshimura and S. Somiya, Hydrothermal reaction sintering of monoclinic HfO_2, J.Am.Ceram.Soc., 65,(9), C-159-C-160 (1982).

9. E. Tani, M. Yoshimura and S. Somiya, Formation of ultrafine tetragonal ZrO_2 powder under hydrothermal conditions, J. Am. Ceram. Soc., 66(1) 11-14 (1983).

10. H. Toraya, M. Yoshimura and S. Somiya, Hydrothermal oxidation of Hf metal chips in the preparation of monoclinic HfO_2 powders, J. Am. Ceram. Soc., 66(2) 148-150 (1983).

11. E. Tani, M. Yoshimura and S. Somiya, Hydrothermal crystallization and crystal growth of un-doped and ZrO_2-CeO_2,Rept. RLEMTIT, Vol. 8, 47-53 (1983).

12. E. Tani, M. Yoshimura and S. Somiya, Revised phase diagram of the system ZrO_2-CeO_2 below 1400°C, J. Am. Ceram. Soc., 66(7) 506-510 (1983).

13. H. Toraya, M. Yoshimura and S. Somiya, Reaction kinetics in the hydrothermal oxidation of Hf, J. Am. Ceram. Soc. 66(11) 818-822 (1983).

14. M. Yoshimura and S. Somiya, Fine Zirconia powders by hydrothermal processing, Rept. Res. Lab. Eng. Met. 9 53-64 (1984).

15. S. Somiya, Hydrothermal preparation and sintering of fine ceramic powders, MRS, Symposium 24 225-271 (1984).

16. H. Toraya, M. Yoshimura and S. Somiya, Preparation of mixed fine Al_2O_3-HfO_2 powders by hydrothermal oxidation, pp. 794-805, in Advances in Ceramics 12, N. Claussen, M. Ruhle A. H. Heuer, editors, American Ceramic Society (1984).

17. S. Somiya, M. Yoshimura and S. Kikugawa, Preparation of Zirconia-Alumina fine powders by hydrothermal oxidation of Zr-Al alloys, pp.155-166 in R. F. Davis, H. Palmour, III and R. L. Porter editors, Emergent Process Method for High Technology Ceramics, Materials Science Research Vol.17, Plenum Publ.,1984.

18. S. Somiya, M. Yoshimura, H. Torava, E. Tani, M. Suzuki, S. Kikugawa and T. Hatori, Hydrothermal processing for ceramics-powder preparation and sintering, pp. 739-752, in S. Somiya, E. Kanai and K. Ando, editors, Proceedings of the First International Symposium on Ceramic Components for Engine, KTK Scientific Publishers,1984.

19. M. Yoshimura, T. Hiuga and S. Somiya, Crystal growth of Yttria stabilized Zirconia (YSZ) under hydrothermal conditions, J. Cryst. Growth. 71 (1) 277-79 (1985).

20. S. Somiya, M. Yoshimura, H. Toraya, Y. Fushii, Cubic Eu-doped
     Hafnia ultrafine particles crystallized under hydrothermal
     conditions, Z. Anorg. Allgem. Chem., 540/541, 251-8 (1986).
21. S. Somiya, M. Yoshimura, Z. Nakai, K. Hishinuma and T. Kumaki,
     Hydrothermal processing of ultrafine single crystalline
     Zirconia and Hafnia powders with homogeneous dopants, pp.43-51
     G.L. Messing, K. S. Mazdiyasni, J. W. McCauley, R.A. Haber,
     editors, Ceramic Powder Science & Technology, Advances in
     Ceramics Vol.21 Amer.Ceram. Soc., 1987.
22. S. Somiya, M. Yoshimura, Zenjiro Nakai and K. Hishinuma, T. Kumai,
     Microstructure development of hydrothermal powders and
     ceramics, pp.465-474 in J. A. Pask and A. G. Evans, editors,
     Ceramic Microstructure, Plenum Press, New York, 1987.
23. K. Hishinuma,T. Kumak, Z. Nakai, M. Yoshimura, S. Somiya,
     Characterization of Y_2O_3- ZrO_2 powders synthesized under
     hydrothermal conditions, pp. 201-209 in S.Somiya, N.Yamamoto,
     H.Yanagida, editors Advances in Ceramics Vol. 24, Science and
     Technology of Zirconia III, Am. Ceram. Soc., 1988.
24. M. Yoshimura, S. E. Yoo, S. Somiya, ZrO_2 formation by anodic
     oxidation of Zr metal under hydrothermal conditions, Rept
     R.L.E.M. Tokyo Inst.Tech. 14 21-32 (1989).

EFFECT OF GREEN COMPACT PORE SIZE DISTRIBUTION

ON THE SINTERING OF α-Fe$_2$O$_3$

C.V. Santilli, S.H. Pulcinelli, J.A. Varela, and
J.P. Bonnet*

Instituto de Quimica, UNESP c.p. 174, Araraquara SP
Brazil, *Laboratoire de Quimie du Solide, Universite
de Bordeaux, Talence, France

INTRODUCTION

The bulk electrical properties of α-Fe$_2$O$_3$ are very useful for many
potential applications.[1,2] The macroscopic properties of sintered α-Fe$_2$O$_3$
are due to inter- and intra-granular features which are highly
dependent on the ceramic microstructure and on the presence of
structural defects.[3] Twinning and dislocations normally present in
α-Fe$_2$O$_3$ ceramics influence the magnetic and electric structure. As a
consequence, electronic transport associated with the varistor effect
in this material is dependent on the concentration of these defects,
which in turn are dependent on the morphological characteristics of the
precursors and on the sintering conditions.[4,5] There is no agreement on
sintering mechanisms of α-Fe$_2$O$_3$ reported in the literature. For low
temperatures between 500 and 700°C Whittemore and Varela[6] have shown
large pore growth during sintering, explained by the coalescence of
grains. Idzikowski[7] observed a fast grain growth in this range of
temperatures, suggesting a mechanism of mass transport due to different
types of necks. Santilli et al[8] showed that grain coalescence occurs
mainly in powders derived from goethite while grain growth does not
occur in compacts made from amorphous iron oxide precursors. Yamaguchi
and Kosha[9] studied the sintering of α-Fe$_2$O$_3$ with elongated and
spherical particle shapes, derived from the calcining of α-FeOOH and
Fe$_2$(SO$_4$)$_3$, respectively. They observed a fast initial densification
with pore growth and grain reorientation of acicular particles. This
behavior was attributed to structural rearrangement.

The sintering of α-Fe$_2$O$_3$ seems to be strongly dependent on the
morphological characteristics of the powders chosen. The relative
arrangement between pores and grains and the solid-solid interface
originated in the initial stage of sintering are factors dependent on
powder morphology. In this work the influence of compact
characteristics of acicular α-Fe$_2$O$_3$ particles was investigated.

EXPERIMENTAL

Powder Preparation and Compaction

The powder used in this work was obtained through calcination of goethite at 600°C for 16 hours using the procedure described in the literature.[8] Figure 1 is a TEM picture of the calcined powder. The particles are elongated in the $\langle 001 \rangle_{hex}$ axis with 0.4 μm as the mean dimension in the $\langle 012 \rangle_{hex}$ direction. The α-Fe_2O_3 powder was compacted in a stainless steel die in cylindrical pellets. Then the compacts were isostatically pressed using the following pressures: 25, 50, 75, 100, 150 and 200 MPa.

Compacts were sintered in dry air atmosphere at 1000°C using an electrical furnace with an alumina tube chamber. The samples put inside of an alumina boat were pre-heated at 600°C for 30 min. Then the boat was pushed to the hot zone of the furnace with help of an alumina rod. Sintering isotherms of 2, 4, 8, 16, 32 and 64 minutes were obtained. The zero time of sintering was defined when the furnace reached the thermal equilibrium after introduction of the boat at hot zone. It took about 3 minutes for thermal equilibration.

Sample Characterization

Pore size distribution (PSD) of green and sintered samples were obtained by mercury intrusion and extrusion with help of the Washburn equation.[10] From PSD curves surface areas and mid-pore sizes were determined. The apparent densities were measured by mercury displacement.

Microstrain in the crystalline lattice was determined from profile analysis of the X-ray diffraction lines by using the method proposed by Riella et al.[11] The X-ray peaks of direction $\langle 012 \rangle_{hex}$ were scanned with 0.05 degrees/min by using cobalt K_{α_1} radiation and iron filter.

RESULTS AND DISCUSSION

Effect of Compaction Pressure

Figures 2a and 2b show the effect of compaction pressure on the PSD of green and sintered samples at 1000°C for 2 minutes, respectively. Solid lines represent intrusion data while dashed lines stand for extrusion data. The observed hysteresis is attributed to cavities interconnected by small channels or necks, forming a tridimensional capillary network.[12] The mid-pore diameter obtained for intrusion, d_i, and extrusion, d_e, depend on the shape of capillary interconnection and consequently on the relative arrangement of grains. The results show that above 50 MPa of compaction pressure there is a continuous change in the pore structure with decreasing of d_i and d_e (Figure 2a). Figure 2b clearly illustrates the effect of green compaction on the sample sintered at 1000°C for 2 minutes. Under all conditions, an increase of D_i and a decrease of both D_e and pore volume with respect to compaction pressure was observed. It should also be noted that there is a large variation of pore sizes in samples formed under low compaction pressure. This indicates that under low pressure forming conditions, there is an increase in pore coordination and a decreasing pore stability, as proposed by Kingery and Francois.[13]

During sintering, a pore can be eliminated thermodynamically if its coordination number is smaller than a critical value R_c. The

Fig. 1. TEM micrograph of α-Fe$_2$O$_3$ particles.

decrease of pore coordination number can be achieved by grain growth and/or structural rearrangement. Grain growth comes about as a consequence of differential sintering of particles, domains and agglomerates. Considering that the starting α-Fe$_2$O$_3$ powder used in this study is already agglomerated (Fig. 1) this mechanism for grain growth is highly probable. In the sample sintered at 1000 °C for 2 minutes there is a large decrease of surface area (Table I) which can be attributed to differential rate of sintering and grain growth.

Densifying by means of structural rearrangement also leads to a decrease in pore coordination.[15] This mechanism is associated with stresses generated by asymmetric neck growth. However, rearrangement associated with differential sintering leads to pore growth and

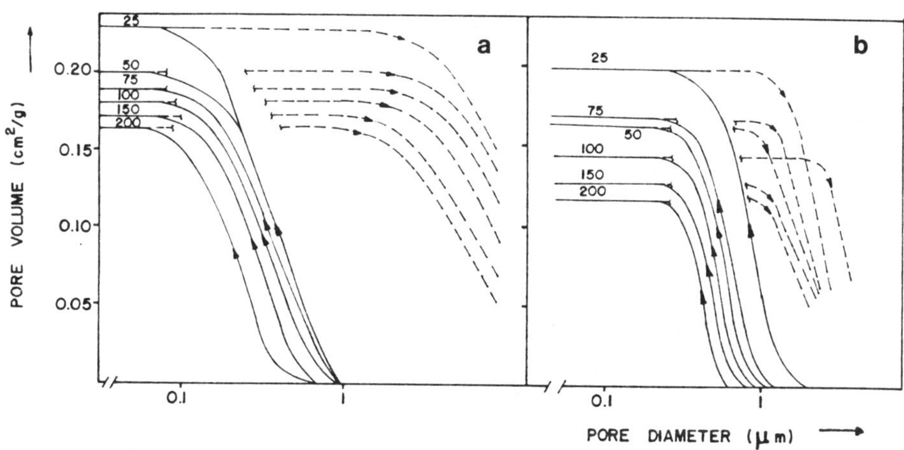

Fig. 2. PSD curves for mercury intrusion (solid lines) and extrusion (dashed lines): (a) green sample compacted at indicated pressures; (b) sample compacted at indicated pressures and sintered at 1000 °C for 2 minutes.

increasing pore coordination.[14] Only in the late stage of sintering does this mechanism lead to a decrease in pore coordination, and thereby become helpful for further densification.

The stresses generated by differential sintering is related with the highest local neck density, n, and few average neck density of the compact, n_a,[14] as is given by:

$$\sigma = (n_a - n)\gamma dA \tag{1}$$

where γ is the free surface energy and dA is the change in area due to a single neck. When $n_a > n_a$ tensile stresses are generated leading to disruptive rearrangement. For $n_a < n$ compressive stresses are generated, thereby favoring rotational rearrangement.

The results of Fig. 2 and Table II shows that rearrangement is more effective in samples with lower and higher compaction pressure. This behavior suggests that rearrangement is controlled by two distinct driving forces: (a) distribution of pore coordination number which increases with porosity, and b) neck asymmetry which increases with the number of contacts of particles in the green body, i.e., with green density.

Weiser and De Jonghe[15] showed that differential densification due to distribution of pore coordination number is the dominant cause for rearrangement in two dimensional arrays of spherical copper particles. They suggested that in three dimensional arrays the same mechanism should be dominant. However in the case of acicular particles, as used in this study, the neck asymmetry can generate high rotational momentum.

<u>Effect of Sintering Time</u>

The rotational momentum and stresses generated by differential densification and/or neck asymmetry can have a significant effect on the sintering kinetics.[16] In this work, the influence of green compaction on the sintering kinetics was analyzed for samples pressed at 25, 75, and 150 MPa, respectively. Figures 3a and 3b show the dependence of sintering time on PSD curves for samples pressed at 25 and 150 MPa. After a fast densification during the first two minutes of sintering, a pronounced increase in the mid-pore size and the total pore volume for samples compacted at 150 MPa was observed. Similar behaviour is observed for samples compacted at 75 MPa. Fast densification occurring in the first two minutes of sintering is attributed to densifying rearrangement processes characteristic of neck asymmetry.[17]

Figure 4 shows relative variation on apparent densities, $\Delta\rho/\rho_0$ mid-pore size $\Delta d/d_0$, and of the microstrain, σ, with sintering time. These figures show that the evolution of pore size distribution, densities and stresses generated can be correlated in a complex manner. The increase of difficulty in rotation of elongated grains as the green density increases should be the cause of the increase of pore volume with sintering time observed in samples compacted at 75 and 150 MPa (Figure 4a). Pore widening is then due to rearrangement resulting from differential densification.

Microstrains generated by microstresses are higher for less dense material in the first two minutes of sintering. In dense compacts, the stresses are relieved by breaking the necks and by consequent densifying rearrangements. For longer times, there is an increase of microstrain

Table I. Physical Characteristics of α-Fe$_2$O$_3$ Compacts Sintered at 1000°C

Compaction Pressure (MPa)	Sintering Time (min.)	Mid-pore Diam. (μm)		Surface Area (m^2/g)	Apparent Density (g/cm^3)
		Intrusion	Extrusion		
25	green	0.39	15.8	3.50	2.40
25	2	0.88	2.35	1.00	2.60
25	4	0.90	3.00	0.93	2.63
25	8	0.90	2.85	0.90	2.67
25	16	0.89	3.00	0.78	2.73
25	32	0.76	2.20	0.84	2.96
25	64	0.67	1.70	0.85	3.20
75	green	0.39	14.0	3.50	2.64
75	2	0.66	2.15	1.19	2.89
75	4	0.66	2.30	1.09	2.85
75	8	0.75	2.40	0.97	2.77
75	16	0.86	2.50	0.77	2.87
75	32	0.80	2.10	0.80	3.02
75	64	0.69	1.90	0.80	3.18
150	green	0.30	8.0	3.50	2.76
150	2	0.48	2.15	1.02	3.22
150	4	0.49	2.20	0.98	3.17
150	8	0.92	3.05	0.82	2.98
150	16	0.93	2.20	0.71	3.06
150	32	0.68	1.58	0.71	3.30
150	64	0.47	1.35	0.66	3.75

Table II. Variation in Physical Characteristics of the α-Fe$_2$O$_3$ Compacts Sintered at 1000°C

Compaction Pressure (MPa)	$\Delta d_i/d_o$	$\Delta d_e/d_o$	$\Delta S/S_o$	$\Delta\rho/\rho_o$
25	1.25	-0.85	-0.72	0.08
50	0.49	-0.86	-0.68	0.12
75	0.51	-0.85	-0.68	0.10
100	0.54	-0.75	-0.71	0.13
150	0.61	-0.73	-0.73	0.16
200	1.00	-0.71	-0.73	0.17

$\Delta d_i/d_o$ — relative variation of intrusion mid-pore size.
$\Delta d_e/d_o$ — relative variation of extrusion mid-pore size.
$\Delta S/S_o$ — relative variation in surface area.
$\Delta\rho/\rho_o$ — relative variation in apparent density.

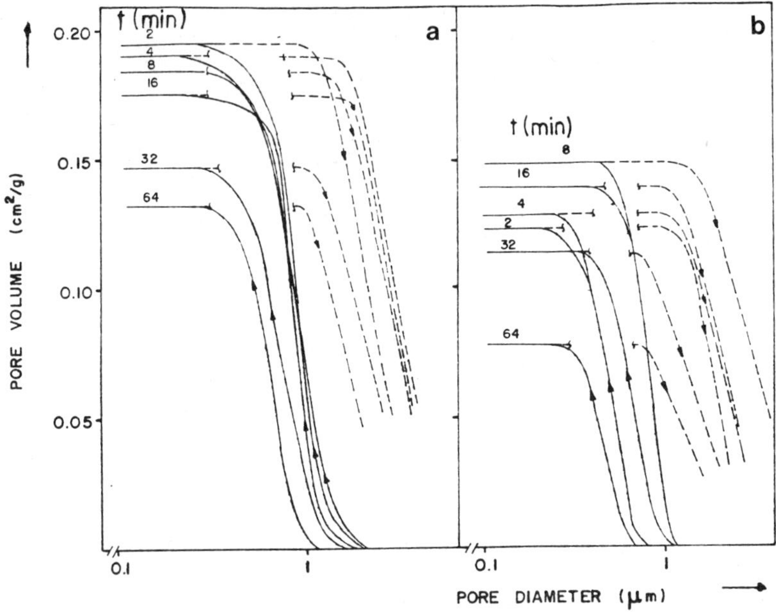

Fig. 3. PSD curves for mercury intrusion (solid line) and extrusion
(dashed line) for samples sintered during the indicated time:
(a) compacted at 25 MPa; (b) compacted at 150 MPa.

for samples compacted at 75 and 150 MPa, respectively. In this case,
the stresses generated by differential densification are higher than
those due to asymmetric neck growth, resulting in pore growth. After
the first two minutes of sintering there is an decrease of density

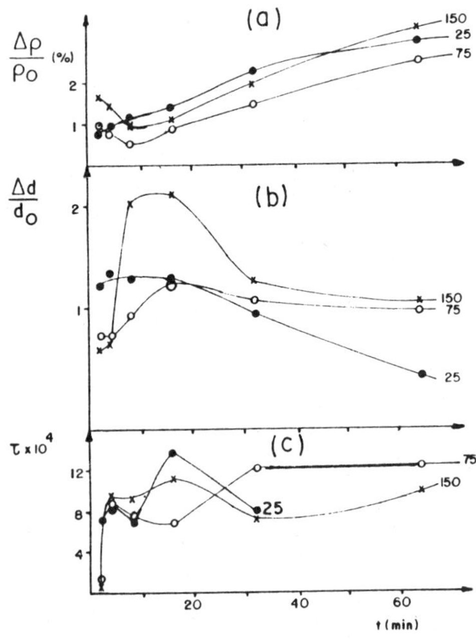

Fig. 4. Relative variations with time of: (a) apparent density; (b) mid
-pore diameter; (c) microstrains.

524

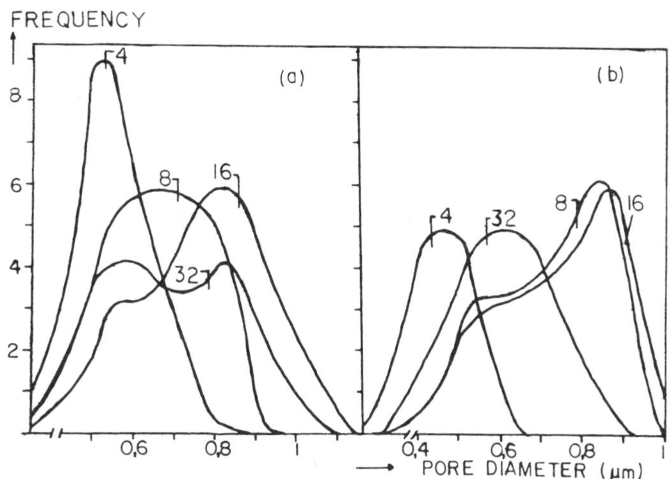

Fig. 5. Volumetric frequency curves (dV/dr) for samples sintered at
indicated times:(a) compacted at 75 MPa; (b) compacted at
150 MPa.

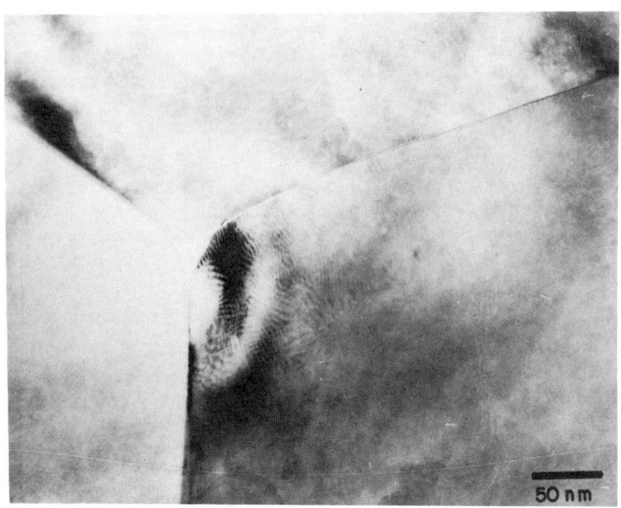

Fig. 6. TEM micrograph of sample compacted at 150 MPa and sintered at
1000 °C for 64 minutes.

followed by densification for long times. It is also observed from Fig.
4 that above the critical value for stresses, in samples compacted at
75 and 150 MPa (Fig. 4b), there is dramatic pore growth. As a
consequence, there is thereafter a decrease of these stresses due to
collapsing necks (Fig. 4c) or decreases in asymmetry.

The disruptive rearrangement leads to separation of certain groups
of particles with formation of large pores between them.[15] As a
consequence there is an increase in the distribution of pore
coordination number, thereby promoting additional increase in stresses,
as given by equation 1. The PSD curves presented in Figure 5 shows the
formation of a second family of pores after a certain sintering time
thus confirming this mechanism.

The two maxima of observed deformation in the initial stage of
sintering (t < 31 minutes) can be understood. For long sintering time
the residual stresses are associated with interactions between grain
boundaries and pores. Dislocations observed near pores at triple points
illustrate this behaviour (Fig. 6).

CONCLUSIONS

Structural rearrangement controls the evolution of microstructure
during the first minutes of sintering of α-Fe_2O_3. The results suggest
that differential densification is the predominant cause of
rearrangement in compacts with low green densities, while neck
asymmetry determines rearrangement for compacts with high apparent
green densities. In all cases there is a bimodal distribution of pores
inherent to such disruptive process. This favors differential
densification, thereby extending the duration of rearrangement.

ACKNOWLEDGEMENT

The authors acknowledge L. G. Martinez and H. G. Riella by
providing computer program for microstrain calculation and the
financial support of CNPq, FAPESP, FINEP and FUNDUNESP.

REFERENCES

1. C. Leygraf, M. Hendewerd and G. Smorgai, J. Solid State Chem. <u>48</u>,
    357(1983).
2. Y. Nakatami, M. Sakai and M. Matsuoka, Jpn J. Appl. Phys <u>22</u>(6),
    912(1983).
3. A. J. Bosmam and H. J. VanDaal, Adv. Phys. <u>19</u>(77), 1 (1970).
4. C. V. Santilli, J. P. Bonnet, P. Dordor, M. Onillon and P.
    Hagenmuller, Ceramics International to be published.
5. C. V. Santilli, M. Onillon and J. P. Bonnet, Mater. Sci. and
    Engr. B, to be published.
6. O. J. Whittemore and J. A. Varela, in <u>Sintering Processes</u>, ed
    G. C. Kuczynski, Plenum Publ. Co. p. 51 (1980).
7. S. Idzikowski, Trans. Brit. Ceram. Soc. <u>76</u>(4), 74 (1977).
8. C. V. Santilli, M. Onillon and J. P. Bonnet, Ceramics
    International, to be published.
9. T. Yamaguchi and H. Kosha, J. Am. Ceram. Soc. <u>66</u>(3), 210 (1983).
10. E. W. Washburn, Proc. Nat. Acad. Sci. <u>7</u>, 115 (1921).
11. H. G. Riella, G. L. Martinez and K. Imakuma, J. of Nuclear Mater.
    <u>153</u>, 71 (1988).

12. S. J. Gregg and K. S. W. Sing, <u>Adsorption, Surface Area and Porosity</u>, Academic Press p. 173 (1982).
13. W. D. Kingery and B. Francois in <u>Sintering and Related Phenomena</u>, ed. G. C. Kuczynski, N. A. Hooton and G. I. Gibbon, Gordon and Breach, N. York, p. 471 (1967).
14. F. F. Lange, J. Am. Ceram. Soc. <u>67</u>(2), 83 (1984).
15. J. A. Varela and O. J. Whittemore, J. Am. Ceram. Soc. <u>66</u>(1), 78 (1983).
16. M. W. Weiser and L. C. De Jonghe, J. Am. Ceram. Soc. <u>69</u>(11), 822 (1986).
17. O. J. Whittemore and J. A. Varela, <u>Advances in Ceramics</u>, vol 10, Ed. W. D. Kingery, Am. Ceram. Soc., p. 583 (1985).

EFFECT OF NIOBIA ON THE SINTERING OF SnO_2

D. Gouvea, J.A. Varela, C.V. Santilli, and E. Longo*

Instituto de Quimica, UNESP, Caixa Postal 174
Araraquara, SP Brazil[1,2,3]
*Departamento de Quimica, UFSCar, CP 676, Sao Carlos
SP Brazil

INTRODUCTION

SnO_2 crystallizes in tetragonal system with structure similar to
rutile and is a n-type semiconductor. The electrical conductivity is
mainly due to the existence of point defects (oxygen vacancies) that
generate donor electron levels. Its electrical and optical properties
have been studied for many technological applications[1,2], and thin
films and compacts are used as electrodes for electrochemical
processes. The grain boundary properties of SnO_2 have been studied with
respect to its use in gas sensors. SnO_2 films are used as electrodes in
electron optical devices and as heating elements.

Some electrical and structural characteristics of SnO_2 are similar
to those found in ZnO and TiO_2. These oxides are normally used in the
fabrication of ceramic varistors.[3,4] Deviations from Ohm's law are
normally observed in a ceramic semiconductor of narrow gap band, i.e.
those with good electrical conductivity at room temperature. For
varistor applications these materials should also have a good thermal
conductivity that will allow dissipation of heat generated by Joule
effect. Thus the residual porosity for this application should be very
low. This latter condition is one of the most difficult achieve to for
utilization of SnO_2 in the fabrication of varistors.

SnO_2 only densifies in the presence of certain additives.[5-9] The
addition of CuO, Bi_2O_3 or ZnO help in the densification of SnO_2
through the formation of small amounts of liquid phase or by creation
of oxygen vacancies. However, these additives create acceptor states
that decrease the electrical conductivity of SnO_2. To avoid this
condition the use of dopants with valence 5 or 6 such as Nb_2O_5, Sb_2O_5
or WO_3 is commonly recomended. The influence of Nb_2O_5 on the
densification and grain growth of SnO_2 was considered in this study.

EXPERIMENTAL

Sample preparation

The niobia used in this study was 99.9% pure and very reactive

(125 m^2/g, obtained from CBMM). SnO_2 (99.99% pure, obtained from
Backer) had a surface area of 7.5 m^2/g. The powders were slurried in
acetone and magnetically stirred for 2 hours. After this period the
acetone was evaporated at 60 ^{o}C with continuous agitation. The powder
was completely dried in an oven at 110 ^{o}C and calcined at 700°C for 4
hours.

Pellets 1.3 cm diameter and 0.3 cm thick were obtained in two
stages. In the first stage the powder was placed in a cylindrical die
and pressed at 15 MPa. No binder was used. In the second stage the
formed pellets were placed in latex balloon and isostatically pressed
at 210 MPa. Samples containing 1.0, 2.0, 3.0, 5.0 and 10.0% Nb_2O_5 were
prepared by this method.

Sintering

Sintering of pellets were carried at 1100 and 1250 ^{o}C in an
alumina tube furnace heated by six silicon carbide heating elements.
The pellets were placed in an alumina boat and pushed slowly to the
furnace hot zone. Sintering atmospheres used were argon, air and oxygen
previously dried in a silica gel column. Sintering at 1500 ^{o}C was
conducted in a muffle furnace with molybdenum silicide heating
elements. In this case sintering was carried out in air for a period of
4 hours.

Physical Characterization

Pore size distributions were determined by using a Micromeretics
mercury porosimeter. Pore diameters were determined by applying
pressure P on mercury and using the equation derived by Washburn[10]

$$P = (4\gamma_m \cos\theta)/d, \tag{1}$$

where γ_m is the mercury surface tension, θ is the contact angle between
mercury and sample surface and d is the pore diameter. This technique
also permits determination of surface area and total open pore volume.

Apparent densities were determined by mercury displacement. With
this technique it is possible to determine apparent densities in
samples with pore diameters smaller than 200 μm, by measuring sample
volume V_a using the following equation:[11]

$$V_a = V_p - (M_{psm} - M_{ps})/\rho_m, \tag{2}$$

where V_p is the internal picnometer volume, M_{psm} is the mass of
picnometer with sample filled with mercury, M_{ps} is the mass of
picnometer with the sample and ρ_m is the Hg density.

DTA was performed on samples containing 10 mol % Nb_2O_5 using a
DuPont Thermal Analyzer. This permits identification of phase
transformations in the SnO_2-Nb_2O_5 system.

SEM was used to characterize the microstructure of the sintered
compacts.

RESULTS AND DISCUSSIONS

According to Isupova et al[12] the SnO_2-Nb_2O_5 system does not form
an intermediate phase. The system has a limited solid solution of Nb_2O_5
in SnO_2 up to 3 mol% at room temperature. Fig. 1 shows the differential

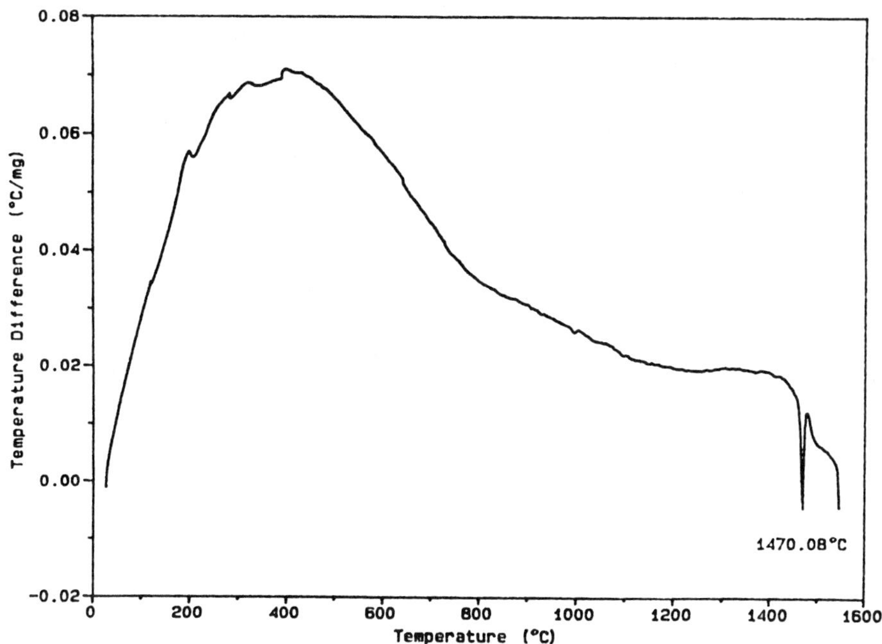

Fig. 1. DTA of SnO$_2$ powder containing 10 mol% Nb$_2$O$_5$.

thermal analysis of samples containing 10 mol% Nb$_2$O$_5$. There is an endothermic peak at 1470 ^{o}C, indicating liquid phase formation. This temperature agrees with the eutectic reported by Isupova et al.[12]

X ray diffraction in samples containing 0, 1, 3, 5 mol% of Nb$_2$O$_5$ and calcined at 700 ^{o}C indicated that SnO$_2$ is the only phase present for concentrations up to 3 mol% of Nb$_2$O$_5$. For large concentrations of dopant, the Nb$_2$O$_5$ peaks are also present (Fig. 2).

In order to verify the influence of niobia on the densification rate of SnO$_2$, samples containing 1 to 10 mol% of niobia were calcined at 700 ^{o}C and sintered at temperatures of 1100 and 1250 ^{o}C in

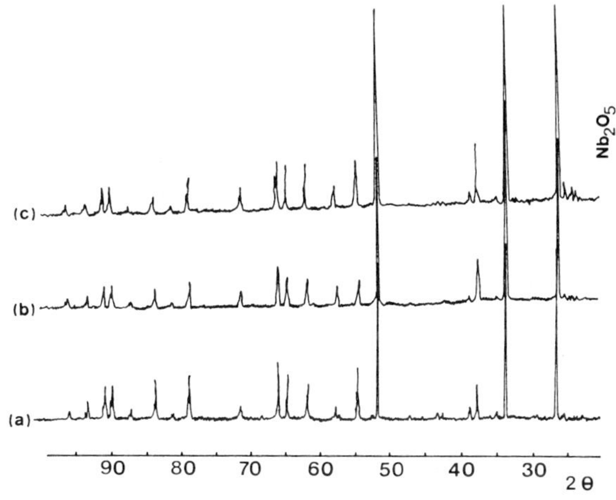

Fig. 2. X-ray diffraction for SnO$_2$ powder containing niobia: a) 1 mol%; b) 3 mol %; c) 5 mol %.

Table I. Physical characteristics of SnO_2 pellets sintered in dry argon

$[Nb_2O_5]$ %	Temp. (^oC)	Time (min.)	Bulk Dens. (g/cm^3)	Midpore dia. (μm)	Surf.Area (m^2/g)
0	1100	30	4.1	0.198	2.51
0	1100	60	4.1	0.205	2.38
0	1100	120	4.1	0.216	2.32
0	1100	240	4.1	0.270	1.75
0	1250	30	4.1	0.435	1.10
0	1250	60	4.1	0.512	0.99
0	1250	120	4.1	0.636	0.91
0	1250	240	4.1	0.804	0.64
1	1100	30	4.0	0.140	3.20
1	1100	60	4.0	0.141	3.00
1	1100	120	4.0	0.156	2.80
1	1100	240	4.0	0.160	2.70
1	1250	30	4.2	0.167	2.70
1	1250	60	4.2	0.175	2.50
1	1250	240	4.2	0.195	2.40

Table II. Physical characteristics of SnO_2 pellets sintered in dry air

$[Nb_2O_5]$ (%)	Temp. (^oC)	Time (min.)	Bulk dens. (g/cm^3)	Midpore dia. (μm)	Surf.Area (m^2/g)
0	1100	240	4.0	0.205	2.02
1	1100	240	4.1	0.130	3.17
3	1100	240	4.1	0.115	3.85
5	1100	240	4.0	0.107	4.19
10	1100	240	3.8	0.116	3.94
0	1200	240	4.0	0.410	1.15
1	1200	240	4.1	0.165	2.71
3	1200	240	4.0	0.145	3.01
5	1200	240	4.0	0.142	3.00
10	1200	240	3.8	0.159	2.94
0	1500	240	4.2	3.210	0.27
1	1500	240	4.6	0.350	0.95
2	1500	240	5.4	0.219	0.84
3	1500	240	5.5	0.209	0.81
5	1500	240	5.1	0.241	0.86
10	1500	240	4.7	0.329	0.84

atmospheres of argon and oxygen, and at 1500 $^{\circ}$C in air. Tables I, II
and III show apparent densities, surface area and midpore diameter,
respectively of samples containing different concentration of niobia
sintered at several temperatures in argon, air and oxygen atmospheres.
The theoretical density of these samples is 7.0 g/cm^3. It is observed
from these tables that 1 mol% of niobia has little effect on
densification of SnO_2. However niobia retards pore and grain growth
when compared with samples without additives.[12]

Pore growth is also inhibited by increasing oxygen partial
pressure.[13] Considering that pore and grain growth of SnO_2 is due to
evaporation-condensation mechanism it seems that niobia will increase
vaporization energy of SnO_2 decreasing the rate of grain growth.

Fig. 3 shows the influence of Nb_2O_5 concentration on the pore size
distributions and densities of SnO_2 samples sintered at 1500 $^{\circ}$C in air
for 4 hours. It is verified that additions of niobia up to 3 mol %
favor the decrease of pore volume and pore diameter (Fig. 3a). However,
additions of more than 3 mol % of niobia inhibit pore shrinkage and
densification. At this sintering temperature there are two distinct
behaviours. For $[Nb_2O_5] \leqslant$ 3 mol % the apparent density increases and
midpore diameter decreases with niobia concentrations. For 3 mol % <
$[Nb_2O_5] \leqslant$ 10 mol % it is verified that a decrease of apparent densities
and an increase of midpore diameter results from increasing niobia
concentrations. These results show that the densifying action of niobia
is limited at values $\leqslant$ 3 mol %. This is approximately the limit of
solid solution of niobia in SnO_2 at 1500 $^{\circ}$C. The solid solubility of
niobia in SnO_2 can be represented by equation:

$$2Nb_2O_5 <-Sn O_2-> 4Nb_{Sn}^{\bullet} + V_{Sn}^{''''} + 10 \; O_o \tag{3}$$

The bulk diffusion coefficient for Sn ions increases with vacancy
formation. One immediate conclusion would be that the rate of
densification increases because tin diffusion is controlling the rate
of sintering. However, no densification was observed for Nb_2O_5 doped
SnO_2 sintered at temperatures up to 1250 $^{\circ}$C, in argon, air and oxygen
(Tables I, II and III).

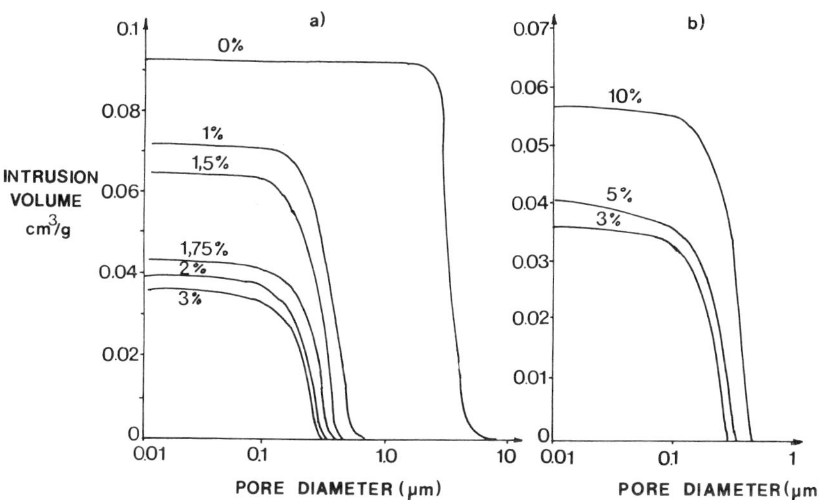

Fig. 3. PSD curves of Nb_2O_5 doped SnO_2 pellets sintered for 4 h at
1500°C in air.

Table III. Physical characteristics of SnO$_2$ pellets sintered in dry oxygen

[Nb$_2$O$_5$] (%)	Temp. (^{O}C)	Time (min)	Bulk dens. (g/cm^3)	Midpore dia. (μm)	Surf.Area (m^2/g)
0	1100	240	4.1	0.150	2.80
1	1100	30	4.1	0.100	4.00
1	1100	60	4.1	0.118	3.70
1	1100	120	4.1	0.119	3.40
1	1100	240	4.1	0.120	3.20
1	1250	240	4.1	0.185	2.20

It was observed in former studies that for CuO-doped SnO$_2$ the rate of densification is very high at 1100 ^{O}C due to increase of oxygen vacancies. Thus oxygen diffusion is considered to be the rate limiting step for densification of SnO$_2$.[9,14] Consequently the inferred increase of Sn vacancies cannot explain the enhanced rate of densification of Nb$_2$O$_5$-doped SnO$_2$ observed at 1500 ^{O}C.

As observed in Fig. 4. and Table II the increase of niobia concentration up to 3 mol % causes a decrease in the rate of pore growth and surface area variation with no densification. This indicates that the rate of evaporation-condensation decreases with niobia concentration. As grain growth caused by evaporation-condensation competis with densification, the high rate of evaporation-condensation at 1500 ^{O}C will establish a negligible densification for undoped SnO$_2$ pellets. However, as the concentration of niobia increases, the rate of evaporation-condensation of SnO$_2$ decreases, and densification increases due to the fairly high oxygen diffusion coefficient obtained at this temperature.

Fig. 5 shows SEM micrographs of undoped and 1% Nb$_2$O$_5$-doped SnO$_2$ sintered at 1500 ^{O}C for 4 hours in air. The mean grain size is on the order of 0.8 μm (Fig. 5a), much smaller than undoped SnO$_2$ sintered at same conditions (Fig. 5b).

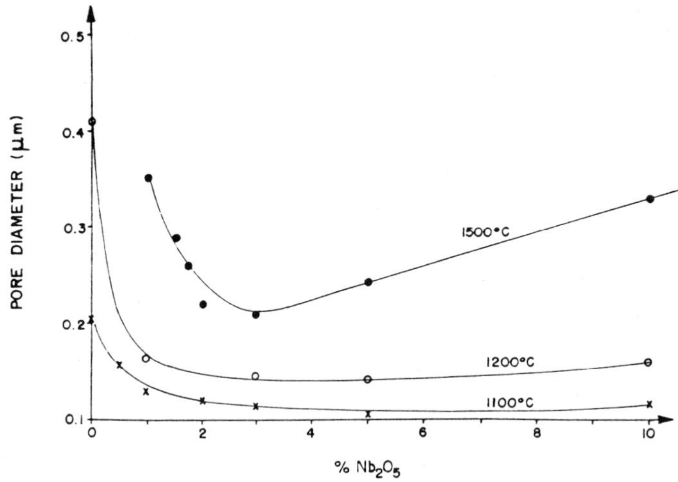

Fig. 4. Pore diameter versus niobia concentration for SnO$_2$ samples sintered for 4 h in air at 1000, 1200 and 1500 ^{O}C.

Fig. 5. SEM micrograph of a) undoped and b) niobia doped SnO_2, sintered at 1500 °C for 4 h in air.

The addition of more than 3 mol % of niobia inhibits densification and pore shrinkage, as observed in Fig. 6. In this case, there is a formation of liquid phase that increases with niobia concentration. It would be expected that the rate of densification would increase with the amount of liquid phase, e.g., through rearrangement and solution-precipitation processes. However, opposite behaviour is observed in this case. Thus the mechanism of solution-precipitation seems to be

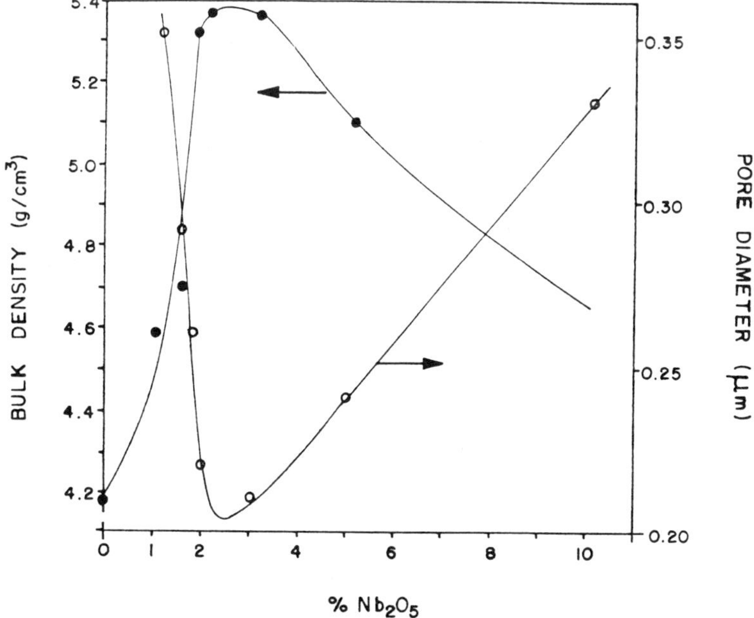

Fig. 6. Apparent bulk density and pore diameter versus niobia concentration in SnO_2 pellets sintered at 1500 °C for 4 h in air.

ineffective for densification in this system, probably because SnO_2 is
not soluble in the liquid formed. The inhibition of pore shrinkage and
densification in the presence of liquid indicates that liquid increases
the rate of evaporation-precipitation thereby enhancing grain and pore
growth. Thus evaporation-condensation would complete with solid state
diffusion. This is so because the latter process depends upon vacancies
for material transport, yet the vacancies sources would be markedly
diminished by pore coarcening caused by the abovementioned evaporation-
condensation processes.

CONCLUSIONS

As results of this study it can be concluded that additions of
niobia up to the limit of solid solution in SnO_2 increases the rate of
densification probably due to niobia segregation at free surfaces and
resultant inhibition of SnO_2 evaporation. Liquid phase formation
inhibits densification due to low solubility of SnO_2 in niobia-rich
liquid, and to high evaporation-condensation rate of SnO_2 in the liquid
formed.

ACKNOWLEDGEMENT

The authors acknowledge FAPESP, CNPq, FINEP and FUNDUNESP for
financial support of this study.

REFERENCES

1. Z. M. Jarzebski and J. P. Morton, J. Electrochem. Soc. (6), 199
    -205C (1976).
2. Z. M. Jarzebski and J. P. Morton, J. Electrochem. Soc. (9), 299
    -310C (1976).
3. L. M. Levinson and N. R. Philipp, J. Appl. Phys. 47(3), 1117-1122
    (1976).
4. M. F. Yann and W. W. Rhotes, Appl. Phys. Lett. 40(6), 536-537
    (1982).
5. P. H. Duvigneaud and D. Reinhard, Science of Ceramics 12, 287-292
    (1980).
6. M. K. Paria and H. S. Maiti, J. Mat. Sci. 17, 3275-3280 (1982).
7. A. Watanabe, T. Kikuchi, M. Tsutsume, S. Takenouchi and K. Uchida,
    J. Am. Ceram. Soc. 66, C 104-C 105 (1983).
8. H. E. Matthews and E. E. Kohnke, J. Phys. Chem. Solids 29, 653-661
    (1968).
9. J. A. Varela, O. J. Whittemore and M. J. Ball, in Sintering '85,
    Ed. G. C. Kuczynski, D. P. Uskokovic, H. Palmour and M. M.
    Ristic, Plenum Press, New York, p.259 (1987).
10. E. W. Washburn, Proc. Nat. Acad. Sci. 7, 115 (1921).
11. J. A. Varela and O. J. Whittemore, Ceramica 28(152), 337 (1982).
12. E. N. Isupova, T. I. Panova and E. P. Savchenko, Izv. Akad. Nauk.
    SSSR, Neorg. Mater. 17(6), 836 (1981).
13. J. A. Varela, E. Longo, N. Barelli, A. S. Tanaka and W. A.
    Mariano, Ceramica 31(191) 241-246 (1985).
14. N. Dolet, J. M. Hentz, E. Longo, J. A. Varela, M. Onillon and J.
    P. Bonnet, submitted to Ceramica (Sao Paulo).

KINETICS AND MECHANISMS OF THE SOLID-STATE SINTERING OF ALKALINE-EARTH
TITANATES

V.I. Lapshin and V.K. Yarmarkin

Positron Res. and Mfg. Corporation
10 Kurchatov St., Leningrad 194021, USSR

INTRODUCTION

Alkaline-earth-titanate-based electronic ceramics are widely used
in the fabrication of modern products such as capacitors, varistors,
thermistors, etc.[1,2] The sintering of electronically active ceramics is
one of most complicated and critical operations that determine the
performance and cost characteristics of such products. To gain a deeper
insight into the mechanisms of the sintering of the above materials and
to explore the ways of promoting this process, we studied both
experimentally and theoretically mass transfer processes in sintering
alkaline-earth titanates.

SAMPLE PREPARATION AND CHARACTERIZATION

The compounds under investigation ($CaTiO_3$, $SrTiO_3$, $BaTiO_3$, and
some others) were prepared by ceramic synthesis starting from
components containing impurities (Fe, Si, Al) in amounts not exceeding
10^{-3}wt.%.

The synthetized products were ground in different types of mills
with a power supplied per unit weight ranging from fractions of a
kilowatt to tens of kilowatts per kilogram of material to be processed.
The particle size distribution was characterized by optical and
electron microscopic techniques using statistical analysis of the
images obtained, by sedimentation technique, Coulter's technique, from
low-temperature adsorption data, and mercury porosimetry, as well as
X-ray techniques.

After being pressed in cylindrical dies at a fixed pressure
(usually 100 MPa), samples were sintered in programmed heat treatments,
or under isothermal conditions. Densification kinetics were studied
using a linear dilatometer with 0.1% accuracy.

The parameters of the self- and hetero-diffusion of atoms in the
materials under investigation (single-crystals and ceramics) were
studied by tracer techniques and electron microprobe analyses.

To obtain information on the nature and mechanisms of the chemical

interactions of a material with additives and atmospheres, we used
complex thermal analyses, micro-calorimetry, gas chromatography, ESCA,
etc.

RESULTS AND DISCUSSION

By varying the synthesis and grinding conditions, powders were
prepared with a particle size ranging from 0.03 to 50 μm. For these
powders as well as for those of a number of model systems of both
metals (Ag, Ta, Pt, etc.) and oxides (TiO_2, Al_2O_3), we investigated the
relationship between the density of compacts, ρ, and the pressure
used, P, as well as the particle size dependence of the relative fill
density, γ, and the density of compacts. It has been found that, in
the specified particle range, $\rho_0 \sim P^{1/4}$ holds true while γ and ρ_0
systematically increase with increasing average particle size, d (Fig.
1, curves 1,2).

Starting from the well-known pressure dependence of density of a
compact of the form

$$\rho_0 = f\ P^{1/n} + \gamma, \tag{1}$$

we obtained[3] a refined expression reflecting (in a hydrostatic
approximation) a decrease in the pressure applied because of
interparticulate friction forces in the powder being pressed,

$$\rho_0 = f(P - K\ \alpha\ P_n\ V/D^2 d)^{1/4} + \gamma, \tag{2}$$

where f is the densification factor, D and V are diameter and volume of
a compact, respectively, α is the interparticulate friction
coefficient, P_n is the normal component of the pressure acting in
particle junctions, and K is a factor taking into account the fraction
of the particle surface area involved in the junction and the average
number of junctions per particle.

Curve 3 of Fig. 1 representing the density of compacts as a
function of the average particle size in $SrTiO_3$ powders processed in a
vibrational activator with a high power per unit weight ($\sim$40 kW/kg)
exhibits a somewhat different character as compared to curves 1 and 3,
which relate to powders processed in an equipment with a lower power
per unit weight, and a higher density of compacts achieved in the
specified powder processing conditions. We have shown that such a
behaviour of compacts is accounted for by hydration of their surface,
during processing, resulting in a decrease in interparticulate friction
and by the formation of aggregates whose size and density depend on
powder processing conditions.

The study of the kinetics of shrinkage, y, in sintering the above
powder materials in the specified wide range of their particle sizes,
with the data obtained being processed on the basis of diffusion
models,[4] has shown that, in the absence of liquid phases whose
appearance in the compounds under investigation may be caused by non-
stoichiometry or special additives, the principal mass transfer
mechanism is the volume self-diffusion of the components of a compound.
Thus, it has been found that the shrinkage at an initial sintering
stage is proportional to the isothermal heat treatment duration to
power 0.5 $\pm$ 0.05 (Fig. 2) while the sintering activation energy for
coarse powders (more than 1 μm particle size) has relatively high
values listed in Table I which also presents, for comparison, our data

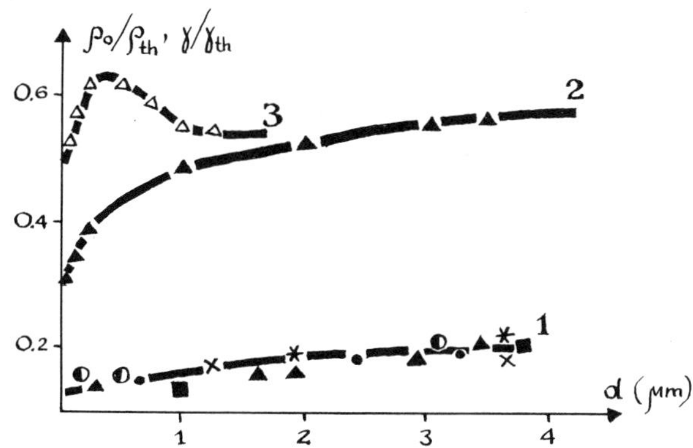

Fig. 1. Apparent fill density (1) and density of compacts
(P=100MPa)(2,3) vs.average size of powder particles
(• Ag *Ta ▪ TiO$_2$ Δ/▲ SrTiO$_3$ x CaTiO$_3$ ○ BaTiO$_3$),▲ conventional
milling Δ vibrational activation.

on the sintering of some other compounds close in structure and
properties to those under study.

Considering an insignificant (less than 10^{-3}) contribution of mass
transfer through grain boundaries as inferred from an analysis of
dependences of the form $y^{2.06} \dot{y} = f(y^{1.03})$, with the straight lines
obtained passing through the origin of coordinates,[4] this suggests that
the volume self-diffusion of components in the compounds under
investigation is the predominant mechanism of the mass transfer in our
case.

It should be noted that our values of the exponent in the time
dependence of shrinkage (0.5 ± 0.05) and of the sintering activation
energy for BaTiO$_3$ (750 ± 100 kJ/mol) and TiO$_2$ (350 ± 40 kJ/mol) and
hence the conclusion that the volume self diffusion of components makes

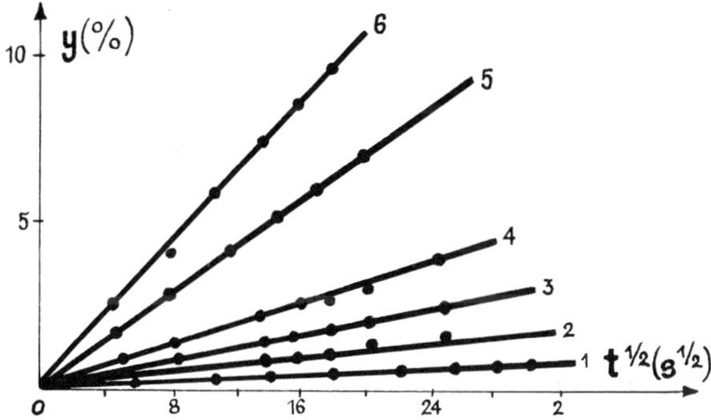

Fig. 2. Sintering kinetics of commercial-grade (1 m^2/g) SrTiO$_3$ at
different sintering temperatures: (1) 1200°C, (2) 1300°C, (3)
1350°C, (4) 1400°C, (5) 1450°C, (6) 1500°C.

Table I. Sintering activation energies of commercial-grade oxide
compounds (more than 1 µm particle size)

Compound	$CaTiO_3$	$SrTiO_3$	$BaTiO_3$	TiO_2	$LiNbO_3$	$NaNbO_3$
E_a (kJ/mol)	610 ± 70	630 ± 80	750 ± 100	350 ± 40	400 ± 50	440 ± 50

a predominant contribution to mass transfer differ from those of
Anderson.[7,8] It is felt that the discrepancies are due to the difference
in temperature range between our study (1200 to 1500°C) and Anderson's
(700 to 1130°C).

Note also that the relatively high E_a's of Table I substantially
exceed the activation energies of the self-diffusion of components
(cations and oxygen) in the compounds under study as known from the
literature[5] and determined in our tracer experiments on single-
crystals. This is accounted for by the high concentration of
dislocations and associated point defects which may be present in
single-crystals.[6] Therefore, in tracer experiments where low
concentrations of diffusing atoms are used, diffusion in the "extrinsic
region" manifests itself so that the activation energy of this process
corresponds to the energy of atom movement to a neighbouring site, E_m,
under the conditions where thermal vacancy generation characterized by
the energy of formation of such defects, E_f, does not restrict the mass
transfer observed. In these experiments, for $SrTiO_3$ and $BaTiO_3$ single-
crystals, cation self-diffusion activation energies were found to be in
the range 240 to 340 kJ/mol. At the same time, at higher concentrations
(above 1%) of a diffusing element that imitates self-diffusion,
e.g.,Zr, in alkaline-earth titanate single-crystals, electron
microprobe analysis yielded, in the sintering temperature range 1400 to
1850 K, much higher E_a's (600 to 700 kJ/mol) that correspond, in our
opinion, to diffusion in the "intrinsic region" where $E_a = E_m + E_f$,
which is in line with the temperature dependence of the densification
process in the materials investigated with particle sizes upwards of
1 µm.

Experiments on the sintering of powders with different particle
sizes have shown that the effective sintering activation energy depends
on the particle size of the powder used. Fig. 3 illustrates a typical
case of such a dependence for the materials investigated. This can be
accounted for by the fact that, for regions confined by a concave
surface of intergranular "necks" the relationship

$$\xi = A \exp[-(E_f - 2\ \gamma\ \Omega/R)/K\ T] \tag{3}$$

holds true, since the vacancy concentration beneath a curved surface
with curvature radius R is given by

$$\xi = \xi_0 \exp(\pm 2\ \gamma\ \Omega/R\ K\ T) = A \exp[-(E_f \pm 2\ \gamma\ \Omega/R)/K\ T],$$

$\xi_0 =$ A $\exp(-E_f/R\ T)$ being the vacancy concentration beneath a plane
surface.

A calculation performed for the case $2\gamma \simeq 5$ J/m^2, $\Omega \simeq 5\times10^{-5}$
m^3/mol, $E_f \simeq 300$ kJ/mol shows that the exponent in Eq.(3) becomes zero
for $R \simeq 10^{-9}$ m. This means that decreasing size of particles and hence
of pores decreases the effective energy of vacancy formation while for

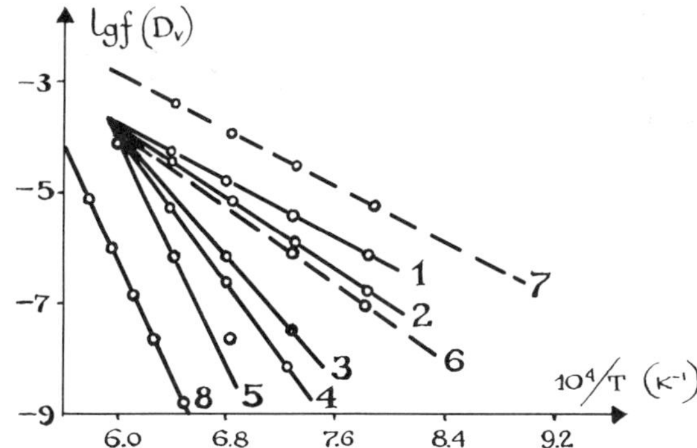

Fig. 3. Temperature dependence of diffusion rate in sintering SrTiO$_3$
with different equivalent diameters of powder particles: (1)
0.05, (2) 0.17, (3) 0.45, (4) 0.53, (5) 0.78, (6) 0.25, (7)
0.04, (8) 1.5 µm. (1) to (4) and (8) are obtained by
decomposition of strontium titanyl oxalate, (5) to (7) by
vibrational grinding.

very small particles (10^{-7} to 10^{-8} m) where particle junctions feature
a curvature on the order of 10^{-8} to 10^{-9} m, the mass transfer process
may not be subjected to restrictions as regards thermal generation of
vacancies, the temperature dependence of the sintering process being
only determined by the energy of their movement, E_m. In the general
case, substitution of the above values of γ, Ω, E_f and R into the
expression

$$E = E_m + (E_f - 2\,\gamma\,\Omega/R) \tag{4}$$

yields $E \simeq E_m$ for $d \leqslant 0.05$ µm and $E \simeq E_m + E_f$ for $d > 1$ µm, which is
confirmed experimentally (Fig. 3). It should be admitted that this
approach to the interpretation of the dependence observed is not
unique. An elevated concentration of defects in ultrafine powders may
be due to the injection of vacancies by a migrating interphase boundary
determining the recrystallization mechanism of the sintering of such
powders.[9] Of importance to the sintering may be the phenomenon of
superplasticity in ultrafine systems.[10] It is essential that a high
concentration of point defects in fine powder systems is observed
experimentally[7] and directly reflected in the character of the
temperature dependence of the sintering process.

With regards to Eq. (4), the expression for the "degree of
activity" of powders determined[11] as

$$g = 1 - \exp(-\Delta E/K\,T), \tag{5}$$

where $\Delta E = E - E'$ is the difference in sintering activation energy
between the starting (E) and activated (E') materials, in the case
where the activity is determine by the grain size, assumes the form

$$g \simeq 1 - \exp(-40\,\gamma\,\Omega/d\,K\,T) \tag{6}$$

where d is the average diameter of powder particles which is related to the radii of curvature of concave surface regions, R, by $d \simeq 20\,R$, as shown by our mercury porosimetric measurements. In this case, in contrast to the findings of other researchers,[11] the sintering may be considered as a process of diffusion transfer in kinetically homogeneous media (E_m = const) and, as a measure of the activation of a material, the quantity g may be taken that is determined by the concentration of excess vacancies associated with the grain size of the material. Clearly, this conclusion is only valid in the absence of other defect sources capable of promoting the sintering. And indeed, we have found that microstrains, dislocations and still less heat-stable nonequilibrium point defects stored during calcination and mechanical treatment are eliminated at an initial stage without substantial contribution to the densification of samples.

Partial aggregation of a product is characteristic of powder materials prepared using vibrational or planetary-centrifugal activators with a high power per unit weight. This results in a two-stage densification in sintering (Fig. 4). At a first stage, sintering takes place inside aggregates, the shrinkage value being independent of the relative content of such aggregates in the powder. At a second stage, sintering takes place amongst aggregates and in separate (initial) grains. This feature of the densification process is important from the stand-point of ceramic structure formation and ceramic properties. In particular, for $BaTiO_3$-based ferroelectric ceramics, this densification character combined with small grain sizes (~ 1 μm) provides for considerable increase in the dielectric constant and dielectric strenght of the ceramics.

We have also studied the effect of additives (ZrO_2, MgO, SrF_2, etc.), non-stoichiometry and atmosphere on the sintering of alkaline-earth titanates.

Taking into account the low solubility of the additives in question as well as the data available on the formation, at their presence, of new phases at grain surfaces and boundaries, the mechanism of the effect of these additives on the solid-state sintering consists in diffusion acceleration (with excess TiO_2 or ZrO_2 additive) or deceleration (with excess alkaline-earth oxides or MgO additive), with the presence of such phases, as compared to a homogeneous material. Similarly, for the materials investigated, sintering in a reducing atmosphere (or when using pre-reduction) is not accompanied by an increase in activity in cases where the homogeneity is not disturbed

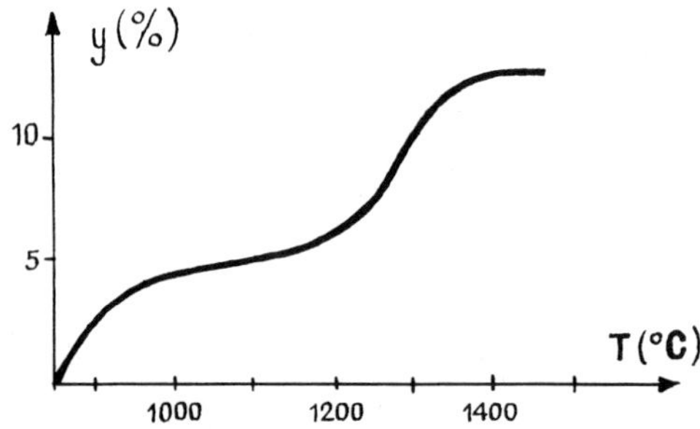

Fig. 4. Densification kinetics at 0.3 K/s for $BaTiO_3$ obtained using a planetary-centrifugal mill.

and no phases are formed with an increased diffusion mobility of
components restricting the sintering rate.

CONCLUSION

In the absence of liquid phases, the principal mechanism of the
mass transfer in sintering alkaline-earth titanates, for a wide range
of particle sizes, is the volume self-diffusion of components, the rate
and the temperature dependence of the self-diffusion being determined
primarily by the particle sizes of the powder used. Additives, non-
stoichiometry and sintering atmosphere affect the sintering process as
new phases are formed with an increased (or decreased) diffusion
mobility of components whose self-diffusion rate restricts the mass
transfer observed in the sintering process.

ACKNOWLEDGEMENTS

Thanks are due to Professors V. V. Boldyrev, P. Yu. Butiagin, P.
S. Kislyi, V. V. Skorohod, and L. I. Trusov for their interest and
assistance in this study.

REFERENCES

1. J. M. Herbert, "Ceramic Dielectrics and Capacitors", Gordon and
   Breach, New York (1985).
2. "Poluprovodniki na Osnove Titanata Bariia" (Semiconducting Barium
   Titanates), Energoizdat, Moscow (1982), translated from the
   Japanese, Gakkensha, Tokyo (1977).
3. V. I. Lapshin, Powder particle size dependence of the density of
   compacts (in Russian), Poroshk. Metallurg.1:23 (1982).
4. D. L. Johnson, Sintering kinetics for combined volume, grain
   boundary and surface diffusion, Phys. Sinter.1:1 (1969).
5. R. Freer, Bibliography: self-diffusion and impurities in oxides,
   J. Mater. Sci.5:803 (1980).
6. W. H. Rhodes and W. D. Kingery, Dislocation dependence of cationic
   diffusion in $SrTiO_3$, J. Am. Cer. Soc.49:521 (1966).
7. H. U. Anderson, Initial sintering of $BaTiO_3$ compacts, J. Am. Cer.
   Soc.48:118 (1965).
8. H. U. Anderson, Initial sintering of rutile, J. Am. Cer.
   Soc.50:235(1967).
9. V. I. Novikov, L. I. Trusov, V. V. Lapovok, and T. V. Geleishvili,
   On the mechanism of low-temperature diffusion activated by a
   migrating boundary (in Russian), Fiz. Tverd. Tela 25:369(1983).
10. Ya. E. Geguzin, Initial stage of active sintering: superplasticity
   of a porous structure (in Russian), Dokl. AN SSSR 229:601
   (1976).
11. P. S. Kislyi and M. A. Kuzenkova, "Spekanie tugoplavkikh
   soedinenii" (Sintering of Refractory Compounds), Naukova Dumka,
   Kiev (1980).

THE INFLUENCE OF THERMAL TREATMENT ON POLARIZATION

BEHAVIOUR OF $Bi_4Ti_3O_{12}$ CERAMICS

Č. Jovalekić, Lj. Atanasoska*, V. Petrović and
M.M. Ristić

Center for Multidisciplinary Study of Belgrade
University, Belgrade, Yugoslavia
*Institute of Technical Sciences of Serbian Academy of
Sciences and Arts, Belgrade, Yugoslavia

ABSTRACT

The influence of thermal treatment (750-950°C) on microstructure
and electric properties of ferroelectric ceramics $Bi_4Ti_3O_{12}$ has been
studied. The shape of monocrystalline grains undergoes changes upon
thermal treatment as observed by optical microscopy. We have found that
the modified dielectric properties of investigated samples are related
to the grain shape transformation. The high value of dielectric
permittivity and the appearence of hysteresis have been correlated to
the presence of oxygen vacancies within the perovskite structure of
$Bi_4Ti_3O_{12}$. The oxygen vacancies are preferentially sited in the
vicinity of bismuth ions as evidenced by x-ray photoemission data.
Variations in the valency state of titanium ions are also possible. The
XPS and AES measurements confirm that the surface elemental composition
of $Bi_4Ti_3O_{12}$ ceramics does not deviate from the nominal bulk
composition.

INTRODUCTION

As a member of a large family of complex mixed Bi_2O_3-MeO
ferroelectrics, bismuth titanate ($Bi_4Ti_3O_{12}$) is of particular interest
due to its unique electrical-optical behaviour. The $Bi_4Ti_3O_{12}$ exists as
a layer-type compound. Its structure comprises the $Bi_4Ti_3O_{12}$ layer,
formed by two $BiTiO_3$ unit cells of hypothetical perovskite structure,
alternating with the single $Bi_2O_2^{+2}$ layer.[1]

At room temperature $Bi_4Ti_3O_{12}$ belongs to crystallographic group of
C_{1h} = m (monoclinic) symmetry and is characterized by ferroelectric
properties. The vector of spontaneous polarization lies in monoclinic
plane <u>ac</u> at an angle of 4°-5° with respect to the main crystal surface.
Along axis <u>a</u> the vector intensity of spontaneous polarization is P_a =
50 μC/cm, while along axis <u>c</u>, P_c = 4μC/cm.[2,3]

At a temperature of 675°C (the Curie temperature), $Bi_4Ti_3O_{12}$ shows
a reversible $\alpha \rightleftarrows \beta$ transfer. Above the Curie temperature, the symmetry is
D_{4h} = 4mmm.[4]

The existance of spontaneous polarization leads to formation of hysteresis loops in alternating fields of high intensity.[5] Apart from the presence of hysteresis loop, anomalous behaviour of dielectric permittivity and specific conductivity below the Curie point[6] was noticed, as well. These anomalous phenomena are supposed to result from both volume and surface charges. Generally speaking, high values of dielectric permittivity (ε) are closely connected with volume charge relaxation and existance of defects that condition the presence of either electron or ionic polarization. The dependence   in $Bi_4Ti_3O_{12}$ results from formation of oxygen defects in the form of anionic vacancies.[7]

The above mentioned properties indicate the possibility of application of $Bi_4Ti_3O_{12}$ as piezo-materials and electrets, and a high phase transformation point allows its application within a wide temperature range what proves to be of particular importance.

EXPERIMENTAL

In our experimental work we used Bi_2O_3 (Bismuth Institute) and TiO_2 powders both at a purity of 99.8% (GmbH Ventron) as initial powders. Mixture of these two powders, being in stoichiometric ratio, was homogenized in agate ball mill for 1440 min, and then cold pressed under a pressure of 50 MPa. Such obtained samples were afterwards sintered at temperature of $1100^{\circ}C$ in air atmosphere held for 240 min. After cooling the samples were re-ground in agate mill and re-compacted. The twice ground powder of sintered $Bi_4Ti_3O_{12}$ was again sintered at $1100^{\circ}C$ in 240 min period. The samples obtained, with a diameter of 9 mm and thickness of 1 mm, were heat treated within the temperature range from 750° to $950^{\circ}C$. Electrodes were put on polished surfaces of samples using method of screen printing, and polymerization of the silver paste was carried out at $600^{\circ}C$ for 30 min.

Microstructural investigations of $Bi_4Ti_3O_{12}$ samples were carried out on chemically etched surfaces by light microscopy ("Reithert").

LCZ automatic bridge (Hewlett-Packard 4276 A) was used to determine, the dielectric permittivity, which was measured at frequency of 1 kHz with a measuring voltage of V_{rms} = 1V.

Polarization, both spontaneous and residual, as well as the coercive field, were determined from hysteresis loops formed by modified Sawyer-Tower circuit.

To determine the relative surface composition we used X-ray photoemission spectra obtained by Perkin-Elmer PHI spectrometer.

RESULTS AND DISCUSSION

a) <u>Microstructure</u>

Microstructural analysis (Fig. 1) shows significant similarities, in the appearance of chemically etched sample surfaces. Chaotic orientations of monocrystal grains that are separated by paraelectric intergranular layers can be easily noticed. During sintering, the growth of grains along the <u>c</u> axis occurred such that the grains become elongated and show prism-like forms. The width of prism-like grains ranges from 5 to 10 µm and their lengths vary from 20 to 50 µm. Depending on the temperature of heat treatment, the length/width ratio

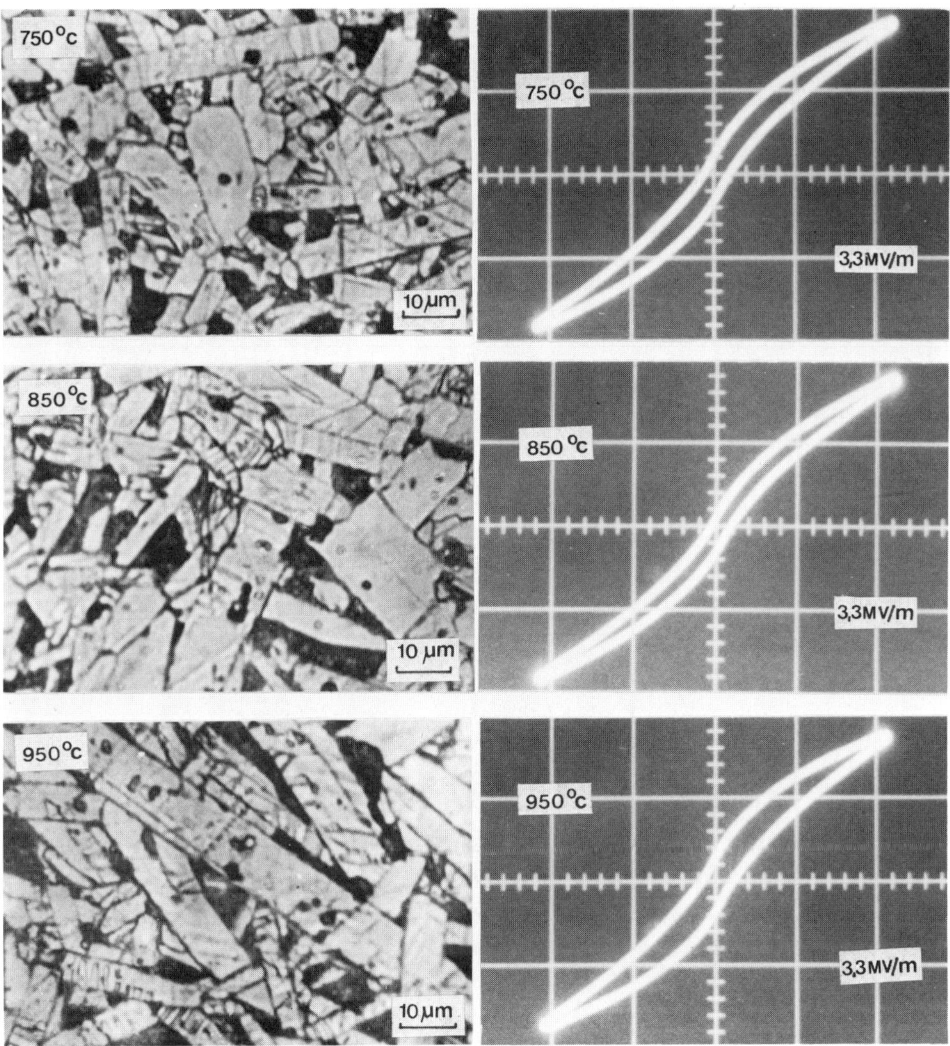

Fig. 1. Optical micrographs of
chemicaly etched surfaces
of $Bi_4Ti_3O_{12}$ ceramic samples.

Fig. 2. Hysteresis loops of ceramic
$Bi_4Ti_3O_{12}$ samples at room
temperature.

of grains changes; it is lowest in the case of heat treatment at a
temperature of 850 °C.

As the dielectric properties of ceramics are functions of the
electrical properties of monocrystal grains along a given axis, thus it
is logical to expect that change in relative dielectric permittivity
will correspond to changes in grain lenght/width ratio (Fig. 3). Thus,
dielectric permittivity is also minimal at a heat treatment temperature
of 850 °C.

b) <u>Dielectric properties</u>

Under the influence of an imposed electrical field, quasi-dipoles
are directed towards the applied field and domains in monocrystal
grains are also reoriented in this direction. On this basis we can
evaluate the dynamic polarizability of $Bi_4Ti_3O_{12}$ ceramics, and estimate

the value of, coercive field E_c, spontaneous polarization P_s and residual polarization, P_r from the obtained hysteresis loops.

At temperatures below the Curie point, ferroelectric grains are divided (grouped) into domains in which the vector of spontaneous polarization is directed towards either the $\underline{a}$ or the $\underline{c}$ axis. The total vector of spontaneous polarization, P_s of the entire sample must be equal to zero. Spontaneous polarization was not detected inside the intergranular layer. When electrical fields exceeding the coercive one are applied, a symmetric hysteresis loop is formed (Fig. 2). In our experimental procedure, a symmetric hysteresis loop was formed at room temperature within the field $E_p = 3,3$ MV/m.

The value P_s could have been expected to exist between the values characterizing spontaneous polarization of P_a and P_c monocrystals, measured in the direction of the $\underline{a}$ and $\underline{c}$ axes, respectively. But, far lower values were actually observed (Table I). These small values resulted from the existence of a paraelectrical intergranular layer, accounting for the lower residual polarization and coercive field.

Table I. Values of both spontaneous and residual polarization and of coercive field estimated from a hysteresis loop

T	(°C)	750	800	850	900	950
P_s	(nC/cm^2)	9.3	11.7	10.1	12.2	12.5
P_r	(nC/cm^2)	28.7	28	26.2	26.5	32.8
E_c	(V/m)	158	198	144	208	189

Dielectric permittivities of $Bi_4Ti_3O_{12}$ ceramics were measured at room temperature. Depending on the temperature of heat treatment (Fig. 3), the values within a temperature interval ranged from 117 to 135. High values of dielectric permittivity were associated with the perovskite structure of $Bi_4Ti_3O_{12}$ compound, and with oxygen defects formed within it during sintering. These defects take part in quasi-dipole formation. Either an increment or a decrement in oxygen defects – anionic vacancies–causes a corresponding increase or decrease of dielectric permittivity. The same conclusion was established by other authors, as well.[6,9]

(c) <u>Auger data</u>

The Auger spectrum recorded for $Bi_4Ti_3O_{12}$ ceramics is shown in Fig. 4. All peaks characteristic for bismuth, titanium and oxygen are clearly visible in the spectrum. The most intense Bi Auger signal, for $N_5N_{6,7}N_{6,7}$ transition, appears at 100 eV. The weaker Bi peaks at 249 eV and 268 eV, for $N_5O_3O_3$ and $N_4O_3O_3$ Auger transitions, respectively, are also easily detected. The intensity ratio of bismuth peaks at 268 eV and 249 eV deviate slightly from the value given in the standard spectra for clean bismuth.[10] The Bi 268 eV peak in Fig. 4 is larger than the Bi 249 eV peak, in contrast to a known relative Auger yield of these two bismuth peaks. This indicates the presence of a trace amount of carbon impurity, because a carbon KLL transition at 270 eV partially overlaps with $Bi(N_4O_3O_3)$ at 268 eV. The prominent Ti signals, for $L_2M_{2,3}M_{2,3}$ and $L_2M_{2,3}M_{4,5}$ Auger transitions, are located at 387 eV and 418 eV, respectively. The amplitude of the oxygen peak at 512 eV, for $KL_{2,3}$

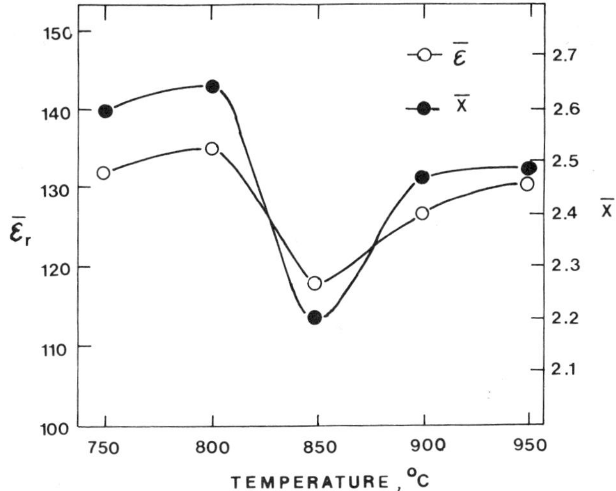

Fig. 3. Change in dielectric permittivity and grain width/length ratio as a function of heat treatment temperature.

$L_{2.3}$ transition,exhibits a considerable intensity as expected for this specimen of an oxide. During the Auger measurements all peaks were shifted by 50 eV due to charging of the dielectric specimen, which required a recalibration of the electron energy scale.

Under our experimental conditions the Auger spectra were not reliable for a quantitive composition evaluation. The stoichiometry information derived from the Auger spectra is based on the measurements of the peak amplitudes corrected with the appropriate sensitivity factors. The Auger sensitivity factors were compiled for the energy of the primary electron beam of 3 KeV.[10] In order to avoid charging problems, we had to reduce the energy of the primary beam from 3 to 2 KeV as well as the adsorbed beam current. These modifications of the primary electron beam characteristics during our AES experiments prevented use of established Auger sensitivity factors to arrive at an elemental composition estimate.

(d) <u>XPS data</u>

The XPS survey spectra of the "as received" and the ion etched $Bi_4Ti_3O_{12}$ ceramics are shown in Fig. 5a and 5b respectively. These survey spectra demonstrate a decrease in carbon contamination after ion sputtering.

Core level photoemission spectra were also collected in order to obtain valence state information for the bismuth titanate compound. All peak positions in the high resolution spectra (x-ray beam was not monochromatic) were equally shifted due to the already-mentioned charging of dielectric specimens. In our XPS spectra (Fig. 7b), the C 1s photoelectron peak was located at 287.4 eV instead at 284.6 eV, the binding energy position common for the carbon impurity peak. Therefore the binding energy scale was calibrated in respect to the C 1s photoemission, taking into account the shift of 2.8 eV determined as a consequence of charging.

The photoemissions obtained for Bi 4f, Bi 5d, Ti 2p and O 1s core levels shown in Figs. 6a-6c are sensitive to the oxidation states of the elements constituting the bismuth titanate compound. The 7/2 and

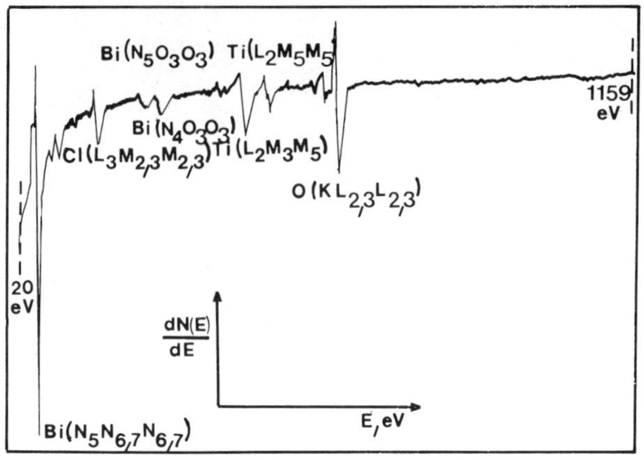

Fig. 4. The Auger electron spectrum for a $Bi_4Ti_3O_{12}$ sample.

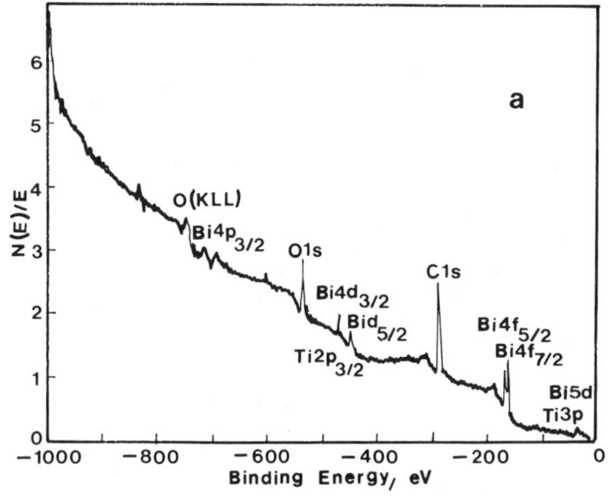

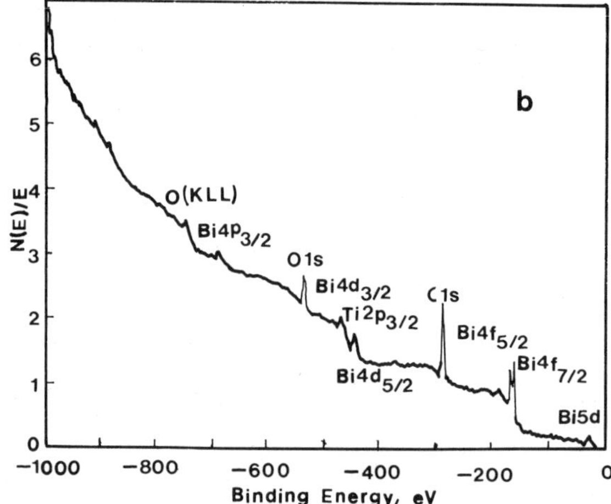

Fig. 5. a) XP spectrum before etching.
b) XP spectrum after etching.

550

5/2 spin-orbit doublet components of the Bi 4f core level photoemission
are located at 158.4 eV and 163.8 eV. The spin-orbit splitting of Bi 4f
core levels in bismuth titanate compound is 5.4 eV. The ion etched
samples exibit a shoulder at low binding energy side of Bi 4f and Bi 5d
photoemission (Fig. 6a and 6b). The shoulder is attributed to the
contribution of bismuth reduced to metallic state due to preferential
sputter-induced oxygen dissosiation. Binding energy locations of the
metallic bismuth $4f_{7/2}$ and $4f_{5/2}$ core levels, determined by line-shape
deconvolution, are at 156.2 eV and 161.6 eV, respectively. The chemical
shift of Bi 4f core levels for $Bi_4Ti_3O_{12}$ ceramics with respect to
unreacted elemental bismuth is 2.2 eV, which is less than for the Bi_2O_3
compound (3.1 eV).[11,12] The observed shift suggests the existence of
bismuth in a valence state of (+3-x). The Bi (+3-x) formal oxidation
state can be generated due to enhanced concentration of oxygen
vacancies in the vicinity of bismuth cations either in the perovskite
lattice structure or in the Bi_2O_2 layer. The case of oxygen vacancies
neighboring bismuth ions is more likely to occur inside the Bi_2O_2 layer
than in the perovskite structure. In the perovskite portion of the
$Bi_4Ti_3O_{12}$ structure, the Bi ion is placed in the center of the unit
cell built by Ti cations. The oxygen anions form octahedra enclosing
titanium ions. In the hypothetical perovskite unit cell, Bi and Ti
cations share oxygen atoms. Strong Ti - O bonds in the perovskite
lattice do not break easily, implying a non-random distribution of
oxygen vacancies, and their confinement inside the Bi_2O_2 layer.

The $Ti2p_{3/2}$ photoemission is partially overlapped by the $Bi4d_{3/2}$
core level peak (Fig. 6b). However the $Ti2p_{3/2}$ photoelectron peak
appears in the XPS spectrum as a distinctively resolved feature. Based
on its binding energy position (459 eV) it can be unambiguously
concluded that the titanium atoms are present in 4+ valence state. This
finding supports our assumption about oxygen vacancies residing inside
the Bi_2O_2 layer. The outstanding affinity of Ti towards oxygen could
further explain the proposed $Bi_4Ti_3O_{12}$ defect structure.

The line-shape of the O 1s core level photoemission reveals two
peaks. The first one at low binding energy side corresponds to the
stronger Ti-O bond while the second peak is ascribed to oxygen attached
to bismuth.

The stoichiometry of the bismuth titanate compound has been
estimated from the XP spectra. The relative surface was calculated from
the measured area of core level peaks corrected for Scofield photoionization cross sections. The atomic concentration ratio of Bi:Ti:O
was determined to be 4:3:14. The Bi:Ti atomic ratio is in good
agreement with the ideal stoichiometry of the bismuth titanate
compound. However, the enhancement in oxygen content is rather
surprising, especially when one takes into consideration the observed
reduction of bismuth to a suboxide state. The increase in oxygen
content by ∼ 20% compared to the stoichiometric value can be explained
by an error in applying Scofield photoionization cross section to
different oxides matrix.

CONCLUSION

Depending on the temperature of heat treatment the microstructure of
$Bi_4Ti_3O_{12}$ changes as do the dielectric and ferroelectric properties. These
changes are correlated with the change in microstructure. On the basis of
x-ray photoemission spectra it was established that oxygen vacancies caused
by such heat treatments gather together in the vicinity of bismuth ions. An
increment in the concentration of vacancies results in a corresponding increment in dielectric permitivity.

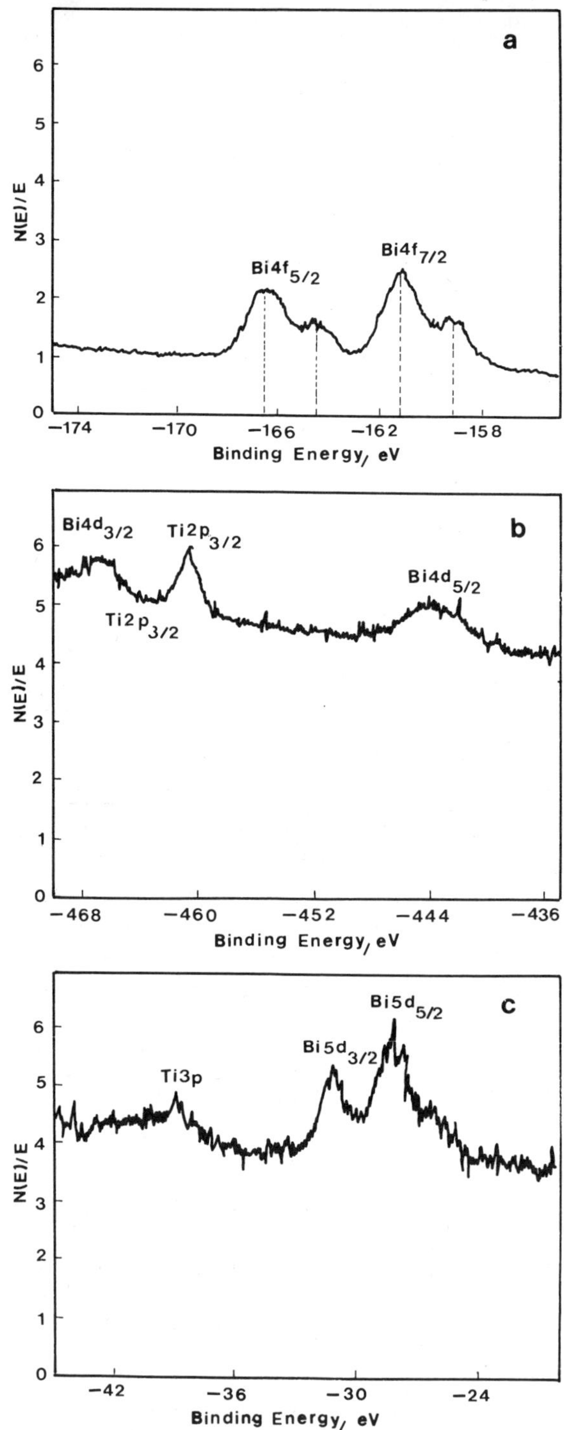

Fig. 6. a) Bi 4f, b) Bi 4d, Ti 2p and c) Ti 3p, Bi 5d region of XP spectrum of $Bi_4Ti_3O_{12}$.

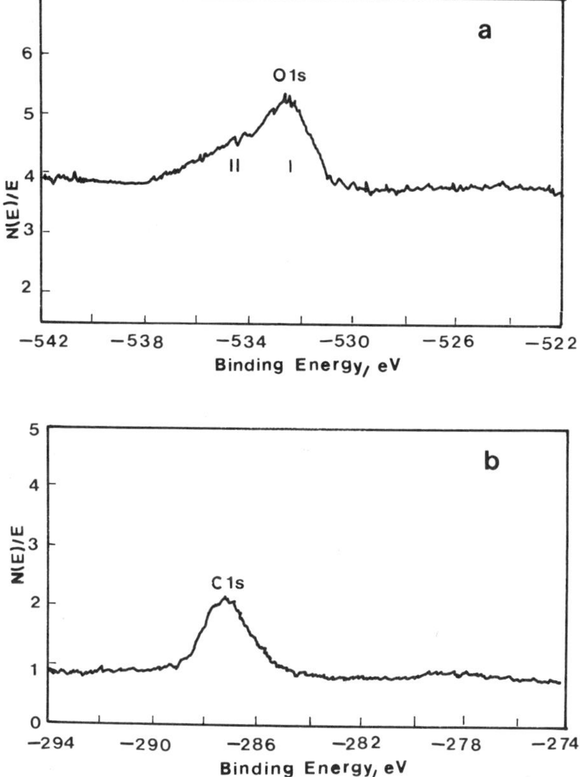

Fig. 7. a) O 1s; and b) C 1s field of XP spectrum of $Bi_4Ti_3O_{12}$.

REFERENCES

1. B. Aurivillius, Mixed Bismuth Oxides with Layer Lattices, Arkiv för kemi 1 (1949) 499.

2. J. F. Dorrian, R. E. Newnham, D. K. Smith, Crystal Structure of $Bi_4Ti_3O_{12}$, Ferroelectrics 3 (1971) 17.

3. G. W. Taylor, S. A. Keneman, A. Miller, Depoling of Single-Domain Bismuth Titanate, Ferroelectrics 2 (1971) 11.

4. E. I. Speranskaya, I. S. Rez, L. V. Kozlova, V. M. Skorikov, V. I. Slovov, Sistema okis vismuta dvuokis titana, Neorganicheskie materiali 1 (1965) 232.

5. L. G. van Uitert, L. Egerton, Bismuth Titanate. A Ferroelectrics, Journal of Applied Physics 32 (1961) 959.

6. A. Fouskova, L. E. Cross, Dielectric Properties of Bismuth Titanate, Journal of Applied Physics 41 (1970) 2834.

7. E. V. Sinjakov, E. F. Dudnik, V. M. Duda, V. A. Podolski, M. A. Gorfunkel, Relaksaciya dielektricheskoi pronicaemosti vismuta, Fizika tverdogo tela 16 (1974) 1515.

8. C. B. Sawyer, C. H. Tower, Rochelle Salt as a Dielectric, Physics Review 35 (1930) 269.

9. S. E. Cummins, L. E. Cross, Electrical and Optical Properties of Ferroelectric $Bi_4Ti_3O_{12}$ Single Crystal, Journal of Applied Physics 39 (1968) 2268.

10. L. E. Davis, N. C. MacDonald, P. W. Palmberg, G. E. Riach, R. E. Weber in: "Handbook of Auger Spectroscopy", PHI, Eden Prairy NM (1976).

11. V. S. Dharmadhikari, S. R. Sainkar, S. Badrinarayan, A. Goswami,
     Characterisation of Thin Films of Bismuth Oxide by X-ray
     Photoelectron Spectroscopy, Journal of Electron Spectroscopy
     and Related Phenomena 25 (1982) 181.

12. V. S. Dharmadhikari, A. Goswami, Efects of Bi_2O_3 Dissociation on
     the Electrical Properties of Thermally Evaporated Films of
     Bismuth Oxide, Journal of Vacuum Sciences and Technology Al
     (1983) 383.

13. C. D. Wagner, W. M. Riggs, L. E. Davis, J. F. Moulder in:
     "Handbook of X-ray Photoelectrons Spectroscopy",
     G.E.Mullinberg, ed: Perkin Elmer Physical Electronic Division,
     Eden Prairy NM (1978).

CHANGES OF ELECTRICAL AND STRUCTURAL CHARACTERISTICS OF COLD SINTERED

POTASSIUM DIHYDROGEN ARSENATE WITH TEMPERATURE

D. Minić, R. Dimitrijević* and M. Šušić

Institute of Physical Chemistry, Faculty of Chemistry
Belgrade University, Belgrade, Yugoslavia
*Faculty of Mining and Geology, Department of Mineralogy
and Crystallography, Belgrade University, Yugoslavia

ABSTRACT

The electrical conductivity of the cold sintered salt KH_2AsO_4 has
been investigated within temperature range from room temperature to
about $500^\circ C$ during multiple heating and cooling cycles. The
reversibility of the dehydration process $KH_2AsO_4 \rightleftharpoons \beta\text{-}KAsO_3$ has been
established. The electrical characteristics were correlated with
thermal and structural transformations. The unit cell dimensions of the
$\beta\text{-}KAsO_3$ phase were determined: $a_0 = 14.092(5)$ Å, $b_0 = 13.099(4)$ Å,
$c_0 = 8.891(2)$ Å, $\beta_0 = 96.37(2)^\circ$ and $V_0 = 1631(1)$ Å^3.

INTRODUCTION

Many solid electrolytes, acid salts and crystallohydrates are
known for their conductivity, which is characteristically based on the
migration of protons.[1-4] The conductivity of these substances at room
temperature ranges within wide limits, from about 10^{-2}S/m for $LiH_2PO_4^2$
up to below 10^{-7}S/m for $MgHPO_4 \cdot 3H_2O^4$ and is closely connected with the
crystal structure of the respective electrolyte. In these structures
hydrogen forms a series of hydrogen bonds along which the protons are
moving by a complex mechanism which also involved a tunnelling step.
Besides the latter, the complex mechanism of transfer includes also the
step in which the charge carrier is prepared as well as the preparation
step of the acceptor group to capture an approaching proton.[5] The
latter two steps require a definite amount of energy and therefore they
determine the total activation energy of the conductivity. In some
crystal structures from the class of crystallohydrates, the
aforementioned steps do not require a significant energy consumption,
and in that case, these systems acquire superconductive characteristics.[1-3]
However, upon heating even the most promising of the hitherto known
systems lose the superconductive characteristic irreversibly. Heating
involves the processes of dehydration and structural transformations,
where the dehydration step most often takes place between 50° and
$200^\circ C$. Therefore, there is still a need to discover a system which
would retain good conductive properties within a wider temperature

Science of Sintering
Edited by D. P. Uskoković *et al.*
Plenum Press, New York

range. From this standpoint, of particular interest are those systems
in which the phase transformations are reversible within the
aforementioned temperature range.

To attain this aim, we studied electrical, thermal and X-ray
crystallographic characteristics of the solid acid salt KH_2AsO_4, within
the range from room temperature to about $500^{\circ}C$, during multiple heating
and cooling cycles. The conductivity of this salt, measured in the
range from room temperature to about $100^{\circ}C$, having an activation
energy of about 92 kJ/mol, classifies this salt within the group of
good solid conductors.[6]

EXPERIMENTAL

The polycrystalline powder of the acid salt KH_2AsO_4 (p.a. Carlo
Erba) was cold sintered under the pressure of 5.1×10^8 Pa into pellets
of 8 mm diameter and 1-3 mm thickness, the density of which was
2.85 g/cm^3.

The electrical conductivity of the sintered sample was measured
during multiple heating within the temperature range from room
temperature to about $500^{\circ}C$ by an alternating current of 1 kHz
frequency, as described previously.[2]

The thermochemical investigations were carried out with a Du Pont
Thermal Analyzer at a heating rate of 20°/min, if not otherwise stated,
using DSC and high-temperature DTA cell ($1200^{\circ}C$). The peak temperature
and the peak area were determined by using INTERACTIVE DSC V2.0
program. The TG curves were recorded on a TG-951 Analyzer in nitrogen
atmosphere.

The X-ray powder diffractograms were obtained using a Philips
PW-1051 diffractometer with $Cu_{k\alpha}$ radiation and graphite monochromator,
at room temperature. Fully automatic programs[7,8] for finding the
crystal symmetry, and a program for refinement[9] of unit cell dimensions
from powder data, were used.

RESULTS AND DISCUSSION

The electrical conductivity measured within the temperature range
from room temperature to $500^{\circ}C$ during two heating cycles, is shown in
Fig. 1. During the first heating, the logarithmic dependence of the
conductivity as a function of reciprocal temperature in the
investigated temperature range, showed two linear regions, namely from
20° to $220^{\circ}C$ and from 360° to $500^{\circ}C$, with a sudden increase and
decrease of the conductivity in the region from 220° to $360^{\circ}C$ and with
a maximum at $305^{\circ}C$. The activation energies of the conductivity in the
regions of these two linear dependences, $\log \chi = f(1/T)$, determined from
the slopes, were found to be 91.9 kJ/mol and 160.8 kJ/mol,
respectively, which is in good agreement with data reported in the
literature for the region of the first linear dependence.[6] The
discontinuity in the dependence shown, line a in Fig. 1, points to
structural changes of the sample within the investigated temperature
range. These changes are also seen on the DTA diagram, Fig. 2, and are
confirmed and followed by recording X-ray powder diffractograms.

The same temperature dependence of the logarithm of the
conductivity on the reciprocal temperature was obtained during the

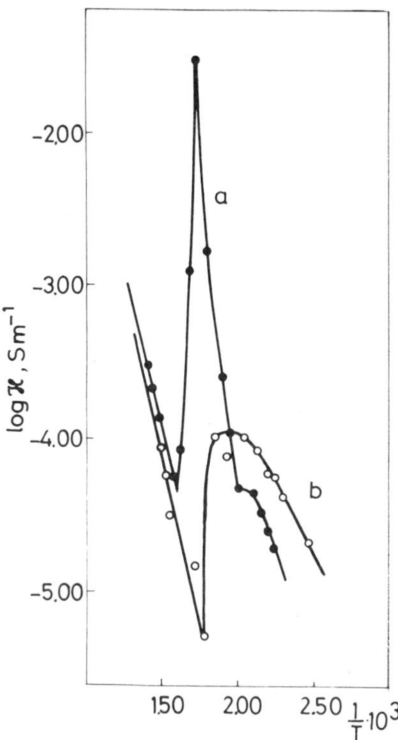

Fig. 1. Change of conductivity with the temperature of cold sintered KH_2AsO_4 during first (a) and second (b) cycle heating

repeated measurement of the conductivity with the same sample, line b in Fig. 1. Such a behaviour indicates the possibility of some reversibility of structural transformations occurring within the investigated temperature range.

The DSC diagram for two successive heatings, curves a and c, and for cooling, curve b, is given in Fig. 3. During the first heating, the DSC diagram, curve a in Fig. 3, in agreement with DTA, Fig. 2, shows several endothermic peaks with maxima at 250°, 280°, 360° and $420^\circ C$, respectively, whereas during the repeated heating only one peak at $280^\circ C$, curve c in Fig. 3, was obtained. According to the TG curve, Fig. 4, recorded in the range from room temperature to about $600^\circ C$, in a temperature range where there appeared poorly separated peaks I and II on the DSC curve, namely between 235.5 and $302.3^\circ C$, the sample lost 9.84% of its original mass. The latter corresponds to loss of one mole of water due to the dehydration of the sample and the formation of polyarsenate $KAsO_3$.

From peak areas on DSC curves the enthalpies of the corresponding endothermic processes were determined. Thus, the total enthalpy of the stepwise dehydration, peak areas I and II was determined to amount to 389 J/g, Fig. 5, curve a. The enthalpies of the corresponding endothermic transformations for peak areas III and IV were found to be 2.15 J/g and 13.8 J/g, respectively, Fig. 5b. In agreement with data

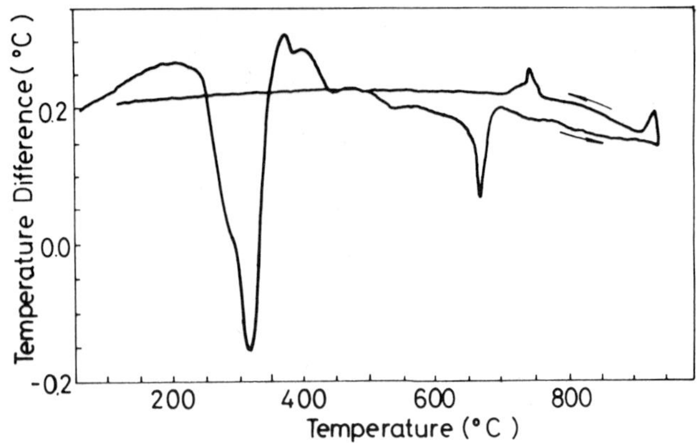

Fig. 2 DTA diagram for cycle heating-cooling of solid KH_2AsO_4; heating
    rate 30 °C/min.

reported so far,[10] the appearance of the peaks III and IV correspond to
gamma $\rightleftarrows$ beta and beta $\rightleftarrows$ alpha structural transformations,
respectively; the aforementioned process can be described by the
following thermochemical reactions:

$$KH_2AsO_4 \xrightarrow[\Delta H=389J/g]{230-390\,^{\circ}C} \gamma KAsO_3 \xrightarrow[\Delta H=2.15J/g]{350\,^{\circ}C} \beta KAsO3 \xrightarrow[\Delta H=13.8J/g]{420\,^{\circ}C} \alpha KAsO_3$$

In a direct repeated measurement the DSC curve, Fig. 6, shows a
repeated dehydration of the part of water which the sample had regained
during standing in the open atmosphere in the cell, between two
successive measurements, peak I, Fig. 6; the enthalpy of 1.49 J/g was
found to correspond to this process. However, in the course of a longer
standing in an open atmosphere the sample regained a larger part of
water lost by the dehydration process as might be seen from the DSC
diagram recorded after standing of the heated sample in an open

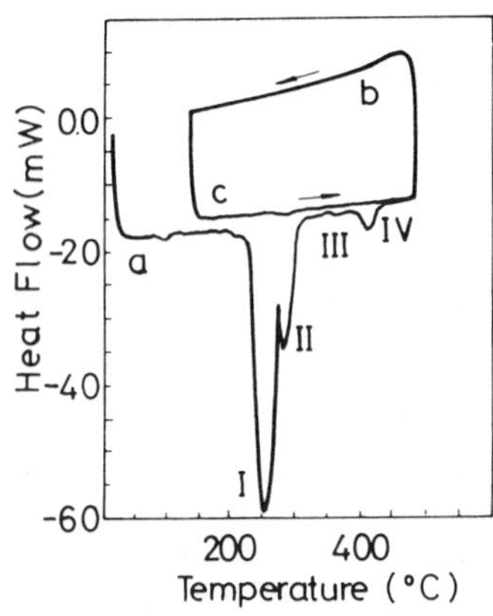

Fig. 3. DSC diagram for multiple heating-cooling cycles of solid KH_2AsO_4
    heating rate 20° C/min.

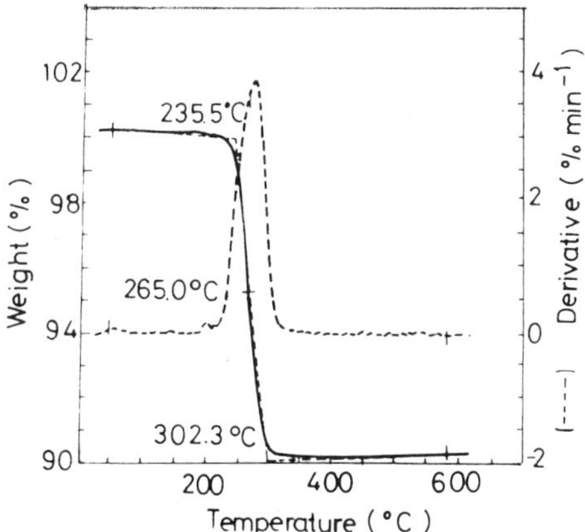

Fig. 4. TGA diagram for solid KH_2AsO_4; heating rate 20°C/min; in
nitrogen atmosphere.

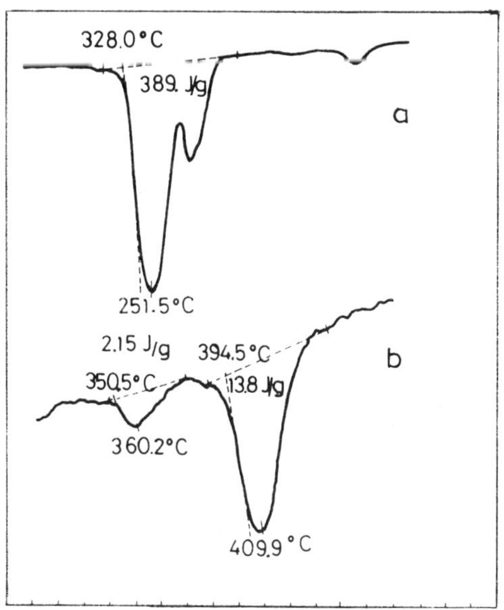

Fig. 5. Determination of the peak temperature and the enthalpies from
peak areas of DSC curves of KH_2AsO_4.

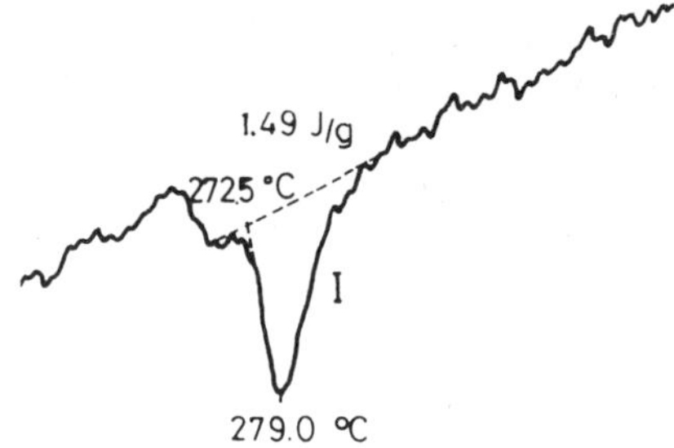

Fig. 6. Determination of the peak temperature and the enthalpies from peak areas in successive heating of KH_2AsO_4.

atmosphere for several hours, Fig. 7; namely, on the DSC diagram the appearance of poorly separated peaks I and II is repeated. The total area of these peaks corresponds to the dehydration enthalpy of 376 J/g, which is very close to the enthalpy of 389 J/g determined during the first heating of the sample.

This interesting phenomenon indicates a definite reversibility of the process: dehydration $\rightleftarrows$ hydration, which can explain to some extent similar electrical behaviour of the investigated sample during the repeated heating.

The dehydrated phase $KAsO_3$ heated at 450°C whose X-ray diffractogram is shown in Fig. 8c, as well as an indexed powder pattern presented in Table I, corresponds to the beta-$KAsO_3$ polyarsenate phase described in the literature.[10] However, X-ray data for the beta $KAsO_3$ phase found in the JSPDS file and which had been taken from the former paper, differ considerably from our results. That was the reason to carry out the calculation of unit cell dimensions of the beta-$KAsO_3$ phase. By using programs VISSER,[7] TREOR-4,[8] and LSUCRIPC,[9] a monoclinic unit cell with a =14.092(5)Å, b =13.099(4)Å, c =8.891(2)Å, β_0=96.37(2)° and V_0=1631(1)Å^3 was found and refined. From Fig. 8b, it can be seen that the transformation from KH_2AsO_4 phase to a dehydrated beta $KAsO_3$ phase at 350°C is a solid-solid transformation, the same being valid

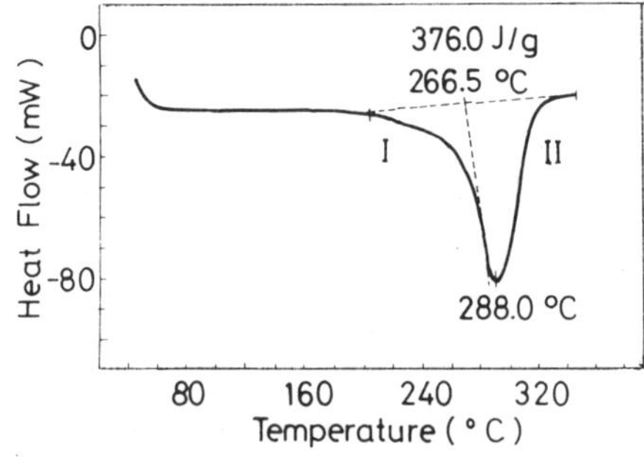

Fig. 7. Determination of the peak temperature and the enthalpies from peak areas of DSC curves after longer standing heated sample in open atmosphere.

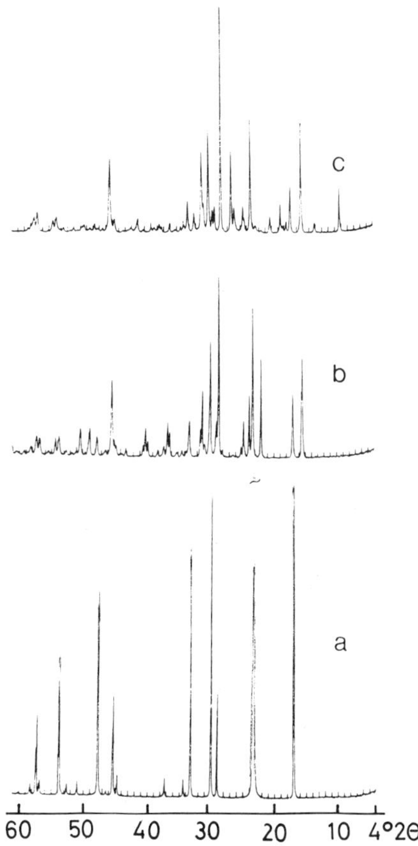

Fig. 8. X-ray powder diffractograms of KH_2AsO_4; at room temperature (a), calcined at $330^{\circ}C$(b), calcined at $450^{\circ}C$ (c).

Table I. Powder diffraction data for β-KAsO$_3$ phase

H K L	I/I$_o$	D$_{obs}$(Å)	D$_{calc}$(Å)	H K L	I/I$_o$	D$_{obs}$(Å)	D$_{calc}$(Å)
-1 1 0	30	9.5534	9.5671	-6 0 1	3	2.3198	2.3212
-1 1 1	6	6.7315	6.7511	-1 5 2	2	2.2455	2.2449
-2 0 1	60	5.8034	5.8109	-6 2 0	7	2.1999	2.1987
2 0 1	20	5.2179	5.2135	6 0 1	6	2.1950	2.1973
-1 2 1	5	5.0271	5.0366	-2 1 4	4	2.1489	2.1476
2 1 1	3	4.8484	4.8440	0 2 4	2	2.0908	2.0931
-2 2 0	15	4.7774	4.7835	-3 0 4			2.0883
-3 1 0	10	4.3987	4.3974	5 4 1	7	2.0304	2.0308
-3 1 1	2	4.1109	4.1120	7 0 0	30	2.0011	2.0007
3 0 1	5	3.9563	3.9504	-7 0 1	28	1.9984	1.9997
-1 3 1	53	3.8140	3.8192	7 0 1	5	1.9063	1.9062
0 2 2	6	3.6661	3.6626	-6 4 0	3	1.9025	1.9008
-1 2 2	12	3.6322	3.6261	-3 3 4	2	1.8833	1.8839
2 1 2	11	3.4356	3.4374	4 5 2	4	1.8481	1.8479
-2 2 2	32	3.3724	3.3756	-5 0 4	4	1.8360	1.8364
-3 3 0	100	3.1858	3.1890	-4 6 1			1.8350
-1 3 2	11	3.0792	3.0832	4 4 3	3	1.7916	1.7914
-3 3 1			3.0750	-8 1 1	5	1.7401	1.7392
3 0 2	11	3.0448	3.0445	-7 4 1	8	1.7063	1.7067
1 3 2	45	2.9822	2.9832	0 2 5			1.7061
-4 0 2	11	2.9101	2.9054	-3 1 5	7	1.7028	1.7024
-1 1 3	32	2.8800	2.8769	3 7 1	6	1.6916	1.6913
2 4 1	10	2.7755	2.7733	-8 2 0			1.6912
-5 1 1	15	2.6966	2.7004	-7 2 3	6	1.6900	1.6885
2 0 3	3	2.6094	2.6132	-6 1 4	5	1.6867	1.6873
4 0 2	3	2.6062	2.6067	-7 3 3	10	1.6217	1.6225
-5 0 2	4	2.4915	2.4041	2 2 5	8	1.6177	1.6182
-1 5 1	3	2.4835	2.4861	1 3 5	7	1.6075	1.6080
2 4 2	4	2.4093	2.4109	3 6 3			1.6073
5 2 1			2.4074	-4 2 5	6	1.6034	1.6029
-2 5 1	4	2.3898	2.3885	-8 0 3	2	1.5833	1.5840
2 5 1	3	2.3399	2.3410	3 1 5			1.5834
3 1 3			2.3367				

for the reverse process. The X-ray diffractogram of the beta-KAsO$_3$ phase, after standing of the latter for several hours in an open atmosphere, is identical with that obtained for the starting KH$_2$AsO$_3$ shown in Fig. 8a. However, it seems that these are not the only phase transformations. The poorly defined endothermic maximum at about 300°C and the "shoulder" at 280°C on the DTA curve in Fig. 2, probably indicate additional interphases in the transformation KH$_2$AsO$_4$ → beta-KAsO$_3$. For checking of this assumption, and for its confirmation, high-temperature X-ray investigations in the range from room temperature to 450°C will need to be performed.

From all our results, it may be concluded that the transformation process: hydrated-dehydrated K-arsenate in the temperature range from 20 to 450°C, is reversible, which, as a phenomenon, by itself improves and prolongs the electrical conductivity. The results obtained also show that the mechanism of phase transformations in this system deserve further investigation.

REFERENCES

1.   L. Glasser, Chem. Rev., 75(1975)21.
2.   M. Šušić and D. Minić, Solid State Ionics, 2(1981)309.
3.   M. Sharon and A. Kalia, J. Solid State Chem., (1977)171.
4.   D. Minić, M. Šušić, N. Petranović and R. Dimitrijević, Materials
     Chemistry and Physics 19(1988)579.
5.   N. Petranović, U. Mioč and D. Minić, Thermochimica Acta,
     116(1987)131.
6.   C. Perrino, B. Lan and R. Alsdorf, Inorg.Chemistry, 11(1972)571.
7.   J. W. Visser, J. Appl. Cryst., 2(1969)89.
8.   P. E. Werner, Z. Kristall., 120(1964)375.
9.   D. E. Appleman and H. T. Evans, Jr., U.S.Dep.Commerce, Nat.
     Techn.Inform.Serv., pb 216188,(1973).
10. E. Thilo und K. Dostal, Z. Anorg.Allge.Chemie, 298(1959)100.

A STUDY OF ISOTHERMAL SINTERING AND PROPERTIES OF MAGNESIUM OXIDE

FROM SEA WATER

N. Petrić, B. Petrić*, V. Martinac, and
M. Mirošević-Anzulović

Faculty of Technology, Split, Yugoslavia
*"Dalmacija", Dugi Rat, Yugoslavia

ABSTRACT

In order to understand the sintering process in magnesium oxide
and to obtain a mathematical model describing behaviour, tests were
carried out with different quantities of SiO_2, Al_2O_3 and TiO_2 added to
magnesium oxide obtained by substoichiometric and overstoichiometric
precipitation of magnesium hydroxide in sea water with dolomite lime.

The isothermal sintering temperature ranged from $1300^{\circ}C$ to $1600^{\circ}C$.
The process of isothermal sintering was examined by determining
densification and sample density, and by mathematical processing of
results.

The B_2O_3 content in samples sintered was examined in forms of its
dependence on the isothermal sintering temperature, on the quantity of
TiO_2 added, and on the way magnesium oxide was obtained. The results
were processed thermodynamically.

INTRODUCTION

Physical and chemical examinations of the magnesium oxide system
have proved that magnesium oxide, when mixed with refractory oxides
such as SiO_2 and Al_2O_3, also yields refractory materials, forming
forsterite and spinel, respectively.[1-3] These oxides being present in
the process of obtaining magnesium oxide from sea water, and especially
when magnesium oxide is obtained from magnesite, it is of special
interest to determine how their presence in individual samples affects
the properties of sintered magnesium oxide, i.e., the quality of the
resultant refractory material. The influence of TiO_2 is also
interesting.

The magnesium oxide used was obtained from sea water by 80% and
120% precipitation of magnesium hydroxide by dolomite lime. The 80%
precipitation of magnesium hydroxide has a number of advantages, and is
a more acceptable method.[4]

EXPERIMENTAL PROCEDURES AND ANALYSES OF RESULTS

The composition of sea water used for precipitation of magnesium hydroxide was:

$$MgO = 2.360 \text{ g dm}^{-3}; \quad CaO = 0.574 \text{ g dm}^{-3}$$

The composition of dolomite used was:

$$0.076\% \text{ } SiO_2, \text{ } 0.064\% \text{ } Fe_2O_3, \text{ } 0.042\% \text{ } Al_2O_3, \text{ } 57.55\% \text{ } CaO, \text{ } 42.27\% \text{ } MgO$$

The magnesium hydroxide obtained was dried at $105^{o}C$ and then calcined at $950^{o}C$. The magnesium oxide obtained had the following chemical composition:

MgO (80% precipitation) contains 99.49% MgO, 0.33% CaO
MgO (120% precipitation) contains 98.25% MgO, 1.36% CaO

The magnesium oxide p.a. (pro analysi) used contained a minimum of 97% MgO.

Specific surface area (according to Blaine) was determined for magnesium oxide samples used:

MgO (80% precipitation)   1.45 m^2g^{-1}
MgO (120% precipitation)   1.31 m^2g^{-1}
MgO p.a.   2.01 m^2g^{-1}

Mixtures with 2%, 5% and 10% of SiO_2, with 2%, 5% and 10% of Al_2O_3 with 0.5%, 1%, 2%, 5% and 10% of TiO_2 were then prepared, with the stoichiometric relation of MgO and SiO_2 being (2:1) and that of MgO and Al_2O_3 (1:1).

SiO_2 and Al_2O_3 were added in the form of α-quartz and α-corundum respectively, and TiO_2 in the form of rutile.

Samples were homogenized by manual mixing in absolute alcohol. Pressing of the samples was carried out in a hydraulic press at a pressure of 625 MPa. The compacts were sintered at $1300^{o}C$, $1400^{o}C$,

Table I. Densities, ρ, of magnesium oxide p.a. samples sintered at different temperatures, $\tau = 5$ hours

Additive	% added	ρ (g cm^{-3})			
		$1300^{o}C$	$1400^{o}C$	$1500^{o}C$	$1600^{o}C$
no additive	0	3.173	3.202	3.314	3.395
SiO_2	2	3.092	3.154	3.225	3.309
SiO_2	5	2.933	3.975	3.042	3.194
SiO_2	10	2.862	2.901	2.966	3.124
SiO_2	stoich.rel.	1.938	1.953	1.987	2.015
Al_2O_3	2	2.994	3.175	3.249	3.300
Al_2O_3	5	2.899	3.053	3.075	3.171
Al_2O_3	10	2.845	2.965	3.002	3.140
Al_2O_3	stoich.rel.	–	–	1.804	1.823

Table II. Densities, ρ, of sintered samples of magnesium oxide obtained from sea water by 80% precipitation, at different temperatures, τ = 5 hours

Additive	% added	ρ(g cm^{-3})			
		1300°C	1400°C	1500°C	1600°C
no additive	0	3.349	3.401	3.426	3.480
SiO_2	2	3.299	3.380	3.407	3.456
SiO_2	5	3.150	3.251	3.296	3.338
SiO_2	10	3.077	3.170	3.220	3.265
SiO_2	stoich.rel.	1.948	1.992	2.186	2.213
Al_2O_3	2	–	3.337	3.365	3.396
Al_2O_3	5	2.979	3.231	3.267	3.315
Al_2O_3	10	2.866	3.079	3.206	3.256
Al_2O_3	stoich.rel.	1.814	1.859	1.924	2.011

1500°C, and 1600°C, with the isothermal heating duration τ = 5 hours. It took about 2 hours to reach the maximum temperature in the furnace.

The density (ρ) of magnesium oxide samples sintered was determined from the volume of water displaced in a calibrated cylinder; before sintering, it was determined from dimensions of compacts.

The values obtained experimentally for density of magnesium samples sintered in the operating conditions described, and with appropriate additives applied, represent the average of a number of measurements (Table I and II).

The experimental data obtained indicate a dependence of sample density on temperature and the quantity of SiO_2, Al_2O_3, and TiO_2 added.

Dependence of function $Z = Z$ (x,y) on variables x and y was analyzed in order to obtain the most suitable expression for the function $Z = Z$ (x,y).

    z = density of sintered samples (ρ)
    y = isothermal sintering temperature
    x = % of additive

Based on this analysis, various expressions were chosen for the function $Z = Z(x,y)$. The most appropriate expressions were selected by determining the mean relative error (r) of deviation of theoretical from experimental data. The expressions which describe the process in the most accurate way are:

- for magnesium oxide p.a.

$$Z = (x + 1)^{-0.04322} y^{0.1639}$$ (for forsterite, i.e., when SiO_2 is added)

$$r = 0.0177$$

$$Z = (x + 1)^{-0.03946} y^{0.1635}$$ (for spinel, i.e., when Al_2O_3 is added)

$$r = 0.0194$$

for magnesium oxide obtained by 80% precipitation:

$$Z = (x + 1)^{-0.03017} \, Y^{0.1699} \qquad \text{(for forsterite, i.e., when } SiO_2 \text{ is added)}$$
$$r = 0.0098$$
$$Z = (x + 1)^{-0.02241} \, Y^{0.1565} \qquad \text{(for spinel, i.e., when } Al_2O_3 \text{ is added)}$$
$$r = 0.0664$$

In order to study the sintering process in magnesium oxide, dependence of densification $[(\Delta V/V)-m]$ on two variables: isothermal sintering temperature (y) and % of SiO_2 or Al_2O_3 added (x), was examined.

$$m = \left(\frac{\Delta V}{V}\right)_{\tau = o} , = \text{the change in volume (\%) until reaching maximum temperature}$$

$$u = \frac{\Delta V}{V} - m = \text{the change in volume (\%) after reaching maximum temperature to the end of the process}$$

Experiments were carried out at temperatures from $1300^{\circ}C$ to $1600^{\circ}C$. The duration of isothermal heating was 5 hours, and the pressure applied in production of compacts was 625 MPa. Table III presents the results obtained.

The experimental data were treated mathematically in order to obtain functions best representative of the densification process in isothermal sintering.[5]

The following function forms were examined:

$$U = y^A \, e^{Bx+C} \tag{1}$$

$$U = y^{Ax+B} \tag{2}$$

$$U = e^{Ax+By+C} \tag{3}$$

Of these functions, the most acceptable one is (3), having the smallest r value.

- for magnesium oxide obtained by 80% precipitation

$$U = e^{-0.0219 \, x - 0.4679 \cdot 10^{-2}y + 9.0259} \qquad (SiO_2 \text{ added})$$
$$r = 0.1487$$

$$U = e^{-0.0232x - 0.4053 \cdot 10^{-2}y + 8.1001} \qquad (Al_2O_3 \text{ added})$$
$$r = 0.176$$

- for magnesium oxide p.a.

$$U = e^{-0.0399x - 0.3466 \cdot 10^{-2} + 7.3334} \qquad (SiO_2 \text{ added})$$
$$r = 0.2430$$

$$U = e^{-0.0173x - 0.2910 \cdot 10^{-2}y + 6.6007} \qquad (Al_2O_3 \text{ added})$$
$$r = 0.1691$$

The influence of TiO_2 added to magnesium oxide in quantities of 0.5%, 1%, 2%, 5% and 10%, on product density and boron content in samples sintered was also examined in terms of its dependence on the way magnesium oxide was obtained (Table IV and V).

B_2O_3 is a very important component of sintered magnesium oxide, as the B_2O_3 content markedly affects the properties of the sintered

Table III. Densification, (%) in dependence on sintering temperature
(y) and quantity of SiO or Al_2O_3 added

Sample	t,°C	no addition	2% SiO_2	5% SiO_2	10% SiO_2	2% Al_2O_3	5% Al_2O_3	10% Al_2O_3
MgO (80%)	1300	17.00	16.76	14.74	13.95	14.78	12.56	10.93
	1400	11.54	11.45	10.88	10.11	12.23	10.70	9.74
	1500	9.71	9.61	8.54	8.33	9.77	8.07	8.45
	1600	4.12	3.79	3.29	2.89	3.91	3.40	3.48
MgO p.a.	1300	14.34	12.52	10.39	10.12	14.24	13.17	12.49
	1400	13.52	12.14	10.15	9.65	13.06	12.53	11.08
	1500	12.35	10.84	9.49	9.01	12.20	11.21	10.03
	1600	5.66	3.84	3.37	2.84	5.48	5.06	5.00

product, i.e., good quality sintered magnesium oxide contains a very
small quantity of B_2O_3.

Examination have shown that, while TiO_2 influences the reduction
of B_2O_3 in the isothermal sintering process in magnesium oxide,
addition of SiO_2 or Al_2O_3 has no such influence. The boron content in
samples examined was determined by potentiometrical analysis. Previous
papers established the presence of dicalcium borate ($Ca_2B_2O_5$) in
sintered magnesium oxide. In this paper ΔG has been calculated from the
experimental data for dicalcium borate formation reaction, i.e. $2CaO +
B_2O_3 = Ca_2B_2O_5$, at different conditions and at isothermal sintering
temperatures of 1300°C and 1500°C, with 1%, 2%, and 5% of TiO_2 added,
and without TiO_2 addition, for magnesium oxide obtained by 80% and by
120% precipitation.

By applying the fundamental equations[6,7] and the following
expressions for dicalcium borate, $Ca_2B_2O_5$:

$$\Delta H_T^O = -291838 + 60.12\ T - 4.52\ 10^{-2}\ T^2 - 1306282\ T^{-1} \quad J/mol$$

$$\Delta S_T^O = -425.47 + 60.12\ lnT - 9.04\ 10^{-3}T - 653141\ T^{-2} \quad J/mol\ K$$

the data presented in Table VI were obtained, where ΔG_T^O is a change of
standard Gibbs free energy, defined by the expression $\Delta G_T^O = \Delta H_T^O - T\Delta S_T^O$,
while K_a is the thermodynamic equilibrium constant.

Table IV. Densities, ρ, of magnesium oxide sintered samples with TiO_2
added, t = 1300°C, τ = 5 hours

Sample	TiO_2					
	no addition	0.5%	1%	2%	5%	10%
MgO p.a.	3.173	3.231	3.256	3.268	3.288	3.315
MgO (120%)	3.320	3.366	3.382	3.388	3.416	3.432
MgO (80%)	3.349	3.367	3.378	3.438	3.446	3.478

Table V. Dependence of % B_2O_3 in magnesium oxide on temperature and % of TiO_2 added

Sample	% B_2O_3 in Mg $(OH)_2$	% B_2O_3 in MgO (950°C)	% B_2O_3 in MgO w/o addit.	% B_2O_3 in MgO+ 1% TiO_2	% B_2O_3 in MgO+ 2% TiO_2	% B_2O_3 in MgO+ 5% TiO_2
			t=1300°C, τ = 5 hours			
MgO (80%)	0.1950	0.1934	0.1192	0.0852	0.0645	0.0586
			t=1500°C, τ = 5 hours			
			0.0690	0.0173	0.0159	0,0130
			t=1500 C, τ = 5 hours			
MgO(120%)	0.0376	0.0376	0.0318	0.0204	0.0050	0.0035

The experimental data, i.e., boron content established by potentiometrical analysis (τ = 5 hours, Table V), were used to calculate the degree of yield at the temperatures of 1300°C and 1500°C, in terms of its dependence on the percentage of boron in magnesium oxide calcined at 950°C.

If the value for ξ is known, the reaction constant, K'_x, can be determined for the dicalcium borate formation reaction according to the expression:

$$K'_x = \frac{x(Ca_2B_2O_5)}{x(B_2O_2)[x(CaO)]^2}$$

where x = the quantity of reactant $[x(B_2O_5); x(CaO)]$, or the product $[x(Ca_2B_2O_5)]$ of the reaction.

The data from Table VI and experimental data (K'_x after 5 hours of reaction) were used to calculate the change in Gibbs free energy ΔG for the reaction examined, according to the relation $\Delta G = \Delta G^O + RT \ln K'_x$, which represents the basic criterion of system equilibrium.

In order to determine the degree of yield of dicalcium borate formation reaction in comparison to theoretical yield ξ_t for a given temperature and quantity of B_2O_3 in the sample, the ξ/ξ_t ratio was calculated. The results obtained are shown in Table VII.

Table VI. Changes in ΔH_T^O, ΔS_T^O, ΔG_T^O in dependence on temperature, and K_a for the reaction $2CaO+B_2O_3 = Ca_2B_2O_5$

t (°C)	ΔH_T^O (J/mol)	ΔS_T^O (J/mol K)	ΔG_T^O (J/mol)	K_a
1300	−209284	2.58	−213342	1.1986 10^7
1400	−204689	5.41	−213740	4.6556 10^6
1500	−200191	8.29	−214889	2.1170 10^6
1600	−195787	9.43	−213449	8.8687 10^5

Table VII. ξ, ξ/ξ_t, $K_x^!$, ΔG for the reaction $2CaO+B_2O_3= Ca_2B_2O_5$, $\tau = 5$ hours, for MgO (80% precip.) and MgO (120% precip.) with 0%, 1%, 2%, and 5% of TiO_2

Sample	$t(^oC)$	%TiO_2	ξ (%)	ξ/ξ_t (%)	$K_x^!$	$\dfrac{\Delta G}{J/mol}$
MgO (80% precip.)	1300	0	61.63	61.80	8.52	−185300
		1	44.05	44.17	2.82	−199763
		2	33.35	33.44	1.54	−207676
		5	30.33	30.41	1.28	−210111
MgO (80% precip.)	1500	0	35.70	35.93	1.75	−206633
		1	8.94	9.00	0.24	−236192
		2	8.22	8.27	0.21	−237914
		5	6.72	6.76	0.17	−241031
MgO (120% precip.)	1500	0	84.57	85.11	98.59	−147157
		1	54.25	54.59	5.19	−190594
		2	13.30	13.38	0.38	−229164
		5	9.30	9.36	0.25	−235341

DISCUSSION

For the dicalcium borate formation reaction described, ΔG^o is negative, and a decrease in temperature favours a better yield. At higher temperatures, a smaller quantity of dicalcium borate remains in the sintered sample (Table V), i.e., during the isothermal sintering process at higher temperatures, a greater quantity disappears from the sample into the atmosphere.8

An increase in the quantity of TiO_2 added, as shown by the experimental data (Table VII) also causes a decrease in the degree of yield ξ, i.e., a larger quantity of boron disappears into the atmosphere.

Under the same operating conditions, the experimental data indicate that ξ is higher in magnesium oxide obtained by 120% precipitation than in that obtained by 80% precipitation. In magnesium oxide (120% precip.) the content of CaO=1.36% is significantly higher than in magnesium oxide (80% precip.) where CaO = 0.33%, i.e. the CaO content is much higher than in 80% precipitation, and thus favours the $Ca_2B_2O_5$ formation reaction, which may be the reason for a higher percent of boron disappearing from magnesium oxide obtained by 80% precipitation.

Thus the addition of TiO_2 has a significant influence on boron disappearing from the samples into the atmosphere. In this paper a mathematical expression for this dependence has been developed. Namely, in order to put into proper relationship all the factors influencing the disappearance of boron during the isothermal sintering process, the functional dependence of boron lost on the temperature of isothermal sintering and the quantity of TiO_2 added was examined for both the 80% and the 120% precipitation. The following expressions were obtained:

$$J = 1.3736\ 10^{-2}\ x + 15.8133\ y \qquad \text{(for 120\% precipitation)}$$

$$J = 4.5854 \cdot 10^{-2}x + 4.4925 \, y \qquad \text{(for 80\% precipitation)}$$

J = % of boron lost (as compared to the content in magnesium oxide calcined at 950 $^{\circ}$C)

x = temperature difference, i.e. isothermal sintering temperature minus room temperature (25 $^{\circ}$C)

y = % of TiO_2 added

CONCLUSIONS

Mathematical and thermodynamical treatments of magnesium oxide sintering process with and without different quantities of SiO_2, Al_2O_3, and TiO_2 added, now make it possible to predict mathematically, without additional experiments, the sample density, densification percentage, and B_2O_3 content in sintered samples.

Densities obtained for sintered magnesium oxide closely approach theoretical densities, especially for magnesium oxide (80% precipitation) (3.349–3.480 g cm^{-3}).

Addition of TiO_2 to magnesium oxide increases the product density in the process of isothermal sintering, and decreases the quantity of B_2O_3 present.

With magnesium oxide produced by 80% precipitation, the influence of TiO_2 is greater in terms of B_2O_3 lost by evaporation than with that obtained by 120% precipitation. This is probably due to different initial content of CaO in the magnesium oxide samples of different chemical origin.

REFERENCES

1. N. Petric, B. Petric, E. Tkalčec, V. Martinac, N. Bogdanić, M. Mirošević-Anzulović, Sci. Sinter., 19(1987)81.
2. S. Yangyun, R. J. Brook, Sci. Sinter., 17(1985)35.
3. A. R. West, Solid State Chemistry and its Applications, John Wiley and Sons, Chichester, New York, Brisbane, Toronto, Singapore, 1984,5.
4. B. Petric, N. Petric, Ind. Eng. Chem. Process Des. Dev., 19(1980)329.
5. N. Petric, B. Petric, N. Bogdanić, M. Mirošević-Anzulović, E. Tkalčec, V. Martinac, J. Chem. Tech. Biotechnol., 43(1988)139.
6. D. D. Wagrnan, W. H. Evans, V. B. Parker, R. H. Schum, I. Halow, S. M. Bailey, K. L. Churney, R. L. Nuttall, J. Phys. Chem. Ref. Data, Vol. 11, Suppl. 2, 1982.
7. O. Kubaševskij, C. B. Alkok, Metalurgicheskaya termohimija, Moskva, Metalurgiya, 1982.
8. H. M. Richardson, M. Lester, F. T. Palin, P. T. A. Hodson, Trans. Brit. Ceram. Soc., 68(1969)29.

Part X. HIGH TEMPERATURE SUPERCONDUCTORS

PROPERTIES OF HIGH T_c SUPERCONDUCTING OXIDES*

Boyd W. Veal and S.-K. Chan

Materials Science Division
Argonne National Laboratory
Argonne, IL 60439, U.S.A.

I. INTRODUCTION

With the recent discovery of superconductivity at temperatures exceeding 30 K[1] and the rapid follow-up discovery of superconductivity at temperatures exceeding 90 K,[2] an unprecedented worldwide effort has been directed to exploring these materials and to search for new superconducting compounds. The effort has been remarkably productive, yielding numerous new superconducting compounds, some with transition temperatures T_c in excess of 120 K.[3] All of these new materials are oxides and those with highest T_c's contain copper, generally in a high oxidation state, bonded to oxygen in a near-planar configuration. [The search has yielded one new copper-free compound, potassium doped $BaBiO_3$,[4,5] with $T_c \sim 30$ K, a transition temperature significantly exceeding that of Nb_3Ge, the long time record holder (at 23 K)].

These superconductors are multicomponent oxides which have extremely complex composition-transition temperature phase fields. In general, sintering techniques are used to fabricate the materials after a calcine step to decompose oxide or carbonate precursors. Because of the high anisotropy of the crystallites comprising the individual grains, electronic conduction through the polycrystalline sintered materials tends to be severely limited, providing a serious impediment to their commercial exploitation. Since current densities can be high in appropriate directions through the crystals, much effort is being directed toward the control of texture and, for thin films, epitaxial growth.[6] Studies of single crystals are needed for determination of intrinsic properties.

With all of the excitement surrounding these new discoveries, however, is the realization that mechanisms responsible for the superconductivity at these remarkably high temperatures are not understood. The Bardeen-Cooper-Schreiffer (BCS) theory[7] provided the essential formalism upon which our understanding of superconductivity is based. In this theory, when two electrons of opposite spin are appropriately mediated by a lattice excitation, or phonon, the electrons experience an attractive force and condense into the superconducting state. However, widespread sentiment has emerged to suggest that the phonon mediated electron pairing proposed in the BCS model cannot provide an adequate explanation and a variety of alternate pairing mechanisms have been proposed (see Section III).

---

*Work supported by the U.S. Department of Energy, Basic Energy Sciences-Materials Science under Contract #W-31-109-ENG-38.

Thus, while dramatic progress has been made in reporting discoveries of new high temperature superconductors, the progress has not resulted from predictive capability that comes from a clear understanding of the phenomena but rather from an intensive and broad ranging effort devoted to oxide, and especially copper oxide compounds. Clearly, a better understanding of superconductivity in these materials is badly needed. Guidance is needed for expanding the search to new materials systems and to solve troublesome problems (especially the problem of low permissible current densities) that limit industrial use of the materials.

The large superconductivity literature must contain essential clues to understanding the mediating mechanisms as well as processes which compete with superconductivity. However, reported experiments have, apparently, not yet provided the key insights to clearly identify the relevant mechanisms. Consequently, the effort must continue to find those correlated properties in the oxide systems that potentially will provide the essential insights for understanding the superconducting behavior. $YBa_2Cu_3O_{7-x}$ (123) is a remarkably good candidate for such a study since it offers easy variation of a multitude of parameters, usually with dramatic effect on the superconductivity. These studies suggest a strong link between atomic structure and superconductivity. They also point to the possible importance of cation (and anion) valence, carrier concentration, bond overlap, defect structures, magnetic order, polar effects, and insulating behavior as properties that influence or compete with the superconductivity. These properties are systematically studied, frequently with wide parameter variability, by controllably varying cation and oxygen vacancy content.

In this paper we shall first briefly consider the history of oxide superconductor discoveries followed by a brief sketch of the current state of theory. We then consider crystallographic structures of the high T_c oxide superconductors. We will then examine some phase fields (T_c versus composition) of the oxide superconductors. Since these phase fields show continuous variation between superconducting (metallic) and insulating behavior, they contain both those elements favorable to superconductivity and those detrimental. Particularly well studied are the remarkably variable physical properties of the 123 superconductor that are associated with varied oxygen stoichiometry and cation substitution. We present selected results to provide a brief overview of this problem. Then, recent work on electronic properties, providing insight into the fundamental nature of electronic conduction will be examined.

These considerations impact on the problem of sintering. Sample density, grain size, grain boundary impurities and texture all affect the physical properties. Of particular importance to superconducting properties is the control of oxygen stoichiometry and sample homogeneity associated with dispersal of dopant ions.

II. OXIDE SUPERCONDUCTORS – A BRIEF HISTORY

Though superconductivity was discovered in 1911, it was not thought to exist in oxides and essentially no effort was expended to look for the phenomenon in such materials. Consequently, it was not until the 1960's that superconductivity was first reported in an oxide material, even though hundreds of superconductors were known by that time. The first reported superconducting oxide[8] was the perovskite $SrTiO_3$, normally an insulator, but now processed to make the material oxygen deficient. This procedure enables the conductivity to be controlled over a wide range. With increasing oxygen vacancy content, the material showed behavior of a wide gap insulator, semiconductor and finally a metal. In this final metallic condition, the material went superconducting at about 0.3 K.[8]

After this discovery, scientific advances in the field of oxide superconductivity came rapidly. Off stoichiometric monoxides TiO and NbO, compounds with the cubic NaCl structure showed transitions at 1-2 K.[9] Other perovskites, the tungsten bronzes (e.g., Na_xWO_3), became superconducting[10] at temperatures as high as 7 K and transition temperatures were further increased to 13 K with the discovery[11] of $BaPb_{1-x}Bi_xO_3$. Only very recently, the perovskite $Ba_{1-x}K_xBiO_3$ has been reported to show superconductivity at about 30 K.[4,5] Another oxide structure also showed superconductivity. The spinel $Li_{1-x}Ti_{2-x}O_4$ became superconducting at about 14 K.[12]

It was the discovery of Ba doped La_2CuO_4, with a transition temperature in excess of 30 K[1] that stirred an unprecedented worldwide interest in superconductivity. That interest was quickly and profoundly intensified by the report of superconductivity at 90 K in the Y-Ba-Cu-O system.[2] This superconductor, subsequently to be identified as $YBa_2Cu_3O_{7-x}$,[13] suggested the possibility of profound technological advances and contained the promise of great economic impact. Suddenly, an old dream was realized: that a superconductor exists with T_c in excess of 77 K, the boiling point of liquid nitrogen, an inexpensive refrigerant. The old dream has become a new dream; that superconductivity might one day be found at room temperature.

And the pace moves rapidly. Another family of superconductors[14] with nominal compositions $Bi_2Sr_2Ca_{n-1}Cu_nO_x$ with $1 \leq n \leq 3$ show transition temperatures as high as 110 K. A similar family,[3,15] $Tl_2Ba_2Ca_{n-1}Cu_nO_x$, can be fabricated with superconducting transitions as high as 125 K. Advances continue to be reported at a very rapid pace with numerous superconducting variants of these systems having been reported.

III. THEORY

In the Fermi liquid description[16] of the normal state of a metal, the interacting electrons may be regarded as quasi-particles that move independently with an effective mass m* that accounts for the effects of interaction. These independent quasi-particles obey Fermi-Dirac statistics and bear a one-to-one correspondence with the noninteracting electrons. In the ground state, they occupy energy levels up to the Fermi energy E_F. In this sea of fermions, only quasi-particles near to E_F are responsible for carrying current. In some materials, an effective attraction between the electrons may exist because of a coupling provided by the intervening medium. For such materials, the quasi-particles can form pairs of opposite spins and momenta known as Cooper pairs. At sufficiently low temperature, a macroscopic number of quasi-particles near to the Fermi energy may form such pairs to achieve a more stable (superconducting) state for the entire system. This macroscopic condensation into a pairing state radically affects the quasi-particle spectrum near the Fermi energy, resulting in the occupied states of the allowed energy spectrum being separated from the empty states by a gap 2Δ. This gap, which corresponds to the binding energy of a Cooper pair, is the energy that must be expended to break up the pair and recover the independent quasi-particles. The presence of this gap in the excitation spectrum leads to a rigidity of the spectrum in its response to external perturbations, contributing to the remarkable properties of the absence of resistance and the expulsion of magnetic field (Meissner effect) from the bulk interior of the material.

In the first successful microscopic theory of superconductivity, due to Bardeen, Cooper and Schrieffer,[7] the attractive interaction between the electrons that leads to the pairing state is provided by lattice vibrations or phonons. Loosely speaking, an electron polarizes the lattice ions and alters their vibrational characteristics, the effect of which is felt by another electron resulting in an effective attraction between the two electrons. This mechanism leads to a transition temperature T_c which was thought to be no higher than about 40 K, based on realistic parameters for the previously known superconducting materials.[17]

The discovery of the high T_c superconductors throws open questions about the nature of the "pair" and also of the "coupling" that provided the attractive interaction needed for the formation of the "pair." Dozens of theories have been put forth. For the most part, they fall into the following categories:

(1) Bipolarons. The pair involves two localized electrons of opposite spin, with an electron on each of two atoms of a unit cell. The electron-phonon interaction provides the attractive coupling.[18]
(2) Resonant bonds. A singlet pair of electrons or holes resides on nearest neighbors of similar atoms with exchange interaction providing the coupling.[19,20]
(3) Pair of ionic states (e.g., opposite spin oxygen p hole states) with antiferromagnetic superexchange providing the coupling.[21]
(4) Pair of covalent bonding orbitals with coupling provided by phonons.[22,23]

(5)   Pairs of electrons, as in BCS theory but with enhanced coupling provided by either (a) a high density of states near the Fermi surface or (b) some novel mechanism such as soft plasmons both of which are associated with the 2-D nature of the materials.[24-26]

(6)   Pairs of electrons with an attraction provided by the exchange of a local exciton.[27]

None of the present theories can account for the many facets of the experimental findings. Some of them may have become irrelevant because they are fundamentally inconsistent with the experimental facts (a) that the carriers are, for most of the high T_c materials, in the oxygen p band and not in the copper d band and (b) there exists a Fermi surface. Probably, it will be some time before a viable theory is evolved that can achieve the kind of success attained previously by the BCS theory.

IV. STRUCTURES

While superconductivity has been found in cubic and spinel oxide structures at 14 K or lower, the large majority of oxide superconductors have the perovskite structure or a structure usually considered to be perovskite related. Figure 1 shows the cubic perovskite structure, an ABO_3 material, that has a large A ion, here pictured at the cube corner, that is 12-fold coordinated to neighboring oxygen atoms. A smaller 6-coordinated B cation resides at the cube center. The superconductors $BaPb_{1-x}Bi_xO_3$, $Ba_{1-x}K_xBiO_3$, $(Na,K,Rb,Cs)_xWO_3$ and $SrTiO_{3-x}$, with transition temperatures between 0 and 30 K, crystallize in this structure or small variants of it.

The compound La_2CuO_4 when optimally doped with the divalent ions Ba, Sr, or Ca, or when prepared with suitable oxygen stoichiometry, exhibits transition temperatures as high as 37 K.[28] Under pressure, T_c can be further increased.[29] This material, which displays a second level of high T_c behavior, after the simple perovskites, also shows a trend toward more structural complexity. This structure is shown in Fig. 2.[30] While it may be viewed as a stacking of distorted perovskite cells, there is considerable variation in Cu-O bond distances (the Cu-O1 distances are 1.89 Å while the Cu-O2 distances are 2.43 Å) which provides the structure with a strong layered or planar character. In this view, 4-coordinated Cu atoms lie in planes separated by La or the dopant (Ba, Sr, Ca, ...) atoms that are bonded to the O2 oxygens. The presence of Cu-O planar regions in the structure would seem to be an important, and perhaps essential, ingredient for high T_c superconductivity. All known systems with superconducting transition temperatures exceeding 30 K display the omnipresent Cu-O planes. This material is p-type, obtained from the insulating stoichiometric La_2CuO_4 compound by doping the La site with a divalent ion or by controlling the oxygen defect concentration.

Recently, a new class of compounds consisting of R_2CuO_4 (R = Nd, Pr, Sm and Eu) with doping on the trivalent R-site by tetravalent Ce or Th ions has been reported.[31,32] Transition temperatures varying from 0 to about 24 K are observed. This class of compounds has attracted considerable interest because they are n-type, at least at temperatures well above T_c. If oxygen is stoichiometric, the tendency of the doping is to reduce Cu^{2+} to Cu^{1+} and create electron carriers. For $La_{2-x}Sr_xCuO_4$, La^{3+} is replaced by Sr^{2+}, tending to drive Cu trivalent and creating a p-type conductor.

The location of oxygen atoms in these n-type compounds differs somewhat from their location in $La_{2-x}Sr_xCuO_4$. In particular, one or both of the O2 atoms of Fig. 2 shift from the apical position leaving an even more clearly defined planer structure than shown in Fig. 2.[31]

Figure 3 shows the structure of the $RBa_2Cu_3O_{7-x}$ (R is rare earth or Y) compound, a 92 K superconductor. Pairs of Cu-O planes (defined by Cu2 and O2, O3 atoms) are found with a Y (or R) atom sandwiched between them. A distinguishing structural feature of this compound is the presence of "chains," consisting of the Cu1 and O1 atoms. In both the plane and chain regions, Cu atoms are 4-coordinated. Recently, discovery of the related $RBa_2Cu_4O_8$ (124) and $R_2Ba_4Cu_7O_{15}$ (247) phases was reported.[33] For these superconductors, T_c's occur at temperatures as high as 82 K. The R-sandwiched double plane structures found in the 123 phase are preserved in these materials but they are

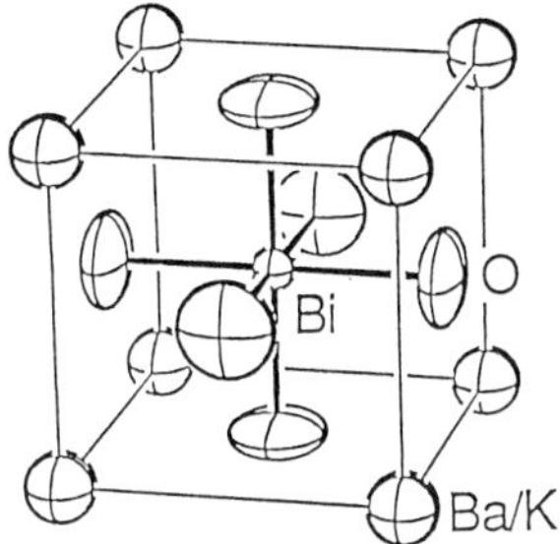

Fig. 1.

The crystal structure of the cubic
perovskite $Ba_{1-x}K_xBiO_3$ (Refs.4,5).

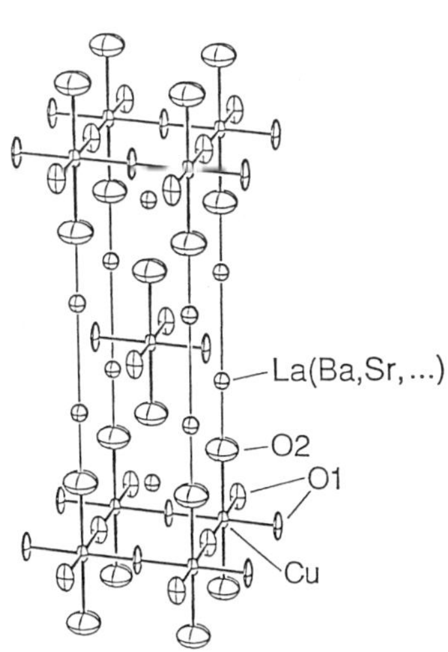

Fig. 2.

The crystal structure of $La_{2-x}m_xCuO_4$
where m = Ba, Ca, Sr. The Cu-O1 atoms
form a plane (Ref. 30).

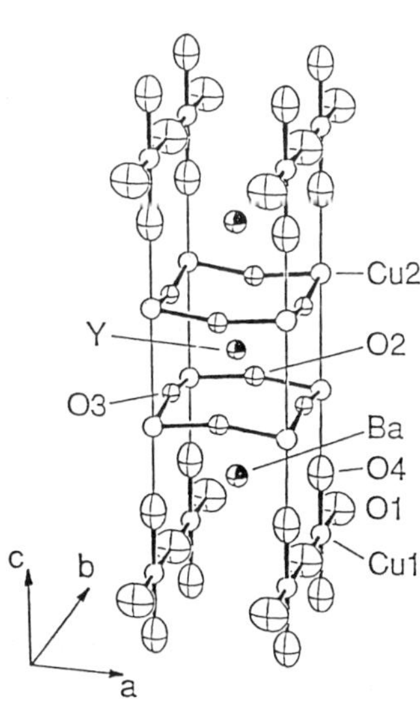

Fig. 3.

The crystal structure of $YBa_2Cu_3O_{7-\delta}$.
Cu2-O2, O3 atoms form planes separated
by a Y atom. Cu 1-O1 atoms form
chains. Oxygen vacancies occur at O1
sites (Ref. 30).

distinguished by the appearance of double chain layers and a consequent increase in the c-axis cell dimension. When R-123 was first discovered, and the highest recorded T_c increased from ~40 K to ~90 K, it was widely felt that the chains provided the essential new ingredient which made the large increase in T_c possible.

With the subsequent discovery of the Bi-Sr-Ca-Cu-O and Tl-Ba-Ca-Cu-O compounds,[34] however, transition temperatures reached new record highs[3] but the chain structures of the 123 compound were not present. It now appeared that multiple layers of the Cu-O planes were most advantageous to high T_c superconductivity. These new families of compounds can be fabricated in several chemical compositions given by the generic formulas $[Bi_2Sr_2Ca_{n-1}Cu_nO_x$ and $Tl_mBa_2Ca_{n-1}Cu_nO_x]$ where, for the Tl compounds, m = 1 or 2. n measures the number of adjacent Cu-O layers (separated by Ca atoms) which appear in these structures. Figure 4 shows the $Tl_2Ba_2Ca_{n-1}Cu_nO_x$ structures[35] when n = 1, 2, and 3. The groups of CuO_2 planes are separated by intercalated layers of BiO and SrO, or TlO and BaO. These materials are difficult to prepare in single phase form and tend to show intergrowths of the different structures. For structures with a given n, transition temperatures may vary considerably, but T_c systematically increases[36] with n, at least until n = 3. The Bi and Tl based superconductors have provided a different perspective for the 123 superconductor; seemingly, the important structural feature of R-123 is that two adjacent Cu-O planes occur, in this case separated by an Y atom. The system is analogous to the n = 2 members of the Bi or Tl based oxide families.

Other, recently discovered, compounds that show superconducting behavior at temperatures $\gtrsim 50$ K include $Pb_2Sr_2LnCu_3O_{8+x}$[37] and $(Ba_{1-x}Ln_x)_2(Ln_{1-y}Ce_y)_2Cu_3O_{8+z}$[38] where Ln is a lanthanide ion. Again Cu-O planes appear but with complex and unusual "interconnects."

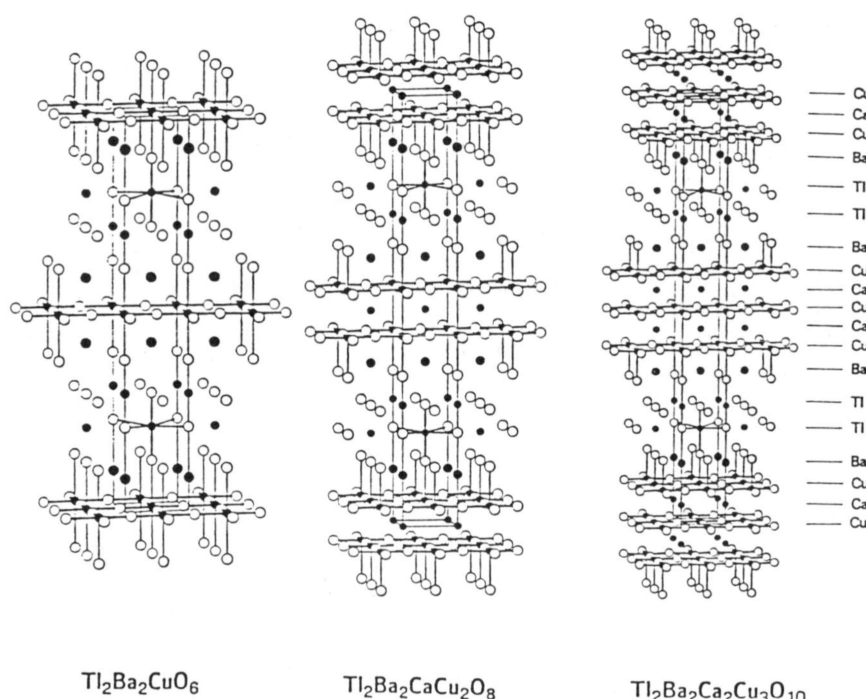

$Tl_2Ba_2CuO_6$ $Tl_2Ba_2CaCu_2O_8$ $Tl_2Ba_2Ca_2Cu_3O_{10}$

Fig. 4. Crystal structures of three members of the family of compounds $Tl_2Ba_2Ca_{n-1}Cu_nO_{2n+4}$ where n = 1, 2, 3. These compounds have n adjacent Cu-O layers separated by Ca atoms (Ref. 35).

In all of the structures discussed above, there are many subtle structural features that have not been considered. For example, the degree of "puckering" or corrugating of the CuO_2 planes varies from structure to structure. The ability of cations to experience site interchange varies in the different structures; the tolerance for oxygen vacancies or cation substitution varies from site to site and structure to structure; the tendency for disorder appears in some regions of some structures. Symmetries may be orthorhombic or tetragonal. The common feature of the high T_c (>30 K) oxides is the appearance of CuO_2 planes that are well ordered and free of observable vacancies.

V. METAL-INSULATOR PHASE FIELDS

One of the remarkable features of the high T_c superconductors is that they exist in phase fields (T_c vs composition) which show behavior ranging from metallic (superconducting) to insulating (and sometimes metallic-nonsuperconducting). For a given superconductor, there may be many composition variations which define phase fields in which this behavior can be seen. Taking the position that a suitable theory of the oxide superconductors should ultimately account for the metal-insulator transition (as well as the superconductivity), we briefly consider this behavior in a variety of systems.

Figure 5 shows transition temperatures for the system $La_{2-x}Sr_xCuO_4$ when the oxygen stoichiometry is maintained at four.[39] Only a relatively small region of this phase field shows superconductivity. In this system, superconductivity is obtained as a consequence of appropriate doping of a parent insulator La_2CuO_4. Figure 6 shows that superconductivity appears in an even smaller dopant concentration[31] range for the "n-type" superconductor $Nd_{2-x}Ce_xCuO_4$. Dabrowski et al.[40] have examined the phase field for $La_2CuO_{4+\delta}$ and similarly discover a small region of oxygen stoichiometry ($0.1 < \delta < 0.2$) where superconductivity occurs.

Phase fields for the complex oxides (with T_c in excess of 100 K) are being investigated. Because of their recent discovery, the great complexity of these systems, and because of their frequent metastable character, the phase fields have not yet been extensively explored. Nonetheless, studies of cation substitution and oxygen stoichiometry variation in $Tl_mBa_2Ca_{n-1}Cu_nO_x$ and $Bi_2Sr_2Ca_{n-1}Cu_nO_x$ show that changes in superconducting behavior are observed. Reportedly, for the Tl-Ba-Ca-Cu-O n = 1 system, with oxygen variation, T_c can be made to vary continuously between 0 and 90 K.[41]

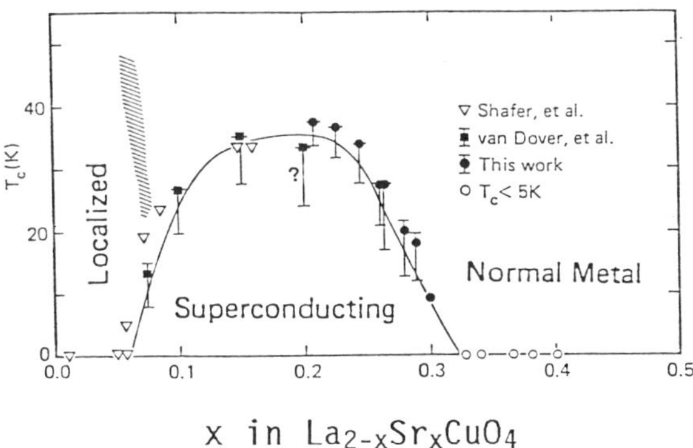

Fig. 5. The superconducting phase field for $La_{2-x}Sr_xCuO_4$ with no oxygen vacancies. Semiconducting, superconducting and normal metallic behavior is observed (Ref. 39).

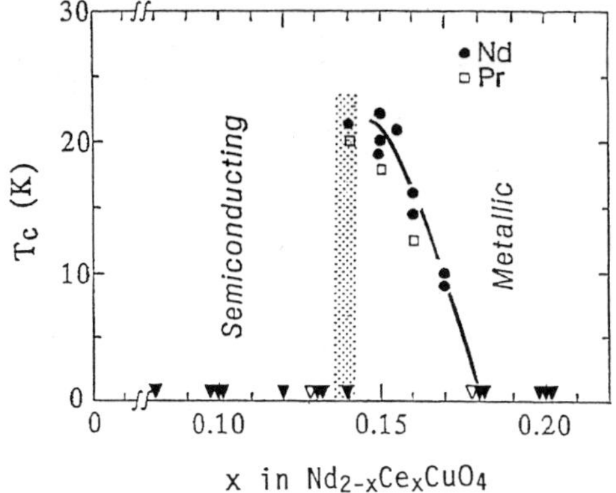

Fig. 6.

The superconducting phase field for $Nd_{2-x}Ce_xCuO_4$, a material believed to show n-type conduction. The superconducting phase exists in a narrow composition range (Ref. 31).

Probably most extensively studied are the phase fields of the system $RBa_2Cu_3O_{7-x}$ when oxygen stoichiometry is varied or substitutions are made on cation sites (see Ref. 42 and references cited therein). The 123 superconductor exhibits a remarkably high tolerance to varied oxygen stoichiometry. Metastable compounds can readily be prepared using a variety of techniques with stoichiometries in the range $6 < 7-x < 7$. Figure 7 shows a plot of transition temperatures versus oxygen content.[42-44] Also shown are the a_0 and b_0 lattice parameters of the orthorhombic structure as they collapse to the tetragonal a_t. As superconductivity is lost, insulating behavior develops along with antiferromagnetic order.[45] Neel temperatures are also shown in Figure 7. The features shown in Fig. 7 ($T_c \rightarrow 0$, orthorhombic→tetragonal transition, metal-insulator transition, development of antiferromagnetic order) are highly correlated. The correlations apparently persist in the series of R-123's where R = Y, Gd, and Nd (Figs. 8 and 9) even though the level of oxygenation where superconductivity is lost systematically varies with ion size.[44,46] Interatomic bond distances also show correlations with loss of superconductivity that may signal important charge transfer effects within the structure.[42,44,47]

Fig. 7.

The superconducting phase field for $YBa_2Cu_3O_{7-\delta}$ as a function of oxygen content. T_c falls to zero as oxygen is depleted; the material then becomes an antiferromagnetic insulator. The crystal symmetry changes from orthorhombic in the superconducting phase to tetragonal in the insulating phase (Refs. 43,45).

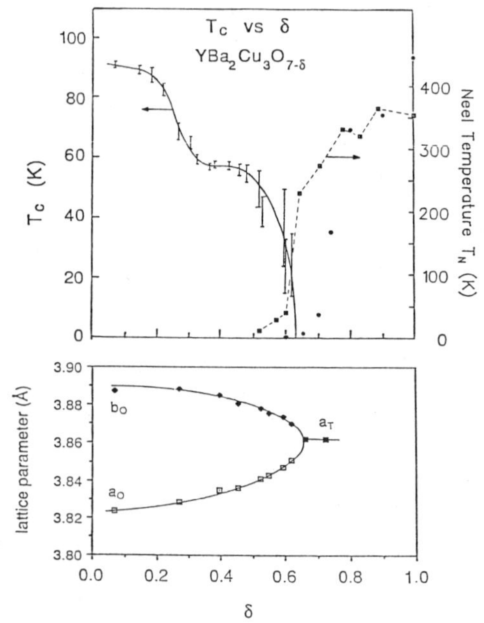

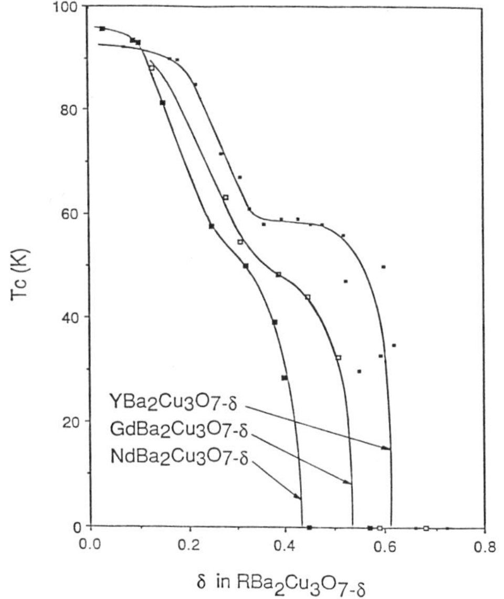

Fig. 8.

Superconducting phase fields for $RBa_2Cu_3O_{7-\delta}$ (R = Y, Gd, Nd) as a function of oxygen content. T_c falls more rapidly as the R-ion size increases (Ref. 46).

Fig. 9.

For the superconductors $RBa_2Cu_3O_{7-\delta}$ (R = Y, Nd) with varied δ, the orthorhombic-to-tetragonal phase transition occurs when superconducting behavior is lost (Ref. 44).
Circles = Nd-123. Triangles = Y-123.

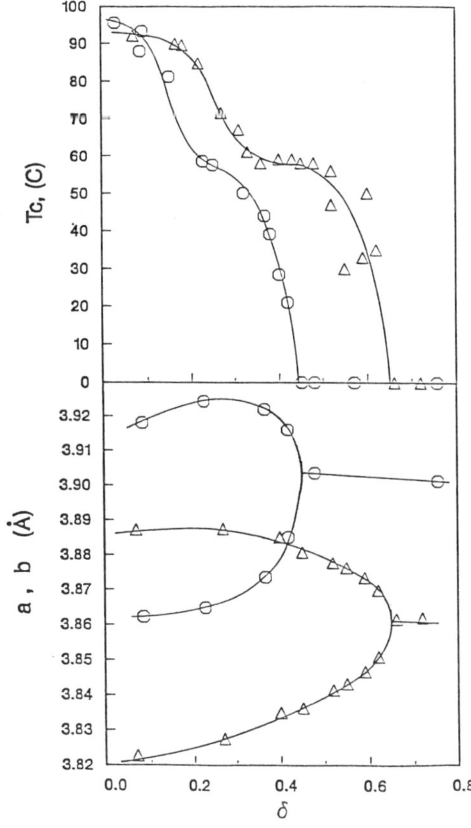

Cation substitution in the R-123 system has been extensively studied. Generally, substitution of trivalent rare earth ions on the R site produces only minor changes in the transition temperature for well oxygenated samples.[48] Since magnetic moments of these substituted ions vary dramatically, it appears that magnetic ions at the R site do not act as magnetic pair breakers. A possible exception is Pr substitution[49] which significantly depresses T_c; Pr-123 does not superconduct. With the exception of the trivalent substituents on the R site, cation replacement almost universally acts to depress T_c. Furthermore, the R-123 phase typically persists (in either orthorhombic or tetragonal symmetry) to sufficiently high levels of substitution such that superconductivity is completely lost. So great is the tolerance for substituted species that several categories of substitution may be considered:

1. Divalent ion on the R-site (e.g., Ca^{2+})[50]
2. Divalent ion on Ba site (e.g., Sr^{2+})[51,52]
3. Trivalent ion on Ba site (e.g., Nd^{3+})[53]
4. Substitution on CuI (chain) site (e.g., Co^{3+})[54,55]
5. Substitution on CuII (plane) site (e.g., Zn^{2+})[55,56]
6. Combinations of substitutions (e.g., La^{3+} on Ba site and Ca^{2+} on R site).[52]

An example of substitution in Y-123, in this case substitution of Fe on the CuI site, is shown in Fig. 10.[57] T_c systematically falls with increasing Fe content, reaching zero when approximately 12% of the Fe is replaced. The $YBa_2(Cu_{1-x}Fe_x)_3O_{7-d}$ system undergoes a transition from orthorhombic to apparent tetragonal symmetry when T_c is higher than 80 K. Perhaps because Fe substitutes in the chains, too remote from the "superconducting planes," Fe does not appear to be a magnetic pair breaker (Fe is no more detrimental to T_c than Al[55] in Y-123). Numerous Mössbauer, diffraction, and X-ray absorption experiments have been conducted to investigate local Fe coordinations. Because of the extreme complexity of the problem, no consensus has been reached to describe the local coordination.

Fig. 10.

The superconducting phase field for $YBa_2Cu_{3-x}Fe_xO_{7-y}$. In this material, Fe substitutes on the Cu 1 (chain) site. Fe depresses T_c and the structure goes tetragonal but apparently not at the composition where superconductivity is lost (Refs. 57).

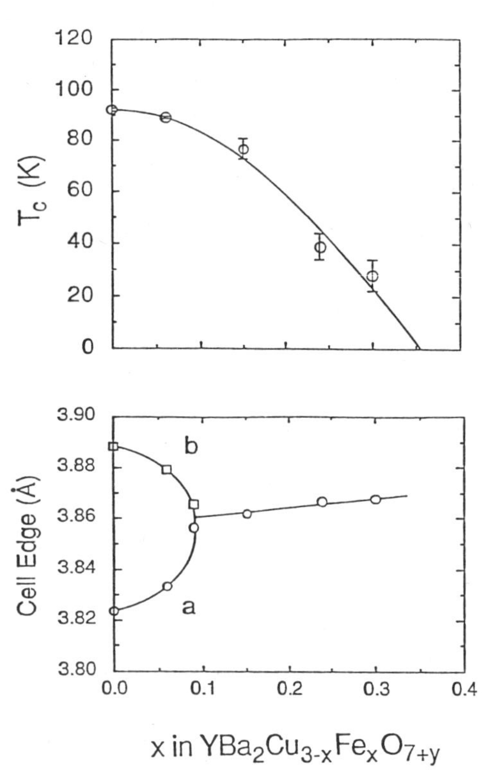

The substitution categories listed above have been extensively studied with emphasis on structural changes, local site symmetries, solubility limits, magnetic properties, charge redistribution, etc. Perhaps the most general characteristic associated with these substitutions (and oxygen depletion) is charge redistribution leading to a reduction of carrier concentration and the eventual onset of insulating behavior. If conduction occurs in the CuO_2 planes, as widely believed, then increased electron density transferred to the planes can reduce the hole concentration and the (hole) density of states at the Fermi level E_F. Reduction of the hole density would be expected for oxygen depletion and for substitution of cations with higher valent species. When hole densities become sufficiently low, insulating behavior develops with the appearance of antiferromagnetic order on the CuI sublattice.

VI. ELECTRONIC PROPERTIES

For conventional metal superconductors, transition temperatures are strongly influenced by the density of electron states at the Fermi level. These conventional metals are treated as Fermi liquids, a sea of interacting electrons contained within a lattice of positive ions. The electronic properties of the metals are nicely described by band theory, a mean-field approach which treats exchange and correlation in an average way. In contrast, oxides which display a substantial degree of ionic or covalent bonding may not be satisfactorily described with band theory. Electrons or holes may be associated with specific atomic sites; their motion through the lattice is highly correlated. In this situation, concepts that arise from Fermi liquid theory (such as the Fermi surface) have no meaning.[58]

While the superconducting oxides frequently display metallic behavior (i.e., the conductivity increases with decreasing temperature), at this time it is not fully resolved whether the superconductors are best described as Fermi liquids or highly correlated electronic systems whose conduction processes are characterized by intersite hopping. Most of the superconductors occur in phase fields which also show insulating behavior and antiferromagnetic order (See Section V). Since the insulators more closely fit the latter (highly correlated) description, and since some vestiges of this correlated behavior show up in the metallic phases (e.g., satellites in photoemission data[59]), the debate continues as to how best to deal with the superconductors and their complex phase fields.

Positron annihilation spectroscopy (PAS) experiments have indicated the presence of a Fermi surface in $YBa_2Cu_3O_{7-x}$ and close agreement with band structure predictions has been reported.[60] The reported Fermi surface topology is consistent with the layered character of Y-123. However, since complex metallic oxide systems have not been extensively studied with PAS, some interpretative complications may remain. For example, it is becoming clear that, for R-123, positrons preferentially sample specific regions of the structure.[61] Trapping by oxygen vacancies may also play a role. Nonetheless, these results have provided strong evidence to support a Fermi liquid description of the electronic properties.

Probably, experiments poised to shed the most direct light on this issue are angle resolved photoelectron spectroscopy (ARPES) experiments. In these experiments, now generally performed on single crystals cleaved at very low temperature ($\lesssim 20$ K),[62] it is possible to monitor the dispersion of electron energy bands in momentum space. The Fermi surface is defined by these bands as they cross E_F throughout the Brillouin zone. ARPES experiments have identified dispersing bands in Y-123 and in $Bi_2Sr_2Ca_1Cu_2O_8$ and have followed them through k-space monitoring their apparent intersection with E_F (hence measuring a portion of a Fermi surface).[63-66] Figure 11 shows an example of band dispersion as observed in an ARPES experiment when an energy band apparently cuts across E_F in midzone.[64] Campuzano et al.[63] have mapped out a substantial portion of this Fermi surface for Y-123. In Fig. 12, these results (circles) are shown superimposed on Fermi surfaces (lines) calculated by band theory.[67] Solid dots represent Fermi surfaces, open dots indicate points in the Brillouin zone where no electrons are found at E_F. While a great deal of detail remains to be garnered from the photoemission experiments (hopefully including measurements with higher energy and momentum resolution), these initial results present a strong case for the validity of Fermi liquid theory as a starting point for understanding the high T_c oxide superconductors.

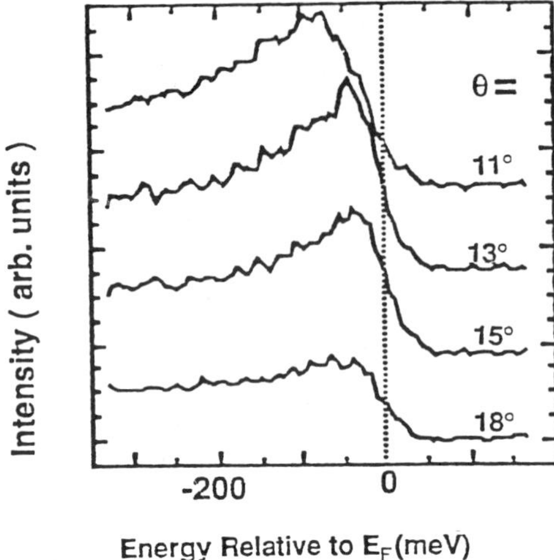

Fig. 11.

Angle resolved photoemission measurements for $Bi_2Sr_2CaCu_2O_8$ taken at different electron pickoff angles θ. These spectra probe electron states at different k-points in the Brillouin zone. These data show that an energy band (below E_F at $\theta = 11°$) moves to E_F with increasing θ (Ref. 64)

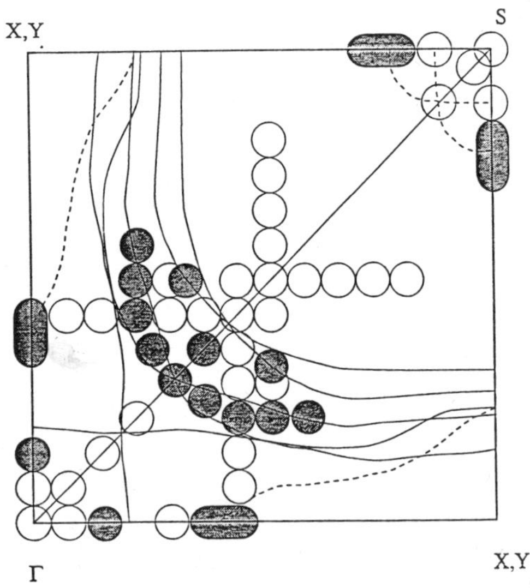

Fig. 12.

Experimentally determined Fermi surfaces of $YBa_2Cu_3O_{7-\delta}$ obtained using angle resolved photoemission (Ref. 63). Filled circles indicate bands crossing the Fermi level. Empty circles are points in the Brillouin zone where bands were not detected at E_F. The lines are theoretical predictions of Yu et al. (Ref. 67).

VII. SUPERCONDUCTING ENERGY GAP

In addition to their great value in providing insight into the fundamental character of the charge carriers in the oxide superconductors, photoemission experiments have provided direct measurement of the superconducting energy gap, Δ, which appears in the electron density of states spectrum as the material becomes superconducting.[64,66,68] Figure 13, taken from Moog et al.,[69] shows the expected behavior of the electron density of states (broadened to account for instrument effects) as the superconducting gap opens. States normally at E_F are piled up below the gap with the magnitude of the effect depending on the size of the gap. Figure 14 shows ARPES data for $Bi_2Sr_2CaCu_2O_8$ taken at 90 K and at 20 K at a selected incident photon energy and electron pickoff angle such that a band apparently intersects the Fermi level.[64] These results show that a superconducting gap, displaying the predicted behavior (Fig. 13), has opened as the temperature is lowered below T_c. The size of the gap (so far reported at limited points of the Brillouin zone), suggests that $2\Delta/kT_c \sim 7\text{-}8$. This is a "strong coupling" result, a deviation from the weak coupling formulation of the original BCS

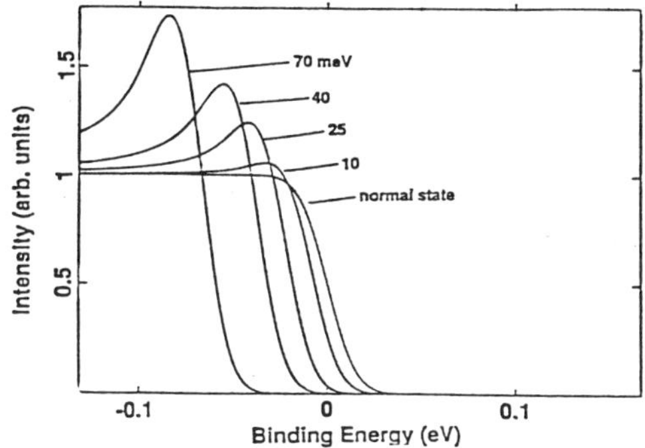

Fig. 13.

Resolution broadened model spectra near the Fermi level calculated for a normal-state $(T > T_c)$ Fermi edge and for BCS superconducting-state spectra $(T \ll T_c)$ for the indicated energy gap values (Ref. 69).

Fig. 14.

Angle resolved photoemission spectra for $Bi_2Sr_2CaCu_2O_8$ ($T_c = 82$ K) taken at hv = 22 eV and an electron pickoff angle of 18°. The open circles are data taken above T_c (90 K) and the crosses are data taken below T_c (20 K). An energy gap (as characterized in Fig. 13) opens as the material goes super-conducting (Ref. 64).

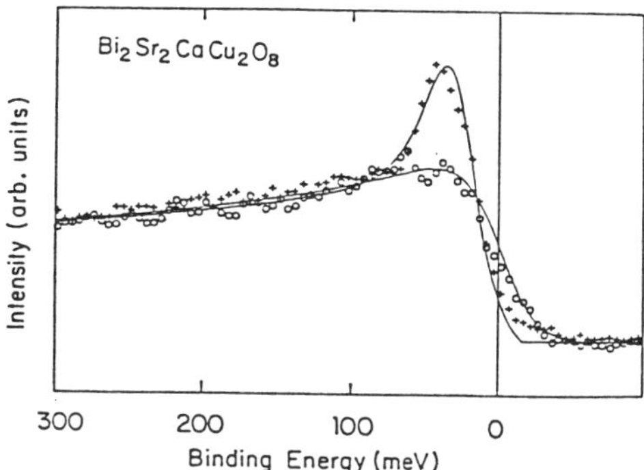

theory. Strong coupling was also implied by PES measurements of Y-123 when oxygen stoichiometries were systematically varied.[70] The conclusion was inferred from the functional dependence of T_c on the electronic density of states.

Other attempts to quantitatively measure 2Δ have met with very mixed results. The classical methods (e.g., tunneling,infrared spectroscopy) have not yet provided a consistent picture though results overlap those reported by the PES measurements.[71]

VIII. SUMMARY

Although a huge worldwide effort has been devoted to the study of oxide superconductors, the mechanism(s) mediating electron pairing have not yet been clearly identified. Consequently the predictive capability afforded by a sound theoretical base remains to be developed. The vast experimental effort has yielded many new compounds and many insights have been garnered from the numerous experiments on oxygen depletion and ionic substitution. The most pervasive characteristic of the high T_c ($\gtrsim$40 K) oxides is the presence of CuO_2 layers. If structural elements in regions external to the planes can produce a suitable charge distribution within those layers, then, presumably, superconductivity at elevated temperatures will occur.

In this paper we have considered a very small subset of experiments chosen to show common trends in a variety of systems or to help clarify critical issues. Apparently essential to such a discussion are considerations of atomic structure and the pervasive CuO_2 layers. Electronic and conduction processes must also be considered. Of particular value are recent photoemission experiments which make a strong case that the superconductors are Fermi liquids. These (and other) experiments observe superconducting gaps of 25-35 meV, indicative of strong coupling behavior. A variety of phase fields were briefly considered, showing that oxide superconductors commonly appear as modifications of insulators with the superconductor and insulator frequently sharing a common (or very similar) crystallographic structure. The data base is extensive; a great theoretical challenge remains.

REFERENCES

1. J. G. Bednorz and K. A. Muller, Z. Phys. B 64:189 (1989).
2. M. K. Wu, J. R. Ashburn, C. J. Tong, P. H. Hor, R. L. Meng, L. Gao, Z. J. Huang, Y. Q. Wang, and C. W. Chu, Phys. Rev. Lett. 58:908 (1987).
3. S.S.P. Parkin, V. Y. Lee, E. M. Engler, A. Nazzal, T. C. Huang, G. Gorman, R. Savoy, and R. Beyers, Phys. Rev. Lett. 60:2539 (1988); S.S.P. Parkin, V. Y. Lee, A. Nazzal, R. Savoy, R. Beyers, and S. J. La Placa, Phys. Rev. Lett. 61:750 (1988).
4. R. J. Cava, B. Batlogg, J. J. Krajewski, R. Farrow, L. W. Rupp, Jr., A. E. White, K. Short, W. F. Peck, and T. Kometani, Nature 332:814 (1988).
5. D. G. Hinks, B. Dabrowski, D. R. Richards, J. D. Jorgensen, Shiyou Pei, and J. F. Zasadzinski, presented at MRS Spring Meeting, Symp. on High T_c Superconductors, San Diego, CA, April 24-28, 1989; Nature 333:836 (1988).
6. P. Chaudhari, J. Mannhart, D. Dimos, C. C. Tsui, J. Chi, M. M. Oprysko and M. Scheuermann, Phys. Rev. Lett. 60:1653 (1988).
7. J. Bardeen, L. N. Cooper and J. R. Schreiffer, Phys. Rev. 106:162 (1957); Phys. Rev. 108:1175 (1957).
8. J. F. Schooley, W. R. Hosler and M. L. Cohen, Phys. Rev. Lett. 12:474 (1964).
9. J. K. Hulm, C. K. Jones, R. Mazelsky, R. C. Miller, R. A. Hein and J. W. Gibson, Low Temperature Physics edited by J. G. Daunt (Plenum, New York, 1965) p. 600; A. M. Okaz and P. H. Keesom, Phys. Rev. B 12:4917 (1975).
10. Ch. J. Raub, A. R. Sweedler, M. A. Jensen, S. Broadston and B. T. Matthias, Phys. Rev. Lett. 13:746 (1964); J. P. Remeika, T. H. Geballe, B. T. Matthias, A. S. Cooper, G. W. Hull and E. M. Kelly, Phys. Lett. 24A:565 (1967).
11. A. W. Sleight, J. L. Gillson and P. E. Bierstadt, Solid State Commun. 17:27 (1975).
12. D. C. Johnston, H. Prakash, W. H. Zachariasen and R. Viswanathan, Mat. Res. Bull. 8:777 (1973).
13. Y. LePage, W. R. McKinnon, J. M. Tarascon, L. H. Greene, G. W. Hull, Phys. Rev. B 35:7115 (1987).
14. H. Maeda, Y. Tanaka, M. Fukutomi, and T. Asano, Jpn. J. Appl. Phys. Lett. 27:L209 (1988); B. W. Veal, H. Claus, J. W. Downey, A. P. Paulikas, K. G. Vandervoort, J. S. Pan and D. J. Lam, Physica C 156:635 (1988).
15. Z. Z. Sheng and A. M. Hermann, Nature 332:138 (1988); Phys. Rev. Lett. 60:937 (1988).
16. E. M. Lifshitz and L. P. Pitaevskii, Stat. Phys., Part 2, Pergamon Press, 1980, Chap. 1.
17. V. L. Ginzburg, Soviet Phys. Uspekhi 13:335 (1970). C. M. Varma, in "Superconductivity in d and f Band Metals," pp. 603-613, ed. W. Buckel and W. Weber, Kernforschungszentrum, Karlsruhe, West Germany (1982).
18. A. Alexandrov and J. Ranninger, Phys. Rev. B 23:1796 (1986).
19. P. W. Anderson and E. Abraham, Nature 327:363 (1986).
20. P. W. Anderson, Science 235:1195 (1987).
21. V. J. Emery, Phys. Rev. Lett. 58:2794 (1987).
22. R. P. Messmer, Solid State Commun. 63:405 (1987).
23. L. Pauling, Phys. Rev. Lett. 59:225 (1987).
24. J. Ashkenazi, C. G. Kuper, and R. Tyk, Solid State Commun. 63:1145 (1987).
25. V. Z. Kresin, Phys. Rev. B 35:8716 (1987).
26. J. Labbe and J. Bok, Euro. Phys. Lett. 3:1225 (1987).
27. C. M. Varma, S. Schmitt-Rink, and E. Abrahams, Solid State Commun. 62:681 (1987).
28. J. D. Jorgensen, B. Dabrowski, Shiyou Pei, D. G. Hinks, L. Soderholm, B. Morosin, J. E. Schirber, E. L. Venturini and D. S. Ginley, Phys. Rev. B 38:11337 (1988).
29. C. W. Chu, P. H. Hor, R. L. Meng, L. Gao and Z. J. Huang, Science 235:567 (1987).
30. J. D. Jorgensen, Jpn. J. Appl. Phys. 26 (Supple. 26-3):2017 (1987).

31. Y. Tokura, H. Takagi, and S. Uchida, Nature 337:345 (1989); H. Takagi, S. Uchida, and Y. Tokura, Phys. Rev. Lett. 62:1197 (1989).

32. J. T. Markert, E. A. Early, T. Bjørnholm, S. Ghamaty, B. W. Lee, J. J. Neumeier, R. D. Price, C. L. Seaman and M. B. Maple, Physica C 158:178 (1989).

33. J. Karpinski, E. Kaldis, E. Jilek, S. Rusiecki, and B. Bucher, Nature 336:660 (1988); P. Fischer, J. Karpinski, E. Kaldis, E. Jilek, and S. Rusiecki, Solid State Comm. 69:531 (1989); D. E. Morris, J. H. Nickel, J.Y.T. Wei, N. G. Asmar, J. S. Scott, V. M. Scheven, C. T. Hultgren, A. G. Markelz, J. E. Post, P. J. Heaney, D. R. Veblen and R. M. Hazen, Phys. Rev. B 39:7347 (1989).

34. Pb may also be substituted on the Bi or Tl site.

35. C. C. Torardi, M. A. Subramanian, J. C. Calabrese, J. Gopalakrishnan, K. J. Morrissey, T. R. Askew, R. B. Flippen, U. Chowdhry and A. W. Sleight, Science 240:631 (1988).

36. For literature, see R. V. Kasowski, W. Y. Hsu and F. Herman, Phys. Rev. B 38:6470 (1988).

37. R. J. Cava, B. Batlogg, J. J. Krajewski, L. W. Rupp, L. F. Shneemeyer, T. Siegrist, R. B. van Dover, P. Marsh, W. F. Peck, Jr., P. K. Gallagher, S. H. Glarum, J. H. Marshall, R. C. Farrow, J. V. Waszczak, R. Hull and P. Trevor, Nature (London) 336:211 (1988).

38. H. Sawa, K. Obara, J. Akimitsu, Y. Matsui and S. Horiuchi, J. Phys. Soc. Jpn. (in press).

39. J. B. Torrence, Y. Tokura, A. I. Nazzal, A. Bezinge, T. C. Huang, and S.S.P. Parkin, Phys. Rev. Lett. 61:1127 (1988); M. W. Shafer, T. Penney and B. L. Olson, Phys. Rev. B 36:4047 (1987); R. B. van Dover, R. J. Cava and E. A. Reitman, Phys. Rev. B 35:5337 (1987).

40. B. Dabrowski, D. G. Hinks, J. D. Jorgensen, and D. R. Richards, presented at MRS Spring Mtg., Symp. on High T_c Superconductors, San Diego, CA, April 24-28, 1989.

41. Y. Kubo, Y. Shimakawa, T. Manako, T. Satoh, S. Iijima, T. Ichahashi and H. Igarashi, to appear in Physica C-Special Issue, Proc. Intl. M^2S-HTSC Conf., Stanford, CA, July 23-28, 1989.

42. J. D. Jorgensen, B. W. Veal, A. P. Paulikas, L. J. Nowicki, G. W. Crabtree, H. Claus and W. K. Kwok, Phys. Rev. B (accepted for publication).

43. J. D. Jorgensen, H. Shaked, D. G. Hinks, B. Dabrowski, B. W. Veal, A. P. Paulikas, L. J. Nowicki, G. W. Crabtree, W. K. Kwok, L. H. Nunez, and H. Claus, Physica C 153-155:578 (1988).

44. H. Shaked, B. W. Veal, J. Faber, Jr., R. L. Hitterman, U. Balachandran, G. Tomlins, H. Shi, L. Morss and A. P. Paulikas, Phys. Rev. B, submitted.

45. J. H. Brewer, E. J. Ansaldo, J. F. Carolan, A.C.D. Chaklader, W. N. Hardy, D. R. Harshman, M. E. Hayden, M. Ishikawa, N. Kaplan, R. Keitel, J. Kempton, R. F. Kiefl, W. J. Kossler, S. R. Kreitzman, A. Kulpa, Y. Kuno, G. M. Luke, H. Miyatake, K. Nagamine, Y. Nakazawa, N. Nishida, K. Nishiyama, S. Ohkuma, T. M. Riseman, G. Roehmer, P. Schleger, D. Shimada, C. E. Stronach, T. Takabatake, Y. J. Uemura, Y. Watanabe, D. Ll. Williams, T. Yamazaki, and B. Yang, Phys. Rev. Lett. 60:1073 (1988).

46. B. W. Veal, A. P. Paulikas, J. W. Downey, H. Claus, K. Vandervoort, G. Tomlins, H. Shi, M. Jensen, and L. Morss, to appear in Physica C-Special Issue, Proc. of the Intl. M^2S-HTSC Conf., Stanford, CA, July 23-28, 1989.

47. P. F. Miceli, J. M. Tarascon, L. H. Greene, P. Barboux, J. D. Jorgensen, J J. Phyne and D. A. Neumann, Proc. MRS Spring Meeting, San Diego, CA, April 24-28, 1989.

48. J. M. Tarascon, W. R. McKinnon, L. H. Greene, G. W. Hull and E. M. Vogel, Phys. Rev. B 36:226 (1987).

49. L. Soderholm, K. Shang, D. G. Hinks, M. A. Beno, J. D. Jorgensen, C. U. Segre and I. K. Schuller, Nature (London) 328:604 (1987); S. P. Goncalves, I. C. Santos, E. B. Lopes, R. T. Henriques, M. Almeida and M. O. Figueiredo, Phys. Rev. B 37:7476 (1988).

50. A. Manthiram, S. J. Lee and J. B. Goodenough, J. Solid State Chem. 73:278 (1988).

51. B. W. Veal, W. K. Kwok, A. Umezawa, G. W. Crabtree, J. D. Jorgensen, J. W. Downey, L. J. Nowicki, A. W. Mitchell, A. P. Paulikas, and C. H. Sowers, Appl. Phys. Lett. 51:279 (1987).

52. A. Manthiram and J. B. Goodenough, Physica C (in press).

53. K. Zhang, B. Dabrowski, C. U. Segre, D. G. Hinks, I. K. Schuller, J. D. Jorgensen and M. Slaski, J. Phys C: Solid State 20:L935 (1987).

54. P. F. Miceli, J. M. Tarascon, L. H. Greene, P. Barboux, F. J. Rotella and J. D. Jorgensen, Phys. Rev. B 37:5932 (1988).

55. J. M. Tarascon, P. Barboux, P. F. Miceli, L. H. Greene, G. W. Hull, M. Eibschutz and S. A. Sunshine, Phys. Rev. B 37:7458 (1988).

56. H. Shaked, J. Faber, Jr., B. W. Veal, R. L. Hitterman and A. P. Paulikas, Phys. Rev. B, submitted.

57. C. W. Kimball, J. L. Matykiewicz, J. Giapintzakis, H. Lee, B. D. Dunlap, M. Slaski, F. Y. Fradin, C. Segre and J. D. Jorgensen, MRS Proc. Vol. 99, High Temperature Superconductors, M. B. Brodsky, R. C. Dynes, K. Kitazawa and H. L. Tuller, eds. (1988) p. 107; B. D. Dunlap, J. D. Jorgensen, C. Segre, A. E. Dwight, J. Matykiewicz, H. Lee, W. Peung and C. W. Kimball, Physica C 158:397 (1989).

58. R. B. Laughlin, Science 242:525 (1988).

59. P. Steiner, S. Hüffner, A. Jungmann, S. Junk, I. Sander, W. R. Thiele, N. Backes and C. Politis, Physica C 156:213 (1988).

60. L. C. Smedskjaer, J. Z. Liu, R. Benedek, D. G. Legnini, D. J. Lam, M. D. Stahulak, H. Claus, and A. Bansil, Physica C 156:269 (1988).

61. D. Singh, W. E. Pickett, R. E. Cohen, H. Krakauer and S. Berko, Phys. Rev. B 39:9667 (1989).

62. A. J. Arko, R. S. List, R. J. Bartlett, S.-W. Cheong, Z. Fisk, J. D. Thompson, C. G. Olson, A.-B. Yang, R. Liu, C. Gu, B. W. Veal, J. Z. Liu, A. P. Paulikas, K. Vandervoort, H. Claus, J. C. Campuzano, J. E. Schirber and N. D. Shinn, Phys. Rev. B 40:2268 (1989).

63. J. C. Campuzano, G. Jennings, M. F aiz, L. Beaulaigue, B. W. Veal, J. Z. Liu, A. P. Paulikas, K. Vandervoort, H. Claus, R. S. List, A. J. Arko and R. J. Bartlett (in preparation).

64. C. G. Olson, R. Liu, A.-B. Yang, D. W. Lynch, A. J. Arko, R. S. List, B. W. Veal, Y. C. Chang, P. Z. Jiang, and A. P. Paulikas, Science 245:731 (1989); Proc. of the Intl. M^{2}S-HTSC Conf., Stanford, CA, July 23-28, 1989, to appear in Physica C-Special Issue.

65. T. Takahashi, H. Matsuyama, H. Katayama-Yoshida, Y. Okabe, S. Hosoya, K. Seki, H. Fujimoto, M. Sato, and H. Inokuchi, Nature (London) 334:691 (1988), Phys. Rev. B 39:6636 (1989).

66. J. M. Imer, F. Pathey, B. Dardel, W.-D. Schneider, Y. Baer, Y. Petroff, and A. Zettl, Phys. Rev. Lett. 62:336 (1989).

67. J. Yu, S. Massidda, A. J. Freeman and D. D. Koelling, Phys. Lett. A 122:203 (1987).

68. R. Manzke, T. Buslaps, R. Claessen and J. Fink, Europhys. Lett. 9:477 (1989).

69. E. R. Moog, S. D. Bader, A. J. Arko, and B. K. Flandermeyer, Phys. Rev. B 36:5583 (1987).

70. B. W. Veal, J. Z. Liu, A. P. Paulikas, K. Vandervoort, H. Claus, J. C. Campuzano, C. Olson, A.-B. Yang, R. Liu, C. Gu, R. S. List, A. J. Arko, and R. Bartlett, Physica C 158:275 (1989).

71. Proc. Intl. M^{2}S-HTSC Conf., Standford, CA, July 23-28, 1989, to appear in Physica C-Special Issue.

THE INFLUENCE OF OXYGEN STOICHIOMETRY ON HIGH-T_c $YBa_2Cu_3O_y$

SUPERCONDUCTOR PROPERTIES

P. Kostić and J.-H. Park

Center for Multidisciplinary Study, Belgrade University
Yugoslavia, and Argonne National Laboratory, Illinois
60439, USA

ABSTRACT

The change of oxygen stoichiometry within $YB_2Cu_3O_y$ ceramics exists
a great influence on electrical properties. This material is p-type,
but under some circumstances (in tetragonal phase, below y = 6.5) it
can be n-type. In the orthorhombic superconducting phase (y between 6.5
and 6.7) the slope of conductivity <u>versus</u> change of stoichiometry is 1,
and above y = 6.7, is 1/4. Frequency measurements of conductivity show
grain boundary properties below y = 6.5, and above that value, some
behaviour characteristic of superconducting material.

INTRODUCTION

After the discovery of superconducting properties in oxides by
Mueller and Bednorz,[1] actually the golden age of superconductivity
started with the emergence of a high-T_c superconducting material
$YBa_2Cu_3O_{7-y}$, wellknown 1-2-3 compound. This material was discovered by
Chu,[2] and it is by now a very well-investigated material. Results
obtained on 1-2-3 compound are in many cases representative for high-T_c
superconductors.[3]

One of the most important issues for this material is the
influence of oxygen stoichiometry on its material properties.[4] It is
well-known that the phase transformation from the high-temperature-
stable tetragonal phase, which shows only non-superconducting
properties, to the low-temperature stable orthorhombic phase which
shows the desired superconducting properties, take place at an oxygen
stoichiometry equivalent to y = 6.5. Below this point exists the
tetragonal state and above it, the orthorhombic state.[5-7] This means
that the oxygen stoichiometry plays a critical role in the needed
crystal transformations and the appearence of the superconducting
properties. But also in the region of stoichiometry where the
superconductivity phenomena are known to exist (6.5 < y < 7.0), there
are also profound stoichiometric influences on electrical properties.
Between y = 6.5 and 6.7, the critical temperature T_c is about 60 K, and
above y = 6.7, the critical temperature is 92 K.[8,9] Therefore, in this
paper the properties of the 1-2-3 material will be analysed from the
viewpoint of the oxygen stoichiometry.

EXPERIMENTAL PROCEDURE

Method

The sample of $YBa_2Cu_3O_y$ ceramic was prepared by sintering a mixture of Y_2O_3, $BaCO_3$ and Cu powder (1/2:2:3). After cutting and polishing, the final sintered specimen had dimensions 0.02x0.2x0.3 cm. Details of sample preparations and experimental techniques have been given in earlier papers.[8,10] The sample was placed in a gas-tight electrochemical cell (Fig. 1). The cell volume was 1 cm^3. The four probe wires were prepared by wrapping the $12.5 \cdot 10^{-6}$ m Pt and $7.5 \cdot 10^{-6}$ m Au wires. The top and the bottom discs of the cell were formed from stabilised zirconia discs (1.25 cm dia. and 0.25 cm thick) that were used as ionic conductors. The upper one was used as an oxygen partial pressure monitor, and the bottom one as an oxygen pump. Electrodes on these discs were made using fired-on platinum paint. By pumping and monitoring partial pressure of oxygen, is one able to determine the change of oxygen stoichiometry inside the sample. The change of oxygen stoichiometry can be controlled by varying the electrical current (i) and the pumping time (t). The equilibrium oxygen concentration (n_{o2}) inside the cell can be determined from the Nernst equation:

$$n_{o2} = it/4F = (w/M)(x/2) \tag{1}$$

or

$$x = (2M/w)\, n_{o2} \tag{2}$$

where F is 1 Faraday, w is the sample weight and M is formula weight $YB_2Cu_3O_{6.5}$, and x in $YBa_2Cu_3O_{y-x}$.

Experimental Measurements

The a.c. electrical conductivity (four probe method, 1592 Hz) and oxygen partial pressure were meeasured simultaneously during oxygen pumping at 500 °C. For coulometric titration (oxygen pumping) measurements, the optimum constant d.c. current was supplied, and at the same time, the pumping current, pumping time and the EMF on the sensor for measuring the oxygen partial pressure were monitored. Conductivity versus a.c. current frequency was measured from 10 to 10^6 Hz. The oxygen stoichiometry was experimentally varied over the range $6.1 < y < 7.0$. Thermodynamical investigations were done in the temperature region 400-900 °C for oxygen stoichiometry level equivalent to y = 6.35, 6.4 and 6.5, respectively.

RESULTS AND DISCUSSION

The thermodynamical investigations were done in the tetragonal non-superconducting phase regime: the resultant conductivity, σ, versus temperature, T, diagram is shown in Fig. 2. The measurements were done while preserving almost the same stoichiometry because both the cell and the oxygen volume inside it were small. There are no significant differences in the shape of the curves between oxygen stoichiometries below 6.5 (see Fig. 2. a,b and c), but with increasing stoichiometry, the magnitude of conductivity decreased. This indicates that n-type conductivity predominates, as compared with the original p-type conductivity of 1-2-3 compound. In Fig. 3. this p-n transition is obvious. This also means that many previously published results were in fact done under conditions where n-type conductivity prevails.

Above the stoichiometry level y = 6.5 exists the stable
orthorhombic superconducting phase. Previously obtained results from
magnetization measurements in this region show two critical
temperatures, the first at about 60 K (6.5<y<6.75) and the second at
92 K (6.75<y<7.0).[11] Continous coulometric conductivity investigations
in this region (6.5<y<7.0) show interesting results (Fig. 4). In the
log-log plot, conductivity <u>versus</u> change of oxygen stoichiometry (x in
$YBa_2Cu_3O_{6.5+x}$) displays a linear dependence for 6.5<y<6.7, and a 1/4
dependence above that range. This suggests that oxygen is incorporated
as the oxygen vacancy in the first region, and also by some other
process in the second region (e.g., incorporation of the uncharged
vacancies).

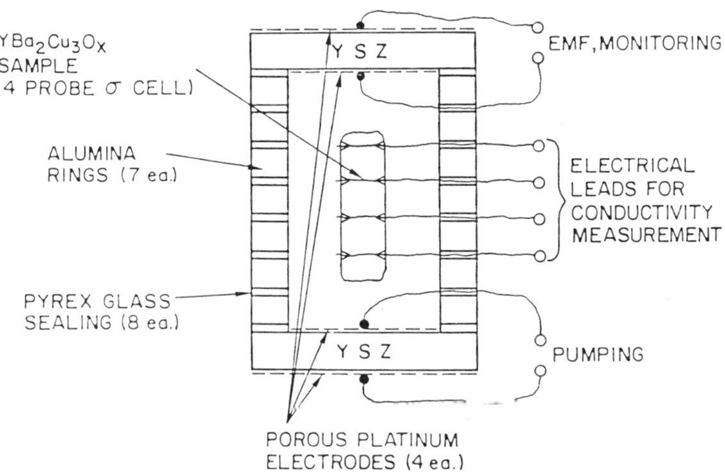

Fig. 1. Gas tight electrochemical cell and assembly.

Measurements of frequency, f, <u>versus</u> conductivity, σ, for
different levels of oxygen stoichiometry are presented in Fig. 5. There
are three different slope shapes: below y=6.5, in the tetragonal phase,
conductivity slowly increases in the second part of frequency region
(slopes 9,10). In the transition region, there exists some frequency
resonance maximum (slopes 6 and 7), and finally, in the orthorhombic
phase, there exist antiresonance minima (slopes 6,5,4,3,2 and 1). In
explaining this behaviour we must consider properties of the ceramic
itself. As the tegtragonal phase has higher conductivity we must
consider the grain boundary which can be represented by parallel
connection of resistivity and capacitance. With the frequency
increases, the current starts to flow by means of capacitance as well
as grain resistivity, and the total grain-grain boundary resistivity
with increase.[12-15] As the oxygen stoichiometry is altered, increasing
oxygen diffusion takes place from the grain boundary to the grain, and
at approximately y=6.5, there probably exists a mixed phase (including
both tetragonal and orthorhombic phases) and therefore slopes 7 and 8
are measured, as a result of surface states on the grain boundary.[16]
Finally, above y=6.5, the resonant observed maxima must be essentially
connected with superconductivity phenomena. Several explanations for
these findings can be suggested e.g., a cation disordering in the

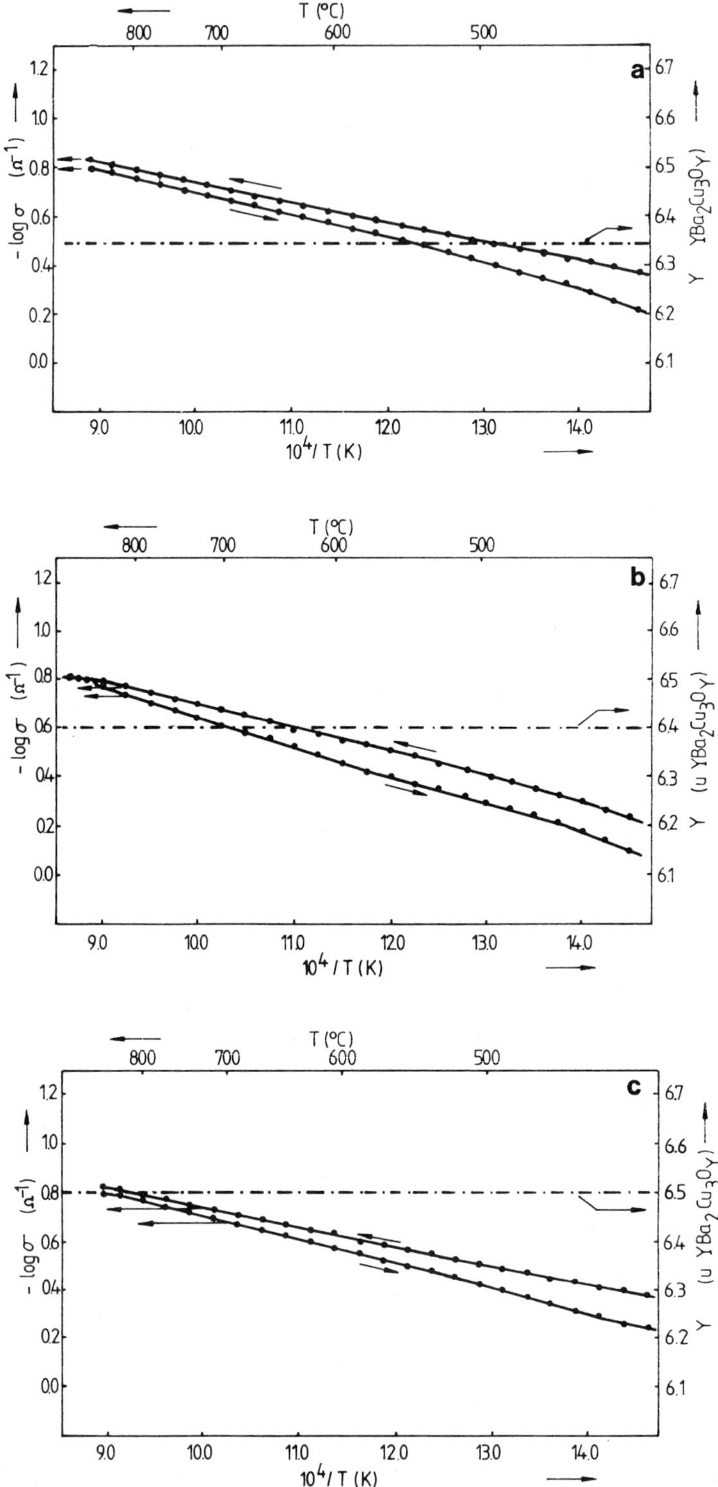

Fig. 2. Electrical conductivity (σ) versus temperature (T) for oxygen
stoichiometry: (a) y = 6.35, (b) y = 6.4 and (c) y = 6.5.

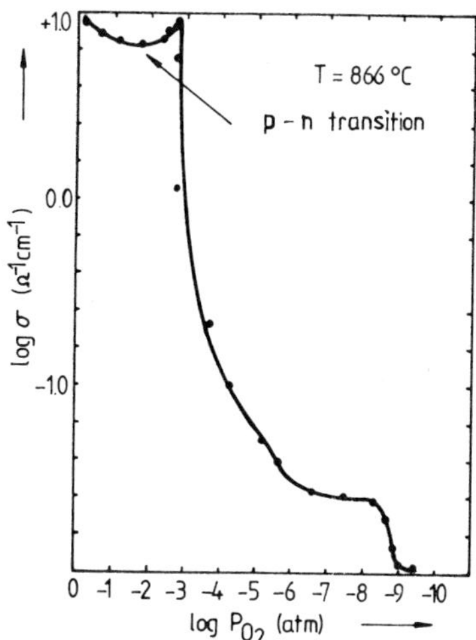

Fig. 3. Electrical conductivity (σ) versus partial pressure of oxygen
        (P_{O2})

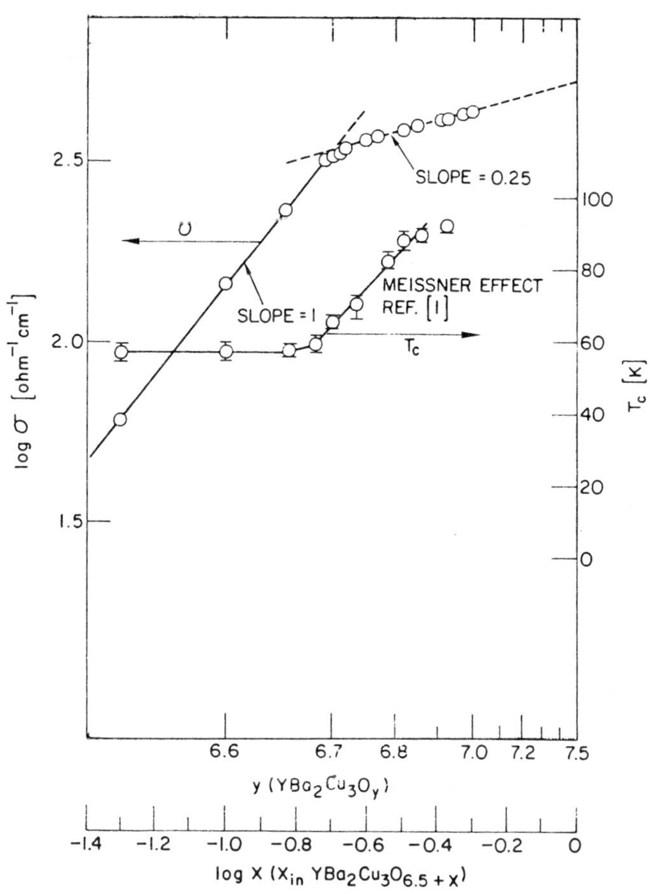

Fig. 4. Electrical conductivity (σ) versus change of oxygen
        stoichiometry x (x in $YBa_2Cu_3O_{6.5+x}$), and magnetization
        measurements of the T_c.

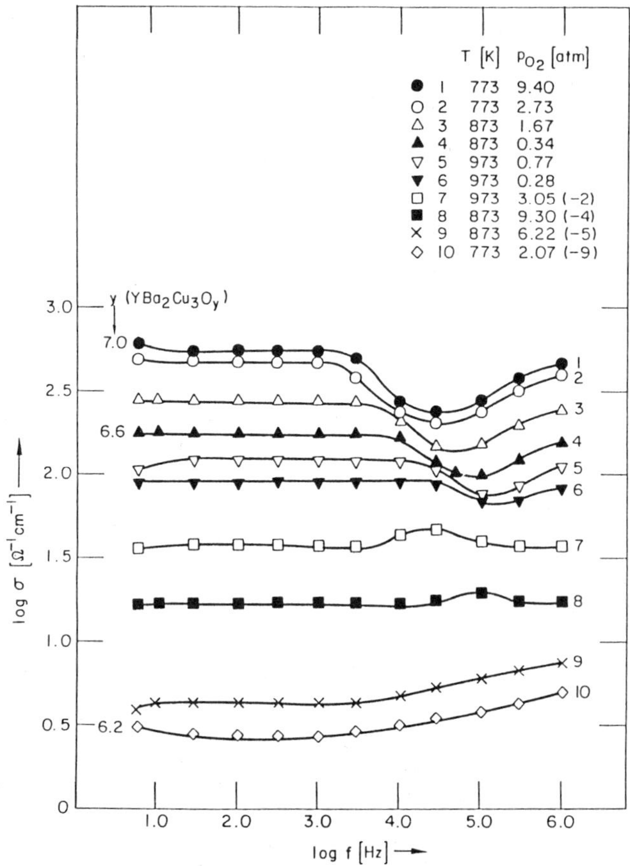

Fig. 5. Frequency dependence of conductivity for different oxygen
stoichiometry (y).

lattice,[17] disconnections of the different valence copper ions,
interaction between copper ion and oxygen vacancies,[12] etc.

CONCLUSION

As in other oxide materials, the oxygen stoichiometry plays a most
important role in the forming process for the superconducting 1-2-3
compound, and also affects its superconducting properties. This
material is p-type, but under some circumstances it can also be n-type.
This happens below stoichiometry y=6.5 in the tetragonal phase. In this
case the measured conductivity is thermally activated.

The coulometric titration measurements show two distinct regions
in the orthorhombic phase. Namely, the slope of conductivity <u>versus</u>
change of stoichiometry equals 1 in the region 6.5<y<6.7 and <u>1/4</u> above
this range. This indicates the presence of oxygen, pressumably
incorporated as (a) ionised and (b) ionised and neutral vacancies,
respectively.

Measurements of conductivity <u>versus</u> frequency serve to illustrate
these different ceramic behaviours. Below stoichiometry y=6.5, the
grain boundary capacitance and resistivity are significant. At the

transition point, y=6.5, a mixed structure of two phases exists, and there are some indications that additional surface states are present. In the superconducting phase, y>6.5, the conductivity <u>versus</u> frequency slopes are directly related to the presence of copper ions and oxygen vacancies, and hence with the intrinsic superconducting phenomenon itself.

ACKNOWLEDGEMENTS

The authors thanks to Drs D. J. Lam, S. J. Rothman and S.- K. Chan for helpful discussion during these studies. This work supported by the U. S. Department of Energy, Basic Energy Sciences-Materials Science, under Contract W-31-109-Eng-38, and by the Serbian Academy of Sciences and Arts Committee for Physical Chemistry.

REFERENCES

1. J. G. Bednorz, K. A. Mueller, K. Takahashi, Science, April (1987) 73.
2. C. W. Chu, P. H. Hor, R. L. Meng, L. Gao, Z. J. Huang, Y. Q. Wang, Phys. Rev. Lett., 58(1987)405.
3. P. Kostić, "Philosophies and Prognosis in High T_c Superconducting Ceramics", "Electroceramic II" Brisel, September 1988. (in press).
4. J. D. Jorgensen, M. A. Bendo, D. G. Hinks, L. Soderholm, K. J. Volin, R. L. Hitterman, J. D. Grace, I. K. Shuller, "Oxygen Ordering and the Orthorhombic-to-Tetragonal Phase Transition on $YBa_2Cu_3O_{7-x}$", Phys Rev. B (in press).
5. M. H. Whangbo, M. Eavin, M. A. Beno, J. M. Williams, "Band Electronic Structure of High-Temperature ($T_c<90$ K.) Superconductor Orthorhombic $YBa_2Cu_3O_7$", Inorg.Chem. (in press).
6. J. F. Maruco, C. Noguera, P. Goroshe, G. Kollin, "Thermodynamic Study at High Temperature of the Superconducting System $YBa_2Cu_3O_z$ with $6<z<7$", J. Mat. Res., (in press).
7. E. G. Derauane, Z. Gabelica, J. L. Bredas, J. M. Andre, Ph. Lambin, A. A. Lucas, J. P. Vigneron, Solid State Comm., 7(1987)1061.
8. J. H. Park, P. Kostić, Mat. Lett., 6(1988)393.
9. M. M. Ristić, P. Kostić, "Materials Science - the Base of Progress in New Electronics", German-Yugoslav Meeting 1989. Stuttgart.
10. J. H. Park, P. Kostić, J. P. Sing, "Electrical Conductivity and Chemical Diffusion in Sintered $YBa_2Cu_3O_y$", Mat.Lett., (in press).
11. B. Veal, a private communication.
12. C. W. Wert, R. M. Thomson, "Physics of Solids", McGraham Hill Book, (1964)495.
13. J. Kawamura, R. Sato, S. Mishina, M. Shimoji, Solid State Ionics, 26(1987)155.
14. S. Geller, "Solid Electrolytes", Topic in Appl. Phys., Springer -Verlag, 21(1977)89.
15. P. Hagenmuller, W. VanGool, "Solid Electrolytes General Principles, Characterisation, Materials, Applications", Mat. Sci. and Tehn., Academic Press, (1978) 127.
16. H. Deuling, K. Klausman, A. Goetzberger, J. State Electronics, 15(1972)559.
17. K. Funke, J. Klaus, R. E. Lechner, Solid. State Comm., 10(1974)1021.

A MODIFIED NONCONTACT METHOD FOR CRITICAL TEMPERATURE MEASUREMENTS OF

HIGH TEMPERATURE SUPERCONDUCTORS

P.M. Nikolić, M. Miletić, M.B. Pavlović,
Z.B. Maričić, and D. Raković

Belgrade University, P.O. Box 816, 11001 Belgrade
Yugoslavia

INTRODUCTION

A significant breakthrough in the synthesis of high-Tc
superconductors has been accomplished during the last three years.[1] As
large numbers of various experimental samples are produced, there is an
increasing need for a rapid, non-contact method for their evaluation.

There are several methods which can be employed to obtain the
critical temperature of high-Tc superconductors: four-contact
resistivity techniques, SQUID dc-magnetization measurements, inductance
bridge ac-susceptibility measurements, eddy current resonant frequency-
shift technique, etc.

The four-contact Van der Pauw method[2] is often used, but it
requires reliable, low resistivity ohmic contacts which cannot be
easily made. Silver paint is usually used for this purpose, although
leads could be better connected with the superconducting sample by
indium ultrasonic soldering, etc.

Non-contact methods for characterization of superconductor
susceptibility and penetration depth have been used for several
decades.[3] They are rather expensive as they need very sensitive SQUIDs,
ac voltmeters, inductance bridges, etc. Recently, a cheaper non-contact
method for Tc measurements, based on a modification of the resonant
frequency of an LC circuit, as changing levels of eddy current modify
the inductance of a sensor-coil arrangement, has been developed.[4]

In this work, we present an inexpensive noncontact method for
critical temperature measurements of high-Tc ceramic superconductors.[5]

EXPERIMENTAL

Our technique is a modified form of an inductance bridge method
(Fig. 1). Current of the desired frequency f is supplied by a 10 Hz –
200 kHz oscillator and fed to inductances, L1 and L2, in series, wound
in the opposite way. The inductances L3 and L4 are inductively coupled
with inductances L1 and L2, respectively. In this bridge, we also use
two resistors, R1 and R2, each of 22 kΩ, and a variable resistor of

10 kΩ. The coils, L2 and L4, have soft ferrite cores, to amplify the
response of the bridge. However, it makes the bridge asymmetric in
principle, displaying the output voltage Uo, even without a
superconducting sample placed between inductors L2 and L4. As
illustrated in Fig. 1, a superconducting sample is placed between
inductors L2 and L4, which remains at room temperature as the sample is
thermally manipulated. Eddy currents are induced in the sample by the
coils, L2 and L4, and produce magnetic fields that modify both the
self-inductances (L2 and L4) and the mutual-inductance (L24) of the
sensor coils. The magnitude of these eddy-current fields, and their
output voltage U, are functions of the temperature-dependent sample
resistivity, $\rho(T)$.

The response of the inductance bridge is measured using a phase
detector (ORTEX 9503). The reference input (U_{ref}) of this detector is
connected to a function generator. The output signal of the phase
detector supplies a personal computer across an AD converter (cf. Fig.
1).

The superconducting sample was an 8 mm diameter pellet, 2 mm
thick, which was put in a sample holder made of a pertinax insulator. A
copper-constantan thermocouple was attached to the sample with
thermoconductive paste. The other end of the thermocouple was held
permanently in liquid nitrogen. The measuring signal for this
thermocouple was connected to the personal computer using the same AD
converter (cf. Fig. 1).

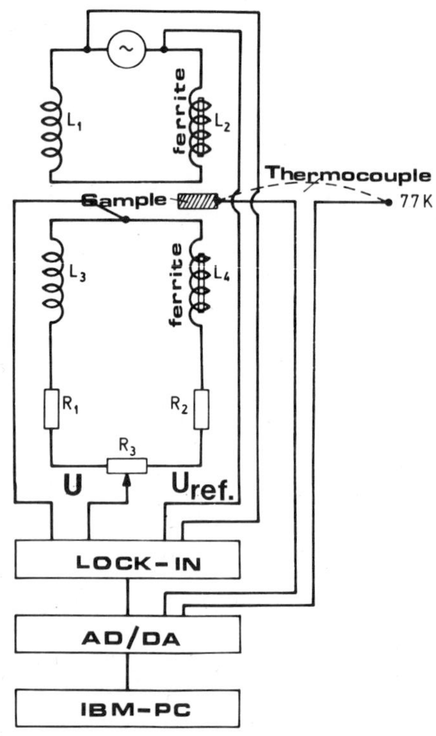

Fig. 1. Modified noncontact inductance bridge for critical temperature
measurements

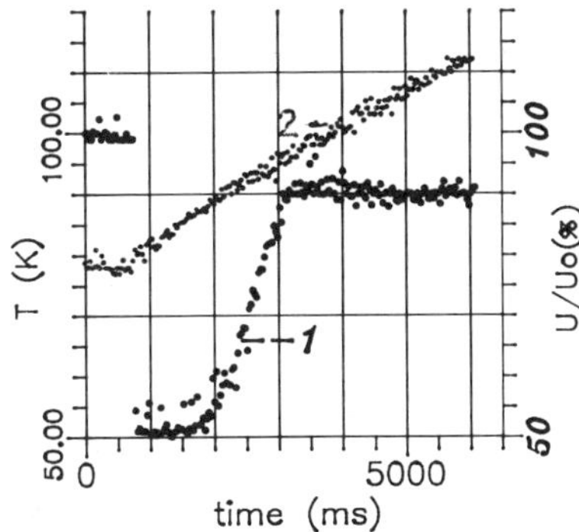

Fig. 2. Temporal changes of relative bridge response (curve 1) and temperature (curve 2) for a superconducting sample (placed between coils L2 and L4) and warmed spontaneously from liquid nitrogen to room temperature (cf. Fig. 1)

The sample holder was always put in the same position in the narrow space between coils L2 and L4 using a specially made lead. Measurements were done with sample-in (U) and sample-out (Uo) positions at room temperature first. Afterwards, the sample was cooled down by immersing the sample holder, with the sample and its thermocouple together, into liquid nitrogen. Then, the sample holder was returned between the coils L2 and L4 as quickly as possible (in 1-2 s). The sample was warmed up slowly and the change of the responses of the bridge and thermocouple were registered by the computer as a function of time.

The time dependencies of the relative response of the bridge (U/Uo) and the sample temperature (T) are given in Fig. 2. The responses of both systems were registered every 30 ms. A sudden change of the bridge response was noticed about 800 ms after the beginning of the experiment. Then, for another 1200 ms the level was almost

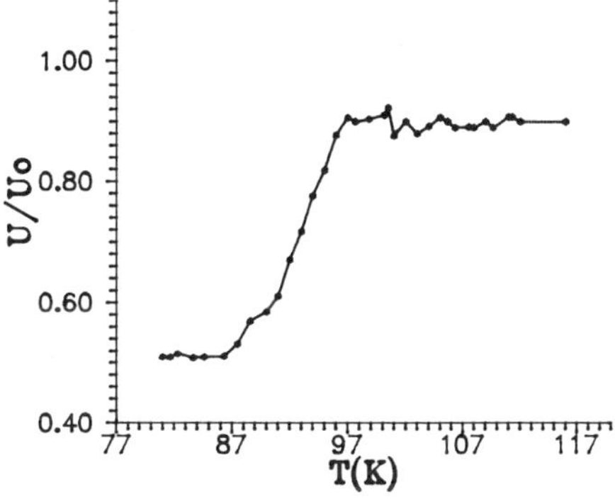

Fig. 3. Calculated relative change of the bridge response versus sample temperature

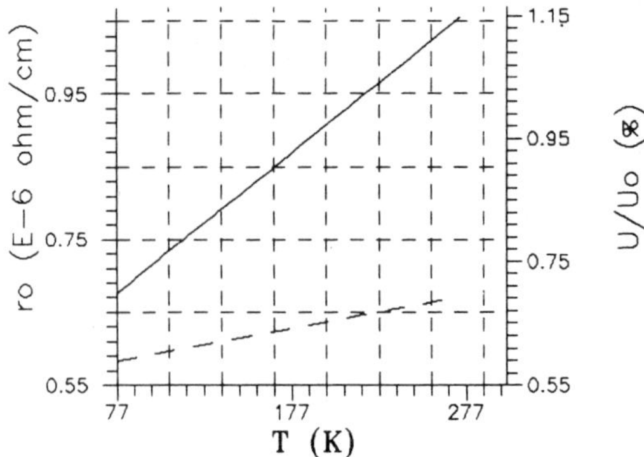

Fig. 4. The temperature changes of relative bridge response (solid
line) and calculated resistivity (dashed line) for a copper
sample

unchanged and then it started to increase almost linearly, reaching the
onset after approximately another 1200 ms. It should be pointed out
that the change of relative response between room and liquid nitrogen
temperatures was about 50%. The change of signal (U) was about 50 mV,
which could be obviously measured by ordinary multimeters, in contrast
to indispensable ac voltmeters with a sensitivity better than 1 μV in
the 1 mV range for the usually applied Tc measurements.

The critical temperature of our superconducting sample could be
easily read using the temporal changes of the relative bridge response
(U/Uo) and sample temperature (T) versus time, displayed in Fig. 2. Its
value was about 87 K and the onset temperature was about 97 K. The
relative change of the bridge response from the bottom to the top level
happened in about 1.5s. The values of critical and onset tempeatures of
the suprconducting sample can be read more accurately from Fig. 3,
where recalculated relative change of the bridge response versus
temperature (obtained from Fig. 2) is displayed.

A similar experiment was done using a copper sample of the same
diameter and thickness. The results obtained during this experiment are
given in Fig. 4 (solid line), together with the calculated resistivity
of the copper sample (dashed line).

DISCUSSION

Our method[5] is a simple, modified inductance bridge method, where
the change of relative response is very high ($\sim$ 50-100 mV), and can be
easily measured using ordinary laboratory multimeters. For the other
techniques, for instance, the usually used inductance bridge and four-
contact resistivity methods, one needs very expensive and much more
accurate instruments. This means that the measured signals are roughly
about a thousand time larger with our method than with currently well
known and used methods.[3]

It is interesting to compare our diagrams, obtained by the
inductance bridge method, with the results of Van der Pauw's resistance
measurements versus temperature, measured in a $YBa_2Cu_3O_{7-x}$ pellet (cf.

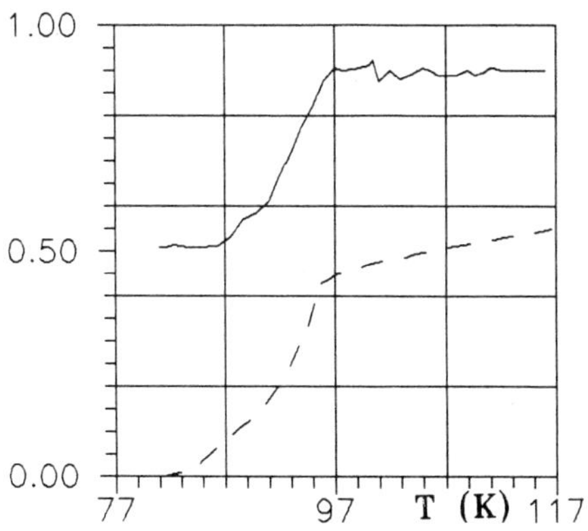

Fig. 5. The temperature changes of relative bridge response (solid
line) and electrical resistance obtained by four-contact
method (dashed line) for $YBa_2Cu_3O_{7-x}$ sample

Fig. 5). They are slightly different, giving the same onset temperature
(~97 K), but a slightly higher critical temperature by inductance
bridge method (~87 K) in comparison with the resistivity method (~ 82 K).

With some modification our method could be used for detection of
the relative proportion of superconducting phase in the high-Tc
samples. To achieve this, the bridge should be made symmetrical by
using soft ferrite cores in all coils (L1, L2, L3 and L4), giving zero
response (Uo=0) without the samples between coils. By placing
successively two ethalonic high-Tc samples (with known relative ratios
of superconducting phase, evaluated by ac susceptibility measurements,
for instance) between coils L2 and L4, and the sample between coils L1
and L3, the obtained bridge responses (U) could serve to obtain the
superconducting relative ratio of the characterized high-Tc sample.

As our inductance bridge method, based on eddy-current inductance
influences, probes a volume of the sample corresponding to the surface
area times a depth determined by either the normal-state skin depth (~ μm)
or superconducting-state London penetration depth (~ few hundred nm),
this approach appears to provide a better measure of the
superconducting quality for thin films than for bulk samples.

For the same reason, our method is generally a better measure of
superconducting quality than the four-contact dc resistance
measurement, which is not a true measure of bulk superconductivity, as
it is only necessary for the current to find the minimum percolative
path for zero resistance to be achieved. In addition, the inductance
bridge method does not require electrical contacts, which are very
difficult to make in a low resistivity form on high-Tc ceramic
superconductors.

CONCLUSION

A modified non-contact inductance bridge method for fast and inexpensive critical temperature measurements of high-T_c superconductors,
by using ordinary ac instruments, has been described.

REFERENCES

1. J. G. Bednorz and K. A. Muller, Z.Phys. B 64, 189 (1986); M. K.
   Wu, J. R. Ashburn, C. T. Torng, P. H. Hor, R. L. Meng, L. Gao,
   Z. J. Huang, Y. Q. Wang, and C. W. Chu, Phys.Rev.Lett. 58, 908
   (1987); C. Michel, M. Hervien, M. M. Borel, A. Grandin, F.
   Deslandes, J. Provost, and B. Raveau, Z.Phys. B 68, 421
   (1987); Z. Z. Sheng and M. A. Hermann, Nature 332, 138 (1988).
2. L. J. van der Pauw: "Measurement of Sheet Resistivities with the
   Four-Point Probe", Bell Syst. Techn. J. 411, 1958.
3. D. Shoenberg, Proc. R. Soc. A 175, 49 (1940); M. Desirant and D.
   Shoenberg, Proc. Phys. Soc. 60, 413 (1948).
4. J. D. Doss, D. W. Cooke, C. W. Mc Cabe, and M. A. Maez, Rev. Sci.
   Instrum. 59, 659 (1988); J. D. Doss, D. W. Cooke, P. N.
   Arendt, M. Nastasi, R. E. Muenchausen, and J. R. Tesmer,
   Supercond Sci. Technol. 2, 63 (1989).
5. P. M. Nikolić, M. Miletić, M. B. Pavlović, Z. B. Maričić, and D.
   I. Raković, Patent application P. 1286/89, Belgrade (Yugoslavia).

Part XI. NON-OXIDE CERAMICS

ON THE KINETICS OF DENSIFICATION DURING LIQUID PHASE SINTERING OF Si_3N_4

J. Pabst and M. Herrmann

Central Institute of Solid-State Physics
and Materials Research
Academy of Sciences of the DDR, Dresden, DDR

INTRODUCTION

The classical model for describing sintering kinetics has already
been developed by Kingery[1] in 1959, who divided the sintering process
into three sequential processes, rearrangement,solution/reprecipitation
and coalescence, and formulated temporal laws for all stages. This
paper will try to investigate the possibility of application of the
Kingery law (three sequential processes, power law for the solution/
reprecipitation step) for describing the densification of Si_3N_4-powders
and reaction bonded silicon nitride (RBSN) during sintering.

First the temporal law of the solution/reprecipitation stage will
be discussed. In the scope of a two particle model, a statement for the
influence of particle shape will be succeeded by a more realistic
description (particles smoothed, with different proportions). Already
small changes in the shape lead to great divergency from the power law.
Therefore, it is necessary to verify through observation the validity
of the power law[2,3] during densification.

Next, there will be discussed the division of the densification
process into sequential processes. The sequential stages of the Kingery
law can be used in order to describe the kinetics during densification
of Si_3N_4-powders.[2] Very important is the densification during sintering
of RBSN, since great deviations occur in the principal behaviour during
densification in contrast to Si_3N_4-powders. In the area of densification
of RBSN, other processes appear than the three sequential processes of
liquid phase sintering. The densification process of RBSN must be
divided into two modified stages.

THE MODEL

As a result of rearrengement, contacts between neighbouring
particles are developed. Such contacts can be described in the two
particle model. Both particles are surrounded by a liquid film. In the
contact region between both, a liquid neck is developed. The capillary
forces of the liquid cause an elastic tension of the particles and a
stress field develops in the contact plane. The contact plane is
characterized by both contact areas with special microstructures and

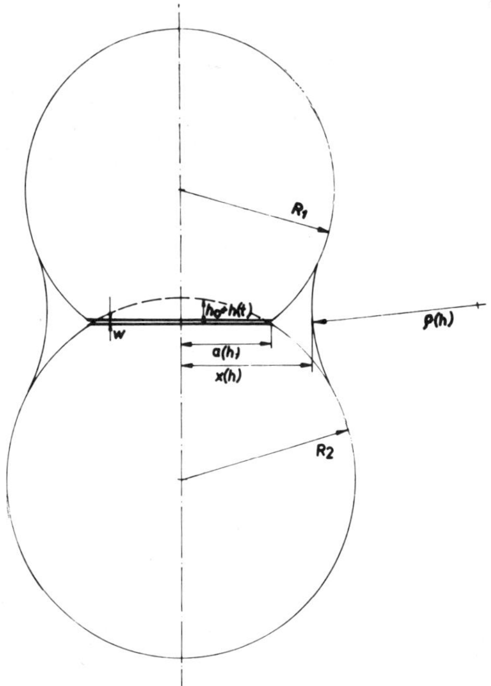

Fig. 1. Geometry model for model calculations.

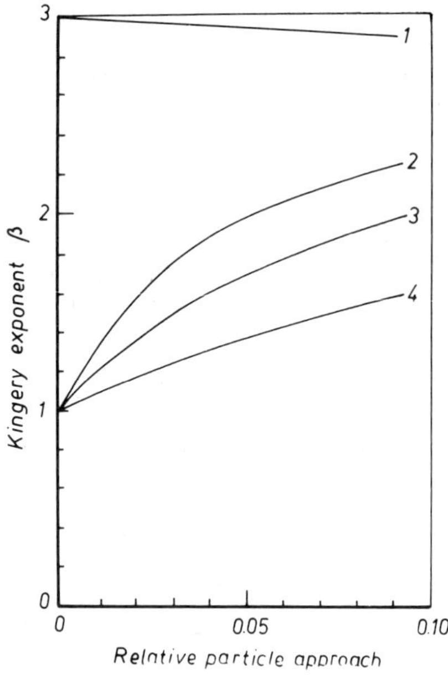

Fig. 2. Kingery exponent as a function of particle approach h for equal
        particles and a volume fraction of 2,5% liquid phase at
        different geometries
        (1)   exact spheres,
        (2) - (4)   spheres with initial flattenings
             ((2)...0.025, (3)...0.05, (4)...0.10 relative initial
             flattenings).

between a thin liquid film, the existence of which is dependent on microstructure and adhesional forces.

The normal stress gradients occurring in the contact region and to which equivalent concentration gradients correspond, cause a mass transport via diffusion from the contact region, which at adjusted quasistationary conditions is characterized by an uniform material friction in each point of the contact plane. The normal stress distribution in the contact plane corresponds to the quasistationary state.

In the liquid shell, the capillary pressure induced by surface geometry dominates. Furthermore, the normal forces are balanced in the contact plane. For the description of the centre approach of the particles, the variable h is used. It represents the "flattening" of the spheres, which takes place during sintering, at the time t=0 also h=0 is valid. Thus it becomes possible to determine the geometry functions x, a, and ρ (see Fig. 1) as a function of h, the initial flattening h_0, the ratio of sphere radii R_1/R_2 as well as the amount of liquid phase numerically exact within the torus approximation. Thus, the stress field as a function of these variables is also known.

To obtain a sintering equation, the relationship between centre approach and the actual sintering time must be known. Such a relationship is obtained by considering the continuity of the process. In order to realize an approximation of the particle centres around dh, the volume $(h + h_0)$ $(2 R - h - h_0)$ dh per particle must be solved and transported out of the process zone. In the following, it is assumed that the rate-determining mechanism is not the solution, but the diffusion through the liquid phase. Then a material amount $2\pi wa$ j(a) dt must be transported out of the process zone of the thickness w within time interval dt. The diffusion flow j (a) is given by the stress gradient at the border of the process zone. The equalization of both contributions and the integration leads to the required relationship between time of sintering and centre approximation:

$$F (h) = k \cdot t \tag{1}$$

By normalization of the specific surface energy, it has been made possible that the left-hand side of equation (1) is a pure geometrical function, whereas k on the right-hand side contains all the material-specific characteristics.

In order to compare the results with the Kingery model, the function F (h) is represented by

$$F (h) = ch^\beta \tag{2}$$

In the classical Kingery law, the value 3 stands for β. The Kingery exponent β has been calculated for equally sized exact spheres and for spheres with different initial flattenings.

The volume amount of liquid phase has been varied. Additionally, calculations for differently sized spheres have been performed.

RESULTS OF THE CALCULATIONS

A very important influence on sintering kinetics is given by the shape of the sintering particles (Fig. 2). If the spheres are exactly spherical, a weakly compaction-depending exponent β results, which is

as great as that predicted by Kingery. Deviations from spherical shape
(initial flattening) have a strong effect on the exponent. In contrast
to the exact spheres, the exponent has the value 1 in the initial phase
of the process and only for small flattenings and great approximations
does n the exponent approach the value of Kingery.

The densification kinetics are determined only by the geometry of
the processing zone within the frame of the model presented here, the
residual shape of the particles is of minor importance. Consequently,
for polyhedrons a comparable kinetics as can be seen in Fig. 2 can be
expected. The initial flattenings in these calculations are chosen in
such a way, that the structures in the ratio of total surface contact
area correspond to regular polyhedrons, (2) ikosahedrons, (3)
dodecahedrons and (4) to an octahedron. Calculations of the shrinkage
exponent for the case of two equally and differently sized spheres
(ratio of the sphere radii $R_1/R_2 = 1.2$) shows that the size of
particles (initial flattenings/polyhedrons) has the greatest effect on
the deviation from the Kingery law.

APPLICATION OF THE RESULTS FOR THE DENSIFICATION OF SILICON NITRIDE

For the densification of silicon nitride, Hampshire and Jack[2] as
well as Hayashi et al.[3] proved the validity of the Kingery law by
experiment (with n = 3 for MgO-doped silicon nitride). The presented
volume shrinkage ($\Delta V/V_o$), however, is composed of two parts: firstly,
the fraction already densified by rearrangement in the beginning of the
resolution step ($\Delta V_R/V_o$, i.e. $\Delta V(t_R)/V_o$) and secondly the fraction
densified by solution/precipitation ($\Delta V_{SP}(t)/V_o$, i.e. $\Delta V(t)/V_o -
\Delta V(t_R)/V_o$ for $t \geq t_R$). Both these fractions represent physically
different effects. The Kingery law, however, contains only one
statement concerning the shrinkages obtained by solution/precipitation

$$\Delta V_{SP}(t_{SP})/V_o = \bar{c} \cdot t_{SP}^{1/3} \tag{3}$$

where the t_{SP} is the difference between actual total time and
rearrangement time (at this moment being constant).

The fraction densified by solution/precipitation is obtained:

$$\Delta V_{SP}(t_{SP})/V_o = \bar{c} \cdot t_R^{1/n} ((1+t_{SP}/t_R)^{1/n} -1) \quad (n = 3) \tag{4}$$

Equation (3) is only contained as a limiting case for long times
($t_{SP} \gg t_R$) or dissapearing rearrangement times (and contributions). It
is also recommended that eq. (4) be fitted to a function

$$\Delta V(t)/V_o = \bar{c} \cdot t_{SP}^{1/\bar{\beta}} \tag{5}$$

for $t = 0, \bar{\beta} = 1$ is valid, for propagating times, $\bar{\beta}$ increases similarly
as the theoretically calculated exponent β above.

A principally analogous behaviour is observed as with the sintered
exponents for polyhedrons determined from the model. This finding will
be used to transform the sintering equation (eq. 2) into a more
convenient form, where the form of (eg. 4) will take over:

$$\Delta V_{SP}/V_o = c (t_{SP}+ t_G)^{1/n} - c \cdot t_G \tag{6}$$

$\Delta V_{SP}/V_o$ is the change of volume caused by the contact flattening
mechanism described. The time t_G is physically evaluated as the
fictitous time, which is necessary to create hypothetically such a

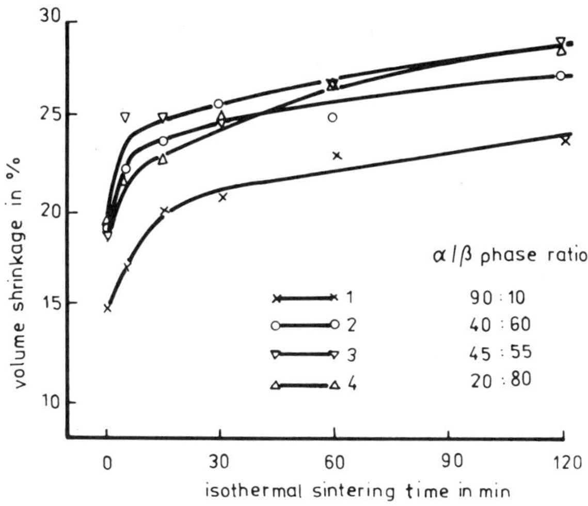

Fig. 3. Isothermal shrinkage of sintered 3%-MgO-Si$_3$N$_4$ (RBSN) at 1720°C with different α/β phase ratio.

geometry corresponding to the real one from ideal spheres by solution/precipitation processes. It is considered to be a characteristic of the given microstructure, since it contains integer findings about the deviations from spherical shape.

To obtain a sintering equation for the solution/precipitation step, it is recommended that the results of the rearrangement process be considered:

$$\Delta V(t)/V_o = r(t-t_R+t_G)^{1/n} + \Delta V_R/V_o - c\, t_G^{1/n}, \quad t > t_R \tag{7}$$

According to this equation, a classical Kingery law can be observed only under two conditions:

1. a system of nearly equal, exact spheres ($t_G=0$), which are packed as dense as possible ($t_R=0$).

2. the rearrangement time corresponds to geometry time $t_R \approx t_G$.

While the first case can be realized only in special model systems, the case $t_R \approx t_G$ is considered to be an occasion, since different physical phenomena contribute to both quantities.

THE DENSIFICATION KINETICS OF REACTION-BONDED SILICON NITRIDE

The shrinkage curves during the isothermal sintering of RBSN (MgO) at a nitrogen pressure of 0.15 MPa and an isothermal holding temperature of 1720°C in a powder bed (Si$_3$N$_4$/BN/MgO mixture) are presented in Fig. 3. Details of preparation of the materials and sintering are given in reference 4. It is possible to describe the densification process of RBSN during the isothermal phase with a law like (7), but t_G is nearly zero and t_R corresponds to the beginning of the isothermal phase. The difference of shrinkage results from the initial phase (non-isothermal range), where at high content a shrinkage lower by about 25% in comparison with the high β content has been attained.[4,5] X-ray phase analysis showed that in the initial phase the α/β phase transformation is finished.

Since the phase transition develops via solution and precipitation
processes, the whole structure during densification of RBSN must have
been dissolved and reprecipitated once. The REM pictures show that the
processes were like this, because grains >1 μm have been found, while
the RBSN structure is characterized by grain sizes of about 100 nm.
These small structures resolve very rapidly because of their high
chemical potential (large curvatures).

In the subsequent isothermal phase, densification increases in
combination with grain growth by solution-reprecipitation processes.

The whole densification process (RBSN) must consequently be
divided into two stages:

1. Initial stage. Primary solution/reprecipitation. Solution of the
   RBSN crystallites and precipitation as β-silicon nitride in
   combination with rearrangement processes.

2. Isothermal phase. Secondary solution/reprecipitation. Further
   densification combined with grain growth.

SUMMARY

It is posible to use the three stages of the Kingery law to
describe the kinetics during densification of Si_3N_4-powders. The
densification process of reaction-bonded Si_3N_4 must be divided,
however, into two modified stages (primary and secondary
solution/reprecipitation).

A sintering equation is given, which contains the original power
law according to Kingery as a special case. This equation describes the
isothermal densification during sintering of Si_3N_4 powders and reaction
bonded Si_3N_4.

REFERENCES

1. W. D. Kingery, Densification during sintering in the presence of a
    liquid phase, J. Appl. Phys. $\underline{30}$ (1959) 301.
2. S. Hampshire, K. H. Jack, The kinetics of densification and phase
    transformation of nitrogen ceramics, Proc. Brit. Ceram. Soc. $\underline{31}$
    (1981) 37.
3. T. Hayashi, H. Munakate, H. Suzuki, H. Saito, Pressureless
    sintering of Si_3N_4 with Y_2O_3 and Al_2O_3, J. Mater. Sci. $\underline{21}$
    (1986) 3501.
4. J. Pabst, M. Herrmann, The kinetic of postsintering of reaction
    -bonded silicon nitride with different α/β phase content, to be
    published Sc. Sinter.
5. W. Hermel, M. Herrmann, J. Pabst, R. Schober, Kinetic
    investigation of nitriding and postsintering of Si/Si_3N_4
    systems, Sintering '87, Tokyo, Japan, November 4-6, 1987.

STABILITY OF PURE AND AlN-ALLOYED Al$_2$OC AND INFLUENCE

ON ABRASIVE PROPERTIES OF Al$_2$O$_3$-Al$_4$C$_3$-AlN MATERIALS

T. Zambetakis, J.M. Lihrmann[*], Y. Larrère and M. Daire

Ecole Européenne des Hautes Etudes des Industries Chimiques
Département Science des Matériaux, F- 67008- Strasbourg, and
* Université Paris XIII, CNRS - LIMHP, 4 Av. J.B.Clément
F- 93430- Villetaneuse

INTRODUCTION

The basic results published by Foster et al. (1) in 1956 allowed to
understand twenty years after many phenomena linked to high temperature
reactivity of alumina against carbon. They called attention to possible
syntheses by melting, in a difficult range of temperature, and likewise to
three compounds: Al$_4$C$_3$ as a carbon bearer in high temperature processes;
Al$_4$O$_4$C as a stable oxycarbide the formation of which is ensured by eutectic crystallization ; and Al$_2$OC as a quite particular phase that many laboratories had expected to find in fumes and condensates rather than in
the body of ceramics.

Otherwise Al$_2$O$_3$ is a thermally stable material with an interesting
hardness (about 2000 daN/mm^2). Although not fashionable at that time, the
idea to harden or toughen ceramic materials by inserting dispersed stable
phases seemed natural for metallurgists. Therefore we suggested as early
as 1976 to use the Al$_2$O$_3$-Al$_4$C$_3$ system as a reference for synthesis of
abrasives, and especially so because it was often difficult to avoid carbon for the temperatures used when melting such ceramic materials.

The industry of abrasives indeed wanted aluminium oxide- or silicium
carbide-based materials, and a diversification was expected from the use
of new iono-covalent compounds as well as from the improvment of existing
processes. It seemed thus interesting to report on the abrasive behaviour
of some melted or sintered alumina-based products, with some contents of
pure or AlN-alloyed Al$_2$OC.

As described elsewhere (2,3), all pellets, disks or cylinders necessary for this study were prepared by mixing powders (molar compositions)
of Al$_2$O$_3$ (mean diameter 0,6 μm), AlN (1 μm), and Al$_4$C$_3$ (50 μm). The two
former sizes were classical. Although coarser the latter proved to be
efficacious, for it was obtained by freshly crushing Al$_4$C$_3$ crystals imme-
diatly before compaction and high temperature exposure, i.e. before a
decomposition by wet air could begin. The powders were cold-pressed or
only compacted without any binder. Melting or sintering were conducted
either in a graphite resistor furnace under Helium protection, or in a
hot-press equipped with a HF-heated graphite matrix under an atmosphere of
Argon.

Science of Sintering
Edited by D. P. Uskoković *et al.*
Plenum Press, New York

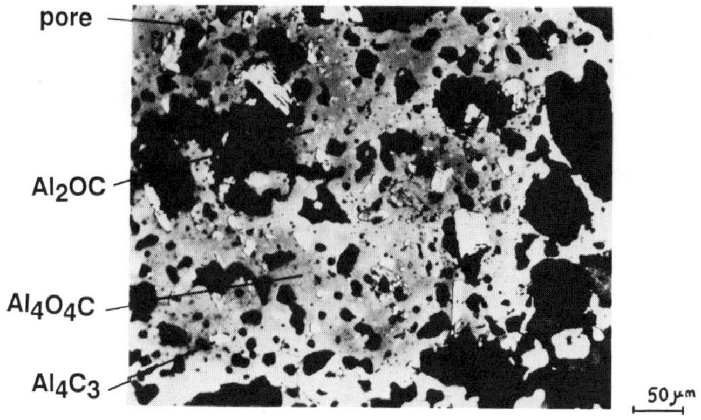

Fig.1 *Microstructure of composition 65 Al$_2$O$_3$-35 Al$_4$C$_3$ sintered at 1820°C under mechanical pressure(300 bars) for one hour.*

SYNTHESIS OF Al$_2$OC BY SINTERING

Reaction-Sintering below 1890°C

Let us see first and foremost the sintering ability of Al$_2$O$_3$-Al$_4$C$_3$ mixtures at temperatures lower than the first peritectic stage (1890°C). Pellets were heated at 1820°C for one hour. Wether they were sintered after previous cold-pressing under 300 bars, or hot-pressed under 300 bars, many pores were irreducible and made the metallographic examination difficult. Micrography (Fig.1) on the molar composition 65-Al$_2$O$_3$-35-Al$_4$C$_3$ shows Al$_4$O$_4$C is the main constituent, with some residual often twinned Al$_4$C$_3$ and a few small Al$_2$OC crystals.

At first, whatever the initial composition may have been, the porosity of samples was high and increased with the Al$_4$C$_3$ residual content. The number and size of pores, as well as the degeneration resulting from destruction of Al$_4$C$_3$ by wet air, made this type of sintering unfit for abrasives containing Al$_4$O$_4$C or Al$_2$OC. Furthermore samples were much too devoid of Al$_2$OC in comparison with expectations from the phase diagram of Foster et al.(1), thereby casting some doubt relating to the complete accuracy of the latter.

Sintering with Carbon

An other way to approach the topic is to sinter powder mixtures of Al$_2$O$_3$ and carbon in a graphite resistor furnace. We established that Al$_2$O$_3$ powder, previously cold-pressed under 4000 daN/cm2, remains nearly insensitive to carbon surroundings up to 1800°C. However a mixture of Al$_2$O$_3$ and C powders begins to react as early as 1500°C. It is therefore not surprising that a cylinder containing 57,0 Al$_2$O$_3$-14,3 Al$_4$C$_3$-28,7 C mole% converted mainly into Al$_4$O$_4$C if sintered at 1700°C for 5 hours, as expected from results of Elyutin et al. (4).

However when a sample of composition 18,8 Al$_2$O$_3$-12,2 C, is hot-pressed under 10 daN/cm2 up to 1900°C, it is possible to melt partially the mixture, i.e. to obtain core-sintered and skin-melted samples (Fig.2). XRD analysis shows the light color crown — which was closest to the hot

614

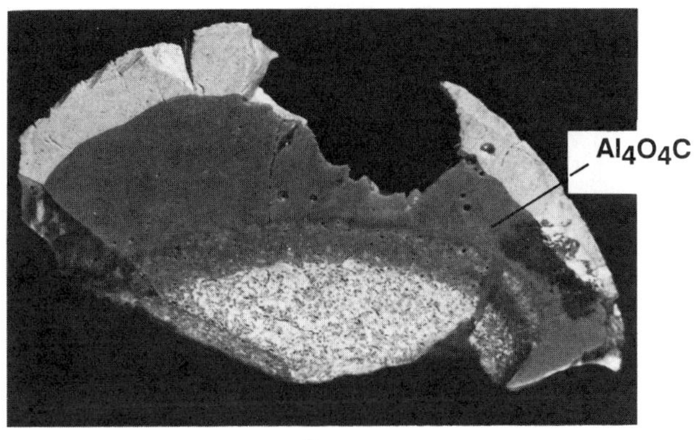

*Fig.2   Macrography of Al_2O_3 sintered with carbon (12,2%)
by hot-pressing at 1900°C for about one minute.*

crucible — is made out of Al_2O_3 and Al_2OC; the dark grey ring, almost pure Al_4O_4C; and the relatively cold center, a mixture of Al_4O_4C, Al_2O_3 and Al_4C_3. This observation led us to believe that Al_2OC could have been formed from Al_4O_4C, according either one of the following two reactions:

$$Al_4O_4C + 3C \; ---> \; 2\ Al_2OC + 2\ CO$$

$$x\ Al_4O_4C + (1-x)\ Al_2OC \; ---> \; Al_2O_3 + x\ Al_2OC$$

The first matches the mechanism reported by Elyutin et al. (4), and is not unlikely to occur near the graphite matrix. The other, a true solid state reaction, is nothing else than the reverse of the decomposition reaction of Al_2OC proposed (3) for Al_2OC at temperatures above 1200°C.

<u>Sintering with Partial Melting</u>

Rising the sintering temperature higher than the peritectic line leads the liquid to take a part in the reaction, and induce the primary crystallisation of Al_2OC in the composition range 15-45 mole% Al_4C_3. With the composition 55 Al_2O_3-45 Al_4C_3, it was shown (3) that Al_2OC was present in high amounts in samples heated at 1910°C, and rapidly cooled. Unreacted Al_4C_3 still was present in the microstructure after 3 minutes of sintering (Fig. 3). But when extending the reaction time to 10 minutes, we obtained samples with very little Al_4C_3 remaining and minor amounts of Al_4O_4C, apparently in good enough agreement with the diagram of Foster et al.(1), but really not yet at equilibrium. Those samples were used by Lihrmann to prove by annealing treatments the instability of Al_2OC below 1715°C (3) in the Al_2O_3-Al_4C_3 stable phase equilibrium diagram.

SYNTHESIS OF Al_2OC BY MELTING

Partial melting beyond the first peritectic stage leads to Al_2OC crystals. At temperatures above the liquidus, experiments may become troublesome due to the possible direct reaction of Al_2O_3 with the graphite crucible, thus resulting in a shift of the initial composition. On the other hand, vaporization of Al_4O_4C and Al_2OC may become intense but balan-

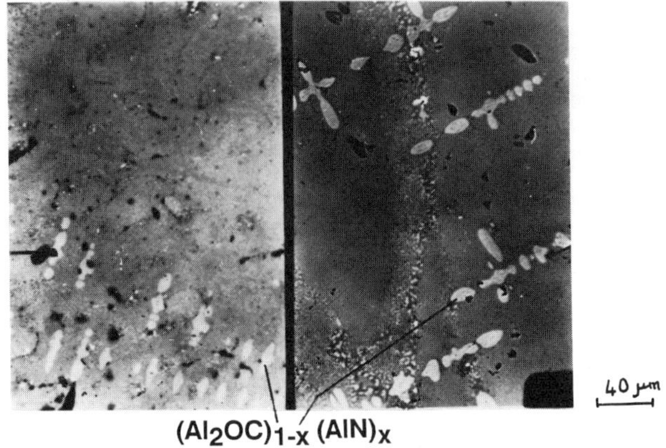

Fig.6 Dendrites of AlN-alloyed Al$_2$OC - Composition 81Al$_2$O$_3$-13Al$_4$C$_3$-06AlN, melted at 1970°C: A= rapidly cooled (70°C/mn); B= slowly cooled (5°C/mn).

near the impressions; that should enhance the abrasion ability by making attrition easier.

Abrasion Test

Evidently a sufficiently high hardness is a necessary condition for the abrasive to penetrate into the metal. Measurement is often obtainable from Knoop indentation under a 4 daN/mm^2 load for 20 sec. However complementary tests need generally to be done on grinding wheels or abrasive grains. Our previously described method (9) consists of machining a flat surface on a cylindrical metallic test tube (XC 38 steel) fixed on the spindle of a modified recti-fying machine, with a triangular base pyramid carved from an abrasive grain. The uniform horizontal displacement of this tool, driven in up to 100 μm , draws a set of grooves. Periodic weighing the gauge measures the volume V of ground or wrested metal. Likewise the wear of the abrasive point is known by measuring the cross section area S with a metallographic microscope. These parameters V and S obey reasonably well the linear relationship : $V = \lambda S + \mu$. So the slope λ characterizes the couple abrasive grain-abraded metal, and allows a first comparison of potentially abrasive materials if a same gauge is used ; especially the factor λ corresponds to the cutting ability of tested materials against a regular carbon steel.

Abrasive Ability of Al$_2$OC-reinforced Al$_2$O$_3$ Materials

Factors λ measured for current alumina-based abrasives rarely exceed 200 mm, especially with sintered products. Introduction of Al$_2$OC into Al$_2$O$_3$ materials is able to increase λ . The hypo-eutectic crystallisation in the metastable Al$_2$O$_3$-Al$_4$C$_3$ system seemed to be very favourable to highest values of λ (Table.I), i.e. to abrasive materials which much tool and rub much metal off, while they wear out little. The formation of Al$_4$O$_4$C, either directly by crystallisation in the stable phase diagram, or by annealing of the metastable monooxycarbide, causes the hardness and the abrasive efficiency to decrease. The (Al$_2$OC)$_{1-x}$(AlN)$_x$ solid solution allowed the oxycarbide to be maintained when slow cooling or long annealing were used ; but abrasive properties did not enhance and rather decreased.

Then we noticed the grinding efficiency of the eutectic microstructures were highly variable : if a high hardness is generally obtained due

Table I - Hardness and abrasive capacity of some Al_2O_3-based materials.

composition Al_2O_3-Al_4C_3-AlN			treatment	main phases	KHN daN/mm^2	λ mm
100				sintered bauxite	905	150
93		07	melted 1910°	eutectic	1200	1270
90	10		melted 1860°	Al_2O_3 + metast. eutectic	1400	7150
90	10		slowly cooled	Al_2O_3 + stable eutectic	1100	890
88	12		melted 1840°	eutectic	1360	1400
80	20		melted 1920°	Al_2OC + Al_2O_3	900	1130
80	20		melted 1960°	Al_2OC + Al_2O3	1220	2080
80	20		anneal.1810°	Al_4O_4C + Al_2O_3	950	1520
81	13	06	melted 1960°	Al_2OCss + Al_2O_3	1290	6700
81	13	06	melted 1920°	Al_4O_4C + AlON	980	1160
81	13	06	anneal.1680°	Al_2OCss + Al_2O_3	1010	1020
81	12	07	melted 1960°	Al_2OCss + eutectic	1070	92
73,5	11.5	15	melted 2010°	Al_2OCss + Al_2O_3	1120	390
64.9	10.1	25	melted 2030°	Al_2O_3 + eutectic	1090	1180
60.5	9.5	30	melted 2040°	Al_2O_3 + AlON	porous	5

to strengthening, this structure could be favourable to fractures when a
load is applied during the abrasion processes, and so caused the material
to be worn out more rapidly. Furthermore these results showed that the
juxtaposition of the two compounds, Al_2O_3 and pure or AlN-alloyed Al_2OC,
led to excellent strength, satisfactory toughness and the best abrasion
ability.

CONCLUSIONS

It is possible to crystallize Al_2OC or the Al_2OC-AlN solid solution
in several Al_2O_3-based materials, either by sintering or melting. However
pure solid state sintering does not sufficiently reduce porosities nor
adequatly strengthen the matrix, unless perhaps through the use of hot-
pressing under severe conditions. Al_4O_4C is obtained more easily than
Al_2OC, especially when sintering is carried out under 1900°C. Al_2OC can
form easily if we pass through the peritectic line, when melting or when
sintering with par-tial melting; the critical cooling velocity of quenching is low enough for this oxycarbide be retained in the metastable
phase equilibrium diagram.

As a true alloying compound, AlN reduces this velocity and therefore
causes hardening. If well dispersed, Al_2OC crystals increase the abrasi-
veness of Al_2O_3-based materials by hardening or toughening, and the cutting capacity increases. But in fact the intrinsic brittleness of Al_2OC
limits the size of useful inserts: if dendritesare too large, they may
lead to many fractures and consecutively to rapid wear of the abrasives.

Moreover the carbide, Al_4C_3, is very sensitive to wet atmospheres,
and therefore must imperatively be avoided. Subject to these limitations
an enhancement of hardness and a promising abrasion behaviour are possible, mainly in samples that are fully melted, and in those that are
sintered with partial melting.

REFERENCES

1. L. M. Foster, G. Long and M.S. Hunter, Reactions between Aluminum Oxide and Carbon. The Al_2O_3-Al_4C_3 diagram, J. Am. Ceram. Soc., 39:1 (1956)

2. Y. Larrère, B. Willer, J. M. Lihrmann and M. Daire, Stable and Metastable Phase Equilibrium Diagrams of the binary System Al_2O_3- Al_4C_3 (in Fr.), Rev. Int. Hautes Tempér. Réfract., 21:3 (1984)

3. J. M Lihrmann, T. Zambetakis and M. Daire, High Temperature Behaviour of the Aluminum Oxycarbide Al_2OC in the system Al_4C_3-AlN and with Additions of Aluminum Nitride, J. Am. Ceram. Soc., 72: (1989)

4. V. P. Elyutin, Y. A. Pavlov and V. S. Chelnokov, High Temperature Reactions of Al_2O_3 with Carbon, Isv. Akad. Nauk. SSSR, Neorg. Mat., 1973:1365 (1973)

5. J. L. Henry, J. H. Russel and H.J. Kelly, The System Al_4C_3-AlN-Al_2O_3. Powder Forming and Sintering Behaviour. Phase Identification and Refractory Composition Properties, Report of investigations 17320, US Bureau of Mines (1969)

6. J. M. Lihrmann, T. Zambetakis and M. Daire, The Aluminum Monooxycarbide Al_2OC in the system Si-C-Al-O-N : some Thermodynamic Properties, 1st Europ. Ceram. Conf., Elsevier Publ., Maastrich (1989)

7. G. M. Zaretskaya, F. I. Eidel'shtein and M. I. Sokhor, Aluminum Monooxycarbide in Oxysulfide slag, Inorg. Mat. (Engl. Transl.), 8(1):70 (1972)

8. I. B. Cutler, P. D. Miller, R. Rafaniello, H. K. Park, D. D. Thompson and K. H. Jack, New Materials in the Si-C-Al-O-N and related Systems, Nature (London) 275:434 (1978)

9. F. Schneider, B. Willer and M. Daire, Method of Study and definition of the Abrasion Capacity (in Fr.), Matériaux et Techniques, 8-9:319 (1984)

PREPARATION AND PROPERTIES OF α/β

SiAlON COMPOSITES

S. Bošković and K.G. Nickel*

"Boris Kidrič" Institute of Nuclear Sciences, POB 522
Mat.Sci.Dept., 11001 Belgrade, Yugoslavia
*Max-Planck Institut fur Metallforschung, PML, 7 Stuttgart
80, DBR

ABSTRACT

The results of a phase relations study in the Y-Si-Al-O-N system
within the Si_3N_4 - $SiAl_2O_2N_2$ - YAl_3N_4 plane are presented. It is shown
that within this plane close to the Si_3N_4 corner α and β SiAlONs coexist
at equilibrium with minor amount of glass phase. The α/β ratio in this
field is controlled by Y concentration. It was found that properties of
α/β composite materials such as density, hardness, toughness and microstructure depend on α/β ratio.

INTRODUCTION

Due to high thermal shock resistance, wear resistance, hardness and
toughness, a commercially successfull application of SiAlONs has been
the use as a cutting tool material for machining metals. Because α-SiAlONs
exceed β-SiAlONs in hardness and thermal shock resistence[1,2], the improvement of these properties can be achieved by using α/β SiAlON composites instead of β-SiAlONs alone.

In this paper attention was paid to a study of phase relations in
the Si_3N_4 - rich corner of Si_3N_4 - $SiAl_2O_2N_2$ - YAl_3N_4 plane of the
Y-Si-Al-O-N system. A two phase (α and β) field is defined which is important for α/β composite preparation. It is shown that phase composition
and α/β ratio in this field is controlled by Y content and is nearly independent of temperature. This points had not been precisely clear from
existing literature.[3,4]

Properties of α/β composites were determined as a function of α phase
amount. It was found that density and hardness increase with increasing
α phase content while toughness is inversly dependent on α phase amount.
Microstructure of α/β composites also changed with variation of α phase
content.

EXPERIMENTAL WORK

The following powders were used for experiments. Si_3N_4 and AlN (H.C.
Starck), Al_2O_3 (Alcoa) and Y_2O_3 (Ventron). The compositions of starting
mixtures (Table I) were chosen to allow for differing α/β ratios at equ-

ilibrium. Characterization of samples hot pressed at 1600-1800°C for 2
hours was performed by X-ray analysis, microstructure analysis, as well
as by density, hardness and toughness measurements.

RESULTS AND DISCUSSION

Phase relations were studied in the plane shown in Fig.1. This field
covers the theoretically possible extent of α-SiAlONs ($Y_x Si_{12-(3+y)} Al_{3x+4y} O_y N_{16-y}$).

According to phase analysis data for samples hot pressed between
1600-1850°C, two phases (α and β SiAlONs) coexist in the field between 1
and 3,1 eq% Y ($\sim$4-15 mol% YAl_3N_4) and up to 5,5 eq.% O ($\sim$15 mol% $SiAl_2O_2N_2$)
as shown in Fig.2.

Table I. Compositions of starting mixtures (wt%)

Sample	Si_3N_4	Y_2O_3	Al_2O_3	AlN	eq% Y
A	82.96	2.5	5.8	8.6	1.0
B	90.62	3.5	0.0	5.9	1.12
C	85.80	3.5	3.5	7.2	1.15
D	82.40	3.5	5.5	8.8	1.15
E	78.90	5.7	4.3	11.1	1.80
F	79.90	7.1	1.6	11.3	2.00
G	73.53	7.1	5.2	14.2	2.00
H	75.70	7.4	3.4	13.5	2.50
I	75.56	9.2	0.8	14.4	3.10

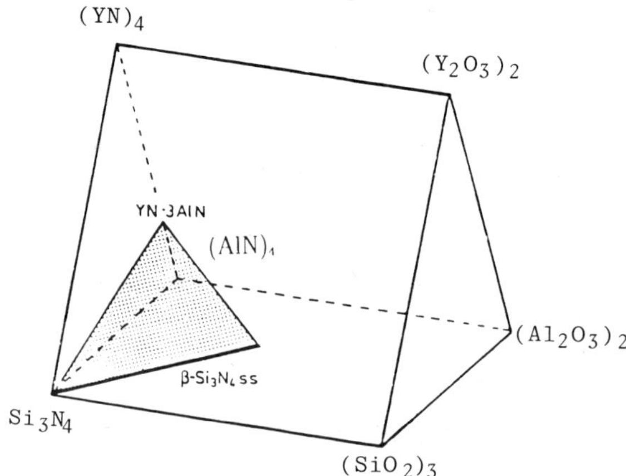

Fig.1. Jänicke prism of the system Si_3N_4-SiO_2-Al_2O_3-AlN-YN-Y_2O_3.

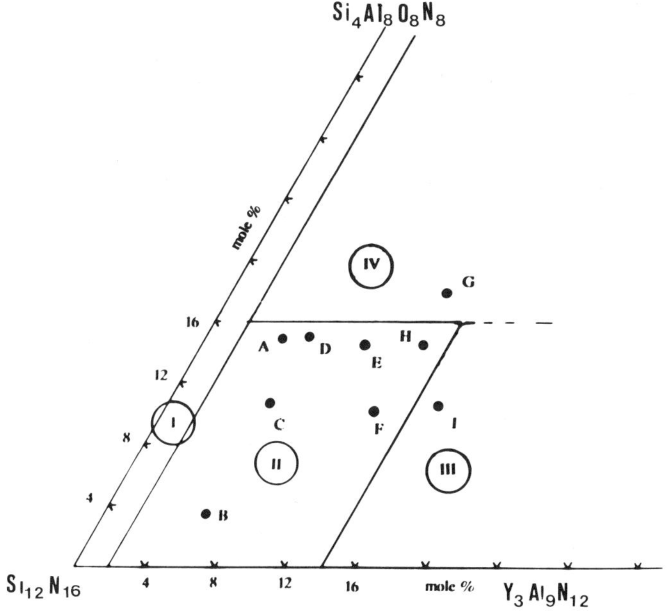

Fig.2. Si$_3$N$_4$ rich corner of shaded plane in Fig.1. I - β–SiAlON, II - α + β–SiAlON, III - α–SiAlON, IV - α + β + phase x.

Equilibrium phase compositions had already been achieved at 1600°C after 4 hours[5] and did not change in the temperature range investigated (up to 1850°C). However, α phase content depended on the Y concentration as illustrated in Fig.3.

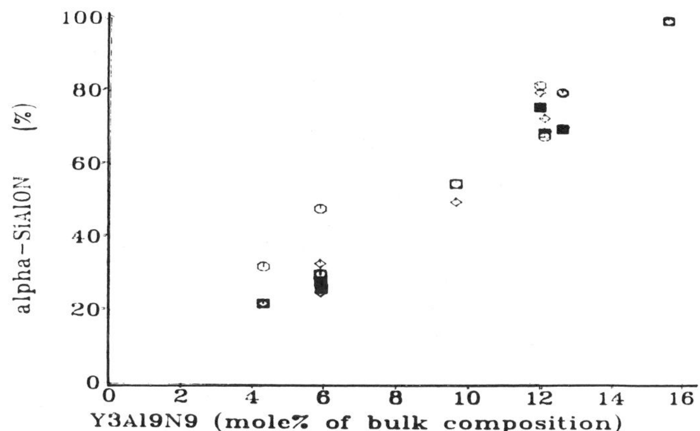

Fig.3. Dependence of α phase amount of Y content at 1600° (circles) 1700°C (diamonds) and 1800°C (squares).

Using these data it is possible to easily read the composition of a starting mixture that will produce a desired α/β ratio at equilibrium.[5]

Densification of investigated compositions takes place in the presence of a liquid phase during heating at about 1500°C.[6] Increasing temperature or run time results in an adjustment of α/β ratio only. In the investigated temperature range no other crystalline phase was detected except unreacted α-Si_3N_4 (for lower temperatures and shorter heating times) and α and β SiAlON as reaction products. Achieved density did not depend on time and temperature above 1500°C, but depended on Y concentration as shown in Fig. 4.

Above 1600°C grain growth can be observed in all investigated compositions, which is illustrated in Fig. 5 (composition B having 1,12 eq.%,

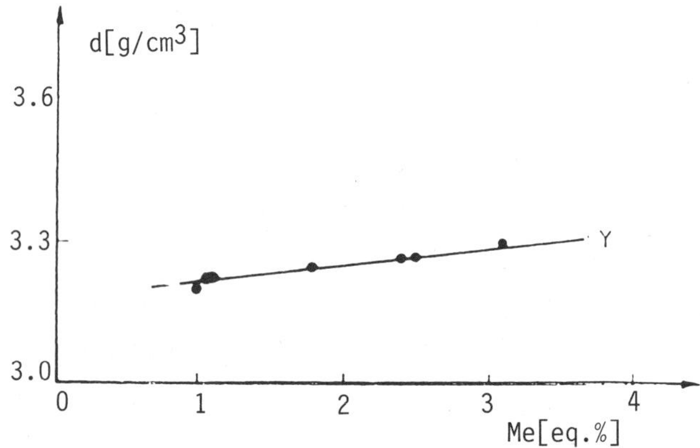

Fig.4. Density dependence on Y concentration.

Fig.5. Etched surfaces of samples with B composition hot pressed at 1600°C (a), 1700°C (b) and 1800°C (c).

Y, i.e. 30% α phase). At 1800°C, coallescence of grains can be observed.
Microstructure changes also with the variation of composition. In
Fig.6 the microstructures of samples having 25 (A), 50 (E) and 100 (I) %
α are given. It can be seen that with increasing additive amount (Table I),

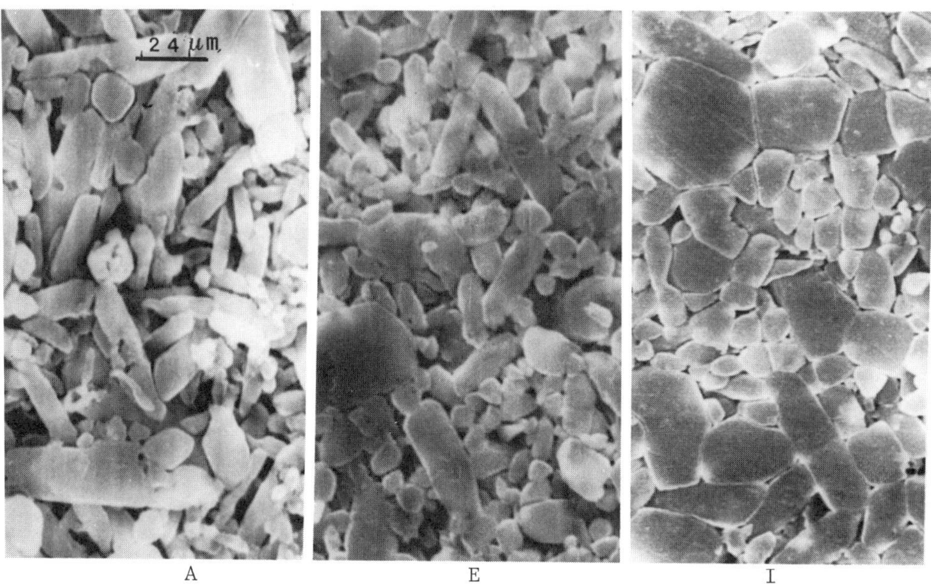

Fig.6. Etched surfaces of samples with 20 (A), 50 (E) and 100 (I) %
α phase 1800°C - 2 hours

i.e. with increasing liquid phase amount during densification, more coarse
grain microstructure is obtained. With increasing additive amount, α phase
content increases, as well.

Hardness was measured as a function of composition, time and temperatures. With increasing α phase amount in the composites, hardness increases (Fig.7), while it decreases with increasing hot-pressing temperature (Fig.8). Each composition defines an individual correlation of hardness with α-phase amount and/or hot pressing temperature. The slopes in
Figs.7 and 8 change with composition. Compositions having high Y content
(F and H) show the lowest slope. The same conclusion can be drawn from
data in Fig.8.

Fracture toughness, measured by indentation method (Fig.9) decreases
with increasing α phase amount, but increases with increasing hot pressing
temperature. It can be observed, as in the case of hardness (Fig.8) that
the composition I exhibits a different temperature dependence. It should
be stressed that composition I is a single α-phase ceramic, and not a two-
phase composite.

The amount of α-phase varies, however, with isothermal heating time
as well, which means that for the same chemical composition, composite
materials with differing α/β ratio can be synthesized.[5] This means that
the properties vary, too. In Fig. 10, the change of hardness and toughness
is presented as a function of α-phase amount achieved for different isothermal heating times at 1600°C.

These data are of importance for practical purposes because it is of
no concequence whether the α/β ratio is obtained as equilibrium or metastable value. These data, along with the former, show that property tailoring seems possible.

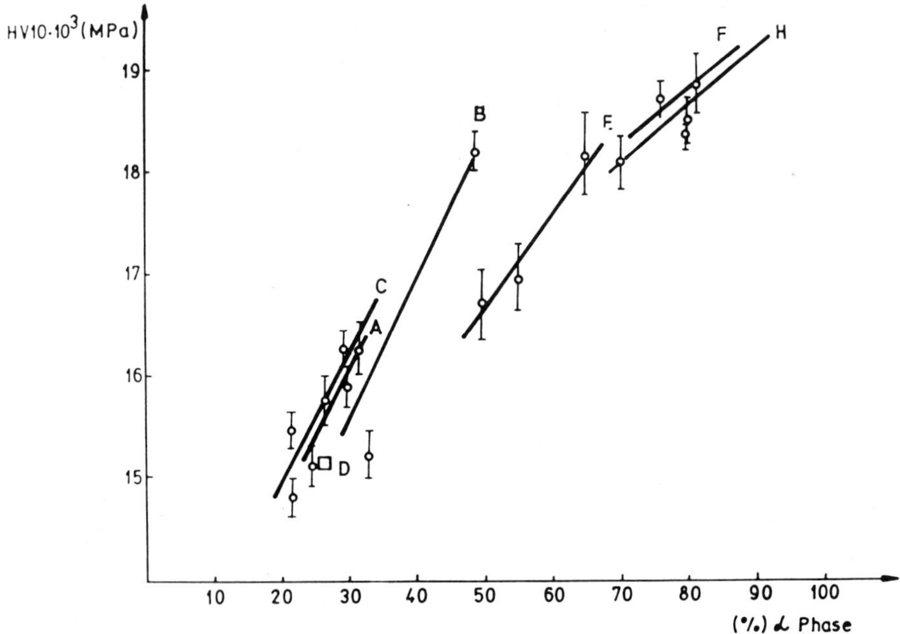

Fig.7. Change in hardness as a function of α phase content.

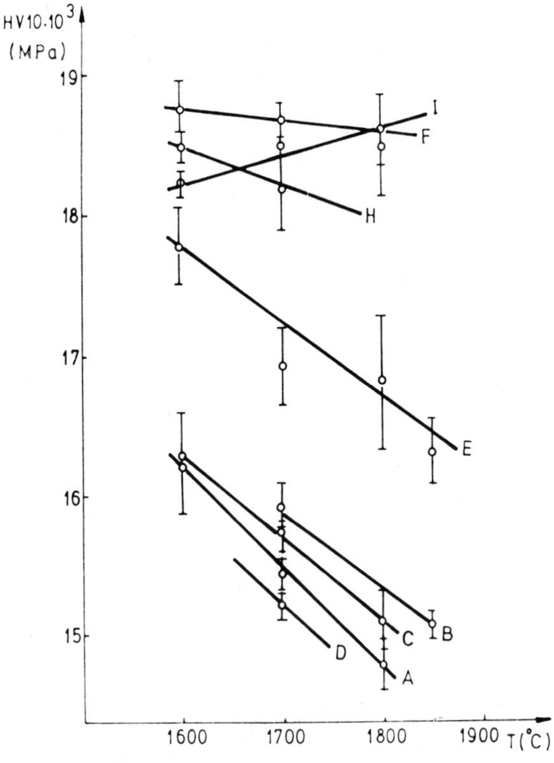

Fig.8. Dependence of hardness on hot-pressing temperature and composition.

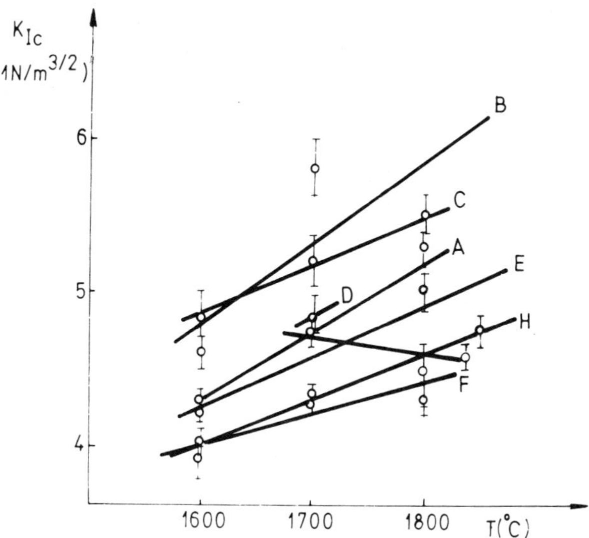

Fig.9. K_{Ic} as a function of temperature and α phase amount.

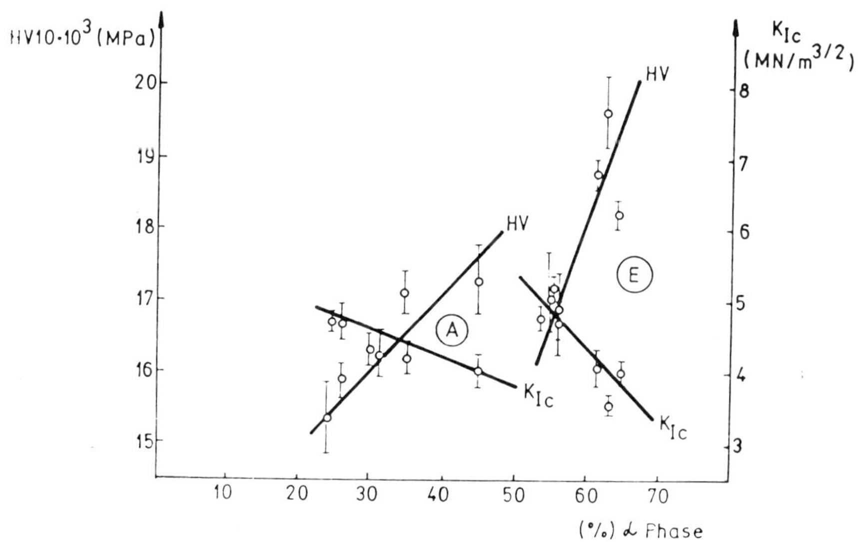

Fig.10. Hardness and toughness as a function of α phase content
achieved for different isothermal heating times at 1600°C
(compositions A and B).

In Fig.11, hardness against K_{Ic} is presented for all compositions
studied. It seems that all obtained data fall upon four different curves,
a,b,c and d. In Fig.11 we compare our data with those of pure β-SiAlONs
and other literature data for α/β composites.[7] Curve e represents the
data for α/β composites, while curve f represents the data for β-SiAlONs
(7).

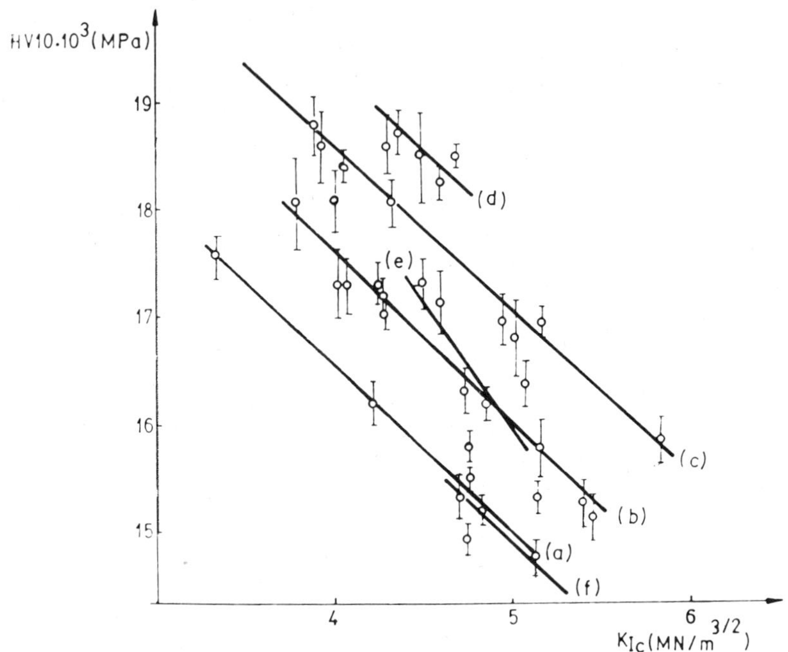

Fig.11. Vickers hardness (HV10) against indentation fracture toughness.

CONCLUSION

Coexistence of α- and β-SiAlONs was found in $Si_{12}N_{16}$ - $Si_4Al_8 N_8$ - $Y_3Al_9N_{12}$ plane in the range of Y concentration between one to 3.1 equiv.% ($\sim$4-15 mole % YAl_3N_9) and up to 5.5 equiv. % O ($\sim$15 mole % $SiAl_2O_2N_2$). The α/β-ratio within the composite materials depends strongly on the Y-content and is nearly independent of temperature in the range 1600° to 1850°C.

Property tailoring of α/β SiAlONs by adjusting composition, hot-pressing temperature and time is possible.

The hot pressed density increases with increasing Y-content.

Grain growth becomes intense above 1600°C.

Hardness increases with increasing α phase amount and decreases with increasing temperature and time.

Toughness increases with increasing temperature and time, but decreases with increasing α phase amount.

<u>Acknowledgements</u>

We greatfully acknowledge support and funding from a bilateral project ("composites") between our institutes, organized by the International Bureau of KFA Julich within the framework of Germany-Yugoslav agreements (YUZAMS).

REFERENCES

1. Jack, K.H. and Wilson, W.I. Ceramics Based on Si-Al-O-N and Related Systems, Nature phys. Sci. 1972, 328, 28-29.

2. Huang, Z.K., Greil, P. and Petzow, G.Formation of α-Si_3N_4 Solid Solutions in the System Si_3N_4-AlN-Y_2O_3, J.Am.Ceram.Soc., 1983, 66, C 96-97.

3. Peuckert, D. Herstellung und Untersuchung von mehrphasigen Si_3N_4-Misckristallkeramiken, Ph. D.Thesis, University of Stuttgart, 1987.

4. Sun, W.Y., Fu, F.Y, and Yan, D.S., Studies of Formation of α and β SiAlON, Mat.Lett, 1987, 6, 11-15.

5. S.Bošković, and K.G.Nickel, Multiphase α/β SiAlON Ceramics, 1st E.Cer. Soc., Conf. Maastricht, June, 1989.

6. Slasor, S. and Thompson D.P., Preparation and Characterization of Yttrium α' SiAlONs, Non-Oxide Tehcnical and Engineering Ceramics, Ed S.Hampshire, Limerick, Ireland, 1986, 223-230.

7. Ekström, T. and Ingelström, N., Characterisation and Properties of SiAlON Materials, Non-Oxide Technical and Engineering Ceramics, Ed. S.Hampshire, Limerick, Ireland, 1986 231-254.

INTERNATIONAL PROGRAM COMMITTEE

R. J. Brook
    Max-Planck-Institute for Metals Research, Stuttgart, F.R.Germany

A. C. D. Chaklader
    University of British Columbia, Vancouver, Canada

R. L. Coble
    Massachusetts Institute of Technology, Cambridge, U.S.A.

K. E. Easterling
    University of Technology, Lulea, Sweden

Ya. E. Geguzin[+]
    Harkov University, Harkov, U.S.S.R.

R. M. German
    Rensselaer Polytechnic Institute, Troy, U.S.A.

N. A. Haroun
    Tablin Institute of Metallurgy, Cairo, Egypt

W. J. Huppmann
    HILTI AG, Schann, Principality of Lichtenstein

D. L. Johnson
    Northwestern University, Evanston, U.S.A.

W. A. Kaysser
    Max-Planck-Institute for Metals Research, Stuttgart, F.R.Germany

S. S. Kiparisov
    Institute of Fine Chemical Technology, Moscow, U.S.S.R.

P. S. Kisliy
    Institute for Superhard Materials AS USSR, Kiev, U.S.S.R.

D. Kolar
    Institute "Jožef Štefan", University of Ljubljana, Yugoslavia

N. C. Kothary
    James Cook University, North Queensland, Townsville, Australia

G. C. Kuczynski (Honorary President)
    University of Notre Dame, Indiana, U.S.A.

E. Labusca
    Institute ISEFIZ/IFTM, Bucharest, Romania

A. Moccelin
     Ecole des Mines, Nancy, France

Li Nan
     Wuhan Institute of Iron and Steel Technology, Wuhan, China

H. Palmour III
     North Carolina State University, Raleigh, U.S.A.

H. Pastor
     Ugicarb Morgon, Grenoble, France

G. Petzow (Vice President)
     Max-Planck-Institute for Metals Research, Stuttgart, F.R.Germany

P. Reynen
     Institut fur Gesteinshuttenkunde RWTH, Aachen, F.R.Germany

W. Schatt
     Technical University, Dresden, D. R. Germany

V. V. Skorohod
     Institute for Problems of Mater. Sci. AS USSR, Kiev, U.S.S.R.

M. Šlesar
     Institute of Experimental Metallurgy, Košice, Č.S.S.R.

S. Somiya (Vice President)
     The Nishi Tokyo University, Tokyo, Japan

M. J. Sparnaay
     Philips Research Laboratories, Eindhoven, The Netherlands

R. M. Spriggs (President)
     Alfred University, Alfred, U.S.A.

S. Stolarž
     Institute of Non-Ferous Metals, Gliwice, Poland

A. Thölen
     Technical University of Denmark, Lungby, Denmark

V. I. Trefilov (Vice President)
     Institute for Problems of Mater. Sci. AS USSR, Kiev, U.S.S.R.

G. S. Upadhyaya
     Indian Institute of Technology, Kanpur, India

D. P. Uskoković (Secretary)
     Institute of Techn. Sci. of the SASA, Belgrade, Yugoslavia

J. A. Varela
     Universidade Estadual Paulista Araraquara, Sao Paulo, Brasil

P. Vincenzini
     National Research Council of Italy, Faenza, Italy

R. Watanabe
     Tohoku University, Sendai, Japan

CONTRIBUTORS

CONFERENCE CHAIRMEN

R. T. DeHoff
        University of Florida, Gainesville, U.S.A.

H. E. Exner
        Max-Planck-Institute for Metals Research, Stuttgart, F.R.Germany

R. M. German
        Rensselaer Polytechnic Institute, Troy, U.S.A.

L. Jakešova
        Institute for Nuclear Research, Rež, Č.S.S.R.

D. L. Johnson
        Northwestern University, Evanston, U.S.A.

W. A. Kaysser
        Max-Planck-Institute for Metals Research, Stuttgart, F.R.Germany

B. F. Kieback
        Institute for Solid State Physics, AS DDR, Dresden, D.R.Germany

P. S. Kisliy
        Institute for Superhard Materials AS USSR, Kiev, U.S.S.R.

D. Kolar
        "Jožef Stefan" Institute, Ljubljana University, Yugoslavia

H. Palmour III
        North Carolina State University, Raleigh, U.S.A.

J. Pask
        University of California, Berkeley, U.S.A.

H. Pastor
        Ugicarb Morgon, Grenoble, France

M. M. Ristić
        Serbian Academy of Sciences and Arts, Belgrade, Yugoslavia

W. Schatt
        Technical University, Dresden, D.R.Germany

V. V. Skorohod
        Institute for Problems of Mater. Sci. AS USSR, Kiev, U.S.S.R.

S. Somiya
        The Nishi Tokyo University, Tokyo, Japan

R. M. Spriggs
        Alfred University, Alfred, U.S.A.

F. Thümmler
        Karlsruhe University, Karlsruhe, F.R.Germany

D. P. Uskoković
        Institute of Technical Sciences of the SASA, Belgrade, Yugoslavia

J. A. Varela
        Instituto de Quimica UNESP, Araraquara, Sao Paulo, Brasil

AUTHORS

P. Arato
        Research Institute for Technical Physics, Budapest, Hungary

Lj. Atanasoska
        Institute of Technical Sciences of the Serbian Academy of
        Sciences and Arts, Belgrade, Yugoslavia

E. Besenyei
        Research Institute for Technical Physics, Budapest, Hungary

J. E. Blendell
        National Inst. of Standard and Technology, Gaithersburg, U.S.A.

J. P. Bonnet
        University of Bordeax, Talence, France

S. Bošković
        "Boris Kidrič" Institute, Belgrade, Yugoslavia

D. Božić
        "Boris Kidrič" Institute, Belgrade, Yugoslavia

M. Buchberger
        Electrotechnical Institute "Rade Končar", Zagreb, Yugoslavia

L. E. G. Cambronero
        School of Mines, Madrid, Spain

S.-K. Chan
        Argonne National Laboratory, Argonne, U.S.A.

M. Clee
        University of Wales, Cardiff, U.K.

R. L. Coble
        Massachusetts Institute of Technology, Cambridge, U.S.A.

M. Daire
        Ecole Europeene des Hautes Etudes des Industries Chimiques,
        Strasbourg, France

R. T. DeHoff
        University of Florida, Gainesville, U.S.A.

S. Dick
    Ecole Europeene des Hautes des Industries Chimiques,
    Strasbourg, France

R. Dimitrijević
    Belgrade University, Belgrade, Yugoslavia

K. Dorfschmidt
    Karlsruhe University, Karlsruhe, F.R.Germany

D. Duževic
    Electrotechnical Institute "Rade Končar", Zagreb, Yugoslavia

H. E. Exner
    Max-Planck-Institute for Metals Research, Stuttgart, F.R.Germany

G. Fischman
    Alfred University, Alfred, U.S.A.

A. Frisch
    Max-Planck-Institute for Metals Research, Stuttgart, F.R.Germany

R. M. German
    Rensselaer Polytechnic Institute, Troy, U.S.A.

E. A. Giess
    IBM Research Division, Yorktown Heights, U.S.A.

R. Gopalakrishnan
    Institute "Jožef Štefan", University of Ljubljana, Yugoslavia

S. Gouvea
    Universidade Estadual Paulista Araraquara, Sao Paulo, Brasil

P. Halary
    Max-Planck Inst. for Metals Research, Stuttgart, F.R. Germany

C. A. Handwerker
    National Inst. of Standard and Technology, Gaithersburg, U.S.A.

K. Hashimoto
    Fujitsu Laboratories Ltd., Yokohama, Japan

W. Hermel
    Institute for Solid State Physics AS DDR, Dresden, D.R. Germany

M. Herrmann
    Institute for Solid State Physics AS DDR, Dresden, D.R. Germany

S. Hess
    Institute for Solid State Physics AS DDR, Dresden, D.R.Germany

M. Hiraishi
    Hiroshima University, Hiroshima, Japan

E. Horikoshi
    Fujitsu Laboratories Ltd., Yokohama, Japan

T. Iikawa
    Fujitsu laboratories Ltd, Yokohama, Japan

Y. Ishiwata
     Toshiba Corp., Yokohama, Japan

Y. Itoh
     Toshiba Corp., Yokohama, Japan

V. A. Ivensen
     Inst. for Refractory Metals and Hard Alloys, Moscow, U.S.S.R.

D. L. Johnson
     Northwestern University, Evanston, U.S.A.

C. Jovalekić
     Center for Multidisciplinary Study, Belgrade, Yugoslavia

Lj. Karanović
     Belgrade University, Belgrade, Yugoslavia

H. Kashiwaya
     Toshiba Corp. Yokohama, Japan

J.Katanić-Popović
     Institute "Boris Kidrič"-Vinča, Belgrade, Yugoslavia

W. A. Kaysser
     Max-Planck-Institute for Metals Research, Stuttgart, F.R.Germany

A. Kele
     Research Institute for Technical Physics, Budapest, Hungary

H. Kessler
     Institute for Solid State Physics AS DDR, Dresden, D.R.Germany

Y.-H. Kim
     Han Yang University, Seoul, Korea

P. Kisliy
     Institute for Superhard Materials AS USSR, Kiev, U.S.S.R.

G. Klassen
     Alfred University, Alfred, U.S.A.

D. Kolar
     Institute "Jožef Štefan", Ljubljana University, Yugoslavia

T. Kosmač
     Institute "Jožef Štefan", Ljubljana University, Yugoslavia

E. Kostić
     "Boris Kidrič" Institute, Belgrade, Yugoslavia

P. Kostić
     Center for Multidisciplinary Studies, Belgrade, Yugoslavia

Lj. Kostić-Gvozdenović
     Faculty of Technology and Metallurgy, Belgrade, Yugoslavia

V. Kraševec
     Institute "Jožef Štefan", Ljubljana University, Yugoslavia

U. Kunaver
    Institute "Jožef Štefan, Ljubljana University, Yugoslavia

H. Kuroki
    Hiroshima University, Hiroshima, Japan

R. Laag
    Max-Planck-Institute for Metals Research, Stuttgart, F.R.Germany

V. N. Lapovok
    NPO "Krasnaya Zvezda", Moscow, U.S.S.R.

V. I. Lapshin
    NPO "Pozitron", Leningrad, U.S.S.R.

Y. Larrere
    Ecole Europeene des Hautes Etudes des Industries Chimiques,
    Strasbourg, France

J. Laughner
    Alfred University, Alfred, U.S.A.

J. M. Lihrmann
    Universite de Paris XIII, Villetaneuse, France

T. Liu
    Karlsruhe University, Karlsruhe, F.R.Germany

E. Longo
    Department of Chemistry UFS Car, S.Carlos, Sao Paolo, Brasil

Z. Maričić
    Faculty of Sciences, Belgrade University, Yugoslavia

V. Martinac
    Faculty of Technological Engineering, Split, Yugoslavia

E. Matijević
    Clarkson University, Potsdam, U.S.A.

R. Maurer
    Max-Planck-Institute for Metals Research, Stuttgart, Yugoslavia

B. Mihelić
    Institute of Technical Sciences of the Serbian Academy of
    Sciences and Arts, Belgrade, Yugoslavia

M. Miletić
    Belgrade University, Belgrade, Yugoslavia

O. Milošević
    Institute of Technical Sciences of the Serbian Academy of
    Sciences and Arts, Belgrade, Yugoslavia

M. Mirošević-Anzulović
    Faculty of Technological Engineering, Split, Yugoslavia

D. Minić
    Faculty of Sciences, Belgrade University, Yugoslavia

M. Mitkov
    Institute "Boris Kidrič"-Vinča, Belgrade, Yugoslavia

I.-H. Moon
    Han Yang University, Seoul, Korea

K. G. Nickel
    Max-Planck-Institute for Metals Research, Stuttgart, Yugoslavia

P. M. Nikolić
    Belgrade University, Belgrade, Yugoslavia

Z. Nikolić
    Faculty of Electronic Engineering, Nish University, Yugoslavia

V. I. Novikov
    NPO "Krasnaya Zvezda", Moscow, U.S.S.R.

R. Oberacker
    Karlsruhe University, Karlsruhe, F.R.Germany

S. Ohno
    National Research Institute for Metals, Tokyo, Japan

H. Okuyama
    National Research Institute for Metals, Tokyo, Japan

E. Ozawa
    K.K.L'Air Liquide Labs., Tokodai, Ibaraki, Japan

M. Ozawa
    National Research Institute for Metals, Tokyo, Japan

C. P. Ostertag
    Nat. Inst. of Standard and Technology, Gaithersburg, U.S.A.

J. Pabst
    Institute for Solid State Physics AS DDR, Dresden, D.R.Germany

J. A. Pask
    University of California, Berkeley, U.S.A.

H. Palmour III
    North Carolina State University, Raleigh, U.S.A.

J.-H. Park
    Argonne National Laboratory, Argonne, U.S.A.

M. B. Pavlović
    Belgrade University, Belgrade, Yugoslavia

B. Petrić
    Faculty of Technological Engineering, Split, Yugoslavia

N. Petrić
    Faculty of Technological Engineering, Split, Yugoslavia

V. Petrović
    Institute of Technical Sciences of the Serbian Academy of
    Sciences and Arts, Belgrade, Yugoslavia

G. Petzow
     Max-Planck-Institute for Metals Research, Stuttgart, F.R.Germany

M. Pijolat
     Ecole des Mines, Saint Etienne, France

D. Poleti
     Belgrade University, Belgrade, Yugoslavia

S. H. Pucinelli
     Department of Chemistry UNESP, Arraquara Sao Paolo, Brasil

R. Pruemmer
     Fraunhofer-Institute fuer Werkstoffmechanik, Freiburg,
     F.R.Germany

T. S. Rakotch
     Institute for Refractory Metals and Hard Alloys VNIITS, Moscow,
     U.S.S.R.

D. Raković
     Belgrade University, Belgrade, Yugoslavia

J. L. Rehspringer
     Ecole Europeene des Hautes Etudes des Industries Chimique,
     Strassbourg, France

K. Rzesnitzek,
     Max-Planck-Institute for Metals Research, Stuttgart, F.R.Germany

M. M. Ristić
     Serbian Academy of Sciences and Arts, Belgrade, Yugoslavia

W. Rossner
     Siemens AG, Research and Development, Muenchen, F.R.Germany

D. M. Rowe
     University of Wales, Cardiff, U.K.

J. M. Ruiz Prieto
     School of Mines, Madrid, Spain

Y. Sakka
     National Research Institute for Metals, Tokyo, Japan

C. V. Santilli
     Department of Chemistry UNESP, Arraquara, Sao Paolo, Brasil

T. Sato
     Fujitsu Laboratories Ltd., Yokohama, Japan

C. Sauer
     Technical University, Dresden, D.R.Germany

W. Schatt
     Technical University, Dresden, D.R. Germany

H. Schubert
     Max-Planck-Institute for Metals Research, Stuttgart, F.R.Germany

V. V. Skorohod
    Inst. for Probles of Materials Science AS USSR, Kiev, U.S.S.R.

S. Somiya
    The Nishi Tokyo University, Tokyo, Japan

M. Soustelle
    Ecole des Mines, Saint Etienne, France

R. M. Spriggs
    Alfred University, Alfred, U.S.A.

M. J. Sook
    Han Yang University, Seoul, Korea

M. Šušić
    Faculty of Sciences, Belgrade University, Yugoslavia

B. Šerbedžija
    Institute of Technical Sciences of the Serbian Academy of
    Sciences and Arts, Belgrade, Yugoslavia

F. Thuemmler
    Karlsruhe University, Karlsruhe, F.R.Germany

J. M. Torralba
    School of Mines, Madrid, Spain

V. I. Trusov
    NPO "Krasnaya Zvezda", Moscow, U.S.S.R.

G. S. Upadhyaya
    Indian Institute of Technology, Kanpur, India

T. Ushikoshi
    National Research Institute for Metals, Tokyo, Japan

D. P. Uskoković
    Institute of Technical Sciences of the Serbian Academy of
    Sciences and Arts, Belgrade, Yugoslavia

D. Vasović
    Belgrade University, Belgrade, Yugoslavia

J. Varela
    Department of Chemistry UNESP, Arraquara, Sao Paolo, Brasil

B. W. Veal
    Argonne National Laboratory, Argonne, U.S.A.

F. Weber
    Research Institute for Technical Physics, Budapest, Hungary

V. K. Yarmarkin
    NPO "Positron", Leningrad, U.S.S.R.

M. Yoshimura
    Tokyo Institute of Technology, Tokyo, Japan

T. Zambetakis
    Ecole Europeene des Hautes Etudes des Industries Chimiques,
    Strasbourg, France

Particle size distribution
    Al_2O_3-ZrO_2,130
    atomized powder,159,160
    mono-sized,5,101
    narrow,5,102,295

Phase stability,215

Pore
    boundary separation,6
    channels,342
    closed,344
    coalescence,12
    curvature,6
    entrapment,351,408
    gas-filled,12
    growth,14
    isolated,11
    mobility,7
    planar,387
    removal,408
    shapes,75
    shrinkage,6
    solid interphase,74
    volume reduction,470

Porosity
    coating,330
    grain size,50
    Kirkendall,440
    measurement,332
    Ni_3Al,444
    time dependence,406

Powders
    aluminide,295,439
    amorphous $Ni_{78}P_{22}$,285
    amorphous spheres,104
    calcination,141
    coated,110
    composite,104
    forced hydrolysis,102
    glass,75
    hydrothermal,507
    inheritance,469
    manganese phosphate,104
    monodispersed,101
    morphology,105,129
    mullite,135
    precipitation,102,118,128,406
    shapes,108
    sintering behaviour,143
    SnO_2,103,106
    sol-gel,45,118,129,132,135
    spherical,102
    surface effect,149
    thermal treatment,179
    ultrafine,185,193,203,204
    well defined,101
    ZnO,107,117

Precision digital dilatometry,353

Rate controlled sintering
    ceramics,337
    cordierite/zirconia,347
    glossary of terms,352
    hot isostatic pressing,344
    injection molded metals,340
    liquid phase, 338
    metals,337
    microstructures,340,346,360,
        365,372,374
    porcelain,339,346
    process limitations,343
    SiC with additions,367
    solid state,338
    strength properties,361
    technology,51
    translucent alumina,348
    ZrO_2(Y)-TZP,357

Recrystallization,249,475

Seebeck coefficient,414,417

Shock
    compaction,267,287
    front,269,270
    matter transformation,276,286
    synthesis,271
    wave,267,269,286

Shrinkage
    aggregate of spheres,386,388
    differential,453
    high rates,505
    isotropic,73
    rate,358
    volume,42

Sinterability,10,44

Sinter-forging,13

Sinter/HIP,311,344

Sintering
    activation energy,8,41,540
    activity,43,541
    Al-bronze,220
    Al_2OC,614
    Al_2OC-AlN,614
    alkaline-earth titanates,537
    alumina,3
    Al-carbides composites,223
    atmosphere,6,15,18,83,444,532,
        542
    brass,221
    $BaTiO_3$,395
    cemented carbides,224
    ceramics,3,497